Handbook of
GROWTH FACTORS
Volume I: General Basic Aspects

HANDBOOK OF GROWTH FACTORS
Enrique Pimentel, M.D.

Volume I
General Basic Aspects

Regulation of Cell Functions
Growth Factor Receptors
Postreceptor Mechanisms of Growth Factor Action
Cyclic Nucleotides
Guanosine Triphosphate-Binding Proteins
The Calcium-Calmodulin System
Phosphoinositide Metabolism
Protein Phosphorylation
Proto-Oncogene and Onco-Suppressor Gene Expression
Role of Growth Factors in Neoplastic Processes

Volume II
Peptide Growth Factors

Insulin
Insulin-Like Growth Factors
Epidermal Growth Factor
Fibroblast Growth Factors
Neurotrophic Growth Factors
Organ-Specific Growth Factors
Cell-Specific Growth Factors
Transforming Growth Factors
Regulatory Peptides with Growth Factor-Like Properties

Volume III
Hematopoietic Growth Factors and Cytokines

Hematopoietic Growth Factors
Interleukins and Cytokines
Colony-Stimulating Factors
Interferons
Tumor Necrosis Factors
Erythropoietic Growth Factors
Platelet-Derived Growth Factor
Transferrins

Handbook of
GROWTH FACTORS
Volume I: General Basic Aspects

Enrique Pimentel, M.D.
National Center of Genetics
Institute of Experimental Medicine
Central University of Venezuela
Caracas, Venezuela

CRC Press
Taylor & Francis Group
Boca Raton London New York

CRC Press is an imprint of the
Taylor & Francis Group, an **informa** business

First published 1994 by CRC Press
Taylor & Francis Group
6000 Broken Sound Parkway NW, Suite 300
Boca Raton, FL 33487-2742

Reissued 2018 by CRC Press

© 1994 by Taylor & Francis
CRC Press is an imprint of Taylor & Francis Group, an Informa business

No claim to original U.S. Government works

A Library of Congress record exists under LC control number: 93040108

Publisher's Note
The publisher has gone to great lengths to ensure the quality of this reprint but points out that some imperfections in the original copies may be apparent.

Disclaimer
The publisher has made every effort to trace copyright holders and welcomes correspondence from those they have been unable to contact.

ISBN 13: 978-1-138-10575-1 (hbk)
ISBN 13: 978-1-138-55951-6 (pbk)
ISBN 13: 978-0-203-71260-3 (ebk)

Visit the Taylor & Francis Web site at http://www.taylorandfrancis.com and the CRC Press Web site at http://www.crcpress.com

CONTENTS

Chapter 3
Postreceptor Mechanisms of Growth Factor Action

Chapter 4
Cyclic Nucleotides

Chapter 5
Guanosine Triphosphate-Binding Proteins

Chapter 8
Protein Phosphorylation

Chapter 9
Proto-Oncogene and Onco-Suppressor Gene Expression

Chapter 10
Role of Growth Factors in Neoplastic Processes

THE AUTHOR

Enrique Pimentel, M.D., is Professor of General Pathology and Pathophysiology at the School of Medicine, Central University of Venezuela, Caracas. He was formerly Director of the Institute of Experimental Medicine at the same university and founded and directed the National Center of Genetics in Venezuela.

Born on April 7, 1928, in Caracas, Venezuela, he obtained an M.D. degree from the Universities of Madrid, Spain, and Caracas, Venezuela, in 1953. He is President of the National Academy of Medicine in Venezuela and an honorary, corresponding, or active member of 32 national and international scientific academies and societies. He is an active member of the New York Academy of Sciences and Vice President of the International Academy of Tumor Marker Oncology (IATMO). He has received several decorations in his own country and the Grosse Verdienstkreuz (Great Cross to the Merit) of the Federal Republic of Germany. In 1982 he received the National Award of Science in Venezuela. On many occasions Dr. Pimentel has been invited to give lectures and seminars at universities and other scientific institutions in North America and Europe.

Dr. Pimentel is the author of more than 100 papers and co-author of seven books on topics related to endocrinology, genetics, and oncology. In addition, he is the author of *Hormones, Growth Factors, and Oncogenes* and *Oncogenes* (first edition in one volume, second edition in two volumes) published by CRC Press. He is editor of the bimonthly journal *Critical Reviews in Oncogenesis*.

Chapter 1

Regulation of Cell Functions

I. INTRODUCTION

Hormones and peptide growth factors are important regulatory substances present in metazoan organisms. They represent biological signals involved in the regulation of cell growth and differentiation as well as in the control of specific metabolic processes during both pre- and postnatal life. Certain hormones and peptide growth factors act in a restricted manner on specific types of cells, whereas others may have a broad, perhaps universal, spectrum of activity for different types of cells and tissues.

Hormones are defined as chemical messengers synthesized in the endocrine glands of multicellular organisms and secreted into the extracellular body fluids, which transport them to more or less distantly located target cells (hormone-responsive cells), where they can exert important regulatory actions. Upon arrival to its target cell, the hormone is recognized by and binded to a specific site, the hormone receptor. The formation of a hormone-receptor complex determines the response that is specific for both the hormone and the cell.[1,2] Hormones are involved in the integrated regulation and modulation of the differentiated functions of multicellular organisms. The neural system and the endocrine system have partially overlapping and complementary functions related to the integration and coordination of complex biological processes. They constitute a functionally important integrative structure of the organism, the neuroendocrine system.

In contrast to the classical hormones, the peptide growth factors (usually called in an abbreviated manner "growth factors") are not necessarily synthesized in specialized endocrine organs but are produced and secreted by cells from a wide variety of tissues, and their target cells are frequently located not far from the site of release (paracrine response). Even the cell producing a growth factor may in some cases (when it is endowed of the specific receptor) respond to the factor (autocrine response).[4-8] Certain growth factors remain anchored to the cell membrane.[9] The distinction between peptide hormones and growth factors may be a subtle one, however. Many growth factors, or proteolytic products of growth factors, circulate in the blood and may act at sites far away from the place of their secretion or may display specific regulatory functions. Growth factors, or substances with growth factor-like activities, have been found not only in vertebrates but also in invertebrates and even in plants, where they are represented by biologically active agents called cytokinins.

Growth factors are produced by normal and neoplastic cells *in vitro* and *in vivo*. They are essential components of the media required for the survival and growth of cells cultured *in vitro* and afford protection of cells against death. Growth factors are involved in cell survival *in vivo* and have a crucial role in the control mechanisms of the development of organs and tissues. In addition to their growth-promoting and differentiation-inducing activities, growth factors are able to elicit a wide variety of effects in their target cells, including diverse metabolic effects. Growth factors are importantly involved in physiological processes such as inflammation, immune reactions, and tissue repair.[10-12] Wound repair requires close control of both degradative and regenerative processes. It involves numerous types of cells and complex interactions between multiple biochemical pathways and growth factors such as epidermal growth factor (EGF), fibroblast growth factor (FGF), platelet-derived growth factor (PDGF), and transforming growth factor (TGF), which when released in the traumatized area have a crucial role in the regulation of these processes.[13] The effects of growth factors in relation to wound healing include promotion of cell migration into the wound area (chemotaxis), stimulation of the proliferation of epithelial cells and fibroblasts (mitogenesis), formation of new blood vessels (angiogenesis), and formation of matrices and remodeling of the affected region. Growth factors produced by a subpopulation of dermal fibroblasts contribute to the excessive accumulation of extracellular matrix that occurs in scleroderma.[14]

Growth factors are importantly implicated in the development of common human diseases such as atherosclerosis and cancer.[15-18] There are numerous structural and functional relationships between extracellular signaling agents, such as hormones, growth factors, and regulatory peptides, and intracellular signaling molecules, such as oncogene products (oncoproteins) and tumor suppressor gene products (onco-suppressor proteins).[19,20] Expression of growth factors, oncoproteins, and onco-suppressor proteins is altered in a diversity of benign and malignant tumors.[21-23] Quantitative and/or qualitative alterations of hormones and growth factors can be profitable as valuable markers for the diagnosis and follow-up of various human tumors.[24,25] In general, tumor growth depends on complex interactions between hormones and growth factors,[26] but malignant cells may depend less than normal cells on the exogenous supply of these agents, and some tumor cells are able to produce and utilize them by an autocrine or paracrine mechanism. The altered growth of tumor cells depends, at least in part, on the action of certain growth factors that are produced and utilized in inappropriate amounts by the tumor cells themselves.

The biological properties of growth factors can be studied either *in vitro* or *in vivo*. The isolation and characterization of growth factors are facilitated by culturing cells in a defined protein-free medium.[27] The undefined serum component of the medium can be replaced by specific mixtures of nutrients, hormones, growth factors, and metal ion transporting proteins.[28] Bioassay can be used as the initial method for the detection and measurement of growth factors. It requires the discovery and characterization of a factor-dependent morphological, physiological, or biochemical event in whole animals, tissue fragments, or cultured cells. The biological activity of growth factors can be examined in organ culture assays or in colony assay systems in agarose culture. Each of these methods has some advantages and disadvantages.[29] While the organ culture may be a better method for screening possible growth factors, the colony assay may provide quantitative results for statistical analysis. Growth factors are defined by their ability to induce stimulation of target cell proliferation, and their activity is measured by assays where the increase of cell number is estimated or the incorporation of radioactively labeled thymidine into DNA is determined by autoradiography and counting of the labeled cells or by determination of radioactivity in liquid scintillation vials. The most suitable routine test for detection of growth factor activity in a biological sample may lie on measuring the incorporation of tritiated thymidine into DNA of target cells with a scintillation counter.[30] After isolation guided by bioassay, the purified growth factor peptide can be used to develop antisera for detection and measurement of the growth factor by immunodiffusion or radioimmunoassay procedures, as well as by enzyme-linked immunoassay (ELISA) and membrane receptor assay. Polyclonal or monoclonal antibodies can be developed by using synthetic oligopeptides corresponding to specific amino acid sequences of the growth factor. A list of growth factors and other peptides with growth factor-like properties appears in Table 1.1 The chromosomal localization of genes coding for hormones and growth factors or their respective receptors is indicated in Table 1.2 and Figure 1.1.

II. THE CELL CYCLE

Depending on the cell type and the physiological conditions, cells may be engaged in proliferation and may continuously traverse the different phases of the cell cycle or may not divide and remain in a quiescent state. In general, there is an antagonism between the rapid proliferation of cells and the expression of highly differentiated functions by the cells. Cultured cells have been extensively used in the past decades for the analysis of the different phases of the cell cycle and the evaluation of exogenous factors capable of altering them. Serum contains a complex mixture of mitogens; cultured fibroblasts from mouse and other rodent species are widely used as a simple system for evaluating the influence of purified hormones, growth factors, and other mitogens on cell proliferation. Fibroblastic cells whose growth has been arrested by serum starvation (quiescent cells) can undergo a synchronous progression through the cycle by the addition of fresh serum or specific components purified from serum.

A. PHASES OF THE CELL CYCLE

The cell cycle comprises four major phases: G_1, S (DNA synthesis period), G_2, and M (mitosis period).[31,32] The phases of the cell cycle are schematically represented in Figure 1.2. The S phase corresponds to the period of DNA replication, a process that depends on the activity of several types of DNA polymerases,[33] as well as on the activity of other enzymes and factors. G_1 is the gap period between M and the initiation of DNA synthesis, and G_2 is the period between S and M. Cells in G_2 contain double the amount of DNA than cells in G_1. For most cells growing exponentially in culture, the interval between cell divisions is between 10 and 30 h. Differences in the duration of the cycle between different types of cells or different environmental conditions are mainly due to variation in the length of G_1, with the duration of S (6 to 8 h) + G_2 (2 to 6 h) + M (1 h) being relatively constant. In addition, there is much variability in the length of G_1 among individual cells in a single population. Animal cells, both *in vivo* and *in vitro*, can also exist in a nongrowing, quiescent state during which they do not divide for long periods. Most frequently, normal cells that have ceased to grow have the G_1 content of DNA. Quiescent cells may be metabolically different from cycling G_1 cells and are considered to be in a distinct state, termed G_0.

Table 1 **Growth factors and other endogenous growth regulatory peptides**

Growth factors	*Partially characterized growth factors*
Insulin	EGF-like mitogens
Insulin-like growth factor I (IGF-I) or somatomedin C	TGF-like growth factors
Insulin-like growth factor II (IGF-II) or somatomedin A	PDGF-like growth factors
	Melanocyte growth factor (MGF)
Epidermal growth factor (EGF)	Mammary-derived growth factor 1 (MDGF-1)
Fibroblast growth factors (acidic FGF and basic FGF)	Prostate growth factors
Nerve growth factor (NGF)	Cartilage-derived growth factor (CDGF)
Brain-derived neurotrophic factor (BDNF)	Chondrocyte growth factor (CGF)
Neurotrophins 3 and 4 (NT-3 and NT-4)	Bone-derived growth factor (BDGF)
Ciliary neurotrophic factor (CNTF)	Osteosarcoma-derived growth factor (ODGF)
Hepatocyte growth factor (HCGF)	Glial growth-promoting factor (GGPF)
Transforming growth factor α (TGF-α)	Colostrum basic growth factor (CBGF)
Transforming growth factor β (TGF-β)	Endothelial cell growth factor (ECGF)
Macrophage colony-stimulating factor (M-CSF or CSF-1)	Tumor angiogenesis factor (TAF)
	Hematopoietic stem cell growth factor (SCGF)
Granulocyte-macrophage colony-stimulating factor (GM-CSF or CSF-2)	B-cell stimulating factor 2 (BSF-2)
	B-cell differentiation factor (BCDF)
Granulocyte colony-stimulating factor (G-CSF or CSF-3)	Leukemia-derived growth factor (LDGF)
	Myelomonocytic growth factor (MMGF)
Platelet-derived growth factor (PDGF)	Macrophage-derived growth factor (MDGF)
Platelet-derived endothelial cell growth factor (PD-ECGF)	Macrophage-activating factor (MAF)
	Erythroid-potentiating activity (EPA)
Interleukins 1 to 13 (IL-1 to IL-13)	
Interferons α, β, and γ (IFN-α, IFN-β, and IFN-γ)	
Tumor necrosis factor α (TNF-α) or cachectin	
Tumor necrosis factor β (TNF-β) or lymphotoxin	
Erythropoietin	

Peptides with growth factor-like activities
Transferrin
Bombesin and bombesin-like peptides
Angiotensin II
Endothelin
Atrial natriuretic factor (ANF) and ANF-like peptides
Vasoactive intestinal peptide (VIP)
Bradykinin

B. REGULATION OF THE CELL CYCLE

The regulation of cell division depends on crucial events that occur during the G_1 phase of the cycle.[34] Two major control points ("checkpoints") have been defined in the G_1 phase of animal cells: a first point which operates in the early G_1 phase and allows the cell to exit the cycle and enter into the nonproliferative G_0 phase, and a second point, the restriction point (R point), which occurs in late G_1 phase. Protein synthesis is required through G_1 phase for cells to pass the R point and become committed to enter S phase and initiate DNA synthesis. However, not only two but three or more restriction control points may be present in the cycle of certain cells. In activated B lymphocytes the cycle would be controlled by three main restriction points.[35] The study of embryonic cell cycles indicates that diverse modes of regulation of the cycle are used during development.[36] Defects in one or more cell cycle checkpoints would be responsible for the genomic instability of cancer cells.[37]

The mechanisms associated with the regulation of cell cycle events had remained little characterized at the biochemical level;[38] however, there have been recent important advances in our knowledge of these mechanisms, as discussed in a number of reviews,[39-44] as well as in a book.[45] Cyclins and the cdc2 kinase and its related family of proteins, as well as certain protein phosphatases, play central roles in the mechanisms associated with the regulation of the cell cycle. Calcium and calmodulin also are importantly

Table 2 **Chromosome localization of the human genes for hormones and growth factors and their cellular receptors**

Chromo-some	Location	Hormone or Growth Factor	Chromo-some	Location	Hormone or Growth Factor
1	1p13-p21	CSF-1	7	7p12-p14	IGF-BP-1
1	1p21-p22.1	NGF	7	7p22	PDGF-A
1	1p32-p34	CSF-3 receptor	7	7q21	Erythropoietin
1	1q21	IL-6 receptor α-chain	7	7q21.1	HCGF
1	1q23-q24	Trk-A NGF receptor	7	7p15	IL-6
2	2p11-p13	TGF-α	8	8p12	Flg FGF receptor
2	2p16-p21	FSH receptor	8	8q12-q13	IL-7
2	2p21	LH/CG receptor	8	8q24	Thyroglobulin
2	2q13	IL-1-α	9	9p22	IFN-α and IFN-β
2	2q13-q21	IL-1-β	10	10p14-p15	IL-2 receptor α-subunit
2	2q14	IL-1 receptor antagonist	11	11p11.21	PTH
2	2q35	IL-8 receptors A and B	11	11p14.1	Insulin
3	3p12-3q13.2	IL-12 A subunit	11	11p15	IGF-II
3	3p21-p25	Thyroid hormone receptor β	11	11q13	β-Adrenergic receptor
3	3p24-p26	IL-5 receptor α-subunit	11	11q22	Progesterone receptor
3	3q15-q25	Transferrin	12	12p13	TNF-R1 receptor
3	3q26.2	Transferrin receptor	12	12q12-q13.1	Calcitriol receptor
4	4q11-q12	PDGF receptor α-subunit	12	12q22	SCGF
4	4q25-q27	EGF	12	12q23	IGF-I
4	4q26-q27	Basic FGF	14	14q23-q24	TGF-β_3
4	4q26-q28	IL-2	14	14q31	TSH receptor
4	4q31.2	Aldosterone receptor	15	15q25-q26	IGF-I receptor
5	5p12-p13.1	Growth hormone receptor	17	17q11	CSF-3
5	5p13-p14	Prolactin receptor	17	17q11.2	Thyroid hormone receptor α
5	5p13	IL-7 receptor	17	17q12-q22	NGF receptor p75
5	5q23-q31	CSF-2	17	17q21.1	RAR-α receptor
5	5q23-q31	Adrenergic receptors	17	17q22-q24	Growth hormone
5	5q23-q31	IL-3	17	17q22-q24	Placental lactogen
5	5q23-q31	IL-13	17	17q23	Bek FGF receptor
5	5q23.3-q31.2	IL-4	19	19p13.2-p13.3	Insulin receptor
5	5q31	IL-5	19	19q13.1-q13.3	TGF-β_1
5	5q31-q33	IL-12 B subunit	19	19q13.3-q13.4	IL-11
5	5q31-q35	IL-9	19	19p	Erythropoietin receptor
5	5q31-q32	PDGF receptor β-subunit	20	20p13	Oxytocin receptor
5	5q31-q32	Glucocorticoid receptor	21	21q	IFN-α receptor
5	5q31.2-q33.2	ECGF	22	22q12-q13	IL-2 receptor β-subunit
5	5q33.2-q33.3	CSF-1 receptor/c-*fms*	22	22q12.2-q13.1	CSF-2 receptor β-chaim
6	6p21.3	TNF-β	22	22q13	PD-ECGF
6	6p21.1-p22	TNF-α	22	22q13.1	PDGF-B/c-*sis*
6	6q16-q22	IFN-γ receptor	X	Xp11.1-p11.4	EPA
6	6q24-q27	Estrogen receptor	X	Xq11-q12	Androgen receptor
7	7p12-p14	EGF receptor			

involved in this regulation.[46] A number of cellular genes, including proto-oncogenes and tumor suppressor genes, are implicated in the regulation of the cell cycle. Arrest of cell growth prior to the R point may be mediated, in part, by the action of the nuclear phosphoprotein p53, which is the product of a tumor suppressor gene. This action appears to occur through repression of the expression of the c-*myb* proto-oncogene and the DNA polymerase-α gene.[47] Other participants in the signal transduction pathway related to cell cycle arrest at the R point are the *GADD45* gene, which is activated by ionizing radiation in lymphoblasts and fibroblasts, and the AT gene, which is involved in the hereditary, cancer-prone disease ataxia-telangiectasia.[48] The nuclear protein RB, which is the product of the *RB1* gene whose

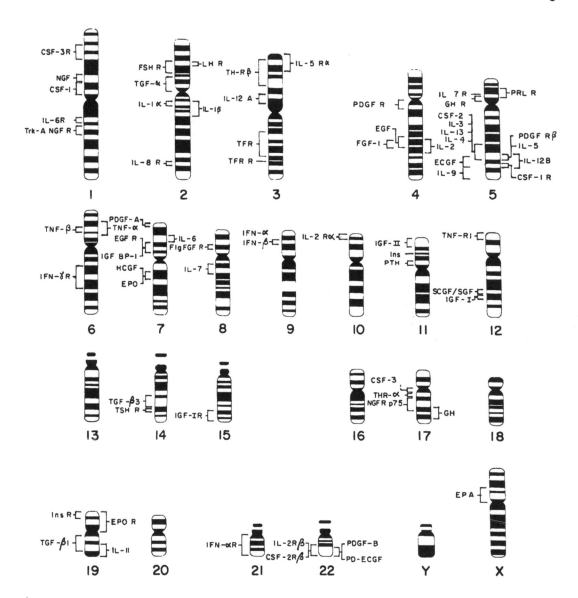

Figure 1.1 Chromosomal localization of the genes encoding growth factors and their receptors.

deletion is associated with human retinoblastoma and other tumors, has an important role in cell cycle regulation and may be involved in cell growth arrest.[49] Expression of the RB protein is crucial for the differentiation of certain types of cells.

Both endogenous and exogenous agents contribute to the regulation of the cell cycle. Hormones and growth factors are involved in the regulation of the cycle in different cell types and may influence, in particular, the operation of the R point. During early G_1 phase, the cells complete the growth factor-dependent processes leading to commitment for cellular proliferation. Thereafter, they enter a growth factor-independent pre-DNA-synthetic part of the G_1 phase. Suboptimal conditions such as the absence of serum may shift normal cells into quiescence (G_0 phase). In contrast, neoplastic cells lack the R point control in whole or in part and may grow in the absence of serum (or growth factors contained in serum).

On the basis of the cell cycle arrest induced by brief serum starvation or transient inhibition of protein synthesis, the G_1 period of exponentially growing murine fibroblasts (3T3 cells) has been subdivided into two discrete subphases: one postmitotic (G_1pm) and one pre-DNA-synthetic (G_1ps).[50] The commitment process of G_1pm may be completed in the presence of PDGF as the only supplied growth factor. EGF and insulin are not sufficient for the completion of the commitment process in G_1pm. Under conditions optimal for proliferation, the cells may complete the commitment process of G_1pm within a constant time

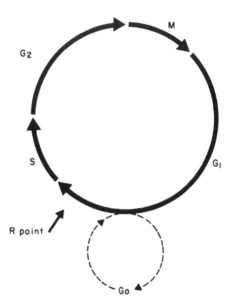

Figure 1.2 The cell cycle.

period (3 to 4 h after mitosis). The duration of G_1ps, on the other hand, shows a large intercellular variability. In fact, G_1ps accounts for most of the variability in cell cycle time. The mechanisms of the progression of cells through the G_1 phase of the cycle are associated with the action of multiple protein kinases, and the activities of these enzymes could be impaired in neoplastic cells.[51]

The initiation of late events in the cycle may depend on the completion of early events, and this process is regulated at the cellular level by mechanisms involving checkpoints that ensure the order of cell cycle events.[42] Progression through each checkpoint in certain cell types (for example, in sea urchin embryo cells) may be triggered by transient increases in the concentration of cytoplasmic calcium ions, termed $[Ca^{2+}]_i$.[53] Specific changes in gene expression are crucially involved in cell cycle-associated events. Studies on Rous sarcoma virus (RSV)-induced growth of avian neuroretina cells suggest that the state of quiescence is actively maintained by a delicate balance between positive and negative regulatory signals that act through alterations in gene expression.[54] The expression of certain cellular genes, which are developmentally regulated, may thus be required for maintaining specific cell types in a nonproliferative state, and suppression of the expression of these genes would lead to an active growth of the cells. In general, the regulation of the cycle of different types of cells in different states of development and different physiologic conditions depends on complex interactions between positive and negative external signals represented by specific hormones, growth factors, growth inhibitors, and regulatory peptides, and internal signals represented by specific proteins, including the products of proto-oncogenes and tumor suppressor genes.[55] In any case, the microenvironment where the cell resides is crucial for the regulation of its functional behavior.

Expression of specific genes is of great importance in relation to regulation of the cell cycle. Genes involved in cell cycle regulation can be identified by the analysis of the specific mRNAs or proteins. The construction of cell cycle phase-specific cDNA libraries represents a valuable tool for the identification of genes specifically involved in the regulation of the cell cycle.[56] Immediate-early growth response genes participate in a crucial way in the transition from quiescence into the G_1 phase. These genes code for transcription factors, secreted proteins with autocrine or paracrine functions, or structural proteins. Expression of the immediate-early genes can be activated by extracellular signaling agents, including hormones, growth factors, and other mitogens. Characterization of cell cycle mutants may contribute to a better knowledge of the biochemical events associated with the cycle in different types of cells, especially those that control or execute cell cycle-specific processes. Unfortunately, many of these mutations are lethal and few of them have been defined at the biochemical level. The study of temperature-sensitive (*ts*) cell cycle mutants in defined cell systems *in vitro* may contribute to the identification of genes involved in the initiation of DNA synthesis or other phases of the cycle.[57] Chinese hamster cells carrying a recessive *ts* mutation, *ts41*, grow normally at the permissive temperature, but exhibit at nonpermissive temperatures a time-dependent lethality and accumulate excess DNA, often to as high as 16°C.[58] The *ts41* mutant cells can replicate their genome normally, but instead of going on into G_2, S, and

G_1, they pass directly into a second S phase without requiring the presence of serum.[59] The *ts41* gene product may participate in two functions in the cell cycle (entry into mitosis and inhibition of entry into S phase), but its precise structure and function remain unknown.

The discontinuous model for regulation of the cell cycle in which a regulatory control of proliferation specifically occurs in the G_1 phase of the cycle has been challenged, and a continuum model has been proposed in which the available data have been reinterpreted as indicating that regulation occurs in all phases of the division cycle.[60] Studies on c-*fos* and c-*myc* proto-oncogene expression in serum-stimulated normal and transformed rat 3Y1 fibroblasts show that the processes leading to the initiation of S phase occur continuously throughout all phases of the cycle.[61] These results support a continuum model of regulation of the cell cycle.

It may be concluded that, although we have a good general picture of the mechanistic operation of the cell cycle, there remain many open questions regarding the biochemical events that are crucially involved in the regulation of the cycle.[62] The regulation of the cell cycle is very complex and further studies are required for a better characterization of the factors involved in this process.

C. CELLULAR SENESCENCE

All normal somatic cells eventually undergo senescence and die. In rapidly renewing tissues such as epithelial tissues, the terminal differentiation of the cells is associated with their destruction or elimination from the body by a process of decimation. Other cells such as hepatocytes usually are not engaged in cell cycling, have a prolonged lifespan, and may divide only occasionally or under the influence of specific mitogenic stimulation. Nondividing cells that are terminally differentiated, such as neurons, persist throughout the life of the organism, although some of them also die.[63]

The process of cell aging also occurs *in vitro*. Cultures of normal (diploid) cells of human or nonhuman origin have a defined lifespan and die out after 40 to 50 replications.[64-66] Starting with a primary culture of cells (phase I), the cells multiply at a constant rate for a number of generations (phase II), but finally enter a senescent condition (phase III) that leads to the death of the cell culture. The final failure in the proliferative response of culture cells is preceded by a lengthening in the cell cycle time, primarily at the expense of the G_1 phase of the cycle.

The mechanisms involved in cell senescence are exceedingly complex and may include the accumulation of genetic and metabolic defects. The hallmark of cell aging *in vitro*, and probably also *in vivo*, is a decreased mitogenic response to growth factors.[67] The diminished response is not due to decreased numbers of cellular receptors or to changes in the affinity of the receptors for their ligands, but may be primarily associated with an altered expression of genes involved in cell cycle regulatory events. However, specific genes responsible for cellular aging remain to be identified. Most of the genes examined in senescent cells are expressed in a normal fashion, including proto-oncogenes such as c-*myc*, c-*jun*, and c-*ras*. Only the c-*fos* gene may not be adequately expressed in serum-stimulated senescent cells to allow them to enter DNA synthesis.[68]

D. APOPTOSIS

Cells may die by the action of injurious exogenous agents such as toxins or viruses through a process of necrosis that involves disruption of membrane integrity, cellular swelling, and lysis. In contrast to this passive pathological process, somatic cells are normally subjected to the occurrence of progressive degenerative changes, and some of these cells are eliminated from the body by a process of active programmed cell death called apoptosis.[69-75] Apoptosis has an important role in normal developmental processes, which include not only the proliferation of specific groups of cells but also the elimination of other cells that are produced in excess to the final requirements of the organism. In contrast to necrosis, there is no inflammatory reaction and organelle swelling in apoptotic cells, which exhibit an early nuclear compaction and cytoplasmic condensation, followed by breakdown of the nucleus into discrete fragments. A typical feature of apoptosis is internucleosomal DNA fragmentation, associated with double-stranded DNA damage that cannot be repaired and is by itself lethal. However, endonucleases involved in this type of damage have not been identified as yet.

Apoptosis is an essential component of the normal function and development of multicellular organisms. It operates in embryonic development and metamorphosis, as well as in normal physiologic processes of adult life such as the deletion of certain lymphocyte clones from the immune system, lactating mammary gland regression, and ovarian follicle atresia. About half of the neurons produced during embryogenesis die before adulthood, and this process may be initiated by selective reduction of neurotrophic growth factors. Apoptosis represents a universal form of programmed cell death and is essential to the normal structure and function of the organism. However, not all programmed cell deaths may occur via apoptosis. Unlike the death of T-lymphoid cells during negative selection in the thymus which occurs by a typically apoptotic process, the loss of intersegmental muscles at the end of metamorphosis in the tobacco hawkmoth, *Manduca sexta*, does not display the features that characterize apoptosis and may take place by a mechanism that would be distinct from apoptosis.[76]

The process of apoptosis in different tissues and developmental stages depends on both the genetic program of the cell and the local and general physiological conditions. In the adult tissues of vertebrates, apoptosis can be regulated by steroid and peptide hormones as well as by growth factors. The regressive effect of glucocorticoids on lymphoid tissues may include apoptosis, and lack of androgens may result in apoptosis of prostate epithelium. TGF-β may induce apoptosis in some cell types, including gastric carcinoma cells.[77] Progesterone withdrawal elicits apoptosis in the epithelial cells of the uterus, which results in menstrual bleeding. Hematopoietic cell survival depends on the availability of specific growth factors. Withdrawal of IL-3 from immature hematopoietic cells has been shown to lead to internucleosomal DNA cleavage and other events characteristic of apoptosis, and lack of erythropoietin accelerates apoptosis in marrow cells.

The process of apoptosis is dependent on RNA and protein synthesis in the dying cell. Specific cellular genes are responsible for programmed cell death, and one of these genes may be the c-*myc* proto-oncogene.[78] The nuclear c-Myc protein is involved in the regulation of cell proliferation and its apoptotic effect would depend on the availability of growth factors and other proteins. The 26-kDa protein Bcl-2, which is the product of the putative proto-oncogene *bcl*-2 and is an integral component of the inner mitochondrial membrane, the endoplasmic reticulum, and the nuclear envelope, is able to prevent c-*myc*-induced apoptosis.[79,80] Bcl-2 bears no resemblance to other cellular proteins and its mechanism of action remains enigmatic. Bcl-2 blocks apoptosis even in cells lacking mitochondrial DNA.[81] Mitogenic stimulation of lymphocytes with lectins or cytokines results in marked increases in the levels of Bcl-2 expression.[82] Bcl-2 is expressed in a wide variety of human fetal tissues and may have a role in morphogenesis.[83] High levels of Bcl-2 expression occur during neurogenesis,[84] and Bcl-2 may be a mediator of the effect of neurotrophic factors on neuronal survival. Overexpression of Bcl-2 can prevent apoptosis of sympathetic neurons.[85] The possible role of the *bcl*-2 gene in oncogenesis is not clear, but the gene was discovered through its involvement in chromosome translocations t(14;18) associated with human follicular non-Hodgkin's lymphomas.[86] The *bcl*-2 gene is expressed at low or undetectable levels in androgen-dependent prostate cancers, and higher levels of *bcl*-2 gene expression are observed in androgen-independent cancers of the prostate.[87] *bcl*-2 expression is augmented following androgen ablation and may be correlated with the progression of prostate cancer from androgen dependence to androgen independence. Normal and neoplastic cells deprived of growth factors may undergo apoptosis, and apoptosis is a regular component of tumor growth. Available evidence suggests that the Bcl-2 protein plays an important role in cell development and maturation.[88] Bcl-2 may be involved in pathways leading to the terminal differentiation of cells.

Tumor suppressor proteins may have a role in apoptosis. The p53 protein can induce apoptosis of myeloid leukemia cells, and this effect is inhibited by IL-6.[89] Apoptosis induced by p53 in murine erythroleukemia cells occurs predominantly in the G_1 phase or the G_1/S boundary of the cycle, but it is not the result of G_1 arrest alone.[90] The p53 protein may be involved in a fail-safe mechanism that would kill cells in which the normal cell cycle-regulatory pathways have been disturbed. p53 can regulate the survival of hematopoietic progenitor cells, thus modulating the production of mature blood cells. Thymocyte apoptosis depends on p53-dependent and -independent pathways.[91] Agents that initiate DNA-strand breakage kill thymocytes by a p53-dependent pathway, whereas the effects of glucocorticoids, as well as those of aging, appear to be independent of p53 expression. Further studies are required for a better characterization of the role of oncoproteins, tumor suppressor proteins, hormones, and growth factors in apoptotic processes occurring in normal or malignant cells.

E. CELLULAR IMMORTALIZATION

Normal diploid somatic cells are characterized by a finite lifespan *in vitro*, which in the case of human cells would correspond to 50 to 60 serial passages of the cultured cells.[92] After this number of replications the doubling time of the cultured cells increases exponentially and the cells die away, even when the environmental conditions are optimal. In contrast, neoplastically transformed heteroploid cells are immortal and can be maintained in culture for an apparently unlimited time span. Neoplastic cells may have the potential for continuous proliferation under adequate environmental conditions (supply of nutrients, renewal of the medium, etc.). Moreover, transformed cells may be less dependent on the exogenous supply of hormones and growth factors than are their normal counterparts. The mechanisms responsible for this relative independence are not understood but may include, at least in certain cases, the synthesis of specific growth factors by the transformed cells. These factors can be utilized by the same cells when they are endowed with the cognate receptors (autocrine mechanism). However, immortality is not unique to transformed cells. Haploid germ cells can be considered as an immortal cell line specifically involved in perpetuating the species. Certain normal, diploid somatic cells (in particular, stem cells and embryonic cells) could have the potential for immortality. Normal mammalian embryonic stem cells, prior to the segregation into germ and somatic elements, are able to divide indefinitely under appropriate conditions *in vitro* without apparent "crisis" or transformation.[93] Thus, embryonic and neoplastic cells may share the same fundamental characteristic of potential immortality.

III. REGULATION OF CELL PROLIFERATION AND DIFFERENTIATION

Growth factors have an essential role in the regulation of cell proliferation and differentiation during both the pre- and postnatal life. In addition, they are involved in the regulation of specialized cellular functions and distinct biochemical pathways. The regulatory actions of growth factors on their target cells are performed through mechanisms that include various intracellular mediators. Some of these signal transduction pathways are able to modify, either positively or negatively, the expression of specific gene sets. Genomic sequences involved in the regulation of gene expression include positive regulatory elements, such as the promoters and enhancers, and negative regulatory elements, called silencers. The promoters are located within or very close to the gene which is regulated, usually at the 5′ side, while the enhancers can be located far from the regulated gene and in either orientation, 5′ or 3′. The activity of certain transcription factors, which include nuclear oncoproteins, may be modified by the cellular action of growth factors and may lead to specific changes in gene expression.

The effects of growth factors on processes associated with cell proliferation and differentiation have been analyzed in various systems *in vitro*. Activation of multiple intracellular signal transduction pathways by growth factors may be involved in the survival of quiescent 3T3 cells in culture.[94] More than one pathway is required for long-term survival, but the survival of a considerable fraction of the cells may be assured by activation of a single pathway. The requirement of different cell types for survival and growth in culture is variable. Fetal hepatocytes have a complex factor requirement for their maintenance and growth in culture.[95] Terminal differentiation of cultured skeletal muscle cells (myogenic differentiation) is regulated by an interaction between differentiation-inducing growth factors such as FGFs and IGFs and the differentiation-inhibiting action of TGFs.[96,97] Certain types of cells exhibit a restricted ability to respond to growth factors, other cell types may respond to a wide spectrum of these factors. For example, quiescent normal human mesothelial cells can enter DNA synthesis when they are incubated in defined medium supplemented with insulin, hydrocortisone, transferrin, and one of the following mitogens: EGF, TGF-β, PDGF, FGF, IL-1, IL-2, IFN-γ, IFN-β, or cholera toxin.[98] The increases in cellular proliferation caused by the different mitogens do not exhibit significant differences, even though these mitogens recognize different cell surface receptors and activate cell proliferation via different transductional pathways. Sustained growth of the mesothelial cells requires, in addition to the above-mentioned factors, the presence in the medium of high density lipoproteins (HDL).

A. ROLE OF GROWTH FACTORS IN DEVELOPMENTAL PROCESSES

The molecular regulatory mechanisms of cell proliferation and cell differentiation are specified and coordinated by the genetic program of the organism. Accomplishment of a normal equilibrium between the proliferation and differentiation of distinct cell populations is essential during the development of multicellular organisms. The genetic information passed on from generation to generation in the different species contains a developmental program that controls ontogeny, and normal development is based on the reading out of this program in a precise spatial and temporal pattern.[99] A strict sequence of morphogenetic events, dictated by a complex interplay between genetic and environmental factors, is accomplished during ontogeny, resulting in the formation of a diversity of organs and tissues with specialized functions. It is important to remember that all factors, either endogenous or exogenous, that are not strictly genetic must be considered as environmental factors. Thus, hormones, growth factors, and other regulatory peptides are environmental factors, capable of regulating in specific manners the expression of the multiple potentialities contained in the genome. These agents have a crucial role in the activation of specific developmental programs, and alterations in the delicate balance existing between them may result in the production of dysmorphogenic abnormalities. Specific signaling agents of both embryonic and maternal origin participate in the embryogenic processes. The placenta, which plays a key role in gestation not only as a barrier between the maternal and fetal circulation but also as an endocrine organ, is a rich source of hormones, growth factors, and regulatory peptides, and it expresses receptors for many of these agents.[100,101] Growth factors have an essential role in the regulation of early embryo-genesis.[102-105] Growth factors of embryonic or maternal origin may participate in preimplantation embryo development and blastocyst functions in an autocrine and/or paracrine way. In species having hemochorial placentation, the formation of decidual tissues involves proliferation, differentiation, and migration of endometrial stromal cells and other cells, and growth factors participate in the regulation of conceptus growth during embryo implantation and may function as inducers or morphogens during embryogenesis.

Single-cell mRNA phenotyping performed by using a combination of reverse transcription and amplification of the transcribed cDNA in a polymerase chain reaction (PCR) has allowed the character-ization of growth factors required for the preimplantation of mouse embryos.[106] TGF-α, TGF-β, and PDGF-A genes are expressed in mouse blastocysts, whereas EGF, basic FGF, NGF, and CSF-3 genes may not be transcribed. Growth factors expressed by mammalian embryos are probably directed at maternal tissue and may be involved in early angiogenesis and decidualization of the uterus. Expression of TGF-β_1, TGF-β_4, PDGF-A, and low levels of PDGF-β occur in the inner mass of preimplantation embryos while TGF-β_2, TGF-β_3, and IGF-II are expressed in early endodermal and mesodermal deriva-tives.[107] Early embryo cells have receptors for a number of growth factors that are expressed in peri-implantation embryonic, extra-embryonic, or maternal tissues, including the receptors for TGF-α/EGF, PDGF, and basic FGF.[108] The functional capacity of these receptors can be assessed by their capacity to induce c-*fos* gene transcripts following the treatment of embryos with different growth factors.

The extracellular environment plays a key role in the growth and differentiation of cells during embryogenesis and tissue morphogenesis, and growth factors are involved in modulating the structure and function of the extracellular matrix, an effect that is partially exerted through the direct or indirect regulation of the activity of a specific set of proteases.[109,110] Localized production of growth factors and regulation of the expression of their receptors and the function of postreceptor signaling mechanisms may play an important role in the development of organs such as the palate.[111] Particular types of growth factors, including factors related to PDGF, basic FGF, and TGF-β, as well as a mesoderm-inducing activity (MIA), are involved in the induction of mesoderm during vertebrate development.[112,113] Secretion of factors such as MIA may be a characteristic of a subset of cells related to a certain stage of development.

B. REGULATION OF HOMEOBOX-CONTAINING GENES

Homeobox-containing genes (homeotic genes or *Hox* genes) have an essential role in the regulation of developmental processes by morphogens and growth factors.[114-120] The homeobox is a sequence of 183 bp that was originally discovered in genes associated with homeotic mutations (i.e., changes in segment identity leading to transformation of one body structure into another) in the fruit fly, *Drosophila melanogaster*. Homeotic genes are activated in a defined spatial and temporal sequence during the process of developmental segmentation of the insect. Homeobox-containing genes are present in verte-brates, including humans.[121] No less than 30 homeobox-containing genes have been identified in the human genome.[122-124] Human homeotic genes are distributed in four main clusters, termed *HOX*1 to *HOX*4, which reside on different human chromosomes: *HOX*1 includes 8 homeoboxes in 90 kb of DNA on chromosome 7, *HOX*2 includes 9 homeoboxes in 180 kb of DNA on chromosome 17, *HOX*3 contains at least 7 homeoboxes in 160 kb of DNA on chromosome 12, and *HOX*4 includes 6 homeoboxes in 70 kb of DNA on chromosome 2. The homeobox encodes a 61-amino acid helix-turn-helix (HTH) protein structure that binds DNA with sequence specificity. Homeobox proteins may function as transcription factors involved in the regulation of gene expression; however, the identity of target genes regulated by homeobox proteins is almost unknown.

Sequential expression of homeotic genes is associated with the early stages of limb bud outgrowth in the mouse, with each gene showing a graded transcript distribution along the antero-posterior axis.[125] A homeodomain protein, GHF1, is required for pituitary cell proliferation in the mouse.[126] Homeobox genes can organize a complete secondary body axis in the dorsal blastopore lip of the early *Xenopus laevis* gastrula.[127] Analysis of *HOX* gene expression in human hematopoietic cell lines suggests that these genes are switched on or off in blocks at various stages of hematopoietic cell differentiation.[128] Lineage- and stage-specific expression of *HOX*1 genes occurs in the human hematopoietic system.[129] Modulation of *HOX*2 gene expression alters the phenotype of human hematopoietic cell lines.[130] A coordinate regulation of *HOX* genes may play a very important role in lineage determination during early stages of hematopoiesis

Homeotic genes may have a role in the regulation of hormone and growth factor secretion. The mammalian homeotic gene *GHF1* is involved in the regulation of growth hormone gene expression in anterior pituitary somatotrophs.[131] Transcription of the *GHF1* gene is highly restricted and depends on positive autoregulation by the GHF1 protein as well as environmental influences acting through regulation of the intracellular levels of cAMP and the activity of the cAMP response element-binding (CREB) protein. Endogenous retinoids (vitamin A derivatives) may function as natural morphogens and may be involved in the regulation of homeotic gene expression. Little is known, however, about the regulatory effects of hormones and growth factors on homeotic gene expression. Some homeotic genes exhibit structural homology to the genes encoding growth factors such as EGF and TGF-α.[132,133]

Homeoboxes may have a role in oncogenesis.[134] The homeobox gene *Hox*-2.4 is constitutively activated in the mouse myeloid leukemia cell line WEHI-3B, which is due to insertion of a transposable DNA element belonging to the family of intracisternal A particles upstream of the gene. Transfection of the altered *Hox*-2.4 gene into NIH/3T3 cells resulted in their tumorigenic conversion.[135] A marked increase in homeobox RNA and protein occurs in some human tumors.[136] Immunohistochemical analysis showed that normal colonic epithelium expresses very little homeobox proteins, but both villous adenomas (precancerous lesions) of the colon and overt colon carcinomas exhibit high levels of these proteins. The human tumor homeobox proteins have an apparent molecular weight of 63 kDa. The homeobox gene *Hox*-1.1 may be involved in translocations associated with childhood T-cell acute leukemias.[137]

C. COMPETENCE AND PROGRESSION FACTORS

Growth factors contained in serum are crucially required for the growth of cells cultured *in vitro*.[138] Serum-deprived fibroblasts cease to grow, and restoration of serum causes the cells to enter S phase quasi-synchronously after a lag of approximately 16 h.[139] Although each serum factor may have a differential effect on DNA, RNA, protein synthesis, and mitosis of particular cell types, the combined action of them is required for the optimal growth of cells.[140-144] Certain growth factors, called competence factors, act on cells that are either in G_0 or early G_1 phase of the cycle, rendering them competent to initiate DNA replication. Other growth factors, called progression factors, would allow the progression of cells through the prereplicative phase of the cycle, inducing cells to traverse into S, G_2, and M phases.[145-148] In cultured mouse fibroblasts, PDGF behaves as a competence factor, rendering G_0-arrested cells competent to progress into S phase in response to progression factors such as EGF and IGFs. Moreover, IGFs may stimulate DNA synthesis specifically in competent cells primed with EGF (primed competent cells). According to these results, competence and progression factors would act sequentially and synergistically in the cell cycle. The similar action of some growth factors in the prereplicative phase suggests the operation of certain common intracellular pathways.[149]

The classification of growth factors in competence and progression factors has several limitations. Depending on the cell type and the intrinsic and extrinsic conditions, a single growth factor can act as either a competence or a progression factor. For example, in the hematopoietic system, IL-3 can act on IL-3-dependent cell lines as a competence factor, capable of inducing G_0 to G_1 transition; if exposure persists, IL-3 acts as a progression factor, capable of inducing the competent cells to move from G_1 to M.[150] A specific combination of hormone and growth factor influences may be required for the induction of DNA in particular types of cells, for example, in immature gonadal cells.[151] The model based on a sequential action of competence and progression factors is based on studies using cultured mouse fibroblasts and its general validity should be considered with caution. The regulatory events exerted by particular growth factors in other cellular systems (for example, in the proliferation of epithelial cells such as hepatocytes) may be more complicated and less well defined.[152] In contrast to the model of mouse fibroblast cell growth based on a competence/progression model, EGF alone is able to induce the signals necessary for the mitogenic stimulation of the EL2 rat embryo fibroblast cell line.[153] Regulation of human fibroblast proliferation by growth factors does not appear to be in accordance to a competence/progression cell cycle model.[154] The human cells, cultured in a serum-free, chemically defined medium, respond to EGF, IGF-I, transferrin, and dexamethasone (without PDGF) with an extensive and rapid growth, which argues that these cells may not have a "competence factor" requirement. PDGF has an effect similar to that of EGF on human cells, with regulation of the fraction of cycling cells, whereas IGF-I exerts its primary effect by regulating the rate of exit from G_1 into S phase without affecting the cycling fraction of cells.

Induction of certain secreted proteins (molecular weights of 29 to 68 kDa) is associated with a marked increase in DNA synthesis induced by EGF in rat EL2 cells.[155] The mechanisms by which growth factors are capable of regulating cell division are unknown, but they do not appear to include direct regulation of enzymes involved in DNA replication such as DNA polymerase-α.[156] These mechanisms are complex in nature and involve components from different cellular compartments, going from the membrane through the cytoplasm and into the nucleus.

D. CELLULAR RESPONSES TO GROWTH FACTORS

The response of cells to extracellular signaling agents such as hormones and growth factors may result in either stimulation or inhibition of specific functions associated with structural components of the membrane, the cytoplasm, the mitochondria, the nucleus, or other cellular compartments. Frequently, the cellular response includes alterations in the regulation of gene expression, and certain cells may show a mitogenic response. Hormone response domains contained in distinct sets of genes are implicated in the regulation of gene expression by hormones and growth factors.[157] The cellular genes may respond to mitogenic signals after variable periods of time. Interest has been focused on the "immediate-early" or "early response" genes that are characterized by their rapid and transient induction, independent of *de novo* protein synthesis. The normal function of many of these genes is unknown, but some of them code for transcription factors or transcription modulators that are involved in the regulation of gene expression through the G_1 phase of the cell cycle.

Most of the studies related to the action of growth factors have been performed by using defined systems *in vitro*. The *in vivo* response of cells to the mitogenic action of growth factors may show great variation, according to the type of cell, its state of differentiation, and the predominant physiological conditions. In general, cell differentiation is associated with a progressive reduction in proliferative potential, but the precise relationship that exists between differentiation and mitogenic responsiveness has not been established. Studies with 3T3 T mesenchymal cells show that cells at a predifferentiation arrest state express similar growth factor requirement to initiate DNA synthesis as do undifferentiated cells at other reversible growth arrest states, i.e., the state induced by growth factor deficiency.[158] In contrast, expression of a nonterminally differentiated phenotype is associated with reduction of mitogenic responsiveness of 3T3 T cells to growth factors. A progressive decrease in the capacity of response of cells to mitogens may occur during differentiation.

Studies performed on cell populations should be complemented with those focused on the response to exogenous signals of individual cells from the population. Individual mouse fibroblasts display marked heterogeneity in proliferation capacity when cultured at low serum concentrations.[159] Heterogeneity in growth factor sensitivity may arise at high frequency within a clone, suggesting that the ability of cells to respond to growth factors may depend on some cellular component that is distributed unevenly within the population, some cells having enough for many consecutive cycles and others having an insufficient amount even for one.

Age-associated factors are of great importance in relation to the cellular responses to hormones and growth factors. It is well known that cells such as human diploid fibroblasts have a limited lifespan in culture.[64-66] Moreover, this span is inversely related to the age of the human donor.[160,161] There is also clear evidence for a relationship between longevity of mammalian species and lifespans of normal fibroblasts *in vitro* and erythrocytes *in vivo*.[162] Such variations may be attributed, at least in part, to age-related differences in the cellular responses to growth factors. In general, cells from newborn animals have greater mitogenic responses to growth factors than do adult cells. The responsiveness of human fibroblasts to hormones and growth factors contained in fetal calf serum declines with age.[163,164] The mechanisms involved in the decreased responsiveness of senescent cells to growth factors are unknown, but they are not associated with a lack of expression of growth factor receptors, which are usually present at approximately normal levels in the aging cells. Growth factors can induce early prereplicative changes even in senescent human fibroblasts.[165] Among human cells capable of cycling in aging cultures there are few changes in the regulation of the growth fraction by PDGF and EGF, but there may be a greatly increased dependence on IGF-I for the regulation of the rate of entry into S phase.[166] The slower growth of the dividing population of cells in aging cultures may be related to a requirement for IGF-I at levels that are above those usually supplied, and at supraphysiologic IGF-I concentrations older cells may be capable of a faster G_1 exit rate than younger cells. An embryonic sheep cell extract may restore the responsiveness of senescent fibroblast cell cultures to growth factors, thus allowing the aging cells to resume the ability to undergo mitotic divisions.[167] The component(s) of the embryonic cell extract responsible for this effect is unknown.

In certain cases the response of cells to hormones and growth factors may persist far beyond the time of exposure to the exogenous signaling agent. For example, a single dose of estradiol is sufficient to elicit the permanent induction of hepatic estrogen receptor mRNA in male *Xenopus laevis*.[168] The transient exposure of the animal to the hormone or growth factor could establish an autocrine loop or irreversible switching-on of specific sets of hormone- or growth factor-responsive genes, but the molecular mechanisms involved in such phenomena are unknown.

Environmental parameters may influence the response of cells to growth factors. Oxygen concentrations within the physiologic range control the growth pattern of cultured human diploid fibroblasts by modulating their response to serum and growth factors such as EGF and PDGF.[169] Exposure of quiescent cells to reduced oxygen concentrations may enhance serum-induced DNA synthesis in a time-dependent manner. Oxygen concentration could control cell growth indirectly by altering the activity of a stable intermediate that regulates the cellular response to growth factors.

As a general rule, tumor cells are less dependent on the exogenous supply of growth factors than are normal cells, and progression of tumor cells to higher levels of malignant behavior is frequently associated with increasing autonomy from this supply.[170] Murine leukemia occurs by a step-wise development in which immortal, growth factor-dependent, and stromal cell-dependent myeloid cell lines of limited tumorigenic potential slowly progress to become an autonomous line, independent of growth factor requirement and interaction with the stroma cell microenvironment.[171] However, tumor cell growth is not always characterized by an autonomy from the exogenous supply of growth factors.[172,173] Moreover, the proliferative response of tumor cells to growth factors may show great variation according to factors such as the microenvironment and the presence of chromosome abnormalities.[174] Not only highly malignant cells but also cells from certain benign tumors (for example, keloid-derived fibroblasts) have reduced growth factor requirements in culture.[175] Thus, a diminished requirement for the exogenous supply of growth factors is not an exclusive property of malignant cells.

E. REGULATION OF DIFFERENTIATED FUNCTIONS BY GROWTH FACTORS

Growth factors are involved not only in the control of cellular proliferation and differentiation but also in the regulation of processes associated with the expression of differentiated functions that may be independent of cell proliferation and DNA synthesis. PDGF, EGF, and FGF regulate the production of IFN-γ by lymphocytes through a mechanism unrelated to their proliferative effects on cells.[176] TGF-β selectively regulates the secretion of three proteins by 3T3 fibroblasts through a mechanism that does not affect DNA synthesis.[177] Other secretory processes may also be regulated by hormones and growth factors by mechanisms independent of DNA synthesis and cell proliferation. Thus, growth factors can stimulate two independent sets of events: one that leads to an increase in DNA synthesis and cell proliferation and the other that results in a selective increase in the production of certain proteins, some of which may be secreted by the cell. In general, growth factors may be involved in controlling the expression of differentiated functions in many types of cells and tissues. These control mechanisms require the operation of specific interactions between the growth factor and the cell.

F. GROWTH FACTOR-LIKE PROTEINS

In addition to growth factors, normal cells and tumor cells are able to produce various substances that may be useful or required for their proliferation. Regulatory peptides such as bombesin, angiotensin, endothelin, atrial natriuretic factor, vasoactive intestinal peptide, and bradykinin exhibit growth factor-like properties. Other putative growth factors have been only partially identified. Immature *Xenopus* oocytes contain a growth factor-like protein with activity in homotypic and heterotypic cells that is distinct from known growth factors.[178] The ascitic fluid from ovarian cancer patients may contain unidentified growth factors capable of stimulating the proliferation of fresh ovarian cancer cells and the ovarian cancer cell line HEY through the activation of phosphoinositide metabolism and intracellular Ca^{2+} mobilization.[179] A rat hepatoma cell line established in serum-free medium secretes into the medium an unidentified growth factor, FF-GF, which exhibits little specificity in both species and organs.[180,181] Many other substances with growth factor-like properties may be produced by tumors, but they have not been well characterized.

Mouse fibroblasts (BALB/c 3T3 cells) stimulated with either fetal calf serum or growth factors such as PDGF and EGF express an RNA coding for proliferin, a protein with structural homology to prolactin.[182] Proliferin, prolactin, growth hormone, and chorionic somatomammotropin constitute a family of related proteins that seems to have evolved from a common precursor by gene duplication mechanisms.[183] Proliferin is a 28-kDa protein composed of 224 amino acids, including a hydrophobic signal sequence of 29 amino acids.[184,185] It is secreted as a glycoprotein lacking the hydrophobic signal peptide.[186] Proliferin is abundantly produced in the mouse placenta.[187] Several copies of proliferin-related DNA sequences have been detected in the mouse genome. These sequences encode distinct forms of proliferin-related proteins (PRPs) that are expressed in tissue culture and placental cells.[188] A PRP detected in mouse placenta is not present in proliferating BALB/c 3T3 cells in culture.[185] PRP synthesis is restricted to the basal zone of the mouse placenta and PRP is present in the maternal plasma of mice through the latter half of gestation, consistent with its acting as a hormone.[189] The normal functions of PRPs are unknown. Proliferin is apparently identical with a protein called mitogen-regulated protein, which is secreted by certain immortal murine cell lines stimulated with serum or growth factors such as EGF or FGF.[190,191] In serum-stimulated BALB/c 3T3 cells, proliferin mRNA appears in the pre-S phase of growth stimulation and reaches its peak level at the start of DNA synthesis. In contrast to the rapid expression of the c-*fos* and c-*myc* genes, the serum-inducible expression of mouse proliferin genes in BALB/C 3T3 cells depends on both protein synthesis and an extended presence of serum in the medium.[192] Transcription of the proliferin gene occurs in serum-starved resting cells, but these transcripts are not processed to mature proliferin mRNA. DNA sequences upstream of the proliferin gene may be responsible for the serum regulation of proliferin gene expression.[193] Proliferin expression is also regulated at the posttranscriptional level in cultured mouse fibroblasts.[194] The high level of expression of the proliferin gene in certain growing cells in culture suggests that it may serve as an autocrine factor in nonplacental cells *in vitro*, perhaps related to the immortality of these cells. Compounds with an ability to promote morphological transformation of mouse C3H/10T1/2 share specific effects on proliferin gene expression.[195] The function of proliferin is not only related to cell proliferation but also to differentiation. The protein is actively synthesized in multipotential 10T1/2-derived myogenic cell lines and in nondifferentiating variants derived from a myoblast cell line, but not in differentiated myoblasts. Introduction of a proliferin expression construct into muscle-and 10T1/2-derived myoblasts results in the generation of cell lines that are no longer myogenic or that only partially differentiate.[196] The mechanism of action of proliferin in cell proliferation and differentiation is unknown. A simple correlation between proliferin protein expression and cell differentiation may not exist. Although expected to function *in vivo* as a hormone due to its homology to prolactin and growth hormone, proliferin does not appear to act through a cell surface receptor.

G. PROLIFERATION-ASSOCIATED CELLULAR PROTEINS

The functions of many intracellular or secreted proteins detected in proliferating cells are little understood. A proliferation-related cytosolic phosphoprotein, which is induced in lymphocytes following mitogenic stimulation, is represented by the 149-amino acid protein, oncoprotein 18 (Op-18), also called stathmin and prosolin.[197] Op-18 may have a role in intracellular signal transduction and is expressed in markedly increased amounts in acute leukemia cells, which is due to increased RNA transcription from a structurally unaltered gene.

A human nuclear basic phosphoprotein of 54 to 55 kDa, called dividin, shows a rate of synthesis that is increased during the S phase of the cycle.[198] In growth-arrested cells, stimulation with serum induces dividin synthesis, which is first detected late in G_1 near the G_1/S transition border, reaches a maximum in mid- to late S phase, and declines thereafter. No detectable synthesis of dividin is observed in growth-arrested cells. The rate of synthesis of dividin correlates directly with the proliferative state of both normal and transformed cells, which suggests a role for this protein in events leading to DNA replication and cell division. The synthesis of dividin, as that of cyclin, may be an obligatory event in the mitogenic response. However, the physiological role of dividin is unknown.

A protein that may have a role in the common pathway leading to DNA replication and cell division has been termed progressin.[199] This 33-kDa protein was identified in proliferating human cells of epithelial, fibroblast, and lymphoid origin. Progressin synthesis occurs almost exclusively during the S phase of the cycle in cells such as transformed human amnion cells. Increased synthesis of progressin is first detected late in G_1, at or near the G_1/S transition border, reaches a maximum in mid- to late S phase, and declines thereafter. Contrary to histones, progressin synthesis is not coupled to DNA replication. No synthesis of progressin was detected in nonproliferating cells such as human MRC-5 fibroblasts and epidermal basal keratinocytes. Elevated, but variable levels of progressin have been found in proliferating normal fibroblasts and transformed cells of fibroblast, epithelial, and lymphoid cell origin. The normal function of progressin is unknown.

Serum stimulation of quiescent 3T3 cells can induce an increased synthesis of a group of secreted polypeptides of 45 kDa (p45 or 45K).[200] Synthesis of 45K proteins depends on transcription, and 45K expression may be controlled by a labile repressor whose activity depends on factors contained in serum. The level of 45K induction corresponds to the mitogenic capacity of the stimulant. Among growth factors, PDGF and FGF act as strong inducers of 45K, whereas EGF is a weaker inducer. The functions of 45K proteins are unknown but they could be components of the extracellular matrix.

A nuclear antigen associated with proliferation, detected by the monoclonal antibody Ki-67, is expressed in cycling cells and is not found in quiescent cells.[201,202] Antigen Ki-67 is distinct from cyclin, and its expression is highest in cells during the G_2 and M phases of the cycle.[203] The protein recognized by the Ki-67 antibody is synthesized predominantly during the S phase of the cycle.[204] Production of Ki-67 is higher in the latter period of the S phase than in the first half. The antigen is expressed in over 95% of the HeLa cells in an exponentially growing condition. Ki-67 can be used for estimation of the growing fraction of cell populations. Immunohistochemical analyses with Ki-67 may be useful for evaluating the proliferative activity of tumor tissues such as breast carcinoma, and the results correlate with prognosis.[205,206] Flow cytometric analysis of staining for another proliferative marker, Ki-S1, may also be useful for the clinical evaluation of human breast cancer.[207]

Serum stimulation of quiescent 3T3 cells immediately induces the synthesis of a set of cytoplasmic basic proteins that are absent in growing cells, including polypeptides of 27 kDa (p27), 35 kDa (p35), 38 kDa (p38), and 69 kDa (p69).[208] The induction of these proteins is a primary effect of serum factors at the level of gene expression. The levels of mRNAs coding for these proteins are markedly increased in serum-stimulated cells. The p27, p35, p38, and p69 proteins are synthesized during the G_0/G_1 transition, but their precise role is unknown. The serum factors responsible for the induction of these proteins have been identified only in part. PDGF and FGF strongly induce p35 and p69, but are only weak inducers of p27 and p38.

H. MECHANISMS OF REGULATION OF CELL PROLIFERATION BY GROWTH FACTORS

The mechanisms of regulation of cell proliferation by growth factors are complex and include the specific interaction of the factor with the receptor on the cell surface and the action of mediators and modulators such as prostaglandins and polyamines, as well as changes in cyclic nucleotides, ion fluxes and subcellular distribution, phosphoinositide metabolism, and cellular protein phosphorylation. Each growth factor may use preferentially one or more of these signaling pathways in different types of cells. The final mitogenic response may depend on the activation of specific gene expression.

Different growth factors may have synergistic, additive, or opposite effects when they act together on a given type of cell. The mitogenic effect of IGF-I, EGF, and PDGF on cultured human fibroblasts is additive, suggesting that the three peptides have unique mechanisms of action and/or that they each act at different points in the cycle.[209] In general, the state of the cell in relation to the phases of the cycle is crucial for its response to growth factor stimulation. However, the mitogenic action of some growth factors and the antiproliferative effects of interferon on vascular smooth muscle (VSM) cells and endothelial cells may be independent of cell cycle events.[210]

The regulation of DNA synthesis and cell proliferation by growth factors is associated with qualitative and/or quantitative changes in certain nuclear proteins, including transcription factors. Such factors regulate cell growth, development, and differentiation through regulation of gene expression by binding to a specific DNA site or set of sites. Transcription factors include proteins containing HTH, leucine zipper, and zinc finger domains.[211] The products of *myc*, *myb*, *fos*, *jun*, *ski*, *erb*-A, and *ets* proto-oncogenes are nuclear proteins and their functions are related to the control of DNA functions such as replication and transcription. Certain nuclear oncoproteins are transcription factors that are regulated through their phosphorylation and/or dephosphorylation at specific serine/threonine and/or tyrosine residues.[212]

1. Gene Expression During the Cell Cycle

The levels of expression of many genes show variation according to the different phases of the cycle, and mitogenic stimulation of cells with serum or growth factors may result in the specific activation of gene expression.[213-217] A number of genes, called immediate-early genes or early growth-regulated genes, are rapidly induced in the G_0/G_1 transition or the G_1 phase of the cell cycle after the stimulation of cell proliferation, independently of new protein synthesis. The proteins encoded by immediate-early genes include factors involved in the regulation of transcription such as the nuclear oncoproteins Myc, Myb, Fos, Jun, and Ets, as well as secreted proteins that are involved in autocrine or paracrine functions and certain structural proteins such as actin. These proteins may play an important role in the mitogenic response to growth factors. Other genes, called late growth-regulated genes, are secondarily expressed at the G_1/S boundary or during the S phase in the stimulated cells after variable periods of time. Prototypes of the latter genes are those coding for proteins of the DNA-synthesizing machinery of the cell such as thymidine kinase and DNA polymerase-α, as well as the genes encoding cyclin and other proteins involved in the regulation of cell division. However, the levels of DNA polymerase-α may not regulate the entry of cells into S phase, because they are similar in the G_1 and S phases of the cell cycle.[218] Another gene that is induced after several hours of serum stimulation in quiescent murine fibroblasts codes for calcyclin, a protein that may function as a calcium-modulated effector regulated by phosphorylation.[219] Concentration of calcyclin mRNA is highest in organs containing proliferating cells. A gene directly involved in the acquisition of growth factor independency in rat cells infected by chronic retroviruses — designated growth factor independence-1 (*Gfi*-1) gene — begins to be expressed 12 h after mitogenic stimulation of lymphoid cells, suggesting that it may be involved in late events occurring in G_1 to S phase transition.[220] The Gfi-1 protein contains six zinc fingers in its carboxy-terminal region, suggesting that it is involved in the regulation of transcription. Expression of Gfi-1 in adult animals is restricted to thymus, spleen, and testis.

The entry of cells into S phase is regulated by multiple molecular mechanisms, but the relative importance of these mechanisms in cell division is not understood. Expression of early or late cell cycle-associated genes in mitogenically stimulated cells does not necessarily mean that these genes are directly involved in the regulation of the cell cycle. Many of the early and late growth response genes are common to all types of eukaryotic cells, but others may depend on the cell type. Mitogen-stimulated cultured cells such as lymphocytes and fibroblasts have been used most frequently for the identification of growth response genes. A convenient system for the identification of these genes *in vivo* is represented by the regenerating liver, because in adult animals the liver cells are quiescent but exhibit a rapid and marked growth response to the stimulus represented by partial hepatectomy.

Differential screening and cross-hybridization of cDNAs from a library prepared from quiescent NIH/ 3T3 mouse fibroblasts showed that at least 82 distinct mRNAs are rapidly induced by serum, thus corresponding to the expression of immediate-early genes.[221] The expressed mRNAs include those encoding Myc, Fos, and Jun proteins. In another study, approximately 100 immediate-early genes were identified in serum-treated NIH/3T3 fibroblasts and over 70 in regenerating rat liver and insulin-treated Reuber H-35 hepatoma cells.[222] It is thus clear that stimulation of cell growth by the complex mixture of hormones, growth factors, and other mitogens contained in serum results in the induction of a complex genetic program involving a large number of genes. The abundance of immediate-early genes and the highly varied pattern of their expression in different types of cells suggest that the tissue specificity of the proliferative response is not due to expression of a few cell type-specific genes but, rather, the response may arise from the particular set of immediate-early genes expressed in a given tissue or in response to a given growth stimulus.

Identification of cell cycle-specific genes can be achieved by means of cell cycle phase-specific cDNA libraries. Only a few genes are known whose cell cycle phase-specific expression is unequivocally proven in mammalian cells. The most prominent cycle phase-specific genes are represented by late growth-regulated genes that encode S-phase enzymes involved in the synthesis of DNA precursors (thymidilate synthase, thymidine kinase, dihydrofolate reductase, thymidilate kinase, deoxycytosine monophosphate deaminase, and ribonucleotide reductase). The level of thymidine kinase is subjected to cell cycle regulation in human cells, with barely detectable levels in G_1 phase and up to 20-fold increases in M phase.[223] The low level of thymidine kinase in G_1 is due primarily to the specific degradation of the protein during cell division, and residues near the carboxyl terminus of the enzyme are essential for its specific degradation at mitosis.

The normal function of cell cycle-associated genes is frequently unknown. A gene, cMG1, whose expression is transiently activated in growth factor-stimulated epithelial cells of the rat intestine, was identified by nucleic acid hybridization screening.[224]

Analysis of cDNAs corresponding to cMG1 indicates that it encodes a protein of 338 amino acids and shows no similarities to other genes with the exception of TIS11, a gene induced by phorbol ester in 3T3 mouse fibroblasts. The two genes, cMG1 and TIS11, appear to code for early response gene products. A member of the set of immediate-early genes expressed in mouse 3T3 cells, *cyr61*, encodes a polypeptide of 379 amino acids which contains several cysteine residues.[225,226] The Cyr-61 protein is expressed minutes after cell stimulation with serum, PDGF, or FGF. Cyr-61 has a short half-life and is a secreted protein that is associated with the extracellular matrix, suggesting that it may function in cell-to-cell communication. The Cyr-61 protein binds heparin and shows similarities with the product of the *int*-1 proto-oncogene. Another immediate-early gene, *3CH134*, whose transcription is rapidly and transiently stimulated by serum growth factors, encodes a 367-amino acid protein of 40 kDa that does not share similarity with any known protein.[227] The products of late growth-regulated genes have varied functions in addition to the participation of some of them in the DNA replication machinery. Two human genes required for G_1 progression were identified by their ability to complement the mutations of *ts* cell cycle mutants isolated from the BHK-21 Syrian hamster cell line.[228] These two genes were assigned to human chromosome regions 7cen-q35 and 8q21, respectively, and one of them corresponds to the gene encoding asparagine synthetase. The function of the other gene is unknown.

Certain genes are preferentially expressed in quiescent cells and are negatively regulated by serum or purified growth factors. The expression of one of them, gas-1, is downregulated by serum in NIH/3T3 cells transfected with an activated H-ras oncogene.[229,230] The amount of gas-1 mRNA is higher in tumor cells than in normal cells, but gas-1 gene expression is not sufficient to maintain the cells in a quiescent state. The normal function of the Gas-1 protein is unknown. Another gene codes for a 57-kDa protein called statin, which is located in the nucleus of nonproliferating and senescent human fibroblasts and is not present in growing or transformed cells.[231,232] Statin disappears when arrested cells are stimulated with serum and re-enter the cell cycle.

The enzyme DNA polymerase-α plays a key role in the replication of the eukaryotic genome. Regulation of human DNA polymerase-α gene expression occurs at the transcriptional level.[233] During the mitogenic activation of quiescent cells, the steady-state RNA levels, rate of synthesis of nascent protein, and enzymatic activity of DNA polymerase-α exhibit a marked increase prior to the peak of DNA synthesis. Transcription of the DNA polymerase-α gene and other DNA replication-associated genes is reduced in quiescent cells but increases markedly upon mitogenic stimulation. However, the constitutive expression of DNA polymerase-α during the cell cycle suggests that a basal level of the enzyme is required for the maintenance of cellular DNA. In addition to DNA replication, DNA polymerase-α may have a role in DNA repair.

Nuclear proteins that are differentially expressed in the cell cycle may play a role in cell cycle control. The genes encoding both basic and acidic nuclear proteins may exhibit differential expression during the cell cycle. Histones are basic chromosomal proteins rich in arginine, lysine, and histidine and are involved in the packaging of newly replicated DNA into chromatin. Histone gene expression is regulated during the cell cycle.[234,235] The histone genes are expressed in a coordinate manner and are tightly coupled with DNA replication. Histone genes whose regulation is associated with the cell cycle include those encoding histones H2A, H2B, H3, and H4. A human H4 gene shows cell cycle-dependent changes in chromatin structure that correlate to its level of expression.[236] Specific histones are involved in the regulation of gene expression. Nucleosomal cores and histone H1 are involved in the regulation of transcription by RNA polymerase II.[237] The linker histone H5 may play a role in the regulation of chromatin function and DNA replication.[238] Histone expression and function are regulated at the transcriptional and posttranscriptional levels. The histones can undergo posttranslational covalent modifications such as acetylation, phosphorylation, methylation, and ADP-ribosylation, and these modifications may be correlated to changes in genomic regulation and chromatin function. Acetylation of the histones that organize the nucleosome core particle (H2A, H2B, H3, and H4) is a ubiquitous posttranslational modification found in all animal and plant species. The acetylation occurs at specific lysine residues located on the amino-terminal region of the core histone molecules and could influence biochemical events such as transcription, replication, and DNA packaging through the cell cycle as well as DNA repair.[239] Induction of differentiation of Friend erythroleukemia cells is associated with altered expression of the amount of histone fractions as well as with posttranslational modifications of histones.[240] Downregulation of the cell cycle-regulated histone genes may be accompanied by a reciprocal increase in the expression of a distinct subset of the histone genes that are not coupled with DNA replication.[241] There are structural differences between cell cycle-regulated and cell cycle-independent histone genes.

Genes encoding nonhistone nuclear proteins can also exhibit cell cycle-associated changes in their levels of expression. The gene for the human nonhistone chromosomal protein HMG-17 shows cell cycle-dependent expression, with a sharp increase in the mRNA levels at the beginning of the S phase, followed by a second wave of transcription that occurs later in the same phase.[242] An acidic nuclear protein of 55 kDa, termed PSL, is synthesized mostly during the S phase of the cycle and is transiently associated with chromatin.[243] The function of PSL is unknown, but it appears to be involved in certain steps of the granulocyte differentiation process. Expression of two nonhistone nuclear proteins of 85 kDa (p85) and 110 kDa (p110) is associated with the cell cycle in stimulated human lymphocytes.[244]

As mentioned earlier, the fact that certain genes are expressed in a cell cycle-dependent manner does not necessarily mean that they are involved in cell cycle regulatory phenomena. Increased levels of G_1-specific RNAs transcribed from one or more genes in growing cells, including tumor cells, may not indicate overexpression of these genes but may instead simply reflect the fraction of proliferating cells.[245] It is thus difficult to recognize which of these genes are directly responsible for the induction of DNA synthesis. Over 3300 proteins can be distinguished in Swiss 3T3 mouse cells by two-dimensional electrophoresis on giant gels; 34 of these proteins are consistently induced more than threefold after stimulation by serum or growth factors.[246] An additional 30 inductions are variably present, and some of these inductions are inhibited by dexamethasone. Only a few of these proteins show a constant linear relationship to DNA synthesis, suggesting that they play an important role in early control of the cell cycle.

Cloning and characterization of genes that complement mutations of *ts* cell cycle mutants may represent a more direct approach to the identification of genes whose expression is essential for DNA synthesis and cell proliferation than the usual approach based on isolating genes whose expression is regulated differentially through the cycle. A human gene that can complement the *ts 11* mutation of BHK 21 Syrian hamster cell line (a mutation in which the progression through the G_1 phase is arrested at 39.5°C) has been characterized.[247] The *ts 11* gene encodes an mRNA of 2 kb that is expressed in all human, hamster, and mouse cell lines tested. A cDNA of this gene was isolated and sequenced. The gene encodes a protein of 540 amino acids with no homology to other proteins. Experiments with serum-synchronized cells indicate that expression of the *ts 11* gene, which is necessary for G_1 phase progression, is itself cell-cycle regulated, being induced in mid-G_1. A human genomic sequence that can correct the *ts* phenotype of a G_1-specific cell cycle mutant (*ts 13*) was isolated and cloned, but the mechanism by which it is able to correct the cell cycle defect could not be characterized.[248] A *ts* hamster cell mutant with a G_1-specific arrest at a nonpermissive temperature exhibited a defect in glycoprotein synthesis that would be associated with an altered Ca^{2+} distribution between endoplasmic reticulum and cytosol.[249] It is clear that further studies are required for a better characterization of cellular genes directly involved in the regulation of the cell cycle. The recent identification of cyclins and cdc2-related protein kinases represents an important advance in our knowledge of the mechanisms directly implicated in the regulation of the cell cycle.

2. Cyclins

Cyclin is an acidic nuclear protein of 36 kDa that is involved in the control of DNA synthesis and cell proliferation. It was first detected in fertilized sea urchin eggs and clam oocytes and was found later also to be present in higher eukaryotes.[250,251] Cyclin is identical with the protein called proliferating cell nuclear antigen (PCNA).[252-254] In embryonic *Xenopus laevis* cells, cyclin expression occurs in the period of the cycle dedicated to DNA synthesis, and it has been suggested that cyclin may be a useful marker of cells engaged in S phase and could be an alternative to thymidine labeling.[255] However, in HeLa cells, cyclin was found to be present in constant amounts during the cycle and its synthesis is not tightly linked to DNA synthesis.[256] Notwithstanding, the amount of cyclin that binds to chromatin increases in HeLa cells to a maximum during the peak of the S phase of the cycle. Studies on frog eggs showed that cyclin synthesis can induce mitosis and allows the progression from mitosis to the next interphase.[257] The activation of a specific protein, the maturation promoting factor (MPF), induces cells to enter mitosis and meiosis, and cyclin is directly required for the generation of MPF.[258] Injection of a human antibody against cyclin into unfertilized *Xenopus* eggs results in inhibition of DNA replication.[259] It may be concluded that cyclin plays a pivotal role in the control of mitosis. In addition, cyclin may be involved in DNA repair, as shown by UV irradiation of quiescent cultured human fibroblasts.[260] The cellular action of cyclin is exerted through its association with the protein kinase cdc2 in the form of the MPF complex, which is required for G_2 to M transition during the cell cycle.

The general biological importance of cyclin is indicated by its conservation in evolution. Mammalian cyclins form a family of proteins that play a key role in the regulation of the cell cycle by functioning as integrators of growth factor-mediated signals during the G_1 phase of the cycle, before the initiation of DNA synthesis.[261] Cyclin-related proteins are present in normal and transformed cells from all animal species.[262-264] A cyclin-related gene is present in higher plants and a 35-kDa protein can be detected with monospecific anticyclin antibodies in the nucleus of plant cells.[265] The importance of a conservation of the DNA replication apparatus is indicated by the functional similarity of human cyclin and the replication factors A and C (RF-A and RF-C) of bacteriophage T4.[266] Biochemical analyses with highly purified RF-C and human cyclin showed functions that are completely analogous to the functions of bacteriophage T4 DNA polymerase accessory proteins.

Human epidermal basal cells synthesize little cyclin, while SV40-transformed keratinocytes synthesize cyclin constitutively.[267] Induction of cyclin in serum-stimulated cells may be independent of DNA synthesis and cell transformation, but it precedes DNA synthesis and correlates with cell proliferation.[268,269] Cyclin synthesis is not triggered by DNA replication, but changes in the nuclear distribution of cyclin occur in relation to the S phase of the cycle.[270] Immunofluorescence analysis of synchronously growing transformed human amnion cells using autoantibodies specific for cyclin show changes in the nuclear distribution of cyclin during the S phase of the cycle.[271] Individual nuclei in polykaryons produced by polyethylene glycol-induced fusion of transformed human amnion cell populations are able to control cyclin distribution and DNA synthesis in spite of the fact that they share a common cytoplasm.[272] Cyclin is required as a cellular factor for SV40 DNA replication *in vitro*.[273] Cyclin remains stable during the transition from a growing to a quiescent state of the cycle, and a fraction of cyclin is tightly associated with DNA replication sites during the S phase.[274]

A gene coding for cyclin has been mapped to human chromosome 20, at region 20p12,[275] and cDNA clones for this gene have been isolated and sequenced.[276,277] The promoter of the human cyclin gene is bidirectional.[278] The predicted amino acid sequence of human cyclin indicates that it is composed of 261 residues with a high content of acidic vs. basic amino acids. Cyclin may function as an auxiliary protein of DNA polymerase-δ and may enable this enzyme to utilize template/primers containing long stretches of single-strand template.[279,280] Studies with polyclonal and monoclonal antibodies to cyclin suggest a central role for DNA polymerase-δ in cellular DNA replication.[281] Exposure of exponentially growing BALB/c 3T3 mouse fibroblasts to antisense oligodeoxynucleotides to cyclin can completely inhibit cellular DNA synthesis and cell proliferation, indicating that cyclin is required for both DNA synthesis and cell cycle progression.[282]

Expression of the cyclin gene is under the control of genes coding for homeobox-containing proteins. Determination of the complete sequence of the *Drosophila* cyclin gene, including its 5'- and 3'-flanking regions, showed the existence of a cluster of putative binding sites for homeodomain-containing gene products.[283] Moreover, footprint analysis indicated that *Drosophila* homeodomain proteins can specifically bind to these sites. These results implicate a close relationship between the regulation of developmentally programmed cell differentiation and that of cell proliferation. However, little is known about the molecular mechanisms involved in controlling cyclin gene expression. The cyclin gene promoter may be active in mammalian serum-deprived G_0 cells, suggesting that the key to serum dependence of cyclin mRNA levels may reside in the exons, introns, or other sequences of the cyclin gene, exclusive of the proximal 5'-flanking sequences.[284] Studies with transfection of the human cyclin gene into mouse fibroblasts suggest that factors involved in serum-induced regulation of cyclin gene expression include both transcriptional and post-transcriptional components.[285]

Cloning, sequence, and expression of a yeast analog of mammalian cyclin have been reported.[286,287] Yeast cyclin interacts with DNA polymerase III, an enzyme that is the yeast counterpart of mammalian DNA polymerase-δ.[288] Cyclin is synthesized at relatively high levels in late G_1 and early S phase of the cell cycle in *Saccharomyces cerevisiae*, but DNA synthesis is not required for its synthesis.

A number of genes showing variable degrees of homology to the classic cyclin/PCNA gene have been detected in mammalian and nonmammalian cells. These genes encode members of the cyclin family of proteins, which include cyclins A, B, C, D1, D3, and E. At least some of these cyclin types (cyclins A and E) may satisfy the criteria for a labile protein involved in the control of the restriction point (R point) of the cell cycle.[289] Two distinct cyclins, A and B, exhibit differential variation during the cycle of human cells.[290] The two types of cyclin have a different location in the cell, with cyclin A being predominantly found in the nucleus and cyclin B in the cytoplasm.[291] Moreover, cyclin A and cyclin B undergo cell cycle-dependent transport from the cytoplasm to the nucleus, suggesting different functions for the two proteins. Human cyclin A mRNA and protein levels vary during the cell cycle and increase and decrease in advance of cyclin B levels.[292] Cyclin A is required for the onset of DNA replication in mammalian fibroblasts as well as during S phase in normal epithelial cells.[293,294] Cyclin B forms a functional complex with the 34-kDa cdc2 kinase (p34^{cdc2}), whereas cyclin A may act in association with a protein kinase of 33 kDa, the cdk2 kinase (p33^{cdk2}).[295] The cdk2/cyclin A complexes may have a unique role in cell cycle regulation.

The putative human proto-oncogene *prad-1/bcl*-1 encodes cyclin D1, which is a protein of 295 amino acids found at high levels in the G_1 phase of the cell cycle.[296] The human gene encoding a form of cyclin closely related to cyclin D1, cyclin D3, was cloned from a placental cDNA library by cross-hybridization with *prad*-1. The two cyclins, D1 and D3, play distinct roles in cell cycle control. Another cyclin, cyclin E, was discovered in the course of a screen for human cDNAs capable of rescuing a deficiency of G_1 cycling function in yeast.[297] The amounts of cyclin E and an associated kinase activity show periodic fluctuations during the human cell cycle, being maximal in late G_1 and early S phases. The complex between cyclin E and the cdk2 kinase is a human G_1-S phase-specific regulatory protein kinase. It is thus clear that there is a multimember family of human cyclins, but the precise role of each one of these members in cell cycle control is unknown.

A cyclin family exists in other mammalian species, including the mouse. Three cyclin-like genes (*CYL*) have been identified in murine macrophages, and two of these genes are regulated by CSF-1 during the G_1 phase of the cycle.[298] CSF-1 deprivation during G_1 leads to rapid degradation of *CYL*-encoded proteins

($p36^{CYL}$) and correlates with failure to initiate DNA synthesis. However, after entering S phase macrophages no longer require CSF-1 and can complete cell division without expressing *CYL* genes. During G_1, $p36^{CYL}$ is phosphorylated and associates with the cdc2 protein kinase. A family of cyclins is also present in amphibians. In addition to cyclin A, *Xenopus laevis* expresses two closely related B types of cyclins, B1 and B2. These two forms of cyclins are phosphorylated on serine residues in the eggs of the animal by the cdc2 kinase or MAP kinase, but such phosphorylations may not be required for cyclin function.[299]

Cyclins and their associated protein kinases may form functional complexes with different cellular proteins and may be involved in the phosphorylation and activation or inactivation of regulators of the cell cycle such as the RB tumor suppressor protein.[300] The E2F transcription factor can form complexes with cyclin A, and E2F-containing complexes are regulated during the cell cycle.[301]

The role of cyclins in oncogenesis is poorly understood but, in general, the steady-state cyclin mRNA level in a tissue is related to the rate of proliferation of the tissue and some cyclins may display oncoprotein-like activities. As mentioned, the bcl-1/prad-1/PRAD1 proto-oncogene encodes cyclin D1. This gene is located on human chromosome region 11q13 and is overexpressed in various types of human tumors with abnormalities at this region.[302] The gene is coamplified with the FGF-related gene hst-1 in some squamous esophageal cancers.[303] The majority of human centrocytic lymphomas (CD5-positive B-cell neoplasms) exhibit rearrangements at the bcl-1/prad-1/cyclin D1 region on chromosome 11q13.[304] Proviral insertions near the bcl-1/prad-1/cyclin D1 gene occur in some mouse lymphomas and may represent a parallel for bcl-1 gene-associated translocations observed in human B-cell neoplasms.[305] Amplification and/or overexpression of the gene encoding cyclins A, B, or E have been detected in some human breast cancer cell lines and primary breast carcinomas, suggesting that cyclins may function redundantly in cancer cells.[306] Amplification of cyclin genes has been detected in some colorectal carcinomas.[307] Altered expression of cyclins may contribute, in principle, to the origin and/or expression of a transformed phenotype. Interactions between different cyclins and the products of tumor suppressor genes may be important for the control of both normal and tumoral cellular proliferation. Suppression of cell growth induced by wild-type p53 tumor suppressor protein may be accompanied by selective downregulation of cyclin expression.[308] Expression of a construct encoding wild-type RB protein in a human osteogenic sarcoma cell line that lacks the full-length nuclear RB product resulted in growth arrest at the G_0 or G_1 phase of the cycle, but this arrest could be overridden by cotransfection of cyclin A or E.[309] Cyclin A and/or E may be able to direct the cell into S phase of the cycle.

3. The cdc2 and cdk2 Protein Kinases

The *cdc2* gene of *Schizosaccharomyces pombe* and its homolog, the *CDC28* gene of *Saccharomyces cerevisiae*, play an important role in cell cycle control in yeast. The two yeast genes are functionally interchangeable. The product of the *cdc2* gene is required in yeast both at G_1 and G_2 phases of the cell cycle for the initiation of mitosis. The yeast *cdc2* gene encodes a 34-kDa protein kinase, $p34^{cdc2}$ or cdc2, whose function is regulated by phosphorylation on tyrosine as well as on serine/threonine residues, in particular the Tyr-15 and Thr-14 residues.[310,311] Entry into mitosis in fission yeast is elicited by an increase in cdc2 kinase activity, which requires the cdc25 component of the mitotic regulatory gene network.[312] At mitosis, the level of cdc2 protein phosphorylation decreases. Phosphorylation of the yeast cdc2 protein occurs in Tyr-15, a residue situated within an ATP-binding domain. Substitution of Tyr-15 by phenylalanine advances the yeast cell prematurely into mitosis. Tyrosine phosphorylation/dephosphorylation is directly involved in regulation of cdc2 function. Activity of the yeast cdc2 kinase depends on a balance between the cdc25 activator protein and an inhibitor, Wee-1, which phosphorylates cdc2 on Tyr-15. The functional importance of the *cdc2* gene product and its auxiliary proteins is indicated by their presence in a wide diversity of animal and plant species. A functional *cdc2* gene is present in higher plants such as the pea, where its 34-kDa cdc2 product is involved in the regulation of cell division and protein kinase activity related to histone H1.[313]

A homolog of the yeast *cdc2* gene exists in vertebrates, including humans, and is involved in cell cycle control.[314,315] Moreover, homologs of the fission yeast mitotic inducer cdc25 are present in vertebrate species and are represented by a family of protein phosphatases, termed CDC25A, -B, and -C, which exhibit dual specificity.[316] The CDC25C protein phosphatase is implicated as a positive regulator of entry into mitosis, whereas CDC25A appears to be required for progression through the M phase of the cycle. The vertebrate CDC25 protein phosphatases regulate the activity of proteins associated with the regulation of the cell cycle, in particular, the activities of cdc2-related kinases. The cdc2 protein kinase is present in the cytoplasm and the nucleus of mouse FM3A cells, and its activity in the nuclear fraction increases in the G_2/M phase of the cycle.[317] Despite a constant level of cdc2 kinase in mouse fibroblasts, translation of the cdc2 protein is activated at the G_1/S transition and is inactivated at the G_2/M boundary.[318] The accumulation of newly synthesized cdc2 protein is accompanied by a concurrent mechanism of degradation, resulting in the old pool of cdc2 being largely replaced each round of the cell cycle. Expression of cdc2 mRNA and protein in quiescent young human and hamster fibroblasts is stimulated in culture by serum, and the stimulated cells go through DNA synthesis and mitosis. In contrast, serum stimulation of senescent cells does not result in increased cdc2 expression.[319,320] The human senescent cells also exhibit a deficiency of cyclin A and cyclin B mRNA. These deficiencies may be relevant to the lack of DNA synthesis and mitosis in the senescent cells.

Analysis of *Xenopus laevis* oocytes has greatly contributed to a better knowledge of the biochemistry of cell division.[321] The cdc2 protein of *Xenopus laevis* eggs is an essential component of the MPF, a cytoplasmic inducer of mitosis that is composed of cdc2 and cyclin.[322] The kinase activity of the cdc2 protein oscillates during the cycle of *Xenopus laevis* cells, and maximal activity correlates with the dephosphorylated state of the cdc2 protein.[323] The protein kinase encoded by the c-*mos* proto-oncogene is required for MPF activation and mouse oocyte maturation.[324] The c-Mos serine/threonine kinase directly phosphorylates the cyclin B2 component of MPF.

The cdc2 protein kinase is a key regulator of the eukaryotic cell cycle, and the cdc2 phosphorylation state varies according to the different phases of the cell cycle.[325] Four major sites of cdc2 protein phosphorylation are present in chicken cells: Thr-14, Tyr-15, Thr-161, and Ser-277. These residues are highly conserved in evolution. Thr-14 and Tyr-15 are phosphorylated maximally during G_2 and are abruptly dephosphorylated at the G_2/M transition. Phosphorylation of the Thr-14 and Tyr-15 residues inhibits cdc2 kinase activity. A specific type of human phosphatase, cdc25B, is expressed at the G_2/M boundary of the cycle and leads to the activation of the human cdc2 kinase by dephosphorylating both the Thr-14 and Tyr-15 cdc2 residues.[326] Thr-161 phosphorylation is required for cdc2 interaction with cyclin and its function during the cell cycle,[327] and dephosphorylation of Thr-161 is required for cdc2 protein kinase inactivation and normal anaphase.[328] Ser-277 dephosphorylation of the cdc2 protein peaks during G_1 and drops markedly as cells progress through the S phase of the cycle.

Certain structural components of cdc2 are essential for its efficient phosphorylation by an Src-related tyrosine kinase.[329] Coupling of mitosis to the completion of the S phase of the cycle occurs via modulation of the tyrosine kinase that phosphorylates the cdc2 protein.[330] Phosphorylation of cdc2 depends on the activity of a tyrosine kinase that is associated with cdc2 in the cytosol and modifies the Tyr-15 residue.[331] A substantial fraction of cytosolic cdc2 is hypophosphorylated, whereas the nuclear cdc2 protein kinase is hyperphosphorylated. The protein kinase involved in cdc2 phosphorylation is a 50-kDa human protein, Wee-1, which is a homolog of the yeast wee1 protein.[332] The human Wee-1 kinase phosphorylates the cdc2 protein on Tyr-15, which results in the inhibition of cdc2 kinase activity and suppression of mitosis.[333] Genes encoding proteins homologous to yeast CDC25 and Wee-1 are localized on human chromosome 11, at region 11p15.1-11p15.3.[334] Dephosphorylation of the cdc2 kinase depends on the activity of specific protein phosphatases, and one of these enzymes is represented by the cdc25 protein, which dephosphorylates tyrosine and threonine residues on cdc2 and regulates MPF activation in vertebrate and nonvertebrate species.[335] Okadaic acid, a potent inhibitor of protein phosphatases, can activate the cdc2 kinase and transiently induces in BHK21 cells a mitosis-like state with premature chromosome condensation.[336]

Freshly isolated human T cells express low levels of cdc2 mRNA and protein, but exposure of the cells to mitogens such as PHA results in increased cdc2 expression, coincident with G_1/S transition.[337] This induction may be interrelated with c-*myc* and c-*myb* gene expression. In human HeLa cells, the *cdc2* gene is maximally active during mitosis and its product is a nuclear and centrosomal protein that is required for mitosis.[338,339] The phosphotyrosine content and kinase activity of cdc2 in HeLa cells is subjected to cell cycle-associated regulation.[340,341] The cdc2 protein is dephosphorylated upon shifting of cells from exponential growth to quiescence and is rephosphorylated late in the G_1 phase, when cells are stimulated to re-enter the cycle.[342] Phosphorylation of cdc2 on tyrosine is regulated in the cycle of mouse fibroblasts, and dephosphorylation of cdc2 accompanies activation during entry into mitosis.[343]

The cdc2 protein forms a complex with cyclin, and this association contributes to the activation of the cdc2 protein.[344] The complex is phosphorylated in HeLa cells by tyrosine kinases such as the v-Fms oncoprotein.[345] Src-related tyrosine kinases may contribute to the regulation of the cell cycle through the phosphorylation of the cdc2 protein. In the fission yeast *S. pombe*, cdc2 is phosphorylated on Tyr-15, and dephosphorylation of this residue by a specific protein phosphatase triggers activation of the cdc2-cyclin complex, thereby regulating the initiation of mitosis.[346] In mammalian cells, microinjection of cdc2 protein induces marked changes in cell shape, cytoskeletal organization, and chromatin structure.[347] The cdc2 protein isolated from mouse ascites tumor cells would be a component of the RNA polymerase II complex, suggesting a role for cdc2 in the regulation of transcription.[348]

Cloning and analysis of the human *cdc2* gene showed that the predicted sequence of its protein product is very similar to that of the yeast *cdc2* gene.[349] Expression of the human cdc2 protein varies in a cell cycle-dependent manner.[350] Available evidence indicates that elements of the mechanism by which the cell cycle is controlled are highly conserved between yeast and humans. The yeast *cdc2* gene and its homologs in higher eukaryotes encode a perfectly conserved sequence of 16 amino acids that has not been found in any other protein. Microinjection of this oligopeptide into *Xenopus laevis* oocytes triggers a specific increase in the $[Ca^{2+}]_i$, suggesting that cdc2 may interact through its conserved peptide domain with some component of the calcium regulatory system.[351] The substrates for the cdc2 kinase include components of the mitotic apparatus as well as proteins involved in the control of cell division. Reorganization of intermediate filaments during mitosis is mediated phosphorylation of vimentin induced by cdc2.[352,353] Casein kinase II is phosphorylated by the cdc2 kinase *in vitro* and at mitosis.[354] Caldesmon, an actin- and calmodulin-binding protein that is dissociated from microfilaments as a consequence of mitosis-specific phosphorylation, is a substrate for the cdc2 kinase.[355,356] Other substrates of the cdc2 kinase are oncoproteins and tumor suppressor proteins. The c-Src kinase is a substrate for cdc2-induced phosphorylation on serine and threonine residues, which may sensitize it to a Tyr-527-specific protein phosphatase and contribute to its mitotic activation.[357] The c-Abl protein is phosphorylated on three sites during interphase and seven additional sites during mitosis, and this differential phosphorylation is determined by an equilibrium between cdc2 kinase and protein phosphatase activities.[358] The p53 tumor suppressor protein is phosphorylated by the cdc2 kinase *in vitro* and coprecipitates with cdc2 *in vitro* as well as *in vivo*.[359,360] The RB tumor suppressor protein, which exhibits cell cycle-associated variations in its state of phosphorylation, may also be a substrate of the cdc2 kinase.[361] Interactions between cdc2 and proto-oncogene and tumor suppressor gene products are probably of functional importance for normal cell growth control.

Protein kinase activity similar to that of cdc2 is shared by other enzymes that are involved in cell cycle control. A family of human cdc2-related kinases is represented by at least ten members.[362] In addition to the 34-kDa cdc2 kinase, which preferentially forms a complex with cyclin B, a protein kinase of 33 kDa, called cyclin-dependent kinase 2 (cdk2), forms a functional complex with cyclin A.[295] The cdk2 kinase forms complexes with cyclin E. The cdc2 and cdk2 kinases share 65% amino acid sequence identity. The protein encoded by the *cdk2* gene, also termed *Eg1* gene, is expressed during the cell cycle independently of cyclin and cdc2 kinase.[363] The cdk2/cyclin E complex operates in the G_1 to S phase transition and may have a unique role in cell cycle regulation of vertebrate cells. The *cdk2* gene is expressed earlier than the *cdc2* gene when cells are stimulated to enter the cycle.[364] In human cells, cdk2 and cyclin form complexes with the transcription factor E2F during the S phase of the cycle.[365] An important substrate for cdk2-cyclin complexes is the tumor suppressor protein RB. Available evidence indicates that the cdk2 protein kinase is involved in the induction of S phase and activation of DNA synthesis in human cells.[366] This effect of cdk2 may be mediated by RB protein phosphorylation and/or its association with the E2F factor.

In conclusion, control of the cell cycle is effected through the combined actions of proteins that are members of the cyclin and cdc2 families. Control of cell division at the G_2 to M phase transition depends on the activation of a protein kinase composed of a catalytic subunit, cdc2, whose levels are not altered during the different phases of the cycle, and a regulatory subunit, cyclin B, whose levels oscillate in synchrony with the cell cycle. Phosphorylation/dephosphorylation of the cdc2 kinase by Wee-1 and other kinases at specific tyrosine and threonine residues contributes to the regulation of its enzymatic activity. While phosphorylation of the cdc2 protein on Thr-14 and Tyr-15 inhibits its kinase activity, dephosphorylation at the same residues results in cdc2 kinase activation, coincident with the initiation of mitosis. The cdc2-cyclin B protein kinase complex remains active until the end of mitosis, when cyclin B is degraded and cdc2 is released as an inactive monomer. A second type of control involving the cdk2 kinase and cyclin E operates in human cells for regulation of the G_1 to S phase transition (Figure 1.3). Substrates for the cdk2 kinase include the RB tumor suppressor protein and the E2F transcription factor. The roles of other members of the cyclin and cdc2-related protein families in the regulation of the cell cycle remain to be elucidated, but multiple interactions between these proteins may be important for cell cycle control. Restimulation of growth factor-deprived cells to enter the cycle results in the accumulation of complexes between cyclin D and several cdk kinases (cdk2, cdk4, cdk5), and these complexes appear earlier in the G_1 phase than do the cdk2-cyclin E complexes, which are expressed mainly before the G_1/S phase transition. Proteins related to cdk2, including one called cdk3, may also participate in cell cycle control, but their precise roles are not understood at present.[362] It is clear that, although in the last few years there have been important advances in our knowledge of the mechanisms that are directly involved in the

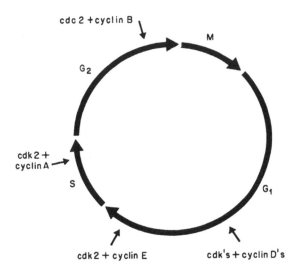

Figure 1.3 Control of the cell cycle by cyclins and cdc/cdk protein kinases.

control of the cell cycle in species ranging from yeast to man, further studies are required for a better characterization of the precise role of each type of control and for the mechanisms by which they are regulated by growth factors.

4. Transcription Factors

The regulation of cellular functions by growth factors, hormones, and other mitogens is exerted frequently at the level of gene expression through modification of the specific activities of transcription factors.[367-369] These factors are nuclear proteins involved in the control transcription initiation by DNA polymerase II through the recognition by them of specific DNA sequences. An enormous number of transcription factors have been isolated and characterized recently.[370] Some of these factors are represented by the products of proto-oncogenes encoding nuclear proteins such as Fos and Jun, which form homo- or heterodimers through leucine zippers and recognize the AP-1 binding site.[371-373] Other nuclear oncoproteins such as Myc, Myb, Ets, and Rel may also function as transcription factors. The specific activities of transcription factors are regulated by phosphorylation, glycosylation, and other posttranslational modifications.[374] Examples of growth factor-regulated transcription factors are mentioned next and others are discussed in chapters dedicated to specific growth factors.

Differential screening of a cDNA library prepared from BALB/c 3T3 murine cells stimulated with serum allowed the identification of a late growth-regulated gene, SC1, which is maximally expressed at the onset of DNA synthesis in human and rodent cells.[375] The SC1 gene encodes a 39-kDa protein that functions as a transcription factor.

The functional activity of the nuclear factor κ B (NF-κB), a ubiquitous transcription factor that can affect the expression of many genes, is regulated by a wide diversity of exogenous stimuli, including growth factors.[376] NF-κB is a homo-or heterodimer of two subunits of 50 kDa (p50) and 65 kDa (p65). The various subunit combinations of NF-κB bind to DNA with an extremely high affinity. A factor termed kBF1, which is identical to the 50-kDa DNA-binding subunit of NF-κB, displays extensive sequence homology with the nuclear v-Rel oncoprotein of the acute avian retrovirus REV-T as well as with the product of maternal effect gene *dorsal* of the fruit fly *Drosophila melanogaster*. The region of sequence similarity between the three proteins (NF-κB, Rel, and dorsal) is termed NRD domain and represents the minimal requirement for DNA binding and dimerization of these proteins. NRD defines a family of transcription factors. The p50 subunit of NF-κB can be considered as a helper subunit imposing a limited regulation of the *trans*-activating p65 subunit by increasing its affinity for DNA. It behaves in a manner similar to that of the Max protein, which is required for the gene *trans*-activating activity of the Myc protein. In nonstimulated cells, NF-κB is not detectable in the nuclear, cytosolic, or membrane fractions due to the presence of a specific inhibitor, I κ B (IκB). The putative proto-oncogene *bcl*-3 encodes an IκB protein.[377] IκB is specialized in negatively controlling the DNA-binding activities of NF-κB and Rel proteins. Expression of both NK-κB and IκB in different types of cells depends, in part, on the action of exogenous stimuli, including growth factors.

An 'immediate-early' gene that is transiently expressed during the G_0/G_1 transition in cells stimulated by serum or purified growth factors has been characterized as a member of a superfamily of ligand-binding transcription factors that includes the steroid and thyroid hormone receptors.[378-384] This gene, called Nur-77, NGFI-B, or N10, is represented by a 601-amino acid nuclear protein, which is rapidly induced in mouse fibroblasts by several mitogens including PDGF, FGF, EGF, and TPA. Expression of Nur-77 protein is regulated at both the transcriptional and posttranscriptional levels, including phosphorylation on serine residues. A human homolog of the rodent Nur-77 gene, ST-59, was cloned from LS-180 colon adenocarcinoma cells.[385] The ST-59 gene is rapidly induced by fetal calf serum through signaling pathways that include activation of both protein kinase C and protein kinase A.

A *ts* cell cycle mutant isolated from the Syrian hamster cell line BHK21 had a defect in G_1 phase progression and was unable to enter S phase at the nonpermissive temperature, but once it entered S phase, DNA replication progressed normally. By using this mutant as the recipient of DNA-mediated gene transfer, a gene located on human chromosome region Xq11-q13 was isolated and cloned.[386-388] The gene, cell cycle gene 1 (*CCG1*), encodes a 210-kDa nuclear DNA-binding protein whose carboxy-terminal part contains the target consensus sequence for casein kinase II and is phosphorylated by this enzyme *in vitro*. The CCG1 protein may have an important role in the transition of cells from the G_1 to S phase of the cycle.

The transcription factor Oct-3 was identified in undifferentiated embryonal carcinoma cells. Oct-3 is the product of a regulatory gene that is expressed in totipotent and pluripotent stem cells before gastrulation as well as in the germ cell lineage. Oct-3 is involved in early mammalian development. Maternally derived Oct-3 is required for mouse development to proceed beyond the one-cell stage. Primordial germ cells express Oct-3 throughout their migration from the allantois to the genital ridges. In the adult animal, Oct-3 is present in the ovary and the testis.[389] In the ovary Oct-3 is confined to oocytes, and maturing oocytes express higher amounts of Oct-3 than resting oocytes. The precise function of Oct-3 is unknown, but it may be involved in the regulation of DNA replication. Another Oct transcription factor, Oct-1 (also termed OTF-1, OBP100, and NF-III), is a homeodomain protein that regulates commonly expressed genes such as those encoding histone H2b and small nuclear RNAs. In addition to these "housekeeping" functions, Oct-1 and an auxiliary 40-kDa protein, the Oct-1-associated protein (OAP[40]), participate in activating IL-2 gene transcriptional expression in T lymphocytes.[390] The Oct-1 protein is phosphorylated in a cell cycle-dependent manner by a cdc2-related protein kinase.[391] The possible role of growth factors in the regulation of Oct factors expression is unknown.

The CREB functions as a Ca^{2+}-regulated transcription factor and is a substrate for the Ca^{2+}/calmodulin dependent (CaM) protein kinases I and II.[392] CREB residue Ser-133 is the major site of phosphorylation by the CaM kinases and mutation of this residue impairs the ability of CREB to respond to Ca^{2+}. CREB is involved in the regulation of gene expression by hormones, growth factors, and other stimuli that modify the $[Ca^{2+}]_i$. The CREB protein appears to be involved in the integration of signal transduction mechanisms involving Ca^{2+} and cAMP as second messengers.

The pituitary transcription factor Pit-1 is a member of the POU family of transcription factors that are involved in regulation of mammalian development.[393] Pit-1 contains two domains, termed POU-specific (POU-S) and POU-homeodomain (POU-HD), which are necessary for high affinity binding on the genes encoding growth hormone and prolactin. Pit-1 is also important for the regulation of the gene encoding the thyroid-stimulating hormone (TSH or thyrotropin) by thyrotropin-releasing hormone (TRH) and cAMP. Point mutation of the human Pit-1 gene POU-HD region results in hypopituitarism with combined deficiencies of growth hormone, prolactin, and TSH.[394,395] The mutant Pit-1 protein may bind DNA response elements normally but acts as a dominant inhibitor of Pit-1 action in the pituitary.

5. Heat Shock Proteins

Heat shock proteins (HSPs) are defined as proteins that can be induced by temperature elevation and other stimuli in a variety of organisms, from bacteria to man.[396-402] HSPs may be necessary for the survival and proliferation of cells under stress conditions and probably also under nonstress conditions. The genes coding for HSPs contain a conserved sequence of 14 bp, the Pelham box, which is located on the coding strand of DNA upstream of the TATA box and serves as the promoter for HSP mRNA transcription. The presence of three inverted repeats of a five-base module, nGAAn, located about 80 to 150 bp upstream of the transcription start site is characteristic of *HSP* genes. This conserved sequence, termed heat-shock element (HSE), is the site of interaction with a transcriptional activator, the heat-shock factor (HSF). The HSF appears in the nucleus as an inactive oligomer that is converted on heat shock to a multimer. In insect polytenic chromosomes, the HSF oligomer transition is accompanied by a relocalization of the factor from a general chromosomal distribution to specific heat-shock puff sites.[403] Other DNA regulatory signals may contribute to the activation of *HSP* gene expression. This expression can be induced by a wide diversity of environmental agents in addition to heat shock, including chemical and physical agents, as well as microbial infections and genetic lesions. The HSPs are involved in the processes of cell proliferation induced by hormones, growth factors, and other mitogens under nonstress conditions. Embryonic development and cellular differentiation are also associated with altered expression of *HSP* gene expression.

The HSPs are termed (according to their molecular weights expressed in kilodaltons) as HSP-100, HSP-90, HSP-70, HSP-60, and HSP-20/30. An HSP of 8 kDa is also known as ubiquitin. HSP 100 is a Ca^{2+}-calmodulin-regulated actin-binding protein.[404] The HSP-60 proteins form complexes with polypeptides and have ATPase activity; whereas the HSP-70 proteins are postulated to be involved in unfolding and disassembly of polypeptides, the HSP-60 proteins may participate in the folding and assembly of polypeptides. The high evolutionary conservation of HSPs indicates their biological importance. The HSPs respond to stress by a dynamic resorting among cellular organelles including the nucleus and the cytoskeleton.[405] The HSPs may be linked with a protective mechanism during and following cellular stress.

In addition to their important role in normal cell proliferation and differentiation, the HSPs may be involved in different ways in the malignant transformation of cells. Expression of HSPs in mammals may be induced by both growth factors and oncoproteins. Stimulation of human cells by either serum or viral oncoproteins may result in a marked induction of *HSP-70* gene transcription.[406] The c-Myc protein may contribute to the regulation of human *HSP-70* expression by binding directly to the promoter region of the gene.[407] HSP-70 mRNA increases transiently in HeLa cells during the S phase at the peak of DNA synthesis, and HSP-70 protein is found in the nucleus of S-phase cells, whereas in G_1 and G_2 the cells show a more weak and generalized distribution of the protein. Induction of a heat-shock response in HeLa cells by diverse stimuli (heat stress, arsenite, heavy metals) is accompanied by an elevation of c-*fos* mRNA, which occurs primarily through regulation at the posttranscriptional level.[408] Expression of both an *HSP-70*-related gene, *HSP-73*, and the c-*fos* gene occurs in rat livers reperfused after non-necrogenic ischemia, although it is not followed by DNA synthesis.[409] The HSP-73 protein is involved in the degradation of cellular proteins in cultured human fibroblasts in response to the serum or growth factor withdrawal.[410] The human genome contains at least 10 genes and pseudogenes with sequence homology to the *HSP-70* gene of *Drosophila*.[411] Inducible *HSP-70*-related DNA sequences have been assigned to human chromosomes 6, 14, and 21, but additional sequences are located on other chromosomes.[412] The expression of a novel gene of the human HSP family, termed *HSP-70* B', and the isolation of its cDNA and genomic DNA have been reported.[413] Human *HSP-70* DNA sequences have clusters of extensive homology with the respective sequences of prokaryotic and eukaryotic species, suggesting that these regions may represent functionally important domains of HSP-70 proteins. However, the precise physiological role of mammalian HSP-70 proteins is not clear. Regulation of *HSP-70* gene expression is very complex and may be dependent on the specificity and magnitude of physiological damage.[414] Expression of the *HSP-70* gene may be induced by serum stimulation or the c-Myc protein.[415,416] Protein HSP-70 could be involved in the control of cell growth, acting during the S phase of the cycle. HSPs usually have a nuclear location, suggesting some role in the regulation of genomic functions. In parasitic protozoa such as *Trypanosoma* and *Leishmania*, expression of *HSP* genes may be responsible for differentiation processes involving extensive morphological and physiological changes upon transmission of the parasite from the insect vector to the mammalian host, which implicates a difference in temperature.[417]

A 2.7-kb transcript of the *HSP-70* gene is expressed in testicular germinal cells, coincidentally with the entry of the cells into the haploid stage.[418] A 4.7-kb transcript of the mouse c-*abl* gene is also expressed in the male germ cells in a haploid stage-specific manner.[419] The biological roles of *HSP-70* and c-*abl* mRNA in male germ cells are not understood, but it is well known that the germinal compartment of the testis can be readily destroyed by external stresses including mildly elevated temperatures.

HSP-70 and other HSPs may have a role in the expression of a transformed phenotype. Tumor cells such as those from Morris hepatoma 3924A may synthesize HSP constitutively, and the effect of heat shock on HSP mRNA is less pronounced than in normal liver cells.[420] Expression of the *HSP-70* gene is high in human tumor cell lines and is augmented by serum in HeLa cells.[421] Increased transcription of two *HSP* genes (*HSP-70* and *HSP-83*), together with increased expression of the c-*myc* and c-H-*ras* proto-oncogenes may occur in rat liver after administration of hepatocarcinogens,[422] but similar changes of gene expression may occur during liver regeneration after partial hepatectomy. Thus, liver damage by hepatocarcinogens induces the expression of genes that are also expressed at increased rates in normal liver growth. Expression of the liver growth factor HCGF is induced during liver regeneration after the action of liver-damaging agents or partial hepatectomy, but the possible influence of HCGF in *HSP* gene expression is not known. In general, the biological significance of changes in *HSP* gene expression during normal and neoplastic hepatic growth is unknown. Decreased HSP-70 protein expression occurs in Friend erythroleukemia cells exposed to differentiation inducers.[423]

HSP-90 is a cytosolic protein abundant in vertebrate cells at physiological temperature. A cDNA sequence of the chick *HSP-90* gene and the complete deduced amino acid sequence of the HSP-90 protein have been reported.[424] The predicted secondary structure of the protein suggests a model where the charged regions of HSP-90 are candidates for ionic interactions with functionally important macromolecules including other proteins and nucleic acids. Human HSP-90 is phosphorylated *in vivo* on serine residues that are a substrate for type II casein kinase *in vitro*.[425,426] The HSP-90 protein exists in cytosol in heteromeric complexes containing HSP-70 and three other proteins with molecular weights of 63, 56, and 50 kDa.[427] HSP-90 forms complexes with steroid hormone receptors as well as with Src-related oncoproteins, including the v-Src, v-Fps, v-Yes, v-Fes, and v-Fgr oncoproteins.[428-433] The physiological role of such complexes is not understood. Formation of a complex between HSP-90, a metal ion (probably Zn^{2+}), and the estradiol receptor may contribute to the regulation of estradiol receptor-DNA interaction.[434] The chicken HSP-90, free or released from the progesterone receptor in the cytosol, occurs in a dimeric form.[435] Both HSP-90 and HSP-100 are actin-binding proteins that can crosslink actin filaments.[436] The interaction between HSP-90 and F-actin is inhibited by calmodulin.[437] The transforming action of some oncoproteins may be exerted, at least in part, through their interaction with particular HSPs, which would involve an alteration of actin components of the cytoskeleton. HSP-90 may be constitutively expressed at high levels in human acute leukemia cells.[438] Further studies are required for a better characterization of the physiological role of HSPs in normal and neoplastic cells. In particular, the possible role of HSPs in the mechanisms of action of growth factor is almost unknown.

I. DIFFERENTIATION-INDUCING FACTORS

An important task in biology is the characterization of proteins involved in the mechanisms of cellular differentiation.[439] Results from many studies suggest that these mechanisms are related to the production of specific endogenous inducers of differentiation, referred to as differentiation-inducing activities (DIAs). Some DIAs are represented by known growth factors. A DIA expressed during rat liver epithelial cell differentiation *in vitro* and *in vivo* is identical with TGF-β.[440] This factor can induce the differentiation of epithelial liver cell progenitors, the oval cells, into the hepatocytic lineage in adult mammalian liver. However, other DIAs are distinct of growth factors and may rather correspond to specific endogenous differentiation-inducing factors (DIFs). Selected examples of such DIFs are discussed next.

1. Osteogenin

An interesting DIF is osteogenin,[441] a bone-inductive protein isolated from bovine bone matrix.[442] Osteogenin activity is represented by a protein of 30 to 40 kDa, and the sequences of a number of tryptic peptides of the gel-eluted material have been determined.[443] Bovine osteogenin amino acid sequences are similar to those deduced from cDNA clones from one of the human bone morphogenetic proteins, BMP-3. Osteogenin is able to stimulate the expression of an osteogenic phenotype in periosteal cells, osteoblasts, and bone marrow stromal cells and the expression of a chondrogenic phenotype in chondroblasts.[444] In the mouse calvaria-derived cell line MC3T3, osteogenin induces alkaline phosphatase activity and the production of calcified bone matrix *in vitro*.[445] The highest concentrations of osteogenin *in vivo*, as determined by autoradiographic localization, occur in skeletal tissues during critical phases of cytodifferentiation and morphogenesis.[446]

2. Preadipocyte-Stimulating Factor

A DIF that promotes the differentiation of rat preadipocytes has been identified in rat serum.[447] This factor, termed preadipocyte-stimulating factor (PSF), is distinct from known hormones and growth factors and is represented by a highly glycosylated protein of 63 kDa. PSF is also present in mouse serum and can promote the adipogenic conversion, biochemical differentiation, and mitogenic activation of cultured rat preadipocytes.

3. Leukemia-Related Differentiation Inducers

HL-60 human promyelocytic leukemia cells can be induced to differentiation by treatment with a diversity of chemical agents including retinoic acid, 1,25-dihydroxyvitamin D_3 (calcitriol), dimethyl sulfoxide (DMSO), and phorbol esters. HL-60 cells initiated to differentiation with either calcitriol or DMSO produce an activity that is capable of inducing differentiation and loss of proliferative capacity in fresh HL-60 cells.[448] An intracellular DIF was detected in HL-60 cells induced to differentiate by treatment with calcitriol.[449] Such DIF is a protein capable of inducing the differentiation not only of HL-60 cells but also of other human leukemia cell lines, including the U-937 monoblastic leukemia cells and KG-1 myeloblastic leukemia cells, but not of K-562 erythroleukemia tumor cells. The relationship between this transacting/interacting DIA (int-DIA) and other differentiation inducers is not clear. The int-DIA could be capable of inducing cells to proceed from a commitment to a phenotype-expression step during the differentiation of cells from the myeloid lineage. Macrophage-like cells, subcloned from a clonal line of mouse myeloid leukemia M1 cells, produce and secrete into the medium a DIF that is capable of inducing differentiation of M1 cells and is also capable of inhibiting the growth of HL-60 cells.[450] This DIF, purified to homogeneity, is a peptide of an approximate molecular weight of 1200. Differentiation of U-937 leukemia cells by phorbol ester is associated with production of a DIF that is present in the conditioned medium.[451] This DIF would be represented by a 67-kDa protein that is distinct from all factors reported to regulate the differentiation of leukemic cell lines. It is not known whether the DIFs produced by HL-60 and U-937 cells are similar or identical. Two DIFs, termed DIF-I and DIF-II, would be involved in a synergistic manner in the erythroid differentiation of mouse erythroleukemia (MEL) cells induced by DMSO or other agents such as hexamethylenebis(acetamide).[452] Although both DIF-I and DIF-II are cytoplasmic proteins, they may be different. Induction of DIF-II is inhibited by phorbol ester and does not take place in a mutant MEL line unable to undergo erythroid differentiation.

4. Leukemia Inhibitory Factor

A DIF, termed D-factor, capable of inducing differentiation of murine myeloid leukemia M1 cells into macrophages, was purified to homogeneity from serum-free conditioned medium of mouse L929 cells.[453] The purified D-factor is a protein of 62 kDa and its half-maximal concentration for inducing M1 cell differentiation is about 1.7×10^{-11} M. It can inhibit pluripotential embryonic stem cell differentiation.[454] The D-factor may be identical to a factor purified from Krebs ascites tumor cell-conditioned medium and called leukemia inhibitory factor (LIF). Such LIF is also capable of inducing the differentiation of M1 cells.[455] M1 cell differentiation into macrophages is also induced by calcitriol, which results in production of IL-6, although this production may not be responsible for the D-factor- or calcitriol-induced M1 cell differentiation.[456] A differentiation inducer, DIF-A, purified from human monocytic leukemia THP-1 cells, is a human counterpart of the murine LIF/D-factor.[457] The cloning and expression of cDNAs encoding murine and human LIF have been reported.[458-460] The gene for LIF/DIF-A resides on human chromosome region 22q12.q12.2.[461]

A factor similar or identical to LIF is present in the serum of patients with chronic myeloid leukemia.[462] Analysis of a LIF cDNA clone showed that it is identical with a myeloid factor termed human interleukin for DA cells (HILDA).[463] LIF is also identical with the hepatocyte-stimulating factor III (HSF-III), which was purified from the conditioned medium of COLO-16 human tumor cells.[464] Furthermore, LIF seems to be identical with a cholinergic neuronal differentiation factor (CNDF), which is secreted by cultured rat heart cells and is capable of directing the choice of neurotransmitter phenotype.[465] CNDF is a 45-kDa glycoprotein that acts on postmitotic sympathetic neurons to induce the expression of acethylcholine synthesis and cholinergic function while suppressing catecholamine synthesis and noradrenergic function.

LIF is normally produced by a variety of cell types and may be involved in regulating the growth, differentiation, and specific functions of a wide diversity of cell types.[466-468] LIF may have a role in the regulation of early developmental processes. It is expressed in mouse blastocytes prior to hematopoiesis and may be involved in the regulation of the growth and development of trophoblasts or embryonic stem cells.[469] LIF is synthesized in the extra-embryonic part of the mouse embryo and acts on the embryonic tissues during early mouse development.[470] It is involved in intertissue relationships during early mouse embryogenesis. LIF is expressed by the implantation uterus and selectively blocks primitive ectoderm formation in vitro.[471] An inhibitor of aortic endothelial cell growth secreted by pituitary follicular cells is identical with LIF.[472] The study of LIF and other similar factors reinforces the concept that the distinction between growth-promoting, growth-inhibiting, and differentiation-inducing activities is largely determined by the target cell type and its stage of differentiation. Because LIF displays cytokine-like functions, it is discussed further in Volume 3.

5. Other Differentiation Inducers

A DIF that is constitutively produced by the malignant T-lymphocyte leukemic cell line HUT-102 is a protein of 50 kDa that is capable of inducing differentiation of HL-60 and U-937 leukemia cell lines.[473] This DIF can also inhibit the proliferation of fresh human AML cells as well as the growth of normal granulocyte-macrophage colonies. The physiologic roles and the mechanisms of action of this DIF are unknown.

The conditioned medium of a CSF-producing tumor (G2T) obtained from a patient with lung cancer contained a DIA, termed G1-DIF, that was capable of inducing differentiation of HL-60 cells into macrophage-like cells.[474,475] G1-DIF has a molecular weight of 33 to 36 kDa and is distinct from CSFs and IFNs.

The cell line CESJ, derived from a murine mammary carcinoma that produces hypercalcemia and granulocytosis, produces a 17-kDa protein that induces osteoclasts to differentiate from bone marrow cells.[476] This protein, designated osteoclast colony-stimulating factor (O-CSF), is distinct from myeloid differentiation factors.

Hematopoietic growth factors (HGFs) and cytokines, either alone or in particular types of combinations, can induce the differentiation of certain types of neoplastic cells *in vitro*. A DIF capable of inducing the differentiation of some human leukemia cells *in vitro* was identified with TNF-α.[477] Recombinant human CSF-3 can induce the *in vitro* terminal differentiation of cultured murine and human leukemic cells into macrophages and granulocytes, including M1 cells.[478,479] However, even under optimum conditions of M1 cell differentiation induced by CSF-3, some blastic cells remain. A combination of IL-1 and IL-6 may act in a synergistic manner to induce the differentiation of mouse and human myeloid leukemic cell lines into macrophage-like cells.[480] Specific HGFs may be capable of inducing differentiation of leukemic cells not only *in vitro* but also *in vivo*. Both CSF-2 and IL-3 injected into mice are capable of inducing differentiation of leukemic cells maintained in diffusion chambers implanted in the animals.[481] Interestingly, one type of leukemia that is resistant to differentiation induced by CSF-2 or IL-3 *in vitro* is sensitive to these factors *in vivo*, which indicates that their action in these cells is indirect and is probably mediated by another factor, probably a species of MGI.

An interesting approach to cancer treatment may consist of the delivery of noncytotoxic DIFs conjugated with anticarbohydrate antibodies. This approach is based on the fact that DIFs are known to modify tumor cell growth *in vitro* and could inhibit malignant tumor cells *in vivo* without damaging normal cells. Sodium butyrate encapsulated in liposomes that were covalently linked to an anticarbohydrate monoclonal antibody were targeted to human colonic adenocarcinoma HRT-18 and HT29 cells expressing the specific antigen *in vitro* as well as in athymic nude mice.[482] The tumor cell growth was inhibited and the tumor cells showed morphological and biochemical signs of differentiation. The major drawback of this approach is that not all tumor cells are susceptible to DIFs.

6. Mechanisms of Cell Differentiation

Cell differentiation during developmental processes is regulated mainly, if not exclusively, at the level of gene expression. An array of factors are involved in both the positive and negative regulation of genes whose expression is related to specific differentiation processes. In turn, the activity of these genes is regulated by other gene products as well as by the intracellular environment, which partially depends on the differentiation-inducing and -inhibiting action of hormones, growth factors, and other extracellular agents. Physical and chemical interactions between neighbor cells, occurring at the level of the cell surface, are of great importance for the operation of cell differentiation processes.

There is a tight coupling between cell cycle regulation and cell differentiation. In general, the terminal differentiation of cells is associated with their withdrawal from the cycle and the cessation of cell proliferation. A system studied in detail is represented by skeletal muscle development in vertebrates which partially depends on growth factor action.[483] During development of muscle tissue, multipotential precursor cells proliferate extensively and progress through a myoblast stage before differentiating to become myocytes which requires withdrawal from the cell cycle. The onset of muscle cell differentiation is characterized by the expression of genes that code for muscle-specific enzymes, contractile proteins, and receptors that are not expressed in myoblasts. Expression of members of the MyoD family of myogenic regulatory genes is crucial for muscle differentiation.[484] This family includes genes encoding MyoD, myogenin, Myf-5, and Myf-6. All of these gene products share a similar 57-amino acid basic helix-loop-helix (B-HLH) domain characterized by a strongly basic region and a potential HLH structure. Similar domains are found in an extended family of nuclear proteins which includes Myc proteins as well as some factors identified in *Drosophila melanogaster*. The structural similarity between the MyoD and Myc families is interesting because the c-Myc protein is implicated in the control of cell differentiation in many different biological systems. Cotransfection experiments have shown that expression of a c-*myc* gene can inhibit the differentiation of mouse NIH/3T3 fibroblasts made myogenic by the expression of a MyoD gene.[485] MyoD and myogenin are not sufficient, acting alone or together, to drive muscle differentiation in the presence of high levels of c-Myc protein. The suppression produced by c-Myc can function independently of the activity of a negative regulator, termed Id, which would act through a different biochemical pathway. The Id protein contains an HLH domain but not an adjacent basic region. The Id protein can associate with MyoD and can inhibit its binding to DNA, and may negatively regulate the differentiation functions of tissue-specific B-HLH complexes by sequestering components of such complexes in inactive hetero-oligomers.[486] These results indicate that whereas myogenic factors such as the members of the MyoD family act as positive regulators of muscle cell differentiation, c-Myc and Id may represent independent negative regulators of this process. It may be concluded that the molecular mechanisms of differentiation are very complex and are based on a balance between the actions of positive and negative regulatory molecules. Although the precise functions of these molecules are unknown, it seems likely that at least some of them act as direct or indirect regulators of the genetic program of the cell associated with the expression of certain differentiated functions.

J. DIFFERENTIATION-INHIBITING FACTORS

The regulation of cell differentiation may occur by mechanisms involving both inducers and inhibitors, but very little is known about the structure and function of endogenous agents specifically involved in inhibiting cell differentiation. However, it is well known that stimulation of cell proliferation by growth factors is usually associated with inhibition of cell differentiation. Some examples of factors with selective differentiation-inhibiting properties are mentioned next.

A differentiation-inhibiting factor was purified from conditioned medium of differentiation-resistant variants of M1 mouse myeloid leukemia cells.[487] The inhibitory factor was represented by a 68-kDa protein that was capable of producing a marked inhibition of glucocorticoid-induced differentiation of M1 cells at a very low concentration without altering the rate of cell proliferation.

A differentiation inhibitory activity was identified as a soluble factor produced by a number of cell types, including rat liver cells.[488] In the absence of this inhibitory activity, embryonic stem cells (the totipotent outgrowths of blastocysts) are able to differentiate into a variety of cell types. LIF can substitute for this inhibitory activity in the maintenance of totipotent embryonic stem cell lines that retain the potential to form chimeric mice.[489] Both LIF and the inhibitory activity are produced by the human bladder carcinoma cell line 5637. They may be represented by the same molecule which would have an important regulatory function in hematopoiesis and early embryogenesis.

A human erythroid differentiation-inhibiting protein (DIP) isolated from the blood of a woman suffering from pure red cell aplasia could specifically block the differentiation of human burst-forming unit-erythroid (BFU-E), but not that of colony-forming unit-erythroid (CFU-E).[490] The DIP also inhibited DMSO-induced differentiation of Friend murine erythroleukemia cells.

K. GROWTH-INHIBITING FACTORS

The control of cell proliferation depends not only on the action of stimulatory factors but also on endogenous and exogenous factors with growth inhibitory properties. Moreover, negative controls would be prevalent in the regulatory mechanisms of normal and neoplastic cell proliferation.[491,492] Unfortunately, most of the growth-inhibiting factors (GIFs) remain poorly characterized and their mechanisms of action are little understood.[493] GIF RNA or protein can be detected, in principle, by the analysis of senescent or growth-arrested cells, where GIFs are expected to be present in increased amounts. The recent identification of tumor suppressor genes (antioncogenes) has opened the way to the characterization of gene products with specific growth inhibitory activity.[494-497] Deletion or inactivation of tumor suppressor genes may contribute, in conjunction or not with proto-oncogene activation, to oncogenesis.

Most frequently, the effect of hormones and growth factors on the target cells and tissues is stimulatory, but in some cases they may have growth inhibitory effects on particular types of cells, thus exhibiting GIF-like activity. Moreover, factors that exert a growth stimulatory action on their normal target cells may have opposite effects when the cells have acquired a malignant behavior. For example, HCGF exerts a stimulatory action on the proliferation of normal hepatocytes but may function as a negative growth regulator for human hepatocellular carcinoma cells.[498] Some growth regulatory factors are bifunctional in nature, displaying stimulatory or inhibitory effects on the growth of target cells depending upon the type of cell and the *in vitro* or *in vivo* conditions. An example of such bifunctional factors is TGF-β, a factor with potent growth inhibitory activity for certain types of cells, in particular for epithelial cells. The growth inhibitory properties of glucocorticoids on lymphoid tissue and other types of cells are widely profitable for the treatment of inflammatory and neoplastic diseases.

A negative control mechanism for cellular proliferation mediated by specific types of GIFs was proposed by Weiss.[499] In general, the proliferation of normal cells is controlled by two types of factors: growth-stimulating factors (positive growth factors) and growth-inhibiting factors (negative growth factors).[500] The altered growth characteristics of malignant cells may be due, in principle, to alterations of the normal balance existing between positive and negative growth factors, with a final predominance of growth stimulation over growth inhibition. The unrestrained growth of malignant cells may result from the abnormal production of growth-stimulating factors in an autocrine or paracrine manner, or from their failure to respond to endogenous or exogenous GIFs. In malignant diseases there could be a defective production of GIFs, or a defective response of the cells to GIFs, or both. However, transformed cells may be able to respond to specific types of GIFs. A GIF identified in rabbit serum was shown to be even more effective on RSV-transformed tumorigenic rat liver epithelial cells than on nontumorigenic cells of similar rat liver origin.[501]

GIFs, or factors with GIF-like properties, are a focus of great interest because of their possible use in cancer therapy. Even GIFs isolated from distantly related species may show antitumor effects. A cytostatic peptide identified in the roe of the shad (*Alosa sapidissima*), a diadromous fish, was found to exert inhibitory effects on the growth of murine L1210 leukemia cells in culture.[502] In addition to the inhibitor, the supernatant fraction of unfertilized ova from shad contains a tumor growth-promoting peptide of similar molecular weight (20 to 30 kDa). The possible homology of the shad growth-inhibiting and -promoting peptides to other known growth-modulating factors remains to be determined.

1. GIFs Produced by Normal Cells

Factors with growth inhibiting properties may be produced by both normal or neoplastic cells. A particular type of GIFs called chalones was defined as nonspecies-specific, low molecular weight peptides with tissue-specific growth inhibitory properties.[503-510] Chalones are produced by the differentiated cells of many tissues and their growth inhibitory effect is exerted on the immature cells of the origin tissue through nontoxic mechanisms. Chalones would contribute to maintaining adequate growth rates within specific tissues. Unfortunately, little progress has occurred recently in the characterization of chalones, which are conceptually most interesting as mitotic inhibitors and as participants in negative feedback regulatory mechanisms. Chalone activity may correspond, at least in part, to factors with known growth inhibitory action such as TGF-β.[511,512] Both TGF-β and the interferons display growth inhibitory activity under certain conditions.[513] TGF-β exerts potent inhibitory effects on the growth of a variety of normal and neoplastic tissues, including the developing mouse mammary gland *in vivo*.[514] TGF-β functions as an inhibitor of hematopoietic progenitor cell proliferation *in vitro* and has a critical role in balancing the development of the hematopoietic system *in vivo*.[515,516] A GIF activity present in the conditioned medium of the mouse embryo fibroblast cell line AKR-MCA treated with *N,N*-dimethylformamide (DMF) is represented mainly, but not completely, by TGF-β.[517] In addition to TGF-β and IFNs, other types of GIFs produced by normal cells have been described, but their structures and functions remain in many cases poorly understood. Examples of such GIFs are mentioned next.

Fibroblasts maintained in culture may produce and secrete GIFs. A 13-kDa protein with GIF-like activity was purified from medium conditioned by dense cultures of 3T3 fibroblasts, and a GIF, termed inhibitory diffusible factor 45 (IDF-45), was purified by similar procedures.[518-523] The IDF-45 was characterized as a monomeric 45-kDa protein that is capable of inhibiting DNA synthesis in chicken embryo fibroblasts (CEF cells) in a reversible and dose-dependent manner. RSV-transformed cells are not sensitive to the growth inhibitory action of IDF-45. Stimulation of RNA synthesis in CEF cells by serum, PMA, and PDGF is partially inhibited by pretreatment with IDF-45, and stimulation induced by IGF-I can be completely inhibited by IDF-45. By contrast, stimulation of RNA synthesis induced by the v-Src oncoprotein or insulin is not inhibited by IDF-45, suggesting that the mechanisms of action of v-Src and insulin may share some similar components.

A GIF called fibroblast-derived growth inhibitor (FDGI) is secreted by human fibroblasts when induced to secrete IFN-β.[524] The FDGI is a peptide of 12 kDa and is different from IFNs and TGF-β. It inhibits the growth of human fibroblasts and mouse L cells as well as the growth of Namalva cells (derived from Burkitt's lymphoma) and HeLa and A549 cells (derived from human solid tumors).

A cell cycle-related protein of 20 kDa (p20K) is synthesized and secreted by nonproliferating CEF cells.[525] Studies with a cDNA clone encoding p20K, isolated from a library representing DNA sequences expressed in contact-inhibited CEF cells, showed that p20K mRNA levels decline rapidly after mitogenic stimulation with insulin or EGF or by expression of the v-Src oncoprotein. Basal levels of p20K are attained during the G_1 phase of the cell cycle, coinciding with the expression of the 'immediate-early' mitogen response gene CEF-4. The amino acid sequence of the p20K protein, as deduced from cDNA clones, exhibited limited similarity to $\alpha_{2\mu}$-globulin, β-lactoglobulin, and plasma retinol-binding protein.

A protein complex detected in mature human granulocytes and monocytes is a GIF that may exert its action through the inhibition of the activities of protein kinase I and II.[526] This GIF, termed PC inhibitor, may be a mediator of growth inhibition during the terminal stage of the differentiation of HL-60 and THP-1 human leukemia cells induced by specific chemical stimuli as well as during physiologic and pathological events associated with the function of mature monomyelocytic cells.[527]

A macromolecule present in the marrow supernatant of C57BL/6J (B6) (Fv-2^{rr}) mice is capable of rapidly shutting down BFU-E DNA synthesis *in vitro*. This compound, termed negative regulatory protein (NRP), appears to be specific to the BFU-E.[528] The NRP would act during the S phase of the cycle and its action on BFU-E can be opposed by IL-3. The action of NRP on BFU-E DNA synthesis is similar to that of TGF-β, but NRP and TGF-β are different molecules.

An inhibitor of hematopoietic stem cell proliferation purified from fetal calf bone marrow was identified as a tetrapeptide that is capable of preventing the recruitment of spleen colony-forming units (CFU-S) when administered to mice after injection of cytosine arabinonucleoside.[529] Since thymosin-β 4 contains the same amino acid sequence of the inhibitor, this could be generated *in vivo* during the proteolytic maturation of thymosin-β 4.

Human pulmonary alveolar macrophages, which serve as the first line of defense against inhaled foreign materials, may secrete both growth stimulatory and growth inhibitory factors. The conditioned medium of human lung tissue cultured for three days promoted DNA synthesis of human alveolar macrophages in a manner similar to that of CSF-2, whereas the medium from human lung cultured for six days did not promote DNA synthesis but strongly suppressed CSF-2-induced DNA synthesis.[530] Partial purification of a growth inhibitor from the medium yielded two distinct peaks of activity with molecular weights of 38 and 110 kDa. The precise structure and function of such putative GIFs were not elucidated.

Platelets contain both growth-stimulating factors and GIFs. A 37-kDa inhibitor of endothelial cell DNA replication and cell growth was isolated from platelets.[531] Two platelet-derived proteins with GIF activity for rat hepatocytes, termed platelet-derived growth inhibitor (PDGI) α and β, were isolated from rat platelets.[532,533] Purification of PDGI-β to homogeneity showed its identity with TGF-β. The PDGI-α, purified to homogeneity, is a homodimeric protein of 26 kDa, distinct from TGF-β and other known growth factors and GIFs. The PDGI-α at a very low concentration can induce a complete inhibition of DNA synthesis in rat hepatocytes stimulated by insulin and EGF.

A GIF purified from African green monkey kidney cells (BSC-1 cells) and human platelets inhibited the proliferation of concanavalin A-stimulated mouse thymocytes; this action was reversed by the addition of IL-2.[534,535] The BSC-1 GIF is identical to a subtype TGF-β (TGF-β$_2$) and can either stimulate or inhibit cell proliferation depending on the particular experimental or natural conditions. It acts as a stimulator of DNA synthesis when added to quiescent Swiss mouse 3T3 fibroblasts in combination with other hormones or growth factors, including bombesin, vasopressin, and PDGF.[536] The patterns of synergistic interactions displayed by the BSC-1 GIF or TGF-β in this assay are indistinguishable from those obtained with insulin, indicating that growth stimulators and GIFs may share some common intracellular pathways.

A hepatic proliferation inhibitor (HPI) was purified from the adult rat liver.[537] HPI activity would be represented by a 17- to 19-kDa protein that is capable of modulating the stimulation of DNA synthesis in hepatocytes by either EGF or a factor activity prepared from serum of hepatectomized rats. The original HPI preparations were not homogeneous and further purification yielded a more potent GIF-like activity. Such purified liver-derived growth inhibitor (LDGI) is a 20-kDa polypeptide that is different from TGF-β and produces, in a reversible way, a growth inhibitory effect on primary cultures of hepatocytes.[538] LDGI also reduced DNA synthesis in AFB$_1$-transformed rat liver cells, MCF-7 human breast carcinoma cells, and Reuber rat hepatoma cells. In contrast, the rat hepatoma cell line UVM 7777 and the human hepatoma cell line HepG2 were resistant to the antiproliferative effects of LDGI, and the proliferation of rat and human mesenchymal cell lines was rather stimulated by LDGI. Thus, in a manner similar to TGF-β and TNF-α, LDGI acts as a bifunctional modulator of cellular proliferation.

Gene transcripts encoding an intracellular protein, prohibitin, were identified through their expression in higher than normal regenerating liver.[539,540] Prohibitin is expressed in different tissues of the adult rat and is a protein highly conserved in evolution. Expression of prohibitin blocks DNA synthesis in both normal and tumor cells, and microinjection of an oligonucleotide containing prohibitin sequences into human fibroblasts stimulates entry of cells into the S phase. The prohibitin gene, *pro*-1, encodes a 30-kDa protein of 272 amino acids which does not appear to contain ATP-binding sites or nuclear localization signals. The human *pro*-1 gene is located on chromosome region 17q21 and is mutated in some sporadic breast cancers.[541] The *pro*-1 gene is the mammalian analog of *Cc*, a *Drosophila* gene that is vital for the normal development of the insect. Prohibitin could have a universal role in the negative control of proliferative activity. However, the level of prohibitin mRNA in different tissues is not necessarily consistent with the growth potential of each tissue, and the product of the prohibitin gene has not been characterized.

GIFs could circulate with the blood or other body fluids in masked forms. Tryptic hydrolysis of serum or α_2-macroglobulin produces a glycopeptide of 3500 to 4600 mol wt capable of inhibiting G_1/S phase transition in rat hepatocytes.[542] α_2-Macroglobulin could be the carrier of the precursor form of the glycopeptide.

A 14.5-kDa polypeptide, the mammary-derived growth inhibitor (MDGI), was isolated from lactating bovine mammary glands on the basis of its ability to inhibit the growth of mammary epithelial cells *in vitro*.[543,544] The MDGI was sequenced and found to belong to a superfamily of lipid-binding proteins.[545] MDGI exhibits extensive homology to several normal proteins, including fatty acid-binding protein from heart, myelin P2 protein, and retinoic acid-binding protein. MDGI is also related to a 70-kDa antigen identified in nuclei of mammary epithelial cells. MDGI mRNA and protein levels are increased in the differentiated mammary gland, as compared with the proliferating gland of pregnant animals. MDGI may be a physiological modulator of β-adrenergic response in cardiac muscle, and this effect is not related to its lipid-binding activity.[546] A protein called cardiac fatty acid-binding protein (FABP) may be identical to MDGI.[547] The growth of various human mammary epithelial cell lines is inhibited by MDGI.[548] The carboxy-terminal sequence of MDGI is crucial for its activity.

Antibodies against MDGI crossreact with a growth inhibitor, the fibroblast growth regulator (FGR), which was purified from the conditioned medium of 3T3 cells.[549] Both MDGI and FGR consist of a single 13-kDa polypeptide chain. Another GIF, mammastatin, with specificity for mammary tissue, was isolated from the conditioned medium of normal human mammary cells.[550] Mammastatin is capable of inhibiting the growth of transformed human mammary cells, both estrogen receptor-positive and -negative, but has no effect on the growth of transformed human cells derived from nonmammary tissues. Mammastatin is composed of polypeptides of 47 and 65 kDa, and is distinct from TGF-β. It is present in normal human mammary cells. Breast tumor cells contain diminished amounts of mammastatin, suggesting that decreased production of mammastatin by transformed mammary cells may contribute to the loss of normal growth control in these cells.

A GIF activity, called decidua inhibitory factor, isolated from the maternal part of second trimester bovine placenta, can inhibit DNA synthesis and proliferation of several transplantation tumors and established tumor cell lines but is ineffective with normal cells.[551] Signal transduction of the decidua inhibitory factor occurs via a pathway that involves its binding to a receptor on the cell surface, the intracellular mobilization of Ca^{2+}, and the downregulation of c-*fos* and c-*myc* proto-oncogene expression.[552] Altered transcription of the c-H-*ras* and N-*ras* genes may also have a role in the mechanism of action of this factor.

GIFs may be synthesized in the central nervous system. An 18-kDa sialoglycopeptide isolated from bovine brain cortex reversibly inhibits protein and DNA synthesis of a variety of normal cell types.[553] The mechanism of action of this GIF at the intracellular level involves changes in Ca^{2+} fluxes, and the GIF is capable of antagonizing the mitogenic activity of EGF. Molecular cloning of a cDNA encoding another GIF that is abundant in normal human brain, but greatly reduced in the brain of patients with Alzheimer's disease, indicated that it is a new member of the metallothionein family.[554] The metallothioneins are ubiquitous, low molecular weight proteins that have a high cysteine content and are capable of binding heavy metal ions such as zinc, copper, and cadmium. As with genes for other members of this family, the gene for the brain-derived GIF is located on human chromosome 16. This gene is expressed only in the central nervous system and, in particular, in astrocytes of the gray matter. The normal functions of brain-derived GIFs remain to be determined.

Supernatants from activated human T lymphocytes are inhibitory for the growth of A375 human melanoma cells.[555] In addition to TGF-β and IFN-γ, these supernatants contain oncostatin M, a glycoprotein produced by phorbol ester-stimulated U-937 human histiocytoma cells that acts as a GIF for A375 melanoma cells and other types of tumor cells, but not for normal human fibroblasts. However, oncostatin M is a multifunctional cytokine and is a potent natural mitogen for rabbit vascular smooth cells in culture.[556] Oncostatin M is a major growth factor for AIDS-associated Kaposi's sarcoma cells.[557,558] Genomic cloning, sequence analysis, and expression of a cDNA coding for oncostatin M has been reported.[559] Purified oncostatin M is a single chain polypeptide of 28 kDa that exerts a reversible cytostatic effect on A375 tumor cells. Binding sites for oncostatin M are present in various human and nonhuman tumor cell lines.[560] Oncostatin M is structurally related to LIF and IL-6, and its physiological effects are mediated by the gp130 IL-6 signal transducer.[561,562] Oncostatin M is internalized and degraded subsequently to binding. It is able to regulate the expression of proto-oncogenes such as c-*myc*, c-*jun*, and *egr*-1.[563] The mechanism of oncostatin M action is associated with tyrosine phosphorylation of cellular proteins.

A factor contained in human plasma, called aproliferin, is capable of inhibiting the terminal differentiation of a murine BALB/c 3T3 mesenchymal stem cell line.[564] Aproliferin is represented by a 45-kDa protein, but the origin and function of aproliferin *in vivo* are unknown.

The mechanisms of action of endogenous GIFs are unknown. A 13-amino acid peptide secreted by yeast, called α-factor, acts as both a GIF and differentiation inducer. The α-factor stimulates yeast cells to prepare for mating by causing them to arrest in G_1 phase, to undergo morphological changes, and to induce expression of many genes involved in mating. The α-factor arrests cells at a point in G_1, called start, that appears to be a key decision-making point in the cell cycle at which cells either commit to a mitotic cycle or undergo differentiation leading to meiosis or mating. A gene, *FAR1*, is necessary for cell cycle arrest in yeast, but not for other responses to α-factor.[565] The FAR1 protein acts by inhibiting CLN2, one of the three G_1 cyclins present in yeast. Other GIFs may also act through the regulation of the functional activity of proteins such as cyclins and the cdc2 and cdk2 kinases associated with the cycle of different types of cells.

2. GIFs Produced by Neoplastic Cells

GIFs, or factors with GIF-like activity, are produced by a wide diversity of malignant cells. Some of these GIFs are capable of inducing the differentiation of neoplastic cell lines *in vitro*. A GIF, termed sarcoma growth inhibitor (SGI), was detected in an ASV-transformed rat cell line that also produced TGFs.[566] The SGI was capable of inhibiting the anchorage-independent growth of transformed cells as well as that of TGF-treated nontransformed BALB/3T3 mouse fibroblasts.

GIFs produced by human monocytic leukemia cells have been detected in the culture medium.[567] A GIF produced by human myeloid leukemia cells, called leukemia-associated inhibitor (LAI), is represented by a large glycoprotein which reversibly reduces the fraction of granulopoietic stem cells in S phase and could function as a normal regulator of the proliferation of granulopoietic stem cells.[568,569] The effect of LAI could be mediated by a trypsin-sensitive receptor on normal marrow cells.[570] Production of LAI by AML cells is inhibited by PGE_1 and glucocorticoids. Leukemic lymphoblasts derived from the marrow of a patient with T-cell acute lymphoblastic leukemia and severe leukopenia released, in the presence of PHA, a protein with GIF-like activity for normal myeloid progenitor cells.[571] Two GIFs capable of suppressing lymphocyte proliferation are produced by the human acute leukemia-derived T-cell line clone GI-CO-T-9.[572]

A GIF was found in rat ascites fluid and a human colon carcinoma cell line.[573,574] This GIF acts in a partially reversible manner and coexists in the same malignant cells with TGF activity, which suggests that control of cell growth, either normal or neoplastic, involves a delicate balance between positive and negative factors. A protein purified from serum-free conditioned medium of the HT29 human colon carcinoma cell line, capable of inhibiting the proliferation of the same cell line, was identified with the IGF-binding protein IGF-BP-4.[575]

A GIF, termed GI, was detected in the conditioned medium of DIF-resistant murine myeloblastic leukemia M1 cells, clone R-1.[576] GI purified to apparent homogeneity is a basic protein of 25 kDa.[577] GI inhibits the growth of other murine leukemia cells but does not affect the growth of monocytic or myeloid human leukemia cell lines or the growth of normal peritoneal macrophages. GI inhibits preferentially the growth of mouse monocytic leukemia cells in intermediate stages of differentiation. The response of malignant cells to GIFs or differentiation inducers may vary depending on the type of cell and its state of differentiation.

The human glioblastoma cell line 308 secretes a 12.5-kDa, 112-amino acid polypeptide, termed glioblastoma-derived T-cell suppressor factor, which has suppressive effects on IL-2-dependent cell growth.[578] The amino acid sequence of the glioblastoma-derived GIF, as deduced from cDNA clones containing the full coding sequence, showed that it shares 71% homology with TGF-β.[579] The fibrosarcoma cell line 8387 produces two peptides (molecular weights 16 and 12 kDa) that can inhibit the soft agar growth of A549 lung adenocarcinoma cells and decrease the secretion of plasminogen activator by human lung fibroblasts.[580] Anti-TGF-β antibodies were found to inhibit the effects of the 16-kDa GIF (but not those of the 12-kDa GIF) in cell culture assays. Thus, 8387 fibrosarcoma cells produce two major GIFs, one of which is closely related to TGF-β. The identity of the 12-kDa GIF was not determined.

An autocrine-secreted melanoma growth-inhibiting activity, MIA, was isolated from supernatants of the human malignant melanoma cell line HTZ-19 dM.[581] The MIA is a peptide of 8 kDa that has potent GIF activity on autologous tumor cells as well as in a number of neuroectoderm-derived tumors of different grades of malignancy. Cell cycle kinetic analysis indicates that MIA induces a prolongation of the S phase and an increased arrest of cells in the G_2 phase.[582] The period during the cell cycle most sensitive to the action of MIA is the G_1 to S traverse.

Two types of tumor growth-inhibiting factors, TIF-1 and TIF-2, were isolated and purified from the A673 human rhabdomyosarcoma cell line.[583,584] TIF-1 is represented by a polypeptide of 10 to 16 kDa that is inactivated by trypsin and dithiothreitol. TIF-1 inhibits the growth of a wide range of human tumor cells and virally transforms human and rat cells in soft agar and monolayer culture but stimulates the growth of normal human fibroblasts and a normal epithelial cell line. TIF-1 blocks the stimulatory effect of both TGF-α and EGF in a soft agar assay with human tumor cells. TIF-2, purified from the rhabdomyosarcoma cell line, shares many properties in common with TIF-1 but can be distinguished by its physicochemical and biological properties. TIF-1 and TIF-2 have no antiviral activity and their growth-inhibiting effects are reversible. The TIFs may be a family of related peptides capable of inhibiting the growth of tumor cells. TIF-2 and IL-1α may be identical molecules.[585]

3. Physiological Role of GIFs

The role of endogenous GIFs in the normal regulation of cellular proliferation is unknown. Regulation of many physiological events and biological activities is governed by processes that can either activate or inhibit specific pathways. In the immune system, for example, the existence of antigen-specific and antigen-nonspecific regulatory proteins that can either promote or inhibit immune functions has been well documented. The soluble immune response suppressor (SIRS) is a protein of 14 kDa produced by mitogen- or IFN-activated suppressor T cells that inhibits antibody secretion by B lymphocytes and cell division by neoplastic cell lines.[586] IL-1, IL-2, and EGF inhibit SIRS-mediated suppression of antibody secretion by cultured mouse spleen cells.[587] Growth factors could act in some systems by interfering with the action of GIFs.

Factors classically considered as growth stimulators may display GIF activity in certain types of cells, or even in the same type of cell under different conditions. EGF, for example, stimulates growth in a wide variety of cells both *in vivo* and *in vitro*, but EGF inhibits the growth of human A431 cells despite the fact that these cells express a very high number of EGF receptors.[570] TGF-β is capable of inducing anchorage-independent growth in mouse embryo AKR-2B cells as well as in normal rat kidney (NRK) cells, although in the latter case either EGF or TGF-α is required in association with TGF-β. However, TGF-β is a bifunctional factor because it is capable of inhibiting the anchorage-independent growth of many human tumor cell lines at concentrations in the same range as those that enhance the anchorage-independent growth of NRK or AKR-2B cells.[588] In rat 3T3 fibroblasts transfected with a c-*myc* gene, TGF-β can function as either an inhibitor or an enhancer of anchorage-independent growth, depending on the particular set of growth factors operant in the cell together with TGF-β.[589] Transformation may result not only from the action of positive, stimulating factors in susceptible cells, but also from the failure of some cells to respond to factors with inhibitory properties. In any case, the general or local environment may critically modify the type of cellular response to growth factors.

4. Role of GIFs in Oncogenesis

The role of GIFs in oncogenic processes occurring in the intact animal is not understood, but studies performed with cells cultured *in vitro* suggest that loss of GIFs may be a component of such complex processes. Studies on primary rat tracheal epithelial cells suggest that two distinct mechanisms are involved in oncogenesis *in vitro*: loss of responsiveness to GIFs in the early stages of transformation and activation of growth stimulatory mechanisms resulting from inappropriate expression of growth factor receptors in late stages of transformation.[590] The precise mechanism of the loss of responsiveness to GIFs by tumor cells is unknown. Further studies will undoubtedly contribute to a better characterization of GIFs and a proper understanding of their role in oncogenesis.

IV. GROWTH FACTORS AS MEDIATORS OF HORMONE ACTION

The cellular actions of hormones may be exerted, at least in part, through the endogenous production of growth factors that may act in autocrine or paracrine fashion. The role of IGF-I as a mediator of growth hormone action (somatomedin) is well known. Available evidence strongly suggests that the actions of thyroid hormone and steroid hormones are also associated with the local production of specific growth factors in the target cells.

A. GROWTH HORMONE

Growth hormone (GH or somatotropin) is a large single-chain polypeptide produced and secreted by the anterior pituitary gland (adenohypophysis). Human GH is a protein of 20 to 22 kDa composed of 191 amino acids, although an array of different molecular forms of the hormone is contained in the circulating blood.[591] Growth hormone exerts varied effects on development, somatic growth, and metabolism.[592-594] The actions of GH are mainly anabolic, favoring the growth of bone and muscles. An excessive production of GH before puberty may result in gigantism, whereas diminished production of the hormone at this time may lead to dwarfism. Hypersecretion of GH after puberty is associated with acromegaly. The effects of GH on development and somatic growth may be mediated by the local production of a "sulfation factor" or somatomedin, which is represented by insulin-like growth factors (IGFs). In particular, IGF-I/somatomedin C may act as a local mediator of GH in peripheral adult tissues. However, other autocrine or paracrine factors in addition to somatomedin C/IGF-I may also be involved in the action of GH in certain tissues.[595]

Production and secretion of GH by the pituitary is controlled by neuroendocrine mechanisms that include both stimulatory and inhibitory factors.[596] A hypothalamic peptide of 40 or 44 amino acids, the growth hormone-releasing hormone (GHRH or somatoliberin), stimulates GH secretion, whereas another hormonal peptide of 14 or 28 amino acids, somatostatin, inhibits GH secretion. Growth hormone itself, as well as adrenergic pathways and other hormones such as glucocorticoids, sex steroids, and thyroid hormone, may exert modulatory effects on GH secretion. Metabolic intermediates also can affect this secretion, with arginine and hypoglycemia inhibiting the secretion and free fatty acids and hyperglycemia stimulating the secretion. An intrinsic hypothalamic-somatotroph rhythm conditions GH secretion in humans. A pulsatile pattern of secretion may be important for the physiologic effects of GH at the peripheral level.

Growth hormone may be synthesized and secreted at extrapituitary sites, in particular by peripheral blood mononuclear cells from normal adult individuals.[597,598] Stimulation of human lymphocytes with mitogens such as PHA results in a marked increase in GH secretion, and addition of GH to the culture augments endogenous secretion of the hormone from the mitogen-stimulated lymphocytes. The growth hormone-releasing factor, GHRF, and a somatostatin analog do not affect the secretion of GH from nonstimulated and PHA-stimulated lymphocytes. Both T and B lymphocytes can secrete GH. The physiological significance of GH secretion by lymphocytes remains to be elucidated, but the inhibitory effects obtained with the use of an antisense oligodeoxynucleotide to GH mRNA suggest that the endogenous hormone may play an autocrine/paracrine role in lymphocyte replication.[599]

Growth hormone circulates with the blood bound to growth hormone-binding proteins (GHBPs).[600,601] Two circulating GHBPs have been characterized, one of high affinity and the other of low affinity, and both are specific for GH. The high affinity GHBP corresponds to the extracellular portion of the hepatic GH receptor, whereas the low affinity does not appear to be related to this receptor. Approximately half of the circulating GH is complexed to the high affinity GHBP. The GHBPs protect GH from degradation and prolong the biological persistence of the hormone *in vivo*. The levels of GHBPs are low in the fetus and neonate, rise rapidly after birth, and stay constant through adult life. The endowment of GHBPs may be relatively fixed for a given individual and may play a pivotal role in somatic growth, determining growth rate and height potential.[602] Certain forms of short stature resistant to GH (pygmies, Laron dwarfism) may have absent or decreased GHPB levels in plasma. However, some cases of Laron syndrome associated with insensitivity to GH are due to mutations in the GH receptor, which lead to defective expression of the receptor on the cell surface.[603]

1. The Growth Hormone Receptor

The physiological effects of GH are initiated by its binding to a membrane receptor. Growth hormone receptors are present not only in liver, bone, and skeletal muscle but also in other organs and tissues, including the heart, the ovary, and the mammary gland.[604] The human placenta also expresses GH receptors.[605] The gene for the GH receptor is located on human chromosome region 5p12-p13.1.[606] The GH receptor, purified from rabbit hepatocytes, is a 130-kDa protein that contains covalently linked ubiquitin and N-linked oligosaccharides.[607] Multiple forms of the GH receptor may exist in the human liver, with two major components of 100 and 50 to 55 kDa and a minor component of 130 kDa, suggesting a possible subunit structure of the receptor.[608] The cloned cDNA sequences of the rabbit and human GH receptor predict a protein of 620 amino acid residues that contains a single transmembrane domain. The extracellular domain of the receptor contains 7 cysteine residues, and 6 of them form 3 disulfide bonds. The receptors for GH and prolactin are members of a cytokine/hematopoietic growth factor (HGF) superfamily of receptors that includes the receptors for several cytokines (IL-2 to IL-7), CSF-2, and erythropoietin.[609] These receptors are structurally related primarily in their extracellular, ligand-binding domains, but the biological significance of these homologies is unknown. The amino-terminal, extracellular sequence of the receptor for GH is identical to that of a 51-kDa GHBP present in serum. Expression in bacteria of a gene fragment encoding the extracellular domain of the human liver GH receptor showed that the unglycosylated domain of the receptor exhibits binding properties identical to those of its natural glycosylated counterpart isolated from human serum.[610]

Binding of GH to its receptor involves specific epitopes on the hormone molecule that have been identified by homolog-scanning mutagenesis.[611] Human GH forms a 1:2 complex with the extracellular domain of its receptor on the cell surface.[612] Analysis of the crystal structure of the complex between human GH and its receptor showed that the complex consists of one molecule of GH per two molecules of the receptor.[613] Dimerization of the receptor may have a crucial role in the signal transduction mechanism of GH. A constructed hybrid molecule containing the extracellular domain of the human GH receptor linked to the transmembrane and intracellular domains of the CSF-3 receptor did undergo dimerization after the addition of GH when expressed in a human myeloid leukemia cell line.[614] Inactive human GH analogs may be useful as antagonists for treating clinical conditions associated with excess of the endogenous GH such as gigantism and acromegaly.

The GH receptor purified from mouse fibroblasts is a transmembrane phosphoprotein of 114 kDa. Binding of the hormone to the receptor stimulates tyrosine phosphorylation of the receptor molecule.[615,616] Tyrosine kinase activity is associated with GH receptors purified from various types of cells from different species.[617] The cloned GH receptor itself, when expressed in GH-responsive and unresponsive cell lines, copurifies with tyrosine-specific kinase activity and is phosphorylated on tyrosine residues both *in vitro* and *in vivo*.[618] Several proteins are phosphorylated on tyrosine in preadipocytes treated with GH *in vitro*.[619] Thus, the GH receptor could possess tyrosine kinase activity. However, the deduced amino acid sequence of human and rabbit GH receptors shows that they are not tyrosine kinases, indicating that their effects on tyrosine phosphorylation are mediated by a distinct tyrosine kinase associated with the receptor molecule. A 121-kDa protein associated with the receptor may be a tyrosine kinase involved in the transduction of growth factor signal.[620] However, more than one tyrosine kinase may be involved in mediating the action of GH after receptor binding.[621] The substrates of these kinases have not been identified, with the exception of two mitogen-activated protein kinases (MAP kinases) of 39 and 42 kDa. The ability of the activated GH receptor to stimulate phosphorylation of tyrosine residues on cellular proteins in spite of the lack of sequence homology to known tyrosine kinases is a property shared with other members of the cytokine/HGF superfamily of cell membrane receptors.

2. Somatomedins

There is much controversy about the role of locally produced autocrine or paracrine growth factors that may function as cellular mediators of GH action in peripheral tissues. It has been postulated that the effects of GH on carbohydrate and lipid metabolism would be direct, but the effects on development and somatic growth would be mediated by the local production of IGFs (somatomedins).[622] Somatomedin-C/IGF-I would act as the major mediator of the actions of GH in peripheral tissues, especially in those associated with somatic growth such as the cartilage plates of long bones. Growth hormone would stimulate the longitudinal growth of bones by inducing the local expression of genes coding for IGF-I.[623-625] The local synthesis of IGF-I in chondrocytes may promote their clonal expansion through paracrine or autocrine mechanisms. The effects of human GH on the proliferation of rat osteoblast-like cells grown in serum-free long-term culture may be mediated by an autocrine mechanism involving IGF-I synthesis,[626] and a similar mechanism occurs in cultured adipocytes, in which the IGF-I gene is expressed relatively early after stimulation with growth hormone.[627] The mitogenic action of GH on human fibroblasts may be mediated by a paracrine and/or autocrine mechanism involving IGF-I production.[628] However, there is some evidence that the effects of GH on somatic growth may be independent of somatomedins. Local administration of human GH to the cartilage growth plate of the proximal tibia of hypophysectomized rats was shown to result in accelerated longitudinal bone growth, and this effect is independent of an increase in the plasma concentration of somatomedins.[629] Thus, GH could stimulate longitudinal bone growth directly. Expression of the human IGF-I gene in GH-deficient transgenic mice is capable of stimulating normal somatic growth in the animals.[630,631] However, skeletal growth is increased in GH, but not IGF-I, transgenic mice; different organs are affected in the two strains. Thus, the effects of GH and IGF-I expression in transgenic mice are not equivalent. IGF-I is a relatively poor growth-promoting agent when administered to hypophysectomized rats.[632]

The differentiation-inducing effects of GH in certain types of cells may be associated, at least in part, with the local production of IGF-I. Growth hormone alone can induce *in vitro* the differentiation of preadipocytes, but the newly differentiated adipocytes are much more sensitive to the mitogenic effect of IGF-I than precursor cells.[633] Thus, the action of IGF-I may be associated with a selective multiplication (clonal expansion) of young differentiated cells induced by GH. IGF-I acts as a mediator of GH negative feedback effects on both the hypothalamus and the pituitary.[634] It may be concluded that many, but perhaps not all, of the physiological effects of GH are mediated by the IGFs/somatomedins. However, further studies are required for a proper evaluation of the role of IGFs and other locally produced growth factors in the peripheral action of GH.

B. THYROID HORMONE

The thyroid hormones are essential for the growth, development, and metabolism of vertebrates. The biosynthesis and secretion of thyroid hormones have been characterized in detail.[635,636] There are two molecular species of thyroid hormones: 3,3′,5-triiodo-L-thyronine (T_3) and L-thyroxine (T_4). Although the main hormone secreted by the thyroid is T_4, T_3 is the main active thyroid hormone at the cellular level. Both T_4 and T_3 circulate in the blood, but the levels of T_4 are two orders of magnitude higher than those of T_3. The T_3 present in blood is originated partially from direct thyroid secretion but peripheral intracellular deiodination processes may contribute to maintain circulating T_3 levels. Thyroglobulin, the protein precursor of thyroid hormones in the thyroid gland, is a dimeric glycoprotein of 2×330 kDa.[637] The human thyroglobulin gene is located on chromosome region 8q24, distal to the c-*myc* proto-oncogene locus.

The biosynthesis and secretion of thyroid hormones depend on a diversity of endogenous and exogenous influences, which may act directly on the thyroid gland but are frequently mediated by the secretion of the TSH (thyrotropin) from the anterior pituitary gland (adenohypophysis).[638-641] TSH secretion is negatively regulated by the circulating levels of thyroid hormones. Suppression of TSH secretion by administration of T_4 to thyroidectomized rats requires the intrapituitary conversion of T_4 into T_3 and the presence of T_3 within the nucleus of pituitary cells. In turn, TSH secretion by the pituitary depends on hypothalamic factors, in particular on the stimulating action of the tripeptide TRH.

Thyroid hormone is crucially involved in the regulation of physiological processes associated with cell proliferation and differentiation as well as with many diverse metabolic reactions. The cellular mechanisms of thyroid hormone action are understood only partially but the available evidence indicates that the nucleus is a major site for the cellular actions of thyroid hormone and that thyroid hormone has an important role in the regulation of gene expression.[642-648] However, the actions of thyroid hormone are pleiotropic and the hormone can exert important physiological effects at the level of the plasma membrane, the mitochondria, and the cytoplasm as well.

Thyroid hormone is essential for the normal development of the brain. It is required briefly after birth to accelerate the rate of microtubule assembly in the brain, thus allowing intensive neurite growth during the critical period of brain development.[649] Axonal and dendritic maturation, which is required for the establishment of the final wiring pattern of neuronal networks, is mediated via tubulin polymerization and neuronal microtubule growth, and these processes are tightly regulated by thyroid hormone.[650] Early treatment of newborns with congenital hypothyroidism with thyroid hormone may prevent irreversible brain damage. The recent advances in molecular biology have a potential impact on diagnosis and treatment of thyroid disorders.[651]

1. Thyroid Hormone Receptors

Multiple molecular forms of thyroid hormone receptors, involved in the regulation of gene expression, have been characterized.[652] Thyroid hormone receptors are present in the nuclei of most mammalian cells and are associated with the chromatin.[653-655] The receptor is a single polypeptide chain that binds either T_3 or T_4,[656] but most of the hormone bound to nuclear receptors is represented by T_3.[657] Thyroid hormone receptors are associated with DNA.[658] The nucleus of most cells contain 4000 to 8000 copies of thyroid hormone receptors. Responsiveness to thyroid hormone varies with receptor content in the nucleus and occupancy of receptor by T_3. Regulation of thyroid hormone receptor expression depends on numerous endogenous and exogenous factors. The receptors are low in human fetal tissues and, in the adult, are markedly reduced by starvation and illness. Developmental processes may be associated with marked changes in thyroid hormone receptor expression. An increase in binding capacity for T_3 is observed in tadpole tail nuclei during metamorphosis.[659]

Thyroid hormone-binding sites have been found in non-nuclear fractions of rodent cells, including mitochondria, cytosol, and plasma membrane.[660-664] A receptor-mediated uptake of T_3 would exist at the level of the cell membrane and would provide an additional regulatory site for thyroid hormone concentration within the cell.[665] A direct action of thyroid hormone on mitochondria through receptors located on the inner membrane of the organelle has been suggested,[666,667] and there is some evidence that thyroid hormone stimulates mitochondrial RNA synthesis in the rat liver.[668,669] However, the precise biological significance of extranuclear-binding sites for thyroid hormone remains to be determined, and the physiological actions of thyroid hormone depend primarily on its binding to nuclear receptors.

The nuclear thyroid hormone receptor purified from rat liver was identified as a protein of 49 kDa.[670] However, there is more than one molecular species of thyroid hormone receptor. A 46-kDa protein with affinity for thyroid hormone detected in rat liver was found to be a degradation product of the nuclear thyroid hormone receptor, whose molecular mass would be 57 kDa.[671] Studies with specific monoclonal antibodies indicate that the 57-kDa protein is a receptor for thyroid hormone.[672] Nuclear thyroid hormone receptors present in malignant cells such as the human hepatoma cell line Hep62 may be structurally identical to the normal receptor.[673] Expression of thyroid hormone receptor genes is regulated at the transcriptional, posttranscriptional, and translational levels in different types of cells, which may result in the production of multiple structurally related proteins.[674] The receptors for steroid hormones and thyroid hormones constitute a family of structurally and functionally related proteins.[675] Both types of receptors are members of a family of DNA-binding finger proteins.[676] These proteins contain "finger" motifs of 29 to 30 amino acids forming a small independent structural domain made of a loop centered around a zinc atom coordinated by two cysteine and two histidine residues located at invariant positions. The finger model postulates that the tip of each loop is in direct contact with the DNA and determines the DNA-binding specificity. Nuclear thyroid hormone receptors are associated with the chromatin and may increase during the S phase of the cycle.[677]

Thyroid hormone receptors are encoded by the c-erb-A proto-oncogenes and are closely related to the v-Erb-A oncoprotein encoded by the avian erythroblastosis virus (AEV).[678] The product of the normal c-erb-A proto-oncogene is a high-affinity nuclear receptor for thyroid hormone.[679,680] The predicted human thyroid hormone receptor is a protein of 52 to 55 kDa with 47% amino acid sequence identity with the human glucocorticoid receptor and 52% identity with the estrogen receptor. The highest degree of structural similarity occurs in a cysteine-rich sequence of 65 amino acids beginning at c-Erb-A amino acid residue 102, and this region would represent the DNA-binding domain of the protein. The c-Erb-A protein is phosphorylated by cellular enzymes in its amino-terminal domain at two distinct serine residues. Phosphorylation of the avian c-Erb-A1/α thyroid hormone receptor protein on Ser-12 is effected by the enzyme casein kinase II.[681]

There is more than one human gene coding for thyroid hormone receptors and multiple transcripts for these receptors have been detected in a variety of cells. A thyroid hormone receptor-encoding gene, termed c-erb-A1, c-erb-Aα, or TRHA, is located on human chromosome region 17q11.2[682,683] and encodes the thyroid hormone receptor protein c-Erb-A1 (TR-α). The TRHA gene is organized in 10 exons distributed along 27 kb and overlaps in its 3′ region a structurally related gene, ear-1.[684] TRHA encodes a polypeptide of 410 amino acids. A second gene, termed c-erb-A2, c-erb-Aβ, or TRHB, encoding the thyroid hormone receptor protein c-Erb-A2 (TR-β), is located on human chromosome region 3p21-p25.[685]

Three forms of nuclear thyroid hormone receptors have been identified and characterized in the rat: TR-α1, TR-β1, and TR-β2. The two β forms (TR-β1 and TR-β2) result from alternative splicing mechanisms. TR-β2 expressed in mouse neuroblastoma cells does not bind thyroid hormone.[686] Another nonthyroid hormone-binding protein, TR-α2, is generated by alternative RNA splicing and differs from the hormone-binding form, TR-α1, at the carboxyl terminus. Thus, the type-2 forms of TR-α and TR-β may not bind thyroid hormone and their function is unknown. Regulation of the biosynthesis of the alternatively spliced TR-α1 and TR-α2 molecules may take place through an antisense mechanism involving an RNA, Rev-ErbAα mRNA, which is transcribed in the opposite direction and is complementary to TR-α1 but not TR-α2 mRNA.[687] The Rev-ErbAα transcript base pairs with the complementary pre-mRNA and blocks formation of TR-α2 but not TR-α1 mRNA.

The different forms of thyroid hormone receptors may have different tissue distribution and functions. While TR-β1 is widely expressed in many different tissues, expression of TR-β2 is restricted mainly to the pituitary. TR-α1 is most abundantly expressed in skeletal muscle and adipose tissue. Expression of different c-erb-A mRNAs varies among different regions of the rat brain and during development, suggesting different roles for thyroid receptor isoforms in brain development and function.[688] The levels of TR-α1 mRNA decrease with T_3 treatment in all tissues except in the brain, where there is no change.[689] However, changes in the mRNAs that code for specific subpopulations of thyroid hormone receptors do not always parallel changes in total nuclear T_3 binding. Expression of the different thyroid hormone receptors is downregulated in a differential fashion by TRH in GH$_3$ rat pituitary cells.[690] TSH and insulin/IGF-I are involved in a differential expression of the thyroid hormone receptor genes in FRTL-5 rat thyroid cells.[691] The levels of TR-α1 and TR-α2 mRNA, but not TR-β1 or TR-β2 mRNA, are negatively regulated by TSH, insulin/IGF-I in FRTL-5 cells. Differential regulation of thyroid hormone receptor isoforms may have important consequences for thyroid hormone action *in vivo*.

2. Growth Factors as Mediators of Thyroid Hormone Action

At least in certain types of cells, the action of thyroid hormone (T_3) may be indirect and may be mediated by the local synthesis of an autocrine growth factor. T_3 stimulates the division of GH$_4$C$_1$ and GC rat pituitary tumor cells by causing the cells to secrete an autocrine factor that seems to be a stable protein of high molecular weight.[692,693] Conditioned medium from the pituitary cells cultured in the presence of physiologic concentrations of T_3 stimulates growth of T_3-depleted pituitary cell cultures even in the presence of anti-T_3 serum. The structure and function of the putative autocrine growth factor induced by thyroid hormone are unknown. In particular, it is not known if this factor acts directly as a growth stimulator, acts permissively to allow expression of another activity, or relieves an inhibition. There is evidence that a thyroid hormone-mediating activity (thyromedin) present in normal serum may be represented by apotransferrin.[694]

C. STEROID HORMONES

Steroid hormones are represented in vertebrates by the androgens, estrogens, gestagens, glucocorticoids, and mineralocorticoids of adrenal or gonadal origin.[695] Steroid hormone biosynthesis in the gonads and adrenals occurs from cholesterol through a series of hydroxylations and other modifications involving the action of various enzymes. The process of steroid hormone biosynthesis is regulated by humoral factors that include the pituitary hormones, adrenocorticotropic hormone (ACTH) and gonadotropins. Insulin may influence steroid hormone metabolism.[696] The active vitamin D derivative 1,25-dihydroxyvitamin D$_3$ (calcitriol), which is synthesized from 2-dehydrocholesterol by successive steps in the skin, the liver, and the kidney, may be properly considered as a steroid hormone.[697]

1. Steroid Hormone Receptors

The classic model proposed for the mechanism of action of steroids proposes the hormones penetrate the cell membrane and bind to specific cytosolic receptors. After binding, the receptor is activated in some manner and the steroid hormone-receptor complex is rapidly translocated into the nucleus, where it may interact with some component of the chromatin, which may result in the stimulation of transcriptional and translational processes leading to the synthesis of specific mRNAs and proteins.[698] The diverse physiologic effects of steroid hormones are exerted primarily at the level of gene transcription.[699,700] Steroid hormone receptors are members of a superfamily of nuclear receptors that also includes the receptors for retinoids and thyroid hormones.[701] The members of this family recognize specific DNA sequences and have an important role in a diversity of normal physiological processes as well as in oncogenesis.[702]

Binding of ligand-activated steroid hormone receptors to specific DNA sequences, called hormone response elements (HREs) or steroid regulatory elements (SREs), is a crucial step in the mechanism of action of steroid hormones. Binding of steroid hormone to the HRE may result in DNA bending,[703] which may facilitate interaction between proteins bound to separate sites on DNA. Steroid hormone-receptor complexes can bind directly to DNA regions that modulate promoter activity and stimulate transcription.[704] The acceptor sites for steroid hormone receptors may be located at positions distant from the genes the steroids ultimately regulate.[705] HREs can affect transcription from distant sites and can act in a cooperative fashion.[706] Sequences required for interactions of steroids with specific binding sites at the DNA level are represented by oligonucleotide segments that are partially symmetric (palindromic). In the intact cell, but not in acellular systems *in vitro*, hormone binding to the receptor is required for the interaction of the receptor with the specific regulatory HRE DNA sequences. The HREs are located on the promoter region of target cells and contain DNA half-sites (core-binding motifs) that are recognized not only by the steroid hormone receptors but also by other members of the nuclear receptor superfamily that are involved in the regulation of gene transcription. The receptors for thyroid hormone and retinoic acid recognize direct repeats in the DNA, rather than palindromic sequences. Differences in both the relative orientation and spacing of core-binding motifs within HREs determine the selective transcriptional responses of the members of the nuclear receptor superfamily.[707,708]

2. Growth Factors as Mediators of Steroid Hormone Action

The physiological effects of steroid hormones, in particular those of androgens and estrogens, may be indirect and may be associated with the local production of growth factors. Characterization of such factors is important because it could provide a possibility for preventing promotion of hormone-responsive cancers.

Androgen-induced growth factors (AIGFs) could be responsible for the physiological effects exerted by testosterone in the prostate and other target tissues. However, the putative AIGFs remain little characterized. The possible role of a factor, the prostate growth factor (PGF), purified from human benign hypertrophic prostates,[709] in prostatic hypertrophy and prostate cancer is not clear at present. A similar or identical factor, called prostate-derived growth factor (PrDGF), was purified from acid extracts of rat prostatic tissue.[710] PrDGF appears to be a 25-kDa glycoprotein that is different from other known growth factors in both chemical composition and biological properties. An autocrine growth factor induced by androgen in SC-3 cells (derived from an androgen-responsive mouse mammary carcinoma, the Shionogi carcinoma 115), is represented by an FGF-like polypeptide.[711-713] The AIGF produced by SC-3 cells may function in an autocrine manner, interacting with the FGF receptor expressed in these cells. Suramin, a polysulfonated naphthylurea, can interrupt the androgen-dependent autocrine loop in Shionogi carcinoma 115 cells, probably by interacting with the FGF receptor.[714] AIGF also stimulates in a paracrine manner the growth of androgen-unresponsive cancer cells derived by progression of the responsive cancer cells. cDNA cloning predicts that the AIGF secreted by SC-3 cells is a protein of 215 amino acids with a putative signal peptide and that it shares 30 to 40% homology with known members of the FGF family. AIGF mRNA is markedly induced by testosterone, and expression of AIGF cDNA in mammalian cells shows that AIGF has a stimulatory effect on cell growth in the absence of androgen. It thus seems likely that androgen-stimulated growth of SC-3 cells is mediated by the AIGF secreted by these cells and that the AIGF functions through an autocrine mechanism.

The mechanism of estrogen action on cell proliferation may also be indirect and may require the operation of specific pathways that would be present only in the estrogen-sensitive cells. Estradiol has no stimulating effect on the growth of mouse 3T3 fibroblasts expressing a human estrogen receptor and may rather exert, at low concentrations, a toxic effect in the transfected cells, inducing a marked inhibition in their proliferation.[715] Estrogens would act on at least some of their target cells through the production of specific mediators (estromedins). A number of studies with both normal and transformed cells suggest the existence of such mediators. Acid homogenates prepared from the uterus of several mammalian species are able to stimulate DNA synthesis in a variety of cell types, including estrogen target and nontarget tissues, and the growth-promoting effects of these homogenates are regulated by estradiol.[716] Stimulation of cell proliferation by estradiol in its target organs such as the uterus would occur only through the secretion of autocrine or paracrine factors (estromedins), which mediate estradiol responses.

Estrogen-induced stimulation of colony formation in NMU-induced rat mammary tumors grown in culture in soft agar is apparently mediated by a secretory protein that is released into the medium and acts as a growth factor.[717] The growth of SC115 mouse mammary carcinoma cells is stimulated by either physiological doses of androgen or pharmacological doses of estrogen *in vivo* but only by androgen in cell culture, suggesting that the action of estrogen on these tumor cells may be indirect.[718] In CAMA mammary carcinoma cells grown in a chemically defined medium with 1% steroid-stripped serum, the estrogen-stimulatory effect is nullified and the cells are more sensitive to antiestrogens, suggesting that some serum component is involved in both estrogen and antiestrogen action.[719]

A number of growth factors and other proteins have been proposed to act as mediators of estrogen action on the target cells. The growth factors produced by mammary tumor cells would not be bloodborne or sufficiently active at a distance, because grafts of hormone-independent mammary tumor tissue in mice cannot replace estrogen in maintaining the growth of hormone-dependent mammary tumors.[720] Endocrine therapy of human breast cancer may inhibit tumor growth by interfering with the expression and secretion of peptide growth factors by the tumor cells.[721]

EGF is a putative estromedin in the uterus. Pre-pro-EGF is present in the mouse uterus and, although the synthesis of this precursor is not estrogen-dependent, its estrogen-stimulated proteolytic processing could be part of a putative autocrine circuit triggered by estrogens.[722] The growth inhibitory action of antiestrogens and progestins on human breast cancer cell lines such as T47D cells can be partially reversed by treatment with EGF or insulin, suggesting that autocrine action of EGF and insulin is partially responsible for the estrogen-stimulated growth of mammary cancer cells.[723,724] TGF-α, which recognizes the EGF receptor, may have a role in the mechanisms of estrogen action in some types of cells.

Differentiated, estrogen-dependent rat mammary adenocarcinomas produce TGF-α during the initial phase of their growth and under estrogen stimulation.[725] By contrast, more aggressive, metastatic rat mammary tumors that are estrogen-independent do not produce TGF-α constitutively, suggesting that TGF-α may be unnecessary for the growth of rat mammary tumors and that escape from hormone dependency may be unrelated to the constitutive production of TGF-α or to an autocrine response to this growth factor. Blockade of the TGF-α/EGF receptor in MCF-7 breast tumor cells with monoclonal or polyclonal antibodies to the receptor does not alter estrogen-regulated growth, suggesting that TGF-α is not a primary mediator of the growth effects of estrogen in these cells.[726] Interaction between EGF/TGF-α and the EGF receptor would not be an obligatory event in the mediation of estrogen-stimulated proliferation of the breast cancer cell line CAMA-1.[727] However, there is evidence that TGF-α may function as a mediator of estrogen action in the mouse uterus.[728] Injection of estrogen to immature female mice results in the expression of TGF-α mRNA in uterine epithelial cells and the secretion of TGF-α into uterine luminal fluid. This induction is specific to estrogen, and an antibody specific to TGF-α reduces estrogen-mediated uterine growth in the treated animals. The action of TGF-α on mouse uterine cells is exerted through EGF receptors that are expressed in these cells. TGF-α could thus function as an autocrine/paracrine mediator of estrogen-induced growth in the normal mouse uterus.

Estrogen stimulation of the proliferation of MCF-7 human breast cancer cells may result in the secretion of peptides which, when partially purified, can replace estrogen as a mitogen.[729-732] The growth factor activities induced by estrogen in MCF-7 cells are sufficient to stimulate MCF-7 tumor growth in ovariectomized athymic mice, thus replacing estradiol. These factors include IGF-I, EGF, TGF-α-like material, TGF-β, a competency factor, and an additional epithelial anchorage-independent growth-promoting activity. While secretion of TGF-α and other factors may have a growth-stimulatory effect on MCF-7 cells, TGF-β has a growth-inhibitory effect, and secretion of TGF-β by MCF-7 cells is stimulated by antiestrogens.[733] The inhibitory effect of TGF-β on MCF-7 cells can be partially blocked by TGF-β-specific antibody. The sensitivity of MCF-7 cells to TGF-β is closely correlated with the cellular response to estradiol: estrogen receptor-positive MCF-7 cells are sensitive to TGF-β, whereas estrogen receptor-defective cells or other clones with altered response to estradiol are not.[734] The repertoire of secreted proteins is markedly altered in the variant MCF-7 cells.

In contrast to MCF-7 cells, estrogen treatment of the T47D human breast cancer cell line does not result in stimulation of cell proliferation.[735] T47D cells show genetic instability and some variants of this cell line are insensitive to growth regulation by estrogen, antiestrogen, and progestin due to mechanisms distal to receptor binding.[736] The sensitivity to treatment with progestins or antiestrogens of a T47D variant that does not express EGF and expresses TGF-α at every low levels is markedly enhanced.[737] The endogenous expression of growth factors in some tumor cells may be associated with decreased sensitivity to growth-inhibitory agents.

TGF-β could act as a mediator for some of the effects of estrogens on some of their target cells. Acute estrogen deficiency induced by ovariectomy in rats reduces the deposition of TGF-β in bone tissue, and the diminished skeletal TGF-β could play a role in the pathogenesis of bone loss, fractures, and microfractures that occur in estrogen-deficient states.[738] By contrast, ovariectomy does not alter the amount of IGF-I or IGF-II in rat bone tissue. The growth inhibitory effects of estrogen on tumor cells could be mediated, at least in part, by autocrine or paracrine mechanisms associated with endogenous production and secretion of TGF-β. Estrogen inhibits the proliferation of UMR106 cells (a clonal cell line derived from a rat osteogenic sarcoma), and this effect is blocked by monoclonal antibodies to TGF-β, suggesting that TGF-β endogenous secretion in UIMR106 cells may be responsible for the inhibitory effect of estrogen on these cells.[739]

IGF-I is a candidate for estromedin function in both normal and neoplastic cells. Expression of IGF-I in the uterus is strongly increased after administration of 17β-estradiol to ovariectomized prepubertal rats.[740] A marked increase in the abundance of uterine IGF-I mRNA occurred after administration of estrogen to the animals. No increase of IGF-I mRNA was observed in the liver or the kidney of the estrogen-treated rats, indicating that the effect is specific for the target organ of estrogen. Estrogen alone is hardly mitogenic for MCF-7 human breast cancer cells, and these cells require insulin or IGFs for their proliferation.[741] Estrogen sensitizes MCF-7 cells to the effects of growth factors that act through the type-1 IGF receptor.[742] The action of estrogen on MCF-7 cells may be indirect and may be mediated, at least in part, by endogenous production of IGF-I. However, MCF-7 cells do not secrete IGFs as a result of estradiol treatment and the results of some studies do not support a possible autocrine function of IGFs

42

in the proliferation of MCF-7 cells.[743] On the other hand, estradiol may stimulate the secretion of IGFs in the rat osteosarcoma-derived clonal cell line UMR106.[744] Expression of a glycoprotein of 52 kDa is regulated by estradiol in MCF-7 human breast cancer cells, and there is evidence that this protein has growth-promoting activity and may act as a mitogen in an autocrine fashion in the same cells.[745] The 52-kDa glycoprotein is found not only in breast cancer cells but is also present in several types of normal cells and fluids as well as in a high percentage of benign mastopathies.[746] The function of this protein is unknown.

Different human cell lines may exhibit different responses in the production of endogenous growth factors when stimulated with estradiol, and in some cell lines no putative estromedins have been identified. Estrogen may act in the MCF-7 human breast cancer cell line through inducing the synthesis and secretion of growth factors such as IGF-I, TGF-α, and/or PDGF, whereas these factors do not appear to play a role in estrogen-induced growth of the human breast cancer cell line CAMA-1.[747] Growth of the mouse osteoblastic cell line MC3T3-E1, maintained in serum-free medium, is directly stimulated by estradiol and is also stimulated by monocyte-conditioned medium. However, conditioned medium prepared by monocytes cultured in the presence of estradiol inhibits the growth of MC3T3-E1 cells.[748] These results show that estradiol can not only exert a direct effect on the growth of these cells, but can also affect their growth through modulating the release of some local regulators of bone metabolism from monocytes. In any case, the microenvironment where estradiol (and other hormones) exerts its action may be of crucial importance in relation to the final effect.

In conclusion, there is evidence that locally produced growth factors may be involved in the action of several hormones in at least some of their target cells through autocrine or paracrine mechanisms, but the possible role of these types of mechanisms in the intact organism remains to be elucidated.

REFERENCES

1. **Pimentel, E.,** Cellular mechanisms of hormone action. I. Transductional events, *Acta Cient. Venez.,* 29, 73, 1978.
2. **Pimentel, E.,** Cellular mechanisms of hormone action. II. Posttransductional events, *Acta Cient. Venez.,* 29, 147, 1978.
3. **Gospodarowicz, D.,** Growth factors and their action *in vivo* and *in vitro, J. Pathol.,* 141, 201, 1983.
4. **Heldin, C.-H. and Westermark, B.,** Growth factors: mechanism of action and relation to oncogenes, *Cell,* 37, 9, 1984.
5. **James, R. and Bradshaw, R. A.,** Polypeptide growth factors, *Annu. Rev. Biochem.,* 53, 259, 1984.
6. **Sporn, M. B. and Roberts, A. B.,** Autocrine, paracrine and endocrine mechanisms of growth control, *Cancer Surv.,* 4, 627, 1985.
7. **Deuel, T. F.,** Polypeptide growth factors: roles in normal and abnormal cell growth, *Annu. Rev. Cell Biol.,* 3, 443, 1987.
8. **Pusztai, L., Lewis, C. E., Lorenzen, J., and McGee, J. O.,** Growth factors: regulation of normal and neoplastic growth, *J. Pathol.,* 169, 191, 1993.
9. **Massagué, J. and Pandiella, A.,** Membrane-anchored growth factors, *Annu. Rev. Biochem.,* 62, 515, 1993.
10. **Canalis, E., McCarthy, T., and Centrella, M.,** Growth factors and the regulation of bone remodeling, *J. Clin. Invest.,* 81, 277, 1988.
11. **Wahl, S. M., Wong, H., and McCartney-Francis, N.,** Role of growth factors in inflammation and repair, *J. Cell. Biochem.,* 40, 193, 1989.
12. **Pierce, G. F., Mustoe, T. A., Lingelbach, J., Masakowski, V. R., Griffin, G. L., Senior, R. M., and Deuel, T. F.,** Platelet-derived growth factor and transforming growth factor-β enhance tissue repair activities by unique mechanisms, *J. Cell Biol.,* 109, 429, 1989.
13. **ten Dijke, P. and Iwata, K. K.,** Growth factors for wound healing, *Bio-Technology,* 7, 793, 1989.
14. **Gay, S., Trabandt, A., Moreland, L. W., and Gay, R. E.,** Growth factors, extracellular matrix, and oncogenes in scleroderma, *Arthritis Rheumat.,* 35, 304, 1992.
15. **Sporn, M. B. and Roberts, A. B.,** Peptide growth factors and inflammation, tissue repair, and cancer, *J. Clin. Invest.,* 78, 329, 1986.
16. **Salomon, D. S. and Perroteau, I.,** Growth factors in cancer and their relationship to oncogenes, *Cancer Invest.,* 4, 43, 1986.
17. **Nilsson, J.,** Growth factors and the pathogenesis of atherosclerosis, *Atherosclerosis,* 62, 185, 1986.
18. **Heldin, C.-H., Betsholtz, C., Claesson-Welsh, L., and Westermark, B.,** Subversion of growth regulatory pathways in malignant transformation, *Biochim. Biophys. Acta,* 907, 219, 1987.

19. **Pimentel, E.,** *Hormones, Growth Factors, and Oncogenes,* CRC Press, Boca Raton, FL, 1987.

20. **Pimentel, E.,** Hormonas, factores de crecimiento y oncoproteínas, *Gac. Med. Caracas,* 100, 93, 1992.

21. **Thompson, T. C.,** Growth factors and oncogenes in prostate cancer, *Cancer Cells,* 2, 345, 1990.

22. **Piccart, M. J., Devleeschouwer, N., Leclercq, G., and L'Hermite, M.,** Les facteurs de croissance locaux du cancer mammaire: revue générale et intérêt pour le clinicien, *Bull. Cancer,* 78, 215, 1991.

23. **Albino, A. P.,** The role of oncogenes and growth factors in progressive melanoma-genesis, *Pigment Cell Res.,* Suppl. 2, 199, 1992.

24. **Pimentel, E.,** Peptide hormone precursors, subunits, and fragments as human tumor markers, *J. Biomed. Lab. Sci.,* 1, 47, 1988.

25. **Pimentel, E.,** Factores de crecimiento como marcadores tumorales, *Gac. Med. Caracas,* 101, 124, 1993.

26. **Sumitani, S., Kasayama, S., Hirose, T., Matsumoto, K., and Sato, B.,** Effects of thyroid hormone on androgen- or basic fibroblast growth factor-induced proliferation of Shionogi carcinoma 115 mouse mammary carcinoma cells in serum-free culture, *Cancer Res.,* 51, 4323, 1991.

27. **Alderman, E. M., Lobb, R. R., and Fett, J. W.,** Isolation of tumor-secreted products from human carcinoma cells maintained in a defined protein-free medium, *Proc. Natl. Acad. Sci. U.S.A.,* 82, 5771, 1985.

28. **Hayashi, I. and Sato, G. H.,** Replacement of serum by hormones permits growth of cells in a defined medium, *Nature,* 259, 132, 1976.

29. **Pyke, K. W. and Gogerly, R. L.,** Murine fetal colon *in vitro*: assays for growth factors, *Differentiation,* 29, 56, 1985.

30. **Plouet, J., Courty, J., Olivié, M., Courtois, Y., and Barritault, D.,** A highly reliable and sensitive assay for the purification of cellular growth factors, *Cell. Mol. Biol.,* 30, 105, 1984.

31. **Pardee, A. B., Dubrow, R., Hamlin, J. L., and Kletzien, R. F.,** Animal cell cycle, *Annu. Rev. Biochem.,* 47, 715, 1978.

32. **Baserga, R.,** Molecular biology of the cell cycle, *Int. J. Radiat. Biol.,* 49, 219, 1986.

33. **Wang, J. C.,** DNA polymerases — why so many, *J. Biol. Chem.,* 266, 6659, 1991.

34. **Pardee, A. B., Coppock, D. L., and Yang, H. C.,** Regulation of cell proliferation at the onset of DNA synthesis, *J. Cell Sci.,* Suppl. 4, 171, 1986.

35. **Melchers, F. and Lernhardt, W.,** Three restriction points in the cell cycle of activated murine B lymphocytes, *Proc. Natl. Acad. Sci. U.S.A.,* 82, 7681, 1985.

36. **O'Farrell, P. H., Edgar, B. A., Lakich, D., and Lehner, C. F.,** Directing cell division during development, *Science,* 246, 635, 1989.

37. **Hartwell, I. L.,** Defects in a cell cycle checkpoint may be responsible for the genomic instability of cancer cells, *Cell,* 71, 543, 1992.

38. **Lloyd, D.,** Biochemistry of the cell cycle, *Biochem. J.,* 242, 313, 1987.

39. **Maller, J. L.,** Mitotic control, *Curr. Opin. Cell Biol.,* 3, 269, 1991.

40. **Nurse, P.,** Eukaryotic cell cycle control, *Biochem. Soc. Trans.,* 20, 239, 1992.

41. **Kidd, V. J.,** Cell division control-related protein kinases: putative origins and functions, *Mol. Carcinogenesis,* 5, 95, 1992.

42. **Norbury, C. and Nurse, P.,** Animal cell cycles and their control, *Annu. Rev. Biochem.,* 61, 441, 1992.

43. **Jacobs, T.,** Control of the cell cycle, *Dev. Biol.,* 153, 1, 1992.

44. **Masui, Y.,** Towards understanding the control of the division cycle in animal cells, *Biochem. Cell Biol.,* 70, 920, 1992.

45. **Marsh, J.,** Ed., *Regulation of the Eukaryotic Cell Cycle, CIBA Found. Symp.,* Vol. 170, John Wiley & Sons, Chichester, England, 1992.

46. **Lu, K. P. and Means, A. R.,** Regulation of the cell cycle by calcium and calmodulin, *Endocrine Rev.,* 14, 40, 1993.

47. **Lin, D., Shields, M. T., Ullrich, S. J., Appella, E., and Mercer, W. E.,** Growth arrest induced by wild-type p53 protein blocks cells prior to or near the restriction point in late G_1 phase, *Proc. Natl. Acad. Sci. U.S.A.,* 89, 9210, 1992.

48. **Kastan, M. B., Zhan, Q., El-Deiry, W. S., Carrier, F., Jacks, T., Walsh, W. V., Plunkett, B. S., Vogelstein, B., and Fornace, A. J., Jr.,** A mammalian cell cycle checkpoint pathway utilizing p53 and *GADD45* is defective in ataxia-telangiectasia, *Cell,* 71, 587, 1992.

49. **Hamel, P. A., Gallie, B. L., and Phillips, R. A.,** The retinoblastoma protein and cell cycle regulation, *Trends Genet.,* 8, 180, 1992.

50. **Zetterberg, A. and Larsson, O.,** Kinetic analysis of regulatory events in G_1 leading to proliferation or quiescence of Swiss 3T3 cells, *Proc. Natl. Acad. Sci. U.S.A.,* 82, 5365, 1985.

51. **Gadbois, D. M., Crissman, H. A., Tobey, R. A., and Bradbury, E. M.,** Multiple kinase arrest points at the G_1 phase of nontransformed mammalian cells are absent in transformed cells, *Proc. Natl. Acad. Sci. U.S.A.,* 89, 8626, 1992.

52. **Hartwell, L. H. and Weinert, T. A.,** Checkpoints: controls that ensure the order of cell cycle events, *Science,* 246, 629, 1989.

53. **Whitaker, M. and Patel, R.,** Calcium and cell cycle control, *Development,* 108, 525, 1990.

54. **Guermah, M., Gillet, G., Michel, D., Laugier, D., Brun, G., and Calothy, G.,** Down regulation by p60^{v-src} of genes specifically expressed and developmentally regulated in postmitotic quail neuroretina cells, *Mol. Cell. Biol.,* 10, 3584, 1990.

55. **Sorrentino, V.,** Growth factors, growth inhibitors and cell cycle control, *Anticancer Res.,* 9, 1925, 1989.

56. **Lu, X., Kopun, M., and Werner, D.,** Cell cycle phase-specific cDNA libraries reflecting phase-specific gene expression of Ehrlich ascites cells growing *in vivo, Exp. Cell Res.,* 174, 199, 1988.

57. **Eki, T., Enomoto, T., Miyajima, A., Miyazawa, H., Murakami, Y., Hanaoka, F., Yamada, M., and Ui, M.,** Isolation of temperature-sensitive cell cycle mutants from mouse FM3A cells. Characterization of mutants with special reference to DNA replication, *J. Biol. Chem.,* 265, 26, 1990.

58. **Hirschberg, J. and Marcus, M.,** Isolation by a replica-plating technique of Chinese hamster temperature-sensitive cell cycle mutants, *J. Cell. Physiol.,* 113, 159, 1982.

59. **Handeli, S. and Weintraub, H.,** The ts41 mutation in Chinese hamster cells leads to successive S phases in the absence of intervening G_2, M, and G_1, *Cell,* 71, 599, 1992.

60. **Cooper, S.,** The continuum model and c-*myc* synthesis during the division cycle, *J. Theor. Biol.,* 135, 393, 1988.

61. **Okuda, A., Masuzaki, A., and Kimura, G.,** Increase in c-*fos* and c-*myc* mRNA levels in untransformed and SV40-transformed 3Y1 fibroblasts after addition of serum: its relationship to the control of initiation of S phase, *Exp. Cell Res.,* 185, 258, 1989.

62. **Baserga, R.,** The cell cycle: myths and realities, *Cancer Res.,* 50, 6769, 1990.

63. **Koli, K. and Keski-Oja, J.,** Cellular senescence, *Ann. Med.,* 24, 313, 1992.

64. **Hayflick, L. and Moorhead, P. S.,** The serial cultivation of human diploid cell strains, *Exp. Cell Res.,* 25, 585, 1961.

65. **Hayflick, L.,** The limited *in vitro* lifetime of human diploid strains, *Exp. Cell Res.,* 37, 614, 1965.

66. **Holliday, R.,** Growth and death of diploid and transformed human fibroblasts, *Fed. Proc.,* 34, 51, 1975.

67. **Matsuda, T., Okamura, K., Sato, Y., Morimoto, A., Ono, M., Kohno, K., and Kuwano, M.,** Decreased response to epidermal growth factor during cellular senescence in cultured human microvascular endothelial cells, *J. Cell. Physiol.,* 150, 510, 1992.

68. **Phillips, P. D. and Pignolo, R. J.,** Altered expression of cell cycle dependent genes in senescent WI-38 cells, *Exp. Gerontol.,* 27, 403, 1992.

69. **Glücksmann, A.,** Cell death in normal vertebrate ontogeny, *Biol. Rev.,* 26, 59, 1951.

70. **Saunders, J. W.,** Death in embryonic systems, *Science,* 154, 604, 1966.

71. **Kerr, J. F. R., Searle, J., Harmon, B. V., and Bishop, C. J.,** Apoptosis, in *Perspectives on Mammalian Cell Death,* Potten, C. S., Ed., Oxford University Press, Oxford, 1987, 93.

72. **Gerschenson, L. E. and Rotello, R. J.,** Apoptosis — a different type of cell death, *FASEB J.,* 6, 2450, 1992.

73. **Vaux, D. L.,** Toward an understanding of the molecular mechanisms of physiological cell death, *Proc. Natl. Acad. Sci. U.S.A.,* 90, 786, 1993.

74. **Wyllie, A. H.,** Apoptosis, *Br. J. Cancer,* 67, 205, 1993.

75. **Schwartzman, R. A. and Cidlowski, J. A.,** Apoptosis — the biochemistry and molecular biology of programmed cell death, *Endocrine Rev.,* 14, 133, 1993.

76. **Schwartz, L. M., Smith, S. W., Jones, M. E. E., and Osborne, B. A.,** Do all programmed cell deaths occur via apoptosis?, *Proc. Natl. Acad. Sci. U.S.A.,* 90, 980, 1993.

77. **Yanagihara, K. and Tsumuraya, M.,** Transforming growth factor beta 1 induces apoptotic cell death in cultured human gastric carcinoma cells, *Cancer Res.,* 52, 4042, 1992.

78. **Shi, Y., Glynn, J. M., Guilbert, L. J., Cotter, T. G., Bissonnette, R. P., and Green, D. R.,** Role for c-*myc* in activation-induced apoptotic cell death in T cell hybridomas, *Science,* 257, 212, 1992.

79. **Bissonnette, R. P., Echeverri, F., Mahboubi, A., and Green, D. R.,** Apoptotic cell death induced by c-*myc* is inhibited by *bcl*-2, *Nature,* 359, 552, 1992.

80. **Wagner, A. J., Small, M. B., and Hay, N.,** Myc-mediated apoptosis is blocked by ectopic expression of Bcl-2, *Mol. Cell. Biol.,* 13, 2432, 1993.

81. **Jacobson, M. D., Burne, J. F., King, M. P., Miyashita, T., Reed, J. C., and Raff, M. C.,** Bcl-2 blocks apoptosis in cells lacking mitochondrial DNA, *Nature,* 361, 365, 1993.

82. **Reed, J. C.,** Regulation of p26-Bcl-2 protein levels in human peripheral blood lymphocytes, *Lab. Invest.,* 67, 443, 1992.

83. **LeBrun, D. P., Warnke, R. A., and Cleary, M. L.,** Expression of *bcl*-2 in fetal tissues suggests a role in morphogenesis, *Am. J. Pathol.,* 142, 743, 1993.

84. **Abedohmae, S., Harada, N., Yamada, K., and Tanaka, R.,** *bcl*-2 gene is highly expressed during neurogenesis in the central nervous system, *Biochem. Biophys. Res. Commun.,* 191, 915, 1993.

85. **Garcia, I., Martinou, I., Tsujimoto, Y., and Martinou, J.-C.,** Prevention of programmed cell death of sympathetic neurons by the *bcl*-2 proto-oncogene, *Science,* 258, 302, 1992.

86. **Tsujimoto, Y. and Croce, C. M.,** Analysis of the structure, transcripts, and protein products of *bcl*-2, the gene involved in human follicular lymphoma, *Proc. Natl. Acad. Sci. U.S.A.,* 83, 5214, 1986.

87. **McDonnell, T. J., Troncoso, P., Brisbay, S. M., Logothetis, C., Chung, L. W. K., Hsieh, J.-T., Tu, S.-M., and Campbell, M. L.,** Expression of the protooncogene *bcl*-2 in the prostate and its association with emergence of androgen-independent prostate cancer, *Cancer Res.,* 52, 6940, 1992.

88. **Lu, Q.-L., Poulsom, R., Wong, L., and Hanby, A. M.,** *bcl*-2 expression in adult and embryonic non-haemotopoietic tissues, *J. Pathol.,* 169, 431, 1993.

89. **Yonish-Rouach, E., Resnitzky, D., Lotem, J., Sachs, L., Kimchi, A., and Oren, M.,** Wild-type p53 induces apoptosis of myeloid leukaemia cells that is inhibited by interleukin-6, *Nature,* 352, 345, 1991.

90. **Ryan, J. J., Danish, R., Gottlieb, C. A., and Clarke, M. F.,** Cell cycle analysis of p53-induced cell death in murine erythroleukemia cells, *Mol. Cell. Biol.,* 13, 711, 1993.

91. **Clarke, A. R., Purdie, C. A., Harrison, D. J., Morris, R. G., Bird, C. C., Hooper, M. L., and Wyllie, A. H.,** Thymocyte apoptosis induced by p53-dependent and independent pathways, *Nature,* 362, 849, 1993.

92. **Hayflick, L.,** The limited *in vitro* lifetime of human diploid strains, *Exp. Cell Res.,* 37, 614, 1965.

93. **Suda, Y., Suzuki, M., Ikawa, Y., and Aizawa, S.,** Mouse embryonic stem cells exhibit indefinite proliferative potential, *J. Cell. Physiol.,* 133, 197, 1987.

94. **Tamm, I. and Kikuchi, T.,** Activation of signal transduction pathways protects quiescent Balb/c 3T3 fibroblasts against death due to serum deprivation, *J. Cell. Physiol.,* 148, 85, 1991.

95. **Hoffmann, B., Piasecki, A., and Paul, D.,** Proliferation of fetal rat hepatocytes in response to growth factors and hormones in primary culture, *J. Cell. Physiol.,* 139, 654, 1989.

96. **Florini, J. R. and Magri K. A.,** Effects of growth factors on myogenic differentiation, *Am. J. Physiol.,* 256, C701, 1989.

97. **Johnson, S. E. and Allen, R. E.,** The effects of bFGF, IGF-I, and TGF-β on RMo skeletal muscle cell proliferation and differentiation, *Exp. Cell Res.,* 187, 250, 1990.

98. **Laveck, M. A., Somers, A. N. A., Moore, L. L., Gerwin, B. I., and Lechner, J. F.,** Dissimilar peptide growth factors can induce normal human mesothelial cell multiplication, *In Vitro Cell. Dev. Biol.,* 24, 1077, 1988.

99. **Gourdeau, H. and Fournier, R. E. K.,** Genetic analysis of mammalian cell differentiation, *Annu. Rev. Cell Biol.,* 6, 69, 1990.

100. **Blay, J. and Hollenberg, M. D.,** The nature and function of polypeptide growth factor receptors in the human placenta, *J. Dev. Physiol.,* 12, 237, 1989.

101. **Ohlsson, R.,** Growth factors, protooncogenes and human placental development, *Cell Differ. Dev.,* 28, 1, 1989.

102. **Mercola, M. and Stiles, C. D.,** Growth factor superfamilies and mammalian embryogenesis, *Development,* 102, 451, 1988.

103. **Melton, D. A. and Whitman, M.,** Growth factors in early embryogenesis, *Annu. Rev. Cell Biol.,* 5, 93, 1989.

104. **Paria, B. C. and Dey, S. K.,** Preimplantation embryo development *in vitro*: cooperative interactions among embryos and role of growth factors, *Proc. Natl. Acad. Sci. U.S.A.,* 87, 4756, 1990.

105. **Simmen, F. A. and Simmen, R. C. M.,** Peptide growth factors and proto-oncogenes in mammalian conceptus development, *Biol. Reprod.,* 44, 1, 1991.

106. **Rappolee, D. A., Brenner, C. A., Schultz, R., Mark, D., and Werb, Z.,** Developmental expression of PDGF, TGF-α, and TGF-β genes in preimplantation mouse embryos, *Science,* 241, 1823, 1988.

107. **Mummery, C. L., van den Eijnden-van Raaij, A. J. M., Feijen, A., Freund, E., Hulskotte, E., Schoorlemmer, J., and Kruijer, W.,** Expression of growth factors during the differentiation of embryonic stem cells in monolayer, *Dev. Biol.,* 142, 406, 1990.

108. **Nielsen, L. L., Werb, Z., and Pedersen, R. A.,** Induction of c-*fos* transcripts in early postimplantation mouse embryos by TGF-α, EGF, PDGF, and FGF, *Mol. Reprod. Dev.,* 29, 227, 1991.

109. **Laiho, M. and Keski-Oja, J.,** Growth factors in the regulation of pericellular proteolysis: a review, *Cancer Res.,* 49, 2533, 1989.

110. **Matrisian, L. M. and Hogan, B. L. M.,** Growth factor-regulated proteases and extracellular matrix remodeling during mammalian development, *Curr. Top. Dev. Biol.,* 24, 219, 1990.

111. **Hollenberg, M. D.,** Growth factors, their receptors and development, *Am. J. Med. Genet.,* 34, 35, 1989.

112. **Kimelman, D. and Kirschner, M.,** Synergistic induction of mesoderm by FGF and TGF-β and the identification of an mRNA coding for FGF in the early *Xenopus* embryo, *Cell,* 51, 869, 1987.

113. **Snoek, G. T., Koster, C. H., de Laat, S. W., Heideveld, M., Durston, A. J., and van Zoelen, E. J. J.,** Effects of cell heterogeneity on production of polypeptide growth factors and mesoderm-inducing activity by *Xenopus laevis* XTC cells, *Exp. Cell Res.,* 187, 203, 1990.

114. **Gehring, W. J.,** Homeoboxes in the study of development, *Science,* 236, 1245, 1987.

115. **Schughart, K., Kappen, C., and Ruddle, F. H.,** Mammalian homeobox-containing genes: genome organization, structure, expression and evolution, *Br. J. Cancer,* 58 (Suppl. 9), 9, 1988.

116. **Scott, M. P., Tamkun, J. W., and Hartzell, G. W., III,** The structure and function of homeodomain, *Biochim. Biophys. Acta,* 989, 25, 1989.

117. **Kessel, M. and Gruss, P.,** Murine development control genes, *Science,* 249, 374, 1990.

118. **Deschamps, J. and Meijlink, F.,** Mammalian homeobox genes in normal development and neoplasia, *Crit. Rev. Oncogenesis,* 3, 117, 1992.

119. **Lawrence, H. J. and Largman, C.,** Homeobox genes in normal hematopoiesis and leukemia, *Blood,* 80, 2245, 1992.

120. **Mavilio, F.,** Regulation of vertebrate homeobox-containing genes by morphogens, *Eur. J. Biochem.,* 212, 273, 1993.

121. **Scott, M. P. and Carroll, S. B.,** The segmentation and homeotic gene network in early *Drosophila* development, *Cell,* 51, 689, 1987.

122. **Logan, C., Willard, H. F., Rommens, J. M., and Joyner, A. L.,** Chromosomal localization of the human homeobox-containing genes, *EN1* and *EN2, Genomics,* 4, 206, 1989.

123. **Boncinelli, E., Acampora, D., Pannese, M., D'Esposito, M., Somma, R., Gaudino, G., Stornaiuolo, A., Cafiero, M., Faiella, A., and Simeone, A.,** Organization of human class I homeobox genes, *Genome,* 31, 745, 1989.

124. **Acampora, D., D'Esposito, M., Faiella, A., Pannese, M., Migliacci, E., Morelli, F., Stornaiuolo, A., Nigro, V., Simeone, A., and Boncinelli, E.,** The human *HOX* gene family, *Nucleic Acids Res.,* 17, 10385, 1989.

125. **Dollé, P., Izpisúa-Belmonte, J.-C., Falkenstein, H., Renucci, A., and Duboule, D.,** Coordinate expression of the murine *HOX-5* complex homeobox-containing genes during limb pattern formation, *Nature,* 342, 767, 1989.

126. **Castrillo, J.-L., Theill, L. E., and Karin, M.,** Function of the homeodomain protein GHF1 in pituitary cell proliferation, *Science,* 253, 197, 1991.

127. **Blumberg, B., Wright, C. V. E., De Robertis, E. M., and Chok, K. W. Y.,** Organizer-specific homeobox genes in *Xenopus laevis* embryos, *Science,* 253, 194, 1991.

128. **Magli, M. C., Barba, P., Celetti, A., De Vita, G., Cillo, C., and Boncinelli, E.,** Coordinate regulation of *HOX* genes in human hematopoietic cells, *Proc. Natl. Acad. Sci. U.S.A.,* 88, 6348, 1991.

129. **Vieille-Grosjean, I., Roullot, V., and Courtois, G.,** Lineage- and stage-specific expression of *HOX*-1 genes in the human hematopoietic system, *Biochem. Biophys. Res. Commun.,* 183, 1124, 1992.

130. **Shen, W.-F., Detmer, K., Mathews, C. H. E., Hack, F. M., Morgan, D. A., Largman, C., and Lawrence, H. J.,** Modulation of homeobox gene expression alters the phenotype of human hematopoietic cell lines, *EMBO J.,* 11, 983, 1992.

131. **McCormick, A., Brady, H., Theill, L. E., and Karin, M.,** Regulation of the pituitary-specific homeobox gene GHF1 by cell-autonomous and environmental cues, *Nature,* 345, 829, 1990.

132. **Bender, W.,** Homeotic gene products as growth factors, *Cell,* 43, 559, 1985.

133. **Burgess, A. W.,** Growth factors and their receptors: specific roles in development, *BioEssays,* 6, 79, 1987.

134. **Blatt, C.,** The betrayal of homeobox genes in normal development: the link to cancer, *Cancer Cells,* 2, 186, 1990.

135. **Aberdam, D., Negreanu, V., Sachs, L., and Blatt, C.,** The oncogenic potential of an activated HOX-2.4 homeobox gene in mouse fibroblasts, *Mol. Cell. Biol.,* 11, 554, 1991.

136. **Wewer, U. M., Mercurio, A. M., Chung, S. Y., and Albrechtsen, R.,** Deoxyribonucleic-binding homeobox proteins are augmented in human cancer, *Lab. Invest.,* 63, 447, 1990.

137. **Kennedy, M. A., Gonzalez-Sarmiento, R., Kees, U. R., Lampert, F., Dear, N., Boehm, T., and Rabbits, T. H.,** *HOX*11, a homeobox-containing T-cell oncogene on human chromosome 10q24, *Proc. Natl. Acad. Sci. U.S.A.,* 88, 8900, 1991.

138. **Holley, R. W. and Kiernan, J. A.,** Control of the initiation of DNA synthesis in 3T3 cells: serum factors, Proc. Natl. Acad. Sci. U.S.A., 71, 2908, 1974.

139. **Brooks, R. F.,** Regulation of the fibroblast cell cycle by serum, *Nature,* 260, 248, 1976.

140. **Olashaw, N. E., Harrington, M., and Pledger, W. J.,** The regulation of the cell cycle by multiple growth factors, *Cell Biol. Int. Rep.,* 7, 489, 1983.

141. **Shipley, G. D., Childs, C. B., Volkenant, M. E., and Moses, H. L.,** Differential effects of epidermal growth factor, transforming growth factor, and insulin on DNA and protein synthesis and morphology in serum-free cultures of AKR-2B cells, *Cancer Res.,* 44, 710, 1984.

142. **McKeehan, W. L., Adams, P. S., and Rosser, M. P.,** Direct mitogenic effects of insulin, epidermal growth factor, glucocorticoid, cholera toxin, unknown pituitary factors and possibly prolactin, but not androgen, on normal rat prostate epithelial cells in serum-free, primary cell culture, *Cancer Res.,* 44, 1998, 1984.

143. **Balk, S. D., Morisi, A., Gunther, S., Svoboda, M. F., Van Wyk, J. J., Nissley, S. P., and Scanes, C. G.,** Somatomedins (insulin-like growth factors), but not growth hormone, are mitogenic for chicken heart mesenchymal cells and act synergistically with epidermal growth factor and brain fibroblast growth factor, *Life Sci.,* 35, 335, 1984.

144. **Ethier, S. P., Kudla, A., and Cundiff, K. C.,** Influence of hormone and growth factor interactions on the proliferative potential of normal rat mammary epithelial cells *in vitro, J. Cell. Physiol.,* 132, 161, 1987.

145. **Pledger, W. J., Stiles, C. D., Antoniades, H. N., and Scher, C. D.,** Induction of DNA synthesis in BALB/c-3T3 cells by serum components: re-evaluation of the commitment process, *Proc. Natl. Acad. Sci. U.S.A.,* 74, 4481, 1977.

146. **Pledger, W. J., Stiles, C. D., Antoniades, H. N., and Scher, C. D.,** An ordered sequence of events is required before BALB/c-3T3 cells become committed to DNA synthesis, *Proc. Natl. Acad. Sci. U.S.A.,* 75, 2839, 1978.

147. **Jimenez de Asua, L., Richmond, K. M. V., and Otto, A. M.,** Two growth factors and two hormones regulate initiation of DNA synthesis in cultured mouse cells through different pathways of events, *Proc. Natl. Acad. Sci. U.S.A.,* 78, 1004, 1981.

148. **O'Keefe, E. J. and Pledger, W. J.,** A model of cell cycle control: sequential events regulated by growth factors, *Mol. Cell. Endocrinol.,* 31, 167, 1983.

149. **Westermark, B. and Heldin, C.-H.,** Similar action of platelet-derived growth factor and epidermal growth factor in the prereplicative phase of human fibroblasts suggests a common intracellular pathway, *J. Cell. Physiol.,* 124, 43, 1985.

150. **London, L. and McKearn, J. P.,** Activation and growth of colony-stimulating factor-dependent cell lines is cell cycle stage dependent, *J. Exp. Med.,* 166, 1419, 1987.

151. **Khan, S., Teerds, K., and Dorrington, J.,** Growth factor requirements for DNA synthesis by Leydig cells from the immature rat, *Biol. Reprod.,* 46, 335, 1992.

152. **Sand, T.-E. and Christoffersen, T.,** Temporal requirement for epidermal growth factor and insulin in the stimulation of hepatocyte DNA synthesis, *J. Cell. Physiol.,* 131, 141, 1987.

153. **Liboi, E., Pelosi, E., Testa, U., Peschle, C., and Rossi, G. B.,** Proliferative response and oncogene expression induced by epidermal growth factor in EL2 rat fibroblasts, *Mol. Cell. Biol.,* 6, 2275, 1986.

154. **Chen, Y. and Rabinovitch, P. S.,** Platelet-derived growth factor, epidermal growth factor, and insulin-like growth factor I regulate specific cell-cycle parameters of human diploid fibroblasts in serum-free culture, *J. Cell. Physiol.,* 140, 59, 1989.

155. **Di Francesco, P., Favalli, C., and Liboi, E.,** Secreted proteins induced by epidermal growth factor and transforming growth factor beta in EL2 rat fibroblasts. Role in the mitogenic response, *Cell Biol. Int. Rep.,* 12, 365, 1988.

156. **Otto, A. M., Smith C., and Jimenez de Asua, L.,** Stimulation of DNA replication by growth factor and hormones in Swiss 3T3 cells: comparison of the rate of entry into S phase with *in vitro* DNA synthesis and DNA polymerase alpha activity, *J. Cell. Physiol.,* 134, 57, 1988.

157. **Lucas, P. C. and Granner, D. K.,** Hormone response domains in gene transcription, *Annu. Rev. Biochem.,* 61, 1131, 1992.

158. **Hoerl, B. J. and Scott, R. E.,** Nonterminally differentiated cells express decreased growth factor responsiveness, *J. Cell. Physiol.,* 139, 68, 1989.

159. **Brooks, R. F. and Riddle, P. N.,** Differences in growth factor sensitivity between individual 3T3 cells arise at high frequency: possible relevance to cell senescence, *Exp. Cell Res.,* 174, 378, 1988.

160. **Martin, G. M., Sprague, C. A., and Epstein, C. J.,** Replicative life-span of cultivated human cells. Effects of donor's age, tissue and genotype, *Lab. Invest.,* 23, 86, 1970.

161. **Schneider, E. L. and Mitsui, Y.,** The relationship between *in vitro* cellular aging and *in vivo* human age, *Proc. Natl. Acad. Sci. U.S.A.,* 70, 1263, 1976.

162. **Röhme, D.,** Evidence for a relationship between longevity of mammalian species and life spans of normal fibroblasts *in vitro* and erythrocytes *in vivo*, *Proc. Natl. Acad. Sci. U.S.A.,* 78, 5009, 1981.

163. **Plisko, A. and Gilchrest, B. A.,** Growth factor responsiveness of cultured human fibroblasts declines with age, *J. Gerontol.,* 38, 513, 1983.

164. **Phillips, P. D., Kaji, K., and Cristofalo, V. J.,** Progressive loss of the proliferative response of senescing WI-38 cells to platelet-derived growth factor, epidermal growth factor, insulin, transferrin, and dexamethasone, *J. Gerontol.,* 39, 11, 1984.

165. **Paulsson, Y., Bywater, M., Pfeiffer-Ohlsson, S., Ohlsson, R., Nilsson, S., Heldin, C.-H., Westermark, B., and Betsholtz, C.,** Growth factors induce early pre-replicative changes in senescent human fibroblasts, *EMBO J.,* 5, 2157, 1986.

166. **Chen, Y. and Rabinovitch, P. S.,** Altered cell cycle responses to insulin-like growth factor I, but not platelet-derived growth factor and epidermal growth factor, in senescing human fibroblasts, *J. Cell. Physiol.,* 144, 18, 1990.

167. **Amtmann, E., Eddé, E., Sauer, G., and Westphal, O.,** Restoration of the responsiveness to growth factors in senescent cells by an embryonic cell extract, *Exp. Cell Res.,* 189, 202, 1990.

168. **Barton, M. C. and Shapiro, D. J.,** Transient administration of extradiol-17 beta establishes an autoregulatory loop permanently inducing estrogen receptor mRNA, *Proc. Natl. Acad. Sci. U.S.A.,* 85, 7119, 1988.

169. **Storch, T. G. and Talley, G. D.,** Oxygen concentration regulates the proliferative response of human fibroblasts to serum and growth factors, *Exp. Cell Res.,* 175, 317, 1988.

170. **Scher, C. D. and Todaro, G. J.,** Selective growth of human neoplastic cells in medium lacking serum growth factor, *Exp. Cell Res.,* 68, 479, 1972.

171. **Boswell, H. S., Srivastava, A., Burgess, J. S., Nahreini, P., Heerema, N., Inhorn, L., Padgett, F., Walker, E. B., and Geib, R. W.,** Cellular control of *in vitro* progression of murine myeloid leukemia: progression accompanies acquisition of independence from growth factor and stromal cells, *Leukemia,* 1, 765, 1987.

172. **Metcalf, D., Moore, M. A. S., Sheridan, J. W., and Spitzer, G.,** Responsiveness of human granulocytic leukemia cell to colony-stimulating factor, *Blood,* 43, 847, 1974.

173. **Greenberg, B. R., Hirasuna, J. D., and Woo, L.,** *In vitro* response to erythropoietin in erythroblastic transformation of chronic myelogenous leukemia, *Exp. Hematol.,* 8, 52, 1980.

174. **Haas, M., Altman, A., Rothenberg, E., Bogart, M. H., and Jones, O. W.,** Mechanism of T-cell lymphomagenesis: transformation of growth-factor-dependent T-lymphoblastoma cells to growth-factor-independent T-lymphoma cells, *Cancer Res.,* 81, 1742, 1984.

175. **Russell, S. B., Trupin, K. M., Rodríguez-Eaton, S., Russell, J. D., and Trupin, J. S.,** Reduced growth-factor requirement of keloid-derived fibroblasts may account for tumor growth, *Proc. Natl. Acad. Sci. U.S.A.,* 85, 587, 1988.

176. **Johnson, H. M. and Torres, B. A.,** Peptide growth factors PDGF, EGF, and FGF regulate interferon-γ production, *J. Immunol.,* 134, 2824, 1985.

177. **Chiang, C.-P. and Nilsen-Hamilton, M.,** Opposite and selective effects of epidermal growth factor and human platelet transforming growth factor-β on the production of secreted proteins by murine 3T3 cells and human fibroblasts, *J. Biol. Chem.,* 261, 10478, 1986.

178. **Théry, C., Jullien, P., and Lawrence, D. A.,** Evidence for a novel growth factor in *Xenopus* oocytes, *Biochem. Biophys. Res. Commun.,* 160, 615, 1989.

179. **Mills, G. B., May, C., McGill, M., Roifman, C. M., and Mellors, A.,** A putative new growth factor in ascitic fluid from ovarian cancer patients: identification, characterization, and mechanism of action, *Cancer Res.,* 48, 1066, 1988.

180. **Matsuda, H., Matsumoto, M., Haraguchi, S., and Kanai, K.,** Partial purification of a growth factor synthesized by a rat hepatoma cell line established in serum-free medium, *Cancer Res.,* 49, 2118, 1989.

181. **Matsuda, H., Matsumoto, M., Kitahara, H., Haraguchi, S., and Kanai, K.,** Purification and characterization of a novel growth factor (FF-GF) synthesized by a rat hepatoma cell line, FF101, *Biochem. Biophys. Res. Commun.,* 189, 654, 1992.

182. **Linzer, D. I. H. and Nathans, D.,** Growth-related changes in specific mRNAs of cultured mouse cells, *Proc. Natl. Acad. Sci. U.S.A.,* 80, 4271, 1983.

183. **Hirt, H., Kimelman, J., Birnbaum, M. J., Chen, E. Y., Seeburg, P. H., Eberhardt, N. L., and Barta, A.,** The human growth hormone gene locus: structure, evolution, and allelic variations, *DNA,* 6, 59, 1987.

184. **Linzer, D. I. H. and Nathans, D.,** Nucleotide sequence of a growth-related mRNA encoding a member of the prolactin-growth hormone family, *Proc. Natl. Acad. Sci. U.S.A.,* 81, 4255, 1984.

185. **Linzer, D. I. H. and Nathans, D.,** A new member of the prolactin-growth hormone gene family expressed in mouse placenta, *EMBO J.,* 4, 1419, 1985.

186. **Lee, S.-J. and Nathans, D.,** Secretion of proliferin, *Endocrinology,* 120, 208, 1987.

187. **Linzer, D. I. H., Lee, S.-J., Ogren, L., Talamantes, F., and Nathans, D.,** Identification of proliferin mRNA and protein in mouse placenta, *Proc. Natl. Acad. Sci. U.S.A.,* 82, 4356, 1985.

188. **Wilder, E. L. and Linzer, D. I. H.,** Expression of multiple proliferin genes in mouse cells, *Mol. Cell. Biol.,* 6, 3283, 1986.

189. **Colosi, P., Swiergiel, J. J., Wilder, E. L., Oviedo, A., and Linzer, D. I. H.,** Characterization of proliferin-related protein, *Mol. Endocrinol.,* 2, 579, 1988.

190. **Parfett, C. L. J., Hamilton, R. T., Howell, B. W., Edwards, D. R., Nilsen-Hamilton, M., and Denhardt, D. T.,** Characterization of a cDNA clone encoding murine mitogen-regulated protein: regulation of mRNA levels in mortal and immortal cell lines, *Mol. Cell. Biol.,* 5, 3289, 1985.

191. **Nilsen-Hamilton, M., Hamilton, R. T., and Alvarez-Azaustre, E.,** Relationship between mitogen-regulated protein (MRP) and proliferin (PLF), a member of the prolactin/growth hormone family, *Gene,* 51, 163, 1987.

192. **Linzer, D. I. H. and Wilder, E. L.,** Control of gene expression in serum-stimulated mouse cells, *Mol. Cell. Biol.,* 7, 2080, 1987.

193. **Linzer, D. I. H. and Mordacq, J. C.,** Transcriptional regulation of proliferin gene expression in response to serum in transfected mouse cells, *EMBO J.,* 6, 2281, 1987.

194. **Edwards, D. R., Parfett, C. L. J., Smith, J. H., and Denhardt, D. T.,** Evidence that post-transcriptional changes in the expression of mitogen regulated protein accompany immortalization of mouse cells, *Biochem. Biophys. Res. Commun.,* 147, 467, 1987.

195. **Parfett, C. L. J.,** Induction of proliferin gene expression by diverse chemical agents that promote morphological transformation in C3H/10T1/2 cultures, *Cancer Lett.,* 64, 1, 1992.

196. **Wilder, E. L. and Linzer, D. I. H.,** Participation of multiple factors, including proliferin, in the inhibition of myogenic differentiation, *Mol. Cell. Biol.,* 9, 430, 1989.

197. **Melhem, R. F., Zhu, X., Hailat, N., Strahler, J. R., and Hanasch, S. M.,** Characterization of the gene for a proliferation-related phosphoprotein (oncoprotein 18) expressed in high amounts in acute leukemia, *J. Biol. Chem.,* 266, 17747, 1991.

198. **Celis, J. E. and Nielsen, S.,** Proliferation-sensitive nuclear phosphoprotein "dividin" is synthesized almost exclusively during S phase of the cell cycle in human AMA cells, *Proc. Natl. Acad. Sci. U.S.A.,* 83, 8187, 1986.

199. **Celis, J. E., Ratz, G. P., and Celis, A.,** Progressin: a novel proliferation-sensitive and cell cycle-regulated human protein whose rate of synthesis increases at or near the G_1/S transition border of the cell cycle, *FEBS Lett.,* 223, 237, 1987.

200. **Santarén, J. F. and Bravo, R.,** Immediate induction of a 45K secreted glycoprotein by serum and growth factors in quiescent mouse 3T3 cells: two-dimensional gel analysis, *Exp. Cell Res.,* 168, 168, 494, 1987.

201. **Gerdes, J., Schwab, U., Lemke, H., and Stein, H.,** Production of a mouse monoclonal antibody reactive with a human nuclear antigen associated with proliferation, *Int. J. Cancer,* 31, 13, 1983.

202. **Gerdes, J., Lemke, H., Baisch, H., Wacker, H. H., Schwab, U., and Stein, H.,** Cell cycle analysis of a cell proliferation-associated human nucelar antigen defined by the monoclonal antibody Ki-67, *J. Immunol.,* 133, 1710, 1984.

203. **Sasaki, K., Murakami, T., Kawasaki, M., and Takahashi, M.,** The cell cycle associated change of the Ki-67 reactive nuclear antigen expression, *J. Cell. Physiol.,* 133, 579, 1987.

204. **Bruno, S. and Darzynkiewicz, Z.,** Cell cycle dependent expression and stability of the nuclear protein detected by Ki-67 antibody in HL-60 cells, *Cell Prolif.,* 25, 31, 1992.

205. **Lelle, R. J., Heidenreich, W., Stauch, G., and Gerdes, J.,** The correlation of growth fractions with histologic grading and lymph node status in human mammary carcinoma, *Cancer,* 59, 83, 1987.

206. **Sahin, A. A., Ro, J., Ro, J. Y., Blick, M. B., El-Naggar, A. K., Ordonez, N. G., Fritsche, H. A., Smith, T. L., Hortobagyi, G. N., and Ayala, A. G.,** Ki-67 immunostaining in node-negative stage I/II breast carcinoma. Significant correlation with prognosis, *Cancer,* 68, 549, 1991.

207. **Camplejohn, R. S., Brock, A., Barnes, D. M., Gillett, C., Raikundalia, B., Kreipe, H., and Parwaresch, M. R.,** Ki-S1, a novel proliferative marker flow cytometric assessment of staining in human breast carcinoma cells, *Br. J. Cancer,* 67, 657, 1993.

208. **Santarén, J. F. and Bravo, R.,** A basic cytoplasmic protein (p27) induced by serum in growth-arrested 3T3 cells but constitutively expressed in primary fibroblasts, *EMBO J.,* 5, 877, 1986.

209. **Conover, C. A., Rosenfeld, R. G., and Hintz, R. L.,** Hormonal control of the replication of human fetal fibroblasts: role of somatomedin C/insulin-like growth factor I, *J. Cell. Physiol.,* 128, 47, 1986.

210. **Heyns, A. du P., Eldor, A., Vlodavsky, I, Kaiser, N., Fridman, R., and Panet, A.,** The antiproliferative effect of interferon and the mitogenic action of growth factors are independent cell cycle events: studies with vascular smooth muscle cells and endothelial cells, *Exp. Cell Res.,* 161, 297, 1985.

211. **Pabo, C. O. and Sauer, R. T.,** Transcription factors: structural families and principles of DNA recognition, *Annu. Rev. Biochem.,* 61, 1053, 1992.

212. **Meek, D. W. and Street, A. J.,** Nuclear protein phosphorylation and growth control, *Biochem. J.,* 287, 1, 1992.

213. **Bravo, R. and Celis, J. E.,** Gene expression in normal and virally transformed mouse 3T3B and hamster BHK21 cells, *Exp. Cell Res.,* 127, 249, 1980.

214. **Kaczmarek, L.,** Protooncogene expression during the cell cycle, *Lab. Invest.,* 54, 365, 1986.

215. **Denhardt, D. T., Edwards, D. R., and Parfett, C. L. J.,** Gene expression during the mammalian cell cycle, *Biochim. Biophys. Acta,* 865, 83, 1986.

216. **Ferrari, S. and Baserga, R.,** Oncogenes and cell cycle genes, *BioEssays,* 7, 9, 1987.

217. **Bravo, R.,** Growth factor-responsive genes in fibroblasts, *Cell Growth Differ.,* 1, 305, 1990.

218. **Stokke, T., Erikstein, B., Holte, H., Funderud, S., and Steen, H. B.,** Cell cycle-specific expression and nuclear binding of DNA polymerase α, *Mol. Cell. Biol.,* 11, 3384, 1991.

219. **Guo, X., Chambers, A. F., Parfett, C. L. J., Waterhouse, P., Murphy, L. C., Reid, R. E., Craig, A. M., Edwards, D. R., and Denhardt, D. T.,** Identification of a serum-inducible messenger RNA (5B10) as the mouse homologue of calcyclin: tissue distribution and expression in metastatic, *ras*-transformed NIH 3T3 cells, *Cell Growth Differ.,* 1, 333, 1990.

220. **Gilks, C. B., Bear, S. E., Grimes, H. L., and Tsichlis, P. N.,** Progression of interleukin-2 (IL-2)-dependent rat T cell lymphoma lines to IL-2-independent growth following activation of a gene (*Gfi*-1) encoding a novel zinc finger protein, *Mol. Cell. Biol.,* 13, 2759, 1993.

221. **Almendral, J. M., Sommer, D., Macdonald-Bravo, H., Burckhardt, J., Perera, J., and Bravo, R.,** Complexity of the early genetic response to growth factors in mouse fibroblasts, *Mol. Cell. Biol.,* 8, 2140, 1988.

222. **Mohn, K. L., Laz, T. M., Hsu, J.-C., Melby, A.E., Bravo, R., and Taub, R.,** The immediate-early growth response in regenerating liver and insulin-stimulated H-35 cells: comparison with serum-stimulated 3T3 cells and identification of 41 novel immediate-early genes, *Mol. Cell. Biol.,* 11, 381, 1991.

223. **Kauffman, M. G. and Kelly, T. J.,** Cell cycle regulation of thymidine kinase: residues near the carboxyl terminus are essential for the specific degradation of the enzyme at mitosis, *Mol. Cell. Biochem.,* 11, 2538, 1991.

224. **Gomperts, M., Pascall, J. C., and Brown, K. D.,** The nucleotide sequence of a cDNA encoding an EGF-inducible gene indicates the existence or a new family of mitogen-induced genes, *Oncogene,* 5, 1081, 1990.

225. **O'Brien, T. P., Yang, G. P., Sanders, L., and Lau, L. F.,** Expression of *cyr61*, a growth factor-inducible immediate-early gene, *Mol. Cell. Biol.,* 10, 3569, 1990.

226. **Yang, G. P. and Lau, L. F.,** Cyr61, product of a growth factor-inducible immediate early gene, is associated with the extracellular matrix and the cell surface, *Cell Growth Differ.,* 2, 351, 1991.

227. **Charles, C. H., Abler, A. S., and Lau, L. F.,** cDNA sequence of a growth factor-inducible immediate early gene and characterization of its encoded protein, *Oncogene,* 7, 187, 1992.

228. **Greco, A., Ittmann, M., Barletta, C., Basilico, C., Croce, C. M., Cannizzaro, L. A., and Huebner, K.,** Chromosomal localization of human genes required for G_1 progression in mammalian cells, *Genomics,* 4, 240, 1989.

229. **Ciccarelli, C., Philipson, L., and Sorrentino, V.,** Regulation of expression of growth arrest-specific genes in mouse fibroblasts, *Mol. Cell. Biol.,* 10, 1525, 1990.

230. **Cairo, G., Ferrero, M., Biondi, G., and Colombo, M. P.,** Expression of a growth arrest specific gene (gas-1) in transformed cells, *Br. J. Cancer,* 66, 27, 1992.

231. **Wang, E.,** A 57,000-mol wt protein uniquely present in nonproliferating cells and senescent human fibroblasts, *J. Cell. Biol.,* 100, 545, 1986.

232. **Wang, E. and Lin, S. L.,** Disappearance of statin, a protein marker for non-proliferating and senescent cells, following serum-stimulated cell cycle entry, *Exp. Cell Res.,* 167, 135, 1986.

233. **Wahl, A. F., Geis, A. M., Spain, B. H., Wong, S. W., Korn, D., and Wang, T. S.-F.,** Gene expression of human DNA polymerase alpha during cell proliferation and the cell cycle, *Mol. Cell. Biol.,* 8, 5016, 1988.

234. **Heintz, N.,** The regulation of histone gene expression during the cell cycle, *Biochim. Biophys. Acta,* 1088, 327, 1991.

235. **Osley, M. A.,** The regulation of histone synthesis in the cell cycle, *Annu. Rev. Biochem.,* 60, 827, 1991.

236. **Chrysogelos, S., Riley, D. E., Stein, G., and Stein, J.,** A human histone H4 gene exhibits cell cycle-dependent changes in chromatin structure that correlate with its expression, *Proc. Natl. Acad. Sci. U.S.A.,* 82, 7535, 1985.

237. **Laybourn, P. J. and Kadonaga, J. T.,** Role of nucleosomal cores and histone H1 in regulation of transcription by RNA polymerase II, *Science,* 254, 238, 1991.

238. **Sun, J.-M., Wiaderkiewicz, R., and Ruiz-Carrillo, A.,** Histone H5 in the control of DNA synthesis and cell proliferation, *Science,* 245, 68, 1989.

239. **Turner, B. M.,** Histone acetylation and control of gene expression, *J. Cell Sci.,* 99, 13, 1991.

240. **Knosp, O., Redl, B., and Puschendorf, B.,** Biochemical characterization of chromatin fractions isolated from induced and uninduced Friend erythroleukemia cells, *Mol. Cell. Biochem.,* 89, 37, 1989.

241. **Shalhoub, V., Gerstenfeld, L. C., Collart, D., Lian, J. B., and Stein, G. S.,** Downregulation of cell growth and cell cycle regulated genes during chick osteoblast differentiation with the reciprocal expression of histone gene variants, *Biochemistry,* 28, 52318, 1989.

242. **Bustin, M., Soares, N., Landsman, D., Srikantha, T., and Collins, J. M.,** Cell cycle regulated synthesis of an abundant transcript for human chromosomal protein HMG-17, *Nucleic Acids Res.,* 15, 3549, 1987.

243. **Barque, J.-P., Lagaye, S., Ladoux, A., Della Valle, V., Abita, J. P., and Larsen, C.-J.,** PSL, a nuclear cell-cycle associated antigen is increased during retinoic acid-induced differentiation of HL-60 cells, *Biochem. Biophys. Res. Commun.,* 147, 993, 1987.

244. **Black, A., Freeman, J. W., Zhou, G., and Busch, H.,** Novel cell cycle-related nuclear proteins found in rat and human cells with mononuclear antibodies, *Cancer Res.,* 47, 3266, 1987.

245. **Calabretta, B., Venturelli, D., Kaczmarek, L., Narni, F., Talpaz, M., Anderson, B., Beran, M., and Baserga, R.,** Altered expression of G_1-specific genes in human malignant myeloid cells, *Proc. Natl. Acad. Sci. U.S.A.,* 83, 1495, 1986.

246. **Levenson, R., Iwata, K., Klagsbrun, M., and Young, D. A.,** Growth factor- and dexamethasone-induced proteins in Swiss 3T3 cells: relationship to DNA synthesis, *J. Biol Chem.,* 260, 8056, 1985.

247. **Greco, A., Ittmann, M., and Basilico, C.,** Molecular cloning of a gene that is necessary for G_1 progression in mammalian cells, *Proc. Natl. Acad. Sci. U.S.A.,* 84, 1565, 1987.

248. **Sauvé, G. J., Shen, Y.-M., Zannis-Hadjopoulos, M., Chang, C.-D., Baserga, R., and Hand, R.,** Isolation of a human sequence which complements a mammalian G_1-specific temperature-sensitive mutant of the cell cycle, *Oncogene Res.,* 1, 137, 1987.

249. **Feige, J.-J. and Scheffler, I. E.,** Analysis of the protein glycosylation of a temperature-sensitive cell cycle mutant by the use of mutant cells overexpressing the human epidermal growth factor receptor after transfection of the gene, *J. Cell. Physiol.,* 133, 461, 1987.

250. **Bravo, R.,** Synthesis of the nuclear protein cyclin (PCNA) and its relationship with DNA replication, *Exp. Cell Res.,* 163, 287, 1986.

251. **Minshull, J., Pines, J., Golsteyn, R., Standart, N., Mackie, S., Colman, A., Blow, J., Ruderman, J. V., Wu, M., and Hunt, T.,** The role of cyclin synthesis, modification and destruction in the control of cell division, *J. Cell Sci.,* Suppl. 12, 77, 1989.

252. **Miyachi, K., Fritzler, M. J., and Tan, E. M.,** Autoantibody to a nuclear antigen in proliferating cells, *J. Immunol.,* 121, 2228, 1978.

253. **Tan, E. M.,** Autoantibodies to nuclear antigens: their immunobiology and medicine, *Adv. Immunol.,* 33, 167, 1982.

254. **Mathews, M. B., Bernstein, R. M., Franza, B. R., Jr., and Garrels, J. I.,** Identity of the proliferating cell nuclear antigen and cyclin, *Nature,* 309, 374, 1984.

255. **Leibovici, M., Monod, G., Géraudie, J., Bravo, R., and Méchali, M.,** Nuclear distribution of PCNA during embryonic development in *Xenopus laevis*: a reinvestigation of early cell cycles, *J. Cell Sci.,* 102, 63, 1992.

256. **Morris, G. F. and Mathews, M. B.,** Regulation of proliferating cell nuclear antigen during the cell cycle, *J. Biol. Chem.,* 264, 13856, 1989.

257. **Murray, A. W. and Kirschner, M. W.,** Cyclin synthesis drives the early embryonic cell cycle, *Nature,* 339, 275, 1989.

258. **Murray, A. W., Solomon, M. J., and Kirschner, M. W.,** The role of cyclin synthesis and degradation in the control of maturation promoting factor activity, *Nature,* 339, 280, 1989.

259. **Zuber, M., Tan, E. M., and Ryoji, M.,** Involvement of proliferating cell nuclear antigen (cyclin) in DNA replication in living cells, *Mol. Cell. Biol.,* 9, 57, 1989.

260. **Toschi, L. and Bravo, R.,** Changes in cyclin/proliferating cell nuclear antigen distribution during DNA repair synthesis, *J. Cell Biol.,* 107, 1623, 1988.

261. **Sherr, C. J.,** Mammalian G_1 cyclins, *Cell,* 73, 1059, 1993.

262. **Bravo, R., Fey, S. J., Bellatin, J., Larsen, P. M., Arevalo, J., and Celis, J. E.,** Identification of a nuclear and of a cytoplasmic polypeptide whose relative proportions are sensitive to changes in the rate of cell proliferation, *Exp. Cell Res.,* 136, 311, 1981.

263. **Celis, J. E., Madsen, P., Nielsen, S., and Rasmussen, H. H.,** Expression of cyclin (PCNA) and the phosphoprotein dividin are late obligatory events in the mitogenic response, *Anticancer Res.,* 7, 605, 1987.

264. **Celis, J. E., Madsen, P., Celis, A., Nielsen, H. V., and Gesser, B.,** Cyclin (PCNA, auxiliary protein of DNA polymerase delta) is a central component of the pathway(s) leading to DNA replication and cell division, *FEBS Lett.,* 220, 1, 1987.

265. **Suzuka, I., Daidoh, H., Matsuoka, M., Kadowaki, K., Takasaki, Y., Nakane, P. K., and Moriuchi, T.,** Gene for proliferating-cell nuclear antigen (DNA polymerase delta auxiliary protein) is present in both mammalian an higher plant genomes, *Proc. Natl. Acad. Sci. U.S.A.,* 86, 3189, 1989.

266. **Tsurimoto, T. and Stillman, B.,** Functions of replication factor C and proliferating-cell nuclear antigen: functional similarity of DNA polymerase accessory proteins from human cells and bacteriophage T4, *Proc. Natl. Acad. Sci. U.S.A.,* 87, 1023, 1990.

267. **Celis, J. E., Fey, S. J., Larsen, P. M., and Celis, A.,** Expression of the transformation-sensitive protein "cyclin" in normal human epidermal basal cells and simian virus 40-transformed keratinocytes, *Proc. Natl. Acad. Sci. U.S.A.,* 81, 3128, 1984.

268. **Bravo, R. and Graf, T.,** Synthesis of the nuclear protein cyclin does not correlate directly with transformation in quail embryo fibroblasts, *Exp. Cell Res.,* 156, 450, 1985.

269. **Macdonald-Bravo, H. and Bravo, R.,** Induction of the nuclear protein cyclin in serum-stimulated quiescent 3T3 cells is independent of DNA synthesis, *Exp. Cell Res.,* 156, 455, 1985.

270. **Bravo, R. and Macdonald-Bravo, H.,** Changes in the nuclear distribution of cyclin (PCNA) but not its synthesis depend on DNA replication, *EMBO J.,* 4, 655, 1985.

271. **Celis, J. E. and Celis, A.,** Cell cycle-dependent variations in the distribution of the nuclear protein cyclin (proliferating cell nuclear antigen) in cultured cells: subdivision of S phase, *Proc. Natl. Acad. Sci. U.S.A.,* 82, 3262, 1985.

272. **Celis, J. E. and Celis, A.,** Individual nuclei in polykaryons can control cyclin distribution and DNA synthesis, *EMBO J.,* 4, 1187, 1985.

273. **Prelich, G., Kostura, M., Marshak, D. R., Mathews, M. B., and Stillman, B.,** The cell-cycle regulated proliferating nuclear antigen is required for SV40 DNA replication *in vitro, Nature,* 326, 471, 1987.

274. **Bravo, R. and Macdonald-Bravo, H.,** Existence of two populations of cyclin/proliferating cell nuclear antigen during the cell cycle: association with DNA replication sites, *J. Cell Biol.,* 105, 1549, 1987.

275. **Rao, V. V. N. G., Schnittger, S., and Hansmann, I.,** Chromosomal localization of the human proliferating cell nuclear antigen (PCNA) gene to or close to 20p12 by *in situ* hybridization, *Cytogenet. Cell Genet.,* 55, 169, 1991.

276. **Almendral, J. M., Huebsch, D., Blundell, P. A., Macdonald-Bravo, H., and Bravo, R.,** Cloning and sequence of the human nuclear protein cyclin: homology with DNA-binding proteins, *Proc. Natl. Acad. Sci. U.S.A.,* 84, 1575, 1987.

277. **Travali, S., Ku, D.-H., Rizzo, M. G., Ottavio, L., Baserga, R., and Calabretta, B.,** Structure of the human gene for the proliferating cell nuclear antigen, *J. Biol. Chem.,* 264, 7466, 1989.

278. **Rizzo, M. G., Ottavio, L., Travali, S., Chang, C., Kaminska, B., and Baserga, R.,** The promoter of the human proliferating cell nuclear antigen (PCNA) gene is bidirectional, *Exp. Cell Res.,* 188, 286, 1990.

279. **Bravo, R., Frank, R., Blundell, P. A., and Macdonald-Bravo, H.,** Cyclin/PCNA is the auxiliary protein of DNA polymerase-delta, *Nature,* 326, 515, 1987.

280. **Prelich, G., Tan, C.-K., Kostura, M., Mathews, M. B., So, A. G., Downey, K. M., and Stillman, B.,** Functional identity of proliferating cell nuclear antigen and a DNA polymerase-δ auxiliary protein, *Nature,* 326, 517, 1987.

281. **Tan, G.-K., Sullivan, K., Li, X., Tan, E. M., Downey, K. M., and So, A. G.,** Autoantibody to the proliferating cell nuclear antigen neutralizes the activity of the auxiliary protein for DNA polymerase delta, *Nucleic Acids Res.,* 15, 9299, 1987.

282. **Jaskulski, D., deRiel, J. K., Mercer, W. E., Calabretta, B., and Baserga, R.,** Inhibition of cellular proliferation by antisense oligodeoxynucleotides to PCNA cyclin, *Science,* 240, 1544, 1988.

283. **Yamaguchi, M., Nishida, Y., Moriuchi, T., Hirose, F., Hui, C.-C., Suzuki, Y., and Matsukage, A.,** *Drosophila* proliferating cell nuclear antigen (cyclin) gene: structure, expression during development, and specific binding of homeodomain proteins to its 5'-flanking region, *Mol. Cell. Biol.,* 10, 872, 1990.

284. **Ottavio, L., Chang, C.-D., Rizzo, M. G., Petralia, S., Travali, S., and Baserga, R.,** The promoter of the proliferating cell nuclear antigen (PCNA) gene is active in serum-deprived cells, *Biochem. Biophys. Res. Commun.,* 169, 509, 1990.

285. **Chang, C.-D., Ottavio, L., Travali, S., Lipson, K. E., and Baserga, R.,** Transcriptional and posttranscriptional regulation of the proliferating cell nuclear antigen gene, *Mol. Cell. Biol.,* 10, 3289, 1990.

286. **Bauer, G. A. and Burgers, P. M. J.,** The yeast analog of mammalian cyclin/proliferating-cell nuclear antigen interacts with mammalian DNA polymerase δ, *Proc. Natl. Acad. Sci. U.S.A.,* 85, 7506, 1988.

287. **Bauer, G. A. and Burgers, P. M. J.,** Molecular cloning, structure and expression of the yeast proliferating cell nuclear antigen gene, *Nucleic Acids Res.,* 18, 261, 1990.

288. **Burgers, P. M. J.,** Mammalian cyclin/PCNA (DNA polymerase delta auxiliary protein) stimulates DNA synthesis by yeast DNA polymerase III, *Nucleic Acids Res.,* 16, 6297, 1988.

289. **Dou, Q.-P., Levin, A. H., Zhao, S., and Pardee, A. B.,** Cyclin E and cyclin A as candidates for the restriction point protein, *Cancer Res.,* 53, 1493, 1993.

290. **Pines, J. and Hunter, T.,** Human cyclin A is adenovirus E1-A-associated protein p60 and behaves differently from cyclin B, *Nature,* 346, 760, 1990.

291. **Pines, J. and Hunter, T.,** Human cyclins A and B1 are differentially located in the cell and undergo cell cycle-dependent nuclear transport, *J. Cell Physiol.,* 115, 1, 1991.

292. **Pines, J. and Hunter, T.,** Human cyclin A is adenovirus E1A-associated protein p60 and behaves differently from cyclin B, *Nature,* 346, 760, 1990.

293. **Girard, F., Strausfeld, U., Fernandez, A., and Lamb, N. J. C.,** Cyclin A is required for the onset of DNA replication in mammalian fibroblasts, *Cell,* 67, 1169, 1991.

294. **Zindy, F., Lamas, E., Chenivesse, X., Sobczak, J., Wang, J., Fesquet, D., Henglein, B., and Brechot, C.,** Cyclin A is required in S-phase in normal epithelial cells, *Biochem. Biophys. Res. Commun.,* 182, 1144, 1992.

295. **Tsai, L.-H., Harlow, E., and Meyerson, M.,** Isolation of the human *cdk2* gene that encodes the cyclin A- and adenovirus E1A-associated p33 kinase, *Nature,* 353, 174, 1991.

296. **Motokura, T., Bloom, T., Kim, H. G., Juppner, H., Ruderman, J. V., Kronenberg, H. M., and Arnold, A.,** A novel cyclin encoded by a *bcl*1-linked candidate oncogene, *Nature,* 512, 1991.

297. **Dulic, V., Lees, E., and Reed, S. I.,** Association of human cyclin E with a periodic G_1-S phase protein kinase, *Science,* 257, 1958, 1992.

298. **Matsushime, H., Roussel, M. F., Ashmun, R. A., and Sherr, C. J.,** Colony-stimulating factor 1 regulates novel cyclins during the G_1 phase of the cell cycle, *Cell,* 65, 701, 1991.

299. **Izumi, T. and Maller, J. L.,** Phosphorylation of *Xenopus* cyclins B1 and B2 is not required for cell cycle transitions, *Mol. Cell. Biol.,* 11, 3860, 1991.

300. **Williams, R. T., Carbonaro-Hall, D. A., and Hall, F. L.,** Co-purification of p34[cdc2]/p58[cyclin A] proline-directed protein kinase and the retinoblastoma tumor susceptibility gene product: interaction of an oncogenic serine/threonine protein kinase with a tumor-suppressor protein, *Oncogene,* 7, 423, 1992.

301. **Mudryj, M., Devoto, S. H., Hiebert, S. W., Hunter, T., Pines, J., and Nevins, J. R.,** Cell cycle regulation of the E2F transcription factor involves an interaction with cyclin A, *Cell,* 65, 1243, 1991.

302. **Motokura, T., Bloom, T., Kim, H. G., Jüppner, H., Ruderman, J. V., Kronenberg, H. M., and Arnold, A.,** A novel cyclin encoded by a *bcl*1-linked candidate oncogene, *Nature,* 350, 512, 1991.

303. **Jiang, W., Kahn, S. M., Tomita, N., Zhang, Y.-J., Lu, S.-H., and Weinstein, I. B.,** Amplification and expression of the human cyclin D gene in esophageal cancer, *Cancer Res.,* 52, 2980, 1992.

304. **Williams, M. E., Swerdlow, S. H., Rosenberg, C. L., and Arnold, A.,** Chromosome 11 translocation breakpoints of the PRAD1/cyclin D1 gene locus in centrocytic lymphoma, *Leukemia,* 7, 241, 1993.

305. **Lammie, G. A., Smith, R., Silver, J., Brookes, S., Dickson, C., and Peters, G.,** Proviral insertions near cyclin D1 in mouse lymphomas: a parallel for BCL1 translocations in human B-cell neoplasms, *Oncogene,* 7, 2381, 1992.

306. **Keyomarsi, K. and Pardee, A. B.,** Redundant cyclin overexpression and gene amplification in breast cancer cells, *Proc. Natl. Acad. Sci. U.S.A.,* 90, 1112, 1993.

307. **Leach, F. S., Elledge, S. J., Sherr, C. J., Willson, J. K. V., Markowitz, S., Kinzler, K. W., and Vogelstein, B.,** Amplification of cyclin genes in colorectal carcinomas, *Cancer Res.,* 53, 1986, 1993.

308. **Mercer, W. E., Shields, M. T., Lin, D., Appella, E., and Ullrich, S. J.,** Growth suppression induced by wild-type p53 protein is accompanied by selective down-regulation of proliferating-cell nuclear antigen expression, *Proc. Natl. Acad. Sci. U.S.A.,* 88, 1958, 1991.

309. **Hinds, P. W., Mittnacht, S., Dulic, V., Arnold, A., Reed, S. I., and Weinberg, R. A.,** Regulation of retinoblastoma protein functions by ectopic expression of human cyclins, *Cell,* 70, 993, 1992.

310. **Simanis, V. and Nurse, P.,** The cell cycle control gene *cdc2*+ of fission yeast encodes a protein kinase potentially regulated by phosphorylation, *Cell,* 45, 261, 1986.

311. **Gould, K. L. and Nurse, P.,** Tyrosine phosphorylation of the fission yeast cdc2+ protein kinase regulates entry into mitosis, *Nature,* 342, 39, 1989.

312. **Moreno, S., Hayles, J., and Nurse, P.,** Regulation of p34[cdc2] protein kinase during mitosis, *Cell,* 58, 361, 1989.

313. **Feiler, H. S. and Jacobs, T. W.,** Cell division in higher plants: a *cdc2* gene, its 34-kDa product, and histone H1 kinase activity in pea, *Proc. Natl. Acad. Sci. U.S.A.,* 87, 5397, 1990.

314. **Draetta, G., Brizuela, L., Moran, B., and Beach, D.,** Regulation of the vertebrate cell cycle by the cdc2 protein kinase, *Cold Spring Harbor Symp. Quant. Biol.,* 53, 195, 1988.

315. **Draetta, G. and Beach, D.,** The mammalian cdc2 protein kinase: mechanisms of regulation during the cell cycle, *J. Cell Sci.,* Suppl. 12, 21, 1989.

316. **Sebastian, B., Kakizuka, A., and Hunter, T.,** cdc25M2 activation of cyclin-dependent kinases by dephosphorylation of threonine-14 and tyrosine-15, *Proc. Natl. Acad. Sci. U.S.A.,* 90, 3521, 1993.

317. **Yasuda, H., Kamijo, M., Honda, R., Nagahara, M., and Ohba, Y.,** The difference in murine cDC2 kinase activity between cytoplasmic and nuclear fractions during the cell cycle, *Biochem. Biophys. Res. Commun.,* 172, 371, 1990.

318. **Welch, P. J. and Wang, J. Y. J.,** Coordinated synthesis and degradation of cdc2 in the mammalian cell cycle, *Proc. Natl. Acad. Sci. U.S.A.,* 89, 3093, 1992.

319. **Richter, K. H., Afshari, C. A., Annab, L. A., Burkhart, B. A., Owen, R. D., Boyd, J., and Barrett, J. C.,** Down-regulation of cdc2 in senescent human and hamster cells, *Cancer Res.,* 51, 6010, 1991.

320. **Stein, G. H., Drullinger, L. F., Robetorye, R. S., Pereira-Smith, O. M., and Smith, J. R.,** Senescent cells fail to express cdc2, cycA, and cycB in response to mitogen stimulation, *Proc. Natl. Acad. Sci. U.S.A.,* 88, 11012, 1991.

321. **Maller, J. L.,** *Xenopus* oocytes and the biochemistry of cell division, *Biochemistry,* 29, 3157, 1990.

322. **Dunphy, W. G., Brizuela, L., Beach, D., and Newport, J.,** The *Xenopus* cdc2 protein is a component of MPF, a cytoplasmic regulator of mitosis, *Cell,* 54, 423, 1988.

323. **Gautier, J., Matsukawa, T., Nurse, P., and Maller, J.,** Dephosphorylation and activation of *Xenopus* p34[cdc2] protein kinase during the cell cycle, *Nature,* 339, 626, 1989.

324. **Zhao, X., Batten, B., Singh, B., and Arlinghaus, R. B.,** Requirement of the c-*mos* protein kinase for murine meiotic maturation, *Oncogene,* 5, 1727, 1990.

325. **Krek, W. and Nigg, E. A.,** Differential phosphorylation of vertebrate p34[cdc2] kinase at the G_1/S and G_2/M transitions of the cell cycle: identification of major phosphorylation sites, *EMBO J.,* 10, 305, 1991.

326. **Honda, R., Ohba, Y., Nagata, A., Okayama, H., and Yasuda, H.,** Dephosphorylation of human p34[cdc2] kinase on both Thr-14 and Tyr-15 by human cdc25B phosphatase, *FEBS Lett.,* 318, 331, 1993.

327. **Ducommun, B., Brambilla, P., Félix, A.-M., Franza, B. R., Jr., Karsenti, E., and Draetta, G.,** cdc2 phosphorylation is required for its interaction with cyclin, *EMBO J.,* 10, 3311, 1991.

328. **Lorca, T., Labbé, J. C., Devault, A., Fesquet, D., Capony, J. P., Cavadore, J. C., Lebouffant, F., and Dorée, M.,** Dephosphorylation of cdc2 on threonine-161 is required for cdc2 kinase inactivation and normal anaphase, *EMBO J.,* 11, 2381, 1992.

329. **Cheng, H., Litwin, C. M. E., Hwang, D. M., and Wang, J. H.,** Structural basis of specific and efficient phosphorylation of peptides derived from p34^{cdc2} by a pp60^{src}-related protein tyrosine kinase, *J. Biol. Chem.,* 266, 17919, 1991.

330. **Smythe, C. and Newport, J. W.,** Coupling of mitosis to the completion of S phase occurs via modulation of the tyrosine kinase that phosphorylates p34^{cdc2}, *Cell,* 68, 787, 1992.

331. **Ferris, D. K., White, G. A., Kelvin, D. J., Copeland, T. D., Li, C.-C. H., and Longo, D. L.,** p34^{cdc2} is physically associated with and phosphorylated by a *cdc2*-specific tyrosine kinase, *Cell Growth Differ.,* 2, 343, 1991.

332. **Honda, R., Ohba, Y., and Yasuda, H.,** The cell cycle regulator, human p50^{wee1}, is a tyrosine kinase and not a serine/tyrosine kinase, *Biochem. Biophys. Res. Commun.,* 186, 1333, 1992.

333. **McGowan, C. H. and Russell, P.,** Human Wee1 kinase inhibits cell division by phosphorylating p34^{cdc2} exclusively on Tyr15, *EMBO J.,* 12, 75, 1993.

334. **Taviaux, S. A. and Demaille, J. G.,** Localization of human cell cycle regulatory genes *CDC25C* and *WEE1* to 15p15.3-11p15.1 by fluorescence *in situ* hybridization, *Genomics,* 15, 194, 1993.

335. **Gautier, J., Solomon, M. J., Booher, R. N., Bazan, J. F., and Kirschner, M. W.,** cdc25 is a specific tyrosine phosphatase that directly activates p34^{cdc2}, *Cell,* 67, 197, 1991.

336. **Yamashita, K., Yasuda, H., Pines, J., Yasumoto, K., Nishitani, H., Ohtsubo, M., Hunter, T., Sugimura, T., and Nishimoto, T.,** Okadaic acid, a potent inhibitor of type 1 and type 2A protein phosphatases, activates cdc2/H1 kinase and transiently induces a premature mitosis-like state in BHK21 cells, *EMBO J.,* 9, 4331, 1990.

337. **Furukawa, Y., Piwnica-Worms, H., Ernst, T. J., Kanakura, Y., and Griffin, J. D.,** *cdc2* gene expression at the G$_1$ to S transition in human T lymphocytes, *Science,* 250, 805, 1990.

338. **Riabowol, K., Draetta, G., Brizuela, L., Vandre, D., and Beach, D.,** The cdc2 kinase is a nuclear protein that is essential for mitosis in mammalian cells, *Cell,* 57, 393, 1989.

339. **Bailly, E., Dorée, M., Nurse, P., and Bornens, M.,** p34^{cdc2} is located in both nucleus and cytoplasm; part is centrosomally associated at G$_2$/M and enters vesicles at anaphase, *EMBO J.,* 8, 3985, 1989.

340. **Draetta, G. and Beach, D.,** Activation of cdc2 protein kinase during mitosis in human cells: cell cycle-dependent phosphorylation and subunit rearrangement, *Cell,* 54, 17, 1988.

341. **Draetta, G., Piwnica-Worms, H., Morrison, D., Druker, B., Roberts, T., and Beach, D.,** Human cdc2 protein kinase is a major cell-cycle regulated tyrosine kinase substrate, *Nature,* 336, 738, 1988.

342. **Lee, M. G., Norbury, C. J., Spurr, N. K., and Nurse, P.,** Regulated expression and phosphorylation of a possible mammalian cell-cycle control protein, *Nature,* 333, 676, 1988.

343. **Morla, A. O., Draetta, G., Beach, D., and Wang, J. Y. J.,** Reversible tyrosine phosphorylation of cdc2: dephosphorylation accompanies activation during entry into mitosis, *Cell,* 58, 193, 1989.

344. **Roy, L. M., Swenson, K. I., Walker, D. H., Gabrielli, B. G., Li, R. S., Piwnica-Worms, H., and Maller, J. L.,** Activation of p34^{cdc2} kinase by cyclin-A, *J. Cell Biol.,* 113, 507, 1991.

345. **Tamura, T., Simon, E., Geschwill, K., and Niemann, H.,** cdc2/pp56-62 are *in vitro* substrates for the tyrosine kinase encoded by the v-*fms* oncogene, *Oncogene,* 5, 1259, 1990.

346. **Gould, K. L., Moreno, S., Tonks, N. K., and Nurse, P.,** Complementation of the mitotic activator, p80^{cdc25}, by a human protein-tyrosine phosphatase, *Science,* 250, 1573, 1990.

347. **Lamb, N. J. C., Fernandez, A., Watrin, A., Labbé, J.-C., and Cavadore, J. C.,** Microinjection of p34^{cdc2} kinase induces marked changes in cell shape, cytoskeletal organization, and chromatin structure in mammalian cells, *Cell,* 60, 151, 1990.

348. **Cisek, L. J. and Corden, J. L.,** Phosphorylation of RNA polymerase by the murine homologue of the cell-cycle control protein cdc2, *Nature,* 339, 679, 1989.

349. **Lee, M. G. and Nurse, P.,** Complementation used to clone a human homologue of the fission yeast cell cycle control gene *cdc2*, *Nature,* 327, 31, 1987.

350. **McGowan, C. H., Russell, P., and Reed, S. I.,** Periodic biosynthesis of the human M-phase promoting factor catalytic subunit p34 during the cell cycle, *Mol. Cell Biol.,* 10, 3847, 1990.

351. **Picard, A., Cavadore, J.-C., Lory, P., Bernengo, J.-C., Ojeda, C., and Dorée, M.,** Microinjection of a conserved peptide sequence of p34^{cdc2} induces a Ca^{2+} transient in oocytes, *Science,* 247, 327, 1990.

352. **Chou, Y.-H., Bischoff, J. K., Beach, D., and Goldman, R. D.,** Intermediate filament reorganization during mitosis is mediated by p34^{cdc2} phosphorylation of vimentin, *Cell,* 62, 1063, 1990.

353. **Chou, Y. H., Ngai, K. L., and Goldman, R.,** The regulation of intermediate filament reorganization in mitosis —p34^{cdc2} phosphorylates vimentin at a unique N-terminal site, *J. Biol. Chem.,* 266, 7325, 1991.

354. **Litchfield, D. W., Luscher, B., Lozeman, F. J., Eisenman, R. N., and Krebs, E. G.,** Phosphorylation of casein kinase II by p34 (cdc2) *in vitro* and at mitosis, *J. Biol. Chem.,* 267, 13943, 1992.

355. **Yamashiro, S., Yamakita, Y., Hosoya, H., and Matsumura, F.,** Phosphorylation of non-muscle caldesmon by p34^{cdc2} kinase during mitosis, *Nature,* 349, 169, 1991.

356. **Mak, A. S., Watson, M. H., Litwin, C. M. E., and Wang, J. H.,** Phosphorylation of caldesmon by cdc2 kinase, *J. Biol. Chem.,* 266, 6678, 1991.

357. **Shenoy, S., Chackalaparampil, I., Bagrodia, S., Lin, P.-H., and Shalloway, D.,** Role of p34^{cdc2}-mediated phosphorylations in two-step activation of pp60^{c-src} during mitosis, *Proc. Natl. Acad. Sci. U.S.A.,* 89, 7237, 1992.

358. **Kipreos, E. T. and Wang, J. Y. J.,** Differential phosphorylation of c-Abl in cell cycle determined by cdc2 kinase and phosphatase activity, *Science,* 248, 217, 1990.

359. **Stürzbecher, H.-W., Maimets, T., Chumakov, P., Brain, R., Addison, C., Simanis, V., Rudge, K., Philp, R., Grimaldi, M., Court, W., and Jenkins, J. R.,** p53 interacts with p34^{cdc2} in mammalian cells: implications for cell cycle control and oncogenesis, *Oncogene,* 5, 795, 1990.

360. **Milner, J., Cook, A., and Mason, J.,** p53 is associated with p34^{cdc2} in transformed cells, *EMBO J.,* 9, 2895, 1990.

361. **Lin, B. T. Y., Gruenwald, S., Morla, A. O., Lee, W. H., and Wang, J. Y. J.,** Retinoblastoma cancer suppressor gene product is a substrate of the cell cycle regulator *cdc2* kinase, *EMBO J.,* 10, 857, 1991.

362. **Meyerson, M., Enders, G. H., Wu, C.-L., Su, L.-K., Gorka, C., Nelson, C., Harlow, E., and Tsai, L.-H.,** A family of human cdc2-related protein kinases, *EMBO J.,* 11, 2909, 1992.

363. **Gabrielli, B. G., Roy, L. M., Gautier, J., Philippe, M., and Maller, J. L.,** A cdc2-related kinase oscillates in the cell cycle independently of cyclins G$_2$/M and cdc2, *J. Biol. Chem.,* 267, 1969, 1992.

364. **Elledge, S. J., Richman, R., Hall, F. L., Williams, R. T., Lodgson, N., and Harper, J. W.,** cdk2 encodes a 33-kDa cyclin A-associated protein kinase and is expressed before cdc2 in the cell cycle, *Proc. Natl. Acad. Sci. U.S.A.,* 89, 2907, 1992.

365. **Pagano, M., Draetta, G., and Jansen-Dürr, P.,** Association of cdk2 kinase with the transcription factor E2F during S phase, *Science,* 255, 1144, 1992.

366. **Pagano, M., Pepperkok, R., Lukas, J., Baldin, V., Ansorge, W., Bartek, J., and Draetta, G.,** Regulation of the cell cycle by the cdck1 protein kinase in cultured human fibroblasts, *J. Cell Biol.,* 121, 101, 1993.

367. **Latchman, D. S.,** Eukaryotic transcription factors, *Biochem. J.,* 270, 281, 1990.

368. **Wolffe, A. P.,** *Xenopus* transcription factors: key molecules in the developmental regulation of differential gene expression, *Biochem. J.,* 278, 313, 1991.

369. **Pabo, C. O. and Sauer, R. T.,** Transcription factors — structural families and principles of DNA recognition, *Annu. Rev. Biochem.,* 61, 1053, 1992.

370. **Faisst, S. and Meyer, S.,** Compilation of vertebrate-encoded transcription factors, *Nucleic Acids Res.,* 20, 3, 1992.

371. **Ransone, L. J. and Verma, I. M.,** Nuclear proto-oncogenes *fos* and *jun, Annu. Rev. Cell Biol.,* 6, 539, 1990.

372. **Angel, P. and Karin, M.,** The role of Jun, Fos and the AP-1 complex in cell proliferation and transformation, *Biochim. Biophys. Acta,* 1072, 129, 1991.

373. **Lucibello, F. C. and Müller, R.,** Proto-oncogenes encoding transcriptional regulators: unravelling the mechanisms of oncogenic conversion, *Crit. Rev. Oncogenesis,* 2, 259, 1991.

374. **Berk, A. J.,** Regulation of eukaryotic transcription factors by post-translational modification, *Biochim. Biophys. Acta,* 1009, 103, 1989.

375. **Ku, D.-H., Chang, D., Koniecki, J., Cannizzarok, L. A., Boghosian-Sell, L., Alder, H., and Baserga, R.,** A new growth-regulated complementary DNA with the sequence of a putative *trans*-activating factor, *Cell Growth Differ.,* 2, 179, 1991.

376. **Grimm, S. and Bauerle, P. A.,** The inducible transcription factor NF-κB: structure-function relationship of its protein subunits, *Biochem. J.,* 290, 297, 1993.

377. **Kerr, L. D., Duckett, C. S., Wamsley, P., Zhang, Q., Chiao, P., Nabel, G., McKeithan, T. W., Bauerle, P. A., and Verma, I. M.,** The proto-oncogene *BCL-3* encodes an IkB protein, *Genes Dev.,* 6, 2352, 1992.

378. **Lau, L. F. and Nathans, D.,** Identification of a set of genes expressed during the G$_0$/G$_1$ transition of cultured mouse cells, *EMBO J.,* 4, 3145, 1985.

379. **Lau, L. F. and Nathans, D.,** Expression of a set of growth-related immediate early genes in BALB/c 3T3 cells: coordinate regulation with c-*fos* or c-*myc, Proc. Natl. Acad. Sci. U.S.A.,* 84, 1182, 1987.

380. **Hazel, T. G., Nathans, D., and Lau, L. F.,** A gene inducible by serum growth factors encodes a member of the steroid and thyroid hormone receptor superfamily, *Proc. Natl. Acad. Sci. U.S.A.,* 85, 8444, 1988.

381. **Ryseck, R.-P., Macdonald-Bravo, H., Mattéi, M.-G., Ruppert, S., and Bravo, R.,** Structure, mapping and expression of a growth factor inducible gene encoding a putative nuclear hormonal binding receptor, *EMBO J.,* 8, 3327, 1989.

382. **Bours, V., Villalobos, J., Burd, P. R., Kelly, K., and Siebenlist, U.,** Cloning of a mitogen-inducible gene encoding a κ B DNA-binding protein with homology to the *rel* oncogene and to cell-cycle motifs, *Nature,* 348, 76, 1990.

383. **Davis, I. J., Hazel, T. G., and Lau, L. F.,** Transcriptional activation bu Nur77, a growth factor-inducible member of the steroid hormone receptor superfamily, *Mol. Endocrinol.,* 5, 854, 1991.

384. **Hazel, T. G., Misra, R., Davis, I. J., Greenberg, M. E., and Lau, L. F.,** Nur77 is differentially modified in PC12 cells upon membrane depolarization and growth factor treatment, *Mol. Cell. Biol.,* 11, 3239, 1991.

385. **Bondy, G. P.,** Phorbol ester, forskolin, and serum induction of a human colon nuclear hormone receptor gene related to the *NUR 77/NGFI-B* genes, *Cell Growth Differ.,* 2, 203, 1991.

386. **Sekiguchi, T., Yoshida, M. C., Sekiguchi, M., and Nishimoto, T.,** Isolation of a human X chromosome-linked gene essential for progression from G_1 to S phase of the cell cycle, *Exp. Cell Res.,* 169, 395, 1987.

387. **Brown, C. J., Sekiguchi, T., Nishimoto, T., and Willard, H. F.,** Regional localization of *CCG1* gene which complements hamster cell cycle mutation BN462 to Xq11-Xq13, *Somat. Cell Mol. Genet.,* 15, 93, 1989.

388. **Kai, R., Ohtsubo, M., Sekiguchi, M., and Nashimoto, T.,** Molecular cloning of a human gene that regulates chromosome condensation and is essential for cell proliferation, *Mol. Cell. Biol.,* 6, 2027, 1986.

389. **Rosner, M. H., Vigano, M. A., Rigby, P. W. J., Arnheiter, H., and Staudt, L. M.,** Oct-3 and the beginning of mammalian development, *Science,* 253, 144, 1991.

390. **Ullman, K. S., Flanagan, W. M., Edwards, C. A., and Crabtree, G. R.,** Activation of early gene expression in T lymphocytes by Oct-1 and an inducible protein, OAP[40], *Science,* 254, 558, 1991.

391. **Roberts, S. B., Segil, N., and Heintz, N.,** Differential phosphorylation of the transcription factor Oct1 during the cell cycle, *Science,* 253, 1022, 1991.

392. **Sheng, M., Thompson, M. A., and Greenberg, M. E.,** CREB: a Ca^{2+}-regulated transcription factor phosphorylated by calmodulin-dependent kinases, *Science,* 252, 1427, 1991.

393. **Verrijzer, C. P. and van der Vliet, P. C.,** POU domain transcription factors, *Biochim. Biophys. Acta,* 1173, 1, 1993.

394. **Radovick, S., Nations, M., Du, Y., Berg, L. A., Weintraub, B. D., and Wondisford, F. E.,** A mutation in the POU-homeodomain of Pit-1 responsible for combined pituitary hormone deficiency, *Science,* 257, 1115, 1992.

395. **Pfäffle, R. W., DiMattia, G. E., Parks, J. S., Brown, M. R., Wit, J. M., Jansen, M., Van der Nat, H., Van den Brande, J. L., Rosenfeld, M. G., and Ingraham, H. A.,** Mutation of the POU-specific domain of Pit-1 and hypopituitarism without pituitary hypoplasia, *Science,* 257, 1118, 1992.

396. **Linquist, S.,** The heat shock response, *Annu. Rev. Biochem.,* 55, 1151, 1986.

397. **Schlessinger, M. J.,** Heat shock proteins: the search for functions, *J. Cell Biol.,* 103, 321, 1986.

398. **Carper, S. W., Duffy, J. J., and Gerner, E. W.,** Heat shock proteins in thermotolerance and other cellular processes, *Cancer Res.,* 47, 5249, 1987.

399. **Schlesinger, M. J.,** Heat shock proteins, *J. Biol. Chem.,* 265, 12111, 1990.

400. **Pechan, P. M.,** Heat shock proteins and cell proliferation, *FEBS Lett.,* 280, 1, 1991.

401. **Morimoto, R. I.,** Heat shock: the role of transient inducible responses in cell damage, transformation, and differentiation, *Cancer Cells,* 3, 295, 1991.

402. **Ang, D., Liberek, K., Skowyra, D., Zylicz, M., and Georgopoulos, C.,** Biological role and regulation of the universally conserved heat shock proteins, *J. Biol. Chem.,* 266, 24233, 1991.

403. **Westwood, J. T., Clos, J., and Wu, C.,** Stress-induced oligomerization and chromosomal relocalization of heat-shock factor, *Nature,* 353, 822, 1991.

404. **Koyasu, S., Nishida, E., Miyata, Y., Sakai, H., and Yahara, I.,** HSP100, a 100-kDa heat shock protein, is a Ca^{2+}-calmodulin-regulated actin-binding protein, *J. Biol. Chem.,* 264, 15083, 1989.

405. **Collier, N. C. and Schlessinger, M. J.,** The dynamic state of heat shock proteins in chicken embryo fibroblasts, *J. Cell Biol.,* 103, 1495, 1986.

406. **Milarski, K. L. and Morimoto, R. I.,** Expression of human HSP70 during the synthetic phase of the cell cycle, *Proc. Natl. Acad. Sci. U.S.A.,* 83, 9517, 1986.

407. **Taira, T., Negishi, Y., Kihara, F., Iguchi-Ariga, S. M. M., and Ariga, H.,** c-Myc protein complex binds to two sites in human hsp70 promoter region, Biochim. Biophys. Acta, 1130, 166, 1992.

408. **Andrews, G. K., Harding, M. A., Calvet, J. P., and Adamson, E. D.,** The heat shock response in HeLa cells is accompanied by elevated expression of the c-*fos* proto-oncogene, *Mol. Cell. Biol.,* 7, 3452, 1987.

409. **Schiaffonati, L., Rappoccolo, E., Tacchini, L., Cairo, G., and Bernelli-Zazzera, A.,** Reprogramming of gene expression in postischemic rat liver: induction of proto-oncogenes and hsp 70 gene family, *J. Cell. Physiol.,* 143, 79, 1990.

410. **Chiang, H.-L., Terlecky, S. R., Plant, C. P., and Dice, J. F.,** A role for a 70-kilodalton heat shock protein in lysosomal degradation of intracellular proteins, *Science,* 246, 382, 1989.

411. **Mues, G. I., Munn, T. Z., and Raese, J. D.,** A human gene family with sequence homology to *Drosophila melanogaster Hsp70* heat shock genes, *J. Biol. Chem.,* 261, 874, 1986.

412. **Harrison, G. S., Drabkin, H. A., Kao, F.-T., Hartz, J., Hart, I. M., Chu, E. H. Y., Wu, B. J., and Morimoto, R. I.,** Chromosomal location of human genes encoding major heat-shock protein HSP70, *Somat. Cell Mol. Genet.,* 13, 119, 1987.

413. **Leung, T. K. C., Rajendran, M. Y., Monfries, C., Hall, C., and Lim, L.,** The human heat-shock protein family. Expression of a novel heat-inducible *HSP70* (*HSP70* B′) and isolation of its cDNA and genomic DNA, *Biochem. J.,* 267, 125, 1990.

414. **Watowich, S. S. and Morimoto, R. I.,** Complex regulation of heat shock- and glucose-responsive genes in human cells, *Mol. Cell. Biol.,* 8, 393, 1988.

415. **Kingston, R. E., Baldwin, A. S., Jr., and Sharp, P. A.,** Regulation of heat shock protein 70 gene expression by c-*myc, Nature,* 312, 280, 1984.

416. **Wu, B. J. and Morimoto, R. I.,** Transcription of the human *HSP70* gene is induced by serum stimulation, *Proc. Natl. Acad. Sci. U.S.A.,* 82, 6070, 1985.

417. **Van der Ploeg, L. H. T., Giannini, S. H., and Cantor, C. R.,** Heat shock genes: regulatory role for differentiation in parasitic protozoa, *Science,* 228, 1443, 1985.

418. **Zakeri, Z. F. and Wolgemuth, D. J.,** Developmental-stage specific expression of the *HSP70* gene family during differentiation of the mammalian male germ line, *Mol. Cell. Biol.,* 7, 1791, 1987.

419. **Ponzetto, C. and Wolgemuth, D. J.,** Haploid expression of a unique c-*abl* transcript in the mouse male germ line, *Mol. Cell. Biol.,* 5, 1791, 1985.

420. **Bardella, L., Schiaffonati, L., Cairo, G., and Bernelli-Zazzera, A.,** Heat-shock proteins and mRNAs in the liver and hepatoma, *Br. J. Cancer,* 55, 643, 1987.

421. **Wu, B. J. and Morimoto, R. I.,** Transcription of the human *HSP70* gene is induced by serum stimulation, *Proc. Natl. Acad. Sci. U.S.A.,* 82, 6070, 1985.

422. **Carr, B. I., Huang, T. H., Buzin, C. H., and Itakura, K.,** Induction of heat shock gene expression without heat shock by hepatocarcinogens and during hepatic regeneration in rat liver, *Cancer Res.,* 46, 5106, 1986.

423. **Hensold, J. O. and Housman, D. E.,** Decreased expression of the stress protein HSP70 is an early event in murine erythroleukemia cell differentiation, *Mol. Cell. Biol.,* 8, 2219, 1988.

424. **Binart, N., Chambraud, B., Dumas, B., Rowlands, D. A., Bigogne, C., Levin, J. M., Garnier, J., Baulieu, E.-E., and Catelli, M.-G.,** The cDNA-derived amino acid sequence of chick heat shock protein M_r 90,000 (HSP90) reveals a "DNA like" structure: potential site of interaction with steroid receptors, *Biochem. Biophys. Res. Commun.,* 159, 140, 1989.

425. **Koyasu, S., Nishida, E., Kadowaki, T., Matsuzaki, F., Iida, K., Harada, F., Kasuga, M., Sakai, H., and Yahara, I.,** Two mammalian heat shock proteins, HSP90 and HSP100, are actin-binding proteins, *Proc. Natl. Acad. Sci. U.S.A.,* 83, 8054, 1986.

426. **Nishida, E., Koyasu, S., Sakai, H., and Yahara, I.,** Calmodulin-regulated binding of the 90-kDa heat shock protein to actin filaments, *J. Biol. Chem.,* 261, 16033, 1986.

427. **Perdew, G. H. and Whitelaw, M. L.,** Evidence that the 90-kDa heat shock protein (HSP90) exists in cytosol in heteromeric complexes containing HSP70 and three other proteins with M_r 63,000, 56,000, and 50,000, *J. Biol. Chem.,* 266, 6708, 1991.

428. **Dougherthy, J. J., Rabideau, D. A., Iannotti, A. M., Sullivan, W. P., and Toft, D. O.,** Identification of the 90 kDa substrate of rat liver type II casein kinase with the heat shock protein which binds steroid receptors, *Biochim. Biophys. Acta,* 927, 74, 1987.

429. **Lees-Miller, S. P. and Anderson, C. W.,** Two human 90-kDa heat shock proteins are phosphorylated *in vivo* at conserved serines that are phosphorylated *in vitro* by casein kinase II, *J. Biol. Chem.,* 264, 2431, 1989.

430. **Oppermann, H., Levinson, W., and Bishop, J. M.,** A cellular protein that associates with the transforming protein of Rous sarcoma virus is also a heat-shock protein, *Proc. Natl. Acad. Sci. U.S.A.,* 78, 1067, 1981.

431. **Brugge, J. S., Erikson, E., and Erikson, R. L.,** The specific interaction of the Rous sarcoma virus transforming protein, pp60^{src}, with two cellular proteins, *Cell,* 25, 363, 1981.

432. **Lipsich, L. A., Cutt, J. R., and Brugge, J. S.,** Association of transforming proteins of Rous, Fujinami and Y73 avian sarcoma viruses with the same two cellular proteins, *Mol. Cell. Biol.,* 2, 875, 1982.

433. **Sanchez, E. R., Toft, D. O., Schlesinger, M. J., and Pratt, W. B.,** Evidence that the 90-kDa phosphoprotein associated with the untransformed L-cell glucocorticoid receptor is a murine heat shock protein, *J. Biol. Chem.,* 260, 12398, 1985.

434. **Schuh, S., Yonemoto, W., Brugge, J., Bauer, V. J., Riehl, R. M., Sullivan, W. P., and Toft, D. O.,** A 90,000-Dalton binding protein common to both steroid receptors and the Rous sarcoma virus transforming protein, pp60^{v-src}, *J. Biol. Chem.,* 260, 14292, 1985.

435. **Ziemicki, A., Catelli, M.-G., Joab, I., and Moncharmont, B.,** Association of the heat shock protein hsp90 with steroid hormone receptors and tyrosine kinase oncogene products, *Biochem. Biophys. Res. Commun.,* 138, 1298, 1986.

436. **Sabbah, M., Redeuilh, G., Secco, C., and Baulieu, E.-E.,** The binding activity of estrogen receptor to DNA and heat shock protein (M$_r$ = 90,000) is dependent on receptor-bound metal, *J. Biol. Chem.,* 262, 8631, 1987.

437. **Radanyi, C., Renoir, J.-M., Sabbah, M., and Baulieu, E.-E.,** Chick heat-shock protein of M$_r$ = 90,000, free or released from progesterone receptor, is in a dimeric form, *J. Biol. Chem.,* 264, 2568, 1989.

438. **Yufu, Y., Nishimura, J., and Nawata, H.,** High constitutive expression of heat shock protein 90 alpha in human acute leukemia cells, *Leukemia Res.,* 16, 587, 1992.

439. **Gabrilove, J. L.,** Differentiation factors, *Semin. Oncol.,* 13, 228, 1986.

440. **Nagy, P., Evarts, R. P., McMahon, J. B., and Thorgeirsson, S. S.,** Role of TGF-beta in normal differentiation and oncogenesis in rat liver, *Mol. Carcinogenesis,* 2, 345, 1989.

441. **Reddi, A. H., Muthukumaran, S., Ma, S., Carrington, J. L., Luyten, F. P., Paralkar, V. M., and Cunningham, N. S.,** Initiation of bone development by osteogenin and promotion by growth factors, *Connect. Tissue Res.,* 20, 303, 1989.

442. **Sampath, T. K., Muthukumaran, N., and Reddi, A. H.,** Isolation of osteogenin, an extracellular matrix-associated, bone-inductive protein, by heparin affinity chromatography, *Proc. Natl. Acad. Sci. U.S.A.,* 84, 7109, 1987.

443. **Luyten, F. P., Cunningham, N. S., Ma, S., Muthukumaran, N., Hammonds, R. G., Nevins, W. B., Wood, W. I., and Reddi, A. H.,** Purification and partial amino acid sequence of osteogenin, a protein initiating bone differentiation, *J. Biol. Chem.,* 264, 13377, 1989.

444. **Vukicevic, S., Luyten, F. P., and Reddi, A. H.,** Stimulation of the expression of osteogenic and chondrogenic phenotypes *in vitro* by osteogenin, *Proc. Natl. Acad. Sci. U.S.A.,* 86, 8793, 1989.

445. **Vukicevic, S., Luyten, F. P., and Reddi, A. H.,** Osteogenin inhibits proliferation and stimulates differentition in mouse osteoblast-like cells (MC3T3-E1), *Biochem. Biophys. Res. Commun.,* 166, 750, 1990.

446. **Vukicevic, S., Paralkar, V. M., Cunningham, N. S., Gutkind, J. S., and Reddi, A. H.,** Autoradiographic localization of osteogenin binding sites in cartilage and bone during rat embryonic development, *Dev. Biol.,* 140, 209, 1990.

447. **Li, Z.-H., Lu, Z., Kirkland, J. L., and Gregerman, R. I.,** Preadipocyte stimulating factor in rat serum: evidence for a discrete 63 kDa protein that promotes cell differentiation of rat preadipocytes in primary cultures, *J. Cell. Physiol.,* 141, 543, 1989.

448. **Djulbegovic, B., Christmas, S. E., and Moore, M.,** Differentiated HL-60 promyelocytic leukaemia cells produce a factor inducing differentiation, *Leukemia Res.,* 11, 259, 1987.

449. **Okazaki, T., Kato, Y., Tashima, M., Sawada, H., and Uchino, H.,** Evidence of intracellular and trans-acting differentiation-inducing activity in human promyelocytic leukemia HL-60 cells: its possible involvement in process of cell differentiation from a commitment step to a phenotype-expression step, *J. Cell. Physiol.,* 134, 261, 1988.

450. **Nakaya, K., Kumakawa, N., Iinuma, H., and Nakamura, Y.,** Purification of a low molecular weight factor that induces differentiation and inhibits growth in myeloid leukemia cells, *Cancer Res.,* 48, 4201, 1988.

451. **Kurata, N., Sawada, M., Ito, Y., and Marunouchi, T.,** A factor inducing differentiation of the human monocytic cell line U-937 produced by 12-*O*-tetradecanoylphorbol 13-acetate-treated U-937, *Jpn. J. Cancer Res.,* 78, 219, 1987.

452. **Watanabe, T. and Oishi, M.,** Dimethyl sulfoxide-inducible cytoplasmic factor involved in erythroid differentiation in mouse erythroleukemia (Friend) cells, *Proc. Natl. Acad. Sci. U.S.A.,* 84, 6481, 1987.

453. **Tomida, M., Yamamoto-Yamaguchi, Y., and Hozumi, M.,** Purification of a factor inducing differentiation of mouse myeloid leukemic M1 cells from conditioned medium of mouse fibroblast L929 cells, *J. Biol. Chem.,* 259, 10978, 1984.

454. **Smith, A. G., Heath, J. R., Donaldson, D. D., Wong, G. G., Moreau, J., Stahl, M., and Roger, D.,** Inhibition of pluripotential embryonic stem cells differentiation by purified polypeptides, *Nature,* 336, 688, 1988.

455. **Hilton, D. J., Nicola, N. A., and Metcalf, D.,** Specific binding of murine leukemia inhibitory factor to normal and leukemic monocytic cells, *Proc. Natl. Acad. Sci. U.S.A.,* 85, 5971, 1988.

456. **Miyaura, C., Jin, C. H., Yamaguchi, Y., Tomida, M., Hozumi, M., Matsuda, T., Hirano, T., Kishimoto, T., and Suda, T.,** Production of interleukin 6 and its relation to the macrophage differentiation of mouse myeloid leukemia cells (M1) treated with differentiation-inducing factor and 1α,25-dihydroxyvitamin D_3, *Biochem. Biophys. Res. Commun.,* 158, 660, 1989.

457. **Abe, T., Murakami, M., Sato, T., Kajiki, M., Ohno, M., and Kodaira, R.,** Macrophage differentiation inducing factor from human monocytic cells is equivalent to murine leukemia inhibitory factor, *J. Biol. Chem.,* 264, 8941, 1989.

458. **Gearing, D. P., Gough, N. M., King, J. A., Hilton, P. J., Nicola, N. A., Simpson, R. J., Nice, E. C., Kelson, A., and Metcalf, D.,** Molecular cloning and expression of cDNA encoding a murine myeloid leukaemia inhibitory factor (LIF), *EMBO J.,* 6, 3995, 1987.

459. **Gough, N. M., Gearing, D. P., King, J. A., Willson, T. A., Hilton, D. J., Nicola, N. A., and Metcalf, D.,** Molecular cloning and expression of the human homologue of the murine gene encoding myeloid leukemia inhibitorty factor, *Proc. Natl. Acad. Sci. U.S.A.,* 85, 2623, 1988.

460. **Lowe, D. G., Nunes, W., Bombara, M., McCabe, S., Ranges, G. E., Henzel, W., Tomida, M., Yamamoto-Yamaguchi, Y., Hozumi, M., and Goeddel, D. V.,** Genomic cloning and heterologous expression of human differentiation-stimulating factor, *DNA,* 8, 351, 1989.

461. **Sutherland, G. R., Baker, E., Hyland, V. J., Callen, D. F., Stahl, J., and Gough, N. M.,** The gene for human leukemia inhibitory factor (*LIF*) maps to 22q12, *Leukemia,* 3, 9, 1989.

462. **Okabe-Kado, J., Honma, Y., Hayashi, M., Hozumi, M., Sampi, K., Sakurai, M., Hino, K., and Tsuruoka, N.,** Induction of differentiation of mouse myeloid leukemia M1 cells by serum of patients with chronic myeloid leukemia, *Jpn. J. Cancer Res.,* 79, 1318, 1988.

463. **Moreau, J.-F., Donaldson, D. D., Bennett, F., Witek-Giannotti, J., Clark, S. C., and Wong, G. G.,** Leukaemia inhibitory factor is identical to the myeloid growth factor human interleukin for DA cells, *Nature,* 336, 690, 1988.

464. **Baumann, H. and Wong, G. G.,** Hepatocyte-stimulating factor III shares structural and functional identity with leukemia-inhibitory factor, *J. Immunol.,* 143, 1163, 1989.

465. **Yamamori, T., Fukada, K., Aebersold, R., Korsching, S., Fann, M.-J., and Patterson, P. H.,** The cholinergic neuronal differentiation factor from heart cells is identical to leukemia inhibitory factor, *Science,* 246, 1412, 1989.

466. **Metcalf, D.,** The leukemia inhibitory factor (LIF), *Int. J. Cell Cloning,* 9, 95, 1991.

467. **Kurzrock, R., Estrov, Z., Wetzler, M., Gutterman, J. U., and Talpaz, M.,** LIF: not just a leukemia inhibitory factor, *Endocrine Rev.,* 12, 208, 1991.

468. **Smith, A. G., Nichols, J., Robertson, M., and Rathjen, P. D.,** Differentiation inhibiting activity (DIA/LIF) and mouse development, *Dev. Biol.,* 151, 339, 1992.

469. **Murray, R., Lee, F., and Chiu, C.-P.,** The genes for leukemia inhibitory factor and interleukin-6 are expressed in mouse blastocysts prior to the onset of hemopoiesis, *Mol. Cell. Biol.,* 10, 4953, 1990.

470. **Conquet, F. and Brûlet, P.,** Developmental expression of myeloid leukemia inhibitory factor gene in preimplantation blastocysts and in extraembryonic tissue of mouse embryos, *Mol. Cell. Biol.,* 10, 3801, 1990.

471. **Shen, M. M. and Leder, P.,** Leukemia inhibitory factor is expressed by the implantation uterus and selectively blocks primitive ectoderm formation *in vitro, Proc. Natl. Acad. Sci. U.S.A.,* 89, 8240, 1992.

472. **Ferrara, N., Winer, J., and Henzel, W. J.,** Pituitary follicular cells secrete an inhibitor of aortic endothelial cell growth — identification as leukemia inhibitory factor, *Proc. Natl. Acad. Sci. U.S.A.,* 89, 698, 1992.

473. **Gullberg, U., Nilsson, E., Sarngadharan, M. G., and Olsson, I.,** T lymphocyte-derived differentiation-inducing factor inhibits proliferation of leukemic and normal hemopoietic cells, *Blood,* 68, 1333, 1986.

474. **Ikeda, K., Motoyoshi, K., Ishizaka, Y., Hatake, K., Kajigaya, S., Saito, M., and Miura, Y.,** Human colony-stimulating activity-producing tumor: production of very low mouse-active colony-stimulating activity and induction of granulocytosis in mice, *Cancer Res.,* 45, 4144, 1985.

475. **Hanamura, T., Motoyoshi, K., Hatake, K., Miura, Y., and Saito, M.,** Human colony-stimulating factor-producing lung cancer tissue releases a differentiation-inducing factor for human leukemic cells, *Leukemia,* 1, 497, 1987.

476. **Lee, M. Y., Eyre, D. R., and Osborne, W. R. A.,** Isolation of a murine osteoclast colony-stimulating factor, *Proc. Natl. Acad. Sci. U.S.A.,* 88, 8500, 1991.

477. **Takeda, K., Iwamoto, S., Sugimoto, H., Takuma, T., Kawatani, N., Noda, M., Masaki, A., Morise, H., Arimura, H., and Konno, K.,** Identity of differentiation inducing factor and tumour necrosis factor, *Nature,* 323, 338, 1986.

478. **Souza, L. M., Boone, T. C., Gabrilove, J., Lai, P. H., Zsebo, K. M., Murdock, D. C., Chazin, V. R., Bruszewski, J., Lu, H., Chen, K. K., Barendt, J., Platzer, E., Moore, M. A. S., Mertelsmann, R., and Welte, K.,** Recombinant human granulocyte colony-stimulating factor: effects on normal and leukemic myeloid cells, *Science,* 232, 61, 1986.

479. **Tomida, M., Yamamoto-Yamaguchi, Y., Hozumi, M., Okabe, T., and Takaku, F.,** Induction by recombinant human granulocyte colony-stimulating factor of differentiation of mouse myeloid leukemic M1 cells, *FEBS Lett.,* 207, 271, 1986.

480. **Onozaki, K., Akiyama, Y., Okano, A., Hirano, T., Kishimoto, T., Hashimoto, T., Yoshizawa, K., and Taniyama, T.,** Synergistic regulatory effects of interleukin 6 and interleukin 1 on the growth and differentiation of human and mouse myeloid leukemic cell lines, *Cancer Res.,* 49, 3602, 1989.

481. **Lotem, J. and Sachs, L.,** *In vivo* control of differentiation of myeloid leukemic cells by recombinant granulocyte-macrophage colony-stimulating factor and interleukin 3, *Blood,* 71, 375, 1988.

482. **Otaka, M., Singhal, A., and Hakomori, S.,** Antibody-mediated targeting of differentiation inducers to tumor cells: inhibition of colonic cancer cell growth *in vitro* and *in vivo*, *Biochem. Biophys. Res. Commun.,* 158, 202, 1989.

483. **Clegg, C. H., Linkhart, T. A., Olwin, B. B., and Hauschka, S. D.,** Growth factor control of skeletal muscle differentiation occurs in G_1-phase and is represented by fibroblast growth factor, *J. Cell Biol.,* 105, 949, 1987.

484. **Olson, E. N.,** MyoD family: a paradigm for development?, *Genes Dev.,* 4, 1454, 1990.

485. **Miner, J. H. and Wold, B. J.,** c-*myc* inhibition of MyoD and myogenin-initiated myogenic differentiation, *Mol. Cell. Biol.,* 11, 2842, 1991.

486. **Benezra, R., Davis, R. L., Lockshon, D., Turner, D. L., and Weintraub, H.,** The protein Id: a negative regulator of helix-loop-helix DNA binding proteins, *Cell,* 61, 49, 1990.

487. **Okabe-Kado, J., Kasukabe, T., Honma, Y., Hayashi, M., and Hozumi, M.,** Purification of a factor inhibiting differentiation from conditioned medium of nondifferentiating mouse myeloid leukemia cells, *J. Biol. Chem.,* 263, 10994, 1988.

488. **Smith, A. G. and Hooper, M. L.,** Buffalo rat liver cells produce a diffusible activity which inhibits the differentiation of murine embryonal carcinoma and embyonic stem cells, *Dev. Biol.,* 121, 1, 1987.

489. **Williams, R. L., Hilton, D. J., Pease, S., Willson, T. A., Stewart, C. L., Gearing, D. P., Wagner, E. F., Metcalf, D., Nicola, N. A., and Gough, N. M.,** Myeloid leukaemia inhibitory factor maintains the developmental potential of embryonic stem cells, *Nature,* 336, 684, 1988.

490. **Durkin, J. P., Biquard, J. M., Whitfield, J. F., Morardet, N., Royer, J., MacDonald, P., Tremblay, R., Legal, J. D., Doyonnas, R., Blanchet, J. P., and Krsmanovic, V.,** The identification and characterization of a novel human differentiation-inhibiting protein that selectively blocks erythroid differentiation, *Blood,* 79, 1161, 1992.

491. **Soto, A. M. and Sonnenschein, C.,** Cell proliferation of estrogen sensitive cells: the case for negative control, *Endocrine Rev.,* 8, 44, 1987.

492. **Sonnenschein, C., Olea, N., Pasanen, M. E., and Soto, A. M.,** Negative controls of cell proliferation: human prostatic cancer cells and androgens, *Cancer Res.,* 49, 3474, 1989.

493. **Miyazaki, K. and Horio, T.,** Growth inhibitors: molecular diversity and roles in cell proliferation, *In Vitro Cell. Dev. Biol.,* 25, 866, 1989.

494. **Geiser, A. G. and Stanbridge, E. J.,** A review of the evidence for tumor suppressor genes, *Crit. Rev. Oncogenesis,* 1, 261, 1989.

495. **Hollingsworth, R. E. and Lee, W.-H.,** Tumor suppressor genes: new prospects for cancer research, *J. Natl. Cancer Inst.,* 83, 91, 1991.

496. **Cavenee, W. K.,** Recessive mutations in the causation of human cancer, *Cancer,* 67, 243, 1991.

497. **Stanbridge, E. J.,** Functional evidence for human tumour suppressor genes — chromosome and molecular genetic studies, *Cancer Surv.,* 12, 5, 1992.

498. **Shiota, G., Rhoads, D. B., Wang, T. G., Nakamura, T., and Schmidt, E. V.,** Hepatocyte growth factor inhibits growth of hepatocellular carcinoma cells, *Proc. Natl. Acad. Sci. U.S.A.,* 89, 373, 1992.

499. **Weiss, P.,** Self-regulation of organ growth by its own products, *Science,* 115, 487, 1952.

500. **Miyazaki, K., Mashima, K., Kimura, T., Huang, W., Yano, K., Ashida, Y., Kihira, Y., Yamashita, J., and Horio, T.,** Growth inhibitors in serum, platelets, and normal and malignant tissues, *Adv. Enzyme Res.,* 26, 225, 1987.

501. **Mashima, K., Kimura, T., Miyazaki, K., Yamashita, J., and Horio, T.,** Growth-inhibitory protein present in rabbit serum, which is more effective on tumorigenic rat liver epithelial cells than on non-tumorigenic ones: its species, and mode of existence, *Biochem. Biophys. Res. Commun.,* 148, 1215, 1987.

502. **Sheid, B., Prat, J. C., and Gaetjens, E.,** A tumor growth inhibitory factor and a tumor growth promoting factor isolated from unfertilized ova of shad (*Alosa sapidissima*), *Biochem. Biophys. Res. Commun.,* 159, 713, 1989.

503. **Bullough, W. S.,** The chalones: a review, *Natl. Cancer Inst. Monogr.,* 38, 5, 1972.

504. **Vogler, W. R. and Winton, E. F.,** Humoral granulopoietic inhibitors: a review, *Exp. Hematol.,* 3, 337, 1975.

505. **Bullough, W. S.,** Mitotic control in adult mammalian tissues, *Biol. Rev.,* 50, 99, 1975.

506. **Finkler, N. and Acker, P.,** Chalones: a mini-review, *Mt. Sinai J. Med.,* 45, 258, 1978.

507. **Rytömaa, T. and Toivonen, H.,** Chalones: concepts and results, *Mechan. Ageing Dev.,* 9, 471, 1979.

508. **Patt, L. M. and Houck, J. C.,** The incredible shrinking chalone, *FEBS Lett.,* 120, 163, 1980.

509. **Fremuth, F.,** Chalones and specific growth factors in normal and tumor growth, *Acta Univ. Carol. Med. Monogr. (Praha),* 110, 1, 1984.

510. **Iversen, O. H.,** What is new in endogenous growth stimulators and inhibitors, *Pathol. Res. Pract.,* 180, 77, 1985.

511. **Strain, A. J.,** Transforming growth factor beta and inhibition of hepatocellular proliferation, *Scand. J. Gastroenterol.,* 23 (Suppl. 151), 37, 1988.

512. **Parkinson, K. and Balmain, A.,** Chalones revisited — a possible role for transforming growth factor β in tumour promotion, *Carcinogenesis,* 11, 195, 1990.

513. **Keski-Oja, J. and Moses, H. L.,** Growth inhibitory polypeptides in the regulation of cell proliferation, *Med. Biol.,* 66, 13, 1987.

514. **Silberstein, G. B. and Daniel, C. W.,** Reversible inhibition of mammary gland growth by transforming growth factor-β, *Science,* 237, 291, 1987.

515. **Ohta, M., Greenberger, J. S., Anklesaria, P., Bassols, A., and Massagué, J.,** Two forms of transforming growth factor-β distinguished by multipotential haematopoietic progenitor cells, *Nature,* 329, 539, 1987.

516. **Sing, G. K., Keller, J. R., Ellingsworth, L. R., and Ruscetti, F. W.,** Transforming growth factor β selectively inhibits normal and leukemic human bone marrow cell growth *in vitro, Blood,* 72, 1504, 1988.

517. **Levine, A. E., Crandall, C. A., and Brattain, M. G.,** Regulation of growth inhibitory activity in transformed mouse embryo fibroblasts, *Exp. Cell Res.,* 171, 357, 1987.

518. **Hsu, Y.-M. and Wang, J. L.,** Growth control in cultured fibroblasts. V. Purification of an M_r 13,000 polypeptide responsible for growth inhibitory activity, *J. Cell Biol.,* 102, 362, 1986.

519. **Blat, C., Villaudy, J., Desauty, G., Goldé, A., and Harel, L.,** Modifications induites par l'expression du gène v-*src* dans la régulation de la prolifération cellulaire. Hypothèse et résultats préliminaires, *C. Rend. Acad. Sci. Paris,* 301, 417, 1985.

520. **Blat, C., Chatelain, G., Desauty, G., and Harel, L.,** Inhibitory diffusible factor IDF45, a G_1 phase inhibitor, *FEBS Lett.,* 203, 175, 1986.

521. **Blat, C., Villaudy, J., Rouillard, D., Goldé, A. and Harel, L.,** Modulation by the src oncogene of the effect of inhibitory diffusible factor IDF45, *J. Cell Physiol.,* 130, 416, 1987.

522. **Blat, C., Bohlen, P., Villaudy, J., Chatelain, G., Goldé, A., and Harel, L.,** Isolation and amino-terminal sequence of a novel cellular growth inhibitor (inhibitory diffusible factor 45) secreted by 3T3 fibroblasts, *J. Biol. Chem.,* 264, 6021, 1989.

64

545. **Boehmer, F. D., Kraft, R., Otto, A., Wernstedt, C., Hellman, U., Kurtz, A., Mueller, T., Rohde, K., Etzold, G., Lehmann, W., Langen, P., Heldin, C.-H., and Grosse, R.,** Identification of a polypeptide growth inhibitor from bovine mammary gland. Sequence homology to fatty acid- and retinoid-binding proteins, *J. Biol. Chem.,* 262, 15137, 1987.

546. **Wallukat, G., Boehmer, F. D., Engstroem, U., Langen, P., Hollenberg, M., Behlke, J., Kuehn, H., and Grosse, R.,** Modulation of the beta-adrenergic response in cultured rat heart cells. II. Mammary-derived growth inhibitor (MDGI) blocks induction of beta-adrenergic supersensitivity. Dissociation from lipid-binding activity of MDGI, *Mol. Cell. Biochem.,* 102, 49, 1991.

547. **Kurtz, A., Vogel, F., Funa, K., Heldin, C.-H., and Grosse, R.,** Developmental regulation of mammary-derived growth inhibitor expression in bovine mammary tissue, *J. Cell Biol.,* 110, 1779, 1990.

548. **Lehmann, W., Widmair, R., and Langen, P.,** Response of different mammary epithelial cell lines to a mammary-derived growth inhibitor (MDGI), *Biomed. Biochim. Acta,* 48, 143, 1989.

549. **Boehmer, F. D., Sun, Q., Pepperle, M., Mueller, T., Eriksson, U., Wang, J. L., and Grosse, R.,** Antibodies against mammary-derived growth inhibitor (MDGI) react with a fibroblast growth inhibitor and with heart fatty acid binding protein, *Biochem. Biophys. Res. Commun.,* 148, 1425, 1987.

550. **Ervin, P. R., Jr., Kaminski, M. S., Cody, R. L., and Wicha, M. S.,** Production of mammastatin, a tissue-specific growth inhibitor, by normal human mammary cells, *Science,* 244, 1585, 1989.

551. **Csaikl, U., Csaikl, F., and Letnansky, K.,** Inhibition of tumor cell proliferation by a novel growth regulator from placenta: modulation of membrane phosphorylation and oncogene expression, *Cancer Lett.,* 44, 227, 1989.

552. **Letnansky, K., Vetterlein, M., and Shi, C.,** Inhibition of DNA synthesis and oncogene expression in tumor cells by the biological factor DIF, *Anticancer Res.,* 11, 981, 1991.

553. **Bascom, C. C., Sharifi, B. G., and Johnson, T. C.,** Inhibition of epidermal growth factor-stimulated DNA synthesis by a bovine sialoglycopeptide inhibitor occurs at an intracellular level, *J. Cell. Biochem.,* 34, 283, 1987.

554. **Tsuji, S., Kobayashi, H., Uchida, Y., Ihara, Y., and Miyatake, T.,** Molecular cloning of human growth inhibitory factor cDNA and its down-regulation in Alzheimer's disease, *EMBO J.,* 11, 4843, 1992.

555. **Zarling, J. M., Shoyab, M., Marquardt, H., Hanson, M. B., Lioubin, M. N., and Todaro, G. J.,** Oncostatin M: a growth regulator produced by differentiated histiocytic lymphoma cells, *Proc. Natl. Acad. Sci. U.S.A.,* 83, 9739, 1986.

556. **Grove, R. I., Eberhardt, C., Abid, S., Mazzucco, C., Liu, J., Kiener, P., Todaro, G., and Shoyab, M.,** Oncostatin M is a mitogen for rabbit vascular smooth muscle cells, *Proc. Natl. Acad. Sci. U.S.A.,* 90, 823, 1993.

557. **Nair, B. C., DeVico, A. L., Nakamura, S., Copeland, T. D., Chen, Y., Patel, A., O'Neil, T., Oroszlan, S., Gallo, R. C., and Sarngadharan, M. G.,** Identification of a major growth factor for AIDS-Kaposi's sarcoma cells as oncostatin M, *Science,* 255, 1430, 1992.

558. **Miles, S. A., Martínez-Maza, O., Rezai, A., Magpantay, L., Kishimoto, T., Nakamura, S., Radka, S. F., and Lindsley, P. S.,** Oncostatin M as a potent mitogen for AIDS-Kaposi's sarcoma-derived cells, *Science,* 255, 1432, 1992.

559. **Malik, N., Kallestad, J. C., Gunderson, N. L., Austin, S. D., Neubauer, M. G., Ochs, V., Marquardt, H., Zarling, J. M., Shoyab, M., Wei, C.-M., Linsley, P. S., and Rose, T. M.,** Molecular cloning, sequence analysis, and functional expression of a novel growth regulator, oncostatin M, *Mol. Cell. Biol.,* 9, 2847, 1989.

560. **Linsley, P. S., Bolton-Hanson, M., Horn, D., Malik, N., Kallestad, J. C., Ochs, V., Zarling, J. M., and Shoyab, M.,** Identification and characterization of cellular receptors for the growth factor regulator, oncostatin M, *J. Biol. Chem.,* 264, 4282, 1989.

561. **Gearing, D. P., Comeau, M. R., Friend, D. J., Gimpel, S. D., Thut, C. J., McGourty, J., Brasher, K. K., King, J. A., Gillis, S., Mosley, B., Ziegler, S. F., and Cosman, D.,** The IL-6 signal transducer, gp130: an oncostatin M receptor and affinity converter for the LIF receptor, *Science,* 255, 1434, 1992.

562. **Liu, J., Modrell, B., Aruffo, A., Marken, J. S., Taga, T., Yasukawa, K., Murakami, M., Kishimoto, T., and Shoyab, M.,** Interleukin-6 signal transducer gp130 mediates oncostatin M signaling, *J. Biol. Chem.,* 267, 16763, 1992.

563. **Liu, J. W., Clegg, C. H., and Shoyab, M.,** Regulation of EGR-1, c-*jun*, and c-*myc* gene expression by oncostatin-M, *Cell Growth Differ.,* 3, 307, 1992.

564. **Wier, M. L. and Scott, R. E.,** Aproliferin — a human plasma protein that induces the reversible loss of proliferative potential associated with terminal differentiation, *Am. J. Pathol.,* 125, 546, 1986.

565. **Chang, F. and Herskowitz, I.,** Identification of a gene necessary for cell cycle arrest by a negative growth control of yeast: FAR1 is an inhibitor of a G₁ cyclin, CLN2, *Cell,* 63, 999, 1990.

566. **Yamaoka, K., Hirai, R., and Hiragun, A.,** Separation of sarcoma growth inhibitor from avian sarcoma virus-transformed rat cells, *Jpn. J. Cancer Res.,* 77, 319, 1986.

567. **Gaffney, E. V., Tsai, S.-C., Dell'Aquila, M. L., and Lingenfelter, S. E.,** Production of growth-inhibitory activity in serum-free medium by human monocytic leukemia cells, *Cancer Res.,* 43, 3668, 1983.

568. **Olofsson, T. and Olsson, I.,** Suppression of normal granulopoiesis *in vitro* by a leukemia associated inhibitor (LAI) derived from a human promyelocytic cell line (HL-60), *Leukemia Res.,* 4, 437, 1980.

569. **Olofsson, T., Cedergren, B. M., and Persson, E.,** Isolation of leukemia associated inhibitor (LAI)-producing cells from normal peripheral blood, *Scand. J. Haematol.,* 35, 511, 1985.

570. **Olofsson, T. and Sallerfors, B.,** Modulation of the production of leukemia associated inhibitor (LAI) and its interaction with granulocyte-macrophage colony-forming cells, *Exp. Hematol.,* 15, 1163, 1987.

571. **Douer, D., Ben-Bassat, I., Froom, P., Shaked, N., and Ramot, B.,** T-cell acute lymphoblastic leukemia with severe leukopenia: evidence for suppressionb of myeloid progenitor cells by leukemic blasts, *Acta Haematol.,* 80, 185, 1988.

572. **Montaldo, P. G., Lanciotti, M., Castagnola, E., Parodi, M. T., Cirillo, C., Cornaglia-Ferraris, P., and Ponzoni, M.,** A human acute leukemia-derived T-cell line produces two inhibitor factors which suppress lymphocyte proliferation: characterization and purification of the molecules, *Lymphokine Res.,* 7, 413, 1988.

573. **Levine, A. E., Hamilton, D. A., Yeoman, L. C., Busch, H., and Brattain, M. G.,** Identification of a tumor inhibitory factor in rat ascites fluid, *Biochem. Biophys. Res. Commun.,* 119, 76, 1984.

574. **Levine, A. E., McRae, L. J., Hamilton, D. A., Brattain, D. E., Yeoman, L. C., and Brattain, M. G.,** Identification of endogenous inhibitory growth factors from a human colon carcinoma cell line, *Cancer Res.,* 45, 2248, 1985.

575. **Culouscou, J.-M. and Shoyab, M.,** Purification of a colon cancer cell growth inhibitor and its identification as an insulin-like growth factor binding protein, *Cancer Res.,* 51, 2813, 1991.

576. **Kasukabe, T., Okabe-Kado, J., Honma, Y., and Hozumi, M.,** Production by undifferentiated myeloid leukemia cells of a novel growth-inhibitory factor(s) for partially differentiated myeloid leukemic cells, *Jpn. J. Cancer Res.,* 78, 921, 1987.

577. **Kasukabe, T., Okabe-Kado, J., Honma, Y., and Hozumi, M.,** Purification of a novel growth inhibitory factor for partially differentiated myeloid leukemic cells, *J. Biol. Chem.,* 263, 5431, 1988.

578. **Wrann, M., Bodmer, S., de Martin, R., Siepl, C., Hofer-Warbinek, R., Frei, K., Hofer, E., and Fontana, A.,** T cell suppressor factor from human glioblastoma cells is a 12.5-kd protein closely related to transforming growth factor-beta, *EMBO J.,* 6, 1633, 1987.

579. **de Martin, R., Haendler, B., Hofer-Warbinek, R., Gaugitsch, H., Wrann, M., Schlüsener, H., Seifert, J.M., Bodmer, S., Fontana, A., and Hofer, E.,** Complementary DNA for human glioblastoma-derived T cell suppressor factor, a novel member of the transforming growth factor-β gene family, *EMBO J.,* 6, 3673, 1987.

580. **Laiho, M.,** Modulation of extracellular proteolytic activity and anchorage-independent growth of cultured cells by sarcoma cell-derived factors: relationships to transforming growth factor-β, *Exp. Cell Res.,* 176, 297, 1988.

581. **Bogdahn, U., Apfel, R., Hahn, M., Gerlach, M., Behl, C., Müller, F., Hoppe, J., and Martin, R.,** Autocrine tumor cell growth inhibiting activities from human malignant melanoma, *Cancer Res.,* 49, 5358, 1989.

582. **Weilbach, F. X., Bogdahn, U., Poot, M., Apfel, R., Behl, C., Drenkard, D., Martin, R., and Hoehn, H.,** Melanoma-inhibiting activity inhibits cell proliferation by prolongation of the S phase and arrest of cells in the G₂ compartment, *Cancer Res.,* 50, 6981, 1990.

583. **Iwata, K. K., Fryling, C. M., Knott, W. B., and Todaro, G. J.,** Isolation of tumor cell growth-inhibiting factors from a human rhabdomyosarcoma cell line, *Cancer Res.,* 45, 2689, 1985.

584. **Fryling, C. M., Iwata, K. K., Johnson, P. A., Knott, W. B., and Todaro, G. J.,** Two distinct tumor cell growth-inhibiting factors from a human rhabdomyosarcoma cell line, *Cancer Res.,* 45, 2695, 1985.

585. **Fryling, C., Dombalagian, M., Burgess, W., Hollander, N., Schreiber, A. B., and Haimovich, J.,** Purification and characterization of tumor inhibitory factor-2: its identity to interleukin 1, *Cancer Res.,* 49, 3333, 1989.

586. **Aune, T. M., Webb, D. R., and Pierce, C. W.,** Purification and initial characterization of the lymphokine soluble immune response suppressor, *J. Immunol.,* 131, 2848, 1983.

587. **Aune, T. M.,** Inhibition of soluble immune response suppressor activity by growth factors, *Proc. Natl. Acad. Sci. U.S.A.,* 82, 6240, 1985.

588. **Gill, G. N. and Lazar, C. S.,** Increased phosphotyrosine content and inhibition of proliferation in epidermal growth factor-treated A431 cells, *Nature,* 293, 305, 1981.

589. **Roberts, A. B., Anzano, M. A., Wakefield, L. M., Roche, N. S., Stern, D. F., and Sporn, M. B.,** Type β transforming growth factor: a bifunctional regulator of cellular growth, *Proc. Natl. Acad. Sci. U.S.A.,* 82, 119, 1985.

590. **Nettesheim, P., Fitzgerald, D. J., Kitamura, H., Walker, C. L., Gilmer, T. M., Barrett, J. C., and Gray, T. E.,** *In vitro* analysis of multistage carcinogenesis, *Environ. Health Perspect.,* 75, 71, 1987.

591. **Baumann, G.,** Growth hormone heterogeneity: genes, isohormones, variants, and binding proteins, *Endocrine Rev.,* 12, 424, 1991.

592. **Raben, M. S.,** Growth hormone, *N. Engl. J. Med.,* 266, 31 and 82, 1962.

593. **Kaplan, S. A.,** Growth hormone, *Am. J. Dis. Child.,* 110, 232, 1965.

594. **Tanner, J. M.,** Human growth hormone, *Nature,* 237, 433, 1972.

595. **Goodman, H. M., Schwartz, Y., Tai, L.R., and Gorin, E.,** Actions of growth hormone on adipose tissue: possible involvement of autocrine or paracrine factors, *Acta Paediatr. Scand.,* Suppl. 367, 132, 1990.

596. **Devesa, J., Lima, L., and Tresguerres, J. A. F.,** Neuroendocrine control of growth hormone secretion in humans, *Trends Endocrinol. Metab.,* 3, 175, 1992.

597. **Weigent, D. A., Baxter, J. B., Wear, W. E., Smith, L. R., Bost, K. L., and Blalock, J. E.,** Production of immunoreactive growth hormone by mononuclear leukocytes, *FASEB J.,* 2, 2812, 1988.

598. **Hattori, N., Shimatsu, A., Sugita, M., Kumegai, S., and Imura, H.,** Immunoreactive growth hormone (GH) secretion by human lymphocytes: augmented release by exogenous GH, *Biochem. Biophys. Res. Commun.,* 168, 396, 1990.

599. **Weigent, D. A., Blalock, J. E., and LeBoeuf, R. D.,** An antisense oligodeoxynucleotide to growth hormone messenger ribonucleic acid inhibits lymphocyte proliferation, *Endocrinology,* 128, 2053, 1991.

600. **Baumann, G.,** Growth hormone binding proteins in plasma — an update, *Acta Paediatr. Scand.,* Suppl. 367, 142, 1990.

601. **Baumann, G.,** Growth hormone-binding proteins, *Trends Endocrinol. Metab.,* 1, 342, 1990.

602. **Martha, P. M., Jr., Reiter, E. O., Dávila, N., Shaw, M. A., Holcombe, J. H., and Baumann, G.,** Serum growth hormone (GH)-binding protein/receptor: an important determinant of GH responsiveness, *J. Clin. Endocrinol. Metab.,* 75, 1464, 1992.

603. **Duquesnoy, P., Sobrier, M. L., Amselem, S., and Goossens, M.,** Defective membrane expression of human growth hormone (GH) receptor causes Laron-type GH insensitivity syndrome, *Proc. Natl. Acad. Sci. U.S.A.,* 88, 10272, 1991.

604. **Jammes, H., Gaye, P., Belair, L., and Djiane, J.,** Identification of growth hormone receptor mRNA in the mammary gland, *Mol. Cell. Endocrinol.,* 75, 27, 1991.

605. **Frankenne, F., Alsat, E., Scippo, M. L., Igout, A., Hennen, G., and Evain-Brion, D.,** Evidence for the expression of growth hormone receptors in human placenta, *Biochem. Biophys. Res. Commun.,* 182, 481, 1992.

606. **Barton, D. E., Foellmer, B. E., Wood, W. I., and Francke, U.,** Chromosome mapping of the growth hormone receptor gene in man and mouse, *Cytogenet. Cell Genet.,* 50, 137, 1989.

607. **Leung, D. W., Spencer, S. A., Cachianes, G., Hammonds, R. G., Collins, C., Henzel, W. J., Barnard, R., Waters, M. J., and Wood, W. I.,** Growth hormone receptor and serum binding protein: purification, cloning and expression, *Nature,* 330, 537, 1987.

608. **Hocquette, J.-F., Postel-Vinay, M.-C., Djiane, J., Tar, A., and Kelly, P. A.,** Human liver growth hormone receptor and plasma binding protein: characterization and partial purification, *Endocrinology,* 127, 1665, 1990.

609. **Cosman, D., Lyman, S. D., Idzerda, R. L., Beckmann, M. P., Park, L. S., Goodwin, R. G., and March, C. J.,** A new cytokine receptor superfamily, *Trends Biochem. Sci.,* 15, 265, 1990.

610. **Fuh, G., Mulkerrin, M. G., Bass, S., McFarland, N., Brochier, M., Bourell, J. H., Light, D. R., and Wells, J. A.,** The human growth hormone receptor. Secretion from *Escherichia coli* and disulfide bonding pattern of the extracellular domain, *J. Biol. Chem.,* 265, 3111, 1990.

611. **Cunningham, B. C., Jhurani, P., Ng, P., and Wells, J. A.,** Receptor and antibody epitopes in human growth hormone identified by homolog-scanning mutagenesis, *Science,* 243, 1330, 1989.

612. Cunningham, B. C., Ultsch, M., de Vos, A. M., Mulkerrin, M. G., Clauser, K. R., and Wells, J. A., Dimerization of the extracellular domain of the human growth hormone receptor by a single hormone molecule, *Science,* 254, 821, 1991.

613. de Vos, A. M., Ultsch, M., and Kossiakoff, A. A., Human growth hormone and extracellular domain of its receptor: crystal structure of the complex, *Science,* 255, 306, 1992.

614. Fuh, G., Cunningham, B. C., Fukunaga, R., Nagata, S., Goeddel, D. V., and Wells, J. A., Rational design of potent antagonists to the human growth hormone receptor, *Science,* 256, 1677, 1992.

615. Foster, C. M., Shafer, J. A., Rozsa, F. W., Wang, X., Lewis, S. D., Renken, D. A., Natale, J. E., Schwartz, J., and Carter-Su, C., Growth hormone promoted tyrosyl phosphorylation of growth hormone receptors in murine 3T3-F442A fibroblasts and adipocytes, *Biochemistry,* 27, 326, 1988.

616. Carter-Su, C., Stubbart, J. R., Wang, X., Stred, S. E., Argetsinger, L. S., and Shafer, J. A., Phosphorylation of highly purified growth hormone receptors by a growth hormone receptor-associated tyrosine kinase, *J. Biol. Chem.,* 264, 18654, 1989.

617. Stred, S. E., Stubbart, J. R., Argetsinger, L. S., Shafer, J. A., and Carter-Su, C., Demonstration of growth hormone (GH) receptor-associated tyrosine kinase activity in multiple GH-responsive cell types, *Endocrinology,* 127, 2506, 1990.

618. Wang, X., Uhler, M. D., Billestrup, N., Norstedt, G., Talamantes, F., Nielsen, J. H., and Carter-Su, C., Evidence for association of the cloned liver growth hormone receptor with a tyrosine kinase, *J. Biol. Chem.,* 267, 17390, 1992.

619. Winston, L. A., and Bertics, P. J., Growth hormone stimulates the tyrosine phosphorylation of 42- and 45-kDa ERK-related proteins, *J. Biol. Chem.,* 267, 4747, 1992.

620. Wang, X., Möller, C., Norstedt, G., and Carter-Su, C., Growth hormone-promoted tyrosyl phosphorylation of a 121-kDa growth hormone receptor-associated protein, *J. Biol. Chem.,* 268, 3573, 1993.

621. Campbell, G. S., Christian, L. J., and Carter-Su, C., Evidence for involvement of the growth hormone receptor-associated tyrosine kinase in actions of growth hormone, *J. Biol. Chem.,* 268, 7427, 1993.

622. Daughaday, W. H., Hall, K., Raben, M. S., Salmon, W. D., Van den Brande, J. L., and Van Wyk, J. J., Somatomedin: proposed designation for sulphation factor, *Nature,* 235, 107, 1972.

623. Nilsson, A., Isgaard, J., Lindahl, A., Dahlström, A., Skottner, A., and Isaksson, O. G. P., Regulation by growth hormone of number of chondrocytes containing IGF-I in rat growth plate, *Science,* 233, 571, 1986.

624. Schlechter, N. L., Russell, S. M., Spencer, E. M., and Nicoll, C. S., Evidence suggesting that the direct growth-promoting effect of growth hormone on cartilage *in vitro* is mediated by local production of somatomedin, *Proc. Natl. Acad. Sci. U.S.A.,* 83, 7932, 1986.

625. Isgaard, J., Möller, C., Isaksson, O. G. P., Nilsson, A., Mathews, L. S., and Norstedt, G., Regulation of insulin-like growth factor messenger ribonucleic acid in rat growth plate by growth hormone, *Endocrinology,* 122, 1515, 1988.

626. Ernst, M. and Froesch, E. R., Growth hormone dependent stimulation of osteoblast-like cells in serum-free cultures via local synthesis of insulin-like growth factor I, *Biochem. Biophys. Res. Commun.,* 151, 142, 1988.

627. Doglio, A., Dani, C., Fredrikson, G., Grimaldi, P., and Ailhaud, G., Acute regulation of insulin-like growth factor-I gene expression by growth hormone during adipose cell differentiation, *EMBO J.,* 6, 4011, 1987.

628. Cook, J. J., Haynes, K. M., and Werther, G. A., Mitogenic effects of growth hormone in cultured human fibroblasts. Evidence for action via local insulin-like growth factor I production, *J. Clin. Invest.,* 81, 206, 1988.

629. Isaksson, O. G. P., Jansson, J.-O., and Gause, I. A. M., Growth hormone stimulates longitudinal bone growth directly, *Science,* 216, 1237, 1982.

630. Mathews, L. S., Hammer, R. E., Behringer, R. R., D'Ercole, A. J., Bell, G. I., Brinster, R. L., and Palmiter, R. D., Growth enhancement of transgenic mice expressing human insulin-like growth factor I, *Endocrinology,* 123, 2827, 1988.

631. Behringer, R. R., Lewin, T. M., Quaife, C. J., Palmiter, R. D., Brinster, R. L., and D'Ercole, A. J., Expression of insulin-like growth factor I stimulates normal somatic growth in growth hormone-deficient transgenic mice, *Endocrinology,* 127, 1033, 1990.

632. Skottner, A., Clark, R. G., Robinson, I. C. A. F., and Fryklund, L., Recombinant human insulin-like growth factor: testing the somatomedin hypothesis in hypophysectomized rats, *J. Endocrinol.,* 112, 123, 1987.

633. **Zezulak, K. M. and Green, H.,** The generation of insulin-like growth factor-1-sensitive cells by growth hormone action, *Science,* 233, 551, 1986.

634. **Berelowitz, M., Szabo, M., Frohman, L. A., Firestone, S., Chu, L., and Hintz, R. L.,** Somatomedin-C mediates growth hormone negative feedback by effects on both the hypothalamus and the pituitary, *Science,* 212, 1279, 1981.

635. **Pitt-Rivers, R.,** The biosynthesis of thyroid hormones, *J. Clin. Pathol.,* 20, 318, 1967.

636. **DeGroot, L. J. and Niepomniszcze, H.,** Biosynthesis of thyroid hormone: basic and clinical aspects, *Metabolism,* 26, 665, 1977.

637. **Christophe, D. and Vassart, G.,** The thyroglobulin gene. Evolutionary and regulatory issues, *Trends Endocrinol. Metab.,* 1, 351, 1990.

638. **Yamada, T.,** Control of thyroid hormone secretion, *Pharmacol. Ther.,* 1, 3, 1976.

639. **Van Herle, A. J., Vassart, G., and Dumont, J. E.,** Control of thyroglobulin synthesis and secretion, *N. Engl. J. Med.,* 301, 239 and 307, 1979.

640. **Cody, V.,** Thyroglobulin and thyroid hormone synthesis, *Endocrine Res.,* 10, 73, 1984.

641. **Lissitzky, S.,** Thyroglobulin entering into molecular biology, *J. Endocrinol. Invest.,* 7, 65, 1984.

642. **Oppenheimer, J. H.,** Thyroid hormone action at the nuclear level, *Ann. Int. Med.,* 102, 374, 1985.

643. **De Nayer, P.,** Thyroid hormone action at the cellular level, *Hormone Res.,* 26, 48, 1987.

644. **Samuels, H. H., Forman, B. M., Horowitz, Z. D. and Ye, Z.-S.,** Regulation of gene expression by thyroid hormone, *Annu. Rev. Physiol.,* 51, 623, 1989.

645. **DeGroot, L. J.,** Thyroid hormone nuclear receptors and their role in the metabolic action of the hormone, *Biochimie,* 71, 269, 1989.

646. **Glass, C. K. and Holloway, J. M.,** Regulation of gene expression by thyroid hormone receptor, *Biochim. Biophys. Acta,* 1032, 157, 1990.

647. **Brent, G. A., Moore, D. D., and Larsen, P. R.,** Thyroid hormone regulation of gene expression, *Annu. Rev. Physiol.,* 53, 17, 1991.

648. **Chatterjee, V. K. K. and Tata, J. R.,** Thyroid hormone receptors and their role in development, *Cancer Surv.,* 14, 147, 1992.

649. **Fellous, A., Lennon, A. M., Francon, J. and Nunez, J.,** Thyroid hormones and neurotubule assembly *in vitro* during brain development, *Eur. J. Biochem.,* 101, 365, 1979.

650. **Laksmanan, J., Mansfield, H., Weichsel, M. E., Jr., Hoath, S., Scott, S., Shapshak, P., and Fisher, D. A.,** Neonatal hypothyroidism — a biochemical disorder of alpha-tubulin metabolism, *Biochem. Biophys. Res. Commun.,* 100, 1587, 1981.

651. **Baxter, J. D.,** Advances in molecular biology — potential impact on diagnosis and treatment of disorders of the thyroid, *Med. Clin. N. Am.,* 75, 41, 1991.

652. **Lazar, M. A.,** Thyroid hormone receptors: multiple forms, multiple possibilities, *Endocrine Rev.,* 14, 184, 1993.

653. **Samuels, H. H. and Tsai, J. S.,** Thyroid hormone action in cell culture: demonstration of nuclear receptors in intact cells and isolated nuclei, *Proc. Natl. Acad. Sci. U.S.A.,* 70, 3488, 1973.

654. **Samuels, H. H., Tsai, J. S., and Casanova, J.,** Thyroid hormone action: *in vitro* demonstration of putative receptors in isolated nuclei and soluble nuclear extracts, *Science,* 184, 1188, 1974.

655. **DeGroot, L. J., Hill, L., and Rue, P.,** Binding of nuclear triiodothyronine (T_3) binding protein-T_3 complex to chromatin, *Endocrinology,* 99, 1605, 1976.

656. **Nikodem, V. M., Chen, S.-Y., and Rall, J. E.,** Affinity labeling of rat liver thyroid hormone nuclear receptor, *Proc. Natl. Acad. Sci. U.S.A.,* 77, 7064, 1980.

657. **Surks, M. I. and Oppenheimer, J. H.,** Concentration of L-thyroxine and L-triiodothyronine specifically bound to nuclear receptors in rat liver and kidney. Quantitative evidence favoring a major role of T_3 in thyroid hormone action, *J. Clin. Invest.,* 60, 555, 1977.

658. **MacLeod, K. M. and Baxter, J. D.,** DNA binding of thyroid hormone receptors, *Biochem. Biophys. Res. Commun.,* 62, 577, 1975.

659. **Yoshizato, K. and Frieden, E.,** Increase in binding capacity for triiodothyronine in tadpole tail nuclei during metamorphosis, *Nature,* 254, 705, 1975.

660. **Sterling, K. and Milch, P. O.,** Thyroid hormone binding by a component of mitochondrial membrane, *Proc. Natl. Acad. Sci. U.S.A.,* 72, 3225, 1975.

661. **Visser, T. J., Bernard, H. F., Docter, R., and Hennemann, G.,** Specific binding sites for L-triiodothyronine in rat liver and kidney cytosol, *Acta Endocrinol.,* 82, 98, 1976.

662. **Pliam, N. B. and Goldfine, I. D.,** High affinity thyroid hormone binding sites on purified rat liver plasma membranes, *Biochem. Biophys. Res. Commun.,* 79, 166, 1977.

663. **Greif, R. L. and Sloane, D.,** Mitochondrial binding sites for triiodothyronine, *Endocrinology,* 103, 1899, 1978.

664. **Maxfield, F. R., Willingham, M. C., Pastan, I., Dragsten, P., and Chen, S.-Y.,** Binding and mobility of the cell surface receptors for 3,3′,5-triiodo-L-thyronine, *Science,* 211, 63, 1981.

665. **Cheng, S.-Y., Maxfield, F. R., Robbins, J., Willingham, M. C. and Pastan, I. H.,** Receptor-mediated uptake of 3,3′,5-triiodo-L-thyronine by cultured fibroblasts, *Proc. Natl. Acad. Sci. U.S.A.,* 77, 3425, 1980.

666. **Sterling, K., Milch, P. O., Brenner, M. A., and Lazarus, J. H.,** Thyroid hormone action: the mitochondrial pathway, *Science,* 197, 996, 1977.

667. **Sterling, K., Lazarus, J. H., Milch, P. O., Sakurada, T., and Brenner, M. A.,** Mitochondrial thyroid hormone receptor: localization and physiological significance, *Science,* 201, 1126, 1978.

668. **Schimmelpfennig, K., Sauerberg, M., and Neubert, D.,** Stimulation of mitochondrial RNA synthesis by thyroid hormone, *FEBS Lett.,* 10, 269, 1970.

669. **Barsano, C. P., DeGroot, L. J., and Getz, G. S.,** The effect of thyroid hormone on *in vitro* rat liver mitochondrial RNA synthesis, *Endocrinology,* 100, 52, 1977.

670. **Ichikawa, K. and DeGroot, L. J.,** Purification and characterization of rat liver nuclear thyroid hormone receptors, *Proc. Natl. Acad. Sci. U.S.A.,* 84, 3420, 1987.

671. **Sheer, D. G., Cahnmann, H. J., and Nikodem, V. M.,** Molecular size of the nuclear thyroid hormone receptor, *Biochim. Biophys. Acta,* 930, 101, 1987.

672. **Schmidt, E. D. L., van Beeren, H. C., Korfage, H., Dussault, J. H., Wiersinga, W. M., and Lamers, W. H.,** Localization of c-ERB A proteins in rat liver using monoclonal antibodies, *Biochem. Biophys. Res. Commun.,* 164, 1053, 1989.

673. **Ichikawa, K. and DeGroot, L. J.,** Thyroid hormone receptors in a human hepatoma cell line: multiple receptor forms on isoelectric focusing, *Mol. Cell. Endocrinol.,* 51, 135, 1987.

674. **Bigler, J. and Eisenman, R. N.,** c-*erb*A encodes multiple proteins in chicken erythroid cells, *Mol. Cell. Biol.,* 8, 4155, 1988.

675. **Evans, R. M.,** The steroid and thyroid hormone receptor superfamily, *Science,* 240, 889, 1988.

676. **Payre, F. and Vincent, A.,** Finger proteins and DNA-specific recognition: distinct patterns of conserved amino acids suggest different evolutionary modes, *FEBS Lett.,* 234, 245, 1988.

677. **Surks, M. I. and Kumara-Siri, M. H.,** Increase in nuclear thyroid and glucocorticoid receptors and growth hormone production during deoxyribonucleic acid synthesis phase of the cell growth cycle, *Endocrinology,* 114, 873, 1984.

678. **Goldberg, Y., Glineur, C., Bosselut, R., and Ghysdael, J.,** Thyroid hormone action and the *erb-A* oncogene family, *Biochimie,* 71, 279, 1989.

679. **Sap, J., Muñoz, A., Damm, K., Goldberg, Y., Ghysdael, J., Leutz, A., Beug, H., and Vennström, B.,** The c-erb-A protein is a high-affinity receptor for thyroid hormone, *Nature,* 324, 635, 1986.

680. **Weinberger, C., Thompson, C. C., Ong, E. S., Lebo, R., Gruol, D. J., and Evans, R. M.,** The c-*erb*-A gene encodes a thyroid hormone receptor, *Nature,* 324, 841, 1986.

681. **Glineur, C., Bailly, M., and Ghysdael, J.,** The c-*erb*-A alpha-encoded thyroid hormone receptor is phosphorylated in its amino terminal domain by casein kinase II, *Oncogene,* 4, 1247, 1989.

682. **Sheer, D., Sheppard, D. M., Le Beau, M., Rowley, J. D., San Roman, C., and Solomon, E.,** Localization of the oncogene c-*erb*A-1 immediately proximal to the acute promyelocytic leukaemia breakpoint on chromosome 17, *Ann. Hum. Genet.,* 49, 167, 1985.

683. **Mitelman, F., Manolov, G., Manolova, Y., Billström, R., Heim, S., Kristofferson, U., Mandahl, N., Ferro, M. T., and San Roman, C.,** High resolution chromosome analysis of constitutional and acquired t(15;17) maps c-*erb*-A to subband 17q11.2, *Cancer Genet. Cytogenet.,* 22, 95, 1986.

684. **Laudet, V., Begue, A., Henry-Duthoit, C., Joubel, A., Martin, P., Stehelin, D., and Saule, S.,** Genomic organization of the human thyroid hormone receptor alpha (c-*erb*A-1) gene, *Nucleic Acids Res.,* 19, 1105, 1991.

685. **Gareau, J. L., Houle, B., Leduc, F., Bradley, W. E. C., and Dobrovic, A.,** A frequent *Hind*III RFLP on chromosome 3p21-25 detected by a genomic c-*erb*-A beta sequence, *Nucleic Acids Res.,* 16, 1223, 1988.

686. **Moeller, M., Rapoport, B., and Gavin, L. A.,** Molecular cloning and characterization of c-*erb*-A mRNA species in mouse neuroblastoma cells, *J. Neuroendocrinol.,* 1, 351, 1989.

687. **Munroe, S. H. and Lazar, M. A.,** Inhibition of c-*erb*-A mRNA splicing by a naturally occurring antisense RNA, *J. Biol. Chem.,* 266, 22083, 1991.

688. **Mellström, B., Naranjo, J. R., Santos, A., Gonzalez, A. M., and Bernal, J.,** Independent expression of the alpha and beta c-*erb*-A genes in developing rat brain, *Mol. Endocrinol.,* 5, 1339, 1991.

689. **Hodin, R. A., Lazar, M. A., and Chin, W. W.,** Differential and tissue-specific regulation of the multiple rat c-erb-A messenger RNA species by thyroid hormone, *J. Clin. Invest.,* 85, 101, 1990.

690. **Jones, K. E. and Chin, W. W.,** Differential regulation of thyroid hormone receptor messenger ribonucleic acid levels by thyrotropin-releasing hormone, *Endocrinology,* 128, 1763, 1991.

691. **Kamikubo, K., Nayfeh, S. N., and Chae, C.-B.,** Differential regulation of multiple c-erb-A expression by thyrotropin, insulin and insulin-like growth factor I in rat thyroid FRTL-5 cells, *Mol. Cell. Endocrinol.,* 84, 219, 1992.

692. **Hinkle, P. M. and Kinsella, P. A.,** Thyroid hormone induction of an autocrine growth factor secreted by pituitary tumor cells, *Science,* 234, 1549, 1986.

693. **Miller, M. J., Fels, E. C., Shapiro, L. E., and Surks, M. I.,** L-triiodothyronine stimulates growth by means of an autocrine factor in a cultured growth hormone-producing cell line, *J. Clin. Invest.,* 79, 1773, 1987.

694. **Sirbasku, D. A., Pakala, R., Sato, H., and Eby, J. E.,** Thyroid hormone regulation of rat pituitary tumor cell growth: a new role for apotransferrin as an autocrine thyromedin, *Mol. Cell. Endocrinol.,* 77, C47, 1991.

695. **Karlson, P.,** Why are so many hormones steroids?, *Hoppe-Seyler's Z. Physiol. Chem.,* 364, 1067, 1983.

696. **Nestler, J. E. and Strauss, J. F., III,** Insulin as an effector of human ovarian and adrenal steroid metabolism, *Endocrinol. Metab. Clin. N. Am.,* 20, 807, 1991.

697. **Bell, N. H.,** Vitamin D-endocrine system, *J. Clin. Invest.,* 76, 1, 1985.

698. **Walters, M. R.,** Steroid hormone receptors and the nucleus, *Endocrine Rev.,* 6, 512, 1985.

699. **King, R. J. B.,** Effects of steroid hormones and related compounds on gene transcription, *Clin. Endocrinol.,* 36, 1, 1992.

700. **Gronemeyer, H.,** Control of transcription activation by steroid hormone receptors, *FASEB J.,* 6, 2524, 1992.

701. **Wahli, W. and Martinez, E.,** Superfamily of steroid nuclear receptors: positive and negative regulators of gene expression, *FASEB J.,* 5, 2243, 1991.

702. **Vedeckis, W. V.,** Nuclear receptors, transcriptional regulation, and oncogenesis, *Proc. Soc. Exp. Biol. Med.,* 199, 1, 1992.

703. **Sabbah, M., Le Ricousse, S., Redeuilh, G., and Baulieu, E.-E.,** Estrogen receptor-induced bending of the *Xenopus* vitellogenin A2 gene hormone response element, *Biochem. Biophys. Res. Commun.,* 185, 944, 1992.

704. **Beato, M., Arnemann, J., Chalepakis, G., Slater, E., and Willmann, T.,** Gene regulation by steroid hormones, *J. Steroid Biochem.,* 27, 9, 1987.

705. **Spelsberg, T. C., Goldberger, A., Horton, M., and Hora, J.,** Nuclear acceptor sites for sex steroid hormone receptors in chromatin, *J. Steroid Biochem.,* 27, 133, 1987.

706. **Schütz, G.,** Control of gene expression by steroid hormones, *Biol. Chem. Hoppe-Seyler,* 369, 77, 1988.

707. **Umesono, K., Murakami, K. K., Thompson C. C., and Evans, R. M.,** Direct repeats as selective response elements for the thyroid hormone, retinoic acid, and vitamin D_3 receptors, *Cell,* 65, 1255, 1991.

708. **Näär, A. M., Boutin, J.-M., Lipkin, S. M., Yu, V. C., Holloway, J. M., Glass, C. K., and Rosenfeld, M. G.,** The orientation and spacing of core DNA-binding motifs dictate selective transcriptional responses to three nuclear receptors, *Cell,* 65, 1267, 1991.

709. **Nishi, N., Matuo, Y., Muguruma, Y., Yoshitake, Y., Nishikawa, K., and Wada, F.,** A human prostatic growth factor (hPGF): partial purification and characterization, *Biochem. Biophys. Res. Commun.,* 132, 1103, 1985.

710. **Maehama, S., Li, D., Nanri, H., Leykam, J. F., and Deuel, T. F.,** Purification and partial characterization of prostate-derived growth factor, *Proc. Natl. Acad. Sci. U.S.A.,* 83, 8162, 1986.

711. **Nonomura, N., Nakamura, N., Uchida, N., Noguchi, S., Sato, B., Sonoda, T., and Matsumoto, K.,** Growth-stimulatory effect of androgen-induced autocrine growth factor(s) secreted from Shionogi carcinoma 115 cells on androgen-unresponsive cancer cells in a paracrine mechanism, *Cancer Res.,* 48, 4904, 1988.

712. **Nonomura, N., Lu, J., Tanaka, A., Yamanishi, H., Sato, B., Sonoda, T., and Matsumoto, K.,** Interaction of androgen-induced autocrine heparin-binding growth factor with fibroblast growth factor receptor on androgen-independent Shionogi carcinoma 115 cells, *Cancer Res.,* 50, 2316, 1990.

713. **Tanaka, A., Miyamoto, K., Minamino, N., Takeda, M., Sato, B., Matsuo, H., and Matsumoto, K.,** Cloning and characterization of an androgen-induced growth factor essential for the androgen-dependent growth of mouse mammary carcinoma cells, *Proc. Natl. Acad. Sci. U.S.A.,* 89, 8928, 1992.

714. **Kasayama, S., Saito, H., Kouhara, H., Sumkitani, S., and Sato, B.,** Suramin interrupts androgen-inducible autocrine loop involving heparin binding growth factor in mouse mammary cancer (Shionogi carcinoma 115 cells), *J. Cell. Physiol.,* 154, 254, 1993.

715. **Gaben, A. M. and Mester, J.,** BALB/c mouse 3T3 fibroblasts expressing human estrogen receptor: effect of estradiol on cell growth, *Biochem. Biophys. Res. Commun.,* 176, 1473, 1991.

716. **Beck, C. A. and Garner, C. W.,** Characterization and estrogen regulation of growth factor activity from uterus, *Mol. Cell. Endocrinol.,* 63, 93, 1989.

717. **Manni, A., Pontari, M., and Wright, C.,** Autocrine control by prolactin of hormone-responsive breast cancer growth in culture, *Endocrinology,* 117, 2040, 1985.

718. **Noguchi, S., Nishizawa, Y., Nakamura, N., Uchida, N., Yamaguchi, K., Sato, B., Kitamura, Y., and Matsumoto, K.,** Growth-stimulating effect of pharmacological doses of estrogen on androgen-dependent Shionogi carcinoma 115 *in vivo* but not in cell culture, *Cancer Res.,* 47, 263, 1987.

719. **Leung, B. S.,** Mode of estrogen action on cell proliferation in CAMA-1 cells: III. Effect of antiestrogen, *Anticancer Res.,* 7, 219, 1987.

720. **Sluyser, M., Moncharmont, B., Ramp, G., de Goeij, C. C. J., and Evers, S. G.,** Hormonal regulation of mouse mammary tumor growth, *J. Steroid Biochem.,* 27, 209, 1987.

721. **Brünner, N., Zugmaier, G., Bano, M., Ennis, B. W., Clarke, R., Cullen, K. J., Kern, F. G., Dickson, R. B., and Lippman, M. E.,** Endocrine therapy of human breast cancer cells: the role of secreted polypeptide growth factors, *Cancer Cells,* 1, 81, 1989.

722. **DiAugustine, R. P., Petrusz, P., Bell, G. I., Brown, C. F., Korach, K. S., McLachlan, J. A., and Teng, C. T.,** Influence of estrogens on mouse uterine epidermal growth factor precursor protein and messenger ribonucleic acid, *Endocrinology,* 122, 2355, 1988.

723. **Koga, M. and Sutherland, R. L.,** Epidermal growth factor partially reverses the inhibitory effects of antiestrogens on T47D human breast cancer cell growth, *Biochem. Biophys. Res. Commun.,* 146, 739, 1987.

724. **Koga, M., Musgrove, E. A., and Sutherland, R. L.,** Modulation of the growth-inhibitory effects of progestins and the antiestrogen hydroxyclomiphene on human breast cancer cells by epidermal growth factor and insulin, *Cancer Res.,* 49, 112, 1989.

725. **Liu, S. C., Sanfilippo, B., Perroteau, I., Derynck, R., Salomon, D. S., and Kidwell, W. R.,** Expression of transforming growth factor α (TGF α) in differentiated rat mammary tumors: estrogen induction of TGF α production, *Mol. Endocrinol.,* 1, 683, 1987.

726. **Arteaga, C. L., Coronado, E., and Osborne, C. K.,** Blockade of the epidermal growth factor receptor inhibits transforming growth factor α-induced but not estrogen-induced growth of hormone-dependent human breast cancer, *Mol. Endocrinol.,* 2, 1064, 1988.

727. **Leung, B. S., Stout, L., Zhou, L., Ji, H. J., Zhang, Q. Q., and Leung, H. T.,** Evidence of an EGF-TGF-α-independent pathway for estrogen-regulated cell proliferation, *J. Cell. Biochem.,* 46, 125, 1991.

728. **Nelson, K. G., Takahashi, T., Lee, D. C., Luetteke, N. C., Bossert, N. L., Ross, K., Eitzman, B. E., and McLachlan, J. A.,** Transforming growth factor-α is a potential mediator of estrogen action in the mouse uterus, *Endocrinology,* 131, 1657, 1992.

729. **Lippman, M. E., Dickson, R. B., Kasid, A., Gelmann, E., Davidson, N., McManaway, M., Huff, K., Bronzert, D., Bates, S., Swain, S., and Knabbe, C.,** Autocrine and paracrine growth regulation of human breast cancer, *J. Steroid Biochem.,* 24, 147, 1986.

730. **Dickson, R. B., McManaway, M. E., and Lippman, M. E.,** Estrogen-induced growth factors of breast cancer cells partially replace estrogen to promote tumor growth, *Science,* 232, 1540, 1986.

731. **Bates, S. E., Davidson, N. E., Valverius, E. M., Freter, C. E., Dickson, R. B., Tam, J. P., Kudlow, J. E., Lippman, M. E., and Salomon, D. S.,** Expression of transforming growth factor α and its messenger ribonucleic acid in human breast cancer: its regulation by estrogen and its possible functional significance, *Mol. Endocrinol.,* 2, 543, 1988.

732. **Eppstein, D. A., Marsh, Y. V., Schryver, B. B., and Bertics, P. J.,** Inhibition of epidermal growth factor/transforming growth factor-α-stimulated cell growth by a synthetic peptide, *J. Cell. Physiol.,* 141, 420, 1989.

733. **Knabbe, C., Lippman, M. E., Wakefield, L. M.,** Flanders, K. C., Kasid, A., Derynck, R., and Dickson, R. B., Evidence that transforming growth factor-β is a hormonally regulated negative growth factor in human breast cancer cells, *Cell,* 48, 417, 1987.

734. **Hagino, Y., Mawatari, M., Yoshimura, A., Kohno, K., Kobayashi, M., and Kuwano, M.,** Estrogen inhibits the growth of MCF-7 cell variants resistant to transforming growth factor-β, *Jpn. J. Cancer Res.,* 79, 74, 1988.

735. **Karey, K. P. and Sibasku, D. A.,** Differential responsiveness of human breast cancer cell lines MCF-7 and T47D to growth factors and 17 β-estradiol, *Cancer Res.,* 48, 4083, 1988.

736. **Reddel, R. R., Alexander, I. E., Koga, M., Shine, J., and Sutherland, R. L.,** Genetic instability and the development of steroid hormone insensitivity in cultured T 47D human breast cancer cells, *Cancer Res.,* 48, 4340, 1988.

737. **Murphy, L. C. and Dotzlaw, H.,** Endogenous growth factor expression in T-47D, human breast cancer cells, associated with reduced sensitivity to antiproliferative effects of progestins and antiestrogens, *Cancer Res.,* 49, 599, 1989.

738. **Finkelman, R. D., Bell, N. H., Strong, D. D., Demers, L. M., and Baylink, D. J.,** Ovariectomy selectively reduces the concentration of transforming growth factor β in rat bone: implications for estrogen deficiency-associated bone loss, *Proc. Natl. Acad. Sci. U.S.A.,* 89, 12190, 1992.

739. **Gray, T. K., Lipes, B., Linkhart, T., Mohan, S., and Baylink, D.,** Transforming growth factor β mediates the estrogen induced inhibition of UMR106 cell growth, *Connect. Tissue Res.,* 20, 23, 1989.

740. **Murphy, L. J., Murphy, L. C., and Friesen, H. G.,** Estrogen induces insulin-like growth factor-I expression in the rat uterus, *Mol. Endocrinol.,* 1, 445, 1987.

741. **van der Burg, B., Rutteman, G. R., Blankenstein, M. A., de Laat, S. W., and van Zoelen, E. J. J.,** Mitogenic stimulation of human breast cancer cells in a growth factor-defined medium: synergistic action of insulin and estrogen, *J. Cell. Physiol.,* 134, 101, 1988.

742. **Stewart, A. J., Johnson, M. D., May, F. E. B., and Westley, B. R.,** Role of insulin-like growth factors and the type I insulin-like growth factor receptor in the estrogen-stimulated proliferation of human breast cancer cells, *J. Biol. Chem.,* 265, 21172, 1990.

743. **van der Burg, B., Isbrücker, L., van Selm-Miltenburg, A. J. P., de Laat, S. W., and van Zoelen, E. J. J.,** Role of estrogen-induced insulin-like growth factors in the proliferation of human breast cancer cells, *Cancer Res.,* 50, 7770, 1990.

744. **Gray, T. K., Mohan, S., Linkhart, T. A., and Baylink, D. J.,** Estradiol stimulates *in vitro* the secretion of insulin-like growth factors by the clonal osteoblastic cell line, UMR1-6, *Biochem. Biophys. Res. Commun.,* 158, 407, 1989.

745. **Vignon, F., Capony, F., Chambon, M., Freiss, G., Garcia, M., and Rochefort, H.,** Autocrine growth stimulation of the MCF 7 breast cancer cells by the estrogen-regulated 52 K protein, *Endocrinology,* 118, 1537, 1986.

746. **Garcia, M., Salazar-Retana, G., Pages, A., Richer, G., Domergue, J., Pages, A. M., Cavalie, G., Martin, J. M., Lamarque, J.-L., Pau, B., Pujol, H., and Rochefort, H.,** Distribution of the M_r 52,000 estrogen-regulated protein in benign breast diseases and other tissues by immunochemistry, *Cancer Res.,* 46, 3734, 1986.

747. **Ji, H., Zhang, Q., and Leung, B. S.,** Survey of oncogene and growth factor/receptor gene expression in cancer cells by intron-differential RNA/PCR, *Biochem. Biophys. Res. Commun.,* 170, 569, 1990.

748. **Kanatani, M., Sugimoto, T., Kano, J., Fukase, M., and Fujita, T.,** Effect of 17 beta-estradiol on the proliferation of osteoblastic MC3T3-E1 cells via human monocytes, *Biochem. Biophys. Res. Commun.,* 178, 866, 1991.

Chapter 2

Growth Factor Receptors

I. INTRODUCTION

The mechanisms of action of growth factors are very complex. The structural and functional effects of these extracellular signaling molecules are intimately related to molecular changes at different subcellular levels.[1,2] The responses of cells stimulated by growth factors comprise two sequential events: transductional and posttransductional. In turn, the transductional events consist of two sequential steps: binding of the signaling agent to a receptor on the cell surface and production or translocation of mediators (second messengers). In the posttransductional events, specific regulatory changes occur at different subcellular sites, including the plasma membrane, the cytoplasm, various cellular organelles, and the nucleus. Alterations in the regulation of gene expression frequently occur in cells stimulated with growth factors.

The proliferation and differentiation of cells are subjected to complex, multifactorial regulatory mechanisms. Each phase of the cell cycle, as well as the transition from one phase to another, is controlled by a complex interaction between intracellular and extracellular factors. Receptors with high or low affinity for several hormones, growth factors, and regulatory peptides are expressed in the cells of different organs and tissues during the intra- and extrauterine life, and the placenta contains receptors for many of these agents.[3] However, the specificity of these endogenous extracellular signaling agents in the regulation of biochemical processes associated with cell proliferation and differentiation is not always clear, because unspecific agents such as extracellular proteases can induce similar or identical processes in some cellular systems.[4] In any case, a delicate balance between the processes associated with cell differentiation and proliferation is continuously operating in metazoan organisms, and a loss or damage in these control mechanisms may result in severe abnormalities, including neoplastic transformation.

II. GROWTH FACTOR RECEPTORS

Endogenous extracellular signaling agents such as hormones, growth factors, regulatory peptides, and neurotransmitters act through their interaction with specific receptors that are capable of discriminating between different types of signaling molecules.[5-9] The receptor and its ligand establish interactions that are specific and reversible and can occur with very high affinity. Interactions between growth factors, protein hormones, and regulatory peptides occur on the cell surface, where the receptors for these signaling agents are located. Nonpeptide hormones, such as the steroid and thyroid hormones, interact with intracellular receptors. In general, the interactions between the physiologic ligand and its receptor exhibit a high degree of specificity, but in certain cases one receptor may be activated by a signal designed for another.[10] This phenomenon of "specificity spillover" is observed mainly in pathologic conditions, when the hormone or growth factor is present in excess.

After formation of the ligand-receptor complex, the receptor undergoes a specific conformational change, which in some manner induces the generation of intracellular transducing signals that lead to a cellular response. For transmembrane receptors such as those of growth factors, the postreceptor signal transduction mechanisms may include the activation of GTP-binding proteins (G proteins), the regulation of plasma membrane ion channels, and the phosphorylation of proteins on tyrosine and/or serine/threonine residues.[11,12] The actions caused by growth factors, hormones, regulatory peptides, and other ligands in their target cells may result, on the one hand, from a given receptor acting selectively on unique substrates in a single cell type and, on the other hand, from the same receptor acting on different substrates in different types of cells.[13] In any case, the genetic program expressed by a particular cell is of crucial importance for the specificity of the physiologic response.

The cellular actions of growth factors are initiated by their binding to transmembrane receptor proteins (Figure 2.1). Cloning strategies for the structural characterization of cell surface receptors for peptide hormone and growth factors have been discussed.[14] The transmembrane receptors include receptors with intrinsic tyrosine kinase activity, receptors coupled to G proteins, and receptors that contain an ion channel as a structural component. Many transmembrane receptors for growth factors and peptide hormones possess intrinsic tyrosine kinase activity.[15-18] G protein-coupled receptors are characterized by

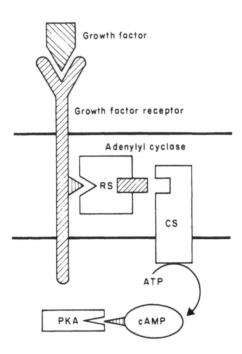

Figure 2.1 Cell surface receptor coupled to adenylyl cyclase.

the presence of seven transmembrane domains and include receptors for pituitary/placental glycoprotein hormones such as thyrotropin and gonadotropins, hypothalamic hormones such as thyrotropin-releasing hormone (TRH) and gonadotropin-releasing hormone (GnRH or LHRH), regulatory peptides such as angiotensin II and bombesin, and the α- and β-adrenergic receptors.[19,20] In addition, to these receptors, there is a family of transmembrane receptors with intrinsic guanylyl cyclase activity which includes the receptors for natriuretic peptides.[21-23] Another important group of surface receptors is represented by the hematopoietin/cytokine receptor superfamily.[24] The mechanism of signal transduction by the members of this superfamily is poorly understood.

There is some evidence that growth factors or their precursors may function in cell communication not only on the same cell where ligand binding occurs but also through a juxtacrine type of mechanism involving the binding of the membrane-anchored factor on one cell to its receptor on the adjacent cell.[25] Depending on the type of receptor, the transmission of its signal can be blocked by the action of various inhibitors, which may be useful in defining the role of the ligand in the different functions of the cell. The polyanionic compound suramin can inhibit the mitogenic effect of several growth factors (PDGF, TGF-β, FGF, and EGF) by altering the interaction between the growth factor and its receptor on the cell surface.[26] It is not known whether suramin binds free ligand and/or disrupts ligand-receptor interactions.

In contrast to the receptors for growth factors, peptide hormones, and regulatory peptides, which are located on the cell surface, the receptors for thyroid hormones, steroids, and retinoids are localized inside the cell. Thyroid hormone receptors are found mainly in the nucleus, although they also may be present in other cellular components including the membrane, the mitochondria, and the cytosol. The receptors for steroid hormones (androgens, estrogens, progestins, glucocorticoids, mineralocorticoids, and calcitriol), as well as the receptors for retinoids such as retinoic acid, are present in the cytoplasm, and upon hormone binding they are rapidly activated and translocated into the nucleus. The nuclear hormone receptor superfamily is characterized by the presence of highly conserved amino acid sequences that are involved in specific DNA binding and gene regulation.[27-31]

A. REGULATION OF RECEPTOR EXPRESSION

Expression of cell surface receptors for peptide hormones, growth factors, and regulatory peptides is subjected to developmental and functional regulatory changes. The ligand is frequently involved in regulating the expression of its receptor (autoregulation or homologous regulation), but other hormones, growth factors, or extracellular signaling agents may also be involved in this regulation (heterologous regulation). As a general rule, exposure of cells to a ligand reduces their response to the subsequent exposure (desensitization), whereas prevention of the interaction of the ligand with its receptor, either by suppressing the ligand or adding an antagonist, enhances the response (hypersensitivity).

Structural modifications of the receptor protein molecule, including glycosylation and phosphorylation, may have an important role in the regulatory mechanisms of receptor expression. The presence of antennary sugars in the cell surface receptors for peptide hormones and growth factors may be essential for signal transduction.[32] Phosphorylation of transmembrane receptors on tyrosine and/or serine/threonine residues is induced by a wide diversity of stimuli and may result in important alterations in the functional properties of the receptor.

Many factors acting at different cellular levels can modify the cellular concentration and/or affinity of receptors in the intact animal.[33-35] Cooperation of different types of cells may be very important in the regulation of receptor expression. Receptor regulation is called homospecific when the physiologic ligand regulates the function and/or number of its own receptor and is heterospecific when the activation of one receptor system leads to changes in a second distinct receptor system. Heterospecific receptor regulation may occur via mechanisms that are localized either at the level of the membrane or at levels comparatively remote from the initial ligand-receptor interaction.

A diversity of exogenous and endogenous factors may be associated with changes in receptor expression. One important factor in the regulation of receptor expression and function on the cell surface is determined by the lipid microenvironment, especially by cell membrane gangliosides,[36] which may contribute to the regulation of receptor-receptor interaction.[37] Cell density, i.e., the total number of cells in a defined space, may be important for the regulation of cell growth *in vitro* and *in vivo*, and this effect is mediated by signals transmitted from cell membrane receptors to the nucleus. The binding of growth factors such as EGF, TGF-β, FGF, and PDGF to their surface receptors is reduced when cell density is increased.[38] The mechanism of such density-induced downregulation of growth factor receptors is unknown, but one function of this phenomenon may be to reduce the responsiveness of cells to specific growth factors, which would result in reduction of growth rate as cell density increases in the culture.

B. RECEPTOR PHOSPHORYLATION

Cell surface receptors for many growth factors display, upon activation by ligand binding, protein kinase activity. Frequently, this activity is specific for tyrosine residues and is crucially required for mediation of the exogenous signal.[39] Ligand binding to the receptor protein kinase may induce autophosphorylation and activates the kinase, which is directed toward substrates located on the membrane, the cytoplasm, or the nucleus. The specificity of the cellular response is determined by the substrates recognized by the receptor kinase, which may be partially different even on a single cell, depending on the receptor and its ligand.[40] Each tyrosine kinase receptor may induce a unique pattern of cellular protein phosphorylation on specific tyrosine residues, although there may be a large overlapping in the substrates recognized by different receptor kinases. Src homology regions (SH regions) contained in certain substrates are important for the specificity and biological effects of activated tyrosine kinase receptors.[41]

Covalent modification of cell surface receptors by specific phosphorylation at tyrosine and nontyrosine residues plays a key role in the regulation of receptor function.[42,43] This regulation is achieved either by the action of G proteins that are involved in the regulation of both cAMP-dependent protein kinase (protein kinase A) and Ca^{2+}-phospholipid-dependent protein kinase (protein kinase C) or by the autophosphorylation of receptors that are themselves protein kinases. Transmembrane receptors with autophosphorylating activity include those for insulin, IGF-I, EGF, and PDGF. Receptor phosphorylation may alter the receptor function in either a positive or negative direction, depending on the cell type as well as on the nature of the receptor and its ligand. Phosphorylation can alter not only the functional activity of the receptor molecule but also its subcellular distribution. Interreceptor transphosphorylation and transactivation may be mechanisms for amplification of signal transduction.[44]

C. MOBILITY AND INTERNALIZATION OF SURFACE RECEPTORS

Upon ligand binding on the cell surface, the ligand-receptor complex is immobilized and undergoes aggregation. Thereafter, these complexes are internalized by a process of endocytosis or other mechanisms and may be degraded in specific intracellular compartments, or the receptor component of the complex may be recycled back to the cell surface.[45]

Lateral mobility and aggregation (clustering) of growth factor-receptor complexes on the cell surface are important for the physiological activities of many of these agents. Plant lectins such as concanavalin A or wheat germ agglutinin are multivalent molecules that bind to specific carbohydrates on the cell surface and can crosslink various cell surface receptors. Lectins may display striking effects at the level of the plasma membrane. They may decrease EGF- and insulin-stimulated DNA synthesis.[46] A close relationship between receptor aggregation/immobilization and response has been observed in a number of cellular systems. For many cell surface receptors, crosslinking by antireceptor antibodies is sufficient for receptor activation. This is not a general rule, however. The NGF receptor is preclustered and immobile on responsive cells, which suggests that immobilization of the receptor prior to ligand binding is required for signal transduction.[47] In general, there are two classes of cell surface receptors: those that are preclustered and immobile and those that only become clustered and immobile upon ligand binding.

The cellular response to hormones and growth factors may be altered, frequently in the form of a desensitization, after the initial exposure of the cell to these agents. This effect is usually correlated with the loss of available receptor sites on the cell surface. The receptors for peptide hormones and growth factors are concentrated on coated pits at the level of the cell membrane and are internalized by endocytosis, a process that involves invagination of the membrane, fusion of the neck of the invagination, and detachment of the fused vesicle with formation of an endosome.[48-53] Internalization of ligands bound to the surface receptors might proceed by one of two mechanisms: in an "escalator model" receptors are continuously being taken into the cell via coated pits and the ligand is a passive passenger, and in an "elevator model" receptors are internalized only after the ligand has given a "signal" for uptake.[54]

Receptor-mediated endocytosis has been classified into four categories, based on the final destination of both receptor and ligand.[55] In the class I category, receptor and ligand dissociate from one another and the receptor is recycled to the cell surface whereas the ligand is degraded in lysosomes. In the class II category, receptor and ligand both recycle to plasma membrane. In the class III category, both receptor and ligand are transported to lysosomes. Finally, in the class IV category the ligand-receptor complexes are delivered to the opposite sides of polarized cells in a process referred to as transcytosis, the ligand is released intact on the opposite side of the membrane, and the receptor is either degraded or recycled. Despite the heterogeneity of receptors, the initial step of endocytosis for all four classes of cell surface receptors is the same: the ligand-receptor complex is internalized via clathrin-coated pits. When clathrin-coated pits are saturated with receptors, they invaginate, forming a clathrin-coated vesicle. These vesicles lose their clathrin and undergo an acidic shift in pH, resulting in an endosome.

Internalization of the ligand-receptor complexes results in downregulation of their expression on the cell surface. Internalized receptors may become associated with different forms of lysosomes as well as with the nuclear membrane, the endoplasmic reticulum, and the Golgi apparatus. Membrane receptors may be degraded by lysosomal proteases or may be recycled back to the cell surface.[49] The endocytosed ligand-receptor complexes may enter an endosomal compartment formed by an extensive network of tubular cisternae. The endocytosed material, identified as multivesicular bodies by electron microscopy, moves along the intracellular reticulum toward the pericentriolar area.[56] The factors that regulate the initiation and subsequent routes of internalization are unknown.

Peptide hormones and growth factors would have, in addition to the cell surface receptors, intracellular binding sites of functional significance.[57,58] Receptors for peptide hormones and growth factors have been found in the nucleus. In cell lines bearing the specific surface receptors, growth factors such as NGF, EGF, and PDGF are associated with the chromatin in a nondegraded form.[59] EGF and NGF binding to isolated chromatin can be inhibited by receptor-specific monoclonal antibodies. Receptor-mediated endocytosis of IL-1 results in its accumulation in the nucleus, and it has been proposed that this mechanism may play a role in mediating some of the actions of IL-1.[60,61] Microinjection of TNF-α into the cytoplasm of target cells, which bypasses the early events of ligand-induced receptor activation at the cell surface, may result in the rapid killing of the cells, suggesting that internalization of TNF-α may have some physiological role.[62] Other factors such as IFN-γ would also have an intracellular activity, as shown by its microinjection into macrophages. However, other cytokines do not mimic their extracellularly initiated actions when microinjected into the respective target cells. The possible physiological significance of intracellular binding sites for peptide hormones and growth factors remains to be characterized. The autocrine function of growth factors such as IGF-I requires the secretion of IGF-I and its interaction with the receptor on the cell surface.[63]

D. RECEPTOR EXPRESSION IN NEOPLASTIC CELLS

Growth factor receptors are expressed by certain tumor cells in normal amounts and with normal affinities. The regulation of receptor expression in normal cells and neoplastic cells may be similar.[64] However, tumor cells frequently exhibit changes in receptor expression, which may be associated with altered response to the ligand. Quantitative and qualitative alterations in the expression of growth factor receptors may contribute to the expression of a transformed phenotype by normal cells or to the growth of transformed cells.[65] The existence of complex positive and negative correlations may become apparent when the receptors for multiple hormones and growth factors are determined in a given type of tumor, for example, in breast cancer. Comparison of the content of EGF, IGF, estrogen, and progesterone receptors in primary human breast cancer tissue and adjacent normal tissue showed that: (1) EGF receptor correlates negatively to estrogen receptor and progesterone receptor; (2) IGF receptor correlates positively to estrogen receptor and progesterone receptor; (3) no correlation between EGF receptor and IGF receptor is apparent; (4) IGF binding is higher in tumor tissues than in adjacent normal tissues; (5) EGF binding in breast cancer tissue may not differ from that in normal breast tissue; (6) the degree of cellular differentiation in ductal breast cancer correlates to EGF receptor but not to IGF receptor; and (7) the binding of EGF and IGF is the same in the primary breast tumor and their metastases.[66] These results suggest that complex interrelations between hormone and growth factor receptors are related to mammary carcinogesis, but the precise role of these complex changes in relation to the origin and development of human breast cancer is unknown.

REFERENCES

1. **James, R. and Bradshaw, R. A.,** Polypeptide growth factors, *Annu. Rev. Biochem.,* 53, 259, 1984.
2. **Rozengurt, E.,** Early signals in the mitogenic response, *Science,* 234, 161, 1986.
3. **Blay, J. and Hollenberg, M. D.,** The nature and function of polypeptide growth factor receptors in the human placenta, *J. Dev. Physiol.,* 12, 237, 1989.
4. **Scher, W.,** The role of extracellular proteases in cell proliferation and differentiation, *Lab. Invest.,* 57, 607, 1987.
5. **Levey, G. S. and Robinson, A. G.,** Introduction to the general principles of hormone-receptor interactions, *Metabolism,* 31, 639, 1982.
6. **Lefkowitz, R. J. and Michel, T.,** Plasma membrane receptors, *J. Clin. Invest.,* 72, 1185, 1983.
7. **Hollenberg, M. D.,** Mechanisms of receptor-mediated transmembrane signalling, *Experientia,* 42, 718, 1986.
8. **Hollenberg, M. D.,** Receptor triggering and receptor regulation: structure-activity relationships from the receptor's point of view, *J. Med. Chem.,* 33, 1275, 1990.
9. **Barnes, P. J.,** Molecular biology of receptors, *Q. J. Med.,* 83, 339, 1992.
10. **Fradkin, J. E., Eastman, R. C., Lesniak, M. A., and Roth, J.,** Specificity spillover at the hormone receptor —exploring its role in human disease, *N. Engl. J. Med.,* 320, 640, 1989.
11. **Hollenberg, M. D.,** Structure-activity relationships for transmembrane signaling: the receptor's turn, *FASEB J.,* 5, 178, 1991.
12. **Pouysségur, J. and Seuwen, K.,** Transmembrane receptors and intracellular pathways that control cell proliferation, *Annu. Rev. Physiol.,* 54, 195, 1992.
13. **Hollenberg, M. D.,** Receptors for insulin and other growth factors: rationale for common and distinct mechanisms of cell activation, *Clin. Invest. Med.,* 10, 475, 1987.
14. **Wright, M. S., Gautvik, V. T., and Gautvik, K. M.,** Cloning strategies for peptide hormone receptors, *Acta Endocrinol.,* 126, 97, 1992.
15. **Ullrich, A. and Schlessinger, J.,** Signal transduction by receptors with tyrosine kinase activity, *Cell,* 61, 203, 1990.
16. **Cadena, D. L. and Gill, G. N.,** Receptor tyrosine kinases, *FASEB J.,* 6, 2332, 1992.
17. **Panayotou, G. and Waterfield, M. D.,** The assembly of signalling complexes by receptor tyrosine kinases, *BioEssays,* 15, 171, 1993.
18. **Fantl, W. J., Johnson, D. E., and Williams, L. T.,** Signalling by receptor tyrosine kinases, *Annu. Rev. Biochem.,* 62, 453, 1993.
19. **Houslay, M. D.,** G-protein linked receptors — a family probed by molecular cloning analysis, *Clin. Endocrinol.,* 36, 525, 1992.
20. **Dias, J. A.,** Recent progress in structure-function and molecular analyses of the pituitary/placental glycoprotein hormone receptors, *Biochim. Biophys. Acta,* 1135, 278, 1992.
21. **Chinkers, M. and Garbers, D. L.,** Signal transduction by guanylyl cyclases, *Annu. Rev. Biochem.,* 60, 553, 1991.
22. **Wong, S. K. F. and Garbers, D. L.,** Receptor guanylyl cyclases, *J. Clin. Invest.,* 90, 299, 1992.
23. **Garbers, D. L.,** Guanylyl cyclase receptors and their endocrine, paracrine, and autocrine ligands, *Cell,* 71, 1, 1992.

24. **Cosman, D.,** The hematopoietin receptor superfamily, *Cytokine,* 5, 95, 1993.

25. **Massagué, J.,** Transforming growth factor-α. A model for membrane-anchored growth factors, *J. Biol. Chem.,* 265, 21393, 1990.

26. **Coffey, R. J., Jr., Leof, E. B., Shipley, G. D., and Moses, H. L.,** Suramin inhibition of growth factor receptor binding and mitogenicity in AKR-2B cells, *J. Cell. Physiol.,* 132, 143, 1987.

27. **Evans, R. M.,** The steroid and thyroid hormone receptor superfamily, *Science,* 240, 889, 1988.

28. **Wahli, W. and Martinez, E.,** Superfamily of steroid nuclear receptor: positive and negative regulators of gene expression, *FASEB J.,* 5, 2243, 1991.

29. **Lazar, M. A.,** Steroid and thyroid hormone receptors, *Endocrinol. Metab. Clin. N. Am.,* 20, 681, 1991.

30. **Vedeckis, W. V.,** Nuclear receptors, transcriptional regulation, and oncogenesis, *Proc. Soc. Exp. Biol. Med.,* 199, 1, 1992.

31. **Laudet, V., Hänni, C., Coll, J., Catzeflis, F., and Stéhelin, D.,** Evolution of the nuclear receptor gene superfamily, *EMBO J.,* 11, 1003, 1992.

32. **Sairam, M. R.,** Complete dissociation of gonadotropin receptor binding and signal transduction in mouse Leydig tumour cells, *Biochem. J.,* 265, 5667, 1990.

33. **Hollenberg, M. D.,** Examples of homospecific and heterospecific receptor regulation, *Trends Pharmacol. Sci.,* 6, 242, 1985.

34. **Hollenberg, M. D.,** Biochemical mechanisms of receptor regulation, *Trends Pharmacol. Sci.,* 6, 299, 1985.

35. **Hollenberg, M. D.,** Pathophysiological and therapeutic implications of receptor regulation, *Trends Pharmacol. Sci.,* 6, 334, 1985.

36. **van Echten, G. and Sandhoff, K.,** Ganglioside metabolism. Enzymology, topology, and regulation, *J. Biol. Chem.,* 268, 5341, 1993.

37. **Bremer, E. G., Schlessinger, J., and Hakomori, S.,** Ganglioside-mediated modulation of cell growth. Specific effects of G_{M_3} on tyrosine phosphorylation of the epidermal growth factor receptor, *J. Biol. Chem.,* 261, 2434, 1986.

38. **Rizzino, A., Kazakoff, P., Ruff, E., Kuszynski, C., and Nebelsick, J.,** Regulatory effects of cell density on the binding of transforming growth factor β, epidermal growth factor, platelet-derived growth factor, and fibroblast growth fact, *Cancer Res.,* 48, 4266, 1988.

39. **Izumi, T., Saeki, Y., Akanuma, Y., Takaku, F., and Kasuga, M.,** Requirement for receptor-intrinsic tyrosine kinase activities during ligand-induced membrane ruffling of KB cells. Essential sites of src-related growth factor receptor kinases, *J. Biol. Chem.,* 263, 10386, 1988.

40. **Kyriakis, J. M. and Avruch, J.,** Insulin, epidermal growth factor and fibroblast growth factor elicit distinct patterns of protein tyrosine phosphorylation in BC_3H1 cells, *Biochim. Biophys. Acta,* 1054, 73, 1990.

41. **Carpenter, G.,** Receptor tyrosine kinase substrates: src homology domains and signal transduction, *FASEB J.,* 6, 3283, 1992.

42. **Sibley, D. R., Benovic, J. L., Caron, M. G., and Lefkowitz, R. J.,** Regulation of transmembrane signaling by receptor phosphorylation, *Cell,* 48, 913, 1987.

43. **Sibley, D. R., Benovic, J. L., Caron, M. G., and Lefkowitz, R. J.,** Phosphorylation of cell surface receptors: a mechanism for regulating signal transduction pathways, *Endocrine Rev.,* 9, 38, 1988.

44. **Lammers, R., Van Obberghen, E., Ballotti, R., Schlessinger, J., and Ullrich, A.,** Transphosphorylation as a possible mechanism for insulin and epidermal growth factor receptor activation, *J. Biol. Chem.,* 265, 16886, 1990.

45. **King, A. C. and Cuatrecasas, P.,** Peptide hormone-induced receptor mobility, aggregation, and internalization, *N. Engl. J. Med.,* 305, 77, 1981.

46. **Kaplowicz, P. B.,** Wheat germ agglutinin and concanavalin A inhibit the response of human fibroblasts to peptide growth factors by a post-receptor mechanism, *J. Cell. Physiol.,* 124, 474, 1985.

47. **Venkatakrishnan, G., McKinnon, C. A., Pilapil, C. G., Wolf, D. E., and Ross, A. H.,** Nerve growth factor receptors are preaggregated and immobile on responsive cells, *Biochemistry,* 30, 2748, 1991.

48. **Gorden, P., Carpentier, J.-L., Fan, J.-Y., and Orci, L.,** Receptor mediated endocytosis of polypeptide hormones: mechanism and significance, *Metabolism,* 31, 664, 1982.

49. **Brown, M. S., Anderson, A. G. W., and Goldstein, J. L.,** Recycling receptors: the round-trip itinerary of migrant membrane proteins, *Cell,* 32, 663, 1983.

50. **Wileman, T., Harding, C., and Stahl, P.,** Receptor-mediated endocytosis, *Biochem. J.,* 232, 1, 1985.

51. **Goldstein, J. L., Brown, M. S., Anderson, R. G. W., Russell, D. W., and Schneider, W. J.,** Receptor-mediated endocytosis: concepts emerging from the LDL receptor system, *Annu. Rev. Cell Biol.,* 1, 1, 1985.

52. **Stahl, P. and Schwartz, A. L.,** Receptor-mediated endocytosis, *J. Clin. Invest.,* 77, 657, 1986.

53. **Carpentier, J.-L., Gorden, P., Robert, A., and Orci, L.,** Internalization of polypeptide hormones and receptor recycling, *Experientia,* 42, 734, 1986.

54. **Larrick, J. W., Enns, C., Raubitschek, A., and Weintraub, H.,** Receptor-mediated endocytosis of human transferrin and its cell surface receptor, *J. Cell. Physiol.,* 124, 283, 1985.

55. **Brown, B. I. and Greene, M. I.,** Molecular and cellular mechanisms of receptor-mediated endocytosis, *DNA Cell Biol.,* 10, 399, 1991.

56. **Hopkins, C. R., Gibson, A., Shipman, M., and Killer, K.,** Movement of internalized ligand-receptor complexes along a continuous endosomal reticulum, *Nature,* 346, 335, 1990.

57. **Goldfine, I. D.,** Interaction of insulin, polypeptide hormones, and growth factors with intracellular membranes, *Biochim. Biophys. Acta,* 650, 53, 1981.

58. **Lipson, K. E. Kolhatkar, A. A., Cherksey, B. D., and Donner, D. B.,** Characterization of glucagon receptors in Golgi fraction of rat liver: evidence for receptors that are uncoupled from adenylate cyclase, *Biochemistry,* 25, 2612, 1986.

59. **Rakowicz-Szulczynska, E. M., Rodeck, U., Herlyn, M., and Koprowski, H.,** Chromatin binding of epidermal growth factor, nerve growth factor, and platelet-derived growth factor in cells bearing the appropriate surface receptors, *Proc. Natl. Acad. Sci. U.S.A.,* 83, 3728, 1986.

60. **Grenfell, S., Smithers, N., Miller, K., and Solari, R.,** Receptor-mediated endocytosis and nuclear transport of human interleukin 1α, *Biochem. J.,* 264, 813, 1989.

61. **Curtis, B. M., Widmer, M. B., DeRoos, P., and Quarnstrom, E. E.,** IL-1 and its receptor are translocated to the nucleus, *J. Immunol.,* 144, 1295, 1990.

62. **Smith, M. R., Munger, W. E., Kung, H.-F., Takacs, L, and Durum, S. K.,** Direct evidence for an intracellular role for tumor necrosis factor-α. Microinjection of tumor necrosis factor kills target cells, *J. Immunol.,* 144, 162, 1990.

63. **Dai, Z., Stiles, A. D., Moats-Staats, B., Van Wyk, J. J., and D'Ercole, A. J.,** Interaction of secreted insulin-like growth factor-I (IGF-I) with cell surface receptors is the dominant mechanism of IGF-I's autocrine actions, *J. Biol. Chem.,* 267, 19565, 1992.

64. **Brizzi, M. F., Avanzi, G. C., Veglia, F. Clark, S. C., and Pegoraro, L.,** Expression and modulation of IL-3 and GF-CSF receptors in human growth factor dependent leukaemic cells, *Br. J. Haematol.,* 76, 203, 1990.

65. **Jasmin, C., Allouche, M., Le Bousse-Kerdiles, C., Smadja-Joffe, F., Krief, P., Georgoulias, V., and Boucheix, C.,** The role of growth-factor receptors (excluding IL-2 receptors) in the proliferation and differentiation of normal and leukemic hematopoietic cells, *Leukemia Res.,* 14, 695, 1990.

66. **Pekonen, F., Partanen, S., Mäkinen, T., and Rutanen, E.-M.,** Receptors for epidermal growth factor and insulin-like growth factor I and their relation to steroid receptors in human breast cancer, *Cancer Res.,* 48, 1343, 1988.

Chapter 3

Postreceptor Mechanisms of Growth Factor Action

I. INTRODUCTION

The transductional events responsible for receptor-operated signaling across the cell membrane involve complex mechanisms that have a remarkable spatial and temporal organization. Binding of the extracellular signaling agent to the specific receptor promotes the generation of early signals that are followed by a sequence of multiple biochemical responses that eventually converge into common final pathways leading to late events that may include DNA synthesis and cell division.[1,2] Binding of the ligand to its receptor on the cell membrane triggers the operation of intracellular signals consisting in the production and/or translocation of second messengers such as cyclic nucleotides, phosphoinositide metabolites, and Ca^{2+}. These alterations may cause the activation of enzymes, including protein kinases and protein phosphatases, which alters phosphorylation/dephosphorylation processes and leads to structural and/or functional changes associated with the specific cellular response. These effects may require only a few seconds to develop and may reach a maximum within one or a few minutes after exposure to the signal molecule.

II. MEDIATORS AND MODULATORS OF GROWTH FACTOR ACTION

A number of mediators or second messengers are associated with the action mechanisms of growth factors, hormones, and other signaling agents. In addition to the second messengers, substances such as prostaglandins and polyamines may act as modulators or regulators of growth factor action at different biochemical levels. For achieving an effective control of cellular processes at every time and site, the production of each mediator and modulator is regulated at different levels.

A. PROSTAGLANDINS

The prostaglandins are oxygenated and unsaturated 20-carbon fatty acids that contain a cyclopentane ring with two aliphatic side chains. They were discovered on the basis of the observation that substances present in human semen can produce contraction and relaxation of the uterus.[3] The effect of semen extracts is due to the presence of lipid-soluble compounds that were believed to have their origin in the prostate.[4,5] These compounds were called "prostaglandins", although it was later demonstrated that they have their origin not in the prostate but in the seminal vesicle fraction of the ejaculate. Isolation and identification of two active compounds of this group in seminal fluid, prostaglandin E_1 (PGE_1) and prostaglandin $F_{1\alpha}$ ($PGF_{1\alpha}$),[6] were followed by the characterization of the biosynthesis, metabolism, and biological properties of other compounds with similar structure and function.[7-10] About 20 natural prostaglandins are known at present. They are divided into the main groups PGE, PGF, PGA, PGB, PGC, and PGD, according to the substituents of the cyclopentane ring. These groups can be divided into subgroups or series on the basis of the number of double bonds present in the side chains (1, 2, or 3), which is indicated in the form of a subscript numeral. In the F series, α and β indicate the stereochemical orientation of the C9-hydroxyl group, but the β variant does not occur naturally. PGEs and PGFs are called primary prostaglandins because they are most abundant in animal tissues. PGA, PGB, and PGC are derivatives of PGE. The thromboxanes differ from the prostaglandins in having an oxane ring structure but share similar side chains and are classified according to a similar nomenclature.

The prostaglandins can be considered as chemical derivatives of a hypothetical 20-carbon prostanoic acid cytoskeleton. Two compounds derived from linoleic acid (an essential fatty acid) are involved in the biosynthesis of prostaglandins; dihomo-γ-linoleic acid (eicosatrienoic acid) and arachidonic acid (eicosatetranoic acid). While dihomo-γ-linoleic acid gives origin to prostaglandins of the 1 series, arachidonic acid is a precursor of the 2 series. The polyunsaturated fatty acid precursors are incorporated into cell membrane phospholipids. Prostaglandins are universally present in mammalian cells and are generated from the unsaturated fatty acid precursors by membrane-bound enzymes called prostaglandin synthetases. The predominant precursor of prostaglandin synthesis in mammalian cells is arachidonic acid, which is split from the cell membrane phospholipids by the action of phospholipase A_2.

The initial step of arachidonic acid metabolism involves site-specific incorporation of oxygen by a lipoxygenase-catalyzed reaction. Following this step, the cyclooxygenase pathway yields unstable products such as thromboxane A_2 and PGI_2 (prostacyclin), as well as the primary prostaglandins, PGE_2, PGD_2, and $PGF_{2\alpha}$, via endoperoxide intermediates. Other lipoxygenase pathways of arachidonic acid metabolism give rise to the leukotrienes, which are noncyclized products that bear three conjugated double bonds. Because the term prostaglandin system may not be appropriate to indicate the entire group of biologically active substances derived from arachidonic acid, the alternative term "eicosanoids" may be more conveniently used. The basic structure of several eicosanoids is indicated in Figure 3.1.

Prostaglandins, or enzyme systems forming or degrading them, have been identified in nearly all animal tissues studied. Various chemical, physical, neural, and hormonal factors can stimulate the synthesis of prostaglandins. Prostaglandins are not stored but are synthesized and released as required. The mechanisms involved in the regulation of prostaglandin production and degradation are little known but prostaglandin biosynthesis is frequently associated with changes in the intracellular concentrations of cAMP, as well as with intracellular mobilization of Ca^{2+}. The metabolism of prostaglandins may be altered in inflammatory processes, associated or not with pain and fever. Acetylsalicylic acid (aspirin) and indomethacin are widely used in clinical practice as anti-inflammatory drugs with potent inhibitory effects on the synthesis of prostaglandins.

1. Physiological Role of Prostaglandins

Prostaglandins exert a wide diversity of metabolic, physiologic, and pharmacologic effects in different tissues. They participate in several ways in the normal functions of the cardiovascular, respiratory, renal, gastrointestinal, hematic, endocrine, reproductive, and neural systems, and may have an important role in pathological processes affecting these systems. The two main products of arachidonic acid via the cyclooxygenase pathway in the large blood vessels, the lung, and other organs are prostacyclin and thromboxane A_2. These two products may display opposing physiologic effects at the sites where they are generated.[11] Prostacyclin is a potent vasodilator and an endogenous inhibitor of platelet aggregation, whereas thromboxane A_2 is a contractor of large blood vessels and induces platelet aggregation. In the lung, there may operate a delicate physiologic balance between the bronchodilator and vasodilator properties of prostacyclin and the bronchoconstrictor and vasoconstrictor properties of thromboxane A_2. The prostaglandins and other products of arachidonic acid metabolism may have a role in regulating the proliferation and differentiation of normal and neoplastic cells. However, the role of these substances in normal cell physiology remains unclear in spite of an enormous volume of basic and applied research.

Prostaglandins, particularly PGEs, are synthesized in many tissues coincident with hormone and growth factor action, and they may act as either positive or negative modulators of these agents. The cyclooxygenase inhibitors, aspirin and indomethacin, increase the growth-stimulating action of EGF and TNF-α on FS-4 human diploid fibroblasts, suggesting that prostaglandins induced by these factors antagonize their growth stimulatory action.[12] The prostaglandins may be involved in several steps of hormone and growth factor action, including the expression of cell surface receptors, but their precise role in the mechanisms of these signaling agents actions remains unclear.

2. Cellular Mechanisms of Action of Prostaglandins

Prostaglandins act as local regulators that are synthesized and degraded near their site of action. The physiologic effects of prostaglandins are exerted through their binding to receptors that are expressed in many types of cells. Binding of prostaglandins to their receptors is a Ca^{2+}-dependent event. Cyclic nucleotides, in particular cAMP, function as second messengers for the cellular actions of prostaglandins. Many cellular functions, including gene expression, can be regulated by prostaglandins. The expression of proto-oncogenes may be modulated by prostaglandins. PGE_2 induces a rapid increase in the intracellular level of cAMP in HL-60 human leukemia cells, and concurrently reduces c-*myc* gene expression, which accompanies the phenotypic differentiation of these cells induced by calcitriol or IFN-γ.[13] EGF-induced mitogenesis in BALB/c 3T3 cells requires metabolism of arachidonic acid to prostaglandins, and this effect is associated with induction of c-*myc* expression.[14] Both cAMP and Ca^{2+} are involved in PGE_1-induced elevation of c-*myc* mRNA levels in 3T3 fibroblasts.[15] EGF-dependent DNA synthesis in Syrian hamster embryo cells by PGE_2 is associated with altered expression of the c-*jun* and *jun*-B genes.[16]

3. Role of Prostaglandins in Tumors

The role of prostaglandins in tumor progression is little known. Incubation of PGA_1 with tumor tissue may result in the generation of polar metabolites possessing growth inhibitory properties.[17] PGA_2 exerts a

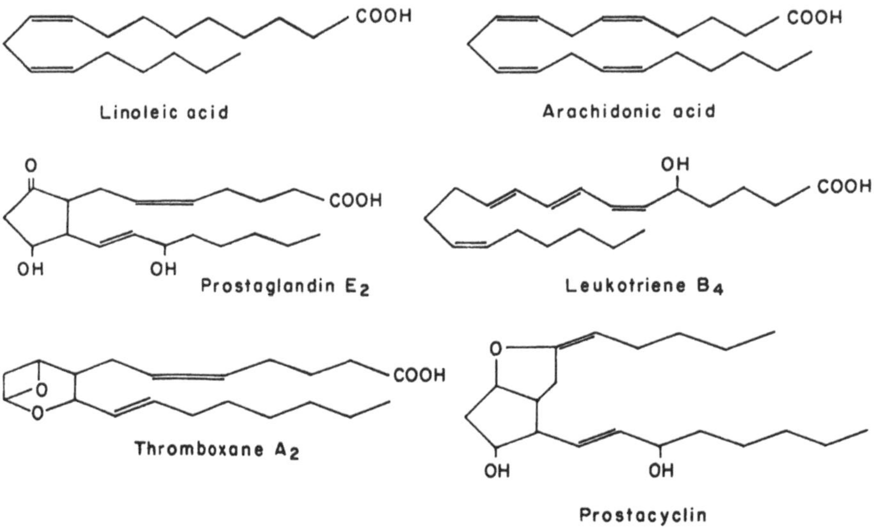

Figure 3.1 Basic structures of eicosanoids.

dose-dependent inhibitory effect on the growth of HL-60 human promyelocytic leukemia cells, with arrest in the G_0-G_1 phase and reduced expression of c-*myc* gene mRNA.[18] In different types of melanomas, including human melanomas, prostaglandins or their analogs exert an inhibitory effect on the growth of the tumor cells, making these agents potential tools for cancer chemotherapy.[19-22] This effect may be exerted independently of alterations in intracellular cAMP concentrations, however, the sensitivity of different tumor cells for the inhibitory effect of prostaglandins on cell proliferation shows wide variation. PGE_2 has little direct effect on the proliferation of murine mammary tumor cell lines.[23] PGE_2 binding is a heterogeneous property of these cells and some of them exhibit an apparent uncoupling of the prostaglandin receptor to the adenylyl cyclase system. The effects of prostaglandins on tumor cell growth may be mediated, at least in part, through their action on effector cells of the host immune system. There is little information about the possible alteration of prostaglandin biosynthesis in human tumors. In a comparative study of matched fresh normal lung and lung cancer tissue samples from lung cancer patients, the levels of PGE_2 and $PGF_{2\alpha}$ were found to be consistently elevated in all of the common histological types of lung cancer, the only exception being an undifferentiated large cell carcinoma.[24]

B. LEUKOTRIENES

The leukotrienes are noncyclized metabolites of arachidonic acid and their biosynthesis is catalyzed by a lipoxygenase specific for C-5 position.[25] These lipid mediators have 20 carbons and 3 conjugated double bonds. The leukotriene A_4 is an unstable epoxide that can be hydrolyzed to generate leukotriene B_4 (LTB_4) or the peptido-leukotrienes C_4, D_4, and E_4 by addition of a peptide group in position C_6. These peptido-leukotrienes are the bioactive components of the slow-reacting substance or anaphylaxis.

Leukotrienes display many biological effects and may have a crucial role in inflammatory and hypersensitivity reactions. LTB_4 stimulates polymorphonuclear leukocytes and leads to their aggregation and degranulation. LTB_4 is also a potent chemotactic and chemokinetic agent for polymorphonuclear leukocytes, eosinophils, and monocytes. LTB_4 can stimulate suppressor T lymphocytes and can enhance natural cytotoxic cell activity. It is also a potent bronchoconstrictor. Bioactive leukotrienes induce vasoconstriction of vascular beds and can produce a negative inotropic effect on cardiac muscle fibers. The precise role of leukotrienes in the mechanisms of action of hormones and growth factors is not clear, but they are members of the eicosanoid family of lipids and have numerous interactions with the prostaglandins. The effects of LTB_4 on leukocyte function are probably mediated by the synthesis of diverse cytokines. There is evidence that the physiological actions of leukotrienes are mediated by a specific receptor located on the cell surface and that the LTB_4 transduction mechanism involves changes in G proteins and alterations in phosphatidylinositol turnover.[26] LTB_4 may regulate the production of different cytokines by modulating the yield or the function of transcription factors such as the Fos and Jun proteins, which recognize AP-1-containing target genes.[27]

C. POLYAMINES

The two polyamines, putrescine and spermidine, and the diamine, spermine, are ubiquitous in vertebrate cells and are implicated as essential factors in the regulation of cell proliferation and differentiation.[28-40] Frequently, the three compounds (spermidine, spermine, and putrescine) are called polyamines (Figure 3.2). In mammalian cells, the polyamines are derived from the amino acids arginine and methionine, and the immediate precursor of the true polyamines is the diamine, putrescine. The polyamines can be interconverted and degraded back to putrescine by the action of specific enzymes. Both acetylated and oxidized forms of polyamines are known to occur naturally, and these polyamine derivatives may be major factors through which the polyamines exert many of their physiological effects.

In spite of their highly conserved and essential nature, the precise physiological role of the polyamines is not clear at present. Stimulation of polyamine synthesis frequently precedes increases in the rates of DNA, RNA, and protein synthesis in different types of cells. The polyamines may play an important role during commitment of serum-stimulated BALB/c 3T3 fibroblasts to DNA replication.[41] Addition of serum to quiescent WI38 human diploid fibroblasts may result in an accumulation of polyamines, especially of putrescine, which is followed by DNA synthesis.[42] Polyamine-depleted cells are arrested in the G_1 phase of the cycle, prior to a point of commitment where their further progression would not require the presence of growth factors. The recruitment of polyamine-depleted, serum-deprived mouse fibroblasts into the cell division cycle does not require PDGF and can be induced by addition of EGF and insulin plus putrescine.[43]

1. Regulation of Ornithine Decarboxylase

The initial and rate-limiting enzyme in the polyamine biosynthetic pathway is ornithine decarboxylase (ODC).[44] The biological importance of this enzyme is indicated by the extensive sequence homology found between ODC proteins from species as distant as the mouse and *Trypanosoma brucei* or even *Saccharomyces cerevisiae*. The human genome contains more than one copy of the ODC gene. ODC-related sequences are present on human chromosomes 2p and 7q.[45] Two ODC-related gene sequences, *ODC1* and *ODC2*, were localized to the human chromosome regions 2p25 and 7q31-qter, respectively.[46] However, the functional ODC locus resides on human chromosome 2p, and the locus on chromosome 7q may be a pseudogene. Isolation and expression of cDNAs encoding human ODC have been reported.[47-49] The predicted enzyme is a 461-residue polypeptide of 51,156 mol wt and has 90% homology with the amino acid sequence of murine ODC. The human ODC gene is divided into 12 exons and spans 8 kb. The murine ODC gene has been cloned from S49 mouse lymphoma cells, in which the ODC gene is amplified.[50] The mouse ODC gene is small in size (6.2 kb) and the transcription start site is located 31 or 32 bp 3′ to a canonical TATA box, which is not usual in a universally expressed gene such as the ODC gene.

The activity of ODC is regulated at both the transcriptional and posttranscriptional levels by many classes of hormones and growth factors acting on their respective target organs and tissues. ODC mRNA levels are regulated transcriptionally in mouse 3T3 cells and posttranscriptionally within the nucleus of T lymphocytes in response to mitogenic stimuli.[51] Serum, phorbol esters, and other mitogenic agents may influence the levels of ODC activity and mRNA.[52,53] Both the proliferative and antiproliferative action of cytokines such as IL-1 and TNF-α may involve regulation of ODC activity as a key component.[54] Expression of the ODC gene, as that of the c-*myc* proto-oncogene, increases rapidly after mitogenic stimulation of quiescent cells, but ODC mRNA is unchanged relative to total RNA during G_1 progression in cycling cells.[55] Two distinct ODC isoforms with different catalytic properties have been found in crude extracts of epidermal papillomas induced by an initiation-promotion protocol consisting of a single treatment with DMBA followed one week later with twice weekly applications of PMA.[56] The biological significance of the two forms of ODC detected in epidermal tumors is not known but only one form of the enzyme was found in PMA-treated normal epidermis.

Alterations in polyamine synthesis, frequently associated with changes in the intracellular concentration of cAMP, may contribute to the regulation of DNA synthesis and cell division. In certain types of cells, insulin and growth factors such as EGF, PDGF, and NGF induce the expression of ODC activity in the presence of asparagine, and this activity is closely related to the synthesis of polyamines and the initiation of DNA synthesis.[57] Inhibition of ODC activity leads to cessation of DNA synthesis and cellular proliferation However, an increase in ODC activity may be necessary but not sufficient by itself for cell division to occur, and an increase in cAMP concentration may be neither necessary nor sufficient for the induction of cell division in particular types of cells.[58] It is clear that polyamines and cAMP must act

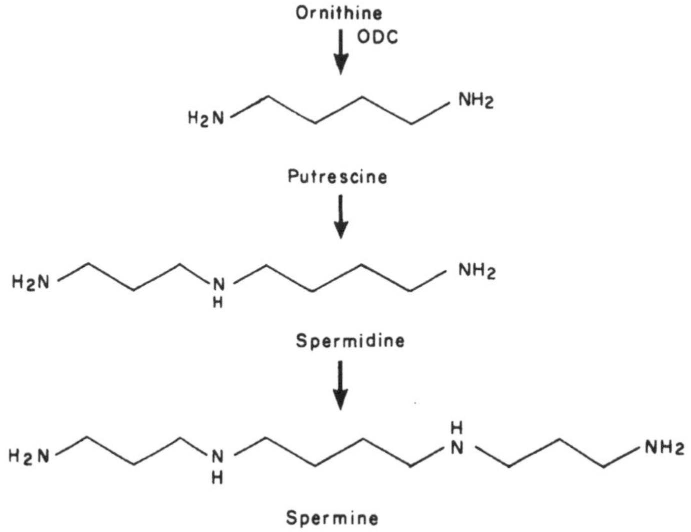

Figure 3.2 Basic structures of polyamines.

synergistically with other substances in the regulation of DNA synthesis and cellular proliferation stimulated by hormones and growth factors.

2. Mechanisms of Action of Polyamines

The mechanisms of action of polyamines in the regulation of DNA, RNA, and protein synthesis are little understood. There is evidence, however, that polyamines contribute to regulation of the phosphorylation of nonhistone proteins in organs such as the liver.[59] Polyamines may have an important influence on membrane functions.[60] They may have a role in the processes of cell differentiation involving phosphorylation and dephosphorylation of certain proteins.[61] These processes are associated with the activation of cAMP-dependent protein kinase and protein kinase C and with increased synthesis of phosphatidylcholine and may depend, at least in part, on the activity of catalase and the generation of H_2O_2. Available evidence suggests that polyamines may increase cAMP levels through a catalase-sensitive mechanism with subsequent changes in protein phosphorylation and phosphoinositide turnover. Polyamines also stimulate the activity of tyrosine kinases.[62] Because high concentrations of polyamines are present in many rapidly proliferating normal cells as well as in tumors, the regulation of tyrosine kinases by polyamines could be a mechanism for the stimulation of normal and tumor cell growth.

The possible role of polyamines in the regulation of proto-oncogene and tumor suppressor gene expression is little known. The levels of expression of the c-*myc*, c-*myb*, and p53 genes may become markedly reduced in cells treated with ODC inhibitors. Polyamine depletion induced in the COLO 320 human colon carcinoma cell line by the suicide inhibitor of ODC activity 2-difluoromethylornithine results in a markedly decreased transcriptional expression of the c-*myc* gene.[63] The biological significance of this alteration is unknown but it does not appear to be simply due to a reduced growth rate of the treated cells.

3. Role of Polyamines in Oncogenesis

Expression of activated oncoproteins may induce alterations in polyamine metabolism.[64] Transfection of a mutant c-H-*ras* gene into Rat-1 fibroblasts results in increased ODC activity, and polyamine uptake is markedly increased in cells transfected with an N-*myc* proto-oncogene. Polyamines could be involved in the growth-promoting effects of oncoproteins, and the polyamine biosynthetic pathway may have an important role in tumorigenesis. High levels of ODC mRNA are found in human colorectal tumors, compared with the normal adjacent mucosa and control mucosa.[46] Overproduction of ODC in mouse myeloma and leukemia cell lines under the pressure of a mechanism-based ODC inhibitor, which induces an amplification of ODC genes, confers a growth advantage to the malignant cells.[65] ODC overexpression can induce transformation in the NIH/3T3 assay system.[66] Application of the tumor promoter phorbol ester TPA to mouse skin leads to a marked increase in epidermal ODC activity.[67]

Polyamines may have a role in the growth of hormone- and growth factor-dependent tumors.[68] The polyamine pathway is important in the expression of an autocrine control of rat mammary tumor growth by prolactin.[69] As a general rule, malignant tumors exhibit higher levels of ODC activity than the respective normal tissues, which could simply reflect an increased rate of cellular proliferation.

Polyamines may have a role in the development of gastrointestinal tumors.[70] Normal-appearing mucosa in the large bowel of patients with multiple large bowel tumors, adenocarcinoma, and adenoma may exhibit elevated levels of ODC activity, suggesting that the mucosa has been exposed to carcinogens or promoters, and/or is highly susceptible to them.[71] A gradient of ODC activity is seen in the human colon mucosa from relatively low levels in the normal mucosa to higher levels in adenomatous polyps and maximum levels in adenocarcinomas.[72] Qualitative alterations of ODC may represent a marker for certain human tumors. A functionally distinct form of ODC, characterized by its activation by GTP, was detected in 13 of 40 human colorectal adenocarcinomas.[73] The GTP-activatable form of ODC was detected mainly in tumors of the cecum and was present in a lower percentage of tumors from other colonic segments.

In addition to a possible direct stimulation of tumor cell growth, polyamines may be involved in tumor-associated angiogenesis, which depends on the action of angiogenic growth factors and is required for tumor progression. The irreversible inhibitor of ODC activity, α-difluoromethylornithine (DFMO), inhibits B16 melanoma-induced angiogenesis in chick embryo chorioallantoic membrane, which may lead to inhibition of the growth of the tumor in the membrane.[74] The inhibitory effects of DFMO in this system can be reversed by exogenous putrescine and spermidine.

III.   ION AND pH CHANGES

The regulation of cell proliferation and differentiation by growth factors is frequently associated with changes in the flux rates and intracellular levels of monovalent ions such as H^+, Na^+, and K^+.[75-78] The divalent cations Ca^{2+} and Mg^{2+} are also involved in the mechanisms of growth factor action and may have a universal role in controlling the cell cycle and the proliferation of both normal and tumor cells.[79,80]

A. ION FLUXES AND CYTOPLASMIC pH

Intracellular (cytoplasmic) pH (pH_i) is strictly regulated in eukaryotic cells by a number of mechanisms that interact in complex ways.[81-83] The membrane-bound sodium-hydrogen exchanger (Na^+/H^+ antiporter), which exchanges internal H^+ for external Na^+, plays a fundamental role in the regulation of pH_i homeostasis in all vertebrate cells.[77,78] The Na^+/H^+ antiporter is unique among transport systems in that it is involved in multiple cellular functions, including not only the regulation of the pH_i but also the control of cell volume and cell proliferation and, in certain epithelia, the transcellular transport of salt, water, and acid-base equivalents. The Na^+/H^+ antiporter is characterized by its sensitivity to the potassium-sparing diuretic amiloride and resides on the plasma membrane; internal cellular membranes do not express amiloride-sensitive Na^+/H^+ exchange activity. Unlike the Na^+/K^+ pump (ATPase), the Na^+/H^+ antiporter does not require input of metabolic energy during the transport cycle. The inward directed Na^+ gradient produced by Na^+/H^+ antiporter activity is used to extrude the intracellular H^+ generated during cellular metabolic processes. In addition to the amiloride-sensitive Na^+/H^+ antiporter, the transport of Na^+ and K^+ into the cell depends on a bumetanide-sensitive $Na^+/K^+/Cl^-$ antiporter that can be stimulated by phorbol ester and different mitogens in cells such as quiescent human skin fibroblasts.[84]

The molecular structure of ion antiporters and channels is only partially understood. Expression of a human genomic DNA was capable of restoring the Na^+/H^+ antiporter defective function in a constructed stable mouse cell line lacking the antiporter activity.[85] Thus, this DNA contains the human gene coding for a protein with Na^+/H^+ antiporter activity. Protein kinases and protein phosphatases may be associated with ion channels and may contribute to regulate the channels on the cell surface. An endogenous protein kinase activity that is closely associated with a Ca^{2+}-activated K^+ channel in rat brain may be involved in regulating the activity of this ion channel.[86] Protein phosphorylation associated with the activity of both protein kinase A and protein kinase C can alter the activity of Ca^{2+}-activated ion channels.

The changes induced by growth factors in cellular pH and ion distribution are complex. Growth factor-induced mobilization of Ca^{2+} from intracellular stores is a pH-dependent process with a pH_i optimum around 7.1.[87] Activation of the Na^+/H^+ antiporter may facilitate the Ca^{2+} mobilization from intracellular stores by shifting the pH_i toward the optimum for the Ca^{2+} release from intracellular compartments. Transient cytoplasmic acidification may occur in association with an early increase in $[Ca^{2+}]_i$ upon cell stimulation, followed by a later cell alkalinization, which reflects activation of the Na^+/H^+ exchanger.[88] Activation of the amiloride-sensitive Na^+/H^+ antiporter with rise in pH_i and cytoplasmic alkalinization may result in profound changes in cells responding to growth factors.[89,90] In turn, phosphorylation of membrane proteins mediated by a spectrum of kinases that may be activated by growth factors and other extracellular stimuli is probably involved in the regulation of the amiloride-sensitive Na^+/H^+ antiporter and the flux of ions across the membrane.

B. MECHANISM OF NA+/H+ EXCHANGE REGULATION

The cellular and molecular mechanisms involved in regulation of Na^+/H^+ exchange through the plasma membrane are only partially understood. In normal NIH/3T3 mouse fibroblasts the exchange of monovalent ions is regulated by at least two mechanisms, one dependent on intracellular Ca^{2+} mobilization and the other on protein kinase C activation.[91] In cells transformed by a mutated c-H-*ras* oncogene, only the protein kinase C-mediated pathway is responsible for modulation of the Na^+/H^+ exchange. A cDNA encoding the Na^+/H^+ antiporter expressed in *Escherichia coli* allowed the characterization of a 110-kDa antiporter glycoprotein, which is phosphorylated by the action of growth factors.[92] Mitogenic activation of hamster and human cells with EGF, thrombin, phorbol esters, or serum-stimulated phosphorylation of the Na^+/H^+ antiporter with a time course similar to that of the rise in pH_i.

C. INFLUENCE OF MITOGENS, HORMONES, AND GROWTH FACTORS

Activation of Na^+/K^+-ATPase-mediated K^+ influx is induced by growth factors in their target cells.[93] The regulatory changes induced by growth factors on membrane ion transport may lead to stimulation of DNA synthesis in particular types of cells and under specific physiological conditions.[94] A growth factor-induced increase in pH_i through stimulation of the Na^+/H^+ antiporter may stimulate the entry of mitogen-stimulated quiescent fibroblasts into the S phase of the cycle. Prevention of mitogen-stimulated Na^+ influx and cytoplasmic pH rise in a fibroblast mutant lacking Na^+/H^+ antiporter activity suppresses growth factor-induced DNA synthesis at neutral and acidic pH.[95]

The relationship between changes in membrane ion fluxes and the mechanisms involved in the control of DNA synthesis and cell proliferation by growth factors is not totally clear. Cytokines such as IL-2 may rapidly induce an increase in pH_i through activation of a Na^+/H^+ antiporter and extrusion of protons, but cytoplasmic alkalinization is not required for lymphocyte proliferation.[96] Studies on the early stages of the proliferative cascade in activated T lymphocytes in which Na^+/H^+ antiporter activity was precluded by omission of Na^+ or by the addition of amiloride analogs indicate that stimulation of the antiporter that accompanies the addition of mitogens is neither sufficient nor necessary for the initiation of cellular proliferation.[97] Activation of the Na^+/H^+ antiporter and cytoplasmic alkalinization may not be required for EGF-dependent changes in gene expression and alteration of cell proliferation.[98] Phorbol ester-induced changes in the regulation of DNA synthesis in human monoblastoid cell lines are not causally associated with cytoplasmic alkalinization.[99] NGF does not activate Na^+/H^+ antiporter in PC12 rat pheochromocytoma cells that possess NGF receptors and respond to the growth factor with dramatic morphological and physiological changes.[100] Glucocorticoids confer normal serum- and growth factor-dependent growth regulation to Fu5 rat hepatoma cells *in vitro*, but DNA synthesis and alterations in proto-oncogene expression in the treated Fu5 cells occur in the absence of any short-term increases of intracellular pH or Ca^{2+}.[101]

A variety of molecular mechanisms may be involved in the activation of Na^+/H^+ exchange by the influence of mitogens, hormones, and growth factors. cAMP receptors may be involved in the regulation of Na^+ channels on the surface of cardiac myocytes through phosphorylation dependent on cAMP-dependent protein kinase A, and this mechanism may include the action of a pertussis toxin-sensitive G protein.[102] The cloning and expression of a cAMP-activated Na^+/H^+ exchanger has been reported, and there is evidence that the cytoplasmic domain of this exchanger mediates regulation by hormones and growth factors.[103] However, in some tissues multiple receptors coupled to the adenylyl cyclase system are capable of regulating Na^+/H^+ exchange independent of changes in cAMP accumulation and GTP regulatory proteins.[104] Changes in $[Ca^{2+}]_i$ and protein kinase C activation may have a role in the regulation of Na^+/H^+ exchange, but other unidentified mechanisms are probably implicated in the activation of Na^+/H^+ exchanges, as shown in T lymphocytes stimulated with concanavalin A.[105] Mechanisms involved in this activation include receptor-associated tyrosine kinase activity and generation of arachidonic acid by either phospholipase A_2 or the combined effect of phospholipase C and diacylglycerol lipase. Cytokines such as IFN-γ and IL-1α can enhance Na^+/H^+ exchange and produce sustained cytoplasmic alkalinization in lymphoid cells.[106] These ion changes are accompanied by a translocation of protein kinase C to the plasma membrane, but the Na^+/H^+ exchange may be independent of protein kinase C activation. Further studies are required for a better characterization of the biochemical mechanisms involved in the regulation of monovalent ion exchange by growth factors.

D. REGULATION OF CELL CYCLE-ASSOCIATED EVENTS

Activation of an Na^+/H^+ antiporter may be closely associated with cell cycle events and may have an important role in the regulation of cellular proliferation. This activation, with its subsequent cytoplasmic alkalinization, is the most common signaling mechanism associated with the stimulation of cell proliferation by HGFs such as CSF-1, CSF-2, CSF-3, and IL-3 in marrow progenitor cells.[107]

The pH_i levels may show variations in relation to the different phases of the cell cycle. During exponential growth of human tumor cell lines, the pH_i is cell cycle-related, being more alkaline during S, G_2, and M phases, whereas cells in the G_1 phase exhibit lower pH_i values.[108] The mechanisms by which pH and ion flux changes regulate cell cycle-associated events and mitogenesis are unknown. Growth factor-induced phosphorylation of the S6 ribosomal protein, as well as synthesis of proteins associated with G_0/G_1 transition of fibroblasts, may depend on changes in the pH_i.[109] Phosphorylation of S6 would be at least one of the pH_i-sensitive limiting steps in growth control. However, studies on mitogen-stimulated murine thymocytes and fibroblasts indicate that there is no requirement for activation of the Na^+/H^+ antiporter for the activation of S6 phosphorylation or protein synthesis by several mitogens.[110] However, activation of the Na^+/H^+ antiporter could be necessary for other cellular responses required for DNA synthesis. Alterations in the local distribution of monovalent ions such as lowering of extracellular Na^+ concentrations may result in the release of growth factors capable of stimulating cell proliferation in an autocrine manner.[111]

The role of monovalent ions and pH_i in the regulation of proto-oncogene expression is little known. Progression, but not competence, to enter the S phase of the cell cycle may be highly dependent on pH_i in growth factor-stimulated mouse fibroblasts, and the levels of expression of two proto-oncogenes, c-*myc* and c-*fos*, which are involved in the events that trigger the competent state and are stimulated after treatment of cultured cells with serum or growth factors, are relatively independent of the external pH and intracellular alkalinization.[112] Reduction of the extracellular Na^+ concentrations may abolish the response of the pH_i to mitogens but does not affect the increase in c-*fos* and c-*myc* gene transcripts stimulated by these mitogens.[113] Activation of Na^+/H^+ exchange and intracellular alkalinization is unnecessary in the induction of c-*fos* mRNA expression in serum-stimulated vascular smooth muscle (VSM) cells.[114] Stimulation of monovalent ion fluxes may be required for the transition of cells into S phase but may not be necessary for the transduction of a mitogenic signal into the nucleus, as evaluated by the transcriptional activation of c-*fos* and c-*myc* genes or the ODC gene in response to serum or growth factors.[115] However, activation of the c-*fos* and c-*myc* genes in response to the calcium ionophore A23187 is inhibited by removal of Na^+ from the incubation medium of cells such as murine thymocytes. Activation of the Na^+/H^+ antiporter, associated or not with changes in $[Ca^{2+}]_i$ protein kinase C activation, may result in induction of c-*fos* expression in U937 human monocyte-like leukemic cells.[116] Further studies are required for a better knowledge of the possible influence of growth factor-induced changes in monovalent ion fluxes and pH_i on proto-oncogene expression.

E. INFLUENCE OF ONCOGENES

Expression of viral oncogenes or activated cellular proto-oncogenes may lead to alterations in the ionic permeability of the plasma membrane. Unlike growth factor receptors with tyrosine kinase activity, the v-Src oncoprotein can induce a change in the membrane permeability to monovalent cations with permanent depolarization of the membrane.[117] Oncogene expression may result in activation of the Na^+/H^+ antiporter and intracellular alkalinization through pathways that are independent, at least in part, of the presence of serum or growth factors. Introduction into NIH/3T3 cells of a v-*mos* oncogene or a mutant c-H-*ras* gene mimics the effect of growth factors and activates the Na^+/H^+ antiporter.[118] The oncogene-induced increase of pH_i would stimulate initiation of DNA synthesis. The normal c-H-*ras* proto-oncogene does not induce cytoplasmic alkalinization and has only a weak mitogenic effect. Activation of the N^+/H^+ antiporter by the mutated c-H-*ras* gene occurs by an unknown protein kinase C-independent mechanism.[119]

F. EXPRESSION OF A TRANSFORMED PHENOTYPE

Changes in pH and Na^+ concentration in the culture medium may have remarkable effects on cell growth and morphology as well as on the expression of a transformed phenotype. The pH level as well as the osmolality and Na^+ concentration of the medium may affect both the clonal growth and morphology of early passage SHE cells.[120] Adjustments in the pH, osmolality, and Na^+ concentration ion in the medium may result in enhancement of the transformation response to agents with oncogenic potential. Neoplastic transformation induced in SHE cells by benzo(a)pyrene is enhanced under low pH culture conditions. A subpopulation of cells may exist, which can reversibly express the altered phenotype depending on medium pH. A role for pH_i itself as a contributory factor in the development of a transformed phenotype is suggested by the results of some studies. In CHE tumorigenic cell lines, the maintenance of an increased pH_i is an early event associated with the acquisition of a tumorigenic phenotype.[121] The pH_i increase is associated with Na^+/H^+ antiporter activity, occurs in the absence of autocrine growth factor production, is not associated with activation of protein kinase C or Ca^{2+}/calmodulin-mediated biochemical pathways, and correlates with an altered proliferative response. Thus an alteration in the functioning of the Na^+/H^+ antiporter itself may be responsible for the aberrant pH_i observed in the tumorigenic hamster cells. Expression of the Na^+/H^+ exchanger is required for tumor growth.[122] Differentiation of tumor cells induced by several agents may be associated with alteration of Na^+/H^+ antiporter activity. For example, induction of HL-60 cell differentiation by retinoic acid or DMSO is accompanied by cell alkalinization related to altered Na^+/H^+ exchanger activity.[123,124] Commitment of MEL cells to differentiation involves a modulated depolarization through Ca^{2+}-activated K^+ channels.[125]

Studies with mutants of hamster lung fibroblast cell lines lacking a functional Na^+/H^+ antiporter indicate that lack of antiporter activity does not preclude cells from acquiring a transformed phenotype induced by transfection of a vector expressing a mutant EJ c-H-*ras* oncogene.[126] However, the antiporter is a limiting factor in cell proliferation by its virtue of maintaining a steady pH_i, and its control applies to both normal and transformed cells. Mutants of hamster lung fibroblasts unable to regulate pH_i through the Na^+/H^+ antiporter system evolve tumors less frequently than the wild-type cells, and transplanted tumors originating in the mutant cells grow at a lower rate than those originating in the wild-type cells.

Embryonal carcinoma cells can undergo differentiation either *in vivo* or *in vitro* and may represent the most appropriate system for the study of the relationship between pH_i changes and neoplastic transformation.[127] While the malignant pluripotent PC19 murine embryonal carcinoma cells, which are autonomous and independent of the exogenous supply of growth factors, have a constitutively activated Na^+/H^+ exchanger, the differentiated MES-1 mesodermal derivatives of PC19 cells, which require the supply of exogenous growth factors for their growth, possess a Na^+/H^+ exchanger that is highly sensitive to growth factors such as EGF and PDGF as well as to TPA. The mechanism underlying the constitutive activation of the Na^+/H^+ antiporter in PC19 cells is unknown, but one possibility is that these cells could produce and utilize endogenous growth factors in an autocrine manner, which would signal the pathways leading to activation of the Na^+/H^+ carrier system. A constitutive activation of Na^+/H^+ exchange is not, however, an exclusive property of malignant cells.

G. ROLE OF MAGNESIUM

Magnesium is the second most abundant cation within animal cells, trailing only potassium, and may have a universal role in regulating the proliferation and differentiation of both normal and tumor cells. Magnesium has an important role in protein synthesis and is necessary for the function of Mg^{2+}-dependent enzymes. It maintains proper conformations of nucleic acids and proteins and regulates the operation of channels, receptors, and intracellular signaling molecules. It modulates photosynthesis, oxidative phosphorylation, muscle contraction, and nerve excitability. Magnesium is indispensable for cell proliferation and cell cycle control. It is contained in all subcellular fractions, including membrane, mitochondria, and nucleus, and is involved in many different metabolic pathways.[128-135] Normal Mg^{2+} concentrations within the cell are necessary for stabilization of the natural conformations of nucleic acids.[136]

The intracellular concentrations of free magnesium ($[Mg^{2+}]_i$) are regulated by several mechanisms and may involve active movements of the cation through the plasma membrane as well as its mobilization from intracellular compartmental stores, which are represented mainly by the mitochondria. Little is known about the influence of extracellular signaling agents such as hormones and growth factors on the regulation of magnesium metabolism. Noradrenaline, through the stimulation of β-adrenergic receptors and an increase in cAMP, stimulates a large efflux of Mg^{2+} from cardiac cells.[137] In a manner similar to Ca^{2+}, Mg^{2+} exchange through the plasma membrane and redistribution of Mg^{2+} among intracellular compartments may have considerable importance on cell structure and function. However, because the $[Mg^{2+}]_i$ is relatively high (0.2 to 1 mM), Mg^{2+} is unlikely to have a function of second messenger in the mechanisms of action of growth factors.

REFERENCES

1. **Rozengurt, E.,** Early signals in the mitogenic response, *Science,* 234, 161, 1986.
2. **Berridge, M. J., Cobbold, P. H., and Cuthbertson, K. S. R.,** Spatial and temporal aspects of cell signalling, *Phil. Trans. R. Soc. London,* B320, 325, 1988.
3. **Kurzrock, R. and Leib, C. C.,** Biochemical studies of human semen, *Proc. Soc. Exp. Biol. Med.,* 28, 268, 1930.
4. **von Euler, U. S.,** Zur Kenntnis der pharmakologischen Wirkungen von Nativsekretion und Extrakten männlicher accessorischer Geschlechtsdrüsen, *Arch. Exp. Pathol. Pharmakol.,* 175, 78, 1934.
5. **von Euler, U. S.,** Über die spezifische blutdrucksenkende Substanz des menschlichen Prostata- und Samenblasensekreten, *Klin. Wochenschr.,* 14, 1182, 1935.
6. **Bergström, S., Carlson, L. A., and Weeks, J. R.,** The prostaglandins, *Pharmacol. Rev.,* 20, 1, 1968.
7. **Bartmann, W.,** The prostaglandins, *Angew. Chem. Int. Ed.,* 14, 337, 1975.
8. **Vapaatalo, H. and Parantainen, J.,** Prostaglandins; their biological and pharmacological role, *Med. Biol.,* 56, 163, 1978.
9. **Olley, P. M.,** The prostaglandins, *Am. J. Dis. Child.,* 134, 688, 1980.
10. **Vane, J. R.,** Prostacyclin: a hormone with therapeutic potential, *J. Endocrinol.,* 95, 3P, 1982.
11. **Moncada, S. and Vane, J. R.,** Pharmacology and endogenous roles of prostaglandin endoperoxides, thromboxane A_2, and prostacyclin, *Pharmacol. Rev.,* 30, 293, 1979.
12. **Hori, T., Kashiyama, S., Hakayama, M., Shibamoto, S., Tsujimoto, M., Oku, N., and Ito, F.,** Possible role of prostaglandins as negative regulators in growth stimulation by tumor necrosis factor and epidermal growth factor in human fibroblasts, *J. Cell. Physiol.,* 141, 275, 1989.
13. **Matsui, T., Nakao, Y., Koizumi, T., Katakami, Y., Takahashi, R., Mihara, K., Sugiyama, T., and Fujita, T.,** Effect of prostaglandin E_2 on gamma-interferon and $1,25(OH)_2D_3$-induced c-*myc* reduction during HL-60 cell differentiation, *Leukemia Res.,* 7, 597, 1988.
14. **Handler, J. A., Danilowicz, R. M., and Eling, T. E.,** Mitogenic signaling by epidermal growth factor (EGF), but not platelet-derived growth factor, requires arachidonic acid metabolism in BALB/c 3T3 cells. Modulation of EGF-dependent c-*myc* expression by prostaglandins, *J. Biol. Chem.,* 265, 3669, 1990.
15. **Yamashita, T., Tsuda, T., Hakamori, Y., and Takai, Y.,** Possible involvement of cyclic AMP and calcium ion in prostaglandin E_1-induced elevation of c-*myc* mRNA levels in Swiss 3T3 fibroblasts, *J. Biol. Chem.,* 261, 16878, 1986.
16. **Cowlen, M. S. and Eling, T. E.,** Modulation of c-*jun* and *jun*-B messenger RNA and inhibition of DNA synthesis by prostaglandin E_2 in Syrian hamster embryo cells, *Cancer Res.,* 52, 6912, 1992.
17. **Gouin, E., Vulliez-Le Normand, B., Gouyette, A., Heidet, V., Nagel, N., and Dray, F.,** Tumor cell biotransformation of prostaglandin A_1 with growth inhibitory activity, *Biochem. Biophys. Res. Commun.,* 141, 1254, 1986.
18. **Ishioka, C., Kanamaru, R., Sato, T., Dei, T., Konishi, Y., Asamura, M., and Wakui, A.,** Inhibitory effects of prostaglandin A_2 on c-*myc* expression and cell cycle progression in human leukemia cell line HL-60, *Cancer Res.,* 48, 2813, 1988.
19. **Bregman, M. D. and Meyskens, F. L., Jr.,** *In vitro* modulation of human and murine melanoma growth by prostanoid analogues, *Prostaglandins,* 26, 449, 1983.
20. **Bregman, M. D., Peters, E., Sander, D., and Meyskens, F. L., Jr.,** Dexamethasone, prostaglandin A, and retinoic acid modulation of murine and human melanoma cells grown in soft agar, *J. Natl. Cancer Inst.,* 71, 927, 1983.
21. **Bregman, M. D., Funk, C., and Fukushima, M.,** Inhibition of human melanoma growth by prostaglandin A, D, and J analogues, *Cancer Res.,* 46, 2740, 1986.
22. **Abdel-Malek, Z. A., Swope, V. B., Amornsiripanitch, N., and Nordlund, J. J.,** *In vitro* modulation of proliferation and melanization of S91 melanoma cells by prostaglandins, *Cancer Res.,* 47, 3141, 1987.
23. **Fulton, A. M., Laterra, J. J., and Hanchin, C. M.,** Prostaglandin E_2 receptor heterogeneity and dysfunction in mammary tumor cells, *J. Cell. Physiol.,* 139, 93, 1989.
24. **McLemore, T. L., Hubbard, W. C., Litterst, C. L., Liu, M. C., Miller, S., McMahon, N. A., Eggleston, J. C., and Boyd, M. R.,** Profiles of prostaglandin biosynthesis in normal lung and tumor tissue from lung cancer patients, *Cancer Res.,* 48, 3140, 1988.
25. **Piper, P. J. and Samhoum, M. N.,** Leukotrienes, *Br. Med. Bull.,* 43, 297, 1987.
26. **Cristol, J.P., Provençal, B., and Sirois, P.,** Leukotriene receptors, *J. Receptor Res.,* 9, 341, 1989.

27. **Stanková, J. and Rola-Pleszczynski, M.,** Leukotriene B4 stimulates c-*fos* and c-*jun* gene transcription and AP-1 binding activity in human monocytes, *Biochem. J., 282,* 625, 1992.

28. **Bachrach, U.,** Metabolism and function of spermine and related polyamines, *Annu. Rev. Microbiol., 24,* 109, 1970.

29. **Jänne, J., Pösö, H., and Raina, A.,** Polyamines in rapid growth and cancer, *Biochim. Biophys. Acta, 473,* 241, 1978.

30. **Williams-Ashman, H. G. and Canellakis, Z. N.,** Polyamines in mammalian biology and medicine, *Perspect. Biol. Med., 22,* 421, 1979.

31. **Heby, O.,** Role of polyamines in the control of cell proliferation and differentiation, *Differentiation, 19,* 1, 1981.

32. **Henningsson, A. C., Henningsson, S., and Persson, L.,** Polyamine metabolism as related to growth and hormones, *Med. Biol., 59,* 320, 1981.

33. **Pegg, A. E. and McCann, P. P.,** Polyamine metabolism and function, *Am. J. Physiol., 243,* C212, 1982.

34. **Cochet, C. and Chambaz, E. M.,** Polyamine-mediated protein phosphorylation: a possible target for intracellular polyamine action, *Mol. Cell. Endocrinol., 30,* 247, 1983.

35. **Tabor, C. W. and Tabor, H.,** Polyamines, *Annu. Rev. Biochem., 53,* 749, 1984.

36. **Pegg, A. E.,** Recent advances in the biochemistry of polyamines in eukaryotes, *Biochem. J., 234,* 249, 1986.

37. **Pegg, A. E.,** Polyamine metabolism and its importance in neoplastic growth and as a target for chemotherapy, *Cancer Res., 48,* 759, 1988.

38. **Canellakis, Z. N., Marsh, L. L., and Bondy, P. K.,** Polyamines and their derivatives as modulators in growth and differentiation, *Yale J. Biol. Med., 62,* 481, 1989.

39. **Heby, O. and Persson, L.,** Molecular genetics of polyamine synthesis in eukaryotic cells, *Trends Biochem. Sci., 15,* 153, 1990.

40. **Davis, R. H.,** Management of polyamine pools and the regulation of ornithine decarboxylase, *J. Cell. Biochem., 44,* 199, 1990.

41. **Schaefer, E. L. and Seidenfeld, J.,** Effects of polyamine depletion on serum stimulation of quiescent 3T3 murine fibroblast cells, *J. Cell. Physiol., 133,* 546, 1987.

42. **Heby, O., Marton, L. J., Zardi, L., Russell, D. H., and Baserga, R.,** Changes in polyamine metabolism in W138 cells stimulated to proliferate, *Exp. Cell Res., 90,* 8, 1975.

43. **Charollais, R. H. and Mester, J.,** Resumption of cell cycle in BALB/c-3T3 fibroblasts arrested by polyamine depletion: relation with "competence" gene expression, *J. Cell. Physiol., 137,* 559, 1988.

44. **Russell, D. H.,** Ornithine decarboxylase: a key regulatory enzyme in normal and neoplastic growth, *Drug Metab. Rev., 16,* 1, 1985.

45. **Winqvist, R., Mäkelä, T. P., Seppänen, P., Jänne, O. A., Alhonen-Hogisto, L., Jänne, J., Grzeschik, K.-H., and Alitalo, K.,** Human ornithine decarboxylase sequences map to chromosome regions 2pter-p23 and 7cen-qter but are not coamplified with the *NMYC* oncogene, *Cytogenet. Cell Genet., 42,* 133, 1986.

46. **Radford, D. M., Nakai, H., Eddy, R. L., Haley, L. L., Byers, M. G., Henry, W. M., Lawrence, D. D., Porter, C. W., and Shows, T. B.,** Two chromosomal locations for human ornithine decarboxylase gene sequences and elevated expression in colorectal neoplasia, *Cancer Res., 50,* 6146, 1990.

47. **Hickok, N. J., Seppänen, P. J., Gunsalus, G. L., and Jänne, O. A.,** Complete amino acid sequence of human ornithine decarboxylase deduced from complementary DNA, *DNA, 6,* 179, 1987.

48. **Fitzgerald, M. C. and Flanagan, M. A.,** Characterization and sequence analysis of the human ornithine decarboxylase gene, *DNA, 8,* 623, 1989.

49. **Moshier, J. A., Gilbert, J. D., Skunca, M., Dosescu, J., Almodovar, K. M., and Luk, G. D.,** Isolation and expression of a human ornithine decarboxylase gene, *J. Biol. Chem., 265,* 4884, 1990.

50. **Brabant, M., McConlogue, L., van Daalen Wetters, T., and Coffino, P.,** Mouse ornithine decarboxylase gene: cloning, structure, and expression, *Proc. Natl. Acad. Sci. U.S.A., 85,* 2200, 1988.

51. **Abrahamsen, M. S. and Morris, D. R.,** Cell type-specific mechanisms of regulating expression of the ornithine decarboxylase gene after growth stimulation, *Mol. Cell. Biol., 10,* 5525, 1990.

52. **Hovis, J. G., Stumpo, D. J., Halsey, D. L., and Blackshear, P. J.,** Effects of mitogens on ornithine decarboxylase activity and messenger RNA levels in normal and protein kinase C-deficient NIH-3T3 fibroblasts, *J. Biol. Chem., 261,* 10380, 1986.

53. **Djurhuus, R., Laerum, O. D., and Lillehaug, J. R.,** Ornithine decarboxylase activity in foetal rat brain cells and in the mouse embryo fibroblasts C3H/10T1/2 CL8 cells: differences in response to medium change and to tumor promoter TPA, *Int. J. Biochem., 19,* 495, 1987.

54. **Endo, Y., Matsushima, K., Onozaki, K., and Oppenheim, J. J.,** Role of ornithine decarboxylase in the regulation of cell growth by IL-1 and tumor necrosis factor, *J. Immunol.,* 141, 2342, 1988.

55. **Stimac, E. and Morris, D. R.,** Messenger RNAs coding for enzymes of polyamine biosynthesis are induced during G_0-G_1 transition but not during traverse of the normal G_1 phase, *J. Cell. Physiol.,* 133, 590, 1987.

56. **O'Brien, T. G., Hietala, O., O'Donnell, K., and Holmes, M.,** Activation of mouse epidermal tumor ornithine decarboxylase by GTP: evidence for different catalytic forms of the enzyme, *Proc. Natl. Acad. Sci. U.S.A.,* 84, 8927, 1987.

57. **Rinehart, C. A., Jr. and Canellakis, E. S.,** Induction of ornithine decarboxylase activity by insulin and growth factors is mediated by amino acids, *Proc. Natl. Acad. Sci. U.S.A.,* 82, 4365, 1985.

58. **Willey, J. C., Laveck, M. A., McClendon, I. A., and Lechner, J. F.,** Relationship of ornithine decarboxylase activity and cAMP metabolism to proliferation of normal human bronchial epithelial cells, *J. Cell. Physiol.,* 124, 1985.

59. **Imai, H., Shimoyama, M., Yamamoto, S., Tanigawa, Y., and Ueda, I.,** Effect of polyamines on phosphorylation of non-histone chromatin proteins from hog liver, *Biochem. Biophys. Res. Commun.,* 66, 856, 1975.

60. **Schuber, F.,** Influence of polyamines on membrane functions, *Biochem. J.,* 260, 1, 1989.

61. **Kiss, Z., Deli, E., and Kuo, J. F.,** Cyclic AMP-like effects of polyamines on phosphatidylcholine synthesis and protein phosphorylation in human promyelocytic leukemia HL60 cells, *FEBS Lett.,* 213, 365, 1987.

62. **Sakai, K., Sada, K., Tanaka, Y., Kobayashi, T., Nakamura, S., and Yamamura, H.,** Regulation of cytosolic protein-tyrosine kinase from porcine spleen by polyamines and negative-charged polysaccharides, *Biochem. Biophys. Res. Commun.,* 154, 883, 1988.

63. **Celano, P., Baylin, S. B., Giardiello, F. M., Nelkin, B. D., and Casero, R. A., Jr.,** Effect of polyamine depletion on c-*myc* expression in human colon carcinoma cells, *J. Biol. Chem.,* 263, 5491, 1988.

64. **Chang, B. K., Libby, P. R., Bergeron, R. J., and Porter, C. W.,** Modulation of polyamine biosynthesis and transport by oncogene transfection, *Biochem. Biophys. Res. Commun.,* 157, 264, 1988.

65. **Polvinen, K., Sinervirta, R., Alhonen, L., and Jänne, J.,** Overproduction of ornithine decarboxylase confers an apparent growth advantage to mouse tumor cells, *Biochem. Biophys. Res. Commun.,* 155, 373, 1988.

66. **Moshier, J. A., Dosescu, J., Skunca, M., and Luk, G. D.,** Transformation of NIH/3T3 cells by ornithine decarboxylase expression, *Cancer Res.,* 53, 2618, 1993.

67. **Verma, A. K., Erickson, D., and Dolnick, B. J.,** Increased mouse epidermal ornithine decarboxylase activity by the tumour promoter 12-*O*-tetradecanoylphorbol 13-acetate involves increased amounts of both enzyme protein and messenger RNA, *Biochem. J.,* 237, 297, 1986.

68. **Manni, A.,** Polyamines and hormonal control of breast cancer growth, *Crit. Rev. Oncogenesis,* 1, 163, 1989.

69. **Manni, A., Wright, C., Hsu, C.-J., and Hammond, J. M.,** Polyamines and autocrine control of tumor growth by prolactin in experimental breast cancer in culture, *Endocrinology,* 119, 2033, 1986.

70. **Sarhan, S., Knodgen, B., and Seiler, N.,** The gastrointestinal tract as polyamine source for tumor growth, *Anticancer Res.,* 9, 215, 1989.

71. **Narisawa, T., Takahashi, M., Niwa, M., Koyama, H., Kotanagi, H., Kusaka, N., Yamazaki, Y., Nagasawa, O., Koyama, K., Wakizaka, A., and Fukaura, Y.,** Increased mucosal ornithine decarboxylase activity in large bowel with multiple tumors, adenocarcinoma, and adenoma, *Cancer,* 63, 1572, 1989.

72. **Porter, C. W., Herrera-Ornelas, L., Pera, P., Petrelli, N. F., and Mittelman, A.,** Polyamine biosynthetic activity in normal and neoplastic human colorectal tissue, *Cancer,* 60, 1275, 1987.

73. **Hietala, O. A., Yum, K. Y., Pilon, J., O'Donnell, K., Holroyde, C. P., Kline, I., Reichard, G. A., Litwin, S., Gilmour, S. K., and O'Brien, T. G.,** Properties of ornithine decarboxylase in human colorectal adenocarcinomas, *Cancer Res.,* 50, 2088, 1990.

74. **Takigawa, M., Enomoto, M., Nishida, Y., Pau, H.-O., Kinoshita, A., and Suzuki, F.,** Tumor angiogenesis and polyamines: α-difluoromethylornithine, an irreversible inhibitor of ornithine decarboxylase, inhibits B16 melanoma-induced angiogenesis *in ovo* and the proliferation of vascular endothelial cells *in vitro, Cancer Res.,* 50, 4131, 1990.

75. **Rozengurt, E.,** Stimulation of Na influx, Na-K pump activity and DNA synthesis in quiescent cultured cells, *Adv. Enzyme Regul.,* 19, 61, 1981.

76. **Rozengurt, E. and Mendoza, S. A.,** Synergistic signals in mitogenesis: role of ion fluxes, cyclic nucleotides and protein kinase C in Swiss 3T3 cells, *J. Cell Sci.,* Suppl. 3, 229, 1985.

77. **Grinstein, S., Rotin, D., and Mason, M. J.,** Na$^+$/H$^+$ exchange and growth factor-induced cytosolic pH changes. Role in cellular proliferation, *Biochim. Biophys. Acta,* 988, 73, 1989.

78. **Barber, D. L.,** Mechanisms of receptor-mediated regulation of Na-H exchange, *Cell. Signal.,* 3, 387, 1991.

79. **Whitfield, J. F.,** The roles of calcium and magnesium in cell proliferation: an overview, in *Ions, Cell Proliferation, and Cancer,* Boynton, A. L., McKeehan, W. L., and Whitfield, J. F., Eds., Academic Press, New York, 1982, 283.

80. **Cameron, I. L. and Smith, N. K. R.,** Cellular concentration of magnesium and other ions in relation to protein synthesis, cell proliferation and cancer, *Magnesium,* 8, 31, 1989.

81. **Seifter, J. L. and Aronson, P. S.,** Properties and physiologic roles of the plasma membrane sodium-hydrogen exchanger, *J. Clin. Invest.,* 78, 859, 1986.

82. **Lagarde, A. E. and Pouysségur, J. M.,** The Na$^+$:H$^+$ antiport in cancer, *Cancer Biochem. Biophys.,* 9, 1, 1986.

83. **Madshus, I. H.,** Regulation of intracellular pH in eukaryotic cells, *Biochem. J.,* 250, 1, 1988.

84. **Panet, R. and Atlan, H.,** Bumetanide-sensitive Na$^+$/K$^+$/Cl$^-$ transporter is stimulated by phorbol ester and different mitogens in quiescent human skin fibroblasts, *J. Cell. Physiol.,* 145, 30, 1990.

85. **Franchi, A., Perucca-Lostanlen, D., and Pouysségur, J.,** Functional expression of a human Na$^+$/H$^+$ antiporter gene transfected into antiporter-deficient mouse L cells, *Proc. Natl. Acad. Sci. U.S.A.,* 83, 9388, 1986.

86. **Chung, S., Reinhart, P. H., Martin, B. L., Brautigan, D., and Levitan, I. B.,** Protein kinase activity closely associated with a reconstituted calcium-activated potassium channel, *Science,* 253, 560, 1991.

87. **Maly, K., Hochleitner, B. W., and Grunicke, H.,** Interrelationship between growth factor-induced activation of the Na$^+$/H$^+$ antiporter and mobilization of intracellular Ca^{2+} in NIH3T3 fibroblasts, *Biochem. Biophys. Res. Commun.,* 167, 1206, 1990.

88. **Ives, H. E. and Daniel, T. O.,** Interrelationship between growth factor-induced pH changes and intracellular Ca^{2+}, *Proc. Natl. Acad. Sci. U.S.A.,* 84, 1950, 1987.

89. **Vicentini, L. M. and Villereal, M. L.,** Activation of Na$^+$/H$^+$ exchange in cultured fibroblasts: synergism and antagonism between phorbol ester, Ca^{2+} ionophore, and growth factors, *Proc. Natl. Acad. Sci. U.S.A.,* 82, 8053, 1985.

90. **Berkman, B. C., Aronow, M. S., Brock, T. A., Cragoe, E., Jr., Gimbrone, M. A., Jr., and Alexander, R. W.,** Angiotensin II-stimulated Na$^+$/H$^+$ exchange in cultured vascular smooth muscle cells: evidence for protein kinase C-dependent and -independent pathways, *J. Biol. Chem.,* 262, 5057, 1987.

91. **Owen, N. E., Knapik, J., Strebel, F., Tarpley, W. G., and Gorman, R. R.,** Regulation of Na$^+$-H$^+$ exchange in normal NIH-3T3 cells and in NIH-3T3 cells expressing the *ras* oncogene, *Am. J. Physiol.,* 254, C756, 1989.

92. **Sardet, C., Counillon, L., Franchi, A., and Pouysségur, J.,** Growth factors induce phosphorylation of the Na$^+$/H$^+$ antiporter, a glycoprotein of 110 kD, *Science,* 247, 723, 1990.

93. **Vairo, G. and Hamilton, J. A.,** CSF-1 stimulates Na$^+$K$^+$-ATPase mediated ^{86}Rb$^+$ uptake in mouse bone marrow-derived macrophages, *Biochem. Biophys. Res. Commun.,* 132, 430, 1985.

94. **Tupper, J. T. and Smith, J. W.,** Growth factor regulation of membrane transport in human fibroblasts and its relationship to stimulation of DNA synthesis, *J. Cell. Physiol.,* 125, 443, 1985.

95. **Pouysségur, J., Franchi, A., L'Allemain, G., and Paris, S.,** Cytoplasmic pH, a key determinant of growth factor-induced DNA synthesis in quiescent fibroblasts, *FEBS Lett.,* 190, 115, 1985.

96. **Mills, G. B., Cragoe, E. J., Jr., Gelfand, E. W., and Grinstein, S.,** Interleukin 2 induces a rapid increase in intracellular pH through activation of a Na$^+$/H$^+$ antiport: cytoplasmic alkalinization is not required for lymphocyte proliferation, *J. Biol. Chem.,* 260, 12500, 1985.

97. **Grinstein, S., Smith, J. D., Onizuka, R., Cheung, R. K., Gelfand, E. W., and Benedict, S.,** Activation of Na$^+$/H$^+$ exchange and the expression of cellular proto-oncogenes in mitogen- and phorbol ester-treated lymphocytes, *J. Biol. Chem.,* 263, 8658, 1988.

98. **Church, J. G., Mills, G. B., and Buick, R. N.,** Activation of the Na$^+$/H$^+$ antiport is not required for epidermal growth factor-dependent gene expression, growth inhibition or proliferation in human breast cancer cells, *Biochem. J.,* 257, 151, 1989.

99. **Forsbeck, K., Nilsson, K., Nygren, P., Larsson, R., Gylfe, E., Skoglund, G., and Ingelman-Sundberg, M.,** Phorbol-ester-induced stable changes in the regulation of DNA synthesis and intracellular pH are accompanied by altered expression of protein kinase C in the monoblastoid cell line U-937, *Int. J. Cancer,* 42, 284, 1988.

100. **Chandler, C. E., Cragoe, E. J., Jr., and Glaser, L.,** Nerve growth factor does not activate Na^+/H^+ exchange in PC12 pheochromocytoma cells, *J. Cell. Physiol.,* 125, 367, 1985.

101. **Cook, P. W., Weintraub, W. H., Swanson, K. T., Machen, T. E., and Firestone, G. L.,** Glucocorticoids confer normal serum/growth factor-dependent growth regulation to Fu5 rat hepatoma cells *in vitro.* Sequential expression of cell cycle-regulated genes without changes in intracellular calcium or pH, *J. Biol. Chem.,* 263, 19296, 1988.

102. **Sorbera, L. A. and Morad, M.,** Modulation of cardiac sodium channels by cAMP receptors on the myocyte surface, *Science,* 253, 1286, 1991.

103. **Borgese, F., Sardet, C., Cappadoro, M., Pouysségur, J., and Motais, R.,** Cloning and expression of a cAMP-activated NA^+/H^+ exchanger: evidence that the cytoplasmic domain mediates hormonal regulation, *Proc. Natl. Acad. Sci. U.S.A.,* 89, 6765, 1992.

104. **Ganz, M. B., Pachter, J. A., and Barber, D. L.,** Multiple receptors coupled to adenylate cyclase regulate Na-H exchange independent of cAMP, *J. Biol. Chem.,* 265, 8989, 1990.

105. **Grinstein, S., Smith, J. D., Rowatt, C., and Dixon, S. J.,** Mechanism of activation of lymphocyte Na^+/H^+ exchange by concanavalin A: a calcium- and protein kinase C-independent pathway, *J. Biol. Chem.,* 262, 15277, 1987.

106. **Ostrowski, J., Meier, K. E., Stanton, T. H., Smith, L. L., and Bomsztyk, K.,** Interferon-γ and interleukin 1α induce transient translocation of protein kinase C activity to membranes in a B lymphoid cell line. Evidence for a protein kinase C-independent pathway in lymphokine-induced cytoplasmic alkalinization, *J. Biol. Chem.,* 263, 13876, 1988.

107. **Cook, N., Dexter, T. M., Lord, B. I., Cragoe, E. J., Jr., and Whetton, A. D.,** Identification of a common signal associated with cellular proliferation stimulated by four haemopoietic growth factors in highly enriched population of granulocyte/macrophage colony-forming cells, *EMBO J.,* 8, 2967, 1989.

108. **Musgrove, E., Seaman, M., and Hedley, D.,** Relationship between cytoplasmic pH and proliferation during exponential growth and cellular quiescence, *Exp. Cell Res.,* 172, 65, 1987.

109. **Chambard, J.-C. and Pouysségur, J.,** Intracellular pH controls growth factor-induced ribosomal protein S6 phosphorylation and protein synthesis in the G_0-G_1 transition of fibroblasts, *Exp. Cell Res.,* 164, 282, 1986.

110. **Pennington, S. R., Moore, J. P., Hesketh, T. R., and Metcalfe, J. C.,** Mitogen-stimulated activation of the Na^+/H^+ antiporter does not regulate S6 phosphorylation or protein synthesis in murine thymocytes or Swiss 3T3 fibroblasts, *J. Biol. Chem.,* 265, 2456, 1990.

111. **Walsh-Reitz, M. M., Gluck, S. L., Waack, S., and Toback, F. G.,** Lowering extracellular Na^+ concentration releases autocrine growth factors from renal epithelial cells, *Proc. Natl. Acad. Sci. U.S.A.,* 83, 4764, 1986.

112. **Bravo, R. and Macdonald-Bravo, H.,** Effect of pH on the induction of competence and progression to the S-phase in mouse fibroblasts, *FEBS Lett.,* 1985, 309, 1986.

113. **Moore, J. P., Todd, J. A., Hesketh, T. R., and Metcalfe, J. C.,** c-*fos* and c-*myc* gene activation, ionic signals, and DNA synthesis in thymocytes, *J. Biol. Chem.,* 261, 8158, 1986.

114. **Nabika, T., Kobayashi, A., Nara, Y., Endo, J., and Yamori, Y.,** Activation of Na^+/H^+ exchange is unnecessary in the induction of c-*fos* mRNA in serum-stimulated vascular smooth muscle cells, *Clin. Exp. Pharmacol. Physiol.,* 18, 543, 1991.

115. **Panet, R., Amir, I., Snyder, D., Zonenshein, L., Atlan, H., Laskov, R., and Panet, A.,** Effect of Na^+ flux inhibitors on induction of c-fos, c-myc, and ODC genes during cell cycle, *J. Cell. Physiol.,* 140, 161, 1989.

116. **Shibanuma, M., Kuroki, T., and Nose, K.,** Inhibition of proto-oncogene c-*fos* transcription by inhibitors of protein kinase C and ion transport, *Eur. J. Biochem.,* 164, 15, 1987.

117. **van der Valk, J., Verlaan, I., de Laat, S. W., and Moolenaar, W. H.,** Expression of pp60[v-src] alters the ionic permeability of the plasma membrane in rat cells, *J. Biol. Chem.,* 262, 2431, 1987.

118. **Doppler, W., Jaggi, R., and Groner, B.,** Induction of v-*mos* and activated Ha-*ras* oncogene expression in quiescent NIH 3T3 cells causes intracellular alkalinisation and cell cycle progression, *Gene,* 54, 147, 1987.

119. **Maly, K., Überall, F., Loferer, H., Doppler, W., Oberhuber, H., Groner, B., and Grunicke, H. H.,** Ha-*ras* activates the Na$^+$/H$^+$ antipaorter by a protein kinase C-independent mechanism, *J. Biol. Chem.,* 264, 11839, 1989.

120. **LeBoeuf, R. A. and Kerckaert, G. A.,** The induction of transformed-like morphology and enhanced growth in Syrian hamster embryo cells grown at acidic pH, *Carcinogenesis,* 7, 1431, 1986.

121. **Ober, S. S. and Pardee, A. B.,** Intracellular pH is increased after transformation of Chinese hamster embryo fibroblasts, *Proc. Natl. Acad. Sci. U.S.A.,* 84, 2766, 1987.

122. **Rotin, D., Steele-Norwood, D., Grinstein, S., and Tannock, I.,** Requirement of the Na$^+$/H$^+$ exchanger for tumor growth, *Cancer Res.,* 49, 205, 1989.

123. **Ladoux, A., Cragoe, E. J., Jr., Geny, B., Abita, J. P., and Frelin, C.,** Differentiation of human promyelocytic HL 60 cells by retinoic acid is accompanied by an increase in the intracellular pH: the role of the Na$^+$/H$^+$ exchange system, *J. Biol. Chem.,* 262, 811, 1987.

124. **Costa-Casnellie, M. R., Segel, G. B., Cragoe, E. J., Jr., and Lichtman, M. A.,** Characterization of the Na$^+$/H$^+$ exchanger during maturation of HL-60 cells induced by dimethyl sulfoxide, *J. Biol. Chem.,* 262, 9093, 1987.

125. **Arcangeli, A., Ricupero, L., and Olivotto, M.,** Commitment to differentiation of murine erythroleukemia cells involves a modulated plasma membrane depolarization through Ca^{2+}-activated K$^+$ channels, *J. Cell. Physiol.,* 132, 387, 1987.

126. **Lagarde, A. E., Franchi, A. J., Paris, S., and Pouysségur, J. M.,** Effect of mutations affecting Na$^+$:H$^+$ antiport activity on tumorigenic potential of hamster lung fibroblasts, *J. Cell. Biochem.,* 36, 249, 1988.

127. **Bierman, A. J., Tertoolen, L. G. J., de Laat, S. W., and Moolenaar, W. H.,** The Na$^+$/H$^+$ exchanger is constitutively activated in P19 embryonal carcinoma cells, but not in a differentiated derivative: responsiveness to growth factors and other stimuli, *J. Biol. Chem.,* 262, 9621, 1987.

128. **Wacker, W. E. C. and Parisi, A. F.,** Magnesium metabolism, *N. Engl. J. Med.,* 278, 712 and 772, 1968.

129. **Ebel, H. and Günther, T.,** Magnesium metabolism: a review, *J. Clin. Chem. Clin. Biochem.,* 18, 257, 1980.

130. **Günther, T.,** Biochemistry and pathobiochemistry of magnesium, *Artery,* 9, 167, 1981.

131. **Juan, D.,** The clinical importance of hypomagnesemia, *Surgery,* 91, 510, 1982.

132. **Günther, T.,** Functional compartmentation of intracellular magnesium, *Magnesium,* 5, 53, 1986.

133. **Walker, G. M.,** Magnesium and cell cycle control: an update, *Magnesium,* 5, 9, 1986.

134. **Cameron, I. L. and Smith, N. K. R.,** Cellular concentration of magnesium and other ions in relation to protein synthesis, cell proliferation and cancer, *Magnesium,* 8, 31, 1989.

135. **Romani, A. and Scarpa, A.,** Regulation of cell magnesium, *Arch. Biochem. Biophys.,* 298, 1, 1992.

136. **Theophanides, T. and Polissiou, M.,** Magnesium-nucleic acid conformational changes and cancer, *Magnesium,* 5, 221, 1986.

137. **Romani, A. and Scarpa, A.,** Hormonal control of Mg^{2+} transport in the heart, *Nature,* 346, 841, 1990.

Chapter 4

Cyclic Nucleotides

I. INTRODUCTION

Cyclic nucleotides are purine derivative compounds that have a key role as second messengers in the mechanism of action of externally signaling agents including hormones, growth factors, regulatory peptides, and neurotransmitters. Two cyclic nucleotides, cyclic adenosine 3′,5′-monophosphate (cyclic AMP or cAMP) and cyclic guanine 3′5′-monophosphate (cyclic GMP or cGMP), are involved in such functions, but the first of them, cAMP, is considered to be the most important cyclic nucleotide second messenger in mammalian cells.[1-6] The possible role of a third cyclic nucleotide, cyclic cytidine 3′,5′-monophosphate (cCMP), in the regulation of cellular functions by external stimuli is controversial.[7]

II. CYCLIC AMP AND THE ADENYLYL CYCLASE SYSTEM

cAMP was discovered during a study on the effects of glycogenolytic hormones (glucagon and epinephrine) in liver slices and homogenates, which were associated with the accumulation of a heat-stable, dialyzable adenosine nucleotide.[8] This molecule, in the presence of ATP, Mg^{2+}, and a cytoplasmic enzyme, was able to convert liver phosphorylase from an inactive precursor to an active form.[9] The active substance, identified as cAMP, was shown to be biosynthesized from ATP by the action of a specific enzyme, adenylyl cyclase, which is an integral component of the plasma membrane.[10] Methyl xanthines such as caffeine, aminophylline, and theophylline, can prevent the conversion of cAMP to an inactive noncyclic metabolite, 5′ adenosine monophosphate (5′ AMP), which is catalyzed by cAMP phosphodiesterase.[11] Thus, by prolonging the survival of cAMP, methyl xanthines produce physiological effects that resemble those of many hormones that activate adenylyl cyclase. It was further demonstrated that the activation of glycogen phosphorylase by cAMP depends on the stimulation of a protein kinase, the cAMP-dependent protein kinase (protein kinase A), which transfers the terminal phosphate of ATP to specific serine and/or threonine residues in the enzyme undergoing activation.[12] Thus, the manyfold cellular effects of cAMP are associated with the phosphorylation of specific proteins and are mediated by activation of a cAMP-dependent protein kinase.

cAMP has a central role in the regulation of cell proliferation and cell differentiation and is a key mediator in the molecular mechanisms of action of hormones, growth factors, regulatory peptides, neurotransmitters, and other external stimuli. The intracellular concentration of cAMP depends on an equilibrium between the activity of adenylyl cyclase and phosphodiesterase. Hormones and growth factors can alter this equilibrium mainly through modification of adenylyl cyclase activity, although influences on phosphodiesterase may also occur. Multiple molecular forms of phosphodiesterases exist, and some of them have been characterized according to substrate specificity, sensitivity to stimulation by cGMP or calmodulin, response to various selective and nonselective phosphodiesterase inhibitors, and tissue and organ distribution.[13] The activity of some cyclic nucleotide phosphodiesterases is regulated by the Ca^{2+}/calmodulin system, depending on the formation of stable complexes between the enzymes and calmodulin.[14,15]

cAMP is a universal second messenger that can influence a wide diversity of biochemical and physiological phenomena. Specificity of the cellular response to the activation of the adenylyl cyclase system and the increased intracellular concentration of cAMP is determined not by the relatively unspecific cAMP molecule but by the type of cell surface receptor involved in the activation of the system and the particular endogenous substrate for cAMP-dependent phosphorylation. In turn, both the receptors expressed by a given cell and the substrates available for phosphorylation depend on the specific genetic program of the cell.

A. THE ADENYLYL CYCLASE SYSTEM

The adenylyl cyclase (adenylate cyclase) system consists of three distinct units: (1) the hormone receptor (R component), which is located in the outer plasma membrane and contains a specific site for a given extracellular signaling agent; (2) the catalytic moiety (C component) of adenylyl cyclase, which is located on the inner face of the plasma membrane and bears the site responsible for catalysis of the cyclizing reaction; and (3) the guanine nucleotide-binding regulatory subunit (G component), which is located mainly within the plasma membrane.[16] The catalytic subunit/C component of adenylyl cyclase purified from rabbit myocardial membranes has a molecular weight of 150 kDa.[17] The C subunit purified from brain has been identified as a calmodulin-binding protein of 135 kDa.[18]

Guanosine triphosphate (GTP) exerts three actions in the adenylyl cyclase system: (1) GTP stimulates basal activity in the absence of hormone; (2) GTP accelerates the rate of hormone binding to and release from its receptor; and (3) GTP promotes coupling of receptors to adenylyl cyclase.[16] The system is regulated by GTPase activity, and hormone-induced stimulation of the system is associated with an exchange process between GTP and guanosine diphosphate (GDP), which promotes dissociation of the inhibitory effects of GDP and allows association of the stimulatory GTP. An intrinsic GTPase activity is responsible for the turn off of adenylyl cyclase activity and the formation of an inactive enzyme-GDP complex.

Analysis of cDNAs derived from a bovine brain library and encoding for adenylyl cyclase indicates that most of the deduced amino acid sequence of 1134 residues is divisible into two alternating sets of hydrophobic and hydrophilic domains and that each of the two large hydrophobic domains contains six transmembrane span regions.[19] The two large hydrophilic domains of adenylyl cyclase contain a sequence that is homologous to a cytoplasmic domain of guanylyl cyclases and that may represent the nucleotide-binding sites of the enzyme. A topographical resemblance was found between adenylyl cyclase and plasma membrane channels and transporters.

The cAMP analog, dibutyryl cAMP, has been extensively used for testing the effects of elevated intracellular levels of cAMP in normal and transformed cells. The diterpene compound, forskolin, increases cAMP levels through the stimulation of adenylyl cyclase activity in different types of cells by a mechanism that is not well understood but is distinct from that of hormones, growth factors, and other regulators of the enzyme. The study of forskolin-resistant mutants isolated from the ACTH-responsive adrenocortical cell line Y1 and other cell types may help to clarify the mechanisms of forskolin action.[20] In the mutant Y1 cells, forskolin-stimulated adenylyl cyclase activity is diminished. A mutant adenylyl cyclase, which is insensitive to ACTH, appears to be responsible for the resistance to forskolin in Y1 cells.

1. The Adenylyl Cyclase Family

There are different isoforms of adenylyl cyclase. Polyclonal antibodies against the C subunit of calmodulin-sensitive adenylyl cyclase from bovine brain distinguish between calmodulin-sensitive and -insensitive forms of the enzyme.[21] Calmodulin-sensitive adenylyl cyclase activity is found predominantly in the brain, whereas liver and other peripheral tissues contain only the calmodulin-insensitive type of activity. Two forms of hormone-responsive adenylyl cyclase have been identified in the rat liver, the less abundant of them being linked to glycogenolysis.[22] Glucagon predominantly stimulates the glycogenolysis-linked form of the enzyme. Recently, six distinct forms of mammalian adenylyl cyclases (types 1 to 6) have been identified by molecular cloning analysis.[23] The activities of these isoforms of adenylyl cyclase are stimulated by the stimulatory guanine nucleotide-binding regulatory protein, G_s. The isoenzymes may be involved in differential modulation of signal transduction in different types of cells. Among the widely distributed forms of the enzyme, type 4 has no predicted protein kinase A sites, whereas types 5 and 6 are not stimulated by $\beta\gamma$ subunits of G proteins. The multiplicity of effectors in a single signaling pathway such as that associated with adenylyl cyclase activity increases the complexity of signal transduction processes at the cell surface.

2. Regulation of the Adenylyl Cyclase System

The guanine nucleotides, GDP and GTP, and the guanine nucleotide-binding proteins (G proteins) have an essential role in the regulation of the activity of the adenylyl cyclase system. G proteins function as transducers of information across the cell membrane by coupling receptors to effectors. They may act as either stimulators or inhibitors of the adenylyl cyclase system, according to the type of α subunit contained in the trimeric G protein. While G_s proteins mediate stimulation of adenylyl cyclase, G_i proteins mediate inhibition of the system. However, the existence of various forms of G protein α subunits, as well as β and γ subunits, gives further diversity to the regulatory action of G proteins in adenylyl cyclase and other effector systems, including enzymes and ion channels.[24-27]

Hormone and growth factor receptors activated by ligand binding on the cell surface function as exchange factors, enabling GDP tightly bound to the G protein to be released, thus permitting the activating GTP to be bound to the α subunit and the dissociation of this subunit from the $\beta\gamma$ subunit complex. G protein activation is terminated by a GTPase intrinsic to the α subunit, which determines the conversion of active GTP to inactive GDP and its binding to the trimeric G protein (Figure 4.1).

In addition to the regulatory action of guanine nucleotides, the activity of the adenylyl cyclase system is regulated by other influences, in particular by the Ca^{2+}/calmodulin system.[28] The Ca^{2+}-calmodulin complex may activate either adenylyl cyclase or phosphodiesterase, thus regulating the intracellular concentration of cAMP. Many actions of the Ca^{2+}-calmodulin system are exerted through activation of phosphorylation/dephosphorylation processes regulated by particular protein kinases, and these modifications may be involved in regulating the activity of the catalytic (C) component of the adenylyl cyclase system.[29]

Regulation of the intracellular levels of cAMP by hormones and growth factors may be exerted not only at the level of its synthesis by adenylyl cyclase but also at the level of its degradation by cAMP phosphodiesterases. Both cAMP itself and hormones regulate the expression of the mRNA for a high-affinity cAMP phosphodiesterase in Sertoli cells and in a glioma cell line.[30] The Ca^{2+}/calmodulin system is also involved in regulation of cAMP phosphodiesterase activity. Changes in $[Ca^{2+}]_i$ levels contribute to regulate cAMP concentrations through activation or inactivation of a calmodulin-sensitive phosphodiesterase.[31] There may be different Ca^{2+}/calmodulin-dependent phosphodiesterases and these enzymes may show differences in tissue distribution. A Ca^{2+}/calmodulin-sensitive phosphodiesterase detected in the RPMI 8392 human lymphoblastoid B-cell line is not present in normal human peripheral blood lymphocytes.[32] Two major Ca^{2+}/calmodulin-dependent phosphodiesterase isozymes of 60 and 63 kDa have been detected in the bovine brain.[15] In turn, the activity of the phosphodiesterases may be mediated by phosphorylation dependent on protein kinases which may result in a decrease in the affinity of the enzyme toward calmodulin. It is thus apparent that complex interactions may occur *in vivo* between the cellular regulatory systems mediated by cyclic nucleotides and Ca^{2+}/calmodulin.

3. Role of cAMP in Hormone and Growth Factor Action

The intracellular levels of cAMP are critically involved in the control of cell proliferation and differentiation. The mitogenic response of cells to hormones and growth factors may be influenced by the intracellular concentrations of cAMP,[33] which results in changes in the activities of cAMP-dependent protein kinases. However, the precise role of cAMP in mitogenic responses is not clear. Depending on the cell type and the particular physiological conditions, the intracellular levels of cAMP may have variable and even opposite effects on cell proliferation and differentiation. Whereas in yeast cells cAMP exerts a positive control on cell proliferation, in multicellular animals there may be an inverse relationship between the levels of cAMP and the rate of cell proliferation. Treatment of the ALL-derived B precursor cell line Reh with forskolin results in greatly augmented concentrations of intracellular cAMP, and this effect is associated with reduced proliferation and accumulation of the Reh cells in the G_1 phase of the cycle, but without appearance of specific differentiation markers.[34] The positive or negative effects of cAMP on the proliferation of normal and transformed cells may be associated with changes in the levels of expression of proto-oncogenes such as c-*myc* and c-H-*ras*. Regulation of hepatocyte proliferation by cAMP may occur by facilitating the traverse of G_0 or early G_1 cells through the prereplicative period and by exerting an inhibitory effect at the R point shortly before the G_1/S border.[35]

cAMP cooperates with a wide variety of hormones and growth factors to synergistically stimulate the proliferation of different types of vertebrate cells. The mechanism of this effect is not clear but may include the induction of growth factor receptors.[36] Another mechanism, associated with the cellular action of hormones such as insulin and glucagon, consists of cAMP-mediated regulatory changes in the expression of genes coding for specific enzymes.[37]

4. Role of cAMP in the Regulation of Gene Expression

cAMP is present in both animals and plants and is involved in the regulation of gene expression.[38] It functions as an important regulator of transcription in prokaryotes, acting through its binding to a specific receptor protein, the catabolite-gene activator protein (CAP). The cAMP-CAP complex binds to specific bacterial DNA sequences and thus can enhance the rate of transcription of several operons. As for bacteria, cAMP is involved in the regulation of gene expression in eukaryotes, where it may act as either an activator or an inhibitor of transcription.[39,40] Expression of cAMP-responsive genes in higher eukaryotes is controlled by proteins that form complexes with cAMP and then display sequence-specific DNA-binding properties. cAMP response elements (CREs) are contained in the promoters of eukaryotic cAMP-inducible genes, and the CREs are recognized by CRE-binding proteins.[41,42] Nuclear extracts from rat liver contain a cAMP-dependent factor capable of specific binding to a synthetic fragment of 46 bp whose

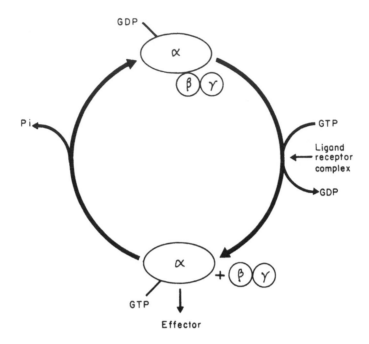

Figure 4.1 Activation and inactivation of a receptor-coupled G protein.

sequence corresponds to that of the 5′ flanking region of a gene known to contain the element that confers regulation of transcription of this gene by cAMP.[43] The 5′-flanking region of the rat corticotropin-releasing hormone (CRH) gene contains a CRE that is localized to the DNA sequence from −238 to −180 bp relative to the putative CRH mRNA cap site.[44] The CRE of the rat CRH gene is conserved at 58/59 bases in the human genome, suggesting an important role for this CRE in the regulation of gene expression.

The CREs are recognized by CRE-binding proteins (CREBPs or CREBs) which are substrates for cAMP-dependent protein kinase (protein kinase A).[42] All CREBs belong to a family of proteins that contain a basic region/leucine zipper that is responsible for DNA binding as well as for heterodimerization with other proteins involved in the control of gene transcription. The CREBs ATF-2, ATF-3, and ATF-4 can heterodimerize with the Fos and Jun oncoproteins. Other members of this class of proteins are the cAMP-responsive element modulators (CREMs), which include negative regulators of cAMP-induced transcription. It is thus clear that transcriptional regulation of cAMP-responsive genes involves a delicate balance between different types of activators and repressors of CRE-containing promoters. All of the proteins involved in these complex regulatory processes show extensive homologies in their basic region/leucine zipper regions.

In addition to protein kinase A, members of the CREB family may be substrates for other kinases such as Ca^{2+}/calmodulin-dependent-kinase II.[45,46] Thus, CREB proteins may represent a convergence point for two different second messenger systems acting on gene transcription: the Ca^{2+}/calmodulin system and the cAMP system. Regulation of the transcription of specific gene subsets by cAMP may occur by two independent mechanisms. Stimulation of growth hormone gene transcription and phosphorylation of a 19-kDa nuclear protein by cAMP occurs through the direct activation of a cAMP-dependent protein kinase, whereas stimulation of prolactin gene transcription by cAMP would reflect the activation of a calcium-dependent event.[47] Recent evidence suggests that a specific type of CREM, CREM-tau, may have an important physiologic role in the regulation of spermatogenesis.[48]

B. cAMP-DEPENDENT PROTEIN KINASES

The manyfold physiologic effects of cAMP in eukaryotic cells are mediated exclusively through activation of the cAMP-dependent protein kinase (protein kinase A or A-kinase).[49-53] The kinase holoenzyme is tetrameric and exists as an inactive complex of two regulatory (R) and two catalytic (C) subunits (R_2C_2). The R subunit contains two cAMP-binding sites, a dimer interaction site, and an autophosphorylation site. In addition, the holoenzyme contains interaction sites between the R and C subunits. Like most protein kinases, the function of the cAMP-dependent protein kinase depends on a regulatory mechanism and the enzyme is maintained in an inactive form in the absence of cAMP. When cAMP is generated by the action of hormones, growth factors, or other stimuli, it binds with a high affinity to the R subunit, and the complex dissociates to liberate active C subunits which appear to be identical. Under normal physiological conditions, this leads to dissociation of the holoenzyme into an R_2-$(cAMP)_4$ dimer and two catalytically active free C subunits. Crystal structure analysis of the C subunit at a resolution of 2.7 Å showed that the enzyme is bilobal with a deep cleft between the lobes.[54] While the smaller lobe is associated with nucleotide binding, the larger lobe is primarily involved in peptide binding and catalysis.

Two distinct isoenzymes of cAMP-dependent protein kinase exist in mammalian cells, the type I and type II kinases, which differ in their R subunits. Differential expression of these isoenzymes may be linked to the growth regulatory effects of cAMP. A mixture of the type I and type II cAMP-dependent protein kinases exists in most tissues, suggesting that selective modulation of the two isoenzymes may be crucial for the different activities of cAMP in normal and transformed cells. However, the relative content of the type I and type II isoenzymes varies among different adult tissues as well as in ontogeny and differentiation, and the general physiologic conditions and hormonal status of the animal may contribute to regulating the expression of the two isoenzymes. The use of site-selective cAMP analogs with specificity for either type I or type II protein kinase A may contribute to a better definition of the role of cAMP in processes related to cell proliferation and differentiation.[55] In contrast to nonsite-specific cAMP analogs such as dibutyryl-cAMP, site-selective cAMP analogs display inhibitory effects at micromolar concentrations and without signs of toxicity.

1. C Subunits of cAMP-Dependent Protein Kinases

The catalytic core of all protein kinases shows extensive sequence homologies that include the ATP-binding domain as well as regions thought to be important in catalysis. The complete amino acid sequence of the C subunit of bovine cardiac muscle protein kinase A has been determined,[56] and a full-length cDNA for the C subunit of the mouse enzyme has been cloned and sequenced.[57] The amino-terminal glycine present in the bovine C subunit is conserved in the mouse sequence. This amino acid residue is myristoylated both in the C subunit of the cAMP-dependent protein kinase and the Src oncoprotein, but, in contrast to Src, myristoylation of this residue in the C subunit of the enzyme does not appear to localize it to the cellular membranes. Certain mutations in the C subunit of cAMP-dependent protein kinase may prevent myristoylation, but the mutant enzyme is fully functional in its ability to interact with specific proteins in different locations of the cell.[58]

Although it has been generally accepted that there is only one type of C subunit of cAMP-dependent protein kinase, screening of a bovine pituitary cDNA library using synthetic oligonucleotides predicted from the known amino acid sequence of the C subunit revealed the existence of two different genes coding for the C subunit.[59] The encoded subunits, C-α and C-β, are 93% identical in primary structure but exhibit differences in tissue distribution. Furthermore, C-α is present in higher quantities than C-β in most tissues, and the substrate specificity or subcellular localization of C-α may differ from that of C-β. The functional differences between protein kinases carrying C-α or C-β are unknown.

2. R Subunits of cAMP-Dependent Protein Kinases

Unique R subunits account for the various holoenzyme forms of cAMP-dependent protein kinase. The type I and type II forms of the kinase are distinguished according to differences in the R components, functional properties, and tissue-specific functional activities.[60,61] Each R subunit of the type I protein kinase A isoenzyme (RI subunit) and the type II protein kinase A isoenzyme (RII subunit) contains two binding sites for cAMP that can be differentiated on the basis of their cAMP dissociation rates and their specificities for cAMP analogs. The RI and RII subunits are phosphorylated *in vivo*. The effect of cAMP on phosphorylation of the R subunit is biphasic, being stimulating at low concentrations and inhibiting at high concentrations.[62] The only established function of the R subunit of cAMP-dependent protein kinase is that of inhibiting the C subunit of the kinase. However, an additional function of R may include inhibition of protein phosphatase activity, especially when cAMP level is elevated.[63]

Expression of the R and C subunits of cAMP-dependent protein kinases depends on the growth state in both normal and tumor cells. The proportion between R and C is usually close to 1:1 in normal tissue. In the rat uterus, as well as in estrogen-dependent DMBA-induced mammary adenocarcinomas, expression of RI, RII, and C subunits depends on the growth state.[64] An increased R/C ratio correlates positively with proliferation of both normal rat uterine cells and mammary carcinoma cells. The turnover rates of the two cAMP-dependent protein kinases are different, the type I isoenzyme being renewed much faster than the type II form, and this difference depends on a much shorter half-life of RI as compared to RII.[65]

The different turnover of the two isoenzymes could be related to different functions, such as modulation of enzyme activity vs. modulation of gene expression. The R subunits of mammalian cAMP-dependent protein kinases show homology in amino acid sequence, as well as in secondary and tertiary structure, with the CAP protein from *Escherichia coli*, a protein which has a DNA-binding domain.[66] Type II protein kinase A may specifically bind single-strand DNA through the R subunit of the kinase, which may be related to regulation of gene expression.[67]

There is evidence for molecular heterogeneity of R subunits. A cDNA clone for the RI-α subunit of cAMP-dependent protein kinase isolated from a human testis cDNA library encoded a protein of 381 amino acids that shows very high homology to bovine skeletal muscle RI and rat brain RI.[68] Two major mRNA species of RI are found in human testis, but only one mRNA species exists in human T lymphocytes. A cDNA encoding the RI-β subunit was also isolated and characterized from a human testis cDNA library.[69] The human RI-β protein is composed of 418 amino acids and has a molecular weight of 53,856. The highest levels of RI-β mRNA were found in the testis, and significant levels were found in fallopian tubes and ovary.

Differentiation of MEL cells, as well as treatment of these cells with a cAMP analog (8-bromo-cAMP), elicit a large and selective increase in the rate of biosynthesis of only one type of RII subunit.[70] The rat has at least three genes coding for R subunits of cAMP-dependent protein kinase, one gene for RI and two for RII (RII-α and RII-β).[71-73] The RII isoform is present and regulated by hormones (FSH and estradiol) in rat ovarian granulosa cells. A cDNA encoding the hormonally regulated RII-β subunit isoform was cloned from these cells and its structure was determined, thus allowing the deduction of the amino acid sequence (415 residues) of the enzyme.[74] A cAMP-resistant CHO cell line containing both wild-type and mutant species of RI has been characterized.[75] It is possible that the two forms of RI detected in these cells occurred by mutation in one of the two alleles of a single gene coding for RI. Studies on the functional organization and activity of cAMP-dependent protein kinase R subunits may be facilitated by the collection of a series of mutants from sublines of S49 mouse lymphoma cells that are hemizygous for expression of the R subunit and possess structural lesions in the gene for the R subunit.[76]

3. Regulatory Actions of cAMP-Dependent Protein Kinases

Activation of cAMP-dependent protein kinases may have important effects at various intracellular levels, including the membrane, the cytoplasm, the mitochondria, the lysosomes, the microtubules, and the nucleus. Unfortunately, the substrates responsible for such pleiotropic effects of the protein kinase remain poorly characterized. At the cell membrane level, some components of the membrane skeleton, acting as accessory proteins of the spectrin-actin complex, may be phosphorylated by cAMP-dependent protein kinases.[77] Protein components of microfilaments may be substrates for cAMP-dependent kinases. Phosphorylation of muscle and nonmuscle actin by cAMP-dependent protein kinase occurs only on serine and decreases actin polymerizability.[78] By contrast, protein kinase C phosphorylates actin filaments on both serine and threonine and increases actin polymerizability. These changes may be associated with important effects in the function of microfilaments. At the nuclear level, phosphorylation and dephosphorylation of histone and nonhistone proteins are most important for the regulation of gene expression.[79] cAMP-dependent protein kinases may have a direct role in the regulation of gene expression in mammalian cells.[80,81] Both RI and RII regulatory subunits of the cAMP-dependent protein kinase isoenzymes are associated with chromatin fractions that are enriched in transcriptionally active DNA. The mechanism of the regulatory action of cAMP-dependent protein kinases on gene expression is associated with the action of CREB-related proteins.[42] The identification of CREs in the promoters of cAMP-inducible genes provided a link between protein kinase A activation and cAMP-induced gene expression. CREBs have been identified as substrates for cAMP-dependent protein kinases, and phosphorylation of CREBs at specific serine residues may result in modification of their DNA-binding activities.

4. Regulation of cAMP-Dependent Protein Kinase Activity

The specific activity of cAMP-dependent protein kinases is regulated mainly through phosphorylation/dephosphorylation at serine/threonine residues. Various types of serine/threonine protein kinases and protein phosphatases may be involved in regulating the state of phosphorylation of protein kinase A. In the human colon carcinoma cell line HT-29, protein kinase C activation results in phosphorylation of the RII-α subunit isoform of cAMP-dependent protein kinase.[82] There is a crosstalk between PKA- and PKC-related intracellular signaling pathways.

Regulation of cAMP-dependent protein kinase is partially due to its specific interaction with high-affinity inhibitors. One of these inhibitors may have an important role in the regulation of transcriptional events associated with cAMP-dependent protein kinase activity in mammalian cells.[83] The use of synthetic genes that code for an active fragment of the kinase inhibitor indicates that an active C subunit of the protein kinase A is a necessary intermediate in the cAMP stimulation of gene transcription. The level of activity of a cAMP-dependent protein kinase inhibitor may be regulated by phosphorylation dependent on tyrosine kinase(s).[84] A potent synthetic peptide inhibitor of cAMP-dependent protein kinase has been constructed,[85] which may contribute to a better knowledge of the enzyme function.

Phosphorylation of different cellular substrates by cAMP-dependent and -independent kinases is associated with cell proliferation and differentiation. The expression of cAMP-dependent protein kinase subunits is differentially regulated during the prereplicative period of liver regeneration after partial hepatectomy in rats.[86] Cellular differentiation, either spontaneous or induced by agents of endogenous or exogenous origin, is associated with altered expression of cAMP-dependent protein kinases. The differentiation of HL-60 and RDFD human leukemia cells induced by retinoids is correlated to elevation of cAMP-dependent protein kinase activity, as well as to a marked stimulation of other kinases (protein kinase C and protamine kinase).[87] However, the precise role of different kinases in cell differentiation is not understood.

C. THE ADENYLYL CYCLASE SYSTEM AND ONCOGENES

There are interesting structural and functional relationships between the adenylyl cyclase system and oncogenes. Structural homology has been detected between the C subunit of mammalian cAMP-dependent protein kinase and the proteins of the *src* gene superfamily, including the v-Src oncoprotein itself. A significant degree of homology exists between the human Abl protein and the C subunit of bovine cAMP-dependent protein kinase, although the tyrosine acceptor residue is replaced by tryptophan in the bovine kinase.[88] This substitution is directly adjacent to a threonine known to represent one of two phosphorylation sites identified in the C subunit of the kinase. The products of other members of the *src* family (the v-Fes, v-Fps, and v-Yes oncoproteins) are more distantly related to the mammalian cAMP-dependent protein kinase.

A distant degree of homology exists between the C subunit of mammalian cAMP-dependent protein kinase and the oncoprotein v-Mos, which possesses kinase activity specific for serine/threonine residues. Both the C subunit of mammalian cAMP-dependent protein kinase and the v-Mos protein are structurally related to the product of the *CDC28* gene of *Saccharomyces cerevisiae*.[89] The product of the *CDC28* gene is involved in the control of yeast cell division. This product is a homolog to that of the *cdc2* gene of the fission yeast, *Schizosaccharomyces pombe*. A human homolog of the yeast *CDC28/cdc2* gene has been cloned and characterized.[90] As discussed in Chapter 1, the mammalian cdc2 protein plays a key role in cell cycle control.

Maturation of *Xenopus* oocytes is stimulated by insulin and IGF-I as well as by progesterone. Mutated c-H-Ras proteins accelerate this maturation through stimulation of cAMP phosphodiesterase.[91] Ras proteins may display a suppressive effect on cAMP-dependent protein kinase activity. Proteolytic cleavage of the R subunit of the kinase is inhibited by Ras proteins, which may be due to thiol proteinase-inhibitor activity of the oncoprotein.[92]

Proto-oncogene expression may be regulated by mechanisms involving cAMP-binding proteins. The c-*fos* gene contains CREs involved in the regulation of its expression.[93] The CRE is required for c-*fos* induction by increases in cAMP levels. Transcriptional activity of the c-*jun* proto-oncogene is repressed by the CREB protein, which binds to the AP-1 site present in the c-*jun* gene promoter.[94,95] The ability of CREB to repress transcription from the c-*jun* promoter extends to both serum and PMA induction. The repression is alleviated when CREB is phosphorylated on a specific serine residue by the activity of protein kinase A. Thus, CREB may have a dual function, one as a repressor of c-*jun* gene expression in the absence of phosphorylation and the other as an activator when it is phosphorylated by protein kinase A. Microinjection of antibody against CREB protein into fibroblasts diminishes gene expression in response to cAMP.[96] Agents that elevate the intracellular concentration of cAMP rapidly and transiently induce expression of the c-*fos* gene in different types of cells. Strong induction of c-*fos* gene expression with stable accumulation of c-*fos* mRNA occurs in macrophages treated with dibutyryl cAMP or exposed to cholera toxin.[97] Treatment of mouse 3T3 fibroblasts with the direct activator of adenylyl cyclase, forskolin, elicits marked increases in c-*fos* mRNA levels.[98] Similar effects can be obtained with receptor-mediated activation of adenylyl cyclase by PGE$_1$ and stimulation with the cAMP analog, 8-bromo-cAMP. Growth factor-induced changes in proto-oncogene expression may depend, at least in part, on activation of the adenylyl cyclase system. Induction of c-*fos* and c-*myc* mRNA by EGF or calcium ionophore is cAMP-dependent.[99] The mouse c-*fos* gene promoter-enhancer region contains multiple CREs that contribute to responsiveness to cAMP.[100] The regulation of c-*fos* gene expression by cAMP depends on cAMP-dependent protein kinase activation. Microinjection of the catalytic subunit of cAMP-dependent protein kinase into fibroblast cell lines of rat, mouse, and human origin results in increased expression of the Fos protein in the nucleus.[101] cAMP-dependent protein kinase is involved in controlling the activity of the c-Fos protein through its phosphorylation on specific serine residues, in particular Ser-362.[102] This phosphorylation would occur in the nucleus.

Synthetic cAMP analogs may exert remarkable effects on proto-oncogene expression. Treatment of PY815 mouse mastocytoma cells with dibutyryl cAMP results in a transient increase in the level of c-*myc* transcripts.[103] Site-selective cAMP analogs, which are potent activators of cAMP-dependent protein kinase, exert a strong inhibition of the estrogen-stimulated growth of MCF-7 human breast cancer cells and reduce the levels of Myc and Ras expression.[104] Treatment of HL-60 human promyelocytic leukemia cells with dibutyryl cAMP induces differentiation along the monocytic pathway, and this is preceded by transcriptional inactivation of the c-*myc* and transferrin receptor genes.[105] The inactivation is followed by loss of c-*myc* and transferrin receptor mRNA and protein. The treated cells complete one round of proliferation, followed by growth arrest and G_1 synchronization.

Elevated intracellular concentrations of cAMP may suppress the transcriptional expression of certain proto-oncogenes. Expression of the c-*sis*/PDGF-B gene in some human glioma and osteosarcoma cell lines is blocked by agents that increase the intracellular levels of cAMP.[106] Thrombin and TGF-β induce expression of c-*sis*/PDGF-B transcripts in human renal microvascular endothelial cells, and this induction can be blocked by adrenergic agonists such as isoproterenol and norepinephrine, as well as by the adenylyl cyclase activator forskolin.[107-109] However, the role of the adenylyl cyclase system in the regulation of proto-oncogene expression in not totally clear. Treatment of HL-60 cells with the β-adrenergic agonist isoproterenol results in increased levels of cAMP as well as in decreased expression of the c-*myc* and c-*fos* genes, but the reduction of c-*myc* and c-*fos* mRNA is not related to cAMP concentrations.[110]

The adenylyl cyclase system may have a role in the neoplastic transformation induced by either v-Ras proteins or mutant c-Ras proteins. The precise nature of this role is not clear, however. While nontransformed NRK cells show increased cAMP levels and decreased growth rate at confluency, the transformation of CEF cells by acute retroviruses such as RSV or K-MuSV is accompanied by a failure of adenylyl cyclase to rise in the growing cells.[111] An increase in adenylyl cyclase activity in early G_1 phase occurs in NRK cells infected with a *ts* mutant of K-MuSV when the temperature is shifted to activate the expression of the v-Ras oncoprotein.[112] In contrast, adenylyl cyclase activity is depressed in C127 mouse fibroblasts infected with wild-type K-MuSV but the activity increases when cells infected with a *ts* mutant of K-MuSV are shifted from the permissive temperature to the nonpermissive temperature.[113] Adenylyl cyclase activity is reduced in NIH/3T3 mouse cells expressing a mutant EJ c-Ras protein.[114] cAMP is able to suppress the expression of a v-H-*ras* oncogene linked to the MMTV promoter.[115] Treatment of H-MuSV-transformed NIH/3T3 cells with cAMP analogs may result in inhibition of both v-Ras protein synthesis and phenotypic transformation.[116] Transformation of BALB/3T3 cells with a mutant c-H-*ras* gene inhibits the adenylyl cyclase response to β-adrenergic agonist while possibly increasing muscarine receptor-dependent hydrolysis of inositol lipids.[117] However, the v-K-Ras oncoprotein can induce changes in growth, morphology, and the rate of collagen production independent of changes in cAMP levels.[118]

D. ROLE OF THE ADENYLYL CYCLASE SYSTEM IN NEOPLASIA

Studies related to cAMP concentrations in transformed cells have yielded conflicting results, with either an increase, a decrease, or no changes in the levels of cAMP in transformed cells, as compared to normal cells.[119-125] cAMP may have a role in the control of density-dependent cell proliferation *in vitro* as well as in the distribution of microfilaments and microtubuli, which are important determinants of the cell shape.[126-128] Loss of contact inhibition and changes in cell shape are among the most characteristic components of the transformed phenotype.

The effects of cAMP in the proliferation of normal and transformed cells may be different. Agents that increase the intracellular concentration of cAMP or calcium increase the proliferation of erythroid progenitors from normal cats *in vitro* but do not enhance or may rather inhibit the growth of progenitors from cats infected with FeLV retrovirus.[179] Treatment of cultured neoplastic cells (Reuber H35 hepatoma cells) or tumors growing *in vivo* (rat mammary carcinomas) with active cAMP derivatives may result in inhibition of the growth of neoplastic cells and regression of tumors.[130-132]

Treatment of a human salivary gland adenocarcinoma cell line with dibutyryl cAMP results in a reversible differentiation of the intercalated duct cells into myoepithelial cells with suppression of both anchorage-independent and -dependent colony growths.[133] Treatment of C1300 mouse neuroblastoma cells with dibutyryl cAMP results in inhibition of cell proliferation and in the appearance of neurite-like cellular processes.[134] Site-selective cAMP analogs can exert a potent inhibitory effect on the growth of a spectrum of human cancer cell lines.[135,136] The tumor cells treated with the cAMP analogs show a slowing down of cell cycle progression, which is paralleled by changes in cell morphology, an increase of the R-II cAMP receptor protein kinase, and a decrease of Ras protein expression.

Changes in cAMP may not be related directly to the origin or maintenance of the transformed phenotype. Some transformed cells may not require a functional endogenous cAMP system for cell cycling.[125] However, decreased levels of cAMP (usually as a consequence of decreased adenylyl cyclase activity rather than increased phosphodiesterase activity) are observed in different types of transformed cells, including epithelial cells or fibroblasts transformed by RNA viruses (K-MuSV and RSV) or DNA viruses (SV40 and polyoma virus).[137] On the other hand, increased intracellular concentrations of cAMP directly correlate with morphological transformation of certain types of cells, for example, in v-*mos*-transformed NRK cells.[138] In these cells, agents that elevate the endogenous levels of cAMP, such as PGE_1 and cholera toxin, stimulate an accumulation of cAMP that is paralleled by morphological signs of transformation.

Transformed cells may exhibit altered response to cAMP-mediated modulation of protein phosphorylation.[139] cAMP metabolism may also be involved in the complex processes of tumor cell spreading *in vivo*. Experimental metastasis from B16 murine melanoma cell clones correlates with higher levels of cAMP accumulation induced by melanocyte-stimulating hormone (MSH) or forskolin.[140] Although the physiological role of cAMP in the regulation of cellular growth remains unresolved,[141] it might be safe to conclude that cAMP can potentiate the effects of mitogenic factors.

Subtle differences may exist in the regulatory subunits of cAMP-dependent protein kinases in tumor tissues.[142] The hormone-dependent DMBA-induced rat mammary carcinoma is especially sensitive to the growth-inhibitory effect of cAMP, because tumor regression occurs under treatment with dibutyryl cAMP. However, a related tumor that is hormone-independent grows autonomously and fails to regress after ovariectomy or treatment with dibutyryl cAMP. Although the mechanisms responsible for such differences are not clear, there are subtle differences in the physicochemical properties of the RII subunits of cAMP-dependent protein kinase II between the two tumors. No such differences were observed in the RI subunits of the enzyme.

Regulation of adenylyl cyclase may be altered in oncogene-transformed cells. The response of adenylyl cyclase to forskolin, pertussis toxin, and cholera toxin is attenuated in rat thyroid cells transformed by the v-K-*ras* oncogene.[143] TSH, which acts on normal thyroid cells in culture as a growth factor by stimulating the adenylyl cyclase activity, is not able to induce DNA synthesis nor does it stimulate adenylyl cyclase in v-K-*ras*-transformed rat thyroid cells. However, it is not known whether the alteration of adenylyl cyclase observed in v-*ras*-transformed cells is a direct effect of the oncogene product or whether it occurs only as a consequence of the process of transformation itself.

Differentiation induced in neoplastic cells *in vitro* may be intimately associated with the generation of cAMP. Induction of differentiation of HL-60 human leukemic cells along the myeloid pathway is associated with a marked increase in adenylyl cyclase activity independent of the inducer used.[144] An increase in the intracellular levels of cAMP occurs after the addition of retinoic acid to HL-60 cells; this increase is not observed in a variant of HL-60 cells that are not able to differentiate. The initial rise in cAMP apparently is related to the initiation of the differentiation process.

Exposure of tumor cells *in vitro* to cAMP analogs may result in inhibition of cell growth and expression of a differentiated phenotype. Reversible differentiation of the HSG human salivary gland adenocarcinoma cell line into myoepithelial cells occurs in a medium containing dibutyryl cAMP.[145] These cells show elevated levels of cAMP, which are inversely related to their content of Ras protein. Treatment of human melanoma cell lines with dibutyryl cAMP also results in reversible induction of differentiation and reduction of the growth rate.[146] Site-selective cAMP analogs can induce growth inhibition and differentiation in the K-562 chronic myeloid leukemia cell line.[147] This effect correlates with differential expression of the genes for the type I and type II regulatory subunits of cAMP-dependent protein kinases as well as with a decrease in c-*myc* mRNA. cAMP receptor proteins may have a major role in regulating the cascade of events that affect expression of the differentiation program in leukemic cells.

III. CYCLIC GMP AND THE GUANYLYL CYCLASE SYSTEM

Cyclic guanosine monophosphate (guanosine 3':5'-monophosphate or cGMP) is generated from GTP by the activity of guanylyl cyclase.[148,149] The concentration of cGMP within the cell depends on the equilibrium existing between guanylyl cyclase and cGMP phosphodiesterase. The role of cGMP in the transductional mechanisms of hormone and growth factor action is not as clear as that of cAMP. In many systems, cGMP is not an essential second messenger in hormone action but may be considered as a modulator of this action.[150] Three major classes of cGMP receptor proteins have been identified in eukaryotic cells: cGMP serine/threonine kinases, cGMP-regulated ion channels, and cGMP-binding cyclic nucleotide phosphodiesterases.[151]

A. GUANYLYL CYCLASE

Guanylyl cyclase activity has been detected in all cell types and phyla. In most cells the enzyme exists as a polymorphic protein, with cytosolic (soluble) and membrane-associated (particulate) forms occurring in the same cell.[152] Guanylyl cyclase activity is an essential enzyme in cell signaling mechanisms, and its activity is regulated by intracellular and extracellular peptides and ions.[153-155] The membrane form of guanylyl cyclase can directly bind peptide ligands present in the extracellular space. Hormones and growth factors may contribute to the regulation of guanylyl cyclase activity through mechanisms that include the generation of cAMP.[156] The activity of cGMP phosphodiesterase is under separate genetic control from cAMP phosphodiesterase.[157,158] The multiple forms of guanylyl cyclase are regulated in a variety of ways, but their relative roles in different cellular systems are still little understood.

The amino acid sequence of guanylyl cyclase, as deduced from a cDNA cloned from sea urchin spermatozoa, predicts an intrinsic membrane protein of 986 amino acids, including an amino-terminal signal sequence.[159] A single transmembrane domain separates the protein into putative extracellular and cytoplasmic-catalytic domains. The cytoplasmic carboxyl terminal of 95 amino acids contains 20% serine, which represents regulatory sites for phosphorylation. The guanylyl cyclase protein exhibits structural homology to some members of the protein kinase family, including the PDGF receptor, the Fes tyrosine kinase, and the Mos serine/threonine kinase.

An increase in cGMP concentrations is one of the earliest events following binding of some peptide hormones and growth factors to their target cells. The recent identification and characterization of transmembrane receptors with intrinsinc guanylyl cyclase activity has shown that cGMP is a second messenger for natriuretic peptides.[160,161] Receptor guanylyl cyclases constitute an important family of proteins involved in the transmission of signals across the cell membrane. An endogenous peptide isolated from the intestine, guanylin, may be recognized by guanylyl cyclase C, a cell surface receptor for bacterially derived peptides known as heat-stable enterotoxins involved in acute secretory diarrheas.[162] The guanylin gene is expressed in various mammalian tissues in addition to intestine, suggesting an important role for the guanylin/guanylyl cyclase C receptor system in epithelial function.

B. cGMP AND CELL PROLIFERATION

The effects of cGMP on the control of the proliferation of normal or transformed cells are, as those of cAMP, difficult to evaluate.[163,164] High levels of guanylyl cyclase activity have been found in nonproliferating, differentiated tissues, including the cerebellum, where it is mainly located within the astrocytes.[165] High concentrations of cGMP have been found in the blood and urine of guinea pigs during the development of a transplantable leukemia.[166,167] In some systems an increase in the intracellular levels of cAMP is associated with inhibition of DNA synthesis and mitosis whereas cGMP may have opposite effects, acting in a positive manner in the mediation of cellular responses to mitogenic agents.[168-171] However, in other systems the situation is less clear and a general conclusion on this subject cannot be reached on the basis of the available data.

C. cGMP-DEPENDENT PROTEIN KINASE

The physiological actions of cGMP are mediated by a cGMP-dependent protein kinase that is present in a diversity of mammalian tissues.[172] Two interconvertible forms of cGMP-dependent protein kinase found in mammalian tissue extracts correspond to the cGMP-free and -bound forms of the enzyme.[173] cAMP- and cGMP-dependent kinases are homologous proteins, having many characteristics in common, but there are some differences between them, especially in the mechanisms of activation of the enzymes by cyclic nucleotides and the effects of different stimulatory and inhibitory modulators on their activity.

Unfortunately, the substrates of cGMP-dependent protein kinases at different cellular sites are, as those of the other protein kinases, little known. Atrial natriuretic factor (ANF), a circulating hormone of atrial origin involved in the regulation of renal excretion of Na^+, acts through a G protein-coupled receptor, which possesses intrinsic guanylyl cyclase activity, and its effects on the renal innermedullary collecting duct may be exerted by the inhibition of a ductal cation channel through a phosphorylation-dependent mechanism involving cGMP and cGMP-dependent protein kinase.[174] cGMP may stimulate the incorporation of phosphate into specific nuclear proteins of peripheral blood lymphocytes subjected to proliferative stimuli, whereas cAMP would exert an opposing effect.[175] cGMP may be associated with the regulation of transcriptional processes such as those occurring in the polytene chromosomes of *Drosophila melanogaster* after heat-shock treatment.[176] It has been suggested that a protein kinase modulator may participate in cGMP stimulation of histone phosphorylation processes occurring in the liver,[177] but the hypothetical modulator was not characterized.

REFERENCES

1. **Sutherland, E. W.,** Studies on the mechanism of hormone action, *Science,* 177, 401, 1972.
2. **Bitensky, M. W., Keirns, J. J., and Freeman, J.,** Cyclic adenosine phosphate and clinical medicine, *Am. J. Med. Sci.,* 266, 320, 1973.
3. **Catt, K. J. and Dufau, M. L.,** Basic concepts of the mechanism of action of peptide hormones, *Biol. Reprod.,* 14, 1, 1976.
4. **Earp, H. S. and Steiner, A. L.,** Compartmentalization of cyclic nucleotide-mediated hormone action, *Annu. Rev. Pharmacol. Toxicol.,* 18, 431, 1978.
5. **Dumont, J. E., Jauniaux, J.-C., and Roger, P. P.,** The cyclic AMP-mediated stimulation of cell proliferation, *Trends Biochem. Sci.,* 14, 67, 1989.
6. **Krebs, E. G.,** Role of the cyclic AMP-dependent protein kinase in signal transduction, *JAMA,* 262, 1815, 1989.
7. **Anderson, T. R.,** Cyclic cytidine 3′,5′-monophosphate (cCMP) in cell regulation, *Mol. Cell. Endocrinol.,* 28, 373, 1982.
8. **Rall, T. W., Sutherland, E. W., and Berthet, I.,** Effect of epinephrine and glucagon on the reactivation of phosphorylase in liver homogenates, *J. Biol. Chem.,* 224, 463, 1957.
9. **Sutherland, E. W. and Rall, T. W.,** Fractionation and characterization of a cyclic adenine ribonucleotide formed by tissue particles, *J. Biol. Chem.,* 232, 1077, 1957.
10. **Rall, T. W. and Sutherland, E. W.,** Formation of a cyclic adenine ribonucleoside by tissue particles, *J. Biol. Chem.,* 232, 1065, 1957.
11. **Butcher, R. W. and Sutherland, E. W.,** Purification and properties of cyclic 3′,5′-nucleotide phosphodiesterase and use of this enzyme to characterize adenosine 3′,5′ phosphate in human urine, *J. Biol. Chem.,* 237, 1244, 1962.
12. **Walsh, D. A., Perkins, J. P., and Krebs, E. G.,** An adenosine 3′,5′-monophosphate dependent protein kinase from rabbit, *J. Biol. Chem.,* 243, 3763, 1968.
13. **Weishaar, R. E., Kobylarz-Singer, D. C., and Kaplan, H. R.,** Subclasses of cyclic AMP phosphodiesterase in cardiac muscle, *J. Mol. Cell. Cardiol.,* 19, 1025, 1987.
14. **Sharma, R. K. and Wang, J. H.,** Purification and characterization of bovine lung calmodulin-dependent cyclic nucleotide phosphodiesterase: an enzyme containing calmodulin as a subunit, *J. Biol. Chem.,* 261, 14160, 1986.
15. **Sharma, R. K. and Wang, J. H.,** Regulation of cAMP concentration by calmodulin-dependent cyclic nucleotide phosphodiesterase, *Biochem. Cell Biol.,* 64, 1072, 1986.
16. **Abramowitz, J., Iyengar, R., and Birnbaumer, L.,** Guanyl nucleotide regulation of hormonally-responsive adenylyl cyclases, *Mol. Cell. Endocrinol.,* 16, 129, 1979.
17. **Pfeuffer, E., Dreher, R.-M., Metzger, H., and Pfeuffer, T.,** Catalytic unit of adenylate cyclase: purification and identification by affinity crosslinking, *Proc. Natl. Acad. Sci. U.S.A.,* 82, 3086, 1985.
18. **Coussen, F., Haiech, J., d'Alayer, J., and Monneron, A.,** Identification of the catalytic subunit of brain adenylate cyclase: a calmodulin binding protein of 135 KDa, *Proc. Natl. Acad. Sci. U.S.A.,* 82, 6736, 1985.
19. **Krupinski, J., Coussen, F., Bakalyar, H. A., Tang, W.-J., Feinstein, P. G., Orth, K., Slaughter, C., Reed, R. R., and Gilman, A. G.,** Adenylyl cyclase amino acid sequence: possible channel- or transporter-like structure, *Science,* 244, 1558, 1989.

20. **Schimmer, B. P., Tsao, J., Borenstein, R., and Endrenyi, L.,** Forskolin-resistant Y1 mutants harbor defects associated with the guanyl nucleotide-binding regulatory protein, G_s, *J. Biol. Chem.,* 262, 15521, 1987.

21. **Rosenberg, G. B. and Storm, D. R.,** Immunological distinction between calmodulin-sensitive and calmodulin-insensitive adenylate cyclases, *J. Biol. Chem.,* 262, 7623, 1987.

22. **Yamatani, K., Sato, N., Wada, K., Suda, K., Wakasugi, K., Ogawa, A., Takahashi, K., Sasaki, H., and Hara,** Two types of hormone-responsive adenylate cyclase in the rat liver, *Biochim. Biophys. Acta,* 931, 180, 1987.

23. **Premont, R. T., Chen, J., Ma, H.-W., Ponnapalli, M., and Iyengar, R.,** Two members of a widely expressed subfamily of hormone-stimulated adenylyl cyclases, *Proc. Natl. Acad. Sci. U.S.A.,* 89, 9809, 1992.

24. **Freissmuth, M., Casey, P. J., and Gilman, A. G.,** G proteins control diverse transmembrane signaling, *FASEB J.,* 3, 2125, 1989.

25. **Birnbaumer, L., Abramowitz, J., and Brown, A. M.,** Receptor-effector coupling by G proteins, *Biochim. Biophys. Acta,* 1031, 163, 1990.

26. **Taylor, C. W.,** The role of G proteins in transmembrane signalling, *Biochem. J.,* 272, 1, 1990.

27. **Spiegel, A. M., Shenker, A., and Weinstein, L. S.,** Receptor-effector coupling by G proteins: implications for normal and abnormal signal transduction, *Endocrine Rev.,* 13, 536, 1992.

28. **Stevens, F. C.,** Calmodulin: an introduction, *Can. J. Biochem. Cell Biol.,* 61, 906, 1983.

29. **Akiyama, T., Gotho, E., and Ogawara, H.,** Alteration of adenylate cyclase activity by phosphorylation and dephosphorylation, *Biochem. Biophys. Res. Commun.,* 112, 250, 1983.

30. **Swinnen, J. V., Joseph, D. R., and Conti, M.,** The mRNA encoding a high-affinity cAMP phosphodiesterase is regulated by hormones and cAMP, *Proc. Natl. Acad. Sci. U.S.A.,* 86, 8197, 1989.

31. **Erneux, C., Van Sande, J., Miot, F., Cochaux, P., Decoster, C., and Dumont, J. E.,** A mechanism in the control of intracellular cAMP level: the activation of a calmodulin-sensitive phosphodiesterase by a rise of intracellular free calcium, *Mol. Cell. Endocrinol.,* 43, 113, 1985.

32. **Epstein, P. M., Moraski, S., and Hachisu, R.,** Identification and characterization of a Ca^{2+}-calmodulin-sensitive cyclic nucleotide phosphodiesterase in a human lymphoblastoid cell line, *Biochem. J.,* 243, 533, 1987.

33. **Ethier, S. P., Kudla, A., and Cundiff, K. C.,** Influence of hormone and growth factor interactions on the proliferative potential of normal rat mammary epithelial cells *in vitro, J. Cell. Physiol.,* 132, 161, 1987.

34. **Blomhoff, H. K., Smeland, E. B., Beiske, K., Blomhoff, R., Ruud, E., Bjoro, T., Pfeiffer-Ohlsson, S., Watt, R., Funderud, S., Godal, T., and Ohlsson, R.,** Cyclic AMP-suppression of normal and neoplastic B cell proliferation is associated with regulation of *myc* and Ha-*ras* protooncogenes, *J. Cell. Physiol.,* 131, 426, 1987.

35. **Thoresen, G. H., Sand, T.-E., Refsnes, M., Dajani, O. F., Guren, T. K., Gladhaug, I. P., Killi, A., and Christoffersen, T.,** Dual effects of glucagon and cyclic AMP on DNA synthesis in cultured rat hepatocytes: stimulatory regulation in early G_1 and inhibition shortly before the S phase entry, *J. Cell. Physiol.,* 144, 523, 1990.

36. **Weinmaster, G. and Lemke, G.,** Cell-specific cyclic AMP-mediated induction of the PDGF receptor, *EMBO J.,* 9, 915, 1990.

37. **Iynedjian, P. B., Jotterand, D., Nouspikel, T., Asfari, M., and Pilot, P.-R.,** Transcriptional induction of glucokinase gene by insulin in cultured liver cells and its repression by the glucagon-cAMP system, *J. Biol. Chem.,* 264, 21824, 1989.

38. **Kolb, A., Busby, S., Buc, H., Garges, S., and Adhya, S.,** Transcriptional regulation by cAMP and its receptor protein, *Annu. Rev. Biochem.,* 62, 749, 1993.

39. **Vaulont, S., Munnich, A., Marie, J., Reach, G., Pichard, A.-L., Simon, M.-P., Besmond, C., Barbry, P., and Kahn, A.,** Cyclic AMP as a transcriptional inhibitor of upper eukaryotic gene transcription, *Biochem. Biophys. Res. Commun.,* 125, 135, 1984.

40. **Nagamine, Y. and Reich, E.,** Gene expression and cAMP, *Proc. Natl. Acad. Sci., U.S.A.,* 82, 4606, 1985.

41. **Roesler, W. J., Vanderbark, G. R., and Hanson, R. W.,** Cyclic AMP and the induction of eukaryotic gene expression, *J. Biol. Chem.,* 263, 9063, 1988.

42. **de Groot, R. P. and Sassone-Corsi, P.,** Hormonal control of gene expression: multiplicity and versatility of cyclic adenosine 3',5'-monophosphate-responsive nuclear regulators, *Mol. Endocrinol.,* 7, 145, 1993.

43. **Lee, C. Q., Miller, H. A., Schlichter, D., Dong, J. N., and Wicks, W. D.,** Evidence for a cAMP-dependent nuclear factor capable of interacting with a specific region of a eukaryotic gene, *Proc. Natl. Acad. Sci. U.S.A.,* 85, 4223, 1988.

44. **Seasholtz, A. F., Thompson, R. C., and Douglass, J. O.,** Identification of a cyclic adenosine monophosphate-responsive element in the rat corticotropin-releasing hormone gene, *Mol. Endocrinol.,* 2, 1311, 1988.

45. **Dash, P. K., Karl, K. A., Colicos, M. A., Prywes, R., and Kandel, E. R.,** cAMP response element-binding protein is activated by Ca^{2+}/calmodulin-dependent as well as cAMP-dependent protein kinases, *Proc. Natl. Acad. Sci. U.S.A.,* 88, 5061, 1991.

46. **Sheng, M., Thompson, M. A., and Greenberg, M. E.,** CREB: a Ca^{2+}-regulated transcription factor phosphorylated by calmodulin-dependent kinases, *Science,* 252, 1427, 1991.

47. **Waterman, M., Murdoch, G. H., Evans, R. M., and Rosenfeld, M. G.,** Cyclic AMP regulation of eukaryotic gene transcription by two discrete molecular mechanisms, *Science,* 229, 267, 1985.

48. **Foulkes, N. S., Mellstrom, B., Benusiglio, E., and Sassone-Corsi, P.,** Developmental switch of CREM function during spermatogenesis: from antagonist to transcriptional activator, *Nature,* 355, 80, 1992.

49. **Taylor, S. S.,** cAMP-dependent protein kinase, *J. Biol. Chem.,* 264, 8443, 1989.

50. **Taylor, S. S., Buechler, J. A., and Yonemoto, W.,** cAMP-dependent protein kinase: framework for a diverse family of regulatory enzymes, *Annu. Rev. Biochem.,* 59, 971, 1990.

51. **Cho-Chung, Y. S.,** Role of cyclic AMP receptor proteins in growth, differentiation, and suppression of malignancy: new approaches to therapy, *Cancer Res.,* 50, 7093, 1990.

52. **Simon, N. M.,** cAMP-dependent protein kinases: structure and role in the control of gene expression, *Bull. Inst. Pasteur (Paris),* 89, 97, 1991.

53. **Cohen, P. and Hardie, D. G.,** The actions of cyclic AMP on biosynthetic processes are mediated indirectly by cyclic AMP-dependent protein kinase, *Biochim. Biophys. Acta,* 1094, 292, 1991.

54. **Knighton, D. R., Zheng, J., Ten Eyck, L. F., Ashford, V. A., Xuong, N., Taylor, S. S., and Sowadski, J. M.,** Crystal structure of the catalytic subunit of cyclic adenosine monophosphate-dependent protein kinase, *Science,* 253, 407, 1991.

55. **Cho-Chung, Y. S., Clair, T., Tagliaferri, P., Ally, S., Katsaros, D., Tortora, G., Neckers, L., Avery, T. L., Crabtree, G. W., and Robins, R. K.,** Site-selective cyclic AMP analogs as new biological tools in growth control, differentiation, and proto-oncogene regulation, *Cancer Invest.,* 7, 161, 1989.

56. **Shoji, S., Parmelee, D. C., Wade, R. D., Kumar, S., Ericsson, L. H., Walsh, K. A., Neurath, H., Long, G. L., Demaille, J. G., Fischer, E. H., and Titani, K.,** Complete amino acid sequence of the catalytic subunit of bovine cardiac muscle cyclic AMP-dependent protein kinase, *Proc. Natl. Acad. Sci. U.S.A.,* 78, 848, 1981.

57. **Uhler, M. D., Carmichael, D. F., Lee, D. C., Chrivia, J. C., Krebs, E. G., and McKnight, G. S.,** Isolation of cDNA clones coding for the catalytic subunit of mouse cAMP-dependent protein kinase, *Proc. Natl. Acad. Sci. U.S.A.,* 83, 1300, 1986.

58. **Clegg, C. H., Ran, W., Uhler, W., and McKnight, G. S.,** A mutation in the catalytic subunit of protein kinase A prevents myristylation but does not inhibit biological activity, *J. Biol. Chem.,* 264, 20140, 1989.

59. **Showers, M. O. and Maurer, R. A.,** A cloned bovine cDNA encodes an alternate form of the catalytic subunit of cAMP-dependent protein kinase, *J. Biol. Chem.,* 261, 16288, 1986.

60. **Corbin, J. D., Keely, S. L., and Park, C. R.,** The distribution and dissociation of cyclic adenosine 3':5'-monophosphate-dependent protein kinases in adipose, cardiac, and other tissues, *J. Biol. Chem.,* 250, 218, 1975.

61. **Hofmann, F., Beavo, J. A., Bechtel, P., and Krebs, E. G.,** Comparison of adenosine 3':5'-monophosphate-dependent protein kinases from rabbit skeletal and bovine heart muscle, *J. Biol. Chem.,* 250, 7795, 1975.

62. **Russell, J. L. and Steinberg, R. A.,** Phosphorylation of regulatory subunit of type I cyclic AMP-dependent protein kinase: biphasic effects of cyclic AMP in intact S49 mouse lymphoma cells, *J. Cell. Physiol.,* 130, 207, 1987.

63. **Khatra, B. S., Printz, R., Cobb, C. E., and Corbin, J. D.,** Regulatory subunit of cAMP-dependent protein kinase inhibits phosphoprotein phosphatase, *Biochem. Biophys. Res. Commun.,* 130, 567, 1985.

64. **Houge, G., Cho-Chung, Y. S., and Doskeland, S. O.,** Differential expression of cAMP-kinase subunits is correlated with growth in rat mammary carcinomas and uterus, *Br. J. Cancer,* 66, 1022, 1992.

65. **Weber, W. and Hilz, H.,** cAMP-dependent protein kinases I and II: divergent turnover of subunits, *Biochemistry,* 25, 5661, 1986.

66. **Weber, I. T., Steitz, T. A., Bubis, J., and Taylor, S. S.,** Predicted structures of cAMP binding domains of type I and II regulatory subunits of cAMP-dependent protein kinase, *Biochemistry,* 26, 343, 1987.

67. **Shabb, J. B. and Miller, M. R.,** Identification of a rat liver cAMP-dependent protein kinase, type II, which binds DNA, *J. Cycl. Nucleot. Protein Phosphoryl. Res.,* 11, 253, 1986.

68. **Sandberg, M., Taskén, K., Oyen, O., Hansson, V., and Jahnsen, T.,** Molecular cloning, cDNA structure and deduced amino acid sequence for a type I regulatory subunit of cAMP-dependent protein kinase from human testis, *Biochem. Biophys. Res. Commun.,* 149, 939, 1987.

69. **Levy, F. O., Sandberg, M., Taskén, K., Eskild, W., Hansson, V., and Jahnsen, T.,** Molecular cloning, complementary deoxyribonucleic acid structure and predicted full-length amino acid sequence of the hormone-inducible regulatory subunit of 3′-5′-cyclic adenosine monophosphate-dependent protein kinase from human testis, *Mol. Endocrinol.,* 2, 1364, 1988.

70. **Schwartz, D. A. and Rubin, C. S.,** Regulation of cAMP-dependent protein kinase subunit levels in Friend erythroleukemic cells, *J. Biol. Chem.,* 258, 777, 1983.

71. **Jahnsen, T., Lohmann, S.M., Walter, U., Hedin, L., and Richards, J. S.,** Purification and characterization of hormone-regulated isoforms of the regulatory subunit of type II cAMp-dependent protein kinase from rat ovaries, *J. Biol. Chem.,* 260, 15980, 1985.

72. **Jahnsen, T., Hedin, L., Lohmann, S.M., Walter, U., and Richards, J. S.,** The neural type II regulatory subunit of cAMP-dependent protein kinase is present and regulated by hormones in the rat ovary, *J. Biol. Chem.,* 261, 6637, 1986.

73. **Jahnsen, T., Hedin, L., Kidd, V. J., Beattie, W. G., Lohmann, S. M., Walter, U., Durica, J., Schulz, T. Z., Schiltz, E., Browner, M., Lawrence, C. B., Goldman, D., Ratoosh, S. L., and Richards, J. S.,** Molecular cloning, cDNA structure, and regulation of the regulatory subunit of type II cAMP-dependent protein kinase from rat ovarian granulosa cells, *J. Biol. Chem.,* 261, 12351, 1986.

74. **Sandberg, M., Levy, F. O., Oyen, O., Hansson, V., and Jahnsen, T.,** Molecular cloning, cDNA structure and deduced amino acid sequence for the hormone-induced regulatory subunit (RII$_\beta$) of cAMP-dependent protein kinase from rat ovarian granulosa cells, *Biochem. Biophys. Res. Commun.,* 154, 705, 1988.

75. **Singh, T. J., Hochman, J., Verna, R., Chapman, M., Abraham, I., Pastan, I. H., and Gottesman, M. M.,** Characterization of a cyclic AMP-resistant Chinese hamster ovary cell mutant containing both wild-type and mutant species of type I regulatory subunit of cyclic AMP-dependent protein kinase, *J. Biol. Chem.,* 260, 13927, 1985.

76. **Steinberg, R. A., Murphy, C. S., Russell, J. L., and Gorman, K. B.,** Cyclic AMP-resistant mutants of S49 mouse lymphoma cells hemizygous for expression of regulatory subunit of type I cyclic AMP-dependent protein kinase, *Somat. Cell Mol. Genet.,* 13, 645, 1987.

77. **Leto, T. L., Marchesi, V. T., Horne, W. C. et al.,** Differential phosphorylation of multiple sites in protein 4.1 and protein 4.9 by phorbol ester-activated and cyclic AMP-dependent protein kinases, *J. Biol. Chem.,* 260, 9073, 1985.

78. **Ohta, Y., Akiyama, T., Nishida, E., and Sakai, H.,** Protein kinase C and cAMP-dependent protein kinase induce opposite effects on actin polymerizability, *FEBS Lett.,* 222, 305, 1987.

79. **Cooper, E. and Spaulding, S. W.,** Hormonal control of the phosphorylation of histones, HMG proteins and other nuclear proteins, *Mol. Cell. Endocrinol.,* 39, 1, 1985.

80. **Sikorska, M., Whitfield, J. F., and Walker, P. R.,** The regulatory and catalytic subunits of cAMP-dependent protein kinases are associated with transcriptionally active chromatin during changes in gene expression, *J. Biol. Chem.,* 263, 3005, 1988.

81. **Day, R. N., Walder, J. A., and Maurer, R. A.,** A protein kinase inhibitor gene reduces both basal and multihormone-stimulated prolactin gene transcription, *J. Biol. Chem.,* 264, 431, 1989.

82. **Taskén, K., Kvale, D., Hansson, V., and Jahnsen, T.,** Protein kinase C activation selectively increases mRNA levels for one of the regulatory subunits (RI$_\alpha$) of cAMP-dependent protein kinases in HT-29 cells, *Biochem. Biophys. Res. Commun.,* 172, 409, 1990.

83. **Grove, J. R., Price, D. J., Goodman, H. M., and Avruch, J.,** Recombinant fragment of protein kinase inhibitor blocks cyclic AMP-dependent gene transcription, *Science,* 238, 530, 1987.

84. **Van Patten, S. M., Heisermann, G. J., Cheng, H.-C., and Walsh, D. A.,** Tyrosine kinase catalyzed phosphorylation and inactivation of the inhibitor protein of the cAMP-dependent protein kinase, *J. Biol. Chem., 262, 3398,* 1987.

85. **Cheng, H.-C., Kemp, B. E., Pearson, R. B., Smith, A. J., Misconi, L., Van Patten, S. M., and Walsh, D. A.,** A potent synthetic peptide inhibitor of the cAMP-dependent protein kinase, *J. Biol. Chem.,* 261, 989, 1986.

86. **Ekanger, R., Vintermyr, O. K., Houge, G., Sand, T.-E., Scott, J. D., Krebs, E. G., Eikhom, T. S., Christoffersen, T., Ogreid, D., and Doskeland, S. O.,** The expression of cAMP-dependent protein kinase subunits is differentially regulated during liver regeneration, *J. Biol. Chem.,* 264, 4374, 1989.

87. **Fontana, J. A., Reppucci, A., Durham, J. P., and Miranda, D.,** Correlation between the induction of leukemic cell differentiation by various retinoids and modulation of protein kinases, *Cancer Res.,* 46, 2468, 1986.

88. **Groffen, J., Heisterkamp, N., Reynolds, F. H., Jr., and Stephenson, J. R.,** Homology between phosphotyrosine acceptor site of human c-*abl* and viral oncogene products, *Nature,* 304, 167, 1983.

89. **Lörincz, A. T. and Reed, S. I.,** Primary structure homology between the product of yeast cell division control gene CDC28 and vertebrate oncogenes, *Nature,* 307, 183, 1984.

90. **Lee, M. G. and Nurse, P.,** Complementation used to clone a human homologue of the fission yeast cell cycle control gene *cdc2*, *Nature,* 327, 31, 1987.

91. **Sadler, S. E., Maller, J. L., and Gibbs, J. B.,** Transforming *ras* proteins accelerate hormone-induced maturation and stimulant cyclic AMP phosphodiesterase in *Xenopus* oocytes, *Mol. Cell. Biol.,* 10, 1689, 1990.

92. **Hiwasa, T., Sakiyama, S., Noguchi, S., Ha, J.-M., Miyazawa, T., and Yokoyama, S.,** Degradation of a cAMP-binding protein is inhibited by human c-Ha-*ras* gene products, *Biochem. Biophys. Res. Commun.,* 146, 731, 1987.

93. **Härtig, E., Loncarevic, I. F., Büscher, M., Herrlich, P., and Rahmsdorf, H. J.,** A new cAMP response element in the transcribed region of the human c-*fos* gene, *Nucleic Acids Res.,* 19, 4153, 1991.

94. **Macgregor, P. F., Abate, C., and Curran, T.,** Direct cloning of leucine zipper proteins: Jun binds cooperatively to the CRE with CRE-BP1, *Oncogene,* 5, 451, 1990.

95. **Lamph, W. W., Dwarki, V. J., Ofir, R., Montminy, M., and Verma, I. M.,** Negative and positive regulation by transcription factor cAMP response element-binding protein is modulated by phosphorylation, *Proc. Natl. Acad. Sci. U.S.A.,* 87, 4320, 1990.

96. **Meinkoth, J. L., Montminy, M. R., Fink, J. S., and Feramisco, J. R.,** Induction of a cyclic AMP-responsive gene in living cells requires the nuclear factor CREB, *Mol. Cell. Biol.,* 11, 1759, 1991.

97. **Bravo, R., Neuberg, M., Burckhardt, J., Almendral, J., Wallich, R., and Müller, R.,** Involvement of common and cell type-specific pathways in c-*fos* gene control: stable induction by cAMP in macrophages, *Cell,* 48, 251, 1987.

98. **Kacich, R. L., Williams, L. T., and Coughlin, S. R.,** Arachidonic acid and cyclic adenosine monophosphate stimulation of c-*fos* expression by a pathway independent of phorbol ester-sensitive protein kinase C, *Mol. Endocrinol.,* 2, 73, 1988.

99. **Ran, W., Dean, M., Levine, R. A., Henkle, C., and Campisi, J.,** Induction of c-*fos* and c-*myc* mRNA by epidermal growth factor or calcium ionophore is cAMP dependent, *Proc. Natl. Acad. Sci. U.S.A.,* 83, 8216, 1986.

100. **Berkowitz, L. A., Riabowol, K. T., and Gilman, M. Z.,** Multiple sequence elements of a single functional class are required for cyclic AMP responsiveness of the mouse c-*fos* promoter, *Mol. Cell. Biol.,* 9, 4272, 1989.

101. **Riabowol, K. T., Gilman, M. Z., and Feramisco, J. R.,** Microinjection of the catalytic subunit of cAMP-dependent protein kinase induces expression of the c-*fos* gene, *Cold Spring Harbor Symp. Quant. Biol.,* 53, 85, 1988.

102. **Tratner, I., Ofir, R., and Verma, I. M.,** Alteration of a cyclic AMP-dependent protein kinase phosphorylation site in the c-Fos protein augments its transforming potential, *Mol. Cell. Biol.,* 12, 998, 1992.

103. **Le Gros, J., De Feyter, R., and Ralph, R. K.,** Cyclic AMP and c-*myc* gene expression in PY815 mouse mastocytoma cells, *FEBS Lett.,* 186, 13, 1985.

104. **Katsaros, D., Ally, S., and Cho-Chung, Y. S.,** Site-selective cyclic AMP analogues are antagonistic to estrogen stimulation of growth and proto-oncogene expression in human breast cancer cells, *Int. J. Cancer,* 41, 863, 1988.

105. **Trepel, J. B., Colamonici, O. R., Kelly, K., Schwab, G., Watt, R. A., Sausville, E. A., Jaffe, E. S., and Neckers, L. M.,** Transcriptional inactivation of c-*myc* and the transferrin receptor in dibutyryl cyclic AMP-treated HL-60 cells, *Mol. Cell. Biol.,* 7, 2644, 1987.

106. **Harsh, G. R., Kavanaugh, W. M., Starksen, N. F., and Williams, L. T.,** Cyclic AMP blocks expression of the c-*sis* gene in tumor cells, *Oncogene Res.,* 4, 65, 1989.

107. **Daniel, T. O., Gibbs, V. C., Milfray, D. F., and Williams, L. T.,** Agents that increase cAMP accumulation block endothelial c-*sis* induction by thrombin and transforming growth factor-beta, *J. Biol. Chem.,* 262, 11893, 1987.

108. **Kavanaugh, W. M., Harsh, G. R. IV, Starksen, N. F., Rocco, C. M., and Williams, L. T.,** Transcriptional regulation of the A and B chain genes of platelet-derived growth factor in microvascular endothelial cells, *J. Biol. Chem.,* 263, 8470, 1988.

109. **Daniel, T. O. and Fen, Z.,** Distinct pathways mediate transcriptional regulation of platelet-derived growth factor B/c-*sis* expression, *J. Biol. Chem.,* 263, 19815, 1988.

110. **Moens, U., Bang, B. E., and Aarbakke, J.,** Physiological cyclic AMP stimulation and oncogene mRNA levels in HL-60 cells, *Life Sci.,* 47, 1555, 1990.

111. **Anderson, W. B., Russell, T. R., Charchman, R. A., and Pastan, I.,** Interrelationship between adenylate cyclase activity, adenosine 3′:5′ cyclic monophosphate phosphodiesterase activity, adenosine 3′:5′ cyclic monophosphate levels, and growth of cells in culture, *Proc. Natl. Acad. Sci. U.S.A.,* 70, 3802, 1973.

112. **Franks, D. J., Whitfield, J. F., and Durkin, J. P.,** Viral *p21* Ki-RAS protein: a potent intracellular mitogen that stimulates adenylate cyclase activity in early G_1 phase of cultured rat cells, *J. Cell. Biochem.,* 33, 87, 1987.

113. **Saltarelli, D., Fischer, S., and Gacon, G.,** Modulation of adenylate cyclase by guanine nucleotides and Kirsten sarcoma virus mediated transformation, *Biochem. Biophys. Res. Commun.,* 127, 318, 1985.

114. **Tarpley, W. G., Hopkins, N. K., and Gorman, R. R.,** Reduced hormone-stimulated adenylate cyclase activity in NIH-3T3 cells expressing the EJ human bladder *ras* oncogene, *Proc. Natl. Acad. Sci. U.S.A.,* 83, 3703, 1986.

115. **Najam, N., Clair, T., Bassin, R. H., and Cho-Chung, Y. S.,** Cyclic AMP suppresses expression of v-*ras*[H] oncogene linked to the mouse mammary tumor virus promoter, *Biochem. Biophys. Res. Commun.,* 134, 431, 1986.

116. **Tagliaferri, P., Clair, T., DeBortoli, M. E., and Cho-Chung, Y. S.,** Two classes of cAMP analogs synergistically inhibit p21 *ras* protein synthesis and phenotypic transformation of NIH/3T3 transfected with Ha-MuSV DNA, *Biochem. Biophys. Res. Commun.,* 130, 1193, 1985.

117. **Chiarugi, V., Porciatti, F., Pasquali, F., and Bruni, P.,** Transformation of Balb/3T3 cells with EJ/T24/H-*ras* oncogene inhibits enylate cyclase response to β-adrenergic agonist while increasing muscarinic receptor-dependent hydrolysis of inositol lipids, *Biochem. Biophys. Res. Commun.,* 132, 900, 1985.

118. **Majmudar, G. and Peterkofsky, B.,** Cyclic AMP-independent processes mediate Kirsten sarcoma virus-induced changes in collagen production and other properties of cultured cells, *J. Cell. Physiol.,* 122, 113, 1985.

119. **Ryan, W. L. and Heidrick, M. L.,** Role of cyclic nucleotides in cancer, *Adv. Cyclic Nucleotide Res.,* 4, 87, 1974.

120. **MacManus, J. P. and Whitfield, J. F.,** Cyclic AMP, prostaglandins, and the control of cell proliferation, *Prostaglandins,* 6, 475, 1974.

121. **Eker, P.,** Inhibition of growth and DNA synthesis in cell cultures by cyclic AMP, *J. Cell Sci.,* 16, 301, 1974.

122. **Minton, J. P., Matthews, R. H., and Wisenbaugh, T. W.,** Elevated adenosine 3′,5′-cyclic monophosphate levels in human and animal tumors *in vivo, J. Natl. Cancer Inst.,* 57, 39, 1976.

123. **Rechler, M. M., Bruni, C. B., Podskalny, J. M., and Carchman, R. A.,** DNA synthesis in cultured human fibroblasts: regulation by 3′:5′-cyclic AMP, *J. Supramol. Struct.,* 4, 199, 1976.

124. **Wang, T., Sheppard, J. R., and Foker, J. E.,** Rise and fall of cyclic AMP required for onset of lymphocyte DNA synthesis, *Science,* 201, 155, 1978.

125. **Gottesman, M. M. and Fleischmann, R. D.,** The role of cAMP in regulating tumour cell growth, *Cancer Surv.,* 5, 291, 1986.

126. **Willingham, M. C. and Pastan, I.,** Cyclic AMP and cell morphology in cultured fibroblasts: effects on cell shape, microfilament and microtubule distribution, and orientation to substratum, *J. Cell. Biol.,* 67, 146, 1975.

127. **Johnson, G. S. and Pastan, I.,** Role of 3′,5′-adenosine monophosphate in regulation of morphology and growth of transformed and normal fibroblasts, *J. Natl. Cancer. Inst.,* 48, 1377, 1972.

128. **Froehlich, J. E. and Rachmeler, M.,** Effect of adenosine 3′,5′-cyclic monophosphate on cell proliferation, *J. Cell Biol.,* 55, 19, 1972.

129. **Zack, P. M. and Kociba, G. J.,** Effects of increasing cyclic AMP or calcium on feline erythroid progenitors *in vitro:* normal cells are stimulated while cells from retrovirus-infected cats are suppressed, *Int. J. Cell Cloning,* 6, 192, 1988.

130. **van Wijk, R., Wicks, W. D., and Clay, K.,** Effects of derivatives of cyclic 3′,5′-adenosine monophosphate on the growth, morphology, and gene expression of hepatoma cells in culture, *Cancer Res.,* 32, 1905, 1972.

131. **Cho-Chung, Y. S. and Redler, B. H.,** Dibutyryl cyclic AMP mimics ovariectomy: nuclear protein phosphorylation in mammary tumor regression, *Science,* 197, 272, 1977.
132. **Huang, F. L. and Cho-Chung, Y. S.,** Dibutyryl cyclic AMP treatment mimics ovariectomy: new genomic regulation in mammary tumor regression, *Biochem. Biophys. Res. Commun.,* 107, 411, 1982.
133. **Yoshida, H., Azuma, M., Yanagawa, T., Yura, Y., Hayashi, Y., and Sato, M.,** Effect of dibutyryl cyclic AMP on morphologic features and biologic markers of a human salivary gland adenocarcinoma cell line in culture, *Cancer,* 57, 1011, 1986.
134. **Prashad, N., Lotan, D., and Lotan, R.,** Differential effects of dibutyryl cyclic adenosine monophosphate and retinoic acid on the growth, differentiation, and cyclic adenosine monophosphate-binding protein of murine neuroblastoma cells, *Cancer Res.,* 47, 2417, 1987.
135. **Katsaros, D., Tortora, G., Tagliaferri, P., Clair, T., Ally, S., Neckers, L., Robins, R. K., and Cho-Chung, Y. S.,** Site-selective cyclic AMP analogs provide a new approach in the control of cancer cell growth, *FEBS Lett.,* 223, 97, 1987.
136. **Tagliaferri, P., Katsaros, D., Clair, T., Ally, S., Tortora, G., Neckers, L., Rubalcava, B., Parandoosh, Z., Chang, Y., Revankar, G. R., Crabtree, G. W., Robins, R. K., and Cho-Chung, Y.,** Synergistic inhibition of growth of breast and colon human cancer cell lines by site-selective cyclic AMP analogues, *Cancer Res.,* 48, 1642, 1988.
137. **Beckner, S. K.,** Decreased adenylate cyclase responsiveness of transformed cells correlates with the presence of a viral transforming protein, *FEBS Lett.,* 166, 170, 1984.
138. **Somers, K. D.,** Increased cyclic AMP content directly correlated with morphological transformation of cells infected with a temperature-sensitive mutant of mouse sarcoma virus, *In Vitro,* 161, 851, 1980.
139. **Rieber, M. S. and Rieber, M.,** Transformed cells exhibit altered response to DB cyclic AMP-mediated modulation of protein phosphorylation and different endogenous phosphoprotein acceptors, *Cancer Biochem. Biophys.,* 5, 163, 1981.
140. **Sheppard, J. R., Koestler, T. P., Corwin, S. P., Buscarino, C., Doll, J., Lester, B., Greig, R. G., and Poste, G.,** Experimental metastasis correlates with cyclic AMP accumulation in B16 melanoma clones, *Nature,* 308, 544, 1984.
141. **O'Keefe, E. J. and Pledger, W. J.,** A model of cell cycle control: sequential events regulated by growth factors, *Mol. Cell. Endocrinol.,* 31, 167, 1983.
142. **Ogreid, D., Cho-Chung, Y. S., Ekanger, R., Vintermyr, O., Haavik, J., and Doskeland, S. O.,** Characterization of the cyclic adenosine 3':5'-monophosphate effector system in hormone-dependent and hormone-independent rat mammary carcinomas, *Cancer Res.,* 47, 2576, 1987.
143. **Colletta, G., Corda, D., Schettini, G., Cirafici, A. M., Kohn, L. D., and Consiglio, E.,** Adenylate cyclase activity of v-*ras-k* transformed rat epithelial thyroid cells, *FEBS Lett.,* 228, 37, 1988.
144. **Fontana, J., Miskis, G., and Durham, J.,** Elevation of adenylate cyclase activity during leukemic cell differentiation, *Exp. Cell Res.,* 168, 487, 1987.
145. **Azuma, M., Yoshida, H., Kawamata, H., Yanagawa, T., Furumoto, N., and Sato, M.,** Cellular proliferation and *ras* oncogene of p21 21,000 expression in relation to the intracellular cyclic adenosine 3':5'-monophosphate levels of a human salivary gland adenocarcinoma cell line in culture, *Cancer Res.,* 48, 2898, 1988.
146. **Giuffrè, L., Schreyer, M., Mach, J.P., and Carrel, S.,** Cyclic AMP induces differentiation in vitro of human melanoma cells, *Cancer,* 61, 1132, 1988.
147. **Tortora, G., Clair, T., Katsaros, D., Ally, S., Colamonici, O., Neckers, L. M., Tagliaferri, P., Jahnsen, T., Robins, R. K., and Cho-Chung, Y. S.,** Induction of megakaryocytic differentiation and modulation of protein kinase gene expression by site-selective cAMP analogs in K-562 human leukemic cells, *Proc. Natl. Acad. Sci. U.S.A.,* 86, 2849, 1989.
148. **Hardman, J. G. and Sutherland, E. W.,** Guanyl cyclase, an enzyme catalyzing the formation of guanosine 3':5'-monophosphate from guanosine triphosphate, *J. Biol. Chem.,* 249, 6363, 1969.
149. **Kimura, H. and Murad, F.,** Subcellular localization of guanylate cyclase, *Life Sci.,* 17, 837, 1976.
150. **Naor, Z.,** Cyclic GMP stimulates inositol phosphate production in cultured pituitary cells: possible implication to signal transduction, *Biochem. Biophys. Res. Commun.,* 167, 982, 1990.
151. **Lincoln, T. M. and Cornwell, T. L.,** Intracellular cyclic GMP receptor proteins, *FASEB J.,* 7, 328, 1993.
152. **Waldman, S. A. and Murad, F.,** Cyclic GMP synthesis and function, *Pharmacol. Rev.,* 39, 163, 1987.
153. **Garbers, D. L.,** Guanylate cyclase, a cell surface receptor, *J. Biol. Chem.,* 264, 9103, 1989.
154. **Garbers, D. L.,** Cyclic GMP and the second messenger hypothesis, *Trends Endocrinol. Metab.,* 1, 64, 1989.

155. **Chinkers, M. and Garbers, D. L.,** Signal transduction by guanylyl cyclases, *Annu. Rev. Biochem.,* 60, 553, 1991.

156. **Earp, H. S.,** The role of insulin, glucagon, and cAMP in the regulation of hepatocyte guanylate cyclase activity, *J. Biol. Chem.,* 255, 8079, 1980.

157. **Russell, T. R. and Pastan, I. H.,** Cyclic adenosine 3′:5′-monophosphate and cyclic guanosine 3′:5′-monophosphate phosphodiesterase activities are under separate genetic control, *J. Biol. Chem.,* 249, 7764, 1974.

158. **Beavo, J. A., Hansen, R. S., Harrison, S. A., Hurwitz, R. L., Martins, T. J., and Mumby, M. C.,** Identification and properties of cyclic nucleotide phosphodiesterases, *Mol. Cell. Endocrinol.,* 28, 386, 1982.

159. **Singh, S., Lowen, D. G., Thorpe, D. S., Rodriguez, H., Kuang, W.-J., Dangott, L. J., Chinkers, M., Goeddel, D. V., and Garbers, D. L.,** Membrane guanylate cyclase is a cell-surface receptor with homology to protein kinases, *Nature,* 334, 708, 1988.

160. **Wong, S. K. F. and Garbers, D. L.,** Receptor guanylyl cyclases, *J. Clin. Invest.,* 90, 299, 1992.

161. **Garbers, D. L.,** Guanylyl cyclase receptors and their endocrine, paracrine, and autocrine ligands, *Cell,* 71, 1, 1992.

162. **Schulz, S., Chrisman, T. D., and Garbers, D. L.,** Cloning and expression of guanylin. Its existence in various mammalian tissues, *J. Biol. Chem.,* 267, 16019, 1992.

163. **Goldberg, N. D. and Haddox, M. K.,** cGMP metabolism and the involvement in biological regulation, *Annu. Rev. Biochem.,* 46, 823, 1977.

164. **Coffey, R. G., Hadden, E. M., Lopez, C., and Hadden, J. W.,** Cyclic GMP and calcium in the initiation of cellular proliferation, *Adv. Cyclic Nucleotide Res.,* 9, 661, 1978.

165. **Bunn, S. J., Garthwaite, J., and Wilkin, G. P.,** Guanylate cyclase activities in enriched preparations of neurones, astroglia and a synaptic complex isolated from rat cerebellum, *Neurochem. Int.,* 8, 179, 1986.

166. **Wood, P. J., Pao, G., and Cooper, A.,** Changes in guinea pig plasma cyclic nucleotide levels during the development of a transplantable leukemia, *Cancer,* 53, 79, 1984.

167. **Williams, A. C. and Light, P. A.,** Alterations in concentrations of cyclic guanosine 3′,5′-monophosphate in guinea pig urine during the development of a transplantable leukaemia, *Cancer Lett.,* 28, 93, 1985.

168. **Hadden, J. W., Hadden, E. M., Haddox, M. K., and Goldberg, N. D.,** Guanosine 3′:5′-cyclic monophosphate: a possible intracellular mediator of mitogenic influences on lymphocytes, *Proc. Natl. Acad. Sci. U.S.A.,* 69, 3024, 1972.

169. **Gillette, R. W., McKenzie, G. O., and Swanson, M. H.,** Modification of the lymphocyte response to mitogens by cyclic AMP and GMP, *J. Reticuloendothelial Soc.,* 16, 289, 1974.

170. **Diamantstein, T. and Ulmer, A.,** Regulation of DNA synthesis by guanosine-5′-monophosphate, cyclic guanosine-3′,5′-monophosphate, and cyclic adenosine-3′,5′-monophosphate in mouse lymphoid cells, *Exp. Cell Res.,* 93, 309, 1975.

171. **Miller, Z., Lovelace, E., Gallo, M., and Pastan, I.,** Cyclic guanosine monophosphate and cellular growth, *Science,* 190, 1213, 1975.

172. **Kuo, J. F.,** Guanosine 3′:5′-monophosphate-dependent protein kinases in mammalian tissues, *Proc. Natl. Acad. Sci. U.S.A.,* 71, 4037, 1974.

173. **Wolfe, L., Francis, S. H., Landiss, L. R., and Corbin, J. D.,** Interconvertible cGMP-free and cGMP-bound forms of cGMP-dependent protein kinase in mammalian tissues, *J. Biol. Chem.,* 262, 16906, 1987.

174. **Light, D. B., Corbin, J. D., and Stanton, B. A.,** Dual ion-channel regulation by cyclic GMP and cyclic GMP-dependent protein kinase, *Nature,* 344, 336, 1990.

175. **Johnson, E. M. and Hadden, J. W.,** Phosphorylation of lymphocyte nuclear acidic proteins: regulation by cyclic nucleotides, *Science,* 187, 1198, 1975.

176. **Spruill, W. A., Hurwitz, D. R., Lucchesi, J. C., and Steiner, A. L.,** Association of cyclic GMP with gene expression of polytene chromosomes of *Drosophila melanogaster, Proc. Natl. Acad. Sci. U.S.A.,* 75, 1480, 1978.

177. **Mackenzie, C. W., III and Donnelly, T. E., Jr.,** Variable dependence on protein kinase stimulatory modulator for cyclic GMP stimulation of histone phosphorylation by rat liver cyclic GMP-dependent protein kinase, *Biochem. Biophys. Res. Commun.,* 88, 462, 1979.

Chapter 5

Guanosine Triphosphate-Binding Proteins

I. INTRODUCTION

The molecular mechanisms associated with the transduction of extracellular signals into intracellular messages frequently involve the guanine nucleotides, guanosine diphosphate (GDP) and guanosine triphosphate (GTP), as well as proteins that bind and hydrolyze GTP (GTPases), as essential components of biochemical pathways that may participate in the regulation of the adenylyl cyclase system, enzymes, or ion channels. A large family of GTP-binding proteins has been identified in eukaryotic cells.[1-10] This superfamily of GTP-binding proteins includes the high molecular weight heterotrimeric GTP-binding proteins involved in transmembrane signaling (G proteins), the 21-kDa (p21) Ras proteins encoded by *ras* proto-oncogenes, translation factors such as the eukaryotic elongation factors 1α and 2 (EF-1α and EF-2), and monomeric low molecular weight GTP-binding proteins encoded by genes such as *rab, rap, ral, rho, sec*-4, and *ypt*-1.[11] Although these proteins have regions exhibiting high sequence similarity, they may be involved in different functions. However, the basic mechanism of the reactions catalyzed by GTP-binding proteins is similar and consists of the conversion of an inactive conformation bound to GDP to an active conformation bound to GTP, which is involved in signal transduction across the cell membrane. The active form is inactivated upon hydrolysis of bound GTP to bound GDP, which results in a shift of the tertiary structure of the protein. Conversion of the inactive to the active form of the protein by nucleotide exchange is regulated by an exchange-promoting protein (EPP), whereas the conversion of active (GTP-bound) to inactive (GDP-bound) is accelerated by a GTPase-activating protein (GAP), which stimulates intrinsic GTPase activity in the protein.

Transductional mechanisms involving GTP-binding proteins depend on the interaction of three essential components on the cell surface: a receptor, a GTP-binding protein, and an effector. In hormonally regulated systems the signal is the hormone, which is recognized by its specific receptor, and the GTP-binding component is a trimeric (α,β,γ) G protein. In the visual system, the exogenous signals are represented by photons, which are captured in the retina by the specific receptor, rhodopsin, and the GTP-binding protein is called transducin.[12,13] The GTP-binding proteins are thus involved in mediating the transduction of both physical and chemical signals across the cell membrane. A growing number of low molecular weight GTP-binding proteins (G proteins) have been identified in eukaryotic cells. G proteins are structurally and functionally related to the 21-kDa (p21) Ras protein products of c-*ras* proto-oncogenes.

II. G PROTEINS

Cell surface receptors in different types of cells are directly coupled to G proteins, which bind and hydrolyze GTP. Active and inactive conformations of the trimeric G proteins are associated with GTP and GDP binding, respectively. G proteins are involved in the transduction of transmembrane signals elicited by receptors for hormones, growth factors, neurotransmitters, and other exogenous stimuli.[14] Receptors involved in the stimulation or inhibition of the adenylyl cyclase system regulate cellular effectors via G proteins. A specific class of receptors, the G protein-coupled receptors, is characterized by containing seven membrane-spanning domains and their cellular actions are associated with phosphoinositide hydrolysis, intracellular Ca^{2+} mobilization, and activation of protein kinase C.[15] G proteins may be targets for bacterial toxins associated with human diseases, including the cholera toxin, produced by *Vibrio cholerae*, and the pertussis toxin, produced by *Bordetella pertussis*. Coupling of cell surface receptors to some G proteins is blocked by pertussis toxin. In addition to stimulation by activated receptors, G proteins may be stimulated by other mechanisms. Arachidonic acid metabolites can stimulate the G protein-gated cardiac K^+ channel in a receptor-independent way.[16] Effectors regulated by G protein-mediated signal transduction include enzymes such as adenylyl cyclase, phospholipase C, and phospholipase A_2, as well as K^+ and Ca^{2+} ion channels. A cell can have multiple G proteins that carry out a similar function such as the activation of phospholipase C, but they may couple to different receptors that determine the cellular response.[17]

The G proteins are heterotrimeric complexes of α, β, and γ subunits. The family of G proteins includes the stimulatory (G_s) and inhibitory (G_i) proteins of the adenylyl cyclase system, the retina-specific protein transducin (G_t), and the protein G_0. The G_s and G_i proteins reside in the cytoplasmic surface of the plasma membrane and are associated with the catalytic subunit of adenylyl cyclase. They act in hormonal signal transduction in a manner similar to "on-off" switches, with activation of the membrane receptors by their ligands resulting in the release of tightly associated GDP and activation of the G protein through binding of GTP. The transducins are present only in retinal photoreceptor cells and are involved in visual transduction. The G_0 protein is particularly abundant in the brain and is involved in signal transduction across the plasma membrane. A G protein, termed G_z, is expressed in neural tissue and platelets, whereas another G protein, G_q, is expressed ubiquitously.

The general biological importance of G proteins is indicated by their high conservation in evolution. G protein homologous genes are present in insects such as *Drosophila melanogaster*, amoebas such as *Dictyostelium discoideum*, yeasts such as *Saccharomyces cerevisiae* and *Schizosaccharomyces pombe*, and plants such as *Arabidopsis thaliana*. G protein α subunits are particularly well conserved in evolution. The primitive eukaryote *D. discoideum*, which is separated from vertebrates by billions of years of evolution, contains two proteins that exhibit functional and physical similarity with the α subunits of vertebrate G proteins.[18] Two genes present in the yeast *S. cerevisiae*, *GPA1* and *GPA2*, encode proteins exhibiting overall structural and functional similarities with α subunits of mammalian G_i and G_s proteins, respectively. Whereas the GP1α protein encoded by the yeast *GAP1* gene is involved in mating factor signal transduction, the GP2α protein encoded by the *GPA2* gene may participate in regulation of intracellular cAMP levels. The yeast GP1α protein is also directly involved in the pheromone signal transduction system.[19] Sterile mutants of the yeast carry a mutation of *GPA1* gene (the *dac1* allele), which makes them unresponsive to mating pheromones. The human genome contains a gene, *GST1-Hs*, which is homologous to the yeast *GST1* gene and encodes a GTP-binding protein that is expressed in a proliferation-dependent manner in mammalian cells.[20] A human homolog of the CDC42 cell division-associated protein of *S. cerevisiae* is represented by the GTP-binding protein G_p, which is a substrate for the EGF receptor kinase.[21]

A. G PROTEIN SUBUNITS

The G proteins involved in hormonal signal transduction are composed of three subunits which are products of distinct genes: the α subunit (39 to 52 kDa), the β subunit (35 to 36 kDa), and the τ subunit (7 to 10 kDa). While the α subunit can be readily dissociated from the heterotrimeric G protein upon its activation in solution, the β and τ subunits are tightly bound and can be dissociated from the G protein only by treatment with denaturants. It is the α subunit which shows structural and functional homology with other members of the GTP-binding protein superfamily. The α subunit binds guanine nucleotide with high affinity, possesses intrinsic GTPase activity, and serves as a substrate for ADP-ribosylation by bacterial toxins (cholera toxin and pertussis toxin). Binding of GTP leads to a change in the conformation of the α subunit, which results in activation of the G protein and dissociation of the α subunit from the $\beta\tau$ complex. A model proposed for the atomic structure of G protein α subunit is based on its close homology to the crystal structure of the human Ras protein.[22] The α subunit plays a critical role for regulation of effectors by G proteins, although there is evidence that the $\beta\tau$ complex may also have a role in the regulation of effectors such as adenylyl cyclase.[23]

Interaction of the G protein with an activated receptor on the cell surface promotes the exchange of GDP, bound to the α subunit, for GTP and the subsequent dissociation of the α-GTP complex from the $\beta\tau$ heterodimer. A single receptor is capable of activating multiple G protein molecules, thus amplifying the ligand binding event. The α subunit with GTP bound and the free $\beta\tau$ unit interact with effectors that further amplify the signal. Such effectors include ion channels and enzymes (adenylyl cyclase, phospholipase C, phospholipase A_2) that generate second messengers. Second messengers such as cAMP or inositol trisphosphate induce various intracellular changes, including protein phosphorylation, gene transcription, cytoskeletal reorganization, and alterations at the level of the plasma membrane including secretion and membrane depolarization. Termination of the signal generated by the receptor-activated G protein occurs when GTP bound by the α subunit of the G protein is hydrolyzed to GDP, and the α subunit then reassociates with the $\beta\tau$ complex (Figure 4.1).

All G proteins are heterotrimeric and consist of α, β, and τ subunits. A relatively high level of molecular heterogeneity exists in these subunits. While the β subunits of G_s and G_i were first considered to be identical in their molecular weights (35 to 36 kDa), amino acid sequences, and proteolytic peptide maps, the α subunits of these two proteins were found to be dissimilar in several aspects, including molecular weight (45 kDa for G_s and 41 kDa for G_i). There are multiple molecular forms of G protein α subunits.[24] Point mutations that replace key conserved amino acids in the α subunits of G_s and G_i can create G proteins that constitutively activate their downstream effector pathways, resulting in increased or decreased production of cAMP, respectively.[25] Myristoylation of the G_i α subunit is essential for its plasma membrane attachment.[26]

A family of G proteins, G_q, characterized by the presence of at least five specific types of pertussis toxin-insensitive α subunits, has been identified recently in both vertebrates and invertebrates.[27] The members of the G_q family possess different subtypes of α subunits that are highly related to one another. The G_q proteins are ubiquitously expressed in murine tissues as well as in stromal and hematopoietic cell lines.[28] However, some members of this family are expressed in a tissue- and development-specific manner. The G_q proteins do not affect adenylyl cyclase activity, but instead mediate phosphoinositide hydrolysis by activation of phospholipase C.[29] The signaling mechanism of the G-coupled family of cell surface receptors involves the specific activation of the phospholipase C isoenzyme, PLC-β1, by G_q proteins. The action of hormones such as TRH and LHRH involves PLC-β1 activation by G_q proteins.[30]

The existence of two distinct types of G protein β subunits in bovine and human cells has been recognized recently.[31-33] These subunits, β_1 and β_2, are encoded by genes that map to different human chromosomes. The 340-residue, 37,329 mol wt β_2 polypeptide is 90% identical with β_1 in predicted amino acid sequence, and both β_1 and β_2 are organized as a series of repetitive homologous segments. Most tissues contain the two types of β subunits, but the functional significance of this heterogeneity is not understood. The γ subunits of G proteins are approximately 70 amino acids long, have a molecular weight of 7 to 10 kDa, and may be all identical.

G proteins can undergo different types of posttranslational modifications. The amino-terminal portion of the α subunit of some G proteins may be myristoylated, i.e., they may contain covalently amide-linked myristic acid.[34] Myristate may play an important role in stabilizing interactions of G proteins with phospholipid or with plasma membrane-bound proteins. Myristoylation may be essential for some α subunits (α_i and α_0) to become membrane-bound as well as for high affinity interaction with the βγ complex. G protein γ subunits undergo a series of posttranslational modifications at their carboxy terminus, including isoprenylation (addition of a farnesyl or geranylgeranyl moiety). The activated insulin receptor kinase may induce phosphorylation of the α subunit of G_i and G_0 proteins.[35] The c-Src kinase is associated with G proteins *in vivo* and may be involved with the phosphorylation of the α subunit of these proteins, which could result in enhanced function of the G proteins.[36] However, the precise physiological significance of G protein phosphorylation is not clear at present.

G protein α subunits control the specificity of interactions with receptor and effector elements of the cell membrane transduction system. Studies with chimeric α subunits of G_i/G_s indicate that carboxy-terminal sequences of the G_s α subunit contain the structural features that are required for specificity of interactions with the effector enzyme adenylyl cyclase, as well as with a hormone receptor such as the β-adrenergic receptor, which stimulates adenylyl cyclase.[37] Activated surface receptors stimulate the α subunit to bind GTP, and the α subunit-GTP complex then interacts with the hormone-sensitive adenylyl cyclase. A mutation that converts a glycine to an alanine in the presumed GDP-binding domain of the G_s α subunit results in interruption of the GTP-driven cycle in the subunit with the consequent alteration in transmembrane signaling.[38] The βγ subunit complex may be responsible for mediating the inhibition of adenylyl cyclase activity.[39] The interaction is terminated when the bound GTP is hydrolyzed to GDP by the α subunit, which is thereafter recoupled with the βγ complex.

In addition to the guanine nucleotide-binding sites, the α subunit of G proteins contains an ADP-ribosylation site. The α subunit of G_s is ADP-ribosylated by pertussis toxin. The pertussis toxin catalyzes ADP-ribosylation of a cysteine located four amino acid residues from the carboxy terminus which prevents the interaction of G proteins with receptors. On the other hand, cholera toxin catalyzes the ADP-ribosylation of the α subunit of G_s at the Arg-201 residue, which results in reduction of the GTPase activity. All G protein α subunits contain an Arg residue at the position corresponding to Arg-201, and mutations replacing the Arg-201 residue may cause constitutive elevation of adenylyl cyclase activity. The conserved Arg residue may be unmasked by interaction with an activated receptor and may play an important role in the regulation of GTP hydrolysis.[9]

Studies at the genomic level have contributed recently to a better molecular characterization of G proteins. Analysis of cDNA clones for G protein α subunits has revealed a great complexity, but the biological significance of this diversity is not understood. The genes for G protein α subunits are not linked and are distributed among different human chromosomes.[40] cDNA clones encoding the bovine G_s α subunit have been isolated and characterized.[41,42] The cloning of cDNAs coding for the bovine and rat

G_s, G_i, and G_0 α subunits also was reported.[43,44] The molecular diversity of the G protein family is indicated by the cloning of cDNAs coding for various subtypes of α subunits in the mouse.[45] cDNA clones for the mouse α chain of G_s and G_i were isolated and sequenced.[46] The cDNA for the α chain of G_s encodes a polypeptide of 377 amino acids (43,856 mol wt) and the cDNA for the α chain of G_i encodes a polypeptide of 355 amino acids (40,482 mol wt). The nucleotide sequence of a cDNA coding for the α subunit of human brain G_i protein was also determined and the deduced amino acid sequences were found to be identical to that of the homologous bovine protein α subunit.[47,48] However, the human and bovine α subunits of G_i brain proteins differ in their primary structure from the respective G protein subunits found in human monocytes, rat glioma, and mouse macrophages, suggesting that there are at least two mRNAs for these subunits. cDNA clones encoding the α subunits of three human G_i proteins, corresponding to three distinct human genes, have been synthesized and sequenced recently.[49] Thus, it is clear that there is a high molecular heterogeneity of G proteins based mainly in the diversity of their α subunits, but the precise biological significance of this heterogeneity is not clear.

B. REGULATION OF G PROTEIN EXPRESSION

G proteins are involved in the transmission of signals elicited by ligand-activated cell surface receptors across the plasma membrane. They have an important role in the mitogenic response of certain types of cells to hormones and growth factors. Incubation of mouse 3T3 cells with bombesin results in a concentration-dependent increase in the level of c-*myc* mRNA and stimulation of DNA synthesis, and these effects are totally abolished by pertussis toxin.[50] By contrast, the mitogenic effects of PDGF on the same cells are not mediated by G proteins because these effects are not abolished by pertussis toxin, although PDGF also stimulates c-*myc* gene expression in the mouse cells. Thus, the mitogenic effects of bombesin-like growth factors appear to be mediated through a pertussis toxin-sensitive step involving G proteins, but the mitogenic effects of other growth factors may follow different pathways, not necessarily involving G proteins.

Hormones and growth factors may be involved in the regulation of G protein expression. Thyroid hormones exert permissive effects on the hormone-sensitive adenylyl cyclase, and these effects may be mediated by regulating the expression of G protein β subunit mRNA.[51] The mechanisms of regulation of G protein expression remain little characterized. The role of proto-oncogenes in the regulation of G proteins is little understood. The amount of 29-kDa G protein that is present in various normal mouse tissues (brain, kidney, lung, and spleen) increases markedly in NIH/3T3 mouse fibroblasts after transformation by the activated human proto-oncogenes c-H-*ras*, c-*raf*, *hst*, and *ret*.[52] The possible role of the 29-kDa G protein in neoplastic transformation is unknown.

C. THE G PROTEIN-COUPLED RECEPTOR FAMILY

The discovery that the β-adrenergic receptor is related to the visual pigment opsin has led to the identification of a growing number of products of a gene family of G protein-coupled receptors that have in common the presence of seven transmembrane segments and the ability to interact with G proteins. The ligands capable of activating these receptors are very diverse and include the classical biogenic amine neurotransmitters such as acetylcholine, noradrenaline, and dopamine, as well as glycoprotein hormones, small peptides, odorants, and light-activated retinal. The use of degenerate primers corresponding to consensus sequences of specific transmembrane segments of available receptors allowed the selective amplification and cloning of several members of this gene family from canine thyroid cDNA libraries.[53] General strategies for cloning of members of the G protein-coupled receptors have been delineated,[54] and the general physiological properties and transductional mechanisms of these receptors have been described.[15,55] Members of the G protein-coupled family of receptors are the adrenergic and serotoninergic receptors, as well as the receptors for substance K, platelet-activating factor (PAF), bombesin, endothelin, bradykinin, vasopressin, and the receptors for the glycoprotein pituitary/placental hormones FSH, LH/CG, and TSH. Proline residues contained within the hydrophobic transmembrane domain of the receptor molecule are highly conserved across the entire superfamily of G protein-coupled receptors and may play key roles in receptor expression, ligand binding, and receptor activation.[56] Some members of this family of receptors, including the muscarinic acetylcholine receptor, have transforming potential when expressed in certain types of cells.[57] The effector mechanisms of G-protein coupled receptors include stimulation of phosphoinositide metabolism, intracellular mobilization of Ca^{2+}, and activation of protein kinase C.

D. ALTERATIONS OF G PROTEINS IN HUMAN DISEASES

G proteins may be altered in neoplastic processes but the data available on this subject are scarce. Distribution of G protein α and β subunits is variable among different clones of the human neuroblastoma cell line SK-N-SH that exhibit different phenotypic characteristics.[58] The G_0 protein is abundant in neural tissues, and neuroendocrine tumors in general, including neuroblastomas, express high amounts of α subunits of G_0.[59] The α subunit of G_0 was detected in the serum from 73% of patients with neuroblastoma at diagnosis, whereas the concentrations of α subunit in the sera of control children or patients with non-neuroendocrine tumors were undetectable. Moreover, the serum concentrations of G_0 were diminished in patients responding to treatment and increased in patients who relapsed. Thus, the G_0 α subunit could be a useful biomarker for neuroendocrine tumors.

The possible role of G proteins in tumor progression is not known. Studies with high and low experimental metastatic B16 melanoma clones suggest that a pertussis-sensitive G protein may regulate second messenger pathways that contribute to the metastatic capacity of the cells.[60] The G protein, probably G_{i2}, would be involved in the regulation matrix-mediated cell adhesion, invasion, and motility. A mutated α subunit of G_q protein can induce transformation in NIH/3T3 cells.[61] Thus, the gene encoding this subunit may be considered a proto-oncogene. The G_q proteins are normally coupled to activation of phospholipase C. Mutations of G-protein coupled receptor genes that activate polyphosphoinositide metabolism, such as the α_{1B}-adrenergic receptor, enhance mitogenesis and can eventually contribute to tumorigenesis.[62]

Somatic mutations of G proteins have been found in a number of tumors, including endocrine tumors, and may have a role in the origin and/or development of these tumors.[10] A subset of growth hormone-secreting human pituitary tumors exhibits high basal adenylyl cyclase activity that responds poorly to GTP analogs and the physiological regulator, the growth hormone-releasing hormone (GHRH).[63] These tumors were found to have somatic mutations within the gene encoding the α_s subunit of G protein (*gsp* mutations).[64] The point mutations contained in the pituitary tumors determined the substitution of Arg-201 (the residue that is ADP-ribosylated by cholera toxin) to Cys or His, or Gln-227 (a residue that is analogous to Gln-61 of Ras proteins) to Arg. Importantly, substitution at position 61 of Ras proteins results in activation of the oncogenic potential of the Ras protein. The amino acid substitutions of α_s contained in the pituitary tumors cause a marked decrease in intrinsic GTPase activity of the G protein, and cell membranes containing the altered G protein produce increased levels of cAMP in the absence of LHRH. Similar mutations of G protein α_s subunits were detected in 18 of 42 growth hormone-secreting pituitary adenomas.[65] Similar mutations of the *gsp* gene have been detected in some thyroid tumors.[66,67] It is thus clear that the *gsp* gene can display some oncogene-like activities.

Activating mutations of the *gsp* gene encoding the α subunit of the G_s protein have been found in the McCune-Albright syndrome.[68,69] This syndrome is characterized by peripheral sexual precocity, polyostotic fibrous dysplasia, café-au-lait pigmentation of the skin, and endocrinopathies, including growth hormone-secreting pituitary adenomas, goitrous hyperthyroidism, and autonomous adrenal hyperplasia. The *gsp* mutations observed in McCune-Albright syndrome are dominant somatic mutations affecting the Arg-201 residue, and germline *gsp* mutations are presumed to be lethal.

III. THE Ras PROTEINS

The mammalian c-*ras* proto-oncogenes encode proteins of 21 kDa, termed p21[ras] or c-Ras, that are the normal counterparts of the the v-Ras oncoproteins encoded by the v-*ras* oncogenes contained in the acute retroviruses Harvey MuSV (H-MuSV) and Kirsten MuSV (K-MuSV). The normal Ras proteins are involved in the control of important cellular functions, including cell proliferation and differentiation as well as specific metabolic pathways.[70-77] The Ras proteins are closely related to the α subunits of GTP-binding proteins (G proteins), and may function as molecular switches in the transduction of signals generating diverse stimuli, including ligand-activated hormone and growth factor receptors that are located on the cell membrane and possess an intrinsic or associated tyrosine kinase activity.[78]

Ras proteins are members of a superfamily of proteins that are present not only in vertebrates but also in invertebrates and even in unicellular organisms. This superfamily can be subdivided into three main groups: Ras-, Rho-, and Rab-like proteins.[79-83] The physiological importance of Ras proteins is indicated by their high conservation in evolution. Genes homologous to mammalian *ras* are present in insects, slime molds, nematodes, and yeasts. The products of *RAS* genes present in yeast participate in important signal transduction pathways.[84,85] The yeast *S. cerevisiae* contains two genes that are structurally and function-ally homologous to the human c-*ras* genes. Genes related to animal c-*ras* genes are active in plant tissues such as hormone-dependent sugarbeet callus and leaves.[86] The dimorphic fungus *Mucor racemosus* expresses three *MRAS* genes which exhibit a striking similarity to human c-*ras* genes and are involved in the regulation of cell proliferation and differentiation.[87]

Ras proteins are composed of 188 or 189 amino acids. Like the G proteins, the products of the three c-*ras* genes (c-H-*ras*, c-K-*ras*, and N-*ras*) contained in the mammalian genome bind guanine nucleotide, hydrolyze GTP, and reside on the inner surface of the plasma membrane. The close structural homology that exists between the amino acid sequences of the α subunit of G proteins and the middle and amino-terminal portions of Ras proteins suggest that both protein families may be derived from a common ancestor molecule. Both the G proteins and the Ras proteins are involved in signal transduction across the plasma membrane.[78]

A. REGULATION OF *ras* GENE EXPRESSION

The c-*ras* genes are expressed at low levels in most tissues and may belong to the class of housekeeping genes. These genes are expressed not only in actively proliferating cells but also in some terminally differentiated cells. Expression of the c-*ras* genes depends on complex regulatory influences originated in promoter and enhancer DNA elements. As other actively transcribed genes, the promoter region of c-*ras* in different human cells of either normal or tumor origin is hypersensitive to DNase I, micrococcal nuclease, endogenous nucleases, and S1 nuclease.[88] The human c-H-*ras* gene promoter corresponds to a 550-bp fragment located about 1 kb upstream from the c-*ras* coding sequence.[89] In addition, regulatory sequences within the variable tandem repeat DNA sequence located downstream of the human c-H-*ras*-1 gene possess an endogenous enhancer activity.[90] A repetitive DNA sequence situated 3′ of the c-H-*ras* polyadenylation site may influence the expression of the proto-oncogene, acting as an enhancer element.[91] Polymorphism of this genetic element in the human population may be responsible for variation in c-H-*ras* expression in normal and transformed cells.

The 5′ end of the mouse c-K-*ras* gene also exhibits features shared by the promoter regions of genes with housekeeping or growth control functions in the cell.[92] In the rat, promoter activity is contained within the 172-bp, 5′-flanking region of the c-H-*ras* gene.[93] A DNA element with suppressive effect on the c-H-*ras* promoter is located on the 5′-flanking region of the rat c-H-*ras* gene.[94]

B. THE MAMMALIAN Ras PROTEIN SUPERFAMILY

Gene sequences closely related to *ras* have been detected in different types of mammalian cells on the basis of nucleic acid homology. These genes are probably derived from a common ancestor and their products form a superfamily of Ras-related proteins including over 35 members that can be classified into three subfamilies according to their structural similarities: Ras-, Rho-, and Rab-related proteins.[79-83] Comparison of the degrees of structural similarities between the members of the Ras superfamily of proteins has allowed the construction of evolutionary trees. Unfortunately, different names have been assigned frequently to the same protein isolated by different investigators using different approaches which has created a high degree of confusion in the nomenclature of these proteins.

The Ras subfamily includes the three mammalian Ras proteins (c-H-Ras, c-K-Ras, and N-Ras), as well as Rap (Rap1A, Rap1B, Rap2A, Rap2B) and Ral (RalA, RalB) proteins. The Rho protein subfamily includes the Rho proteins (RhoA, RhoB, RhoC), as well as Rac proteins (Rac1, Rac2) and other members. The Rab subfamily includes the Rab proteins (Rab1A, Rab1B, Rab2, Rab3A, Rab3B, Rab3C, Rab4A, Rab4B, and Rab5 to Rab11), the Ypt proteins (Ypt1, Ypt2, Ypt3), and the Sas, Sec, and Ara proteins. The members of the Ras subfamily of proteins are located, with few exceptions, on the cytoplasmic face of the plasma membrane in mammalian cells. The members of the Rho subfamily are associated mainly with the cytoplasm and the Golgi complex and are involved in actin organization and control of cell shape, probably by interacting with the cytoskeleton and intracellular membranes. The Rab-related proteins are involved in vesicular traffic and are associated with a variety of cellular compartments. Expression of the members of the Ras protein superfamily in mammalian cells is ubiquitous, although Rab3A is found only in cells of neural origin.

The number of Ras-related proteins identified in the last few years has been growing continuously. Multiple Ras-like guanine nucleotide-binding proteins have been detected in the plasma membranes of mouse fibroblasts.[95] Some of these proteins may be involved in the transmission of mitogenic signals elicited by growth factors and other mitogenic signaling agents, but different Ras-related proteins may have different functions. The Rap1A protein, also called Krev-1, is closely associated with the cytochrome *b* of human neutrophils, suggesting its involvement in the superoxide generating system.[96] The Rap1A protein is associated with the Golgi complex in rat and human cells.[97] The putative effector domains shared between Ras and Rap proteins are functionally similar and interact with their respective GTPase-activating proteins.[98] Expression of the *rap*-1 gene is capable of suppressing *ras*-induced oncogenic transformation.[99]

Expression of the *rho*-B gene is rapidly and transiently induced at the transcriptional level in rat fibroblasts by EGF and PDGF as well as by the oncoproteins v-Fps and v-Src, which possess tyrosine kinase activity.[100] Thus, *rho*-B has the characteristics of a mitogen-inducible, immediate-early gene. The gene *rac*, which exhibits homology to *rho*, is expressed specifically in cells of hematopoietic lineages.[101] The GTP-binding protein G25K, which is closely related to Rho and Rac, is the human homolog of the yeast cell cycle-related gene *CDC42* and is widely expressed in mammalian tissues.[102]

The 23.4-kDa (p23) protein product of a *ras*-related gene, R-*ras*, detected in the human genome by virtue of its homology to v-H-*ras*, may have biochemical properties similar to those of the H-*ras*, K-*ras*, and N-*ras* proto-oncogenes.[103,104] The R-*ras* p23 protein is 218 amino acids long, and its structure is similar to that of p21^{ras} and other members of the GTP-binding protein superfamily. However, p21 and p23 Ras proteins may have different functions. Further studies are required for a better characterization of effectors and functions of the different members of the Ras-related super-family of proteins and the role of growth factors and other stimuli in the regulation of their expression in different types of cells.

C. STRUCTURAL AND FUNCTIONAL HOMOLOGIES OF Ras PROTEINS

The primary, secondary, and tertiary structure of v-Ras and c-Ras proteins have been determined by using a diversity of analytical methods. The primary structure of Ras has been deduced from the nucleotide sequences of the respective cDNA clones. Two-dimensional proton nuclear magnetic resonance analysis was applied to study the molecular interactions of purified recombinant c-H-Ras protein.[105] Three-dimensional models for the structure of the catalytic domain of human c-H-Ras have been proposed.[106,107] Important results for the understanding of Ras protein structure and function have been obtained with crystallographic studies and the elaboration of three-dimensional models of the protein.[108-111] The crystal structure of the guanine nucleotide-binding domain of Ras (residues 1-166) complexed to a GTP analog was determined at a resolution of 2.6 Å. Time-resolved X-ray crystallographic study of c-H-Ras showed the occurrence of structural changes in specific parts of the molecule during GTP hydrolysis. The "on" (GTP-complexed) and "off" (GDP-complexed) forms of c-Ras proteins can be distinguished by confor-mational differences that span a length of more than 40 Å and are induced by the γ-phosphate. Radiation inactivation analysis suggests that Ras proteins exist in the cell as oligomers, which may be important for the performance of their biological roles.[112]

Structural homologies of Ras proteins can be studied by direct comparison of amino acid sequences or by mutational analysis. However, structural homologies do not necessarily implicate the existence of functional homologies in the respective proteins. The 23-kDa protein product (p23) of the human R-*ras* gene exhibits conservation of amino acids in the region corresponding to the proposed Ras effector region (amino acids 35 to 42). However, the p23 Ras protein cannot be considered a proto-oncogene product, because it has a function distinct from that of p21^{ras} and does not possess oncogenic potential.[104] Analysis of constructed mutant c-H-Ras molecules and the p23 molecules that form the R-*ras* gene suggests that individual region sequence divergence between the p23 and p21 products should be considered within the context of the entire molecule and not just as isolated segments of a primary structure.[113]

Comparison of amino acid sequences has revealed the presence of multiple homologous regions common to all members of the human Ras family and the bacterial translation elongation factors Tu (EF-Tu) and G (EF-G), which also contain in their amino-terminal region a binding site for GTP.[114] During protein biosynthesis in bacteria (*Escherichia coli*), EF-Tu recognizes, transports, and positions the codon-specified aminoacyl-tRNA onto the A site of the ribosome. In this role, EF-Tu interacts with GDP and GTP, which act as allosteric effectors to control the protein conformation required during the elongation cycle.[115] High-resolution X-ray diffraction analysis of EF-Tu revealed that four regions of the amino acid sequence that are homologous to human p21^{ras} are located in the vicinity of the GDP-binding site, and most of the invariant amino acids shared by the two proteins interact directly with the GDP ligand.

Mutational analysis may contribute to the definition of the functional domains of Ras proteins.[116,117] The analysis indicates that the basic structure of the GTP-binding site is conserved between Ras and EF-Tu and that this binding site is crucial for the function of Ras proteins.[118] A GTP-binding protein possibly involved in bacterial protein secretion, LepA, shows sequence homology to initiation factor 2 (IF-2) as well as to EF-Tu and EF-G.[119] The mammalian polypeptide chain elongation factor EF-2 shows homology with human c-H-Ras protein.[120]

The Ras proteins exhibit homology with the α subunits of GTP-binding proteins such as G proteins and transducin.[121-123] Homology also exists between the transducin α subunit, c-Ras proteins, and the bacterial translation initiation factor, IF-2. The homology existing between the α subunit of bovine adrenal G_s and the yeast Ras homologous proteins is limited to very short regions of these proteins that appear to be involved in GTP binding and hydrolysis.[42] Ras proteins also show homology to the β subunit of bacterial and mitochondrial ATP-synthase in the region of the protein that contributes to nucleotide binding.[124]

The structural similarities existing between mammalian and yeast *ras* gene products, bacterial translation elongation and initiation factors, G proteins, and the transducin α subunit may reflect their involvement in guanine nucleotide binding and hydrolysis and the requirements for an alternation between GDP-and GTP-bound conformations. The regions of these proteins that are not directly involved in nucleotide binding or hydrolysis may govern the specificity for interaction with different subcellular components.[122] The carboxy-terminal region of G protein α subunits and Ras proteins may represent a functional domain involved in receptor-coupling, thus conferring relative functional specificity to the respective proteins.[125] A model for the Ras tertiary structure predicted sequences of the protein responsible for guanine nucleotide specificity and for GTP binding and hydrolysis.[126] In spite of the striking structural similarities existing between the Ras proteins and the G proteins, there is no evidence that Ras proteins regulate adenylyl cyclase activity in a manner similar to that of G proteins.[127]

D. REGULATION OF Ras PROTEIN ACTIVITY

The functional properties of Ras proteins are similar but not identical to those of other members of the Ras superfamily of proteins involved in GTP binding and hydrolysis. However, as other members of the superfamily, the Ras proteins are biologically active when they are in the GTP-bound form and inactive when bound to GDP. The molecular switching from the GDP-bound to the GTP-bound form of Ras is accompanied by conformational alterations in two parts of the protein, termed switch I and switch II regions.[128] Activation of c-Ras proteins is promoted by extracellular stimuli, including the binding of various growth factors (insulin, IGF-I, EGF, PDGF, NGF) to their cell surface receptors, which are represented by transmembrane proteins with tyrosine kinase activity. A critical step in the activation of c-Ras proteins by growth factors and other extracellular signaling agents is the release of bound GDP and the binding of GTP. This event is facilitated by a specific guanine nucleotide-releasing factor (GNRF or GRF), which in human cells is represented by a 55-kDa protein of 488 amino acids.[129] The human protein is homologous to the CDC25 gene product of the yeast *S. cerevisiae*, which also enhances GDP release, and its gene is localized in human chromosome band 15q2.4.

In contrast to the positive regulation of Ras protein activity by GNRF, Ras activity is negatively regulated by a GTPase-activating protein (GAP), which is localized in the cytoplasm and enhances the GTPase activity of c-Ras proteins.[130] Thus, Ras activity is determined by a balance between the positive action of GNRF and the negative action of GAP, as depicted in Figure 5.1. GAP is ubiquitous in higher eukaryotes and is essential for proper Ras biological activity.[131] GAP interacts with the effector binding domain of Ras and catalyzes the conversion of Ras-GTP to Ras-GDP so that this reaction proceeds at rates more than 100 times higher than intrinsic rates. Binding of GAP to GTP induces translocation of the Ras-GTP complex from the cytosol to the plasma membrane and stimulates the weak intrinsic GTPase activity of normal c-Ras proteins, thereby promoting the return of Ras to an inactive, GDP-bound state. Thus, GAP is a negative regulator of normal Ras proteins.[132] Expression of GAP in NIH/3T3 cells can suppress neoplastic transformation induced by normal c-H-*ras* gene, but does not inhibit transformation caused by the v-*ras* oncogene.[133] The carboxyl-terminal domain of the GAP protein is responsible and sufficient for the interaction of GAP with normal cellular proteins and for stimulating their GTPase activities. The amino-terminal region of GAP contains SH2 and SH3 domains that allow tight binding of GAP to the receptors for PDGF, EGF, and insulin, and the SH3 domain of GAP is essential for signal transduction.[134] The EGF receptor phosphorylates GAP at one site *in vitro*, and GAP remains firmly bound to the receptor at physiological salt concentrations.[135] The EGF receptor phosphorylates the human GAP protein at Tyr-460, a residue adjacent to GAP SH2 domains.[136] The fact that GAP is phosphorylated by diverse tyrosine kinases reinforces the concept of the existence of important biochemical linkages between the Ras and tyrosine kinase signaling pathways. A phospholipid-associated cytoplasmic protein may be involved in counteracting GAP activity, thus increasing the biological activity of Ras and stimulating cell proliferation.[137] The gene of the GAP protein is located on human chromosome 5q13-q15.[138]

The product of a distinct gene, the human gene *NF1*, which is associated with the genetic disorder von Recklinghausen's disease or neurofibromatosis type 1, encodes a protein, NF-1, which shares a region

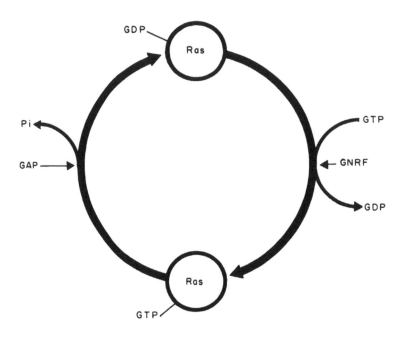

Figure 5.1 Regulation of Ras protein activity.

of homology with GAP. In a way similar to GAP, NF-1 can stimulate the GTPase activity of normal Ras protein, thus inactivating the Ras protein.[139,140] However, the precise role of the NF-1 protein in the regulation of Ras GTPase activity in normal cells remains to be evaluated.

Mutational analysis of Ras has identified a region (residues 32 to 44 and surrounding residues) which may be required for proper signaling to the downstream effector of the protein. Ras mutant-carrying lesions in this "effector domain" may be defective in their ability to transform cells, yet they retain GTP binding and localize to the plasma membrane. In a mutant affecting this region there is a dissociation between the stimulation of Ras GTPase activity by GAP and NF-1 and their potential target function.[141]

Tyrosine kinases may regulate the intrinsic GTPase activity of the GAP protein. GAP forms heteromeric complexes with several tyrosine kinases, including the v-Src oncoprotein, the EGF and PDGF receptors, and two tyrosine phosphoproteins of 62 kDa (p62) and 190 kDa (p190).[142] cDNA cloning revealed that p62 shows extensive homology to a putative hnRNP protein, GRP33, and binds to single-stranded DNA and to RNA, suggesting its involvement in RNA processing and/or utilization.[143] Binding to GAP may allow p62 and p190 to access Ras. GAP is found predominantly as a monomer in the cytosol of unstimulated cells and in the form of heteromeric complexes with p62 and p190 in cells exposed to either growth factors or oncoproteins with tyrosine kinase activity.[144] Phosphorylation of the GAP, p62, and p190 proteins correlates with transformation in cells expressing the v-Src and v-Fps oncoproteins. GAP, p62, and p190 are rapidly phosphorylated on tyrosine in fibroblasts stimulated with either EGF or the c-Fms/CSF-1 receptor kinase.[145] The v-Src protein phosphorylates bovine GAP at Tyr-457 (a residue equivalent to Tyr-460 in the human protein).[146] GAP may play a role in mediating normal functions of c-Src as well as oncogenic activities of v-Src.[147,148] Tyrosine kinases would modify Ras protein function and implicate GAP and its associated proteins as targets of both oncoproteins and growth factor receptors with tyrosine kinase activity. Ras binding induces a conformational change in GAP, which allows the SH2 and SH3 regions contained at the amino-terminal region of GAP to function by relieving a negative constraint that normally prevents GAP from interacting with its target.[149]

The functional properties of Ras proteins may be affected by phosphorylation/dephosphorylation reactions. Transforming mutations of c-Ras proteins at positions 12, 59, and 61 (the phosphoryl-binding region) abolish GTPase stimulation by the GAP protein.[130] The alanine residue at position 59 is changed to a threonine in the v-H-Ras and v-K-Ras oncoproteins, which endows them with transforming activity, and phosphorylation of Thr-59 inhibits biological activity, although Ras proteins phosphorylated at Thr-59 have the same levels of GTPase activity and nucleotide exchange rates for both GDP and GTP as the nonphosphorylated form.[150] v-Ras is susceptible to autophosphorylation by transferring the γ-phosphate of GTP (but not ATP) to Thr-59.[151] Autophosphorylation of Ras at Thr-59 is an intramolecular reaction that takes place as a side reaction of the GTPase reaction.[152] Thr-59 is close to the γ-phosphate group of GTP. In contrast to v-Ras, normal cellular Ras proteins are not phosphorylated due to the presence of an alanine residue at position 59. The v-Ras oncoprotein can be phosphorylated by protein kinase C *in vitro* at serine sites distinct from the site of autophosphorylation.[153] The ligand-activated EGF receptor kinase stimulates guanine nucleotide-binding activity and phosphorylation of Ras products.[154]

The three functional activities displayed by v-Ras proteins (GTP/GDP binding, autophosphorylation, and GTPase) are related and may depend on a single active center within the Ras molecule because they are specifically affected by a monoclonal antibody. The Ras-specific monoclonal antibody Y13-259 hampers the dissociation rate of cellular guanine nucleotide exchange reactions, underscoring the critical importance of these reactions in Ras function.[155] Normal and mutant c-H-Ras proteins undergo ADP-ribosylation in the presence of ADP-ribosyltransferase,[156] and this modification decreases Ras-associated GTPase activity and Ras ability to bind GTP.[157] Glycerol and Mn^{2+} strongly stimulate the GTPase activity of normal c-Ras.[158]

E. FUNCTIONAL ACTIVITIES OF Ras PROTEINS

Ras proteins are involved in the complex mechanisms of signal transduction across the membrane in many different biological systems. They are associated with various cellular proteins of 50 to 150 kDa.[159] c-Ras proteins are synthesized as cytosolic precursors that are proteolytically processed to mature forms that localize to the plasma membrane. Association of Ras proteins with the membrane require posttranslational modifications at the carboxyl-terminal region, including prenylation, removal of three carboxyl-terminal residues and carboxyl methylation, and palmitoylation (addition of palmitic acid residues).[160] Anchorage of Ras to the membrane is associated with prenylation, the covalent addition of a farnesyl moiety (polyisoprenoid group) to the sulfur of a cysteine in the carboxyl-terminal part of the Ras molecule.[161-165] The donor of farnesyl residues required for prenylation is farnesyl pyrophosphate, and the farnesyl residue transfer is effected by the heterodimeric enzyme, farnesyltransferase.[166,167] Farnesyl, a polyisoprenoid derived from mevalonic acid, is an intermediate of the cholesterol biosynthetic pathway. Isoprenoid addition to the Ras protein is critical for both Ras membrane association and transforming activity.[168] By contrast, nonpalmitoylated Ras derivatives containing myristic acid at their amino termini show efficient membrane association and display transforming activity even in their nonmutated, normal forms.[169]

The v-Ras protein contains a functional domain that specifies the guanine nucleotide-binding activity and another domain that is involved in membrane association in transformed cells.[170] The Ras guanine nucleotide-binding domain comprises the amino-terminal region of the protein, and the membrane-binding domain is located at the carboxyl-terminal part. In addition, Ras contains a heterogeneous domain located between the two aforementioned domains.[171] The functional domains of Ras can be characterized by using chimeric genes or molecular vectors containing deletion mutants of *ras* oncogene sequences.[171-173] Residue 12 of Ras proteins may be involved in GTP binding, as suggested by the effects of an antibody that recognizes a Ras epitope that includes residue 12.[174] Microinjection of an antibody specific for residue 12 of the v-K-Ras oncoprotein into cells transformed by the v-K-*ras* oncogene causes a transient reversion of the cells to a normal phenotype.[175] The fact that this antibody inhibits GTP binding to v-K-Ras supports the notion that GTP binding is essential to the transforming function of the v-K-Ras oncoprotein. Substitutions at position 12 of Ras, although impairing GTPase activity, have no effect on GTP binding.[176] GTPase activity of normal c-Ras is about tenfold higher than that of the oncogenically activated forms of the proteins containing amino acid substitutions at position 12.[177] A cytoplasmic protein stimulates GTP hydrolysis by normal c-Ras protein but not by Ras-containing mutations at position 12.[178] In contrast to the normal protein, the mutant Ras protein remains in the active GTP-bound state.

The human c-K-*ras* gene contains two alternative fourth codons (4A and 4B) and can express two distinct Ras proteins ($p21_{4A}$ and $p21_{4B}$), which differ only in their carboxyl termini.[179] Protein $p21_{4B}$ has a basic carboxyl terminus that is not closely related to the K-MuSV-encoded protein. In contrast, the $p21_{4A}$ carboxyl terminus is virtually identical to the carboxyl terminus encoded by K-MuSV. However, the biochemical and transforming properties of $p21_{4A}$ and $p21_{4B}$, carrying a mutation at codon 12, are similar or identical.[180] These results suggest that the effector functions important for transformation reside in the amino-terminal portion of the c-Ras molecule.

The products of constructed *ras* mutant genes containing amino acid substitutions at codon 83, 119, or 144 show decreased affinity for GTP binding as a consequence of increased rates of dissociation of GTP from Ras.[181] Nevertheless, the mutant genes induce neoplastic transformation of NIH/3T3 cells with efficiencies comparable to that of the v-H-*ras* oncogene. A role for GTP binding in Ras-induced transformation is not readily apparent on the basis of the results obtained in these experiments. Both the carboxyl-terminal part of Ras molecules, involved in the location of the protein on the inner aspect of the cell surface, and the amino-terminal part of the same protein, involved in guanine nucleotide binding, may be necessary for the expression of transforming capacity. However, the precise relationship between Ras-associated GTP binding and hydrolysis and the oncogenic transformation induced by mutant versions of Ras proteins is unknown.

The study of human hematopoietic cell lines has suggested that c-Ras proteins are equally expressed in all phases of the cell cycle.[182] However, other studies indicate that normal c-*ras* gene activity may increase specifically in late G_1 and that Ras proteins are required at that time to stimulate cellular events that are needed to initiate DNA replication in serum-stimulated murine fibroblasts.[183,184] The v-Ras oncoprotein can stimulate Ca^{2+}-and serum-deprived NRK cells to transit G_1, replicate DNA, and ultimately divide without help from exogenous growth factors.[185] The activated c-H-Ras protein can induce meiotic maturation of *Xenopus laevis* oocytes, and this effect is associated with phosphorylation of the S6 ribosomal protein and increased levels of M-phase promoting factor (MPF), which is composed of cdc2 and cyclin.[186,187] Induction of oocyte maturation by Ras protein occurs, at least in part, via activation of protein kinase C.[188] Progesterone releases the arrest of *Xenopus* oocytes in prophase of the first meiotic division, which results in activation of MPF, germinal vesicle breakdown, completion of mieiosis I, and production of an unfertilized egg arrested at metaphase II of meiosis. Thus, at least in some cellular systems, the Ras proteins exhibit mitogenic potential.

There is evidence that the three mammalian Ras proteins may not be functionally equivalent. In particular, H-Ras may have biochemical functions that are different from those of K-Ras and N-Ras proteins in relation to the induction of c-*fos* gene expression by TPA in proliferating mouse fibroblasts.[189] In any case, it is clear that interactions of Ras with other cellular proteins are central for their biological activity. Mutational analysis shows that amino acid residues 32 to 40 are required for the interaction of Ras with cellular proteins and for the biological activity of Ras.[190]

F. MUTANT c-Ras PROTEINS

Mutant c-Ras proteins containing amino acid substitutions at positions either 12, 13, or 61 exhibit oncogenic potential. These mutations may not affect Ras localization at the cell membrane or the ability of Ras to bind guanine nucleotides.[191] The mechanism responsible for the oncogenic activation of Ras proteins remains a subject of controversy, but there is some evidence that the signal transduction pathway of the mutated forms of Ras proteins is distinct from that used by normal Ras proteins.[192] The oncogenic activation of Ras may be a consequence of alterations in Ras interaction with other cellular proteins.[193] The level of intrinsic GTPase activity distinguishes normal and mutant forms of c-Ras molecules. A severalfold reduction of GTPase activity is present in the activated, mutant versions of c-Ras, which may determine a persistent activation of the GTP-associated transductional system, leading to uncontrolled cell proliferation and neoplastic transformation of susceptible cells.[194-196] A single point mutation of c-*ras* at either codon 12 or 61 could be enough to reduce the GTPase activity of Ras and activate the oncogenic potential of the gene.[197] Crystallographic studies with refined methodology suggest that the mutant c-H-Ras protein is characterized by an enlargement in the loop that binds the β-phosphate of the guanine nucleotide.[198] Such a change in the conformation of the catalytic site could explain the reduced GTPase capacity of the mutant, which keeps the protein in the GTP-bound "signal on" state for a prolonged period of time. Studies using a single fluorescent probe at the catalytic site of the N-Ras product suggest that mutated Ras proteins may stay in a "GTP-like" conformation throughout the GTPase cycle.[199] However, the role of this alteration in the transforming potential of Ras proteins is not totally clear. There is evidence that Ras proteins with either high or low GTPase activity can efficiently transform NIH/3T3 cells.[200] Amino acid substitutions at position 12 of Ras do not modify the GTP- or GDP-binding ability of the protein, making the possibility of a direct contribution of this residue to the binding site unlikely. Activation of efficient transforming properties by Ras proteins can occur by mechanims not involving reduced GTPase activity. The mitogenic activities of normal and mutant c-H-Ras proteins are equal in human cells, suggesting that these activities do not correlate with alterations in GTPase activity.[201] In general, there is a lack of correlation between guanine nucleotide binding or GTPase activity of Ras proteins and their transforming potential.[202] A constructed Ras mutant that had lost its ability to bind GTP and that did not exhibit autokinase activity was fully capable of inducing transformation of NIH/3T3 cells.[203]

The possible role of the GAP protein in Ras-induced transformation is not understood. GAP stimulates GTP hydrolysis by normal Ras, but has no effect on the impaired GTPase activity of Ras proteins containing mutation at position-12.[178] The major effect of mutations at position 12 would be to prevent stimulation of Ras GTPase activity by GAP, thereby allowing these mutants to remain at the active GTP-bound state, which could result in mitogenic stimulation of the cell.

Ras proteins may be involved in the regulation of G proteins. Ras-activating mutations can produce an altered regulation of cellular systems associated with G protein activity. Hormone-stimulated adenylyl cyclase activity is reduced in NIH/3T3 cells expressing the N-Ras protein, which may result in altered cellular proliferation.[204] Microinjection of the mutated c-K-Ras protein into quiescent rat fibroblasts results in activation of protein kinase C and induction of c-*fos* gene expression.[205] It has been suggested that mutant Ras proteins may induce constitutive activation of phospholipase C, thus enhancing the basal level of inositol phospholipid breakdown.[206] Increased phosphoinositide metabolism is observed in v-H-*ras*-transformed rat fibroblasts, which may lead to increased levels of inositol phosphates, enhanced production of diacylglycerol, and activation of protein kinase C.[207] Elevated levels of diacylglycerol are observed in *ras*-transformed neonatal liver and pancreas of transgenic mice.[208] G proteins regulate the activity of phospholipases A2 and C, and the activities of these enzymes also may be deregulated by the presence of mutated Ras which may result in reduced production of PGE_2 and arachidonic acid.[209] However, the possible role of an altered phospholipid metabolism in Ras-induced oncogenic transformation is not clear. An elevation of diacylglycerol is apparently not necessary for the development of *ras*-induced lung tumors in transgenic animals. Highly tumorigenic Chinese hamster lung fibroblasts expressing either T24 c-H-*ras* or v-K-*ras* oncogenes do not exhibit a constitutive activation of protein kinase C.[210] The enzyme is downregulated in mouse fibroblasts transfected with an activated human c-H-*ras* gene.[211] On the other hand, protein kinase C is able to phosphorylate c-K-Ras and may increase the transforming ability of the protein.[212] Cell lines that stably express high levels of protein kinase C are extremely susceptible to transformation by an activated c-H-*ras* gene.[213] Protein kinase C and the v-K-Ras oncoprotein can cooperate to increase the responsiveness of adenylyl cyclase to agonists.[214] However, the exact role of altered phosphoinositide metabolism and protein kinase C activity in transformation induced by mutant c-*ras* genes or v-*ras* oncogenes remains controversial. According to results obtained with thyroid-derived epithelial cell lines, K-*ras*-induced transformation is associated with activation of phospholipase A2.[215] However, the specificity of such changes is not totally clear. Comparative studies between *ras*-transformed cells and cells transformed by oncoproteins associated with the membrane, the cytoplasm, or the nucleus indicate the existence of common changes in inositol phospholipid metabolism in cells transformed by different oncogenes; these studies do not support the hypothesis that Ras oncoproteins are direct regulatory elements of the enzymes related to the inositol phospholipid pathway.[216]

Structural and/or functional alterations of the cell genome are probably necessary for the expression of a transformed phenotype. Deregulated expression of cellular genes could be a mechanism for the oncogenicity of mutated c-Ras proteins. Expression of a mutant c-H-*ras* gene in NIH/3T3 cells causes a rapid, marked, and transient enhancement in the levels of c-*jun* mRNA, thus delivering a growth factor-like signal to the expression of immediate-early genes.[217] Expression of the mutant c-H-*ras* gene may also induce altered c-*fos* expression and, secondarily, gene deregulations that are characteristic of the transformed phenotype, including increased levels of mRNAs for ODC, transin, and the glucose transporter. However, since expression of the *ras* oncogene may not be required for maintenance of a tumorigenic phenotype once this has been stabilized by selection, the cellular change produced by its expression can only be an early event of low specificity.[218] The results of studies showing the suppression of tumorigenicity in human cell hybrids derived from cell lines expressing different activated c-*ras* genes provide evidence that mutational activation of these genes may be insufficient for tumorigenic conversion.[219] Additional genetic changes, probably including chromosome aberrations, may be required for the expression of a fully transformed phenotype in cells expressing a mutant *ras* gene.

In conclusion, the molecular mechanisms responsible for the acquisition of increased transforming ability by mutant c-Ras proteins are not understood. The role of activated c-Ras proteins in tumorigenic processes occurring *in vivo* is also unknown but Ras mutations may be involved in the origin and/or development of some experimental and natural tumors, including human tumors. Point mutations of c-*ras* genes may contribute to the progression of tumor cell populations in selective environmental conditions.

G. ROLE OF Ras PROTEINS IN CELL PROLIFERATION AND DIFFERENTIATION

The structural and functional homologies existing between G proteins and Ras proteins suggest that they may play similar roles in the control of cell proliferation and/or cell differentiation. Ras proteins are expressed in proliferating cells, but high levels of Ras are also expressed in some nondividing, terminally differentiated cells. In normal human skin cells and human skin tumors, Ras expression occurs in differentiating cells.[220] Expression of c-H-Ras protein appears to be cell cycle-dependent in primary cultures of rat hepatocytes and in some human tumor cell lines, with the levels increasing manifold in the G_1 phase and remaining high during the S phase.[221,222] It has been suggested that c-Ras levels are much lower at the onset of the cell cycle than at its end; hence a drop in c-Ras expression may occur during or immediately after mitotic division. However, the possible role of Ras proteins in the control of cell cycle progression is not understood. Infection of NRK cells with a *ts* mutant of the acute retrovirus K-MuSV indicates that the v-Ras oncoprotein not only is sufficient to induce the G_1 progression but also induces G_2 transit and cell multiplication.[223] The c-Ras proteins may be involved in controlling some critical events in both the G_1 and G_2 phases of the cell cycle.

The role of c-Ras proteins in cell differentiation processes is unclear. Normal and mutant c-Ras proteins may induce or inhibit or may have no effect on cellular differentiation, according to the type of cell, its state of development, and the prevailing environmental conditions. The oncogenic form of c-H-Ras protein blocks the differentiation of the mouse myogenic cell line 23A2, and this effect is associated with inhibition of the expression of the myogenic regulatory gene *MyoD1* and relies on a transduction pathway dependent on protein kinase C activity.[224] In contrast, transfection of a mutant c-H-*ras* gene into P19 multipotential embryonal carcinoma cells does not interfere with differentiation induced by retinoic acid.[225] The transfected cells develop in response to retinoic acid into the same spectrum of neuroectodermal cell types as the parental P19 cells, suggesting that the Ras protein is probably not a component of the cellular machinery involved in the initiation of differentiation processes or the choice between different cell lineages.

Microinjection of the mutated forms of Ras proteins, but not of normal Ras proteins, into rat pheochromocytoma PC12 cells may result in their morphological differentiation into neuron-like cells.[226,227] This differentiation occurs in the absence of NGF, which is required for PC12 cell differentiation. Apparently, the mutated c-Ras protein is able to eliminate the need for an exogenous signal represented by NGF. Retroviruses carrying ras oncogenes also can induce PC12 cell differentiation.[228] In contrast, an irreversible blockade of differentiation is induced by K-MuSV in cultured rat thyroid cells, with suppression of thyroglobulin synthesis and iodide uptake in spite of reversion of the transformed phenotype as assessed by using *ts* mutants of the virus.[229] While the malignant properties of rat thyroid cells are suppressed at the nonpermissive temperature, the blockade of differentiated thyroid functions persists at this temperature. The latter is due to an alteration at the level of gene transcription, but the mechanism of the dissociation between expression of transformation and expression of differentiated functions is not understood. The v-Ras oncoprotein is sufficient to induce a transformed phenotype and to stimulate cell proliferation when it is administered by microinjection into quiescent NIH/3T3 cells.[230] The normal Ras protein can induce similar actions, but only at higher concentrations and the effects are not as pronounced as those of the v-H-Ras protein. Revertants of H-MuSV-transformed cells express reduced levels of v-Ras and exhibit a more normal phenotype.[231] Reversion of the transformed phenotype is observed when the cells are treated with a monoclonal antibody against v-H-Ras oncoprotein.[232] This antibody recognizes both the normal and the oncogenic Ras proteins.

Induction of c-*ras* gene expression may be dissociated from the induction of DNA synthesis and cell proliferation in particular systems. In the uterus of ovariectomized mice, estradiol induces luminal epithelial DNA synthesis, which is followed by a mitogenic response. Progesterone treatment completely suppresses this proliferative response. At the peak of DNA synthesis, the levels of c-H-*ras* and ODC mRNAs are increased but progesterone does not greatly influence the levels of these mRNAs.[233] Thus, in the uterine luminal epithelium estradiol regulates the levels of c-H-*ras* gene expression independently of cell proliferation. The c-H-Ras protein cannot act as a positive regulator of DNA synthesis and cell proliferation independently of other control mechanisms.

A role for c-Ras proteins in differentiated functions is suggested by their presence in fully differentiated tissues. Expression of c-Ras may be more tightly associated with cell differentiation than with cell proliferation. Although c-*ras* gene expression may be associated with cell proliferation in particular tissues, high levels of c-*ras* gene expression occur in nonproliferating tissues. The levels of c-H-*ras*, c-K-*ras*, and N-*ras* gene expression in different mouse tissues are regulated in a complex manner but the three genes are expressed in all tissues.[234] Differences in expression of c-*ras* occur through mouse pre- and postnatal development, and certain adult tissues preferentially express one member of the family over the others. The gene c-H-*ras* is expressed at it highest levels in mouse brain and muscle, where cell division is minimal or nonexistent. The highest levels of c-*ras* gene expression among mammalian tissues are found in the brain, especially in mature neurons.[236-237]

H. Ras PROTEINS AND THE MECHANISMS OF GROWTH FACTOR ACTION

Ras proteins have an important role in the cellular mechanisms of action and mitogenic signal transduction of growth factors, in particular of growth factors whose receptors possess tyrosine kinase activity, including the receptors for insulin, IGF-I, EGF, NGF, and PDGF.

Insulin induces expression of the c-K-*ras* proto-oncogene in rat liver cells *in vivo* and *in vitro*,[238] and the c-K-Ras protein may play a role in the mechanism of insulin action.[239] The carboxy-terminal segment of c-K-Ras is able to mediate insulin-stimulated phosphorylation of calmodulin and stimulates insulin-independent autophosphorylation of the insulin receptor. Insulin stimulation of gene expression may be mediated, at least in part, by Ras activation.[240] Fibroblasts of the 3T3-L1 line differentiate into adipocytes after continuous exposure to high doses of insulin or physiological doses of IGF-I, and expression of a transfected c-H-*ras* oncogene in these cells induces their differentiation in the absence of insulin or IGF-I.[241] Exposure of untransfected 3T3 L1 cells to insulin results in the formation of active Ras-GTP complexes.[242] However, tyrosine phosphorylation of GAP may not be an upstream regulatory event in the activation of Ras by insulin. In any case, as discussed in Volume 2, Ras proteins participate in signal transduction pathways initiated by insulin, IGF-I, and other growth factors whose receptors possess tyrosine kinase activity. Expression of the EGF and PDGF receptors, as well as expression of oncoproteins with a similar activity such as Neu/Erb-B2 and v-Src, may cause an increase in the formation of active Ras-GTP complexes.[243] Several growth factors increase the proportion of Ras in the active GTP-bound form.[244] The activated EGF receptor is much more potent than the insulin receptor for stimulating the formation of Ras-GTP complexes in a variety of cell lines.[245]

The cellular actions of growth factors such as EGF and PDGF, whose receptors possess tyrosine kinase activity, are mediated by Ras proteins through a mechanism involving the growth factor receptor-bound protein 2 (Grb2).[246] The Grb2 protein is composed of a single SH2 domain flanked by two SH3 domains and is associated with the ligand activated EGF and PDGF receptors. EGF can regulate Ras function through the formation of a complex of EGF receptor, Grb2 adapter protein, and the Sos nucleotide exchange factor.[247-251] The human Sos (Sos1) guanine nucleotide exchange factor for Ras that binds to Grb-2 has been characterized.[252] Mammalian Sos proteins are homologous to some yeast nucleotide exchange factors that induce guanine nucleotide exchange by Ras proteins contained in yeast. In *D. melanogaster*, Sos is the product of the *Son of Sevenless* gene, which has been shown to function downstream the EGF receptor tyrosine kinase receptor (DER) of the insect. Sos and DER use the Ras pathway in *Drosophila* to transduce growth and differentiation signals. A similar system is involved in governing vulval development in the nematode *Caenorhabditis elegans*, where the locus *let*-23 encodes a homolog of the EGF receptor, a second gene, *let*-60 specifies a Ras protein, and a third gene encodes a small protein, Sem-5, which consists entirely of two SH3 domains flanking an SH2 domain. Sem-5 is homologous to the mammalian protein Grb2. In mammalian systems, Grb2 plays a central role in signal transmission by coupling growth factor receptor tyrosine kinases to the Sos protein, which is capable of activating the Ras signaling pathway (Figure 5.2). Another phosphotyrosine-containing protein that can associate with Grb2 is Shc, which is the product of the *shc* oncogene and contains a carboxy-terminal SH2 domain that can bind to activated growth factor receptors.[253] Shc can couple tyrosine kinases to Grb2 through formation of an Shc-Grb2 complex that is able to regulate the mammalian Ras signaling pathway. In hematopoietic cells, the Vav protein would play a function similar to that of Grb2.[254] Vav is a 95-kDa protein that is expressed exclusively in hematopoietic cells and contains a guanine nucleotide-binding motif, a cysteine-rich region with similarity to protein kinase C, and an SH2 domain flanked by two SH3 domains. Both Grb2 and Vav may be capable of coupling receptor tyrosine kinases to the Ras signaling pathway.

Interaction between Ras and growth factors or their receptors may contribute to stimulate DNA synthesis and cell proliferation. Expression of the v-H-*ras* oncogene in rat fibroblasts leads to an increased sensitivity to growth factors and tumor promoters.[255] EGF and TGF-α may influence the cellular effects of Ras proteins. Expression in NIH/3T3 cells of a vector containing N-*ras* gene leads to stimulation of DNA synthesis, and this effect can be augmented by EGF.[256] Ras proteins can inhibit the activity of cathepsins, the major enzymes involved in degradation of EGF receptors, and may thus contribute to a more prolonged effect of EGF in its target cells.[257,258] The v-Ras protein can block calcium-induced terminal differentiation of BALB/MK epidermal keratinocytes and can abrogate completely their requirement for EGF.[259] v-Ras can confer to epidermal cells the rapid acquisition of EGF-independent growth. A similar effect is observed when the cells are infected with acute retroviruses of the *src* oncogene family.[260] However, the presence of a functional EGF receptor may not be required for *ras*-mediated neoplastic transformation.[261]

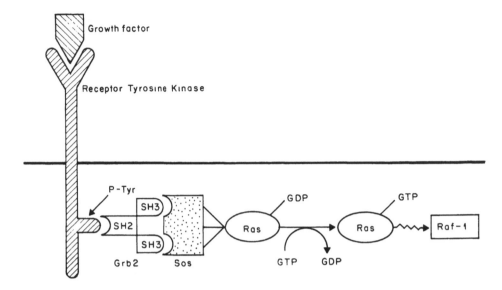

Figure 5.2 Coupling of receptor tyrosine kinase to Ras signaling pathway.

Ras proteins may be involved in mediating the responses of specific organs to hormones and growth factors. Expression of c-K-Ras protein in the rat exhibits variations according to the physiological status of the organ.[262] c-K-Ras is at it lowest level in the immature ovary of the rat and its level of expression increases with development of the corpora lutea to reach high levels at day 16 of pregnancy, after which the levels decline, rising again during lactation. This pattern of Ras expression, which mimics that of progesterone levels in the blood, suggests that ovarian c-K-Ras may play a role in the differentiated functions of the ovary. In contrast, the levels of N-Ras do not show variation with changes in the physiological function of the ovary but may be constitutive.

Ras proteins may act as mediators of factor-induced neuronal differentiation. Introduction of Ras into the cytoplasm of chick embryonic neurons can promote cell survival and fiber outgrowth, mimicking the effects of neurotrophic growth factors such as NGF, BDNF, and CNTF.[263] Ras proteins are required for the induction of neuronal differentiation of PC12 cells by NGF, FGF, IL-6, and other factors with neurotrophic activity. These factors can induce an accumulation of active Ras-GTP complexes.[264] Ras mediates modulation of signal-transducing kinases by the NGF receptor in PC12 cells, inducing MAP kinase, Raf-1, and ribosomal S6 kinases.[265] Ras-GAP complexes can activate the MAP kinase, an enzyme implicated in the regulation of cell proliferation by various stimuli, including hormones and growth factors.[266] The cellular regulatory effects of MAP kinase take place, at least in part, at the nuclear level and may lead to changes in gene expression. The N-Ras protein interacts with metabolic pathways involved in the induction of c-*fos* gene expression by NGF and basic FGF in PC12 cells.[267]

Ras proteins may have a role in the mechanisms associated with the control of cell proliferation by growth factors in hematopoietic tissues.[268] Ras may also be involved in the mechanisms of action of hematopoietic growth factors and cytokines. Ligand-induced activation of IL-2, IL-3, IL-5, and CSF-2 receptors results in Ras protein activation in various hematopoietic cell lines.[269] The upstream signal of Ras in these cells involves tyrosine kinase activity. GTP-bound c-Ras proteins are increased in lympho-cytes and myeloid cell lines after the addition of IL-2, IL-3, or CSF-2.[270,271] Because these factors are strictly required for the survival and growth of these cells, these results suggest that Ras transduces signals controlling cell growth in lymphoid and myeloid cells. In contrast, IL-4 does not seem to transmit signals through c-Ras, although it acts as an indispensable factor for the survival or growth of lymphoid and myeloid cell lines. Ras may induce the expression of cytokines or growth factors in specific types of cells. Antigen triggering of the T-cell receptor results in an accumulation of GTP-bound Ras protein, which may be due to activation of the IL-2 promoter by a mechanism mediated by protein kinase C.[272] The CSF-1 receptor is a transmembrane protein with intrinsic tyrosine kinase activity and, as it is true for other receptors of this type, Ras proteins may have an important role in the transduction of CSF-1-specific signals after CSF-1 binding to its receptor on the cell surface.

Ras proteins may have a role in the processes associated with tissue regeneration, which involves functional interactions between hormones, growth factors, and proto-oncogene products. Liver regeneration after partial hepatectomy is associated with a complex pattern of changes that include alterations in the expression of Ras proteins and α_1-adrenergic receptors.[273]

Ras proteins have an important role in hormone-induced maturation of amphibian oocytes.[274] Micro-injection of a monoclonal antibody against Ras into *Xenopus laevis* oocytes results in acceleration of progesterone-induced germinal vesicle breakdown.[275] On the other hand, microinjection of the antibody reduces the maturation of oocytes after insulin induction.[276] These results suggest that c-Ras proteins have a role in the events associated with amphibian oocyte maturation and that insulin induces maturation of oocytes by a pathway different from that of steroid hormones. However, the oocyte maturation effect of insulin and steroid hormones may be mediated by the appearance of a common cytoplasmic meiosis- or maturation-promoting activity. Oncogenic Ras protein and insulin induction *Xenopus* oocyte maturation is mediated by overlapping yet distinct mechanisms.[277]

Expression of activated c-Ras proteins may result in altered growth factor action. In the rat intestinal epithelial cell line IEC-18, transformed to tumorigenicity by expression of a transfected c-H-*ras* gene, there is a general relationship between the loss of TGF-β-mediated growth regulation and the levels of c-H-*ras* expression.[278] The mechanism responsible for *ras*-induced changes in TGF-β responsiveness is unknown. Microinjection of cloned genes of the *ras* family can stimulate DNA synthesis in quiescent mammalian cells.[279,280] The Ras-specific monoclonal antibody Y13-259 blocks the synthesis of DNA and the proliferation of serum-stimulated mouse NIH/3T3 cells.[281] It may be concluded that Ras proteins are a common element in the events associated with the mechanism of action of growth factors, particularly growth factors whose receptors possess tyrosine kinase activity.

I. OTHER FUNCTIONS OF Ras PROTEINS

Ras proteins have important functions not only in the processes directly associated with cell proliferation and differentiation but also in relation to specialized cellular functions. They are involved in signal transduction at the level of the cell membrane and may induce functional changes in the membrane. Voltage-dependent K^+ currents across the cell membrane may be targets for the action of both v-Ras and c-Ras proteins.[282] In differentiated neurons of the snail *Hermissenda*, expression of either the v-Ras oncoprotein or the normal c-H-Ras protein induces an increase in voltage-sensitive, inward Ca^{2+} current across the soma membrane, although the effect of the viral protein is more rapid and sustained than that of the normal protein.[283] In hybrids between neuroblastoma and glioma cells (N × 108CC15 cells), which are commonly used as a model for neuronal cells, the normal c-H-Ras protein and its mutated (Val-12) counterpart induce enlarged Ca^{2+} currents.[284] Thus, calcium channels may be a target for Ras protein action. Ras proteins may also be involved in the stimulation of cell surface ruffling and fluid phase pinocytosis, as shown by the specific inhibition of these processes in cells microinjected with anti-Ras antibodies.[285] Ras protein activity is essential for T-cell antigen receptor signal transduction as well as for the control of IL-2 gene expression in T cells.[286,287] Thus, Ras proteins may be pivotal signaling molecules in T lymphocyte activation.

The process of fertilization in mammals requires a spermatozoon to undergo a series of events before it can fuse with the egg plasma membrane. Ras proteins are present in mature sperm cells and may have a role in capacitation and/or acrosome reaction of human sperm cells.[288] An H-Ras-specific monoclonal antibody can reduce the capacitation of human sperm cells for egg fertilization.

Ras proteins are indirectly involved in the regulation of protein phosphorylation. There is evidence that Ras proteins may either stimulate or inhibit the phosphorylation of mitochondrial membrane proteins.[289] A 38-kDa protein present in rat liver plasma membrane is phosphorylated *in vitro* by Ras of either cellular or viral origin.[290] The mechanisms by which Ras proteins regulate protein phosphorylation remains to be elucidated, but Ras activity is required for the formation of functional complexes containing serine/threonine kinases such as Raf-1 and MAP kinases.[291]

Ras proteins may have a role in the regulation of gene expression, but this effect is exerted by indirect mechanisms. The c-H-Ras protein induces high levels of expression of metallothionein mRNA in T24 bladder carcinoma and HS578 mammary carcinoma cells.[292] The precise biological significance of this induction is unknown. Metallothioneins are ubiquitous, cysteine-rich proteins that bind heavy metals such as zinc, copper, and cadmium, but their levels are regulated in relation to cell growth and they may have a role in developmental processes. Ras proteins are implicated in the transmission, from the cell surface to the nucleus, of mitogenic signals elicited by growth factor receptors with tyrosine kinase activity.

J. ROLE OF GROWTH FACTORS IN *ras*-INDUCED TRANSFORMATION

Mutant c-*ras* genes are frequently associated with tumors occurring under both experimental and natural conditions, including human tumors.[293,294] Moreover, mutant c-H-*ras*, c-K-*ras*, and N-*ras* genes display oncogenic potential upon their transfection into rodent fibroblasts.[295-297] A correlation may exist in these systems between the expression of mutant c-Ras protein and the expression of a transformed phenotype in the transfected cells. Reversion of these changes may occur when the production of the altered Ras protein is interrupted. It is unlikely, however, that mutations of c-*ras* genes are universally and critically involved in the origin of tumors, especially of the common types of human tumors. A high incidence of mutant c-*ras* genes has been found in some types of human tumors such as carcinomas of the colon and pancreas,[298-300] but in other types of human tumors the incidence of c-*ras* mutations may be rather low. In contrast to the susceptibility of rodent cell lines to transformation by mutant c-*ras* genes, normal human cells are resistant to neoplastic transformation induced by these genes and remain dependent on the exogenous supply of growth factors for their proliferation *in vitro*.[301] Notwithstanding, c-*ras* mutations may participate, in concert with other genetic changes, in certain human tumorigenic processes, especially in the progression of some tumor cell populations.

The mechanisms by which mutated c-Ras proteins induce neoplastic transformation remain a subject of controversy. Point mutations at positions 12, 13, or 61 of the c-Ras protein molecule can activate its oncogenic potential. Because these hot spots are localized at the GTP-binding site of Ras, it has been assumed that GTP binding and hydrolysis to GDP and phosphate are critically involved in the oncogenic action of mutant Ras proteins. However, a constructed mutant with a substitution of valine for glycine at position 10, which had lost its ability to bind GTP and did not show autokinase activity, was fully capable of transforming NIH/3T3 cells to a tumorigenic phenotype.[203] It has been proposed that a common effect of transformation by mutated Ras products and other cytoplasmic and membrane-associated oncoproteins is a functional alteration of G protein coupling and that a similar modification of different kinds of G proteins may account for the pleiotropic alterations of signal transduction observed in the transformed cells.[302] Studies with transfection into the C2 murine myoblast cell line of a vector containing a mutationally activated N-*ras* gene suggest that Ras oncoproteins can reversibly activate an intracellular cascade that prevents the establishment of a differentiated phenotype.[303] It is well known that a blockade of differentiation may be associated in certain types of cells with the expression of a transformed phenotype. Transformation of rat fibroblasts by transfection of mutant c-*ras* genes is a gradual but reversible process that depends on the relative abundance of *ras* gene sequences and their corresponding transcripts.[295] A minimum level of the mutant Ras protein may be of critical importance for the expression of a transformed phenotype in some systems. Microinjection of a monoclonal antibody against Ras proteins can inhibit the proliferation of normal cells, while tumor cells may continue to proliferate following injection.[304] However, tumor cells containing a mutant c-*ras* gene are only partially inhibited to proliferate by the injected antibody. These results suggest that genes which act independently of Ras proteins are in large part responsible for tumor cell proliferation.

Growth factors may have a role in *ras*-induced transformation. Rat-1 cells induced to express high levels of the normal c-H-Ras product may exhibit a transformed phenotype, but this change is dependent on the presence of growth factors in the culture medium.[305] Clonal cell lines of mouse NIH/3T3 cells transformed by the EJ c-H-*ras* oncogene secrete a variety of autocrine growth factors into the medium, including EGF, TGF-α, PDGF, and basic FGF.[306] Growth factors contained in serum, probably including IGF-I and IGF-II, are required for v-*ras*-induced transformation of BALB/c 3T3 cells.[307] Mutations in *ras* genes may alter expression of several cytokine genes in cultured human cells through both transcriptional and posttranscriptional mechanisms.[308] However, the precise role of growth factors in *ras*-induced transformation is not understood.

REFERENCES

1. **Gilman, A. G.,** G proteins: transducers of receptor-generated signals, *Annu. Rev. Biochem.,* 56, 615, 1987.
2. **Casey, P. J. and Gilman, A. G.,** G protein involvement in receptor-effector coupling, *J. Biol. Chem.,* 263, 2577, 1988.
3. **Neer, E. J. and Clapham, D. E.,** Roles of G protein subunits in transmembrane signalling, *Nature,* 333, 129, 1988.

4. **Lochrie, M. A. and Simon, M. I.,** G protein multiplicity in eukaryotic signal transduction systems, *Biochemistry,* 27, 4957, 1988.

5. **Freissmuth, M., Casey, P. J., and Gilman, A. G.,** G proteins control diverse pathways of transmembrane signaling, *FASEB J.,* 3, 2125, 1989.

6. **Birnbaumer, L., Abramowitz, J., and Brown, A. M.,** Receptor-effector coupling by G proteins, *Biochim. Biophys. Acta,* 1031, 163, 1990.

7. **Taylor, C. W.,** The role of G-proteins in transmembrane signalling, *Biochem. J.,* 272, 1, 1990.

8. **Simon, M. I., Strathmann, M. P., and Gautam, N.,** Diversity of G-proteins in signal transduction, *Science,* 252, 802, 1991.

9. **Kaziro, Y., Itoh, H., Kozasa, T., Nakafuku, M., and Satoh, T.,** Structure and function of signal-transducing GTP-binding proteins, *Annu. Rev. Biochem.,* 60, 349, 1991.

10. **Spiegel, A. M., Shenker, A., and Weinstein, L. S.,** Receptor-effector coupling by G proteins: implications for normal and abnormal signal transduction, *Endocrine Rev.,* 13, 536, 1992.

11. **Sanders, D. A.,** A guide to low molecular weight GTPases, *Cell Growth Differ.,* 1, 251, 1990.

12. **Khorana, H. G.,** Rhodopsin, photoreceptor of the rod cell. An emerging pattern for structure and function, *J. Biol. Chem.,* 267, 1, 1992.

13. **Nathans, J.,** Rhodopsin. Structure, function, and genetics, *Biochemistry,* 31, 4923, 1992.

14. **Shamay, A., Pines, M., Waksman, M., and Gertler, A.,** Proliferation of bovine undifferentiated mammary epithelial cells *in vitro* is modulated by G-proteins, *Mol. Cell. Endocrinol.,* 69, 217, 1990.

15. **Moolenaar, W. H.,** G-protein-coupled receptors, phosphoinositide hydrolysis, and cell proliferation, *Cell Growth Differ.,* 2, 359, 1991.

16. **Kurachi, Y., Ito, H., Sugimoto, T., Shimizu, T., Miki, I., and Ui, M.,** Arachidonic acid metabolites as modulators of the G protein-gated cardiac K^+ channel, *Nature,* 337, 555, 1989.

17. **Ashkenazi, A., Peralta, E. G., Winslow, J. W., Ramachandran, J., and Capon, D. J.,** Functionally distinct G proteins selectively couple different receptors to PI hydrolysis in the same cell, *Cell,* 56, 487, 1989.

18. **Snaar-Jagalska, B. E., Kesbeke, F., Pupillo, M., and Van Haastert, P. J. M.,** Immunological detection of G-protein α-subunits in *Dictyostelium discoideum, Biochem. Biophys. Res. Commun.,* 156, 757, 1988.

19. **Fujimara, H.,** The yeast G-protein homolog is involved in the mating pheromone signal transduction system, *Mol. Cell. Biol.,* 9, 152, 1989.

20. **Hoshino, S., Miyazawa, H., Enomoto, T., Hanaoka, F., Kikuchi, Y., Kikuchi, A., and Ui, M.,** A human homologue of the yeast *GST1* gene codes for a GTP-binding protein as is expressed in a proliferation-dependent manner in mammalian cells, *EMBO J.,* 8, 3807, 1989.

21. **Shinjo, K., Koland, J. G., Hart, M. J., Narasimhan, V., Johnson, D. I., Evans, T., and Cerione, R. A.,** Molecular cloning of the gene for the human placental GTP-binding protein Gp (G25K): identification of this GTP-binding protein as the human homolog of the yeast cell-division-cycle protein CDC42, *Proc. Natl. Acad. Sci. U.S.A.,* 87, 9853, 1990.

22. **Holbrook, S. R. and Kim, S.-H.,** Molecular model of the G protein subunit based on the crystal structure of the HRAS protein, *Proc. Natl. Acad. Sci. U.S.A.,* 86, 1751, 1989.

23. **Tang, W.-J. and Gilman, A. G.,** Type-specific regulation of adenylyl cyclase by G protein βγ subunits, *Science,* 254, 1500, 1991.

24. **Kim, S., Ang, S.-L., Bloch, D. B., Bloch, K. D., Kawahara, Y., Tolman, C., Lee, R., Seidman, J. G., and Neer, E. J.,** Identification of cDNA encoding an additional subunit of a human GTP-binding protein: expression of three α_i subtypes in human tissues and cell lines, *Proc. Natl. Acad. Sci. U.S.A.,* 85, 4153, 1988.

25. **Wong, Y. H., Conklin, B. R., and Bourne, H. R.,** Gz-mediated hormonal inhibition of cyclic AMP accumulation, *Science,* 255, 339, 1992.

26. **Jones, T. L. Z., Simonds, W. F., Merendino, J. J., Brann, M. R., and Spiegel, A. M.,** Myristoylation of an inhibitory GTP-binding protein α subunit is essential for its membrane attachment, *Proc. Natl. Acad. Sci. U.S.A.,* 87, 568, 1990.

27. **Strathmann, M. and Simon, M. I.,** G-protein diversity: a distinct class of α subunits is present in vertebrates and invertebrates, *Proc. Natl. Acad. Sci. U.S.A.,* 87, 9113, 1990.

28. **Wilkie, T. M., Scherle, P. A., Strathmann, M. P., Slepak, V. Z., and Simon, M. I.,** Characterization of G-protein α-subunits in the G_q class: expression in murine tissues and in stromal and hematopoietic cell lines, *Proc. Natl. Acad. Sci. U.S.A.,* 88, 10049, 1991.

29. **Smrcka, A. V., Hepler, J. R., Brown, K. D., and Sternweis, P. C.,** Regulation of polyphosphoinositide-specific phospholipase C activity by purified G_q, *Science,* 251, 804, 1991.

30. **Hsieh, K.-P. and Martin, T. F. J.,** Thyrotropin-releasing hormone and gonadotropin-releasing hormone receptors activate phospholipase C by coupling to the guanosine triphosphate-binding proteins G_q and G_{11}, *Mol. Endocrinol.,* 6, 1673, 1992.

31. **Fong, H. K. W., Amatruda, T. T., III, Birren, B. W., and Simon, M. I.,** Distinct forms of the β subunit of GTP-binding regulatory protein identified by molecular cloning, *Proc. Natl. Acad. Sci. U.S.A.,* 84, 3792, 1987.

32. **Gao, B., Gilman, A. G., and Robishaw, J. D.,** A second form of the β subunit of signal-transducing G proteins, *Proc. Natl. Acad. Sci. U.S.A.,* 84, 6122, 1987.

33. **Amatruda, T. T., III, Gautam, N., Fong, H. K. W., Northup, J. K., and Simon, M. I.,** The 35- and 36-kDa β subunits of GTP-binding regulatory proteins are products of separate genes, *J. Biol. Chem.,* 263, 5008, 1988.

34. **Buss, J. E., Mumby, S. M., Casey, P. J., Gilman, A. G., and Sefton, B. M.,** Myristoylated β subunits of guanine nucleotide-binding regulatory proteins, *Proc. Natl. Acad. Sci. U.S.A.,* 84, 7493, 1987.

35. **O'Brien, R. M., Houslay, M. D., Milligan, G., and Siddle, K.,** The insulin receptor tyrosyl kinase phosphorylates homomeric forms of the guanine nucleotide regulatory proteins G_i and G_0, *FEBS Lett.,* 212, 281, 1987.

36. **Hausdorff, W. P., Pitcher, J. A., Luttrell, D. K., Linder, M. E., Kurose, H., Parsons, S. J., Caron, M. G., and Lefkowitz, R. J.,** Tyrosine phosphorylation of G protein α subunits by pp60[c-src], *Proc. Natl. Acad. Sci. U.S.A.,* 89, 5720, 1992.

37. **Masters, S. B., Sullivan, K. A., Miller, R. T., Beiderman, B., Lopez, N. G., Ramachandran, J., and Bourne, H. R.,** Carboxyl terminal domain of $G_s\alpha$ specifies coupling of receptors to stimulation of adenylate cyclase, *Science,* 241, 448, 1988.

38. **Miller, R. T., Masters, S. B., Sullivan, K. A., Beiderman, B., and Bourne, H. R.,** A mutation that prevents GTP-dependent activation of the α chain of G_s, *Nature,* 334, 712, 1988.

39. **Cerione, R. A., Staniszewski, C., Gierschik, P., Codina, J., Somers, R. L., Birnbaumer, L., Spiegel, A. M., Caron, M. C., and Lefkowitz, R. J.,** Mechanism of guanine nucleotide regulatory protein-mediated inhibition of adenylate cyclase: studies with isolated subunits of transducin in a reconstituted system, *J. Biol. Chem.,* 261, 9514, 1986.

40. **Neer, E. J., Michel, T., Eddy, R., Shows, T., and Seidman, J. G.,** Genes for two homologous G-protein α subunits map to different human chromosomes, *Hum. Genet.,* 77, 259, 1987.

41. **Harris, B. A., Robishaw, J. D., Mumby, S. M., and Gilman, A. G.,** Molecular cloning of complementary DNA for the α subunit of the G protein that stimulates adenylate cyclase, *Science,* 229, 1274, 1985.

42. **Robishaw, J. D., Russell, D. W., Harris, B. A., Smigel, M. D., and Gilman, A. G.,** Deduced primary structure of the α subunit of the GTP-binding stimulatory protein of adenylate cyclase, *Proc. Natl. Acad. Sci. U.S.A.,* 83, 1251, 1986.

43. **Itoh, H., Kozasa, T., Nagata, S., Nakamura, S., Katada, T., Ui, M., Iwai, S., Ohtsuka, E., Kawasaki, H., Suzuki, K., and Kaziro, Y.,** Molecular cloning and sequence determination of cDNAs for α subunits of the guanine nucleotide-binding proteins G_s, G_i, and G_0 from rat brain, *Proc. Natl. Acad. Sci. U.S.A.,* 83, 3776, 1986.

44. **Van Meurs, K. P., Angus, C. W., Lavu, S., Kung, H.-F., Czarnecki, S. K., Moss, J., and Vaughan, M.,** Deduced amino acid sequence of bovine retinal $G_0\alpha$: similarities to other guanine nucleotide-binding proteins, *Proc. Natl. Acad. Sci. U.S.A.,* 84, 3107, 1987.

45. **Strathmann, M., Wilkie, T. M., and Simon, M. I.,** Diversity of the G-protein family: sequences from five additional α subunits in the mouse, *Proc. Natl. Acad. Sci. U.S.A.,* 86, 7407, 1989.

46. **Sullivan, K. A., Liao, Y.-C., Alborzi, A., Beiderman, B., Chang, F.-H., Masters, S. B., Levinson, A. D., and Bourne, H. R.,** Inhibitory and stimulatory G proteins of adenylate cyclase: cDNA and amino acid sequences of the α chains, *Proc. Natl. Acad. Sci. U.S.A.,* 83, 6687, 1986.

47. **Didsbury, J. R., Ho, Y.-S., and Snyderman, R.,** Human G_i protein subunit: deduction of amino acid structure from a cloned cDNA, *FEBS Lett.,* 211, 160, 1987.

48. **Bray, P., Carter, A., Guo, V., Puckett, C., Kamholz, J., Spiegel, A., and Nirenberg, M.,** Human cDNA clones for an α subunit of G_i signal-transduction protein, *Proc. Natl. Acad. Sci. U.S.A.,* 84, 5115, 1987.

49. **Itoh, H., Toyama, R., Kozasa, T., Tsukamoto, T., Matsuoka, M., and Kaziro, Y.,** Presence of three distinct molecular species of G_i protein α subunit: structure of rat cDNAs and human genomic DNAs, *J. Biol. Chem.,* 263, 6656, 1988.

50. **Letterio, J. J., Coughlin, S. R., and Williams, L. T.,** Pertussis toxin-sensitive pathway in the stimulation of c-*myc* expression and DNA synthesis by bombesin, *Science,* 234, 1117, 1986.

51. **Rapiejko, P. J., Watkins, D. C., Ros, M., and Malbon, C. C.,** Thyroid hormones regulate G-protein β-subunit mRNA expression *in vivo, J. Biol. Chem.,* 264, 16183, 1989.

52. **Nagahara, H., Nishimura, S., Sugimura, T., and Obata, H.,** A 29 kDa GTP-binding protein expressed in mouse brain, lung, kidney, spleen and transformed NIH3T3 cells, *Biochem. Biophys. Res. Commun.,* 149, 686, 1987.

53. **Libert, F., Parmentier, M., Lefort, A., Dinsart, C., Van Sande, J., Maenhaut, C., Simons, M.-J., Dumont, J. E., and Vassart, G.,** Selective amplification and cloning of four new members of the G protein-coupled receptor family, *Science,* 244, 569, 1989.

54. **Wright, M. S., Gautvik, V. T., and Gautvik, K. M.,** Cloning strategies for peptide hormone receptors, *Acta Endocrinol.,* 126, 97, 1992.

55. **Houslay, M. D.,** G-protein linked receptors: a family probed by molecular cloning mutagenesis, *Clin. Endocrinol.,* 36, 25, 1992.

56. **Wess, J., Nanavati, S., Vogel, Z., and Maggio, R.,** Functional role of proline and tryptophan residues highly conserved among G protein-coupled receptors studied by mutational analysis of the m3 muscarinic receptor, *EMBO J.,* 12, 331, 1993.

57. **Gutkind, J. S., Novotny, E. A., Brann, M. R., and Robbins, K. C.,** Muscarinic acetylcholine receptor subtypes as agonist-dependent oncogenes, *Proc. Natl. Acad. Sci. U.S.A.,* 88, 4703, 1991.

58. **Klinz, F.-J., Yu, V. C., Sadée, W., and Costa, T.,** Differential expression of α-subunits of G-proteins in human neuroblastoma-derived cell clones, *FEBS Lett.,* 224, 43, 1987.

59. **Kato, K., Asano, T., Kamiya, N., Haimoto, H., Hosoda, S., Nagasaka, A., Ariyoshi, Y., and Ishiguro, Y.,** Production of the α subunit of guanine nucleotide-binding protein G_0 by neuroendocrine tumors, *Cancer Res.,* 47, 5800, 1987.

60. **Lester, B. R., McCarthy, J. B., Sun, Z., Smith, R. S., Furcht, L. T., and Spiegel, A. M.,** G-protein involvement in matrix-mediated motility and invasion of high and low experimental metastatic B16 melanoma clones, *Cancer Res.,* 49, 5940, 1989.

61. **Kalinec, G., Nazarali, A. J., Hermouet, S., Xu, N., and Gutkind, J. S.,** Mutated α subunit of the G_q protein induces malignant transformation in NIH 3T3 cells, *Mol. Cell. Biol.,* 12, 4687, 1992.

62. **Allen, L. F., Lefkowitz, R. J., Caron, M. G., and Cotecchia, S.,** G-protein coupled receptor genes as protooncogenes: constitutively activating mutations of the α_{1B}-adrenergic receptor enhance mitogenesis and tumorigenicity, *Proc. Natl. Acad. Sci. U.S.A.,* 88, 11354, 1991.

63. **Vallar, L., Spada, A., and Giannatasion, G.,** Altered G_s and adenylate cyclase activity in human GH-secreting pituitary adenomas, *Nature,* 330, 566, 1987.

64. **Landis, C. A., Masters, S. B., Spada, A., Pace, A. M., Bourne, H. R., and Vallar, L.,** GTPase inhibiting mutations activate the α chain of G_s and stimulate adenylyl cyclase in human pituitary tumours, *Nature,* 340, 692, 1989.

65. **Lyons, J., Landis, C. A., Harsh, G., Vallar, L., Grünewald, K., Feichtinger, H., Duh, Q.-Y., Clark, O. H., Kawasaki, E., Bourne, H. R., and McCormick, F.,** Two G protein oncogenes in human endocrine tumors, *Science,* 249, 655, 1990.

66. **Suarez, H. G., du Villard, J. A., Caillou, B., Schlumberger, M., Parmentier, C., and Monier, R.,** *gsp* mutations in human thyroid tumors, *Oncogene,* 6, 677, 1991.

67. **O'Sullivan, C., Barton, C. M., Staddon, S. L., Brown, C. L., and Lemoine, N. R.,** Activating point mutations of the *gsp* oncogene in human thyroid adenomas, *Mol. Carcinogenesis,* 4, 345, 1991.

68. **Weinstein, L. S., Shenker, A., Gejman, P. J., Merino, M. J., Friedman, E., and Spiegel, A. M.,** Activating mutations of the stimulatory G protein in the McCune-Albright syndrome, *N. Engl. J. Med.,* 325, 1688, 1991.

69. **Schwindinger, W. F., Francomano, C. A., and Levine, M. A.,** Identification of a mutation in the gene encoding the α subunit of the stimulatory G protein of adenylyl cyclase in McCune-Albright syndrome, *Proc. Natl. Acad. Sci. U.S.A.,* 89, 5152, 1992.

70. **Barbacid, M.,** *ras* genes, *Annu. Rev. Genet.,* 56, 779, 1987.

71. **Santos, E. and Nebreda, A. R.,** Structural and functional properties of Ras proteins, *FASEB J.,* 3, 2151, 1989.

72. **Dar-Sagi, D.,** Ras proteins: biological effects and biochemical targets, *Anticancer Res.,* 9, 1427, 1989.

73. **Haubruck, H. and McCormick, F.,** Ras p21: effects and regulation, *Biochim. Biophys. Acta,* 1072, 215, 1991.

74. **Grand, R. J. M. and Owen, D.,** The biochemistry of Ras p21, *Biochem. J.,* 279, 609, 1991.

75. **Bollag, G. and McCormick, F.,** Regulation and effectors of Ras proteins, *Annu. Rev. Cell Biol.,* 7, 601, 1991.

76. **Downward, J.,** Regulatory mechanisms for Ras proteins, *BioEssays,* 14, 177, 1992.

77. **Lowy, D. R. and Willumsen, B. M.,** Function and regulation of Ras, *Annu. Rev. Biochem.,* 62, 851, 1993.

78. **Satoh, T., Nakafuku, M., and Kaziro, Y.,** Function of Ras as a molecular switch in signal transduction, *J. Biol. Chem.,* 267, 24149, 1992.

79. **Chardin, P.,** The Ras superfamily proteins, *Biochimie,* 70, 865, 1988.

80. **McCormick, F.,** Biochemical properties of mammalian Ras proteins and their relatives, *Adv. Regul. Cell Growth,* 1, 233, 1989.

81. **Downward, J.,** The Ras superfamily of small GTP-binding proteins, *Trends Biochem. Sci.,* 15, 469, 1990.

82. **Chardin, P.,** Small GTP-binding proteins of the Ras family: a conserved functional mechanism?, *Cancer Cells,* 3, 117, 1991.

83. **Valencia, A., Chardin, P., Wittinghofer, A., and Sander, C.,** The Ras protein family — evolutionary tree and role of conserved amino acids, *Biochemistry,* 30, 4637, 1991.

84. **Tamanoi, F.,** Yeast *ras* genes, *Biochim. Biophys. Acta,* 948, 1, 1988.

85. **Broach, J. R.,** *ras* genes in *Saccharomyces cerevisiae*: signal transduction in search of a pathway, *Trends Genet.,* 7, 28, 1991.

86. **Hagège, D., Andeol, Y., Boccara, M., Schmitt, P., Jeltsch, J.-M., Barrientos, E., Signoret, J., and Gaspar, T.,** *ras*-related proto-oncogenes are transcribed in leaves and callus from sugarbeet, *J. Plant Physiol.,* 139, 509, 1992.

87. **Casale, W. L., McConnell, D. G., Wang, S.-Y., Lee, Y.-J., and Linz, J. E.,** Expression of a gene family in the dimorphic fungus *Mucor racemosus* which exhibits striking similarity to human *ras* genes, *Mol. Cell. Biol.,* 10, 6654, 1990.

88. **Jordano, J. and Perucho, M.,** Chromatin structure of the promoter region of the human c-K-*ras* gene, *Nucleic Acids Res.,* 14, 7361, 1986.

89. **Trimble, W. S. and Hozumi, N.,** Deletion analysis of the c-Ha-*ras* oncogene promoter, *FEBS Lett.,* 219, 70, 1987.

90. **Spandidos, D. A. and Holmes, L.,** Transcriptional enhancer activity in the variable tandem repeat DNA sequence downstream of the human Ha-*ras*-1 gene, *FEBS Lett.,* 218, 41, 1987.

91. **Cohen, J. B., Walter, M. V., and Levinson, A. D.,** A repetitive sequence element 3′ of the human c-Ha-*ras*1 gene has enhancer activity, *J. Cell. Physiol.,* Suppl. 5, 75, 1987.

92. **Hoffman, E. K., Trusko, S. P., Freeman, N., and George, D. L.,** Structural and functional characterization of the promoter region of the mouse c-Ki-*ras* gene, *Mol. Cell. Biol.,* 7, 2592, 1987.

93. **Damante, G., Filetti, S., and Rapoport, B.,** Studies on the promoter region of the c-Ha-*ras* gene in FRTL5 rat thyroid cells, *Mol. Endocrinol.,* 1, 729, 1987.

94. **Damante, G. and Rapoport, B.,** A suppressor of transcriptional activity is present upstream from the rat c-Ha-*ras* promoter, *J. Mol. Biol.,* 200, 213, 1988.

95. **Wolfman, A., Moscucci, A., and Macara, I. G.,** Evidence for multiple, *ras*-like, guanine nucleotide-binding proteins in Swiss 3T3 plasma membranes. Stimulation of GTPase activity by cytosolic factors, *J. Biol. Chem.,* 264, 10820, 1989.

96. **Quinn, M. T., Parkos, C. A., Walker, L., Orkin, S. H., Dinauer, M. C., and Jesaitis, A. J.,** Association of a Ras-related protein with cytochrome *b* of human neutrophils, *Nature,* 342, 198, 1989.

97. **Béranger, F., Goud, B., Tavitian, A., and de Gunzburg, J.,** Association of the Ras-antagonistic Rap1/Krev-1 proteins with the Golgi complex, *Proc. Natl. Acad. Sci. U.S.A.,* 88, 1606, 1991.

98. **Quilliam, L. A., Der, C. J., Clark, R., O'Rourke, E. C., Zhang, K., McCormick, F., and Bokoch, G. M.,** Biochemical characterization of baculovirus-expressed *rap*1A/Krev-1 and its regulation by GTPase activating proteins, *Mol. Cell. Biol.,* 10, 2901, 1990.

99. **Kitayama, H., Sugimoto, Y., Matsuzaki, T., Ikawa, Y., and Noda, M.,** A *ras*-related gene with transformation suppressor activity, *Cell,* 56, 77, 1989.

100. **Jähner, D. and Hunter, T.,** The *ras*-related gene *rhoB* is an immediate-early gene inducible by v-Fps, epidermal growth factor, and platelet-derived growth factor in rat fibroblasts, *Mol. Cell. Biol.,* 11, 3682, 1991.

101. **Shirsat, N. V., Pignolo, R. J., Kreider, B. L., and Rovera, G.,** A member of the *ras* gene superfamily is expressed specifically in T, B and myeloid hemopoietic cells, *Oncogene,* 5, 769, 1990.

102. **Munemitsu, S., Innis, M. A., Clark, R., McCormick, F., Ullrich, A., and Polakis, P.,** Molecular cloning and expression of a G25K cDNA, the human homolog of the yeast cell cycle gene *CDC42, Mol. Cell. Biol.,* 10, 5977, 1990.

103. **Lowe, D. G., Capon, D. J., Delwart, E., Sakaguchi, A. Y., Naylor, S. L., and Goeddel, D. V.,** Structure of the human and murine R-*ras* genes, novel genes closely related to *ras* protein-oncogenes, *Cell,* 48, 137, 1987.

104. **Lowe, D. G. and Goeddel, D. V.,** Heterologous expression and characterization of the human R-*ras* gene product, *Mol. Cell. Biol.,* 7, 2845, 1987.

105. **Schlichting, I., Wittinghofer, A., and Rösch, P.,** Proton NMR studies of the GDP.Mg^{2+} complex of the Ha-*ras* oncogene product p21, *Biochem. Biophys. Res. Commun.,* 150, 444, 1988.

106. **de Vos, A. M., Tong, L., Milburn, M. V., Matias, P. M., Jancarik, J., Noguchi, S., Nishimura, S., Miura, K., Ohtsuka, E., and Kim, S.-H.,** Three dimensional structure of an oncogene protein: catalytic domain of human c-H-*ras* p21, *Science,* 239, 888, 1988.

107. **Jurnak, F.,** The three-dimensional structure of c-H-*ras* p21: implications for oncogene and G protein studies, *Trends Biochem. Sci.,* 13, 195, 1988.

108. **Pai, E. F., Kabsch, W., Krengel, U., Holmes, K. C., John, J., and Wittinghofer, A.,** Structure of the guanine-nucleotide-binding domain of the Ha-*ras* oncogene product p21 in the triphosphate conformation, *Nature,* 341, 209, 1989.

109. **Milburn, M. V., Tong, L., de Vos, A. M., Brünger, A., Yamaizumi, Z., Nishimura, S., and Kim, S.-H.,** Molecular switch for signal transduction: structural differences between active and inactive forms of protooncogeneic Ras proteins, *Science,* 2447, 939, 1990.

110. **Schlichting, I., Almo, S. C., Rapp, G., Wilson, K., Petratos, K., Lentfer, A., Wittinghofer, A., Kabsch, W., Pai, E. F., Petsko, G. A., and Goody, R. S.,** Time-resolved X-ray crystallographic study of the conformational change in Ha-Ras p21 protein on GTP hydrolysis, *Nature,* 345, 309, 1990.

111. **Brünger, A. T., Milburn, M. V., Tong, L., deVos, A. M., Jancarik, J., Yamaizumi, A., Nishimura, S., Ohtsuka, E., and Kim, S.-H.,** Crystal structure of an active form of Ras protein, a complex of a GTP analog and the HRas p21 catalytic domain, *Proc. Natl. Acad. Sci. U.S.A.,* 87, 4849, 1990.

112. **Santos, E., Nebreda, A. R., Bryan, T., and Kempner, E. S.,** Oligomeric structure of p21 Ras proteins as determined by radiation inactivation, *J. Biol. Chem.,* 263, 9853, 1988.

113. **Lowe, D. G., Ricketts, M., Levinson, A. D., and Goeddel, D. V.,** Chimeric proteins define variable and essential regions of Ha-*ras*-encoded protein, *Proc. Natl. Acad. Sci. U.S.A.,* 85, 1015, 1988.

114. **Halliday, K. R.,** Regional homology in GTP-binding proto-oncogene products and elongation factors, *J. Cyclic Nucleot. Prot. Res.,* 9, 435, 1983.

115. **Jurnak, F.,** Structure of the GDG domain of EF-Tu and location of the amino acids homologous to Ras oncogene proteins, *Science,* 230, 32, 1985.

116. **Sigal, I. S., Gibbs, J. B., D'Alonzo, J. S., and Scolnick, E. M.,** Identification of effector residues and a neutralizing epitope of Ha-*ras*-encoded p21, *Proc. Natl. Acad. Sci. U.S.A.,* 83, 4725, 1986.

117. **Willumsen, B. M., Papageorge, A. G., Kung, H.-F., Bekesi, E., Robins, T., Johnsen, M., Vass, W. C., and Lowy, D. R.,** Mutational analysis of a *ras* catalytic domain, *Mol. Cell. Biol.,* 6, 2646, 1986.

118. **Clanton, D. J., Hattori, S., and Shih, T. Y.,** Mutations of the *ras* gene product p21 that abolish guanine nucleotide binding, *Proc. Natl. Acad. Sci. U.S.A.,* 83, 5076, 1986.

119. **March, P. E. and Inouye, M.,** GTP-binding membrane protein of *Escherichia coli* with sequence homology to initiation factor 2 and elongation factors Tu and G, *Proc. Natl. Acad. Sci. U.S.A.,* 82, 7500, 1985.

120. **Kohno, K., Uchida, T., Okhubo, H., Nakanishi, S., Nakanishi, T., Fukui, T., Ohtsuka, E., Ikehara, M., and Okada, Y.,** Amino acid sequence of mammalian elongation factor 2 deduced from the cDNA sequence: homology with GTP-binding proteins, *Proc. Natl. Acad. Sci. U.S.A.,* 83, 4978, 1986.

121. **Hurley, J. B., Simon, M. I., Teplow, D. B., Robishaw, J. D., and Gilman, A. G.,** Homologies between signal transducing G proteins and *ras* gene products, *Science,* 226, 860, 1984.

122. **Lochrie, M. A., Hurley, J. B., and Simon, M. I.,** Sequence of the alpha subunit of photoreceptor G protein: homologies between transducin, *ras*, and elongation factors, *Science,* 228, 96, 1985.

123. **Tanabe, T., Nukuda, T., Nishikawa, Y., Sugimoto, K., Suzuki, H., Takahashi, H., Noda, M., Haga, T., Ichiyama, A., Kangawa, K., Minamino, N., Matsuo, H., and Numa, S.,** Primary structure of the α-subunit of transducing and its relationship to *ras* proteins, *Nature,* 315, 242, 1985.

124. **Gay, N. J. and Walker, J. E.,** Homology between human bladder carcinoma oncogene product and mitochondrial ATP-synthase, *Nature,* 301, 262, 1983.

125. **Pines, M., Gierschik, P., Milligan, G., Klee, W., and Spiegel, A.,** Antibodies against the carboxyl-terminal 5-kDa peptide of the α subunit of transducin crossreact with the 40-kDa but not the 39-kDa guanine nucleotide binding protein from brain, *Proc. Natl. Acad. Sci. U.S.A.,* 82, 4095, 1985.

126. **McCormick, F., Clark, B. F. C., La Cour, T. F. M., Kjeldgaard, M., Norskov-Lauritsen, L., and Nyborg, J.,** A model for the tertiary structure of p21, the product of the *ras* oncogene, *Science,* 230, 78, 1985.

127. **Beckner, S. K., Hattori, S., and Shih, T. Y.,** The *ras* oncogene product p21 is not a regulatory component of adenylate cyclase, *Nature,* 317, 71, 1985.

128. **Privé, G. G., Milburn, M. V., Tong, L., de Vos, A. M., Yamaizumi, Z., Nishimura, S., and Kim, S.-H.,** X-ray crystal structures of transforming p21 *ras* mutants suggest a transition-state stabilization mechanism for GTP hydrolysis, *Proc. Natl. Acad. Sci. U.S.A.,* 89, 3649, 1992.

129. **Schweighofer, F., Faure, M., Fath, I., Chevallier-Multon, M.-C., Apiou, F., Dutrillaux, B., Sturani, E., Jacquet, M., and Tocque, B.,** Identification of a human guanine nucleotide-releasing factor (H-GRF55) specific for Ras proteins, *Oncogene,* 8, 1477, 1993.

130. **Adari, H., Lowy, D. R., Willumsen, B. M., Der, C. J., and McCormick, F.,** Guanosine triphosphatase activating protein (GAP) interacts with the p21 *ras* effector binding domain, *Science,* 240, 518, 1988.

131. **McCormick, F.,** *ras* GTPase activating protein: signal transmitter and signal terminator, *Cell,* 56, 5, 1989.

132. **Schweighoffer, F., Barlat, I., Chevallier-Multon, M.-C., and Tocque, B.,** Implication of GAP in Ras-dependent transactivation of a polyoma enhancer sequence, *Science,* 256, 825, 1992.

133. **Zhang, K., DeClue, J. E., Vass, W. C., Papageorge, A. G., McCormick, F., and Lowy, D. R.,** Suppression of c-*ras* transformation by GTPase-activating protein, *Nature,* 346, 754, 1990.

134. **Duchesne, M., Schweighoffer, F., Parker, F., Clerc, F., Frobert, Y., Thang, M. N., and Tocqué, B.,** Identification of the SH3 domain of GAP as an essential sequence for Ras-GAP-mediated signaling, *Science,* 259, 525, 1993.

135. **Serth, J., Weber, W., Frech, M., Wittinghofer, A., and Pingoud, A.,** Binding of the H-Ras GTPase activating protein by the activated epidermal growth factor receptor leads to inhibition of the p21 GTPase activity *in vitro, Biochemistry,* 31, 6361, 1992.

136. **Liu, X. Q. and Pawson, T.,** The epidermal growth factor receptor phosphorylates GTPase-activating protein (GAP) at Tyr-460, adjacent to the GAP SH2 comains, *Mol. Cell. Biol.,* 11, 2511, 1991.

137. **Tsai, M.-H., Yu, C.-L., and Stacey, D. W.,** A cytoplasmic protein inhibits the GTPase activity of H-Ras in a phospholipid-dependent manner, *Science,* 250, 982, 1990.

138. **Lemons, R. S., Espinosa, R., III, Rebentisch, M., McCormick, F., Ladner, M., and LeBeau, M. M.,** Chromosomal localization of the gene encoding GTPase-activating protein (*RASA*) to human chromosome 5, bands q13-q15, *Genomics,* 6, 383, 1990.

139. **Xu, G. F., Lin, B., Tanaka, K., Dunn, D., Wood, D., Gesteland, R., White, R., Weiss, R., and Tamanoi, F.,** The catalytic domain of the neurofibromatosis type-1 gene product stimulates *ras* GTPase and complements IRA mutants of *S. cerevisiae, Cell,* 63, 835, 1990.

140. **Martin, G. A., Viskochil, D., Bollag, G., McCabe, P. C., Crosier, W. J., Hausbruck, H., Conroy, L., Clark, R., O'Connell, R., Cawthon, R. M., Innis, M. A., and McCormick, F.,** The GAP-related domain of the neurofibromatosis type 1 gene product interact with *ras* p21, *Cell,* 63, 843, 1990.

141. **DeClue, J. E., Stone, J. C., Blanchard, R. A., Papageorge, A. G., Martin, P., Zhang, K., and Lowy, D. R.,** A *ras* effector domain mutant which is temperature sensitive for cellular transformation: interactions with GTPase-activating protein and NF-1, *Mol. Cell. Biol.,* 11, 3132, 1991.

142. **Ellis, C., Moran, M., McCormick, F., and Pawson, T.,** Phosphorylation of GAP and GAP-associated proteins by transforming and mitogenic tyrosine kinases, *Nature,* 343, 377, 1990.

143. **Wong, G., Müller, O., Clark, R., Conroy, L., Moran, M. F., Polakis, P., and McCormick, F.,** Molecular cloning and nucleic acid binding properties of the GAP-associated tyrosine phosphoprotein p62, *Cell,* 69, 551, 1992.

144. **Moran, M. F., Polakis, P., McCormick, F., Pawson, T., and Ellis, C.,** Protein-tyrosine kinases regulate the phosphorylation, protein interactions, subcellular distribution, and activity of p21*ras* GTPase-activating protein, *Mol. Cell. Biol.,* 11, 1804, 1991.

145. **Heideran, M. A., Molloy, C. J., Pangelinan, M., Choudhury, G. G., Wang, L.-M., Fleming, T. P., Sakaguchi, A. Y., and Pierce, J. H.,** Activation of colony-stimulating factor 1 receptor leads to the rapid tyrosine phosphorylation of GTPase-activating protein and activation of cellular p21*ras*, *Oncogene,* 7, 147, 1992.

146. **Park, S., Liu, X., Pawson, T., and Jove, R.,** Activated Src tyrosine kinase phosphorylates Tyr-457 of bovine GTPase-activating protein (GAP) *in vitro* and the corresponding residue of rat GAP *in vivo*, *J. Biol. Chem.*, 267, 17194, 1992.

147. **Brott, B. K., Decker, S., Shafer, J., Gibbs, J. B., and Jove, R.,** GTPase-activating protein interactions with the viral and cellular Src kinases, *Proc. Natl. Acad. Sci. U.S.A.*, 88, 755, 1991.

148. **Nori, M., Vogel, U. S., Gibbs, J. B., and Weber, M. J.,** Inhibition of v-*src*-induced transformation by a GTPase-activating protein, *Mol. Cell. Biol.*, 11, 2812, 1991.

149. **Martin, G. A., Yatani, A., Clark, R., Conroy, L., Polakis, P., Brown, A. M., and McCormick, F.,** GAP domains responsible for Ras p21-dependent inhibition of muscarinic atrial K$^+$ channel currents, *Science*, 255, 192, 1992.

150. **Chung, H.-H., Kim, R., and Kim, S.-H.,** Biochemical and biological activity of phosphorylated and non-phosphorylated *ras* p21 mutants, *Biochim. Biophys. Acta*, 1129, 278, 1992.

151. **Shih, T. Y., Stokes, P. E., Smythers, G. W., Dhar, R., and Oroszlan, S.,** Characterization of the phosphorylation and surrounding amino acid sequences of the p21 transforming proteins coded for by the Harvey and Kirsten strains of sarcoma viruses, *J. Biol. Chem.*, 257, 11767, 1982.

152. **John, J., Frech, M., and Wittinghofer, A.,** Biochemical properties of Ha-*ras* encoded p21 mutants and mechanism of the autophosphorylation reaction, *J. Biol. Chem.*, 263, 11792, 1988.

153. **Jeng, A. Y., Srivastava, S. K., Lacal, J. C., and Blumberg, P. M.,** Phosphorylation of *ras* oncogene product by protein kinase C, *Biochem. Biophys. Res. Commun.*, 145, 782, 1987.

154. **Kamata, T. and Feramisco, J. R.,** Epidermal growth factor stimulates guanine nucleotide binding activity and phosphorylation of *ras* oncogene proteins, *Nature*, 310, 147, 1984.

155. **Hattori, S., Clanto, D. J., Satoh, T., Nakamura, S., Kaziro, Y., Kawakita, M., and Shih, T. Y.,** Neutralizing monoclonal antibody against *ras* oncogene product p21 which impairs guanine nucleotide exchange, *Mol. Cell. Biol.*, 7, 1999, 1987.

156. **Kawamitsu, H., Miwa, M., Tanigawa, Y., Shimoyama, M., Noguchi, S., Nishimura, S., Ohtsuka, E., and Sugimura, T.,** A hen enzyme ADP ribosylates normal human and mutated c-Ha-*ras* oncogene products synthesized in *Escherichia coli*, *Proc. Jpn. Acad.*, B62, 102, 1986.

157. **Tsai, S.-C., Adamik, R., Moss, J., Vaughan, M., Manne, V., and Kung, H.-F.,** Effects of phospholipids and ADP-ribosylation on GTP hydrolysis by *Escherichia coli*-synthesized Ha-*ras*-encoded p21, *Proc. Natl. Acad. Sci. U.S.A.*, 82, 8310, 1985.

158. **Manne, V. and Kung, H.,** Effect of divalent metal ions and glycerol on the GTPase activity of H-*ras* proteins, *Biochem. Biophys. Res. Commun.*, 128, 1440, 1985.

159. **Kaplan, S. and Bar-Sagi, D.,** Association of p21^{ras} with cellular polypeptides, *J. Biol. Chem.*, 266, 18934, 1991.

160. **Finegold, A. A., Schafer, W. R., Rine, J., Whiteway, M., and Tamanoi, F.,** Common modifications of trimeric G proteins and Ras protein: involvement of polyisoprenylation, *Science*, 249, 165, 1990.

161. **Hancock, J. F., Magee, A. I., Childs, J. E., and Marshall, C. J.,** All *ras* proteins are polyisoprenylated but only some are palmitoylated, *Cell*, 57, 1167, 1989.

162. **Casey, P. J., Solsky, P. A., Der, C. J., and Buss, J. E.,** p21 *ras* is modified by a farnesyl isoprenoid, *Proc. Natl. Acad. Sci. U.S.A.*, 86, 8323, 1989.

163. **Kim, R., Rine, J., and Kim, S.-H.,** Prenylation of mammalian Ras protein in *Xenopus* oocytes, *Mol. Cell. Biol.*, 10, 5945, 1990.

164. **Schafer, W. R., Trueblood, C. E., Yang, C.-C., Mayer, M. P., Rosenberg, S., Poulter, C. D., Kim, S.-H., and Rine, J.,** Enzymatic coupling of cholesterol intermediates to a mating pheromone precursor and to the Ras protein, *Science*, 249, 1133, 1990.

165. **Hancock, J. F., Cadwallader, K., and Marshall, C. J.,** Methylation and proteolysis are essential for efficient membrane binding of prenylated p21$^{K-ras(B)}$, *EMBO J.*, 10, 641, 1991.

166. **Reiss, Y., Stradley, S. J., Gierasch, L. M., Brown, M. S., and Goldstein, J. L.,** Sequence requirement for peptide recognition by rat brain p21^{ras} protein farnesyltransferase, *Proc. Natl. Acad. Sci. U.S.A.*, 88, 732, 1991.

167. **Chen, W. J., Andres, D. A., Goldstein, J. L., and Brown, M. S.,** Cloning and expression of a cDNA encoding the α subunit of rat p21^{ras} protein farnesyltransferase, *Proc. Natl. Acad. Sci. U.S.A.*, 88, 11368, 1991.

168. **Kato, K., Cox, A. D., Hisaka, M. M., Graham, S. M., Buss, J. E., and Der, C. J.,** Isoprenoid addition to Ras protein is the critical modification for its membrane association and transforming activity, *Proc. Natl. Acad. Sci. U.S.A.*, 89, 6403, 1992.

169. **Buss, J. E., Solski, P. A., Schaeffer, J. P., MacDonald, M. J., and Der, C. J.,** Activation of the cellular proto-oncogene product p21 *ras* by addition of a myristoylation signal, *Science,* 243, 1600, 1989.

170. **Weeks, M. O., Hager, G. L., Lowe, R., and Scolnick, E. M.,** Development and analysis of a transformation-defective mutant of Harvey murine sarcoma *tk* virus and its gene product, *J. Virol.,* 54, 586, 1985.

171. **Willumsen, B. M., Papageorge, A. G., Hubbert, N., Bekesi, E., Kung, H.-F., and Lowy, D. R.,** Transforming p21 Ras protein: flexibility in the major variable region linking the catalytic and membrane-anchoring domains, *EMBO J.,* 4, 2893, 1985.

172. **Schejter, E. D. and Shilo, B.-Z.,** Characterization of functional domains of p21 *ras* by use of chimeric genes, *EMBO J.,* 4, 407, 1985.

173. **Lacal, J. C., Anderson, P. S., and Aaronson, S. A.,** Deletion mutants of Harvey Ras p21 protein reveal the absolute requirement of at least two distant regions for GTP-binding and transforming activities, *EMBO J.,* 5, 679, 1986.

174. **Clark, R., Wong, G., Arnheim, N., Nitecki, D., and McCormick, F.,** Antibodies specific for amino acid 12 of the *ras* oncogene product inhibit GTP binding, *Proc. Natl. Acad. Sci. U.S.A.,* 82, 5280, 1985.

175. **Feramisco, J. R., Clark, R., Wong, G., Arnheim, N., Milley, R., and McCormick, F.,** Transient reversion of *ras* oncogene-induced cell transformation by antibodies specific for amino acid 12 of Ras protein, *Nature,* 314, 639, 1985.

176. **McGrath, J. P., Capon, D. J., Goeddel, D. V., and Levinson, A. D.,** Comparative biochemical properties of normal and activated human Ras p21 protein, *Nature,* 310, 644, 1984.

177. **Satoh, Nakamura, S., Nakafuku, M., and Kaziro, Y.,** Studies on Ras proteins. Catalytic properties of normal and activated Ras proteins purified in the absence of protein denaturants, *Biochim. Biophys. Acta,* 949, 97, 1988.

178. **Trahey, M. and McCormick, F.,** A cytoplasmic protein stimulates normal N-*ras* p21 GTPase, but does not affect oncogenic mutants, *Science,* 238, 542, 1987.

179. **McCoy, M. S., Bargmann, C. I., and Weinberg, R. A.,** Human colon carcinoma Ki-*ras*2 oncogene and its corresponding proto-oncogene, *Mol. Cell. Biol.,* 4, 1577, 1984.

180. **McCoy, M. S. and Weinberg, R. A.,** A human Ki-*ras* oncogene encodes two transforming p21 proteins, *Mol. Cell. Biol.,* 6, 1326, 1986.

181. **Feig, L. A., Pan, B.-T., Roberts, T. M., and Cooper, G. M.,** Isolation of *ras* GTP-binding mutants using an *in situ* colony-binding assay, *Proc. Natl. Acad. Sci. U.S.A.,* 83, 4607, 1986.

182. **Andreeff, M., Slater, D. E., Bressler, J., and Furth, M. E.,** Cellular *ras* oncogene expression and cell cycle measured by flow cytometry in hematopoietic cell lines, *Blood,* 67, 676, 1986.

183. **Campisi, J., Gray, H. E., Pardee, A. B., Dean, M., and Sonenshein, G. E.,** Cell-cycle control of c-*myc* but not c-*ras* expression is lost following chemical transformation, *Cell,* 36, 241, 1984.

184. **Mulcahy, L. S., Smith, M. R., and Stacey, D. W.,** Requirement for *ras* proto-oncogene function during serum-stimulated growth of NIH 3T3 cells, *Nature,* 313, 241, 1985.

185. **Durkin, J. P. and Whitfield, J. F.,** Characterization of G_1 transit induced by the mitogenic-oncogenic viral Ki-*ras* gene product, *Mol. Cell. Biol.,* 6, 1386, 1986.

186. **Kamata, T. and Kung, H.-F.,** Modulation of maturation and ribosomal protein S6 phosphorylation in *Xenopus* oocytes by microinjection of oncogenic Ras protein and protein kinase C, *Mol. Cell. Biol.,* 10, 880, 1990.

187. **Daar, I., Nebreda, A. R., Yew, N., Sass, P., Paules, R., Santos, E., Wigler, M., and Vande Woude, G. F.,** The Ras oncoprotein and M-phase activity, *Science,* 253, 74, 1991.

188. **Chung, D. L., Brandt-Rauf, P. W., Weinstein, I. B., Nishimura, S., Yamaizumi, Z., Murphy, R. B., and Pincus, M. R.,** Evidence that the *ras* oncogene-encoded p21 protein induces oocyte maturation via activation of protein kinase C, *Proc. Natl. Acad. Sci. U.S.A.,* 89, 1993, 1992.

189. **Carbone, A., Gusella, G. L., Radzioch, D., and Varesio, L.,** Human Harvey-*ras* is biochemically different from Kirsten- or N-*ras*, *Oncogene,* 6, 731, 1991.

190. **Marshall, M. S., Davis, L. J., Keys, R. D., Mosser, S. D., Hill, W. S., Scolnick, E. M., and Gibbs, J. B.,** Identification of amino acid residues required for Ras p21 target activation, *Mol. Cell. Biol.,* 11, 3997, 1991.

191. **Finkel, T., Der, C. J., and Cooper, G. M.,** Activation of *ras* genes in human tumors does not affect localization, modification, or nucleotide binding properties of p21, *Cell,* 37, 151, 1984.

192. **Pincus, M. R., Chung, D., Dykes, D. C., Brandt-Rauf, P., Weinstein, I. B., Yamaizumi, Z., and Nishimura, S.,** Pathways for activation of the *ras*-oncogene-encoded p21 protein, *Ann. Clin. Lab. Sci.,* 22, 323, 1992.

193. **Der, C. J., Finkel, T., and Cooper, G. M.,** Biological and biochemical properties of human ras^H genes mutated at codon 61, *Cell,* 44, 167, 1986.

194. **Gibbs, J. B., Sigal, I. S., Poe, M., and Scolnick, E. M.,** Intrinsic GTPase activity distinguishes normal and oncogenic ras p21 molecules, *Proc. Natl. Acad. Sci. U.S.A.,* 81, 5704, 1984.

195. **Sweet, R. W., Yokoyama, S., Kamata, T., Feramisco, J. R., Rosenberg, M., and Gross, M.,** The product of *ras* is a GTPase and the T24 oncogenic mutant is deficient in this activity, *Nature,* 311, 273, 1984.

196. **Manne, V., Bekesi, E., and Kung, H.-F.,** Ha-*ras* proteins exhibit GTPase activity: point mutations that activate Ha-*ras* gene products result in decreased GTPase activity, *Proc. Natl. Acad. Sci. U.S.A.,* 82, 376, 1985.

197. **Sekiya, T., Tokunaga, A., and Fushimi, M.,** Essential region for transforming activity of human c-Ha-*ras*-1, *Jpn. J. Cancer Res.,* 76, 787, 1985.

198. **Tong, L., de Vos, A. M., Milburn, M. V., Jancarik, J., Noguchi, S., Nishimura, S., Miura, K., Ohtsuka, E., and Kim, S.-H.,** Structural differences between a *ras* oncogene protein and the normal protein, *Nature,* 337, 90, 1989.

199. **Neal, S. E., Eccleston, J. F., and Webb, M. R.,** Hydrolysis of GTP by p21^{NRAS}, the *NRAS* protooncogene product, is accompanied by a conformational change in the wild-type protein: use of a single fluorescent probe at the catalytic site, *Proc. Natl. Acad. Sci. U.S.A.,* 87, 3562, 1990.

200. **Lacal, J. C., Srivastava, S. K., Anderson, P. S., and Aaronson, S. A.,** Ras p21 proteins with high or low GTPase activity can efficiently transform NIH/3T3 cells, *Cell,* 44, 609, 1986.

201. **Lumpkin, C. K., Knepper, J. E., Butel, J. S., Smith, J. R., and Pereira-Smith, O. M.,** Mitogenic effects of the proto-oncogene and oncogene forms of c-H-*ras* DNA in human diploid fibroblasts, *Mol. Cell. Biol.,* 6, 2990, 1986.

202. **Trahey, M., Milley, R. J., Cole, G. E., Innis, M., Paterson, H., Marshall, C. J., Hall, A., and McCormick, F.,** Biochemical and biological properties of the human N-Ras p21 protein, *Mol. Cell. Biol.,* 7, 541, 1987.

203. **Clanton, D. J., Lu, Y., Blair, D. G., and Shih, T. Y.,** Structural significance of the GTP-binding domain of Ras p21 studied by site-directed mutagenesis, *Mol. Cell. Biol.,* 7, 3092, 1987.

204. **Davies, S.-A., Houslay, M. D., and Wakelam, M. J. O.,** The effects of p21^{N-ras} in NIH-3T3 cells upon cyclic AMP metabolism, *Biochim. Biophys. Acta,* 1013, 173, 1989.

205. **Gauthier-Rouvière, C., Fernandez, A., and Lamb, N. J. C.,** *ras*-induced c-*fos* expression and proliferation in living rat fibroblasts involves C-kinase activation and the serum response element pathway, *EMBO J.,* 9, 171, 1990.

206. **Wakelam, M. J. O., Houslay, M. D., Davies, S. A., Marshall, C. J., and Hall, A.,** The role of N-Ras p21 in the coupling of growth factor receptors to inositol phospholipid turnover, *Biochem. Soc. Trans. (London),* 15, 45, 1987.

207. **Huang, M., Chida, K., Kamata, N., Nose, K., Kato, M., Homma, Y., Takenawa, T., and Kuroki, T.,** Enhancement of inositol phospholipid metabolism and activation of protein kinase C in *ras*-transformed rat fibroblasts, *J. Biol. Chem.,* 263, 17975, 1988.

208. **Wilkison, W. O., Sandgren, E. P., Palmiter, R. D., Brinster, R. L., and Bell, R. M.,** Elevation of 1,2-diacylglycerol in *ras*-transformed neonatal liver and pancreas of transgenic mice, *Oncogene,* 4, 625, 1989.

209. **Benjamin, C. W., Tarpley, W. G., and Gorman, R. R.,** Loss of platelet-derived growth factor-stimulated phospholipase activity in NIH-3T3 cells expressing the EJ-*ras* oncogene, *Proc. Natl. Acad. Sci. U.S.A.,* 84, 546, 1987.

210. **Seuwen, K., Lagarde, A., and Pouysségur, J.,** Deregulation of hamster fibroblast proliferation by mutated *ras* oncogenes is not mediated by constitutive activation of phosphoinositide-specific phospholipase C, *EMBO J.,* 7, 161, 1988.

211. **Weyman, C. M., Taparowsky, E. J., Wolfson, M., and Ashendel, C. L.,** Partial down-regulation of protein kinase C in C3H 10T1/2 mouse fibroblasts transfected with the human Ha-*ras* oncogene, *Cancer Res.,* 48, 6535, 1988.

212. **Ballester, R., Furth, M. E., and Rosen, O. M.,** Phorbol ester- and protein kinase C-mediated phosphorylation of the cellular Kirsten *ras* gene product, *J. Biol. Chem.,* 262, 2688, 1987.

213. **Hsiao, W.-L. W., Housey, G. M., Johnson, M. D., and Weinstein, I. B.,** Cells that overproduce protein kinase C are more susceptible to transformation by an activated H-*ras* oncogene, *Mol. Cell. Biol.,* 2641, 1989.

214. **Franks, D. J., Durkin, J. P., and Whitfield, J. F.,** Protein kinase C and a viral K-Ras protein cooperatively enhance the response of adenylate cyclase to stimulators, *J. Cell Physiol.,* 140, 409, 1989.

215. **Valitutti, S., Cucchi, P., Colletta, G., Di Filippo, C., and Corda, D.,** Transformation by the K-*ras* oncogene correlates with increases in phospholipase A2 activity, glycerophosphoinositol production and phosphoinositide synthesis in thyroid cells, *Cell. Signal.,* 3, 321, 1991.

216. **Alonso, T., Morgan, R. O., Marvizon, J. C., Zarbl, H., and Santos, E.,** Malignant transformation by *ras* and other oncogenes produces common alterations in inositol phospholipid signaling pathways, *Proc. Natl. Acad. Sci. U.S.A.,* 85, 4271, 1988.

217. **Sistonen, L., Hölttä, E., Mäkelä, T. P., Keski-Oja, J., and Alitalo, K.,** The cellular response to induction of the p21[c-Ha-*ras*] oncoprotein includes stimulation of *jun* gene expression, *EMBO J.,* 8, 815, 1989.

218. **Gilbert, P. X. and Harris, H.,** The role of the *ras* oncogene in the formation of tumours, *J. Cell Sci.,* 90, 433, 1988.

219. **Geiser, A. G., Anderson, M. J., and Stanbridge, E. J.,** Suppression of tumorigenicity in human cell hybrids derived from cell lines expressing different activated *ras* oncogenes, *Cancer Res.,* 49, 1572, 1989.

220. **Kikuchi, A., Amagai, M., Hayakawa, K., Ueda, M., Hirohashi, S., Shimizu, N., and Nishikawa, T.,** Association of EGF receptor expression with proliferating cells and of Ras p21 expression with differentiating cells in various skin tumours, *Br. J. Dermatol.,* 123, 49, 1990.

221. **Czerniak, B., Herz, F., Wersto, R. P., and Koss, L. G.,** Expression of Ha-*ras* oncogene p21 protein in relation to the cell cycle of cultured human tumor cells, *Am. J. Pathol.,* 126, 411, 1987.

222. **Ikeda, T., Sawada, N., Fujinaga, K., Minase, T., and Mori, M.,** c-H-*ras* gene is expressed at the G1 phase in primary cultures of hepatocytes, *Exp. Cell Res.,* 185, 292, 1989.

223. **Durkin, J. P. and Whitfield, J. F.,** The viral Ki-*ras* gene must be expressed in the G_2 phase if *ts* Kirsten sarcoma virus-infected NRK cells are to proliferate in serum-free medium, *Mol. Cell. Biol.,* 7, 444, 1987.

224. **Vaidya, T. B., Weyman, C. M., Teegarden, D., Ashendel, C. L., and Taparowsky, E. J.,** Inhibition of myogenesis by the H-*ras* oncogene: implication of a role for protein kinase C, *J. Cell Biol.,* 114, 809, 1991.

225. **Bell, J. C., Jardine, K., and McBurney, M. W.,** Lineage-specific transformation after differentiation of multipotential murine stem cells containing a human oncogene, *Mol. Cell. Biol.,* 6, 617, 1986.

226. **Bar-Sagi, D. and Feramisco, J. R.,** Microinjection of the *ras* oncogene protein into PC12 cells induces morphological differentiation, *Cell,* 42, 841, 1985.

227. **Guerrero, I., Wong, H., Pellicer, A., and Burstein, D. E.,** Activated N-*ras* gene induces neuronal differentiation of PC12 rat pheochromocytoma cells, *J. Cell. Physiol.,* 129, 71, 1986.

228. **Noda, M., Ko, M., Ogura, A., Liu, D., Amano, T., Takano, T., and Ikawa, Y.,** Sarcoma viruses carrying *ras* oncogenes induce differentiation-associated properties in a neuronal cell line, *Nature,* 318, 73, 1985.

229. **Colletta, G., Pinto, A., Di Fiore, P. P., Fusco, A., Ferrentino, M., Avvedimento, V. E., Tsuchida, N., and Vecchio, G.,** Dissociation between transformed and differentiated phenotype in rat thyroid epithelial cells after transformation with a temperature-sensitive mutant of the Kirsten murine sarcoma viruses, *Mol. Cell. Biol.,* 3, 2099, 1983.

230. **Stacey, D. W. and Kung, H.-F.,** Transformation of NIH 3T3 cells by microinjection of Ha-*ras* p21 protein, *Nature,* 310, 508, 1984.

231. **Darfler, F. J., Shih, T. Y., and Lin, M. C.,** Revertants of Ha-MuSV-transformed MDCK cells express reduced levels of p21 and possess a more normal phenotype, *Exp. Cell Res.,* 162, 335, 1986.

232. **Kung, H.-F., Smith, M. R., Bekesi, E., Manne, V., and Stacey, D. W.,** Reversal of transformed phenotype by monoclonal antibodies against Ha-*ras* p21 proteins, *Exp. Cell Res.,* 162, 363, 1986.

233. **Cheng, S. V. Y. and Pollard, J. W.,** c-*ras*[H] and ornithine decarboxylase are induced by oestradiol-17β in the mouse uterine luminal epithelium independently of the proliferative status of the cells, *FEBS Lett.,* 196, 309, 1986.

234. **Leon, J., Guerrero, I., and Pellicer, A.,** Differential expression of the *ras* gene family in mice, *Mol. Cell. Biol.,* 7, 1535, 1987.

235. **Scheinberg, D. A. and Strand, M.,** A brain membrane protein similar to the rat *src* gene product, *Proc. Natl. Acad. Sci. U.S.A.,* 78, 55, 1981.

236. **Furth, M. E., Aldrich, T. H., and Cordon-Cardo, C.,** Expression of *ras* proto-oncogene proteins in normal human tissues, *Oncogene,* 1, 47, 1987.

237. **Tanaka, T., Ida, N., Waki, C., Shimoda, H., Slamon, D. J., and Cline, M. J.,** Cell type-specific expression of c-*ras* gene products in the normal rat, *Mol. Cell. Biochem.,* 75, 23, 1987.

238. **Chan, S. O., Wong, S. S. C., and Yeung, D. C. Y.,** Insulin induction of c-Ki-*ras* in rat liver and in cultured normal rat hepatocytes, *Comp. Biochem. Physiol.,* B104, 341, 1993.

239. **Sacks, D. B., Glenn, K. C., and McDonald, J. M.,** The carboxyl terminal segment of the c-Ki-*ras* 2 gene product mediates insulin-stimulated phosphorylation of calmodulin and stimulates insulin-independent autophosphorylation of the insulin receptor, *Biochem. Biophys. Res. Commun.,* 161, 399, 1989.

240. **Burgering, B. M. T., Medema, R. H., Maassen, J. A., Van de Wetering, M. L., Van der Eb, A. J., McCormick, F., and Bos, J. L.,** Insulin stimulation of gene expression mediated by p21^{ras} activation, *EMBO J.,* 10, 1103, 1991.

241. **Benito, M., Porras, A., Nebreda, A. R., and Santos, E.,** Differentiation of 3T3-L1 fibroblasts to adipocytes induced by transfection of *ras* oncogenes, *Science,* 253, 565, 1991.

242. **Porras, A., Nebreda, A. R., Benito, M., and Santos, E.,** Activation of Ras by insulin in 3T3 L1 cells does not involve GTPase-activating protein phosphorylation, *J. Biol. Chem.,* 267, 21124, 1992.

243. **Satoh, T., Endo, M., Nakafuku, M., Akiyama, T., Yamamoto, T., and Kaziro, Y.,** Accumulation of p21^{ras}-GTP in response to stimulation with epidermal growth factor and oncogene products with tyrosine kinase activity, *Proc. Natl. Acad. Sci. U.S.A.,* 87, 7926, 1990.

244. **Zhang, K., Papageorge, A. G., and Lowy, D. R.,** Mechanistic aspects of signaling through Ras in NIH 3T3 cells, *Science,* 257, 671, 1992.

245. **Osterop, A. P. R. M., Medema, R. H., v.d. Zon, G. C. M., Bos, J. L., Möller, W., and Maassen, J. A.,** Epidermal-growth-factor receptors generate Ras-GTP more efficiently than insulin receptors, *Eur. J. Biochem.,* 212, 477, 1993.

246. **Lowenstein, E. J., Daly, R. J., Batzer, A. G., Li, W., Margolis, B., Lammers, R., Ullrich, A., Skolnik, E. Y., Bar-Sagi, D., and Schlessinger, J.,** The SH2 and SH3 domain-containing protein GRB2 links receptor tyrosine kinases to Ras signaling, *Cell,* 70, 431, 1992.

247. **Egan, S. E., Giddings, B. W., Brooks, M. W., Buday, L., Sizeland, A. M., and Weinberg, R. A.,** Association of Sos Ras exchange protein Grb2 is implicated in tyrosine kinase signal transduction and transformation, *Nature,* 363, 45, 1993.

248. **Rozaki-Sadcock, M., Fernley, R., Wade, J., Pawson, T., and Bowtell, D.,** The SH2 and SH3 domains of mammalian Grb2 couple the EGF receptor to the Ras activator Msos1, *Nature,* 363, 83, 1993.

249. **Li, N., Batzer, A., Dali, R., Yajnik, V., Skolnik, E., Chardin, P., Bar-Sagi, D., Margolis, B., and Schlessinger, J.,** Guanine nucleotide-releasing factor hSos1 binds to Grb2 and links receptor tyrosine kinases to Ras signalling, *Nature,* 363, 85, 1993.

250. **Gale, N. W., Kaplan, S., Lowenstein, E. J., Schlessinger, J., and Bar-Sagi, D.,** Grb2 mediates the EGF-dependent activation of guanine nucleotide exchange on Ras, *Nature,* 363, 88, 1993.

251. **Buday, L. and Downward, J.,** Epidermal growth factor regulates p21^{ras} through the formation of a complex of receptor, Grb2 adapter protein, and Sos nucleotide exchange factor, *Cell,* 73, 611, 1993.

252. **Chardin, P., Camonis, J. H., Gale, N. W., Van Aelst, L., Schlessinger, J., Wigler, M. H., and Bar-Sagi, D.,** Human Sos1: a guanine nucleotide exchange factor for Ras that binds to GRB2, *Science,* 260, 1338, 1993.

253. **Rozakis-Adcock, M., McGlade, J., Mbamalu, G., Pelicci, G., Daly, R., Li, W., Batzer, A., Thomas, S., Brugge, J., Pelicci, P. G., Schlessinger, J., and Pawson, T.,** Association of the Shc and Grb2/Sem5 SH2-containing proteins is implicated in activation of the Ras pathway by tyrosine kinases, *Nature,* 360, 689, 1992.

254. **Hu, P., Margolis, B., and Schlessinger, J.,** *vav*: a potential link between tyrosine kinases and *ras*-like GTPases in hematopoietic signaling, *BioEssays,* 15, 179, 1993.

255. **Huang, M., Kamata, N., Nose, K., and Kuroki, T.,** Modified responsiveness of v-Ha-*ras*-transfected rat fibroblasts to growth factors and a tumor promoter, *Mol.Carcinogenesis,* 1, 109, 1988.

256. **McKay, I. A., Marshall, C. J., Calés, C., and Hall, A.,** Transformation and stimulation of DNA synthesis in NIH-3T3 cells are a titratable function of normal p21^{N-ras} expression, *EMBO J.,* 5, 2617, 1986.

257. **Hiwasa, T., Sakiyama, S., Yokoyama, S., Ha, J.-M., Fujita, J., Noguchi, S., Bando, Y., Kominami, E., and Katunuma, N.,** Inhibition of cathepsin L-induced degradation of epidermal growth factor receptors by c-Ha-*ras* gene products, *Biochem. Biophys. Res. Commun.,* 151, 78, 1988.

258. **Hiwasa, T., Sakiyama, S., Yokoyama, S., Ha, J.-M., Noguchi, S., Bando, Y., Kominami, E., and Katunuma, N.,** Degradation of epidermal growth factor receptors by cathepsin L-like protease: inhibition of the degradation by c-Ha-*ras* gene products, *FEBS Lett.,* 233, 367, 1988.

259. **Weissman, B. E. and Aaronson, S. A.,** BALB and Kirsten murine sarcoma viruses alter growth and differentiation of EGF-dependent BALB/c mouse epidermal keratinocyte lines, *Cell,* 32, 599, 1983.

260. **Weissman, B. and Aaronson, S. A.,** Members of the *src* and *ras* oncogene families supplant the epidermal growth factor requirement of BALB/MK-2 keratinocytes and induce distinct alterations in their terminal differentiation program, *Mol. Cell. Biol.,* 5, 3386, 1985.

261. **McKay, I. A., Malone, P., Marshall, C. J., and Hall, A.,** Malignant transformation of murine fibroblasts by a human c-Ha-*ras*-1 oncogene does not require a functional epidermal growth factor receptor, *Mol. Cell. Biol.,* 6, 3382, 1986.

262. **Palejwala, S. and Goldsmith, L. T.,** Ovarian expression of cellular Ki-Ras p21 varies with physiological status, *Proc. Natl. Acad. Sci. U.S.A.,* 89, 4202, 1992.

263. **Borasio, G. D., John, J., Wittinghofer, A., Barde, Y.-A., Sendtner, M., and Heumann, R.,** Ras p21 protein promotes survival and fiber outgrowth of cultured embryonic neurons, *Neuron,* 2, 1087, 1989.

264. **Nakafuku, M., Satoh, T., and Kaziro, Y.,** Differentiation factors, including nerve growth factor, fibroblast growth factor, and interleukin-6, induce an accumulation of an active Ras-GTP complex in rat pheochromocytoma PC12 cells, *J. Biol. Chem.,* 267, 19448, 1992.

265. **Wood, K. W., Sarnecki, C., Roberts, T. M., and Blenis, J.,** Ras mediates nerve growth factor receptor modulation of three signal-transducing protein kinases: MAP kinase, Raf-1, and RSK, *Cell,* 68, 1041, 1992.

266. **Pomerance, M., Schweighoffer, F., Tocque, B., and Pierre, M.,** Stimulation of mitogen-activated protein kinase by oncogenic Ras p21 in *Xenopus* oocytes. Requirement for Ras p21-GTPase-activating protein interaction, *J. Biol. Chem.,* 267, 16155, 1992.

267. **Thomson, T. M., Green, S. H., Trotta, R. J., Burstein, D. E., and Pellicer, A.,** Oncogene N-*ras* mediates selective inhibition of c-*fos* induction by nerve growth factor and basic fibroblast growth factor in a PC12 cell line, *Mol. Cell. Biol.,* 10, 1556, 1990.

268. **Scolnick, E. M., Weeks, M. O., Shih, T. Y., Ruscetti, S. K., and Dexter, T. M.,** Markedly elevated levels of an endogenous *sarc* protein in a hemopoietic precursor cell line, *Mol. Cell. Biol.,* 1, 66, 1981.

269. **Duronio, V., Welham, M. J., Abraham, S., Dryden, P., and Schrader, J. W.,** p21^{ras} activation via hemopoietin receptors and c-kit requires tyrosine kinase activity but not tyrosine phosphorylation of p21^{ras} GTPase-activating protein, *Proc. Natl. Acad. Sci. U.S.A.,* 89, 1587, 1992.

270. **Satoh, T., Nakafuku, M., Miyajima, A., and Kaziro, Y.,** Involvement of Ras p21 protein in signal-transduction pathways from interleukin 2, interleukin 3, and granulocyte/macrophage colony-stimulating factor, but not from interleukin 4, *Proc. Natl. Acad. Sci. U.S.A.,* 88, 3314, 1991.

271. **Satoh, T., Minami, Y., Kono, T., Yamada, K., Kawahara, A., Taniguchi, T., and Kaziro, Y.,** Interleukin 2-induced activation of Ras requires two domains of interleukin 2 receptor β subunit, the essential region for growth stimulation and Lck-binding domain, *J. Biol. Chem.,* 267, 25423, 1992.

272. **Baldari, C. T., Macchia, G., and Telford, J. L.,** Interleukin-2 promoter activation in T-cells expressing activated Ha-*ras*, *J. Biol. Chem.,* 267, 4289, 1992.

273. **Cruise, J. L., Muga, S. J., Lee, Y.-S., and Michalopoulos, G. K.,** Regulation of hepatocyte growth: α-1 adrenergic receptor and Ras p21 changes in liver regeneration, *J. Cell. Physiol.,* 140, 195, 1989.

274. **Birchmeier, C., Broek, D., and Wigler, M.,** RAS proteins can induce meiosis in *Xenopus* oocytes, *Cell,* 43, 615, 1985.

275. **Sadler, S. E., Schechter, A. L., Tabin, C. J., and Maller, J. L.,** Antibodies to the *ras* gene product inhibit adenylate cyclase and accelerate progesterone-induced cell division in *Xenopus laevis* oocytes, *Mol. Cell. Biol.,* 6, 719, 1986.

276. **Deshpande, A. K. and Kung, H.-F.,** Insulin induction of *Xenopus laevis* oocyte maturation is inhibited by monoclonal antibody against p21 Ras proteins, *Mol. Cell. Biol.,* 7, 1285, 1987.

277. **Chung, D. L., Joran, A., Friedman, F., Robinson, R., Brandt-Rauf, P. W., Weinstein, I. B., Ronal, Z., Baskin, L., Dykes, D. C., Murphy, R. B., Nishimura, S., Yamaizumi, Z., and Pincus, M. R.,** Evidence that oocyte maturation induced by oncogenic Ras-p21 protein and insulin is mediated by overlapping yet distinct mechanisms, *Exp. Cell Res.,* 203, 329, 1992.

278. **Zhao, J. and Buick, R. N.,** Relationship of levels and kinetics of H-*ras* expression to transformed phenotype and loss of TGF-β1-mediated growth regulation in intestinal epithelial cells, *Exp. Cell Res.,* 204, 82, 1993.

279. **Feramisco, J. R., Gross, M., Kamata, T., Rosenberg, M., and Sweet, R. W.,** Microinjection of the oncogene form of the human H-*ras* (T24) protein results in rapid proliferation of quiescent cells, *Cell,* 38, 109, 1984.

280. **Hyland, J. K., Rogers, C. M., Scolnick, E. M., Stein, R. B., Ellis, R. B., and Baserga, R.,** Microinjected *ras* family oncogenes stimulate DNA synthesis in quiescent mammalian cells, *Virology,* 141, 333, 1985.

281. **Mulcahy, L. S., Smith, M. R., and Stacey, D. W.,** Requirement for *ras* proto-oncogene function during serum-stimulated growth of NIH 3T3 cells, *Nature,* 313, 241, 1985.

282. **Collin, C., Papageorge, A. G., Sakakibara, M., Huddie, P. L., Lowie, D. R., and Alkon, D. L.,** Early regulation of membrane excitability by Ras oncogene proteins, *Biophys. J.,* 58, 785, 1990.

283. **Collin, C., Papageorge, A. G., Lowy, D. R., and Alkon, D. L.,** Early enhancement of calcium currents by H-*ras* oncoproteins injected into *Hermissenda* neurons, *Science,* 250, 1743, 1990.

284. **Hescheler, J., Klinz, F.-J., Schultz, G., and Wittinghofer, A.,** Ras proteins activate calcium channels in neuronal cells, *Cell. Signal.,* 3, 127, 1991.

285. **Bar-Sagi, D., McCormick, F., Milley, R. J., and Feramisco, J. R.,** Inhibition of cell surface ruffling and fluid-phase pinocytosis by microinjection of anti-*ras* antibodies into living cells, *J. Cell. Physiol.,* 5, 69, 1987.

286. **Rayter, S. I., Woodrow, M., Lucas, S. C., Cantrell, D. A., and Downward, J.,** p21^{ras} mediates control of *IL-2* gene promoter function in T cell activation, *EMBO J.,* 11, 4556, 1992.

287. **Baldari, C. T., Heguy, A., and Telford, J. L.,** Ras protein activity is essential for T-cell antigen receptor signal transduction, *J. Biol. Chem.,* 268, 2693, 1993.

288. **Naz, R., Ahmad, K., and Kaplan, P.,** Expression and function of *ras* proto-oncogene proteins in human sperm cells, *J. Cell Sci.,* 102, 487, 1992.

289. **Backer, J. M. and Weinstein, I. B.,** Proteins encoded by *ras* oncogenes stimulate or inhibit phosphorylation of specific mitochondrial membrane proteins, *Biochem. Biophys. Res. Commun.,* 135, 316, 1986.

290. **Hedge, A. N. and Das, M. R.,** *Ras* proteins enhance the phosphorylation of a 38 kDa protein (p38) in rat liver plasma membrane, *FEBS Lett.,* 217, 74, 1987.

291. **Moodie, S. A., Willumsen, B. M., Weber, M. J., and Wolfman, A.,** Complexes of Ras-GTP with Raf-1 and mitogen-activated protein kinase, *Science,* 260, 1658, 1993.

292. **Schmidt, C. J. and Hamer, D. H.,** Cell specificity and an effect of *ras* on human metallothionein gene expression, *Proc. Natl. Acad. Sci. U.S.A.,* 83, 3346, 1986.

293. **Bos, J. L.,** *ras* oncogenes in human cancer: a review, *Cancer Res.,* 49, 4682, 1989.

294. **Barbacid, M.,** *ras* oncogenes: their role in neoplasia, *Eur. J. Clin. Invest.,* 20, 225, 1990.

295. **Winter, E. and Perucho, M.,** Oncogene amplification during tumorigenesis of established rat fibroblasts reversibly transformed by activated human *ras* oncogenes, *Mol. Cell. Biol.,* 6, 2562, 1986.

296. **Sistonen, L., Keski-Oja, J., Ulmanen, I., Hölttä, E., Wikgren, B.-J., and Alitalo, K.,** Dose effects of transfected c-Ha-*ras*$^{Val\ 12}$ oncogene in transformed cell clones, *Exp. Cell Res.,* 168, 518, 1987.

297. **Reynolds, V. L., Lebovitz, R. M., Warren, S., Hawley, T. S., Godwin, A. K., and Lieberman, M. W.,** Regulation of a metallothionein-*ras*T24 fusion gene by zinc results in graded alterations in cell morphology and growth, *Oncogene,* 1, 323, 1987.

298. **Bos, J. L., Fearon, E. R., Hamilton, S. R., Verlaan-de Vries, M., van Boom, J. H., van der Eb, A. J., and Vogelstein, B.,** Prevalence of *ras* gene mutations in human colorectal cancers, *Nature,* 327, 293, 1987.

299. **Forrester, K., Almoguera, C., Han, K., Grizzle, W. E., and Perucho, M.,** Detection of high incidence of K-*ras* oncogenes during human colon tumorigenesis, *Nature,* 327, 298, 1987.

300. **Almoguera, C., Shibata, D., Forrester, K., Martin, J., Arnheim, N., and Perucho, M.,** Most human carcinomas of the exocrine pancreas contain mutant c-K-*ras* genes, *Cell,* 53, 549, 1988.

301. **Tubo, R. A. and Rheinwald, J. G.,** Normal human mesothelial cells and fibroblasts transfected with the *EJras* oncogene become EGF-independent, but are not malignantly transformed, *Oncogene Res.,* 1, 407, 1987.

302. **Alonso, T., Srivastava, S., and Santos, E.,** Alterations of G-protein coupling function in phosphoinositide signaling pathways of cells transformed by *ras* and other membrane-associated and cytoplasmic oncogenes, *Mol. Cell. Biol.,* 10, 3117, 1990.

303. **Gossett, L. A., Zhang, W., and Olson, E. N.,** Dexamethasone-dependent inhibition of differentiation of C2 myoblasts bearing steroid-inducible N-*ras* oncogenes, *J. Cell Biol.,* 106, 2127, 1988.

304. **Stacey, D. W., DeGudicibus, S. R., and Smith, M. R.,** Cellular *ras* activity and tumor cell proliferation, *Exp. Cell Res.,* 171, 232, 1987.

305. **Burgering, B. M. T., Snijders, A. J., Maasen, J. A., van der Eb, A. J., and Bos, J. L.,** Possible involvement of normal p21 H-*ras* in the insulin/insulinlike growth factor 1 signal transduction pathway, *Mol. Cell. Biol.,* 9, 4312, 1989.

306. **Pironin, M., Clément, G., Benzakour, O., Barritault, D., Lawrence, D., and Vigier, P.,** Growth in serum-free medium of NIH3T3 cells transformed by the EJ-H-*ras* oncogene: evidence for multiple autocrine growth factors, *Int. J. Cancer,* 51, 980, 1992.

307. **Harada, S. and Nishimoto, I.,** Possible requirement of serum progression factors for transformation of BALB/c 3T3 fibroblasts by v-*ras* p21, *FEBS Lett.,* 295, 59, 1991.

308. **Demetri, G. D., Ernst, T. J., Pratt, E. S., II, Zenzie, B. W., Rheinwald, J. G., and Griffin, J. D.,** Expression of *ras* oncogenes in cultured human cells alters the transcriptional and posttranscriptional regulation of cytokine genes, *J. Clin. Invest.,* 86, 1261, 1990.

Chapter 6

The Calcium-Calmodulin System

I. INTRODUCTION

Calcium is an essential regulatory element for cellular structures and functions, including those related to cell proliferation and cell differentiation.[1] An increase in intracellular concentration of free calcium ions, $[Ca^{2+}]_i$, caused by hormonal or electrical signals, can trigger in different cell types a diversity of responses, including contraction and secretion, as well as changes in metabolism and gene expression.[2] Calcium is involved in the regulation of intercellular communication through gap junctions and can trigger the terminal differentiation program of cells. It has an important role in the programmed cell death (apoptosis) of differentiated senescent or functionally superfluous cells that is required for the maintenance of body homeostasis.[3]

The $[Ca^{2+}]_i$ and the intracellular Ca^{2+} distribution are regulated in eukaryotic cells by various calcium transporting systems.[4-6] These systems maintain a steep concentration gradient between extracellular calcium, which is approximately $10^{-3} M$, and the $[Ca^{2+}]_i$, which is in the order of 10^{-7} to $10^{-8} M$, depending on the physiological state of the cell. The calcium transport systems serve to maintain a low $[Ca^{2+}]_i$ by transporting Ca^{2+} either out of the cell through the plasma membrane, or into intracellular storage sites, which are represented by the sarcoplasmic reticulum in muscle cells and the endoplasmic reticulum in nonmuscle cells. In most cell types in which hormones or growth factors release Ca^{2+} from intracellular stores, the release is triggered by the production of inositol 1,4,5-trisphosphate generated by phospholipase C-mediated breakdown of phosphatidylinositol 4,5-bisphosphate. The membrane receptors that trigger these reactions may be coupled to G proteins and may possess seven membrane-spanning regions. A particular member of the G protein family, G_q, is involved in stimulating the activity of polyphosphoinositide-specific phospholipase C in cells exposed to hormones or growth factors.[7] When microinjected into mammalian cells, inhibitory antibodies to the α subunit of G_q can block both release of Ca^{2+} from intracellular stores and Ca^{2+} influx in response to hormones such as bradykinin and vasopressin.[2]

The major entry pathway for calcium from the extracellular space into cells is via plasma membrane calcium channels. These channels are normally closed, and their opening allows passive flow of calcium through the channels along the calcium electrochemical gradient.[6] Calcium channels are involved in supplying calcium to many types of cells. In muscle and nerve cells, calcium channels play an important role in excitation-contraction coupling and neurotransmitter release, respectively.[8] Calcium channels are present in secretory cells and other cells where their function can be modulated by extracellular signals such as hormones and growth factors. Calcium channels can be classified in two main groups: voltage-dependent channels, which are open in response to an appropriate membrane depolarization, and receptor-operated channels, which are regulated through receptor-dependent mechanisms. Agents acting on voltage-dependent (voltage-gated) calcium channels can affect the signaling mechanisms regulated by growth factors, which act through variations in the $[Ca^{2+}]_i$.[9] Two types of voltage-sensitive Ca^{2+} have been characterized in the membrane of mouse 3T3 fibroblasts, and only one of these two types is selectively suppressed in 3T3 cells transformed by activated oncogenes including c-H-*ras*, EJ-*ras*, and v-*fms*, or by the gene encoding polyoma virus middle-T antigen.[10] The receptor-operated calcium channels (second messenger-operated calcium channels) have an important role in the mechanism of signal transduction across the membrane, and their expression can be dissociated from the formation of inositol phosphates.[11] Protein phosphorylation may have a role in the regulation of calcium channel function.[12] Both cAMP-dependent protein kinases and calcium-dependent phosphatases may regulate the function of voltage-activated calcium channels.[13] However, the substrates of such phosphorylations have not been characterized and it is not known if all cases of calcium current modulation involve phosphorylation.

The effects of calcium on cellular functions are determined not only by the total $[Ca^{2+}]_i$ but also by the distribution of Ca^{2+} among different compartments within the cells. It is known that mammalian cells possess two major types of intracellular Ca^{2+} stores: nonmitochondrial pools, which reside in the endoplasmic reticulum and exhibit high affinity and low capacity for Ca^{2+}, and mitochondrial pools, which represent a low-affinity, high-capacity Ca^{2+} store. The mitochondrial pool can sequester Ca^{2+} under certain physiological conditions. The non-phorbol-ester-type tumor promoter thapsigargin (a sesquiterpene lactone) discharges Ca^{2+} from nonmitochondrial stores because it inhibits ATP-dependent Ca^{2+} channels in the endoplasmic reticulum. Diverse stimuli, including hormones, growth factors, regulatory peptides, and neurotransmitters, can activate channels that release Ca^{2+} from the intracellular stores. One class of intracellular Ca^{2+} channel, the inositol trisphosphate-operated Ca^{2+} channel, is present in most types of cells. A second class of intracellular Ca^{2+} channel, the ryanodine receptor, is sensitive to the plant alkaloids ryanodine and caffeine and is present in the sarcoplamic reticulum of striated muscle as well as in smooth muscle, neurons, eggs, epithelial cells, and secretory cells. The $\beta4$ gene, which encodes the ryanodine receptor, is active in most types of mammalian tissues, and its expression is stimulated by TGF-β in cultured mink lung epithelial cells.[14]

The important role of calcium ions in cell proliferation and differentiation was established many years ago, although the transducing mechanisms responsible for the generation of a Ca^{2+}-mediated signal have been characterized only in the last few years.[15-23] In particular, it has been recognized that calcium ions act through their binding to specific cellular proteins.[24,25] The best characterized and the most widely distributed of these proteins is calmodulin.[26-35] The calcium/calmodulin system plays a key role in the regulation of the cell cycle.[36] Other calcium-binding proteins include parvalbumin, oncomodulin, caldesmon, calcineurin, calpactin, calcimedins, and many other cellular proteins. The metabolic actions of calcium are frequently mediated by the activation of calcium-dependent protein kinases.

The levels of calcium in the plasma are tightly regulated by hormones, in particular by parathormone and calcitonin. Growth factors may contribute to this regulation, and both EGF and TGF-α are capable of causing elevation of plasma calcium in mice.[37] The intracellular distribution of calcium shows marked variation both in the basal state and according to different external stimuli, including the action of hormone and growth factors. Mobilization of intracellular Ca^{2+} by serum growth factors is a pH-dependent process with a pH_i optimum around 7.1.[38] Inhibition of the Na^+/H^+ antiporter and subsequent cytoplasmic acidification attenuates the growth factor-induced Ca^{2+} release from intracellular stores. In the basal state, free Ca^{2+} concentrations within the nucleus are several orders of magnitude higher than those in the cytoplasm.[24] Cycling of mitochondrial Ca^{2+} stores provides a versatile regulation for the intracellular distribution of Ca^{2+}.[39]

II. CALCIUM IONS

Calcium ions play an important role in the regulation of cellular activity and metabolism, and calcium is considered, in conjunction with cyclic nucleotides, as a second messenger to extracellular signaling agents such as hormones, growth factors, regulatory peptides, and neurotransmitters.[40] An increase in the $[Ca^{2+}]_i$ may be caused by Ca^{2+} entry into the cells via ion channels in the plasma membrane or by Ca^{2+} release from intracellular stores. Ca^{2+} entry into the cell may occur by four major types of mechanisms: (1) a receptor-mediated channel coupled to intracellular second messengers; (2) a Ca^{2+} leak channel dependent on the Ca^{2+} electrochemical gradient; (3) a stretch-activated nonselective cation channel; and (4) internal Na^+-dependent Ca^{2+} entry (Na^+/Ca^{2+} exchange).[41] Both Ca^{2+} entry and intracellular Ca^{2+} mobilization can be observed in stimulated cells. Treatment of Ehrlich ascites tumor cells with micromolar concentrations of ATP induces a rapid and transient increase in $[Ca^{2+}]_i$ by mobilizing an intracellular, nonmitochondrial pool of $[Ca^{2+}]_i$ and increasing Ca^{2+} permeability at the level of the plasma membrane.[42] Other nucleotide triphosphates, including GTP, induce Ca^{2+} transient increases that are identical to those produced by ATP.

Growth factors can induce important changes in $[Ca^{2+}]_i$. The addition of serum to serum-deprived, quiescent mouse fibroblasts results in an immediate and transient elevation of $[Ca^{2+}]_i$, and a similar change may be observed by the addition of purified growth factors such as PDGF and FGF, but not EGF.[43] Addition of PDGF, EGF, or fetal calf serum to quiescent cultures of human fibroblasts may cause an immediate rise in $[Ca^{2+}]_i$.[44] cAMP may have an important role in regulating $[Ca^{2+}]_i$.[45] However, at least in certain systems, a hormone-induced rise in $[Ca^{2+}]_i$ and increased levels of intracellular cAMP may be independent from each other.[46]

Growth factors may act by triggering the release of calcium from intracellular stores (e.g., mitochondria, endoplasmatic reticulum, and plasma membrane) rather than by stimulating Ca^{2+} influx into the cell. The Ca^{2+} growth requirement of prostatic epithelial and foreskin fibroblastic cells is affected by EGF. A reduction in the half-maximal requirement for Ca^{2+} occurs in epithelial cells stimulated by EGF.[47,48] However, EGF can induce a rapid but transient rise in $[Ca^{2+}]_i$ which would depend on stimulation of Ca^{2+} entry through a voltage-independent Ca^{2+} channel in the plasma membrane.[49] Entry of extracellular Ca^{2+}

into the cells may be required for a sustained hormonal stimulation of cellular metabolic activities.[50] It is clear that Ca^{2+} is an important intracellular second messenger in the signaling mechanisms of many growth factors.

The transmission of the Ca^{2+} signal requires intracellular Ca^{2+} receptors to trigger the cellular response.[51] The Ca^{2+} receptors, with the exception of the annexins, share a common Ca^{2+}-binding motif, which is represented by a 29-residue HLH structure called an EF hand. Upon binding Ca^{2+}, the intracellular Ca^{2+} receptors undergo conformational changes which allow the activation of target proteins.

Extracellular Ca^{2+} may have an important role in controlling the proliferation and differentiation of both normal and neoplastic cells.[18,21,52-58] The transit of cells through the G_1 phase of the cycle is dependent on extracellular Ca^{2+} at two distinct points: the first point is the G_0-G_1 transition that occurs immediately after stimulation of the cells with serum or growth factors, and the second point is the prereplicative G_1-S period, immediately before DNA synthesis.[22] The events responsible for the two Ca^{2+}-dependent transitions, G_0-G_1 and G_1-S, implicate increased phosphoinositide turnover, protein kinase C activation and/or translocation, and intracellular Ca^{2+} release. Changes in the intracellular levels of cAMP and the activities of cAMP-dependent protein kinases are also involved in these transitions. The functional properties of many enzymes are regulated, either directly or indirectly, by calcium. The activity of DNA polymerase-α, a key enzyme in the process of DNA replication, is partially controlled by calcium, and cAMP-dependent protein kinases may participate in this regulatory mechanism.[59] The activity of the other two enzymes involved in DNA replication, ribonuclease reductase and thymidilate synthase, is also regulated by calcium and the active vitamin D derivative, calcitriol.

Changes in the extracellular concentrations of Ca^{2+} may have important influence on the proliferation and differentiation of cells. Elevating the extracellular level of Ca^{2+} enhances the proliferation of normal human WI-38 diploid cells either when the cultures are dense or are becoming dense.[60] When the cells are cycling, elevation of extracellular Ca^{2+} delays their exit from the cell cycle, and when they are quiescent and are stimulated with serum, it enhances their entry into the cell cycle. WI-38 cells have a limited lifespan and these calcium effects occur throughout the proliferative lifespan of the cells. Young and senescent WI-38 cells may exhibit differences in cell cycle-dependent regulation of calcium.[61]

Calcium-dependent and -independent signals are involved in the stimulation of cell proliferation induced by hormones, growth factors, and other mitogens in many types of cells, including human T lymphocytes.[62] The calcium ionophore A23187, which produces an increase in the levels of $[Ca^{2+}]_i$ via an influx of extracellular Ca^{2+}, can induce blast transformation, DNA synthesis, and mitosis in human lymphocytes.[63] Nontransformed cells require rather large amounts of Ca^{2+} in the media, and Ca^{2+} is necessary for progression of the cells through the cycle from G_1 into S phase as well as for re-entry of G_0 cells into the cycle and for mitosis. In contrast, transformed cells, especially those transformed by oncogenic viruses, require much less Ca^{2+} for growing.[30,64,65]

The mechanism by which Ca^{2+} regulates G_1-S transition of normal cells is not understood but may involve activation of Ca^{2+}-dependent protein kinases located on the cell surface, which would result in altered phosphorylation of particular proteins.[66] Calcium influx may be required for the synthesis and expression of transferrin receptors, and there may be a crucial requirement for transferrin receptors for traversal of the G_1-S boundary and initiation of DNA synthesis.[67] However, the precise role of calcium ions in mitogenic processes remains to be elucidated. In certain cells, DNA synthesis and the proliferative response may be independent of calcium mobilization. For example, calcium is involved in the production of IL-2 and the expression of IL-2 receptors in activated lymphocytes, but the proliferation of these cells may be independent of calcium mobilization.[68] Analysis of $[Ca^{2+}]_i$ in individual cells showed that an increase in $[Ca^{2+}]_i$ is necessary for mitogenesis in BALB/c 3T3 cells stimulated by PDGF, but not that stimulated by FGF.[69] Fetal calf serum plus insulin induces an ordered sequence of proto-oncogene expression and DNA synthesis in Fu5 rat hepatoma cells, but the induced growth of these cells occurs in the absence of any changes in $[Ca^{2+}]_i$ or pH_i.[70] It is thus clear that intracellular calcium increases are not part of a common metabolic pathway used by all competence factors and that there are multiple biochemical pathways associated with the stimulation of cell proliferation.

The precise role of calcium in the biochemical processes leading to cell proliferation remains to be characterized. Studies with strontium, a divalent cation with ionic radius (1.12 Å) similar to that of calcium (0.99 Å), indicate that it can closely mimic calcium in its stimulatory effects on DNA synthesis in CEF cells.[71] Strontium was not capable, however, of stimulating terminal differentiation in cultures of murine and human keratinocytes that are highly responsive to calcium, suggesting that extracellular calcium may separately stimulate the proliferation and terminal differentiation of keratinocytes.[72] Increased extracellular concentrations of calcium or establishment of intercellular communication by cell-cell contact during postconfluent growth of human keratinocytes may regulate the process of differentiation of these cells by influencing the levels of $[Ca^{2+}]_i$.[73]

Calcium may have a role in the initiation of apoptosis, as suggested by the occurrence of sustained increases in the $[Ca^{2+}]_i$ in thymocytes induced to undergo apoptosis by the action of glucocorticoids and the induction of Ca^{2+}-dependent endonuclease activity in the nuclei of cells treated with glucocorticoids. Disturbances in the regulation of cellular Ca^{2+} may occur in IL-3-dependent cells induced to undergo apoptosis by withdrawal of the cytokine.[74] Depletion of IL-3 in the IL-3-dependent hematopoietic cell line 32D results in marked changes in the partitioning of intracellular Ca^{2+} that precede cell death and that are completely reversible by restoration of the growth factor.[75]

III. CALMODULIN

The cellular actions of Ca^{2+} are mediated by a specific calcium-binding protein, calmodulin.[76] This protein is widely distributed in nature and has been found in all eukaryotic cells thus far examined, including plant cells,[77] but it is not present in prokaryotes. Calmodulin is highly conserved in evolution, and calmodulins isolated from diverse organisms are remarkably similar in biological, chemical, and physical properties. Signals such as light and the plant hormone, auxin, can affect the level of calmodulin mRNA in plants.[78]

A. CALMODULIN GENES

The structural organization and nucleotide sequence of the entire rat calmodulin gene has been determined.[79,80] The gene is about 9000 bases long and consists of six exons interrupted by introns of variable size. A segment with nucleotide sequence homologous to that of a rat-specific middle repetitive sequence is contained in the middle of the third exon of the rat calmodulin gene. A segment of the chicken calmodulin gene promoter was sequenced and assayed for promoter activity.[81] The GC-rich proximal 617 bp of this region is required to maximally activate transcription in CEF cells as well as in the mouse myoblast cell line BC3G-1. In proliferating BC3H-1 cells, differentiation induced by serum reduction decreases calmodulin mRNA levels, and the activity of the calmodulin promoter constructs also decreases by a similar extent with differentiation.

Further analyses indicated the existence of at least three calmodulin genes in vertebrate species including rodents. The rat genome contains three legitimate calmodulin genes (CaMI, CaMII, and CaMIII) that can be distinguished by the different sizes of their transcribed mRNAs but which are translated into a 148-residue calmodulin of identical amino acid sequence. The three calmodulin genes of the rat are differentially expressed in fetal and adult tissues as well as in neoplastic tissues.[82] CaMIII is the major calmodulin gene expressed in chemically or virally induced rat tumors and in metastatic lung nodules of mouse mammary carcinoma. The biological significance of differential calmodulin gene expression is not understood.

The human genome contains three divergent calmodulin genes (CaMI, CaMII, and CaMIII) that are under selective pressure to encode an identical protein while maintaining maximally divergent nucleotide sequences.[83] Only 81 to 82% identity exists in the coding sequence of the human calmodulin genes, a value close to the calculated minimum for two nucleotide sequences specifying an identical protein. Codon usage is different for the three calmodulin-coding DNA sequences, further reflecting their evolutionary divergence. The time of separation of the individual human calmodulin genes from their ancestor must lie far back in evolution. The advantage for an organism to possess several expressible calmodulin genes may reside in increased possibilities for differential regulation at the transcriptional level.

B. STRUCTURE AND FUNCTION OF CALMODULIN

Calmodulin is a 17-kDa monomeric protein composed of a single peptide chain of 148 amino acids containing four Ca^{2+}-binding domains. The three-dimensional structure of the calmodulin molecule has been determined by crystallography at 3.0 Å resolution.[84] The molecule consists of two globular lobes connected by a long exposed α-helix. Each lobe binds two calcium ions through helix-loop-helix domains. The long helix between the lobes may be involved in interactions of calmodulin with drugs and various proteins. The molecule is stabilized by multiple interactions between the helices. It is possible that the regulation of different enzymes by calmodulin is partially a function of the number of Ca^{2+} bound, but controversy exists regarding the order in which the four sites are filled and the discrete structural changes associated with Ca^{2+} binding. The three-dimensional solution structure of the complex between calcium-bound calmodulin and a 26-residue synthetic peptide comprising the calmodulin-binding domain (residues 577 to 602) of skeletal muscle myosin light chain kinase has been determined by nuclear magnetic resonance spectroscopy.[85] Calmodulin-binding peptides may share common structural features.

The general biological importance of calmodulin is indicated by the extremely high degree of its evolutionary conservation. All mammalian, chicken, and frog calmodulins have identical amino acid sequences, and even protozoan calmodulin differs from vertebrate calmodulin in only 12 out of 148 amino acids.[79] In plants, calmodulin is involved in a diversity of essential functions through its effects on hormone-regulated processes related to ion fluxes and enzyme activities.[86] Calmodulin exhibits a high degree of homology with other calcium-binding proteins, including parvalbumin and troponin-C. Calmodulin is the intermediate for calcium in many processes that are essential to the functional activity of cells. The kinetics of calcium binding to the calcium-binding domains of the calmodulin molecule and the subsequent activation of calmodulin-dependent enzymes are very complex. A protein of 75 kDa (p75) may represent a link between the protein kinase C and Ca^{2+}/calmodulin pathways.[87] Inhibition of protein kinase C-catalyzed phosphorylation of p75 by calmodulin in A10 smooth-muscle cells may be due to its interaction with the substrate, rather than a direct inhibitory effect on the enzyme, and this inhibition may be regulated by the $[Ca^{2+}]_i$. In addition, there is evidence that calmodulin may regulate cGMP-dependent protein kinase activity in calcium-dependent and -independent forms,[88] suggesting that calmodulin may not always act as a Ca^{2+} receptor but may also have biological actions that are independent of Ca^{2+} binding.

C. REGULATION OF CALMODULIN ACTIVITY

In addition to the regulation by calcium binding, calmodulin activity is regulated by phosphorylation. Insulin increases the phosphorylation of calmodulin and calmodulin-dependent protein kinase.[89-93] The insulin receptor contains a calmodulin-binding domain, and insulin stimulates the phosphorylation of calmodulin in intact cells such as rat adipocytes. This phosphorylation takes place exclusively on tyrosine residues,[94,95] and only the Tyr-99 residue is phosphorylated by the action of the insulin receptor kinase on the calmodulin molecule.[96] Calmodulin phosphorylation could have a role in the signaling mechanisms of growth factors.

IV. THE CALCIUM/CALMODULIN SYSTEM IN THE CONTROL OF CELL PROLIFERATION AND DIFFERENTIATION

Calcium has an important role in the regulation of cell growth and differentiation. The $[Ca^{2+}]_i$ may be directly linked to the control of cell proliferation.[97] Neoplastic cells may be less dependent on calcium and may require lower concentrations of extracellular calcium to proliferate. In serum-deprived human fibroblasts IMR-90 and WI-38 cells, the addition of fetal calf serum stimulates DNA synthesis in a manner that depends on the concentration of Ca^{2+} in the culture medium.[98] By contrast, DNA synthesis and division of SV40-transformed WI-38 cells are relatively independent of the extracellular concentration of Ca^{2+} as well as on the presence of growth factors. These differences my be related to modifications of the RB tumor suppressor protein, whose phosphorylation is necessary for the entry of cells into S phase of the cycle. In nontransformed cells RB protein phosphorylation is strictly dependent on the extracellular concentration of Ca^{2+}, whereas in transformed cells RB phosphorylation is little affected by this concentration. The effects of calcium on RB phosphorylation may be mediated by the cdc2 protein kinase.

Growth factors may evoke different types of responses in the calcium transductional pathways. Two patterns of calcium responses have been observed in myeloid leukemic cell lines stimulated to proliferate by different stimuli.[99] The pattern evoked by serum is characterized by an immediate response in $[Ca^{2+}]_i$ which precedes the onset of cell proliferation, while the pattern elicited by transferrin, insulin, or CSF-2 is characterized by a delayed increase in calcium influx rate which occurs concomitantly with an increase in DNA synthesis.

Changes in the concentration and intracellular distribution of calmodulin during the cell cycle are consistent with its role in the complex processes of cell division. Calmodulin is present in the nucleus and is indirectly bound to DNA through the interaction with specific nuclear protein fractions.[100] Calmodulin increases about threefold in rat liver nuclei after partial hepatectomy, the increase being maximal after 24 h, when DNA synthesis is also maximal.[101] During the same time, redistribution of calmodulin within the nucleus takes place, leading to its association with the nuclear matrix, a site tightly associated with DNA replication. Ca^{2+}-calmodulin complexes may be active in the processes leading to the initiation of DNA synthesis as well as in the regulation of gene expression. Addition of calcium stimulates DNA synthesis in calcium-deprived rat liver cells through the formation of Ca^{2+}-calmodulin complexes which undergo a brief intracellular redistribution in the later prereplicative phase of the cell cycle.[102] Calmodulin acts mainly at two points in the cell cycle: the G_1/S boundary and the metaphase transition.[103] An additional Ca^{2+}-associated event that is calmodulin-independent occurs at mitotic prophase. Studies with mouse C127 cells stably transformed by virus (BPV) expression vectors capable of inducibly synthesizing calmodulin sense or antisense RNA (which results in production of increased or decreased intracellular calmodulin levels, respectively) indicate that intracellular calmodulin levels may limit the rate of cell cycle progression under normal conditions of growth.[104] Variations of intracellular calmodulin levels are regulated mainly by transcriptional mechanisms, i.e., by changes in the synthesis of calmodulin mRNA. Calmodulin levels are elevated twofold at the late G_1 and/or early S phase during the cell cycle in actively growing cells.[105] A rise to a threshold calmodulin intracellular concentration occurring at the G_1-S boundary of the cycle may be a general phenomenon required for cycle progression and may be linked to processes associated with the initiation of DNA synthesis and cell division.[106] Changes in intracellular calmodulin levels occur in the re-entry of growth-arrested cells into the cycle, when the cells have been rendered competent by growth factors or other mitogens.[107] Complex interactions between cyclic nucleotides and calmodulin-dependent enzymes, including cyclic nucleotide phosphodiesterases, may contribute to the regulation of a host of cellular functions.[108]

Calmodulin is involved in at least two distinct phases of the cell cycle. First, release from plateau is associated with a decrease in calmodulin content within the first hour, and second, a rapid increase in calmodulin levels is associated with the entry into S phase 5 to 8 h after release.[107] Specific elevation of intracellular calmodulin levels can affect the rate of cell proliferation. However, elevated levels of calmodulin do not markedly affect overall gene expression or the cell cycle changes observed in mRNAs.[103] Expression of a cloned chicken calmodulin gene in mouse C127 cells using a BPV-based vector resulted in a reduction in the length of the cell cycle due to reduction in the length of the G_1 phase.[109] Mitogenic stimulation of human lymphocytes by lectins results in an increase in the calmodulin content of the blast cells, and treatment of the cells with the anticalmodulin compound, trifluoroperazine, inhibits their ability to undergo blastogenic stimulation.[110] Treatment with anticalmodulin drugs produces important effects on the growth of normal or malignant cells, eliciting specific and reversible delays in the progression into and through S phase of the cycle.[105,111] Ca^{2+} and calmodulin are involved in the processes conferring stability to DNA in proliferating and neoplastic cells.[112] In certain systems *in vitro*, the cellular responses to mitogenic stimuli may be greater at high external Ca^{2+} concentrations, which is probably due to alterations in the state of cell differentiation.

A. REGULATION OF PROTO-ONCOGENE EXPRESSION BY CALCIUM

The effects of calcium on cell proliferation and differentiation are exerted, in part, at the level of the regulation of gene expression.[113] Prominent among the calcium-regulated genes are some proto-oncogenes. Basic calcium phosphate crystals can display some properties similar to those of competence growth factors, controlling the traverse of cells from G_0 or G_1 to S phase of the cycle and can initiate cellular proliferation by rendering density-arrested BALB/c 3T3 fibroblasts competent to respond to growth factors such as the IGFs.[114] These effects of calcium phosphate are associated with increased expression of the c-*fos* and c-*myc* proto-oncogenes. Both extracellular and intracellular Ca^{2+} may participate in the regulation of proto-oncogene expression. Mobilization of intracellular Ca^{2+} is required for the induction of c-*fos* transcription in mouse peritoneal macrophages following activation of phagocytosis by different agents, including phorbol esters, cholera toxin, and dexamethasone.[115] The tumor promoter thapsigargin, which discharges intracellular Ca^{2+} stores by inhibition of endoplasmic reticulum Ca^{2+}-ATPase, induces c-*fos* and c-*jun* expression in murine fibroblasts.[116] Treatment of human lymphocytes with the calcium ionophore A23187 induces a transient accumulation of c-*fos* and c-*myc* mRNA.[117] A23187 also induces c-*fos* expression in the human leukemia cell line U-937.[118] The A23187-induced accumulation of c-*fos* transcripts is blocked by the chelating agent EGTA, which inhibits the calcium influx. Addition of EGTA to the incubation medium of murine thymocytes results in complete abolishment of the transcriptional activation of c-*fos* and c-*myc* genes by mitogenic concentrations of A23187.[119] Increased $[Ca^{2+}]_i$ is necessary for maximal expression of the proto-oncogene c-*jun* in the Jurkat T-cell line.[120] Expression of c-*fos* is induced in PC12 rat pheochromocytoma cells by interaction of NGF with its receptor or by agents that affect voltage-dependent calcium channels.[121,122] Depolarization of PC12 cells with elevated K^+ or veratridine results in prolonged opening of the channels and induction of c-*fos* expression, which is dependent on the exogenous supply of extracellular Ca^{2+}. Opening of the calcium channels appears to be the common denominator in the cascade of events that result in c-*fos* induction in PC12 cells. This induction is associated with the expression of cellular differentiation (neurite formation).

The mechanisms by which calcium induces or modulates gene expression are little known but may include the activation of Ca^{2+}/calmodulin-dependent protein kinases and phosphorylation of specific nuclear proteins. Intracellular redistribution of Ca^{2+} precedes PDGF-stimulated c-*fos* gene induction in NIH/3T3 cells.[123] Expression of c-*fos* mRNA and protein induced by treatment of the cells with PDGF is increased by inhibitors of Ca^{2+}/calmodulin-dependent protein kinase, apparently by inhibition of c-*fos* mRNA degradation. Calcium-inducible proteins and specific DNA sequences may be implicated in cell- and tissue-specific regulation of gene expression by calcium.[124] The c-*fos* gene contains an intragenic Ca^{2+}-responsive element, and mobilizable $[Ca^{2+}]_i$ is required for induction of c-*fos* gene transcription in macrophages.[125] Transcription of the c-*fos* gene in macrophages is modulated by a calcium-dependent block to elongation in intron 1 of the gene. Regulation of c-*fos* and c-*jun* expression by activation of the muscarinic receptor in glial cells depends on protein kinase A activation and $[Ca^{2+}]_i$ increase.[126] Protein kinase C may have a role in the regulation of gene expression by calcium. Oxidative stress induces sequential expression of the c-*fos*, c-*jun*, and c-*myc* genes, which is associated with protein kinase C activation and mobilization of intracellular Ca^{2+}.[127] However, the effects of calcium and protein kinase C on c-*fos* and c-*jun* expression may be different, and calcium induces c-*fos* and prolactin gene expression in cultured rat pituitary cells by mechanisms that are independent of protein kinase C.[128] Expression of the c-*myc* gene is regulated at multiple levels by the calcium ionophore A23187 in the HL-60 human leukemia cell line.[129] An early increase in c-*myc* mRNA steady-state levels seen after treatment with A23187 appears to result from the combined effects of an increase in the initiation of transcription and an enhancement of mRNA stability. A later decline in c-*myc* mRNA level in this system appears to result from an inhibition of elongation of the nascent c-*myc* mRNA.

In certain systems, proto-oncogene expression is independent of changes in $[Ca^{2+}]_i$. Stimulation of c-*myc* expression, DNA synthesis, and mitosis in the Nb2 rat lymphoma cell line by prolactin occurs independently of extracellular calcium and is not accompanied by movement of Ca^{2+} through the plasma membrane.[130] Calcium ionophores and activators of protein kinase C are not mitogenic for Nb2 cells and have no effect on c-*myc* expression in this system. Induction of c-*myc* and c-*fos* gene expression in C3H10T1/2 mouse fibroblasts by EGF or TGF-α may be independent of the $[Ca^{2+}]_i$.[131] It is thus clear that calcium may affect gene expression through different mechanisms, and that in some cases gene activation appears to be independent of changes in the intracellular concentration and/or distribution of calcium.

The intracellular signaling mechanisms mediated by calcium and tyrosine kinases are interconnected, and calcium may play an important role in the regulation of cellular tyrosine kinases. The differentiation of cultured human epidermal keratinocytes induced by calcium is accompanied by increased activity of tyrosine phosphatases, which results in a rapid activation of the c-Src tyrosine kinase and inactivation of the c-Yes tyrosine kinase.[132] These changes are associated with altered phosphorylation of several cellular proteins on tyrosine and correlate with a redistribution of phosphotyrosine from membrane and adhesion sites to the nucleus. The tyrosine kinase activity of the HCGF receptor, which is encoded by the c-*met* proto-oncogene, is negatively modulated in GTL-16 cells by protein kinase C-catalyzed phosphorylation of serine residues in the receptor as well as by an increase in the $[Ca^{2+}]_i$.[133] The mechanism of the latter effect is unknown but may involve the action of Ca^{2+}-dependent protein kinase(s). Calmodulin can inhibit the EGF receptor kinase encoded by the c-*erb*-A1 proto-oncogene.[134]

B. THE CALCIUM/CALMODULIN SYSTEM IN MALIGNANT TRANSFORMATION

Calmodulin levels are usually higher in transformed cells and rapidly dividing normal cells than in their quiescent normal counterparts, suggesting that calmodulin levels are involved in the regulation of cell division.[135,136] Drugs that inhibit the biological activity of calmodulin are capable of killing tumor cells with a potency that correlates with their potency as calmodulin antagonists.[137] However, transformed cells do not obligatorily contain elevated levels of calmodulin and many factors, intrinsic or extrinsic to these cells, may contribute to produce wide variation in the calmodulin levels present in either normal or transformed cells.

Oncogene expression may alter the intracellular distribution of Ca^{2+} and the response of the cell to growth factors. Expression of a mutated c-H-*ras* oncogene in NIH/3T3 cells leads to a rapid desensitization of the intracellular Ca^{2+}-mobilizing system to bombesin and growth factors.[138] The desensitization of the Ca^{2+}-mobilizing system may be caused either by inhibition of inositol 1,4,5-trisphosphate-regulated Ca^{2+} channels or by interference of the c-H-Ras protein with Ca^{2+} translocation between intracellular calcium compartments. However, oncogenic activation of c-*ras* genes is probably not solely responsible for the altered regulation of intracellular calcium observed in neoplastic cells.[139]

Expression of the v-Src oncoprotein in Rat-1 cells induces Ca^{2+} redistribution among different intracellular compartments.[140] Calmodulin is an exogenous substrate for phosphorylation on tyrosine by the purified v-Src kinase.[141] The possible biological significance of this phenomenon is not understood because different exogenous polypeptides can be phosphorylated by the v-Src kinase.[142] However, there is evidence that phosphorylation of calmodulin on tyrosine by the v-Src kinase is inhibited by Ca^{2+} and that the structure of the calmodulin-Ca^{2+} complex may be altered by phosphorylation on tyrosine residues.[143] In contrast, serine and threonine phosphorylation of calmodulin occurs in both normal and RSV-transformed cells.

Myeloid differentiation can be induced in HL-60 leukemia cells with naphthalene sulfonamide calmodulin antagonists, suggesting that inhibition of Ca^{2+}/calmodulin-dependent functions may lead to, or augment, differentiation of certain malignant cells.[144] However, a general role for critical calmodulin changes in the processes of differentiation of either normal or neoplastic cells is not in accordance with some experimental observations. The retinoic acid-induced precommitment state and subsequent myeloid differentiation of HL-60 cells apparently are not regulated by cytosolic calmodulin levels.[145]

C. REQUIREMENT OF NEOPLASTIC CELLS FOR EXTRACELLULAR CALCIUM

The requirement of neoplastic cells for calcium may be highly variable. The proliferation of some tumor cells may be less sensitive than that of normal cells to extracellular calcium deprivation.[146] While the growth of normal cells *in vitro* is tightly controlled by extracellular calcium, the growth of tumor cells may be relatively independent of calcium in the medium. However, there are remarkable exceptions to this general rule, and tumor cells may exhibit large variations in their dependence on extracellular calcium. Initiation of carcinogenesis in the mouse skin may be associated with the evolution of cells that resist calcium-induced terminal differentiation.[147] Proliferation of the estrogen-insensitive human breast cancer cell line HT-39 is more sensitive to changes in the levels of extracellular calcium than is the proliferation of the estrogen-responsive human breast cancer cell line MCF-7.[148] Moreover, inhibition of the growth of HT-39 and MCF-7 cells by calcitriol depends on the concentration of extracellular calcium. Highly malignant leukemia L1210 cells show a marked dependency of the free extracellular Ca^{2+} for proliferation.[149] Fresh leukemic cells from patients with adult T-cell leukemia (ATL) do not grow well in a low calcium medium, and their growth is enhanced by the addition of calcium to the medium in a dose-dependent manner.[150] In contrast, cells from leukemias other than ATL grow well in the low-calcium medium, and their growth is not enhanced by the addition of calcium. Furthermore, calcium antagonists and calmodulin inhibitors inhibit the growth of ATL cells at concentrations that do not inhibit the growth of other leukemic cells. Thus, calcium and calmodulin may play a critical role in regulation of ATL cells, but other leukemic cells may be relatively independent of exogenous calcium supply.

The mechanisms of reduced requirement for exogenous calcium supply for proliferation of neoplastically transformed cells are not understood but may involve an alteration of Ca^{2+}-dependent protein kinases or other cellular enzymes. Human ovarian tumor cells show altered cellular calcium regulatory processes associated with defective downregulation of protein kinase C, and this defect may confer on these cells the ability to proliferate independently of the external calcium concentration.[151] Deregulation of $[Ca^{2+}]_i$ may be an important common component of carcinogenic processes.[152]

D. ROLE OF CALCIUM IN NEOPLASTIC CELL DIFFERENTIATION

Physiological Ca^{2+} concentrations are probably required for the maintenance of a differentiated state of cells in the tissues of adult animals.[153] Several studies suggest a role for calcium in the differentiation of cells transformed by chemical agents *in vitro*. Differentiation of tumor cell lines induced by exogenous agents may be associated with changes in the concentration and/or distribution of intracellular calcium. Differentiation of Friend murine erythroleukemia (MEL) cells induced by DMSO would be associated with an early increase in Ca^{2+} influx.[154] However, the use of a fluorescent indicator shows that, contrary to expectation, a small decrease of $[Ca^{2+}]_i$ occurs upon treatment of Friend MEL cells with DMSO.[155] Both calcium deprivation and the addition of a calcium antagonist (verapamil) in the culture medium enhance the differentiation of HL-60 cells induced by DMSO, retinoic acid, or calcitriol.[156] By contrast, the induction of differentiation of HL-60 cells by phorbol ester is not enhanced by calcium deprivation or verapamil. However, the results from many studies suggest the existence of a relationship between inhibition of intracellular calcium mobilization and cell differentiation. This concept is supported by the fact that addition of an intracellular calcium antagonist (TMB-8) also enhances the differentiation of

HL-60 cells induced by DMSO, retinoic acid, or calcitriol. A cellular calcium pool supplied by ionophoretic increases in the $[Ca^{2+}]_i$ is required for phorbol ester-promoted transformation of the JB6 mouse epidermal cell line.[157] Neither proliferation nor differentiation is affected by calcium deprivation in MEL cells induced to differentiation by treatment with DMSO, but calcium-chelating agents or agents blocking intracellular calcium uptake induce a marked inhibition of differentiation in this cellular system.[158] By contrast, MEL cell differentiation is unaffected by agents that increase intracellular calcium concentration. These results indicate that intracellular calcium mobilization is indispensable for eliciting full differentiation response of MEL cells, but the increase of intracellular Ca^{2+} level is not sufficient for complete signal transduction.

The precise role of calcium in the complex mechanisms associated with cell differentiation is unknown, and a general role for calcium in the differentiation of normal or neoplastic cells is rather unlikely. An increase in the $[Ca^{2+}]_i$ was observed during the process of serum-induced squamous differentiation in normal human bronchial epithelial cells or Ad12-SV40 hybrid virus-transformed human bronchial epithelial cells, but an early calcium signal was not generated during squamous differentiation of these cells induced by either TGF-β or TPA.[159] Therefore, early changes in $[Ca^{2+}]_i$ would not be necessary to initiate the process of squamous differentiation in normal or transformed human bronchial epithelial cells.

E. PRESENCE OF CALMODULIN IN RETROVIRUSES

A calcium-binding site is present in retroviruses, and this site, as determined by a specific radioimmunoassay, is represented by calmodulin or a similar molecule.[160] Retroviruses of different origins, including murine, feline, and primate retroviruses, as well as the human retrovirus HTLV-I, may contain calmodulin. The function of calmodulin in retroviruses is unknown but calcium is involved in the assembly and disassembly of a number of viruses, and calmodulin may be important in maintaining the structural integrity of retroviruses. Further studies are required for a proper evaluation of the possible role of calmodulin in the assembly and function of viral particles.

V. CALCIUM/CALMODULIN-DEPENDENT PROTEIN KINASES

The calcium-calmodulin complex can function either as an integral subunit of an enzyme or through direct interaction with specific enzymes. Signals induced by growth factors and other extracellular agents may result in the activation of Ca^{2+}-dependent enzymes such as phospholipase C and protein kinase C, but there are other intracellular target systems for the fluctuations of the $[Ca^{2+}]_i$ caused by mitogens.[161] An important class of enzymes that can be activated by specific extracellular stimuli is represented by the Ca^{2+}/calmodulin-dependent protein kinases (CaM kinases).[162] The CaM kinases may be involved in mediating the action of hormones, growth factors, regulatory peptides, and neurotransmitters that induce alterations in $[Ca^{2+}]_i$ levels.[163] The calmodulin subunit of CaM kinases is the Ca^{2+} receptor that transduces changes in the $[Ca^{2+}]_i$ into specific biological responses by modulating the activity of the kinase catalytic subunit.[166] This subunit shows substrate specificity for serine/threonine residues. Calmodulin remains bound to the CaM kinase even when the $[Ca^{2+}]_i$ is basal, which provides a potentiating mechanism for brief repetitive calcium signals.[164] The cAMP-response element (CRE)-binding (CREB) protein, which recognizes the consensus DNA sequence TGACGTCA and is involved in the regulation of gene expression by exogenous stimuli, is a substrate for CaM kinases.[165] CaM kinases may transduce signals to the nucleus and the CREB protein may function to integrate Ca^{2+} and cAMP intracellular signals.

Mammalian tissues contain a family of multifunctional CaM kinases that are involved in the selective expression of distinct gene products in different tissues.[167] Four well-characterized members of this family are phosphorylase kinase, myosin light chain kinase, CaM kinase II, and the Ca^{2+}/phospholipid-dependent enzyme, protein kinase C.[168] Two other kinases, the CaM kinases I and III, have been purified from mammalian cells.[169,170] The activity of another enzyme, the CaM kinase IV, is modulated through phosphorylation by cAMP-dependent protein kinase.[171] While phosphorylase kinase and myosin light chain kinase may be specific for their designated substrates *in vivo*, CaM kinase II and protein kinase C have more general substrate specificities.

CaM kinase I is involved in the phosphorylation of a specific serine residue in a collagenase-resistant domain of synapsin I, a synaptic vesicle associated protein that is involved in the regulation of neurotransmitter release. The enzyme was purified from bovine brain using synapsin I as substrate.[169] CaM kinase I contains sequences that are autophosphorylated on threonine in the presence of Ca^{2+} and calmodulin, but the possible role of these autophosphorylations is not known. Also unknown is the function of the phosphorylation of synapsin I as well as the phosphorylation of another substrate of the same kinase, protein III.

CaM kinase II (multifunctional CaM) is involved in the cellular responses to hormones, growth factors, regulatory peptides, and neurotransmitters.[170,173] The enzyme purified from rat brain cytosol is composed of two subunits of 50 and 60 kDa, both of which are autophosphorylated by an intramolecular reaction.[174] Such autophosphorylation is important for modulating the enzyme activity as well as its dependence on the Ca^{2+}-calmodulin system.[175] At low but saturating ATP levels, autophosphorylation causes a 75% reduction in protein kinase II activity, with the residual activity still retaining a dependence on Ca^{2+} and calmodulin, whereas at higher but still physiological levels of ATP the kinase is converted by autophosphorylation to a form that is autonomous of Ca^{2+} and calmodulin, with no accompanying reduction in activity.[176] Autophosphorylation of only one subunit of the enzyme generates Ca^{2+}/calmodulin-independent kinase activity.[177] CaM kinase II is inactive in the absence of Ca^{2+}/calmodulin because an inhibitory domain within residues 281 to 309 of the enzyme interacts with the catalytic domain and blocks ATP binding.[178] Autophosphorylation of the Thr-286 residue results in a Ca^{2+}/calmodulin-independent form of the kinase by disrupting the inhibitory interaction with the catalytic domain. Mutagenesis of Thr-286 in the monomeric form of CaM kinase eliminates the Ca^{2+}/calmodulin-independent activity.[179] A cDNA clone encoding a portion of the 50-kDa subunit of rat brain CaM kinase II was sequenced, the calmodulin-binding region was identified, and a synthetic analog was prepared that bound calmodulin with high affinity in the presence of calcium.[180] CaM kinase II binds to actin filaments under physiologic ionic conditions, and calmodulin in the presence of Ca^{2+} inhibits binding of the enzyme to actin filaments.[181] CaM kinase II may be involved in controlling physiological processes such as cell secretion and intracellular movement of macromolecules through the phosphorylation-induced modification of actin substrate contained in microfilaments. Recently, the enzyme has been localized in the interphase nucleus and the mitotic apparatus of mammalian cells.[182] CaM kinase II activity is much higher in cerebrum than any other rat tissue tested, followed by spleen and lung.[168] The ratios of CaM kinase II to protein kinase C activity are different in various rat tissues as well as between particulate and supernatant subcellular fractions, suggesting different functions for these two Ca^{2+}-regulated protein kinases.

CaM kinase III is involved in the phosphorylation of a 100-kDa protein substrate that is not phosphorylated by other CaM kinases. High levels of this substrate have been found in the pancreas as well as in several mammalian cell lines, including the PC12 rat pheochromocytoma cell line, in which treatment with NGF, forskolin, or other agents that elevate the $[Ca^{2+}]_i$ leads to a marked and rapid decrease in the activity of CaM kinase III.[183] The 100-kDa substrate of CaM kinase III has been identified with EF-2, a factor involved in protein synthesis by catalyzing the translocation of peptidyl-tRNA on the ribosome.[170] Intracellular Ca^{2+} may inhibit protein synthesis in mammalian cells via CaM kinase III-catalyzed phosphorylation of EF-2.

Calmodulin regulates the activity of the adenylyl cyclase system in a wide range of structurally different tissues and species.[184] Unfortunately, the molecular mechanisms by which calmodulin activates adenylyl cyclase are not yet clear. Other questions to be answered are whether both calmodulin-dependent and -independent forms of adenylyl cyclase exist within any one tissue and whether some tissues lack a calmodulin-dependent enzyme entirely. Available evidence indicates that the calcium-calmodulin system modulates, either directly or indirectly, most if not all of the protein phosphorylation occurring in eukaryotic cells. CaM kinase activities contribute to the regulation of cytosolic and nuclear protein phosphorylation.[185-188]

VI. OTHER CALCIUM-BINDING PROTEINS

In addition to calmodulin, a number of proteins with calcium-binding properties have been described in different types of cells. The function of these proteins remains, for most of them, only partially characterized. In particular, the possible role of these proteins in the mechanism of action of growth factors is little understood. Some of these proteins are mentioned next.

Oncomodulin is a calcium-binding protein present in rodent tumors induced by carcinogens as well as in the blood of the tumor-bearing animals.[189] It is also found in a variety of human tumors but is not present in normal fetal or adult human tissues. Oncomodulin is not, however, a tumor-specific protein but is present in human and rodent extra-embryonic tissues descended from both lineages of the blastocyst, including placenta, parietal yolk sac, trophectoderm, parietal endoderm, and amnion.[190,191] In the human placenta, oncomodulin is present in the cytotrophoblastic cell population during all stages of gestation, suggesting that it may have a role in cytotrophoblastic function.[192] Expression of the oncomodulin gene is regulated at the level of transcription. The richest known source of oncomodulin is the rat Morris hepatoma 5123. A bacterial expression system capable of yielding high amounts of oncomodulin has been described.[193] Sequencing of cDNA clones containing the entire coding sequence of oncomodulin from rat hepatoma tissue revealed a single ORF encoding a peptide of 108 amino acids.[194] The oncomodulin-coding DNA sequence does not exhibit significant homology with other members of the calcium-binding family of proteins, including calmodulin, but the oncomodulin and parvalbumin cDNA sequences show 59% homology. Oncomodulin is different from oncodevelopmental proteins such as α-fetoprotein (AFP) and carcinoembryonic antigen (CEA) in several ways, the most important one being that oncomodulin expression is not associated with cellular proliferation but may be linked to the invasive properties of either extra-embryonic tissues or tumor tissues. Studies with chemical transformation of rat fibroblasts suggest that oncomodulin may represent a suitable marker for malignant transformation.[195]

Calcium binds to complexes of calmodulin and several classes of calmodulin-binding proteins,[196] and a prominent member of such proteins is caldesmon,[197,198] a calmodulin-binding and F-actin-binding protein that is involved in the mechanism by which the Ca^{2+}-dependent regulatory action of calmodulin is transmitted to F-actin filaments.[199,200] Caldesmon was first isolated from smooth muscle of chicken gizzard as a 155-kDa protein that binds to actin-tropomyosin filaments, but the molecular weight of monomeric caldesmon in solution is 93 kDa.[201] The actin cross-linking, bundling, and polymerizing properties of caldesmon *in vitro* suggest that its function may be related to cytoskeletal structures *in vivo*. Direct interaction between tropomyosin and caldesmon may contribute to regulate smooth muscle contraction upon binding of Ca^{2+} to the actin-bound calsdesmon.[202] Caldesmon is present in a diversity of cells. Two forms of caldesmon (molecular weight in the range of 120 to 150 and 70 to 80 kDa, respectively) have been detected in a wide variety of smooth muscle and nonmuscle cells. The high molecular weight form of caldesmon is more specific for smooth muscle, whereas the low molecular weight form is widely distributed in tissues and cells with the exception of smooth muscle.[203] Caldesmon may play a major role in the regulation of smooth muscle and nonmuscle contractile events, which implicates its involvement in a wide variety of biological phenomena including contraction, cell movement, cell shape change, and endocytosis. Caldesmon is found in bovine aorta and uterus, in adrenal medulla, and in human platelets as well as in cultured fibroblasts, where it is present as a 77-kDa protein and is localized in the cellular stress fibers and leading edges in close association with actin filaments. In nonmuscle cells, the calmodulin-caldesmon system may play a regulatory role in cellular functions associated with actin-containing microfilaments, such as cellular morphology and locomotion, membrane ruffling, and cell adhesion to a substratum. Caldesmon is present in intact platelets and may be phosphorylated by protein kinase C.[204,205] The possible physiological significance of caldesmon phosphorylation is still uncertain but nonmuscle caldesmon is phosphorylated by the cdc2 kinase during mitosis, which reduces its binding affinity for both actin and calmodulin, suggesting that cdc2 may cause microfilament reorganization during mitosis.[206,207] RSV-transformed cells may contain decreased caldesmon concentrations, and the intracellular distribution of caldesmon in these cells is changed to a diffuse and blurred appearance.[200] These alterations may be important in relation to the morphological and functional expression of a transformed phenotype.

Calcineurin is a major calmodulin-binding protein in bovine brain and human placental membranes.[208-210] Calcineurin is a Ca^{2+}/calmodulin-dependent phosphoprotein phosphatase and is composed of two polypeptide subunits, A and B, with molecular weight of 58 to 61 and 19 kDa, respectively. The A subunit of calcineurin contains the calmodulin-binding and catalytic domains, while the B subunit binds 4 atoms of Ca^{2+} with high affinity. Calcineurin catalyzes the dephosphorylation of both phosphoserine/phosphothreonine and phosphotyrosine residues in various cellular substrates, including the EGF receptor. Calcineurin appears to be a key signaling enzyme in lymphocyte activation.[211] However, the precise physiological role of calcineurin is unknown. The phosphotyrosyl protein phosphatase activity associated with calcineurin can be modulated by calmodulin and protease treatment.[212] CaM kinase II and protein kinase C modulate the enzymatic activity of calcineurin through phosphorylation at regulatory sites of the protein.[213,214]

The calcimedins are another family of calcium-binding proteins. They have been isolated from smooth, cardiac and skeletal muscle and have molecular weights between 30 and 67 kDa.[215] The 67-kDa calcimedin has been purified to homogeneity and is distinct from other Ca^{2+}-binding proteins, but its precise physiological role is unknown.

Other calcium-binding proteins, the calpactins, are proteins of 34 to 39 kDa that undergo a Ca^{2+}-dependent association with the membrane and the cytoskeleton.[216] The calpactins have a particular affinity for phospholipid vesicles and actin filaments. They may play a role in membrane movement or cell attachment or both by interacting with the cytoskeleton and overlaying plasma membrane. Calpactin-like proteins are present in human spermatozoa.[217] Two distinct calpactins, calpactin I and calpactin II, have been detected in lung and placenta tissue as well as in human diploid fibroblasts and the epidermoid carcinoma cell line A431.[218] Calpactin I is identical with a 36-kDa protein called p36, lipocortin II, or annexin II, and calpactin II is identical to a 35-kDa protein called p35, lipocortin I, or annexin I. The term lipocortin may be a misnomer because glucocorticoids have no effect on the synthesis or secretion of calpactins *in vivo*.[219] Calpactin I is a major substrate for oncoproteins with tyrosine kinase activity such as the Src kinase. PDGF induces the phosphorylation of calpactin I on specific tyrosine residues.[220] Calpactin II is a substrate for the activated EGF receptor kinase.[221] In addition to the two calpactins, I and II, a calpactin-related protein of 70 kDa is present in the human placenta. The calpactins can be phosphorylated on their amino-terminal domains by protein kinase C, but this modification occurs only in restricted intracellular compartments where $[Ca^{2+}]_i$ transiently reaches high levels.[222]

The parvalbumins are Ca^{2+}-binding acidic proteins whose expression is limited mainly to muscle and brain. S-100 proteins are a family of acidic, low molecular weight Ca^{2+}-binding proteins originally isolated from nervous tissue, but they have also been found in a number of other tissues. Both calmodulin and the S-100b protein interact with microtubule components in a calcium-dependent manner and inhibit the phosphorylation of these components by protein kinase C.[223] A structural homology was detected between the amino acid sequences of S-100b protein and the regulatory chain of the 36-kDa (p36) substrate of tyrosine kinase oncoproteins.[224] The biological significance of this homology is unknown.

The physiological activities of calmodulin are regulated by calmodulin-binding proteins that are frequently associated with particular cytoskeletal components. These regulatory proteins include caldesmon, calcineurin, calspectin and spectin, as well as the microtubule-associated protein 2 (MAP-2) and cytosynalin. The 35-kDa protein, cytosynalin, purified from synaptosomal membranes, has a broad range of interactions with cytoskeletal elements.[225] Cytosynalin is linked to calspectin on the inner aspect of the membrane and may interact with actin filaments or microtubules.

The calpains are neutral proteases that have an absolute Ca^{2+} dependence for activity.[226] They have been isolated from the cytosolic fraction of animal tissues or cells, but may be translocated to the cell membrane under certain conditions. The two calpains, calpain I and calpain II, have an indentical 30-kDa regulatory subunit but two distinct 80-kDa catalytic subunits. Both types of calpain subunits contain calmodulin-like Ca^{2+}-binding domains at their carboxy-terminal ends, and these domains confer Ca^{2+} sensitivity to the calpains. The activity of calpains is also regulated by an endogenous calpain inhibitor protein, termed calpastatin. The biological functions of calpains are little understood; however, it is clear that they do not have general proteolytic activity, but rather catalyze specific and limited cleavage of specific substrates. Calpain I and calpain II appear to have identical substrate specificities. Endogenous substrates for calpains include enzymes such as kinases as well as myofibrilar, membrane, cytoskeletal, and receptor proteins. More than half of calmodulin-binding proteins are calpain substrates *in vitro*.[226] These proteins are recognized by calpains through the presence of specific amino acid sequences in them, called PEST sequences, which contain proline (P), glutamic acid (E), serine (S), and threonine (T). The Jun and Fos oncoproteins, which contain PEST sequences and form heterodimers with binding activity for AP-1 DNA sites, are substrates for calpain.[227] Some small proteins not containing PEST sequences may also be calpain substrates.

REFERENCES

1. **Williams, R. J. P.,** Calcium and calmodulin, *Cell Calcium,* 13, 355, 1992.
2. **Davis, T. N.,** What's new with calcium?, *Cell,* 71, 557, 1992.
3. **Whitfield, J. F.,** Calcium signals and cancer, *Crit. Rev. Oncogenesis,* 3, 55, 1992.
4. **Anghileri, L. J. and Tuffet-Anghileri, A. M.,** Eds., *The Role of Calcium in Biological Systems,* CRC Press, Boca Raton, FL, 1982.
5. **Carafoli, E.,** Intracellular calcium homeostasis, *Annu. Rev. Biochem.,* 56, 395, 1987.
6. **Hosey, M. M. and Lazdunski, M.,** Calcium channels: molecular pharmacology, structure and regulation, *J. Membrane Biol.,* 104, 81, 1988.
7. **Smrcka, A. V., Hepler, J. R., Brown, K. O., and Sternweis, P. C.,** Regulation of polyphosphoinositide-specific phospholipase C activity by purified G_q, *Science,* 251, 804, 1991.
8. **Miller, R. J.,** Receptor-mediated regulation of calcium channels and neurotransmitter release, *FASEB J.,* 4, 3291, 1990.
9. **Olsen, R., Seewald, M., and Powis, G.,** Contribution of external and internal Ca^{2+} to changes in intracellular free Ca^{2+} produced by mitogens in Swiss 3T3 fibroblasts: the role of dihydropyridine sensitive Ca^{2+} channels, *Biochem. Biophys. Res. Commun.,* 162, 448, 1989.

10. **Chen, C., Corbley, M. J., Roberts, T. M., and Hess, P.,** Voltage-sensitive calcium channels in normal and transformed 3T3 fibroblasts, *Science,* 239, 1024, 1988.

11. **Ng, J., Gustavsson, J., Jondal, M., and Andersson, T.,** Regulation of calcium influx across the plasma membrane of the human T-leukemic cell line, JURKAT: dependence on a rise in cytosolic free calcium can be dissociated from formation of inositol phosphates, *Biochim. Biophys. Acta,* 1053, 97, 1990.

12. **Byerly, L., Leung, H.-T., and Yazejian, B.,** Cellular control of calcium currents, *Biomed. Res.,* 9 (Suppl. 2), 1, 1988.

13. **Armstrong, D. L.,** Calcium channel regulation by protein phosphorylation in a mammalian tumor cell line, *Biomed. Res.,* 9 (Suppl. 2), 11, 1988.

14. **Giannini, G., Clementi, E., Ceci, R., Marziali, G., and Sorrentino, V.,** Expression of ryanodine receptor-Ca^{2+} channel that is regulated by TGF-β, *Science,* 257, 91, 1992.

15. **Rasmussen, H. and Goodman, D. B.,** Relationships between calcium and cyclic nucleotides in cell activation, *Physiol. Rev.,* 57, 421, 1977.

16. **Whitfield, J. F., Boynton, A. L., MacManus, J. P., Sikorska, M., and Tsang, B. K.,** The regulation of cell proliferation by calcium and cyclic AMP, *Mol. Cell. Biochem.,* 27, 155, 1979.

17. **Williamson, J. R., Cooper, R. H., and Hoek, J. B.,** Role of calcium in the hormonal regulation of liver metabolism, *Biochim. Biophys. Acta,* 639, 243, 1981.

18. **Veigl, M. L., Vanaman, T. C., and Sedwick, W. D.,** Calcium and calmodulin in cell growth and transformation, *Biochim. Biophys. Acta,* 738, 21, 1984.

19. **Rasmussen, H.,** The calcium messenger system, *N. Engl. J. Med.,* 314, 1094, 1986.

20. **Metcalfe, J. C., Moore, J. P., Smith, G. A., and Hesketh, T. R.,** Calcium and cell proliferation, *Br. Med. Bull.,* 42, 405, 1986.

21. **Boynton, A. L.,** Calcium and epithelial cell proliferation, *Mineral Electrolyte Metab.,* 14, 86, 1988.

22. **Whitaker, M. and Patel, R.,** Calcium and cell cycle control, *Development,* 108, 525, 1990.

23. **Lu, K. P. and Means, A. R.,** Regulation of the cell cycle by calcium and calmodulin, *Endocrine Rev.,* 14, 40, 1993.

24. **Moore, P. B. and Dedman, J. R.,** Calcium binding proteins and cellular regulation, *Life Sci.,* 31, 2937, 1982.

25. **Heizmann, C. W. and Berchtold, M. W.,** Expression of parvalbumin and other Ca^{2+}-binding proteins in normal and tumor cells: a topical review, *Cell Calcium,* 8, 1, 1987.

26. **Cheung, W. Y.,** Calmodulin plays a pivotal role in cellular regulation, *Science,* 207, 19, 1980.

27. **Means, A. R. and Dedman, J. R.,** Calmodulin in endocrine cells and its multiple roles in hormone action, *Mol. Cell. Endocrinol.,* 19, 215, 1980.

28. **Scharff, O.,** Calmodulin and its role in cellular activation, *Cell Calcium,* 2, 1, 1981.

29. **Means, A. R., Tash, J. S., and Chafouleas, J. G.,** Physiological implications of the presence, distribution, and regulation of calmodulin in eukaryotic cells, *Physiol. Rev.,* 62, 1, 1982.

30. **Means, A. R., Lagace, L. , Guerriero, V., Jr., and Chafouleas, J. G.,** Calmodulin as a mediator of hormone action and cell regulation, *J. Cell. Biochem.,* 20, 317, 1982.

31. **Oldham, S. B.,** Calmodulin: its role in calcium-mediated cellular regulation, *Mineral Electrolyte Metab.,* 8, 1, 1982.

32. **Cheung, W. Y.,** Calmodulin: an overview, *Fed. Proc.,* 41, 2253, 1982.

33. **Lin, Y. M.,** Calmodulin, *Mol. Cell. Biochem.,* 45, 101, 1982.

34. **Stevens, F. C.,** Calmodulin: an introduction, *Can. J. Biochem. Cell Biol.,* 61, 906, 1983.

35. **Johnson, J. D. and Mills, J. S.,** Calmodulin, *Med. Res. Rev.,* 6, 341, 1986.

36. **Lu, K. L. and Means, A. R.,** Regulation of the cell cycle by calcium and calmodulin, *Endocrine Rev.,* 14, 40, 1993.

37. **Tashjian, A. H., Voelkel, E. F., Lloyd, W., Derynck, R., Winkler, M. E., and Levine, L.,** Actions of growth factors on plasma calcium — epidermal growth factor and human transforming growth factor-α cause elevation of plasma calcium in mice, *J. Clin. Invest.,* 78, 1405, 1986.

38. **Maly, K., Hochleitner, B. W., and Grunicke, H.,** Interrelationship between growth factor-induced activation of the Na^+/H^+-antiporter and mobilization of intracellular Ca^{2+} in NIH3T3-fibroblasts, *Biochem. Biophys. Res. Commun.,* 167, 1206, 1990.

39. **Nicholls, D. G. and Crompton, M.,** Mitochondrial calcium transport, *FEBS Lett.,* 111, 261, 1980.

40. **Rebhum, L. I.,** Cyclic nucleotides, calcium and cell division, *Int. Rev. Cytol.,* 49, 1, 1977.

41. **Adams, D. J., Barakeh, J., Laskey, R., and van Breemen, C.,** Ion channels and regulation of intracellular calcium in vascular endothelial cells, *FASEB J.,* 3, 2389, 1989.

42. **Dubyak, G. R. and De Young, M. B.,** Intracellular Ca^{2+} mobilization activated by extracellular ATP in Ehrlich ascites tumor cells, *J. Biol. Chem.,* 260, 10653, 1985.

43. **McNeil, P. L., McKenna, M. P., and Taylor, D. L.,** A transient rise in cytosolic calcium follows stimulation of quiescent cells with growth factors and is inhibitable with phorbol myristate acetate, *J. Cell Biol.,* 101, 372, 1985.

44. **Moolenaar, W. H., Tertoolen, L. G. J., and de Laat, S. W.,** Growth factors immediately raise cytoplasmic free Ca^{2+} in human fibroblasts, *J. Biol. Chem.,* 259, 8066, 1984.

45. **Mollard, P., Zhang, Y., Rodman, D., and Cooper, D. M. F.,** Limited accumulation of cyclic AMP underlies a modest vasoactive-intestinal-peptide-mediated increase in cytosolic $[Ca^{2+}]$ transients in GH3 pituitary cells, *Biochem. J.,* 284, 637, 1992.

46. **Kurstjens, N. P., Heithier, H., Cantrill, R. C., Hahn, M., and Boege, F.,** Multiple hormone actions: the rises in cAMP and Ca^{++} in MDCK-cells treated with glucagon and prostagandin E_1 are independent processes, *Biochem. Biophys. Res. Commun.,* 167, 1162, 1990.

47. **McKeehan, W. L. and McKeehan, K. A.,** Epidermal growth factor modulates extracellular Ca^{2+} requirement for multiplication of normal human skin fibroblasts, *Exp. Cell Res.,* 123, 397, 1979.

48. **Lechner, J. F.,** Interdependent regulation of epithelial cell replication by nutrients, hormones, growth factors, and cell density, *Fed. Proc.,* 43, 116, 1984.

49. **Moolenaar, W. H., Aerts, R. J., Tertoolen, L. G. J., and de Laat, S. W.,** The epidermal growth factor-induced calcium signal in A431 cells, *J. Biol. Chem.,* 261, 279, 1986.

50. **Joseph, S. K., Coll, K. E., Thomas, A. P., Rubin, R., and Williamson, J. R.,** The role of extracellular Ca^{2+} in the response of the hepatocyte to Ca^{2+}-dependent hormones, *J. Biol. Chem.,* 260, 12508, 1985.

51. **Strynadka, N. C. J. and James, M. N. G.,** Crystal structures of the helix-loop-helix calcium-binding proteins, *Annu. Rev. Biochem.,* 58, 951, 1989.

52. **Boynton, A. L., Whitfield, J. F., Isaacs, R. J., and Morton, H. J.,** Control of 3T3 cell proliferation by calcium, *In Vitro,* 10, 12, 1974.

53. **Dulbecco, R. and Elkington, J.,** Induction of growth in resting fibroblastic cell cultures by Ca^{++}, *Proc. Natl. Acad. Sci. U.S.A.,* 72, 1584, 1975.

54. **Swierenga, S. H. H., Whitfield, J. F., and Gillan, D. J.,** Alteration by malignant transformation of the calcium requirements for cell proliferation *in vitro, J. Natl. Cancer Inst.,* 57, 125, 1976.

55. **Hennings, H., Michael, D., Cheng, C., Steinert, P., Holbrook, K., and Yuspa, S. H.,** Calcium regulation of growth and differentiation of mouse epidermal cells in culture, *Cell,* 19, 245, 1980.

56. **Boynton, A. L., McKeehan, W. L., and Whitfield, J. F.,** Eds., *Ions Cell Proliferation and Cancer,* Academic Press, New York, 1982.

57. **Lichtman, A. H., Segel, G. B., and Lichtman, M. A.,** The role of calcium in lymphocyte proliferation (an interpretative review), *Blood,* 61, 413, 1983.

58. **Miyaura, C., Abe, E., and Suda, T.,** Extracellular calcium is involved in the mechanism of differentiation of mouse myeloid leukemia cells (M1) induced by $1\alpha,25$-dihydroxyvitamin D_3, *Endocrinology,* 115, 1891, 1984.

59. **Rixon, R. H., Isaacs, R. J., and Whitfield, J. F.,** Control of DNA polymerase-α activity in regenerating rat liver by calcium and $1\alpha,25(OH)_2D_3$, *J. Cell. Physiol.,* 139, 354, 1989.

60. **Praeger, F. C. and Cristofalo, V. J.,** Modulation of WI-38 cell proliferation by elevated levels of $CaCl_2$, *J. Cell. Physiol.,* 129, 27, 1986.

61. **Brooks-Frederich, K. M., Cianciarulo, F. L., Rittling, S. R., and Cristofalo, V. J.,** Cell cycle-dependent regulation of Ca^{2+} in young and senescent WI-38 cells, *Exp. Cell Res.,* 205, 412, 1993.

62. **Gelfand, E. W., Cheung, R. K., Grinstein, S., and Mills, G. B.,** Characterization of the role for calcium influx in mitogen-induced triggering of human T cells: identification of calcium-dependent and calcium-independent signals, *Eur. J. Immunol.,* 16, 907, 1986.

63. **Luckasen, J. R., White, J. G., and Kersey, J. H.,** Mitogenic properties of a calcium ionophore, A23187, *Proc. Natl. Acad. Sci. U.S.A.,* 71, 5088, 1974.

64. **Durkin, J. P. and Whitfield, J. F.,** Transforming NRK cells with avian sarcoma virus reduces the extracellular Ca^{2+} requirement without affecting the calcicalmodulin requirement for the G_1/S transition, *Exp. Cell Res.,* 157, 544, 1985.

65. **Parsons, P. G., Moss, D. J., Morris, C., Musk, P., Maynard, K., and Partridge, R.,** Decreased calcium dependence of lymphoblastoid cell lines compared with Burkitt lymphoma cell lines, *Int. J. Cancer,* 35, 743, 1985.

66. **Kleine, L. P., Whitfield, J. F., and Boynton, A. L.,** Ca^{2+}-dependent cell surface protein phosphorylation may be involved in the initiation of DNA synthesis, *J. Cell. Physiol.,* 129, 303, 1986.

67. **Neckers, L. M., Bauer, S., McGlennen, R. C., Trepel, J. B., Rao, K., and Greene, W. C.,** Diltiazem inhibits transferrin receptor expression and causes G_1 arrest in normal and neoplastic T cells, *Mol. Cell. Biol.,* 6, 4244, 1986.

68. **Koyasu, S., Suzuki, G., Asano, Y., Osawa, H., Diamantstein, T., and Yahara, I.,** Signals for activation and proliferation of murine T lymphocyte clones, *J. Biol. Chem.,* 262, 4689, 1987.

69. **Tucker, R. W., Chang, D. T., and Meade-Cobun, K.,** Effects of platelet-derived growth factor and fibroblast growth factor on free intracellular calcium and mitogenesis, *J. Cell. Biochem.,* 39, 139, 1989.

70. **Cook, P. W., Weintraub, W. H., Swanson, K. T., Machen, T. E., and Firestone, G. L.,** Glucocorticoids confer normal serum/growth factor-dependent growth regulation to Fu5 rat hepatoma cells *in vitro.* Sequential expression of cell cycle-regulated genes without changes in intracellular calcium or pH, *J. Biol. Chem.,* 263, 19296, 1988.

71. **Rubin, H.,** Specificity of the requirements for magnesium and calcium in the growth and metabolism of chick embryo fibroblasts, *J. Cell. Physiol.,* 91, 449, 1977.

72. **Praeger, F. C., Stanulis-Praeger, B. M., and Gilchrest, B. A.,** Use of strontium to separate calcium-dependent pathways for proliferation and differentiation in human keratinocytes, *J. Cell. Physiol.,* 132, 81, 1987.

73. **Pillai, S., Bikle, D. D., Mancianti, M.-L., Cline, P., and Hincenbergs, M.,** Calcium regulation of growth and differentiation of normal human keratinocytes: modulation of differentiation competence by stages of growth and extracellular calcium, *J. Cell. Physiol.,* 143, 294, 1990.

74. **Rodriguez-Taduchy, G., Malde, P., Lopez-Rivas, A., and Collins, M. K. L.,** Inhibition of apoptosis by calcium ionophores in IL-3-dependent bone marrow cells is dependent upon production of IL-4, *J. Immunol.,* 148, 1416, 1992.

75. **Baffy, G., Miyashita, T., Williamson, J. R., and Reed, J. C.,** Apoptosis induced by withdrawal of interleukin-3 (IL-3) from an IL-3-dependent hematopoietic cell line is associated with repartitioning of intracellular calcium and is blocked by enforced Bcl-2 oncoprotein production, *J. Biol. Chem.,* 268, 6511, 1993.

76. **Cohen, P. and Klee, C. B.,** Eds., *Calmodulin,* Elsevier, Amsterdam, 1988.

77. **Dieter, P.,** Calmodulin and calmodulin-mediated processes in plants, *Plant Cell Environ.,* 7, 371, 1984.

78. **Jena, P. K., Reddy, A. S. N., and Poovaiah, B. W.,** Molecular cloning and sequencing of a cDNA for plant calmodulin: signal-induced changes in the expression of calmodulin, *Proc. Natl. Acad. Sci. U.S.A.,* 86, 3644, 1989.

79. **Nojima, H. and Sokabe, H.,** Structure of a gene for rat calmodulin, *J. Mol. Biol.,* 193, 439, 1987.

80. **Sherbany, A. A., Parent, A. S., and Brosius, J.,** Rat calmodulin DNA, *DNA,* 6, 267, 1987.

81. **Epstein, P. N., Christenson, M. A., and Means, A. R.,** Chicken calmodulin promoter activity in proliferating and differentiated cells, *Mol. Endocrinol.,* 3, 193, 1989.

82. **MacManus, J. P., Gillen, M. F., Korczak, B., and Nojima, H.,** Differential calmodulin gene expression in fetal, adult, and neoplastic tissues of rodents, *Biochem. Biophys. Res. Commun.,* 159, 278, 1989.

83. **Fischer, R., Koller, M., Flura, M., Mathews, S., Strehler-Page, M.-A., Krebs, J., Penniston, J. T., Carafoli, E., and Strehler, E. E.,** Multiple divergent mRNAs code for a single human calmodulin, *J. Biol. Chem.,* 263, 17055, 1988.

84. **Babu, Y. S., Sack, J. S., Greenhough, T. J., Bugg, C. E., Means, A. R., and Cook, W. J.,** Three-dimensional structure of calmodulin, *Nature,* 315, 37, 1985.

85. **Ikura, M., Clore, G. M., Gronenborn, A. M., Zhu, G., Klee, C. B., and Bax, A.,** Solution structure of a calmodulin-target peptide complex by multidimensional NMR, *Science,* 256, 632, 1992.

86. **Elliott, D. C.,** Calmodulin inhibitor prevents plant hormone response, *Biochem. Int.,* 1, 290, 1980.

87. **Zhao, D., Hollenberg, M. D., and Severson, D. L.,** Calmodulin inhibits the protein kinase C-catalysed phosphorylation of an endogenous protein in A10 smooth-muscle cells, *Biochem. J.,* 277, 445, 1991.

88. **Yamaki, T. and Hidaka, H.,** Ca^{2+}-independent stimulation of cyclic GMP-dependent protein kinase by calmodulin, *Biochem. Biophys. Res. Commun.,* 94, 727, 1980.

89. **Graves, C. B., Gale, R. D., Laurino, J. P., and McDonald, J. M.,** The insulin receptor and calmodulin. Calmodulin enhances insulin-mediated receptor kinase activity and insulin stimulates phosphorylation of calmodulin, *J. Biol. Chem.,* 261, 10429, 1986.

90. **Sacks, D. B. and McDonald, J. M.,** Insulin-stimulated phosphorylation of calmodulin by rat liver insulin receptor preparations, *J. Biol. Chem.,* 263, 2377, 1988.

91. **Wong, E. C. C., Sacks, D. B., Laurino, J. P., and McDonald, J. M.,** Characteristics of calmodulin phosphorylation by the insulin receptor kinase, *Endocrinology,* 123, 1830, 1988.

92. **Sacks, D. B., Fujita-Yamaguchi, Y., Gale, R. D., and McDonald, J. M.,** Tyrosine-specific phosphorylation of calmodulin by the insulin receptor kinase purified from human placenta, *Biochem. J.,* 263, 803, 1989.

93. **Sacks, D. B., Davis, H. W., Crimmins, D. L., and McDonald, J. M.,** Insulin-stimulated phosphorylation of calmodulin, *Biochem. J.,* 286, 211, 1992.

94. **Graves, C. B., Goewert, R. R., and McDonald, J. M.,** The insulin receptor contains a calmodulin-binding domain, *Science,* 230, 827, 1985.

95. **Colca, J. R., DeWald, D. B., Pearson, J. D., Palazuk, B. J., Laurino, J. P. and McDonald, J. M.,** Insulin stimulates the phosphorylation of calmodulin in intact adipocytes, *J. Biol. Chem.,* 262, 11399, 1987.

96. **Laurino, J. P., Colca, J. R., Pearson, J. D., DeWald, D. B., and McDonald, J. M.,** The *in vitro* phosphorylation of calmodulin by insulin receptor tyrosine kinase, *Arch. Biochem. Biophys.,* 265, 8, 1988.

97. **Short, A. D., Bian, J. H., Ghosh, T. K., Waldron, R. T., Rybak, S. L., and Gill, D. L.,** Intracellular Ca^{2+} pool content is linked to control of cell growth, *Proc. Natl. Acad. Sci. U.S.A.,* 90, 4886, 1993.

98. **Takuwa, N., Zhou, W., Kumada, M., and Takuwa, Y.,** Ca^{2+}-dependent stimulation of retinoblastoma gene product phosphorylation and p34cdc2 kinase activation in serum-stimulated human fibroblasts, *J. Biol. Chem.,* 268, 138, 1993.

99. **Rephaeli, A., Aviram, A., Rabizadeh, E., and Shaklai, M.,** Proliferation-associated changes of Ca^{2+} transport in myeloid leukemic cell lines, *J. Cell. Physiol.,* 143, 154, 1990.

100. **Bachs, O. and Carafoli, E.,** Calmodulin and calmodulin-binding proteins in liver cell nuclei, *J. Biol. Chem.,* 262, 10786, 1987.

101. **Serratosa, J., Pujol, M. J., Bachs, O., and Carafoli, E.,** Rearrangement of nuclear calmodulin during proliferative liver cell activation, *Biochem. Biophys. Res. Commun.,* 150, 1162, 1988.

102. **Boynton, A. L., Whitfield, J. F., and MacManus, J. P.,** Calmodulin stimulates DNA synthesis by rat liver cells, *Biochem. Biophys. Res. Commun.,* 95, 745, 1980.

103. **Means, A. R. and Rasmussen, C. D.,** Calcium, calmodulin and cell proliferation, *Cell Calcium,* 9, 313, 1988.

104. **Rasmussen, C. D. and Means, A. R.,** Calmodulin is required for cell-cycle progression during G_1 and mitosis, *EMBO J.,* 8, 73, 1989.

105. **Chafouleas, J. G., Bolton, W. E., Hidaka, H., Boyd, A. E., III, and Means, A. R.,** Calmodulin and the cell cycle: involvement in regulation of cell-cycle progression, *Cell,* 28, 41, 1982.

106. **Feinberg, J., Capeau, J., Picard, J., and Weinman, S.,** Calmodulin in Zajdela hepatoma cell growth, *Exp. Cell Res.,* 168, 265, 1987.

107. **Chafouleas, J. G., Lagace, L., Bolton, W. E., Boyd, A. E., III, and Means, A. R.,** Changes in calmodulin and its mRNA accompany reentry of quiescent (G_0) cells into the cell cycle, *Cell,* 36, 73, 1984.

108. **Sharma, R. K. and Wang, J. H.,** Regulation of cAMP concentration by calmodulin-dependent cyclic nucleotide phosphodiesterase, *Biochem. Cell Biol.,* 64, 1072, 1986.

109. **Rasmussen C. D. and Means, A. R.,** Calmodulin is involved in regulation of cell proliferation, *EMBO J.,* 6, 3961, 1987.

110. **Rainteau, D., Sharif, A., Bourrillon, R., and Weinman, S.,** Calmodulin in lymphocyte mitogenic stimulation and in lymphoid cell line growth, *Exp. Cell Res.,* 168, 546, 1987.

111. **Kikuchi, Y., Iwano, I., and Kato, K.,** Effects of calmodulin antagonists on human ovarian cancer cell proliferation *in vitro,* *Biochem. Biophys. Res. Commun.,* 123, 385, 1984.

112. **Lönn, U. and Lönn, S.,** Ca^{2+} and calmodulin are involved in the processes conferring stability to DNA in proliferating neoplasti cells, *Int. J. Cancer,* 37, 891, 1986.

113. **Resendez, E., Jr., Ting, J., Kim, K. S., Wooden, S. K., and Lee, A. S.,** Calcium ionophore A23187 as a regulator of gene expression in mammalian cells, *J. Cell Biol.,* 103, 2145, 1986.

114. **Cheung, H. S., Mitchell, P. G., and Pledger, W. J.,** Induction of c-*fos* and c-*myc* protooncogenes by basic calcium phosphate crystal: effect of β-interferon, *Cancer Res.,* 49, 134, 1989.

115. **Collart, M. A., Belin, D., Briottet, C., Thorens, B., Vassalli, J.-D., and Vasalli, P.,** Receptor-mediated phagocytosis by macrophages induces a calcium-dependent transient increase in c-*fos* transcription, *Oncogene,* 4, 237, 1989.

116. **Schönthal, A., Sugarman, J., Brown, J. H., Hanley, M. R., and Feramisco, J. R.,** Regulation of c-*fos* and c-*jun* protooncogene expression by the Ca^{2+}-ATPase inhibitor thapsigargin, *Proc. Natl. Acad. Sci. U.S.A.,* 88, 7096, 1991.

117. **Grausz, J. D., Fradelizi, D., Dautry, F., Monier, R., and Lehn, P.,** Modulation of c-fos and c-myc mRNA levels in normal human lymphocytes by calcium ionophore A23187 and phorbol ester, *Eur. J. Immunol.,* 16, 1217, 1986.

118. **Shibanuma, M., Kuroki, T., and Nose, K.,** Inhibition of proto-oncogene c-*fos* transcription by inhibitors of protein kinase C and ion transport, *Eur. J. Biochem.,* 164, 15, 1987.

119. **Moore, J. P., Todd, J. A., Hesketh, T. R., and Metcalfe, J. C.,** c-*fos* and c-*myc* gene activation, ionic signals, and DNA synthesis in thymocytes, *J. Biol. Chem.,* 261, 8158, 1986.

120. **Vandenplas, M. L., Mouton, W. J., Vandenplas, S., Bister, A. J., and Ricketts, M. H.,** Increased intracellular Ca^{2+} is necessary for maximal expression of the proto-oncogene c-*jun* in the Jurkat T-cell line, *Biochem. J.,* 267, 349, 1990.

121. **Morgan, J. I. and Curran, T.,** Role of ion flux in the control of c-*fos* expression, *Nature,* 322, 552, 1986.

122. **Curran, T. and Morgan, J. I.,** Barium modulates c-*fos* expression and post-translational modification, *Proc. Natl. Acad. Sci. U.S.A.,* 83, 8521, 1986.

123. **Bravo, R., Neuberg, M., Burckhardt, J., Almendral, J., Wallich, R., and Müller, R.,** Involvement of common and cell type-specific pathways in c-*fos* gene control: stable induction by cAMP in macrophages, *Cell,* 48, 251, 1987.

124. **Huff, C. A., Yuspa, S. H., and Rosenthal, D.,** Identification of control elements 3′ to the human keratin 1 gene that regulate cell type and differentiation-specific expression, *J. Biol. Chem.,* 268, 377, 1993.

125. **Collart, M. A., Tourkine, N., Belin, D., Vassalli, P., Jeanteur, P., and Blanchard, J.-M.,** c-*fos* gene transcription in murine macrophages is modulated by a calcium-dependent block to elongation in intron 1, *Mol. Cell. Biol.,* 11, 2826, 1991.

126. **Trejo, J. and Brown, J. H.,** c-*fos* and c-*jun* are induced by muscarinic receptor activation of protein kinase C but are differentially regulated by intracellular calcium, *J. Biol. Chem.,* 266, 7876, 1991.

127. **Maki, A., Berezesky, I. K., Fargnoli, J., Holbrook, N. J., and Trump, B. F.,** Role of $[Ca^{2+}]_i$ in induction of c-*fos*, c-*jun*, and c-*myc* mRNA in rat PTE after oxidative stress, *FASEB J.,* 6, 919, 1992.

128. **Bandyopadhyay, S. K. and Bancroft, C.,** Calcium induction of the mRNAs for prolactin and c-*fos* is independent of protein kinase C activity, *J. Biol. Chem.,* 264, 14216, 1989.

129. **Salehi, Z. and Niedel, J. E.,** Multiple calcium-mediated mechanisms regulate c-*myc* expression in HL-60 cells, *J. Immunol.,* 145, 276, 1990.

130. **Murphy, P. R., DiMattia, G. E., and Friesen, H. G.,** Role of calcium in prolactin-stimulated c-*myc* gene expression and mitogenesis in Nb2 lymphoma cells, *Endocrinology,* 122, 2476, 1988.

131. **Cutry, A. F., Kinniburgh, A. J., Krabak, M. J., Hui, S.-W., and Wenner, C. E.,** Induction of c-*fos* and c-*myc* proto-oncogene expression by epidermal growth factor and transforming growth factor α is calcium-independent, *J. Biol. Chem.,* 264, 19700, 1989.

132. **Zhao, Y., Sudol, M., Hanafusa, H., and Krueger, J.,** Increased tyrosine kinase activity of c-Src during calcium-induced keratinocyte differentiation, *Proc. Natl. Acad. Sci. U.S.A.,* 89, 8298, 1992.

133. **Gandino, L., Munaron, L., Naldini, L., Ferracini, R., Magni, M., and Comoglio, P. M.,** Intracellular calcium regulates the tyrosine kinase receptor encoded by the *MET* oncogene, *J. Biol. Chem.,* 266, 16098, 1991.

134. **Jose, E. S., Benguria, A., Geller, P., and Villalobo, A.,** Calmodulin inhibits the epidermal growth factor receptor tyrosine kinase, *J. Biol. Chem.,* 267, 15237, 1992.

135. **MacManus, J. P., Braceland, B. M., Rixon, R. H., Whitfield, J. F., and Morris, H. P.,** An increase in calmodulin during growth of normal and cancerous liver *in vivo*, *FEBS Lett.,* 133, 99, 1981.

136. **Epstein, P. M., Moraski, S., Jr., and Hachisu, R.,** Identification and characterization of a Ca^{2+}-calmodulin-sensitive cyclic nucleotide phosphodiesterase in a human lymphoblastoid cell line, *Biochem. J.,* 243, 533, 1987.

137. **Hait, W. N. and DeRosa, W. T.,** Calmodulin as a target for new chemotherapeutic strategies, *Cancer Invest.,* 6, 499, 1988.

138. **Oberhuber, H., Maly, K., Überall, F., Hoflacher, J., Kiani, A., and Grunicke, H. H.,** Mechanism of desensitization of the Ca^{2+}-mobilizing system to bombesin by Ha-*ras*. Independence from down-modulation of agonist-stimulated inositol phosphate production, *J. Biol. Chem.,* 266, 1437, 1991.

139. **Kruszewski, F. H., Hennings, H., Tucker, R. W., and Yuspa, S. H.,** Differences in the regulation of intracellular calcium in normal and neoplastic keratinocytes are not caused by ras gene mutations, *Cancer Res.,* 51, 4206, 1991.

140. **Seuwen, K. and Adam, G.,** Calcium compartments and fluxes are affected by the *src* gene product of Rat-1 cells transformed by temperature-sensitive Rous sarcoma virus, *Biochem. Biophys. Res. Commun.,* 125, 337, 1984.

141. **Fukami, Y. and Lipmann, F.,** Purification of the Rous sarcoma virus *src* kinase by casein-agarose and tyrosine-agarose affinity chromatography, *Proc. Natl. Acad. Sci. U.S.A.,* 82, 321, 1985.

142. **Wong, T. W. and Goldberg, A. R.,** Kinetics and mechanism of angiotensin phosphorylation by the transforming gene product of Rous sarcoma virus, *J. Biol. Chem.,* 259, 3127, 1984.

143. **Fukami, Y., Nakamura, T., Nakayama, A., and Kanehisa, T.,** Phosphorylation of tyrosine residues of calmodulin in Rous sacroma virus-transformed cells, *Proc. Natl. Acad. Sci. U.S.A.,* 83, 4190, 1986.

144. **Veigl, M. L., Sedwick, W. D., Nieldel, J., and Branch, M. E.,** Induction of myeloid differentiation of HL-60 cells with naphthalene sulfonamide calmodulin antagonists, *Cancer Res.,* 46, 2300, 1986.

145. **Yen, A., Freeman, L., Powers, V., Van Sant, R., and Fishbaugh, J.,** Cell cycle dependence of calmodulin levels during HL-60 proliferation and myeloid differentiation, *Exp. Cell Res.,* 165, 139, 1986.

146. **Whitfield, J. F., Boynton, A. L., Healy, G. M., and Sun, A. M.,** Oncogenic transformation and reductions of insulin secretion and proliferative calcium dependence during repeated passage of pancreatic islet cells *in vitro, Cancer Lett.,* 14, 159, 1981.

147. **Yuspa, S. H. and Morgan, D. L.,** Mouse skin cells resistant to terminal differentiation associated with initiation of carcinogenesis, *Nature,* 293, 72, 1981.

148. **Simpson, R. U. and Arnold, A. J.,** Calcium antagonizes 1,25-dihydroxyvitamin D_3 inhibition of breast cancer cell proliferation, *Endocrinology,* 119, 2284, 1986.

149. **Cory, J. G., Carter, G. L., and Karl, R. C.,** Calcium ion-dependent proliferation of L1210 cells in culture, *Biochem. Biophys. Res. Commun.,* 145, 556, 1987.

150. **Shirakawa, F., Yamashita, U., Oda, S., Chiba, S., Eto, S., and Suzuki, H.,** Calcium dependency in the growth of adult T-cell leukemia cells *in vitro, Cancer Res.,* 46, 658, 1986.

151. **Chan, T. C. K.,** Calcium-independent growth of human ovarian carcinoma cells, *J. Cell. Physiol.,* 141, 461, 1989.

152. **Trump, B. F. and Berezesky, I. K.,** Ion regulation, cell injury and carcinogenesis, *Carcinogenesis,* 8, 1027, 1987.

153. **Eckl, P. M., Whitcomb, W. R., Michalopoulos, G., and Jirtle, R. L.,** Effects of EGF and calcium on adult parenchymal hepatocyte proliferation, *J. Cell. Physiol.,* 132, 363, 1987.

154. **Levenson, R., Housman, D., and Cantley, L.,** Amiloride inhibits erythroleukemia cell differentiation: evidence for a Ca^{2+} requirement for commitment, *Proc. Natl. Acad. Sci. U.S.A.,* 77, 5948, 1980.

155. **Faletto, D. L. and Macara, I. G.,** The role of Ca^{2+} in dimethyl sulfoxide-induced differentiation of Friend erythroleukemia cells, *J. Biol. Chem.,* 260, 4884, 1985.

156. **Okazaki, T., Mochizuki, T., Tashima, M., Sawada, H., and Uchino, R.,** Role of intracellular calcium ion in human promyelocytic leukemia HL-60 cell differentiation, *Cancer Res.,* 46, 6059, 1986.

157. **Smith, B. M., Gindhart, T. D., and Colburn, N. H.,** Extracellular calcium requirement for promotion of transformation in JB6 cells, *Cancer Res.,* 46, 701, 1986.

158. **Supino, R., Gibelli, N., Galatulas, I., and Zunino, F.,** Influence of calcium and calcium-modulating agents on differentiation of murine erythroleukaemia cells, *Cell Biol. Int. Rep.,* 9, 1059, 1985.

159. **Miyashita, M., Smith, M. W., Willey, J. C., Lechner, J. F., Trump, B. F., and Harris, C. C.,** Effects of serum, transforming growth factor type β, or 12-*O*-tetradecanoylphorbol-13-acetate on ionized cytosolic calcium concentration in normal and transformed human bronchial epithelial cells, *Cancer Res.,* 49, 63, 1989.

160. **Lewis, M. G., Chang, J. Y., Olsen, R. G., and Fertel, R. H.,** Identification of calmodulin activity in purified retroviruses, *Biochem. Biophys. Res. Commun.,* 141, 1077, 1986.

161. **Takuwa, N., Iwamoto, A., Kumada, M., Yamashita, K., and Takuwa, Y.,** Role of Ca^{2+} influx in bombesin-induced mitogenesis in Swiss 3T3 fibroblasts, *J. Biol. Chem.,* 266, 1403, 1991.

162. **Kuo, J. F., Schatzman, R. C., Turner, R. S., and Mazzei, G. J.,** Phospholipid-sensitive Ca^{2+}-dependent protein kinase: a major protein phosphorylation system, *Mol. Cell. Endocrinol.,* 35, 65, 1984.

163. **Connelly, P. A., Sisk, R. B., Schulman, H., and Garrison, J. C.,** Evidence for the activation of the multifunctional Ca^{2+}/calmodulin-dependent protein kinase in response to hormones that increase intracellular Ca^{2+}, *J. Biol. Chem.,* 262, 10154, 1987.

164. **Lukas, T. J., Haiech, J., Lau, W., Craig, T. A., Zimmer, W. E., Shattuck, R. L., Shoemaker, M. O., and Watterson, D. M.,** Calmodulin and calmodulin-regulated protein kinases as transducers of intracellular calcium signals, *Cold Spring Harbor Symp. Quant. Biol.,* 53, 185, 1988.

165. **Meyer, T., Hanson, P. I., Stryer, L., and Schulman, H.,** Calmodulin trapping by calcium-calmodulin-dependent protein kinase, *Science,* 256, 1199, 1992.

166. **Sheng, M., Thompson, M. A., and Greenberg, M. E.,** CREB: a Ca^{2+}-regulated transcription factor phosphorylated by calmodulin-dependent kinases, *Science,* 252, 1427, 1991.

167. **Shenolikar, S., Lickteig, R., Hardie, D. G., Soderling, T. R., Hanley, R. M., and Kelly, P. T.,** Calmodulin-dependent multifunctional protein kinase: evidence for isoenzyme forms in mammalian tissues, *Eur. J. Biochem.,* 161, 739, 1986.

168. **Hashimoto, Y. and Soderling, T. R.,** Calcium-calmodulin-dependent protein kinase II and calcium-phospholipid-dependent protein kinase activities in rat tissues assayed with a synthetic peptide, *Arch. Biochem. Biophys.,* 252, 418, 1987.

169. **Nairn, A. C. and Greengard, P.,** Purification and characterization of Ca^{2+}/calmodulin-dependent protein kinase I from bovine brain, *J. Biol. Chem.,* 262, 7273, 1987.

170. **Nairn, A. C. and Palfrey, H. C.,** Identification of the major M_r 100,000 substrate for calmodulin-dependent protein kinase III in mammalian cells as elongation factor-2, *J. Biol. Chem.,* 262, 17299, 1987.

171. **Kameshita, I. and Fujisawa, H.,** Phosphorylation and functional modification of calmodulin-dependent protein kinase-IV by cAMP-dependent protein kinase, *Biochem. Biophys. Res. Commun.,* 180, 191, 1991.

172. **Schulman, H. and Lou, L. L.,** Multifunctional Ca^{2+}/calmodulin-dependent protein kinase: domain structure and regulation, *Trends Biochem. Sci.,* 14, 62, 1989.

173. **Colbran, R. J., Schworer, C. M., Hashimoto, Y., Fong, Y.-L., Rich, D. P., Smith, M. K., and Soderling, T. R.,** Calcium/calmodulin-dependent protein kinase II, *Biochem. J.,* 258, 313, 1989.

174. **Kuret, J. and Schulman, H.,** Mechanism of autophosphorylation of the multifunctional Ca^{2+}/calmodulin-dependent protein kinase, *J. Biol. Chem.,* 260, 6427, 1985.

175. **Lai, Y., Nairn, A. C., and Greengard, P.,** Autophosphorylation reversibly regulates the Ca^{2+}/calmodulin-dependence of Ca^{2+}/calmodulin-dependent protein kinase II, *Proc. Natl. Acad. Sci. U.S.A.,* 83, 4253, 1986.

176. **Lou, L. L., Lloyd, S. J., and Schulman, H.,** Activation of the multifunctional Ca^{2+}/calmodulin-dependent protein kinase by autophosphorylation: ATP modulates production of an autonomous enzyme, *Proc. Natl. Acad. Sci. U.S.A.,* 83, 9497, 1986.

177. **Lickteig, R., Shenolikar, S., Denner, L., and Kelly, P. T.,** Regulation of Ca^{2+}/calmodulin-dependent protein kinase II by Ca^{2+}/calmodulin-independent autophosphorylation, *J. Biol. Chem.,* 263, 19232, 1988.

178. **Colbran, R. J., Smith, M. K., Schworer, C. M., Fong, Y.-L., and Soderling, T. R.,** Regulatory domain of calcium/calmodulin-dependent protein kinase II. Mechanism of inhibition and regulation by phosphorylation, *J. Biol. Chem.,* 264, 4800, 1989.

179. **Waxham, M. N., Aronowski, J., Westgate, S. A., and Kelly, P. T.,** Mutagenesis of Thr-286 in monomeric Ca^{2+}/calmodulin-dependent protein kinase II eliminates Ca^{2+}/calmodulin-independent activity, *Proc. Natl. Acad. Sci. U.S.A.,* 87, 1273, 1990.

180. **Hanley, R. M., Means, A. R., Ono, T., Kemp, B. E., Burgin, K. E., Waxham, N., and Kelly, P. T.,** Functional analysis of a complementary DNA for the 50-kilodalton subunit of calmodulin kinase II, *Science,* 237, 293, 1987.

181. **Ohta, Y., Nishida, E., and Sakai, H.,** Type II Ca^{2+}/calmodulin-dependent protein kinase binds to actin filaments in a calmodulin-sensitive manner, *FEBS Lett.,* 208, 423, 1986.

182. **Ohta, Y., Ohba, T., and Miyamoto, E.,** Ca^{2+}/calmodulin-dependent protein kinase II: localization in the interphase nucleus and the mitotic apparatus of mammalian cells, *Proc. Natl. Acad. Sci. U.S.A.,* 87, 5341, 1990.

183. **Nairn, A. C., Nichols, R. A., Brady, M. J., and Palfrey, H. C.,** Nerve growth factor treatment of cAMP elevation reduces Ca^{2+}/calmodulin-dependent protein kinase III activity in PC12 cells, *J. Biol. Chem.,* 262, 14265, 1987.

184. **Mac Neil, S., Lakey, T., and Tomlinson, S.,** Calmodulins regulation of adenylate cyclase activity, *Cell Calcium,* 6, 213, 1985.

185. **Wolff, D. J., Ross, J. M., Thompson, P. N., Brostrom, M. A., and Brostrom, C. O.,** Interaction of calmodulin with histones: alteration of histone dephosphorylation, *J. Biol. Chem.,* 256, 1846, 1981.

186. **Iwasa, Y., Iwasa, T., Higashi, K., Matsui, K., and Miyamoto, E.,** Modulation by phosphorylation of interaction between calmodulin and histones, *FEBS Lett.,* 133, 95, 1981.

187. **Maizels, E. T. and Jungmann, R. A.,** Ca^{2+}-calmodulin-dependent phosphorylation of soluble and nuclear proteins in the rat ovary, *Endocrinology,* 112, 1895, 1983.

188. **Pearson, R. B., Woodgett, J. R., Cohen, P., and Kemp, B. E.,** Substrate specificity for a multifunctional calmodulin-dependent protein kinase, *J. Biol. Chem.,* 260, 14471, 1985.

189. **MacManus, J. P., Whitfield, J. F., Boynton, A. L., Durkin, J. P., and Swierenga, S. H. H.,** Oncomodulin: a widely distributed tumour-specific, calcium-binding protein, *Oncodev. Biol. Med.,* 3, 79, 1982.

190. **MacManus, J. P., Brewer, L. M., and Whitfield, J. F.,** The widely-distributed tumour protein, oncomodulin, is a normal constituent of human and rodent placentas, *Cancer Lett.,* 27, 145, 1985.

191. **Brewer, L. M. and MacManus, J. P.,** Localization and synthesis of the tumor protein oncomodulin in extraembryonic tissues of the fetal rat, *Dev. Biol.,* 112, 49, 1985.

192. **Brewer, L. M. and MacManus, J. P.,** Detection of oncomodulin, an oncodevelopmental protein in human placenta and choriocarcinoma cell lines, *Placenta,* 8, 351, 1987.

193. **MacManus, J. P., Hutnik, C. M. L., Sykes, B. D., Szabo, A. G., Williams, T. C., and Banville, D.,** Characterization and site-specific mutagenesis of the calcium-binding protein oncomodulin produced by recombinant bacteria, *J. Biol. Chem.,* 264, 3470, 1989.

194. **Gillen, M. F., Banville, D., Rutledge, R. G., Narang, S., Seligy, V. L., Whitfield, J. F., and MacManus, J. P.,** A complementary DNA for the oncodevelopmental calcium-binding protein, oncomodulin, *J. Biol. Chem.,* 262, 5308, 1987.

195. **Sommer, E. W. and Heizmann, C. W.,** Expression of the tumor-specific and calcium-binding protein oncomodulin during chemical transformation of rat fibroblasts, *Cancer Res.,* 49, 899, 1989.

196. **Olwin, B. B. and Storm, D. R.,** Calcium binding to complexes of calmodulin and calmodulin binding proteins, *Biochemistry,* 24, 8081, 1985.

197. **Pritchard, K. and Moody, C. J.,** Caldesmon: a calmodulin-binding actin-regulatory protein, *Cell Calcium,* 7, 309, 1986.

198. **Sobue, K., Kanda, K., Tanaka, T., and Ueki, N.,** Caldesmon: a common actin-linked regulatory protein in the smooth muscle and nonmuscle contractile system, *J. Cell. Biochem.,* 37, 317, 1988.

199. **Kakiuchi, R., Inui, M., Morimoto, K., Kanda, K., Sobue, K., and Kakiuchi, S.,** Caldesmon, a calmodulin-binding F actin-interacting protein, is present in aorta, uterus and platelets, *FEBS Lett.,* 154, 351, 1983.

200. **Owada, M. K., Hakura, A., Iida, K., Yahara, I., Sobue, K., and Kakiuchi, S.,** Occurrence of caldesmon (a calmodulin-binding protein) in cultured cells: comparison of normal and transformed cells, *Proc. Natl. Acad. Sci. U.S.A.,* 81, 3133, 1984.

201. **Graceffa, P., Wang, C.-L. A., and Stafford, W. F.,** Caldesmon: molecular weight and subunit composition by analytical ultracentrifugation, *J. Biol. Chem.,* 263, 14196, 1988.

202. **Graceffa, P.,** Evidence for interaction between smooth muscle tropomyosin and caldesmon, *FEBS Lett.,* 218, 139, 1987.

203. **Ueki, N., Sobue, K., Kanda, K., and Hada, T., and Hagashino, K.,** Expression of high and low molecular weight caldesmons during phenotypic modulation of smooth muscle cells, *Proc. Natl. Acad. Sci. U.S.A.,* 84, 9049, 1987.

204. **Umekawa, H. and Hidaka, H.,** Phosphorylation of caldesmon by protein kinase C, *Biochem. Biophys. Res. Commun.,* 132, 56, 1985.

205. **Lichtfield, D. W. and Ball, E. H.,** Phosphorylation of caldesmon[77] by protein kinase C *in vitro* and in intact human platelets, *J. Biol. Chem.,* 262, 8056, 1987.

206. **Yamashiro, S., Yamakita, Y., Hosoya, H., and Matsumura, F.,** Phosphorylation of non-muscle caldesmon by p34[cdc2] kinase during mitosis, *Nature,* 349, 169, 1991.

207. **Mak, A. S., Watson, M. H., Litwin, C. M. E., and Wang, J. H.,** Phosphorylation of caldesmon by cdc2 kinase, *J. Biol. Chem.,* 266, 6678, 1991.

208. **Pallen, C. J., Valentine, K. A., Wang, J. H., and Hollenberg, M. D.,** Calcineurin-mediated dephosphorylation of the human placental membrane receptor for epidermal growth factor urogastrone, *Biochemistry,* 24, 4727, 1985.

209. **Chan, C. P., Gallis, B., Blumenthal, D. K., Pallen, C. J., Wang, J. H., and Krebs, E. G.,** Characterization of the phosphotyrosyl protein phosphatase activity of calmodulin-dependent protein phosphatase, *J. Biol. Chem.,* 261, 9890, 1986.

210. **Tokuda, M., Khanna, N. C., and Waisman, D. M.,** Identification of bovine brain calcium binding protein, *Cell Calcium,* 8, 229, 1987.

211. **Clipstone, N. A. and Crabtree, G. R.,** Identification of calcineurin as a key signalling enzyme in lymphocyte activation, *Nature,* 357, 695, 1992.

212. **Kincaid, R. L., Martensen, T. M., and Vaughan, M.,** Modulation of calcineurin phosphotyrosyl protein phosphatase activity by calmodulin and protease treatment, *Biochem. Biophys. Res. Commun.,* 140, 320, 1986.

213. **Hashimoto, Y. and Soderling, T. R.,** Regulation of calcineurin by phosphorylation. Identification of the regulatory site phosphorylated by Ca^{2+}/calmodulin-dependent protein kinase II and protein kinase C, *J. Biol. Chem.,* 264, 16524, 1989.

214. **Martensen, T. M., Martin, B. M., and Kincaid, R. L.,** Identification of the site on calcineurin phosphorylated by Ca^{2+}/CaM-dependent kinase II: modification of the CaM-binding domain, *Biochemistry,* 28, 9243, 1989.

215. **Moore, P. B.,** 67 kDa calcimedin, a new Ca^{2+}-binding protein, *Biochem. J.,* 238, 49, 1986.

216. **Glenney, J.,** Two related but distinct forms of the M_r 36,000 tyrosine kinase substrate (calpactin) that interact with phospholipid and actin in a Ca^{2+}-dependent manner, *Proc. Natl. Acad. Sci. U.S.A.,* 83, 4258, 1986.

217. **Berruti, G.,** Calpactin-like proteins in human spermatozoa, *Exp. Cell Res.,* 179, 374, 1988.

218. **Glenney, J. R., Jr., Tack, B., and Powell, M. A.,** Calpactins: two distinct Ca^{++}-regulated phospholipid- and actin-binding proteins isolated from lung and placenta, *J. Cell Sci.,* 104, 503, 1987.

219. **Isacke, C. M., Lindberg, R. A., and Hunter, T.,** Synthesis of p36 and p35 is increased when U-937 cells differentiate in culture but expression is not inducible by glucocorticoids, *Mol. Cell. Biol.,* 9, 232, 1989.

220. **Brambilla, R., Zippel, R., Sturani, E., Morello, L., Peres, A., and Alberghina, L.,** Characterization of the tyrosine phosphorylation of calpactin-I (annexin-II) induced by platelet-derived growth factor, *Biochem. J.,* 278, 447, 1991.

221. **Blay, J., Valentine-Braun, K. A., Northup, J. K., and Hollenberg, M. D.,** Epidermal-growth-factor-stimulated phosphorylation of calpactin II in membrane vesicles shed from cultured A-431 cells, *Biochem. J.,* 259, 577, 1989.

222. **Barnes, J. A., Michiel, D., and Hollenberg, M. D.,** Simultaneous phosphorylation of three human calpactins by kinase C, *Biochem. Cell Biol.,* 69, 163, 1991.

223. **Baudier, J., Mochly-Rosen, D., Newton, A., Lee, S.-H., Koshland, D. E., Jr., and Cole, R. D.,** Comparison of S100b protein with calmodulin: interactions with melittin and microtubule-associated tau proteins and inhibition of phosphorylation of tau proteins by protein kinase C, *Biochemistry,* 26, 2886, 1987.

224. **Weber, K. and Johnsson, N.,** Repeating sequence homologies in the p36 target protein of retroviral protein kinases and lipocortin, the p37 inhibitor of phospholipase A_2, *FEBS Lett.,* 203, 95, 1986.

225. **Sobue, K., Okabe, T., Kadowaki, K., Itoh, K., Tanaka, T., and Fujio, Y.,** Cytosynalin: a M_r 35,00 cytoskeleton-interactin and calmodulin-binding protein, *Proc. Natl. Acad. Sci. U.S.A.,* 84, 1916, 1987.

226. **Wang, K. K. W., Villalobo, A., and Roufogalis, B. D.,** Calmodulin-binding proteins as calpain substrates, *Biochem. J.,* 262, 693, 1989.

227. **Hirai, S., Kawasaki, H., Yaniv, M., and Suzuki, K.,** Degradation of transcription factors, c-Jun and c-Fos, by calpain, *FEBS Lett.,* 287, 57, 1991.

Chapter 7

Phosphoinositide Metabolism

I. INTRODUCTION

Phospholipids are essential components of the cell membranes. They were usually considered as relatively inert substances, but in the past few years it has been recognized that inositol phospholipids (phosphoinositides) have an essential role in the transductional mechanisms related to the action of hormones, growth factors, and other chemical messengers involved in signal transduction across the cell membrane.[1-10] The transductional mechanism represented by phosphoinositide metabolism may interact with mechanisms involving tyrosine phosphorylation of cellular proteins.[11] It may also interact with signaling pathways involving the generation of cAMP.[12] The general biological importance of phosphoinositides is suggested by the fact that they are present in vertebrates such as *Xenopus laevis*,[13] as well as in invertebrates. In the amoebal organism *Dictyostelium discoideum*, two pathways of signal transduction have been found to operate, one involving adenylate cyclase for signal relay, and the other involving inositol phosphates.[14] Phosphoinositides are present even in plants,[15-18] where they are involved in the transduction of signals across membranes.[19] In addition to their role in transduction mechanisms, phosphoinositides are involved in other cellular functions. Many proteins in both prokaryotic and eukaryotic membranes contain a covalently attached lipid and in some of these proteins the lipid moiety consists of glycosylated phosphatidylinositol, which may be responsible for membrane anchoring.[20]

II. PHOSPHATIDYLINOSITOL METABOLISM

Phosphatidylinositol and its phosphorylated derivatives, which include mono- and polyphosphates, represent less than 6 to 8% of the components of the membranes of eukaryotic cells. However, they are important components of cell membranes and are crucially involved in receptor-mediated activation of intracellular signaling mechanisms. Various enzymes are involved in the regulation of phospholipid biosynthesis in eukaryotic cells,[21] as well as in the phosphorylation and hydrolysis of cellular phospholipids.[22] The hydrolytic enzymes involved in such processes include several phospholipases. The phosphorylation of phosphatidylinositol depends on the activity of phosphatidylinositol kinases with different substrate specificities. While the generation of phosphatidylinositol 4-monophosphate depends on the activity of phosphatidylinositol kinase type II, the type I enzyme is involved in the generation of phosphatidylinositol 3-monophosphate.[23] Further phosphorylation of the monophosphate gives origin to phosphatidylinositol 4,5-bisphosphate, a compound that has been found to be involved in the activation of a low affinity form of human DNA polymerase-α, suggesting that it may function as a second messenger during the initiation of mitosis.[24]

A. PHOSPHOLIPASE C

Phospholipase C (phosphoinositidase C) has a key role in the initial hydrolysis of phosphoinositides.[25] The enzyme, purified from sheep seminal vesicles or other sources, is capable of mediating the hydrolysis of phosphodiester bonds of all three polyphosphoinositides (phosphatidylinositol and its mono- and bisphosphate derivatives) *in vitro*.[26] However, it hydrolyzes the mono- and bisphosphate phosphoinositide forms more rapidly than the nonphosphorylated phosphatidylinositol, especially at low Ca^{2+} concentrations.[27] In intact cells, the enzyme is more specifically involved in the hydrolysis of inositol 4,5-bisphosphate and the generation of two second messengers with defined sites of action, inositol 1,4,5-trisphosphate and 1,2-diacylglycerol (Figure 7.1). These changes may lead to the activation of the Ca^{2+}/phospholipid-dependent enzyme, protein kinase C.[28] Phospholipase C may be activated when growth factors or regulatory peptides occupy their receptors on the cell surface. G protein-coupled receptors may mediate their actions through the production of second messengers via activation of particular phospholipase C isoenzymes.[27,29,30]

There are several molecular forms of phospholipase C, and some of them may respond in a specific manner to agonist occupancy of cell surface receptors and may have different substrate specificities along the phosphoinositide metabolic pathway.[31] Two phospholipase C isoenzymes found in human platelets show substrate specificity and are distinct from the nonspecific enzyme that acts on all three types of

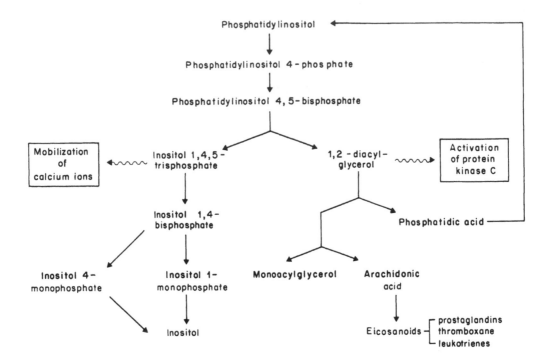

Figure 7.1 Phosphoinositide metabolic pathways.

phosphoinositides.[32] One isoform of phospholipase C is a polyphosphoinositide phosphodiesterase that is involved in the hydrolysis of phosphatidylinositol 4,5-bisphosphate, resulting in the generation of the two second messengers, inositol 1,4,5-trisphosphate and 1,2-diacylglycerol. Nuclear diacylglycerol concentrations may be increased during cellular proliferation *in vivo*.[33] A soluble form of phospholipase C purified from rat liver is specific for both phosphatidylinositol 4-phosphate and phosphatidylinositol 4,5-bisphosphate.[34] A 140-kDa isoform of phospholipase C that acts specifically on polyphosphoinositides but not on phosphatidylinositol was purified from human platelets.[35] This isoform shows maximal activity in the presence of physiological concentrations of Mg^{2+}.

Two forms of phospholipase C with different requirements for calcium were isolated from rat brain and liver,[36] and three forms (PLC-I, PLC-II, and PLC-III) were purified from bovine brain.[37] The molecular weights of these isozymes are different (150,000 for PLC-I, 145,000 for PLC-II, and 85,000 for PLC-III), but they may be similar in their catalytic properties, exhibiting preferential specificity for phosphatidylinositol 4,5-bisphosphate as substrate. The cloning and sequence analysis of phospholipase C isozymes have been reported.[38,39] A cDNA for PLC-II, also termed PLC-γ, was derived from a human endothelial cell library.[40] The enzyme, as deduced from a cDNA clone, is composed of 1289 amino acids and exhibits extensive structural homology to nonreceptor members of the *src* gene family. Two regions of the PLC-II/PLC-γ molecule (residues 555 to 598 and 668 to 705) show significant amino acid similarities to the products of the oncoproteins Yes, Src, Fgr, Abl, Fps, Fes, and Tck. A form of rat phospholipase C, PLC-IV, also has a sequence homologous to the amino-terminal regulatory domains of nonreceptor tyrosine kinases of the src gene family.[41] The *src* homologous domain of PLC-II and PLC-IV is located on a region that is not essential for kinase activity but that may be involved in an interaction with cellular components that modulate kinase function. Phospholipase C shows significant homology to the product of the v-*crk* oncogene contained in the avian sarcoma virus CT10.[42,43] The 47-kDa hybrid oncoprotein Gag-Crk contains 208 amino acids of the viral *gag* sequence fused to 232 amino acids of a proto-oncogene sequence, c-*crk*. The domain common to v-Crk and PLC-II is located in a region that is not essential for tyrosine kinase activity and is likely to be involved in an interaction with cellular components that modulate kinase function.

PLC-I, expressed in the rat uterus, has little similarity with other phospholipase C isozymes, but exhibits a high degree of structural similarity to thioredoxins, which are small proteins that act as cofactors in important oxidation-reduction reactions, including ribonucleotide reduction and photosynthesis.[44] Phospholipase C has oncogenic potential. Introduction of exogenous phospholipase C into quiescent NIH/3T3 mouse fibroblasts induces DNA synthesis and morphologic transformation.[45]

Various mechanisms are involved in the regulation of phospholipase C activity. Ca^{2+} and G proteins participate in this regulation.[46,47] The phospholipase isoform PLC-β, but not PLC-γ or PLC-δ, may be phosphorylated on serine residues by protein kinase C.[48] The PLC-β1 isoform is specifically, but not predominantly, located in the nucleus,[49] and its activity is regulated by G protein-coupled cell surface receptors. In particular, the G_q family of pertussis toxin-insensitive G proteins is involved in the regulation of PLC-β1 activity. Phospholipase C phosphorylation on tyrosine may have an important role in the regulation of phospholipase activity. PLC-γ1 and PLC-γ2 are phosphorylated on tyrosine *in vitro* by members of the Src family of tyrosine kinases, including the Src, Lck, Lyn, Hck, and Fyn kinases.[50] Various growth factors are capable of inducing this phosphorylation through activation of their receptors on the cell surface. Heparin-binding growth factors (HBGFs) such as acidic FGF can induce phosphorylation of the enzyme on tyrosine.[40] Treatment of A431 human epidermoid carcinoma cells with EGF induces an increase in phospholipase C activity, which may be related to tyrosine phosphorylation, as shown by using an antibody to phosphotyrosine.[51] The EGF receptor phosphorylates PLC-II/PLC-γ both *in vitro* and *in vivo*.[52-54] A rapid phosphorylation of PLC-II/PLC-γ on tyrosine is induced by PDGF in BALB/c 3T3 cells.[55] However, the physiologic role of PLC-II/PLC-γ phosphorylation is not clear. Other enzymes may limit PDGF-induced Ca^{2+} signaling and DNA synthesis in NIH/3T3 cells.[56] CSF-1 stimulation of NIH/3T3 cells expressing a human c-*fms*/CSF-1 receptor gene does not result in phosphorylation of PLC-II/PLC-γ, although the PDGF receptor and the CSF-1 receptor express tyrosine kinase activity upon ligand-induced activation.[57] The insulin receptor does not induce phosphorylation of PLC-II/PLC-γ to a significant extent.[58] EGF-stimulated phosphorylation of phospholipase C *in vivo* occurs on both tyrosine and serine residues and is independent of receptor internalization and extracellular Ca^{2+}.[59] EGF and PDGF promote translocation of PLC-II/PLC-γ from the cytosol to the plasma membrane.[60]

A form of PLC-II/PLC-γ, PLC-γ1, is phosphorylated in cells exposed to various external stimuli. NGF rapidly stimulates tyrosine phosphorylation of PLC-γ1 by a kinase activity associated with the product of the c-*trk* proto-oncogene, which functions as an NGF receptor.[61,62] Functional activation of the T-cell antigen receptor induces tyrosine phosphorylation of PLC-γ1.[63,64] Human mammary carcinomas and fibroadenomas have an elevated content of PLC-γ1, and two tyrosine kinases, the EGF receptor and the Erb-B2 protein, are also expressed in these tissues.[65] PLC-γ1 may be a substrate for tyrosine kinases in tumor tissues.

B. PRODUCTS OF PHOSPHATIDYLINOSITOL HYDROLYSIS

Inositol mono- and polyphosphate derivatives are produced during the hydrolysis of phosphatidylinositol. One of these products is 1,4,5-trisphosphate, which is generated through the metabolite intermediates, phosphatidylinositols 4-monophosphate and 4,5-bisphosphate. Inositol 1,4,5-trisphosphate may mobilize calcium ions from intracellular reservoirs and may regulate enzymic reactions or the permeability of the plasma membrane to monovalent cations. Although phosphoinositide turnover and rise in the $[Ca^{2+}]_i$ is partly dependent on extracellular Ca^{2+}, at least in some cellular systems there is an early Ca^{2+}-independent rise in inositol 1,4,5-trisphosphate which precedes the rise in $[Ca^{2+}]_i$.[66]

High-affinity receptors for inositol 1,4,5-trisphosphate have been found in particulate subcellular fractions of the liver, the anterior pituitary gland, and other organs.[67-70] Subtypes of such receptors are expressed in a tissue-specific and developmentally specific manner, suggesting an important role for the distinct subtypes.[71] Hematopoietic growth factors such as IL-3 and GM-CSF/CSF-2 induce in K562 human myeloblast cells the expression of inositol trisphosphate receptors.[72] These receptors can be phosphorylated by protein kinase C, which results in increased potency of inositol 1,4,5-trisphosphate for the release of calcium from nuclei.[73]

The distribution of inositol 1,4,5-trisphosphate receptor mRNA in mouse tissues has been characterized.[74] Cloning and expression of functional inositol 1,4,5-trisphosphate-binding protein, P_{400}, has been reported.[75] The receptor exists in the rat liver in two states displaying different affinities for the ligand.[76] The low affinity state of the receptor is coupled to Ca^{2+} release from the intracellular stores. The ratio of the high to the low affinity state of the receptor is increased by hormones whose effects depend mainly on phosphatidylinositol turnover and this effect is reversed by cAMP-dependent hormones. Inositol 1,4,5-trisphosphate is dephosphorylated in a step-wise manner by the action of two or more enzymes in the liver and other organs and tissues.[77]

An important primary product of phosphodiesteric cleavage of phosphoinositides breakdown is 1,2-diacylglycerol, which functions as a second messenger in the regulation of cellular proliferation. Diacylglycerol is also produced through other pathways, including neosynthesis from a glycolytic intermediate, dihydroxyacetone phosphate and production from phosphatidylcholine turnover, as well as its synthesis from monoacylglycerol.[78] Diacylglycerol is generated from glycerol 3-phosphate through a pathway involving step-wise acylation to lysophosphatidic acid and phosphatidic acid. Diacylglycerol is an endogenous activator of protein kinase C, and its concentration is increased by the action of a wide diversity of hormones and growth factors. Attenuation of diacylglycerol is carried out by phosphorylation by diacylglycerol kinase or degradation by diacylglycerol lipase. Diacylglycerol kinase is a ubiquitous enzyme involved in diacylglycerol remotion by its phosphorylation to phosphatidic acid. Phorbol esters may cause a redistribution of diacylglycerol kinase from the cytosol to the plasma membrane.[79] The phorbol ester-induced translocation of diacylglycerol kinase is relatively small in cells transformed by acute retroviruses carrying the v-*erb*-B or v-*src* oncogenes, which may contribute to altered cellular proliferation.[80]

Choline phospholipids may function as important mediators and modulators of signal transduction.[81] The quarternary amine, choline, is an essential nutrient for humans. Choline is present in cell membranes in the form of ubiquitous phospholipids such as phosphatidylcholine, lysophosphatidylcholine, PAF, choline plasmalogen, and sphingomyelin. Calcium-mobilizing agonists can induce the hydrolysis of phosphatidylcholine by phospholipase C and phospholipase D, thus generating 1,2-diacylglycerol and phosphatidic acid, respectively.[82,83] Phosphatidic acid may function as an intracellular mediator by itself and is able to activate protein kinase C in many different types of cells, including endothelial cells.[84] Phosphatidic acid is degraded by a specific phosphohydrolase to diacylglycerol.

A phosphoinositide pathway characterized in calf brain involves the conversion of inositol 1,4,5-trisphosphate to inositol 1,3,4,5-tetrakisphosphate by a specific kinase and the conversion of this tetrakisphosphate to inositol 1,3,4-trisphosphate by a 5-phosphomonoesterase. Another enzyme, inositol polyphosphate 1-phosphatase, converts inositol 1,3,4-trisphosphate to inositol 3,4-bisphosphate, which in turn is converted to inositol 3-phosphate, inositol 4-phosphate, or inositol 1-phosphate by specific phosphatases.[85,86] The three inositol monophosphates (inositols 1-phosphate, 3-phosphate, and 4-phosphate) are converted to inositol by the enzyme, inositol monophosphate phosphatase. Studies with RBL-2H3 rat basophilic leukemia cells showed that the kinetics of response of different phosphoinositide hydrolytic pathways, including the conversion of inositol 1,4,5-trisphosphate to inositol 1,3,4,5-tetrakisphosphate, may vary according with the type of stimulant used.[87]

Both inositol 1,4,5-trisphosphate and 1,2-diacylglycerol are generated at the membrane level by activation of the enzyme, polyphosphoinositide phosphodiesterase. Physiological levels of Ca^{2+} are required for an optimal activation of this Ca^{2+}-dependent enzyme.[88] The coupling factor linking surface receptors and polyphosphoinositide phosphodiesterase is a G protein.[89] Inositol 1,4,5-trisphosphate and diacylglycerol may function as second messengers to activate signaling pathways that may also be responsible for the release of arachidonic acid and for the activation of guanylyl cyclase. Degradation of 1,2-diacylglycerol by the activity of diacylglycerol lipase results in production of arachidonic acid and monoacylglycerol. Arachidonic acid is a polyunsaturated fatty acid that is oxygenated to produce eicosanoids, a family of compounds that includes prostaglandins, thromboxane, and leukotrienes. Release of arachidonic acid is the principal mechanism for stimulation of eicosanoid biosynthesis, but studies using the tumor cell line Lu-65, derived from human nonsmall cell lung cancer indicate that arachidonate release and stimulation of prostaglandin synthesis can proceed via either calcium/phospholipase A_2-dependent and -independent pathways.[90]

C. PHOSPHOLIPASE D

Hormones and growth factors that do not stimulate inositol phospholipid hydrolysis may be able to increase the cellular content of 1,2-diacylglycerol from another source involving the hydrolysis of phosphatidylcholine.[91] Phospholipase D is involved in the degradation of phosphatidylcholine and the formation of phosphatidic acid and 1,2-diacylglycerol in cells stimulated by a diversity of receptor agonists.[92] Phosphatidic acid and diacylglycerol may function as second messengers in signal transduction, thus providing an alternative pathway for the generation of messengers derived from phospholipids associated with the cell membrane. Phosphatidylcholine degradation stimulated by phorbol esters is mediated principally, if not exclusively, by phospholipase D. The hydrolysis of phosphatidylcholine is a major source of diacylglycerol in stimulated cells. Treatment of cells with serum, vasopressin, or phorbol ester induces degradation of phosphatidylcholine associated with the initial activation of phospholipase D, which results in the release of choline and generation of diacylglycerol.[93] Protein kinase C may be involved in phorbol ester-induced phosphatidylcholine hydrolysis and diacylglycerol generation by the phospholipase D pathway.[94] Formation of diacylglycerol derived from phosphatydylcholine can occur by an indirect pathway dependent on phospholipase D and a direct pathway catalyzed by phospholipase C.[95] The relative importance of phospholipase C and phospholipase D activation in relation to the generation of diacylglycerol depends on the cell type and the acting agonist. In certain cells, diacylglycerol is generated from phosphatidylcholine in the absence of significant phosphatidylinositol turnover.[96] Diacylglycerol provides a backbone for resynthesis of phosphatidylinositol via activity of diacylglycerol kinase, an enzyme that is present in various subcellular fractions from apparently all tissues. Membrane-bound diacylglycerol kinase phosphorylates diacylglycerol and is an arachidonoyl-diacylglycerol kinase that may participate in the formation of arachidonoyl-enriched molecular species of phosphatidylinositol.[97,98] Diacylglycerol phosphorylation by membrane-bound diacylglycerol kinase may result in an increased formation of phosphatidic acid, which can be converted further to phosphatidylinositol. The precise function of diacylglycerol and phosphatidic acid produced from phosphatidylcholine in mitogenic signaling occurring in different types of cells is unknown, but studies with PDGF-stimulated mouse fibroblasts indicate that phosphatidic acid accumulation, rather than diacylglycerol increase, correlates with mitogenesis.[99]

Although the anchoring of proteins to the lipid bilayer of plasma membranes was initially thought to be mediated mainly by hydrophobic amino acid sequences, there is evidence that some cell surface proteins are covalently linked to a phosphatidylinositol-glycan molecule in the lipid bilayer. Phosphatidylinositol-glycan-specific phospholipase D, an enzyme that is abundant in serum and has been purified and characterized, selectively hydrolyzes the inositol phosphate linkage of phosphatidylinositol-glycans, which may result in the secretion of the anchored protein.[100] However, the physiological significance of phosphatidylinositol-glycan-specific phospholipase D has not been determined.

D. PHOSPHOLIPASE A$_2$

Phospholipase A$_2$ can mediate the receptor-induced release of arachidonic acid regulated by a G protein and the subsequent generation of prostaglandins and leukotrienes.[101] The enzyme hydrolyzes arachidonic acid-containing phospholipids, such as phosphatidylcholine, phosphatidylethanolamine, phosphatidylserine, phosphatidic acid, phosphatidylinositides, and plasmalogens to generate free arachidonic acid and lysophospholipids. Arachidonic acid is further metabolized through different pathways: by cyclo-oxygenase to prostaglandins and thromboxanes, by lipoxygenase to leukotrienes and hydroxyeicosatetraenoic acids, and by epoxygenase to epoxides. Arachidonic acid and its metabolites may function not only as second messengers within the cell but, due to their permeability, they can exit through the plasma membrane to the extracellular fluid and can act as first messengers by activating specific receptors on neighboring cells. They may also activate or inhibit adenylyl cyclase, phospholipase C, and ion channels, or may effect other functions on neighboring cells.

Phospholipase A$_2$ purified to homogeneity from the synovial fluid of arthritis patients is identical to that of human placental origin.[102] Two groups of phospholipases A$_2$, I and II, have been described based on their primary structure. Mammalian group I phospholipases A$_2$ are present mainly in the pancreas, whereas the group II enzymes are found in some inflammatory regions.

Phospholipase A$_2$ activation may depend on the interactions of regulatory factors including inositol 1,4,5-trisphosphate and $[Ca^{2+}]_i$. Endogenous generation of diacylglycerol contributes to the modulation of phospholipase A$_2$ activation.[103] Protein kinase C may be involved in the regulation of phospholipase A$_2$ activity and prostaglandin production in some types of cells.[104] The activity of phospholipase A$_2$ is inhibited by the calpactins/lipocortins, from which two forms have been identified, calpactin I or p36 and calpactin II or p35. Calpactin I is a substrate for oncoproteins with tyrosine kinase activity such as the v-Src kinase, as well as for the EGF receptor kinase.[105-107] Differentiation of HL-60 cells induced by various agents is accompanied by an increase in the cellular content of calpactin II.[108] The calpactins/lipocortins are present at high concentrations in many cell types; however, it is not clear whether they function as specific inhibitors of phospholipase A$_2$ or as ubiquitous phospholipid-binding proteins. The inhibitory effects of glucocorticoids on phospholipase A$_2$ activity and prostanoid synthesis are independent of calpactins and may be due to a direct effect of the hormone on phospholipase A$_2$ gene expression.[109] Activation of phospholipase A$_2$ may partly depend on phosphorylation processes. The tyrosine kinase activity of the EGF receptor is necessary for EGF-induced activation of phospholipase A$_2$.[110]

E. PHOSPHATIDYLINOSITOL KINASES

Phosphoinositide kinases catalyze the phosphorylation of the inositol ring of phosphatidylinositol at different positions including the D3 hydroxyl position (phosphatidylinositol 3-kinase), the D4 position (phosphatidylinositol 4-kinase), and the D5 position (phosphatidylinositol 5-kinase).[111]

Phosphoinositides phosphorylated at the D3 position of the inositol ring represent branch points from the conventional polyphosphoinositide pathway.[112] At least two of these compounds, phosphatidylinositol 3,4-bisphosphate and phosphatidylinositol 3,4,5-trisphosphate, may function as intracellular messengers involved in signal transduction. Phosphatidylinositol 3-kinase (type I phosphatidylinositol kinase or PI 3-kinase) phosphorylates the inositol ring of phosphatidylinositol, phosphatidylinositol 4-monophosphate, and phosphatidylinositol 4,5-bisphosphate at the D3 position. PI 3-kinase is found in immune complexes containing polyoma virus middle-T antigen and c-Src protein. The CSF-1 receptor associates with and activates PI 3-kinase in the P388 D1 macrophage cell line and BALB/c fibroblasts made to express the CSF-1 receptor.[113] The IL-7 receptor mediates activation of PI-3 kinase in human B-cell precursors.[114] Activation of the PI 3-kinase occurs in PC12 rat pheochromocytoma cells exposed to EGF, basic FGF, or NGF.[115] The mechanism of PI 3-kinase activation may be associated with phosphorylation of the enzyme. Insulin stimulates the tyrosine phosphorylation of the α-type 85-kDa subunit of PI 3-kinase *in vivo*.[116] However, the physiological significance of the PI 3-kinase remains to be determined, because none of the products of this kinase are in the pathway for generating inositol 1,4,5-trisphosphate. The PI 3-kinase may have a unique role in mitogenic signaling. It is found in direct association with the IL-2 receptor in mitogenically stimulated human and murine lymphocytes.[117] The cloning and expression of a cDNA coding for PI 3-kinase has been reported.[118] The gene of the p85 α, which is the 85-kDa catalytic subunit of PI 3-kinase, has been mapped to human chromosome region 5q12-q13.[119]

Phosphatidylinositol 4-kinase (type II phosphatidylinositol kinase or PI 4-kinase) is the most abundant form of these kinases. It is a membrane-bound enzyme that is activated by divalent cations such as Ca^{2+} and Mn^{2+} and phosphorylates phosphatidylinositol on the 4-hydroxyl position of the inositol ring. The PI 4-kinase is a component of vesicles containing the glucose transporter GLUT-4.[120] Activated tyrosine kinase receptors of growth factors such as EGF may stimulate phosphorylation of phosphatidylinositol kinases on tyrosine residues, suggesting that PI 4-kinase activity may be regulated by tyrosine phosphorylation.[121] However, the possible biological significance of this modification is unknown and there is no clear demonstration that tyrosine phosphorylation of PI 4-kinases is sufficient to induce their activation in intact cells.

F. CYCLIC ESTERS OF INOSITOL PHOSPHATES

Cyclic esters of inositide phosphates may be produced in animal tissues, in addition to noncyclic products, as a result of phosphoinositide degradation by phospholipase C.[122,123] Cyclic inositol trisphosphate and bisphosphate can be generated as products of inositol 1,4,5-trisphosphate metabolism. The possible role of these cyclic phosphate intermediates is unknown, but the cyclic phosphate esters contain a reactive bond that could play some role in phosphoinositide-derived signal transduction.

G. PLATELET-ACTIVATING FACTOR

A very active factor formed in platelets and other types of cells during phospholipid metabolism is a phosphoglyceride called platelet-activating factor (PAF).[124-126] PAF was discovered through its antihypertensive action and was chemically identified as 1-*O*-alkyl-2-acetyl-*sn*-glycero-3-phosphocholine. PAF is released from stimulated basophils, macrophages, polymorphonuclear neutrophils, and platelets and can produce marked cardiovascular effects such as peripheral vasodilatation with hypotension, decrease in cardiac output, increase in cardiovascular permeability, coronary vasoconstriction, and arrhythmias. PAF exerts one of its biological activities (aggregation and secretion in rabbit platelets in bioassay) at a level (1×10^{-10} M) that corresponds to only a few molecules on the cell membrane, which ranks it as one of the most potent naturally produced lipid chemical mediators known to date.

PAF is produced by either a remodeling or a *de novo* biosynthetic pathway.[125] The primary route for the metabolic inactivation of PAF is catalyzed by a cytosolic acetylhydrolase. The activity of this enzyme is regulated by a phosphorylation/dephosphorylation mechanism. PAF is formed in the kidney almost exclusively by the *de novo* biosynthetic pathway. It may contribute to the regulation of blood pressure through its hypotensive activity. In addition, PAF has many important actions, such as those related to the activation of polymorphonuclear leukocytes, monocytes and macrophages, increasing vascular permeability, decreasing cardiac output, stimulation of hepatic glycogenolysis, and stimulation of uterine contraction. It may represent an important signal in reproduction, fetal development, and the initiation of parturition. PAF may contribute to the mediation of many pathological responses, particularly those related to inflammation and allergy.

The biological actions of PAF are initiated by its binding to specific sites on the cell surface. cDNA clones encoding the human PAF receptor protein of 342 amino acids have been isolated and characterized.[127,128]

Binding of PAF to its receptor induces a rapid activation of phospholipase C and increased phosphoinositide turnover with production of inositol 1,4,5-trisphosphate and 1,2-diacylglycerol.[129] The PAF-induced signal transduction process is modulated by a G protein, which couples the PAF receptor to phospholipase C.[130] The stimulatory signals of PAF are blocked in the platelets treated with TPA as well as with agents that increase cellular cAMP concentration. Extracellular Ca^{2+} is required for the action of PAF, which is associated with the phosphorylation of two proteins with approximate molecular weights of 20 and 40 kDa. An early response to the action of PAF (or synthetic derivatives of PAF) on platelets is reflected in the breakdown of phosphatidylcholine, yielding phosphatidic acid and diacylglycerol. In the EBV-transformed human B-cell line SKW6.4, PAF alters membrane phospholipid metabolism, as indicated by the increased incorporation of ^{32}P into phosphatidylcholine, phosphatidylinositol, and phosphatidic acid. In addition, PAF induces elevation in $[Ca^{2+}]_i$, release of arachidonic acid and 5-hydroxyeicosatetraenoic acid, and upregulation of c-*fos* and c-*jun* proto-oncogene expression.[131] In human A431 epidermoid carcinoma cells, PAF induces c-*fos* gene expression through pathways involving activation of tyrosine kinase and protein kinase C.[132] These results suggest that PAF is a potent regulator of cell proliferation and differentiation. In addition to calcium and phospholipid metabolism, the physiologic effects of PAF are associated with protein phosphorylation. Addition of PAF to human neutrophils induces the phosphorylation on tyrosine residues of several proteins with molecular weights ranging from 41 to 116 kDa.[133] The effects of PAF on protein phosphorylation are mediated by PAF receptors and may be due to direct or indirect activation of specific kinases and/or phosphatases. PAF stimulates phosphorylation of the c-Src tyrosine kinase and causes its rapid translocation from cytosol to membranes in platelets.[134]

A Ca^{2+}/phospholipid (PAF)-stimulated protein kinase has been characterized in plants.[135] The plant enzyme is not activated by phorbol esters and is distinct from protein kinase C. It phosphorylates plant microsomal proteins and is involved in the regulation of ATP-dependent H^+ transport. In particular, the PAF-activated protein kinase is implicated in the regulation of two H^+-ATPases located on the plasma membrane of plant cells. The possible role of growth factors in the regulation of PAF activity in animal and plant systems merits further study.

III. PROTEIN KINASE C

Growth factors and other extracellular agents can stimulate in their target cells the production of 1,2-diacylglycerol, a second messenger that functions within the plane of the plasma membrane to induce an increase in the affinity for Ca^{2+} and an activation of the Ca^{2+}/calmodulin-dependent enzyme, protein kinase C.[136-139] The enzyme is involved in the phosphorylation of specific cellular proteins on serine and threonine, and its activity is associated with transmembrane signaling and other important physiological processes, including the regulation of cell proliferation and differentiation.[140-149] Protein kinase C activity is ubiquitous in nature, being present in yeasts and plants.[150,151] The preferred phosphodonor substrate of protein kinase C is ATP.

Protein kinase C has been isolated from several animal sources, including the chick oviduct.[152] The enzyme and its fragments have been quantitated by immunological methods, which allow an estimation of tissue levels, subcellular distribution, and ontogenic changes of the enzyme in brain and heart.[153] The rat brain is one of the richest sources of the 80-kDa native enzyme, but high levels of a 50-kDa species of protein kinase C are contained in HL-60 and K562 leukemic cells. Variation in the levels of protein kinase C is observed in mouse strains, showing different susceptibility to tumor-promoting agents.[154] Studies with a cDNA clone from protein kinase C isolated from a rat brain cDNA library indicated that there are two predominant mRNAs (3.5 and 8 kb), and that both are more prevalent in brain and a neural-derived cultured cell line than in other tissues.[155] Protein kinase C activity shows a considerable variation among different cell lines of human and nonhuman origin.[156]

A. STRUCTURE OF PROTEIN KINASE C AND THE PROTEIN KINASE C GENE

Protein kinase C consists of a single polypeptide chain of 77 to 80 kDa, which can be proteolyzed into two fragments of 32 and 51 kDa.[157] The complete primary structure of the enzyme purified from bovine brain was determined by sequence analysis and recombinant DNA techniques.[158] The enzyme is composed of 672 amino acids arranged in distinct domains and is phosphorylated at specific sites.[159] The regulatory domain occupies the amino-terminal region and the catalytic domain in the carboxy-terminal region. The two domains are joined by a proteolytically sensitive hinge region. Protein kinase C is considered as a cellular receptor for phorbol esters; these compounds and other allosteric cofactors bind to the regulatory domain of protein kinase C. The structural basis for protein kinase C activation by diglycerides and structurally diverse tumor promoters has been discussed.[160] The various tumor promoters and active diglycerides may all be accommodated by the same binding site. Limited proteolysis of protein kinase C generates an active catalytic fragment, indicating that the regulatory domain of the enzyme inhibits its catalytic activity. The amino terminus contains a tandem repeat with a series of six cysteine residues, followed by a calcium-binding domain and a catalytic domain which shares significant homology to other serine/threonine- and tyrosine-specific protein kinases. In particular, the cysteine-rich regions of protein kinase C exhibit structural homologies to the Raf family of proteins, which are the products of the v-*raf/mil* oncogene and the c-*raf/mil* proto-oncogene.[161,162] These homologies suggest the existence of functional and phylogenetical relationships between protein kinase C and the protein products of the c-*raf/mil* proto-oncogene family.

A protein kinase C gene cloned from *Drosophila melanogaster* spans approximately 20 kb and contains at least 14 exons.[163] It encodes a 639-amino acid protein of 75 kDa, which shows extensive homology with bovine, human, and rat protein kinase C as well as with the kinase domains of other protein kinases with specificities for serine/threonine or tyrosine residues. The *Drosophila* protein gene is localized to region 53E of chromosome 2 and transcribes RNAs that are present in the adult insect.

B. PROTEIN KINASE C ISOENZYMES

Protein kinase C activity is present in a family of proteins that comprises several isoforms of the enzyme.[142-149] Analyses of human and bovine cDNA clones indicate the existence of a protein kinase C gene family involved in various cellular signaling pathways.[164-167] No less than eight distinct human protein kinase C isoenzymes, designated using greek letters, have been identified in mammalian tissues. These isoenzymes are serine/threonine kinases of 77 to 83 kDa which may be the product of genes located on different human chromosomes (PKC-α, PKC-β, PKC-γ), but others are generated by alternative RNA splicing mechanisms (the βI and βII isoforms of PKC-β). The precise biological significance of the high number of protein kinase C isoenzymes remains to be elucidated.

Distinct isoforms of protein kinase C have been characterized in mammalian cells.[168] The gene of protein kinase C has been cloned from rat brain tissue.[169,170] Two cDNA clones were isolated, encoding 671- and 673-amino acid polypeptides that differ from each other only in the carboxyl-terminal regions. In another study, three different protein kinase C-related cDNA clones were isolated from a rat brain cDNA library.[171] Three distinct forms of the enzyme were separated from the rat by chromatography.[172] These forms, termed type I (PKC-I), type II (PKC-II), and type III (PKC-III) protein kinase C, have an approximate molecular weight of 82 kDa, are activated by phosphatidylserine, and undergo autophosphorylation in the presence of Ca^{2+}. Four active forms of protein kinase C were isolated from the rat brain with high resolution systems,[173] but additional forms of the enzyme exist in the organ.[174] At least some forms of rat brain protein kinase C isoenzymes are generated by alternative splicing of a single gene.[175] The different rat protein kinase C species have distinct, but partially overlapping, tissue and organ distributions that may reflect important differences in the functions of the respective isoenzymes.[176]

In the rabbit, two forms of protein kinase C (PKC-α and PKC-β) are predominantly present in the brain, while a third form (PKC-γ) is found ubiquitously among various tissues.[177] In addition, protein kinase C-related cDNA clones derived from rabbit brain encode a distantly related member of the protein kinase C family, termed nPKC.[178,179] As the classical form of the enzyme, nPKC behaves as a phorbol ester receptor and possesses a kinase activity that is regulated by phospholipid and diacylglycerol but is independent of Ca^{2+}. The precise physiological role of rabbit nPKC is unknown.

Protein kinase C-related genes coding for three forms of the enzyme (PKC-α, PKC-β, and PKC-γ) have been identified in the bovine and human genome and the corresponding genes were assigned to human chromosomes 17, 16, and 19, respectively. The gene for a new member of the protein kinase C-related family, PKC-L, is expressed in lung, skin, and heart.[180] Alternative RNA splicing of coding exons further increases the structural complexity of the protein kinase family.[181] The physiological roles of different protein kinase C isoenzymes are not known, but each isoenzyme may have distinct tissue distribution and may perform specific roles in signaling. Different protein kinase C isoforms may recognize distinct cellular substrates. The rat embryo fibroblast cell line R6 expresses high levels of the β1 isoform of protein kinase C, and a major substrate of the enzyme in this cell line is a protein of 80 to 87 kDa, called MARCKS.[182] The cellular function of MARCKS is unknown, but MARCKS is cotranslationally myristoylated at its amino terminus, and is phosphorylated and translocated from the membrane to the cytosol upon cell stimulation. MARCKS may be involved in the regulation of calcium/calmodulin-dependent processes and cytoskeletal functions. MARCKs may be a member of a small family of protein kinase C substrates.[183] A MARCKS-related protein, MRP, is encoded by a distinct gene in mouse and human.

Individual cell lines may show different patterns of PKC-α, PKC-β, and PKC-γ mRNA and protein expression. Both normal and *ras* oncogene-transformed NIH/3T3 mouse fibroblasts contain only the α isoenzyme form of protein kinase C.[184] Primary structures may confer type-specific biochemical properties to the different protein kinase C species — for example, in relation to the dependence on Ca^{2+}.[185] Differential coexpression of protein kinase C subspecies by human leukemia-lymphoma cell lines may be related to the state of cell differentiation.[186]

C. SUBCELLULAR LOCALIZATION OF PROTEIN KINASE C

Protein kinase C activity is regulated by its translocation from the cytosol to the plasma membrane and its subsequent activation by 1,2-diacylglycerol. The enzyme is a membrane-associated protein involved in the phosphorylation of serine and threonine residues in specific cellular proteins, including hormone and growth factor receptors as well as the glucose transporter.[187,188] The activity of protein kinase C is subjected to growth state-dependent regulation in normal and transformed cells.[189] Stimulation of sensitive cells with extracellular signaling agents such as hormones and growth factors may result in dramatic changes in the translocation capacity of protein kinase C from the cytosol to the cell membrane.[190] Insertion of protein kinase C into the plasma membrane induces a long-term activation of the enzyme.[191] There is a significant correlation between translocation of protein kinase C activity from the cytoplasm to the plasma membrane and cell proliferation. In confluent cultured mouse fibroblasts, the majority of protein kinase C activity is associated with the cytoplasm and only a minority (5 to 10%) is detected at the level of the membrane.[192] In contrast, most of the enzyme activity in fast growing cells is associated with the membrane.

Various alterations at the level of the cell membrane may result in changes in the association of protein kinase C with the membrane. Binding of hormones, growth factors, and other signaling agents to their receptors on the cell surface may result in rapid translocation or protein kinase C to the cell membrane. The lectin concanavalin A prevents translocation of protein kinase C to the membrane in phorbol ester-treated rat glioma C6 cells.[193] The mechanism by which concanavalin A interferes with the attachment of the enzyme to the membrane is unknown but the treatment blocks phorbol ester-induced desensitization of adenylyl cyclase and sequestration of cell surface β-adrenergic receptors in C6 cells. Bridging of cell-bound IgE antibody molecules on the growth factor (CSF)-dependent cell line PT-18 by multivalent antigen induces an increase in intracellular cAMP concentration, phospholipid methylation, intracellular mobilization and uptake of Ca^{2+}, and formation of 1,2-diacylglycerol.[194] These changes are associated with a rapid increase in protein kinase C activity associated with the cell membrane which precedes the release of chemical mediators such as histamine.

In addition to the plasma membrane, protein kinase C activity has been detected in a variety of intracellular compartments including the nucleus. In rat liver homogenates, maximal activity of protein kinase C was found at two calcium concentrations (1.75 and 3.5 mM), and the activity requiring the higher calcium concentration was associated mainly with the nuclei preparations.[195] The nuclear location of protein kinase C may be related to certain actions of the enzyme, including the phosphorylation of nuclear proteins, which may be relevant for the regulation of transcriptional processes. A rapid and several-hundredfold activation of protein kinase C was observed in isolated rat liver nuclei in response to prolactin, a pituitary hormone thought to exert its effects on macromolecular synthesis at the level of the nucleus and Golgi complex after receptor-mediated endocytosis.[196] Although prolactin is primarily a lactogenic hormone it can exert mitogenic effects on the liver. Another hepatic mitogen, EGF, causes a lesser but significant activation of protein kinase C in the nucleus of liver cells. Prolactin, which exerts comitogenic effects on lymphocytes, stimulates up to 140-fold protein kinase C activity in purified rat splenocyte nuclei, suggesting that this may be a general mechanism for prolactin action. Subcellular structures other than the plasma membrane and the nucleus may possess receptors or binding sites for protein kinase C. Two intracellular receptor proteins for protein kinase C, termed receptors for activated C kinases (RACKs), have been identified in the particular fraction from neonatal rat heart, and an amino acid sequence contained in the RACKs appears to correspond to the binding site for the enzyme.[197] RACKs are associated with a cytoskeletal-rich cell fraction of the heart.

D. REGULATION AND FUNCTION OF PROTEIN KINASE C

Complex endogenous and exogenous factors are involved in the regulation of protein kinase C activity.[198] Phospholipids, in particular phosphatidylserine, are essential cofactors for the enzyme, while *sn*-1,2-diacylglycerol acts as an activator and stimulates translocation of the enzyme to the plasma membrane. Phosphatidylinositol 4,5-bisphosphate may contribute to a positive regulation of protein kinase C activity.[199] Guanine-binding nucleotides may exert regulatory actions on protein kinase C activity.[200] Interaction of protein kinase C with membranes is regulated by Ca^{2+}, ATP, and phorbol esters.[201] Certain Ca^{2+}-binding proteins may be involved in regulating the activity of protein kinase C. According to a model proposed for regulation of the enzyme, the second messengers Ca^{2+} and diacylglycerol may function by causing insertion of protein kinase C into the cell membrane, which would result in constitutive activation of the enzyme until its degradation by protein turnover processes.[202]

Many different exogenous or endogenous agents may be capable of regulating protein kinase C activity in different cell types. Both cytoplasmic and membrane-bound protein kinase C activities are increased in rat thyroid epithelial cells expressing the v-H-*ras* oncogene.[203] Iron transferrin induces a 10-fold increase in the activity of protein kinase C in the human lymphoblastoid cell line CCRF-CEM.[204] This increase is time- and concentration-dependent, is specific for iron transferrin, and may be mediated by increased transcription of the protein kinase C gene and *de novo* synthesis of the enzyme. Endogenous phospholipid cofactors of protein kinase C have not been identified, but there is evidence that the small amounts of polyphosphoinositides that are present in cellular membranes may play a direct role in the activation of the enzyme by serving as phospholipid cofactors.[205] Activation of protein kinase C may occur through the formation of a quarternary complex on the membrane involving the enzyme, Ca^{2+}, phosphatidylserine, and diacylglycerol.[206] Protein kinase C also can be activated independently of phospholipid and Ca^{2+} by *cis* unsaturated fatty acids (oleate and arachidonate).[207] Although the physiological role of this type of activation of protein kinase C is unknown, unsaturated fatty acids are normally present at very small concentrations and are liberated from membrane phospholipids by phospholipase A_2, which may serve as a signal for protein kinase C activation.

E. PROTEIN KINASE C INHIBITORS

Negative regulation of protein kinase C activity may depend on the presence of endogenous inhibitors of the enzyme in specific organs and tissues, but the precise structure and possible physiological role of these inhibitors are little known. The pseudopregnant rat ovary contains an endogenous inhibitor of protein kinase C whose activity predominates over that of the enzyme in the cytosol.[208] The endogenous inhibitor is specific for protein kinase C and does not inactivate the enzyme through proteolysis. A protein from bovine brain is the most potent inhibitor of protein kinase C among this group of proteins.[209,210] The inhibitor, termed PKCI-1, was characterized as a protein of 13,690-Da and 125 amino acid residues with Zn^{2+}-binding properties. The primary structure of PKCI-1 does not exhibit homology with other known proteins. PKCI-1 shows widespread tissue and phylogenetic distribution. It could play an important role in the control of the many physiological processes that depend on protein kinase C activity. The mechanism of action of PKCI-1 is not understood, but it is independent of Ca^{2+} and phospholipids. Endogenous protein kinase C inhibitors have been purified from other tissues in addition to ovary and brain, but the possible role of such inhibitors in the regulation of the enzyme *in vivo* by growth factors, hormones, and other signaling agents is unknown.

A number of exogenous inhibitors of protein kinase C have been described but many of them may be unspecific and their action mechanism is frequently unknown. Staurosporine, an alkaloid purified from *Streptomyces actuosus*, is a potent inhibitor of protein kinase C, but it may also inhibit protein kinase A and tyrosine kinases at similar concentrations. The 7-hydroxy derivative of staurosporine, UCN-10, is a more selective and potent inhibitor of protein kinase C activity. Calphostin C, isolated from the fungus *Cladosporium cladosporioides*, is potent and more selective for protein kinase C due to its interaction with the regulatory domain of the enzyme.[211] The protein K252a, isolated from a microbe (*Nocardipsis* sp.), inhibits protein kinase C activity *in vivo* by competing with ATP.[212] The application of protein kinase C inhibitors may contribute to the elucidation of the precise functions of the enzyme in normal and tumor cells. The discovery that sangivamycin, a purine nucleotide analog with antitumor activity, is an inhibitor of protein kinase C could open new possibilities for the design of antitumor compounds.[213] However, protein kinase C inhibitors may have toxic effects on a number of metabolic processes. The study of mutant cell clones that lack functional protein kinase C molecules may give clues to make clearer the role of the enzyme in physiological processes occurring in different types of cells.[214]

F. PHOSPHORYLATION AND DOWNREGULATION OF PROTEIN KINASE C

Phosphorylation of protein kinase C may contribute to regulate its activity. The enzyme can undergo autophosphorylation at multiple sites,[215] and the resulting autophosphorylated kinase has a lower K_a for Ca^{2+} and a higher affinity for phorbol ester than the nonphosphorylated enzyme, but still requires Ca^{2+} and phospholipid for maximal activity. Compounds that interact with the catalytic site of protein kinase

C, competing with ATP, act as inhibitors of the enzyme in a concentration-dependent manner.[216] Casein kinase I, but not casein kinase II, can phosphorylate protein kinase C in the absence of Ca^{2+} and phospholipids.[217] The possible role of autophosphorylation or *trans*-phosphorylation in the regulation of protein kinase activity in intact cells is not understood.

Phorbol ester binding to its cellular receptor, protein kinase C, is associated with translocation of an inactive cytosolic form of the enzyme to the cell membrane, followed by downregulation (disappearance or diminished binding capacity) of the enzyme.[218] Studies with a constructed mutation at the putative ATP-binding site of protein kinase C, which is associated with abolishment of its kinase activity and insensitivity to downregulation, suggest that intramolecular phosphorylation of the enzyme may be a prerequisite for the proteolytic cleavage associated with its downregulation.[219]

G. PROTEIN KINASE P

A phospholipid-stimulated protein kinase, distinct from protein kinase C, was identified in the murine myeloid leukemia cell line Da-1.[220] This kinase, termed protein kinase P, is activated by phosphatidylglycerol or phosphatidylinositol in the presence of Mn^{2+} or Mg^{2+} and is not influenced by cyclic nucleotides or phosphatidylserine. It is specific for the phosphorylation of serine and threonine residues and is present in murine and human cells. A high level of protein kinase P activity exists in spleen and bone marrow, suggesting that the enzyme may have some role in hematopoiesis. Stimulation of DA-1 cells with IL-3 results in subcellular redistribution of protein kinase P between the cytosol and the particulate fraction.[221] DA-1 cells possess IL-3 receptors and are exceptionally sensitive to IL-3. The possible relationship between protein kinases C and P remains to be ascertained.

H. MUTANT PROTEIN KINASE C

The possible role of qualitatively altered protein kinase C in oncogenic processes is unknown. Four point mutations were detected in cDNA derived from the protein kinase C molecule contained in a murine UV-induced fibrosarcoma cell line which exhibited an unusual subcellular distribution of the enzyme, with most of the enzyme activity associated with the plasma membrane.[222] Three of these mutations were located on the highly conserved regulatory domain of the enzyme, and the other was found in the conserved region of the catalytic domain. Expression of the mutant protein kinase C gene in BALB/c 3T3 cells resulted in the expression of a transformed phenotype and tumorigenicity. Thus, the mutant protein kinase C molecule displays oncoprotein-like activities.

IV. PHOSPHOINOSITIDE METABOLISM AND CELL PROLIFERATION

Phosphoinositide breakdown, generation of 1,2-diacylglycerol, and activation of protein kinase C are important in the control mechanisms of many physiological phenomena, including DNA synthesis, cell proliferation, wound healing, and tissue and organ regeneration.[223] Activation of protein kinase C mediates the action of hormones, growth factors, and exogenous mitogens in many cellular systems.[224] Activation of resting human T lymphocytes requires prolonged stimulation of protein kinase C.[225] Stimulation of serum-deprived mammary tumor cells in culture with insulin, EGF, or thrombin results in changes in the activity of protein kinase C.[190] Human α-thrombin initiates the growth of fibroblasts and other types of cells cultured in serum-free medium; this effect is mediated by activation of protein kinase C.[226] Increased phosphoinositide turnover is crucial for the mitogenic action of PDGF and bombesin in NIH/3T3 cells, as shown by using a monoclonal antibody specific to phosphatidylinositol 4,5-bisphosphate.[227] However, this antibody does not inhibit mitogenesis induced in the same system by FGF, EGF, insulin, or serum. It is thus clear that phosphoinositide breakdown does not represent a universal pathway for mitogenic events elicited by hormones and growth factors. Moreover, receptor-mediated phosphoinositide hydrolysis would not, by itself, serve a major signaling function for driving some quiescent cells into DNA synthesis.[228] G protein-coupled receptors rely on additional signaling pathways for evoking a mitogenic response. Different cellular mechanisms are involved in the activation or modulation of the mitogenic action of hormones and growth factors in relation to phosphoinositide turnover.

A. ACTIVATION OF DNA POLYMERASE

DNA polymerase-α is importantly involved in the replication of DNA in eukaryotic cells. The activity of this enzyme is increased by a number of agents, including treatment with phosphoinositides such as phosphatidylinositol 4-phosphate. This effect is blocked by a monoclonal antibody that does not neutralize polymerase activity, but blocks activation of the enzyme by the phosphoinositide compound.[229] The action of the antibody appears to be directed to a 50-kDa regulatory subunit of DNA polymerase-α rather than to the catalytic subunit. Phosphorylation of the 50-kDa subunit of the enzyme would play a critical role in its binding to DNA prior to DNA replication; this process may depend, at least in part, on some steps of phosphoinositide metabolism, probably including protein kinase C activation.

B. INFLUENCE OF CALCIUM IONS

Calcium ions may influence cell proliferation, at least in part, through changes in the activity of protein kinase C. Extracellular Ca^{2+} deprivation inhibits the growth of mouse BALB/c 3T3 cells, and this effect correlates with a loss of protein kinase C activity from the cell particulate fraction.[230] Addition of calcium to the calcium-deprived cells stimulates DNA synthesis and proliferation of the cells with a parallel increase in the activity of protein kinase C. However, the precise role of calcium in the stimulation of cell proliferation by external signaling agents is not clear. Some mitogens can induce elevation in the $[Ca^{2+}]_i$ levels without requiring the presence of extracellular Ca^{2+}. For example, ligand-induced activation of bombesin receptors present in human small cell lung cancer cells results in an increase in $[Ca^{2+}]_i$ levels, which are not dependent on the presence of extracellular Ca^{2+}.[231] Extracellular Ca^{2+} is not required for the action of angiotensin II, which induces an elevation in the $[Ca^{2+}]_i$ and enhances phosphatidylinositol turnover.[232] However, this independence from the $[Ca^{2+}]_i$ is observed only with high doses of angiotensin II, and responses to low doses of the hormone are dependent on the supply of external Ca^{2+}. Calcium release from intracellular pools may be important for the regulation of protein kinase C. It has been shown that there are at least two distinct intracellular calcium pools, one that is released by inositol 1,4,5-trisphosphate and the other that is released by GTP.[233] These two calcium pools may reside in different intracellular compartments and may have different functions, including their relation to protein kinase C activation. Elevation of the $[Ca^{2+}]_i$ may have a synergistic effect on protein kinase C activity in some cellular systems by acting on the enzyme itself.[234,235] Moreover, activation of protein kinase C may be involved in the regulation of the $[Ca^{2+}]_i$ as well as in the control of cell cycle transitions from G_0 to S and from G_1 to S.[236,237] These mechanisms may lead to induction of DNA synthesis and cell proliferation. Sea urchin eggs can be activated by microinjection of inositol 1,4,5-trisphosphate, which can produce intracellular Ca^{2+} mobilization.[238] Activators of protein kinase C can stimulate meiotic maturation of rat oocytes.[239] In contrast to the uniformity in the tissue concentrations of cAMP-dependent protein kinase and calmodulin, the levels of protein kinase C activity among different organs or the adult rat range over orders of magnitude.[240] The mechanism of inositol 3,4,5-trisphosphate-induced Ca^{2+} mobilization is not known but may be mediated through high affinity binding sites specific for the inositol compound whose activation triggers Ca^{2+} release from intracellular stores. Studies with calmodulin antagonists indicate that calmodulin is an integral component of the inositol 3,4,5-trisphosphate receptor-activated Ca^{2+} release channels.[241]

C. CHANGES IN ION FLUXES AND DISTRIBUTION

Ion fluxes and intracellular redistribution may play an important role in mediating the action of mitogenic agents. Many mitogens stimulate rapid increases in the rate of H^+, Na^+, and K^+ fluxes across the plasma membrane and induce alterations in the levels of H^+, Na^+, K^+, and Ca^{2+} within the cell.[242-244] Activation of phosphatidylinositol turnover and protein kinase C by agents such as LPS and phorbol esters may lead, either directly or indirectly, to enhanced rates of monovalent cation fluxes in quiescent cultured cells.[245,246] Protein kinase C increases the activity of the Na^+/H^+ antiport system, which in turn promotes Na^+ influx, enhances the pH_i, and stimulates Na^+/K^+ pump activity. In turn, protein kinase C may inhibit the activity of phospholipase C, thus reducing the rate of phosphoinositide metabolism turnover.

Activation of protein kinase C may not be a sufficient signal to activate the Na^+/H^+ exchange in some types of cells, which would require some cooperative effects of Ca^{2+} elevation.[246] Activation of Ca^{2+} influx and Na^+/H^+ exchange by EGF in A431 human cells is independent of phosphatidylinositol turnover and is inhibited by phorbol ester and synthetic diacylglycerol.[247] Similarly, activation of Na^+/H^+ exchange, increased entry of Ca^{2+} into the cell, phosphorylation of S6 ribosomal protein, expression of the c-*myc* proto-oncogene, and reinitiation of DNA synthesis can occur in Chinese hamster lung fibroblasts stimulated by the addition of FGF into the culture medium independently of any increase in phosphoinositide breakdown and activation of protein kinase C.[248] Further studies are required to make more clear the precise relationships between phosphatidylinositol turnover, changes in the fluxes and subcellular distribution of monovalent and divalent cations, and initiation of DNA synthesis.

D. RELEASE AND ACTIVATION OF ORNITHINE DECARBOXYLASE

Ornithine decarboxylase (ODC) is the first and rate-limiting enzyme involved in the biosynthetic pathway of polyamines; these compounds are required for cellular proliferation as well as for many growth-related cellular functions. ODC is covalently linked to the membrane by inositol, and release and activation of ODC may involve phosphoinositide breakdown following mitogenic signals such as the interaction of some ligands with their receptors.[249] However, the possible role of ODC activation in the mitogenic effects of hormones and growth factors capable of inducing alterations in phosphoinositide metabolism is unknown.

E. CELLULAR SUBSTRATES FOR PROTEIN KINASE C

Protein kinase C has an important role in the transduction of transmembrane signals involved in the regulation of many cellular functions. Proteins located at different sites within the cell may be substrates for protein kinase C or may be phosphorylated by protein kinase C-activated enzymes. A number of membrane receptors for hormones and growth factors, including the EGF receptor, are substrates for protein kinase C activity. The ubiquitous and amiloride-sensitive Na^+/H^+ exchanger, a plasma membrane phosphoglycoprotein that is involved in the regulation of the pH_i, is rapidly activated by protein kinase C-dependent pathways. Phosphorylation of certain proteins located at the level of the cell surface by protein kinase C may be required for the G_1-S transition of the cell cycle, which is a Ca^{2+}-dependent process.[250] The catalytic subunit of adenylyl cyclase is phosphorylated by protein kinase C, which provides a mechanism for the interaction of the adenylyl cyclase and phosphatidylinositol transmembrane signaling systems.[251]

Several cytoplasmic proteins are phosphorylated by protein kinase C. One of these substrates is the ribosomal protein S6,[252] whose function is related in some manner to the control of cellular DNA synthesis. The cytoplasmic protein p36, which is a substrate for tyrosine kinases, is also a substrate for phosphorylation by protein kinase C and other kinases with specificity for serine and threonine residues *in vitro* and *in vivo*.[253-255] A protein of 87 kDa, widely distributed in different tissues, is also a major specific substrate for protein kinase C.[256] The 80- to 87-kDa MARCKS protein, which is translocated from the membrane to the cytosol upon cell stimulation and may be associated with the cytoskeleton, is a substrate of protein kinase C.[182,183] However, the precise role of cytoplasmic protein kinase C substrates is unknown.

Some cytoskeletal proteins may be substrates for protein kinase C-induced phosphorylation. One of these substrates is vinculin,[257] a cytoskeletal protein that is phosphorylated in response to Ca^{2+} and phorbol esters in intact cells.[258] The microtubule-associated protein 2 (MAP-2) is a substrate for protein kinase C.[259] Protein kinase C phosphorylates MAP-2 exclusively on serine residues which inhibits its ability to cross-link actin filaments. Other cytoskeletal proteins, including tubulin, are poor substrates for protein kinase C. Neomycin, a drug that has a high affinity for phosphoinositides and interferes *in vivo* with phosphatidylinositol turnover, inhibits the polymerization of actin induced by thrombin or ADP in platelets.[260] Phosphatidylinositol turnover may have a role in the control of microfilament-based cell motility.

Nonhistone nuclear proteins present in the testis are substrates for protein kinase C.[261] This phosphorylation is inhibited by Adriamycin, an inhibitor of spermatogenesis. RNA polymerase II, the key enzyme of mRNA synthesis, is also an important substrate for protein kinase C.[262] Phosphorylation of nuclear proteins may represent a mechanism by which protein kinase C is able to modulate gene expression.

Oncoproteins may be substrates for protein kinase C. The v-H-Ras oncoprotein is phosphorylated by protein kinase C on serine residues, at sites distinct from that of autophosphorylation.[263] The possibility that oncoproteins with tyrosine kinase activity, such as the products of the Src family, may be substrates for protein kinase C-mediated phosphorylation is suggested by the fact that tyrosine phosphorylation of a 42-kDa cytosolic protein, which is a common substrate for hormone-, growth factor-, and mitogen-stimulated phosphorylation, is mediated by protein kinase C.[264] The c-Raf protein kinase is activated by phosphorylation dependent on protein kinase C activity.[265]

F. REGULATION OF RECEPTORS FOR ENDOGENOUS MEDIATORS

Protein kinase C may govern some cellular responses through the regulation of receptor-linked endogenous mediators. An example of this type of effect is the negative regulation of leukotriene B_4 (LTB_4) receptors in human neutrophils by protein kinase C.[266] LTB_4 is a metabolite of arachidonic acid in the lipoxygenase pathway of neutrophils and is a potent chemotactic factor that induces adhesion, aggregation, and secretion in the neutrophils. LTB_4 may play an important role in diseases associated with inflammatory processes.[267] LTB$_4$ interacts with specific receptors on the surface of human neutrophils, and one of the earliest events after this interaction is an elevation of the [Ca^{2+}]$_i$, which may be a critical signal in eliciting neutrophil responses. The rise of [Ca^{2+}]$_i$ in LTB$_4$-stimulated neutrophils is associated with Ca^{2+} mobilization from intracellular stores, which results from the formation of the second messenger inositol 1,4,5-trisphosphate as a consequence of the interaction of LTB$_4$ with its cell surface receptor. Activators of protein kinase C, including phorbol ester, inhibit human neutrophil binding of LTB$_4$ by reducing the number of high-affinity receptors available for LTB$_4$.

G. REGULATION OF CELL PROLIFERATION AND DIFFERENTIATION

Activation of phosphoinositide metabolism may play an important role in the control of cellular growth under the influence of hormones, growth factors, and mitogens. It has been proposed that phosphoinositide breakdown, intracellular mobilization of Ca^{2+}, activation of protein kinase C, and phosphorylation of distinct cell surface proteins are required for the initiation of the G$_1$-S transition of the cell cycle and subsequent cell proliferation.[250] However, several lines of evidence indicate that activation of this system does not represent a universal pathway leading to the stimulation of cell growth. Blocking of phosphoinositide turnover may not prevent specific mitogens to act through receptor tyrosine kinases to trigger DNA synthesis and cell division.[268] The mechanism of action of potent mitogens such as FGF in different types of cells may be independent of phosphoinositide turnover and [Ca^{2+}]$_i$.[248,269] Studies on the neuronal cell line NG115-401L-C3, which expresses an endogenous bradykinin receptor and was transfected with the *mas*/angiotensin receptor gene, indicate that neither mitogenic (angiotensin, IGF-I, TGF-β) nor nonmitogenic (bradykinin, NGF, IL-1) receptor activation may correlate with changes in the levels of phosphatidylinositol trisphosphate or with the activity of PI 3-kinase.[270] Phosphatidylinositol turnover may not play a universal role in the postreceptor mechanisms of growth factor action.

The role of Ca^{2+} fluxes and intracellular Ca^{2+} distribution in the processes leading to cellular proliferation through activation of phosphoinositide metabolism is not clear. Addition of thrombin to hamster fibroblasts cultured in serum-free medium stimulates phosphoinositide turnover, but inositol 1,4,5-trisphosphate-induced intracellular mobilization of Ca^{2+} may not be required for thrombin-induced mitogenesis because neomycin blocks inositol 1,4,5-trisphosphate release without inhibiting the initiation of DNA synthesis.[271] Addition of thrombin or ADP to platelets may lead to increased [Ca^{2+}]$_i$ independently of phosphatidylinositol 4,5-bisphosphate degradation.[272] In a PHA-stimulated T-cell leukemia line, this degradation does not depend on Ca^{2+} mobilization.[273] A correlation between PDGF-stimulated inositol trisphosphate generation, an increase in [Ca^{2+}]$_i$, and mitogenesis were not observed in NIH/3T3 cells expressing a transfected vector containing phospholipase C-II/γ.[56] PDGF-induced calcium signaling and DNA synthesis may use biochemical pathways other than phosphoinositide metabolism for transduction.

Phosphoinositide metabolism may be involved in the processes of cellular differentiation occurring in various systems. Increased expression of the inositol 1,4,5-trisphosphate receptor occurs in HL-60 human myeloid leukemia cells induced to differentiation with retinoic acid or DMSO.[274] It has been suggested that modulation of phosphatidylinositol turnover may be even more important than that of tyrosine kinase activity for the differentiation of myeloid leukemia cells, including HL-60 cells.[275] A sustained activation of protein kinase C may be essential to the differentiation of HL-60 to macrophages.[276] Growth and differentiation of murine epidermal cells may be influenced by changes in phosphoinositide turnover.[277] During the proliferative phase of mouse keratinocytes in culture, the cells rapidly incorporate free inositol and synthesize inositol phospholipids. Hydrolytic release of inositol 1,4,5-trisphosphate from phosphoinositides may represent a signal before the onset of differentiation. These changes may be induced by variation in the extracellular Ca^{2+} concentrations. An increase of Ca^{2+} may favor the cessation of keratinocyte proliferation and the induction of terminal differentiation.

H. REGULATION OF PHOSPHOINOSITIDE TURNOVER BY GROWTH FACTORS

A number of hormones, growth factors, and other extracellular signaling agents can stimulate phosphatidylinositol breakdown with the consequent generation of 1,2-diacylglycerol, intracellular mobilization of Ca^{2+}, and protein kinase C activation. Receptor-coupled G proteins activate phospholipase C and favor protein kinase C activation through phosphatidylinositol 4,5-bisphosphate degradation. Growth factors such as IGF-I, EGF, PDGF, FGF, and NGF, whose receptors possess tyrosine kinase activity, transduce their mitogenic signals through pathways that may be either dependent or independent of protein kinase C.[278] A single growth factor may act through multiple signaling pathways to induce cell proliferation, and the action through the pathway involving phosphoinositide turnover and protein kinase C activation does not exclude other mitogenic mechanisms of action for the same factor. However, some growth factors may depend mainly on the activation of protein kinase C for their mitogenic signaling mechanism. The receptors for several growth factors are substrates for protein kinase C, including the receptors for insulin, IGF-I, EGF, and PDGF. Changes in phosphoinositide turnover may play an important role in regulating the functional activity of these receptors by altering their state of phosphorylation through changes in protein kinase C activity.[279] Such changes may precede, follow, or amplify changes in the adenylyl cyclase and calcium-calmodulin systems.

Phosphatidylcholine biosynthesis from phosphatidic acid in Swiss mouse 3T3 fibroblasts is increased in varying degrees by several types of mitogens, including bovine serum, TPA, EGF, FGF, IGF-I, IGF-II, and vasopressin.[280] There may be a synergy between different growth factors in the stimulation of phosphatidylcholine biosynthesis, and the stimulation is associated with an increase in choline kinase activity.[281] Serum, PDGF, and phorbol esters stimulate the hydrolysis of phosphatidylcholine, which results in the generation of 1,2-diacylglycerol and phosphocholine.[282] The production of 1,2-diacylglycerol from phosphatidylcholine may provide an additional mechanism for the regulation of protein kinase C activity.

It is clear that activation of phosphoinositide metabolism does not represent a universal pathway for the cellular actions of growth factors. There is a lack of correlation between EGF-, insulin-, and serum-induced mitogenesis and stimulation of phosphoinositide degradation in BALB/c 3T3 fibroblasts.[283] EGF, alone or in combination with insulin, is capable of stimulating G_0-arrested Chinese hamster lung fibroblasts to undergo DNA replication and to divide without activating phosphatidylinositol turnover.[284] Neither insulin nor EGF stimulates the synthesis of phosphatidylinositol, phosphatidylinositol-4-phosphate, or phosphatidylinositol-4,5-bisphosphate in rat liver plasma membranes or intact hepatocytes, suggesting that the insulin- and EGF-stimulated receptor kinases do not act on phosphoinositides in the liver.[285] Insulin and EGF do not stimulate the translocation of protein kinase C from the soluble to the particulate fraction of the cells.[286] The tyrosine kinase and phosphatidylinositol kinase activity which copurify with the EGF receptor reside on different molecules and the EGF receptor is devoid of the latter type of activity.[287] The role of phosphatidylinositol turnover in relation to the mitogenic effects of several growth factors remains a subject of controversy.

V. PHOSPHOINOSITIDE METABOLISM AND NEOPLASIA

The metabolism of phosphoinositides may have an important role in oncogenesis. It has been suggested that bypassing or subversion of the receptor-operated phospholipid breakdown/protein kinase C signaling mechanism may be the basis of the freeing of cellular proliferation from external controls that characterizes neoplastic transformation.[224] Enhanced phosphatidylinositol kinase activity is associated with early stages of liver carcinogenesis.[288] Activation of the phosphorylation of phosphoinositides, with increased phosphatidylinositol 4,5-bisphosphate pool, precedes the growth of DMBA-induced rat mammary tumors stimulated by prolactin.[289] Transformation by oncogenes such as v-*fms* and v-*fes*, which code for tyrosine kinases, may be associated with enhanced production of 1,2-diacylglycerol and inositol 1,4,5-trisphosphate second messengers.[290]

Protein kinase C may have a role in the complex mechanisms leading to transformation. Phosphorylation of specific cellular substrates by protein kinase C may be relevant during the preneoplastic progression and neoplastic conversion of mouse epidermal cells.[291] Constitutive expression of protein kinase C in NIH/3T3 fibroblasts transfected with plasmids containing rat brain protein kinase C-I cDNA controlled by strong viral promoter/enhancer elements leads to altered growth and tumorigenicity upon inoculation of the transfected cells into nude mice.[292] Inhibition of the activity of protein kinase C by 1-(5-isoquinolinylsulfonyl)-2-methylpiperazine (H7) induces morphological and functional differentiation of Neuro-2a cells.[293] This effect is prevented or reversed by the specific activator of protein kinase C, PMA, and is not produced by an inhibitor that acts more selectively on cAMP- and cGMP-dependent protein kinases. Modulation of protein kinase C activity may also have a role in the mechanisms of transformation induced by chronic retroviruses such as M-MuLV.[294]

The levels of protein kinase C activity, as well as those of other kinases, are frequently altered in human tumor cells. However, there is no general rule for these alterations, and while in some tumors the levels are high, in other tumors they may be normal or low, compared to the respective normal tissue. Immunological quantitation of protein kinase C in various human mammary tumor cell lines showed that estrogen receptor-negative cell lines express higher levels of the enzyme than their estrogen receptor-positive counterparts.[295] Elevated levels of protein kinase C activity have been found in the majority of human breast tumor biopsies, as compared to normal breast tissue specimens from the same patients.[296]

In contrast, the activities of both protein kinase C and Ca^{2+}-dependent protein kinase were found to be reduced in primary human colon carcinomas when compared to the normal adjacent colon mucosa.[297] No difference in the levels of protein kinase C activity was found in esophageal cancers in comparison to the adjacent normal mucosa.[298] It is not known whether altered levels of protein kinase C activity in human tumors reflect an increased turnover of the enzyme or changes in its synthesis at the transcriptional or translational level.

Homozygous deletion of the first exon of the protein kinase C gene was detected in 1 of 19 human melanoma cell lines.[299] Normal human melanocytes, unlike malignant melanoma cells, require the presence of phorbol ester for growth in culture and do not express protein kinase C and protein kinase A-specific subunits.[300] The expression of protein kinase A subunits in human melanomas does not appear to be a requisite for the sustained growth of these cells in culture.

Alterations in phosphoinositide metabolism occurring in some tumor cells may be associated with distinct morphological and functional changes, including invading and metastasizing potential. A strong correlation has been found between the levels of protein kinase C activity and the hematogenous metastasizing abilities of mouse B16 melanoma sublines.[301] Tumor promoter-induced membrane association of protein kinase C additionally stimulates hematogenous metastasis of B16 cells. The cytoskeleton has a crucial role in neoplastic transformation, and phosphoinositides may be involved in maintaining the normal structure and function of this system. Exposure of elongating or assembled microtubules to certain amounts of phosphatidylinositol may alter their interaction with specific proteins, which may result in polymerization arrest or disassembly of the microtubules.[302] Further studies are required for a better characterization of the complex interrelationships that can occur between phosphoinositide metabolism and the expression of a transformed phenotype.

VI. ONCOPROTEINS AND PHOSPHOINOSITIDE METABOLISM

Alterations of phospholipid metabolism are observed in oncogene-transformed cells, and these alterations may have a role in the mechanisms of action of oncoproteins.[303,304] Transformed cells frequently show increased glucose transport and high rates of glycolysis, and diacylglycerol neosynthesis via glycolytic pathway may be constitutively active in cells transformed by several oncogenes, including *ras*, *src*, *sis*, and *abl*.[305] This alteration may result in a permanent activation and downregulation of protein kinase C, with the subsequent deregulation of cellular protein phosphorylation and altered operation of mitogenic signaling mechanisms.

The v-Src, v-Abl, and v-H-Ras oncoproteins contain tightly bound lipids.[306] Turnover of phosphoinositides is stimulated in RSV-transformed cells and the v-Src oncoprotein phosphorylates glycerol.[307,308] Evidence that phosphoinositides are involved in oncogene-induced transformation is supported by the observation that injection of a monoclonal antibody to phosphatidylinositol 4,5-bisphosphate into cells transformed by the viral oncogenes v-*src*, v-*erb*-B, or v-H-*ras* or by an activated c-K-*ras* gene, but not the v-*myc* oncogene, leads to inhibition of cell proliferation.[309] NIH/3T3 cells transformed by various oncogenes (*ras*, *src*, *sis*, and *abl*) exhibit constitutively activated synthesis of diacylglycerol, which may result in permanent activation of protein kinase C, with alteration in the phosphorylation of cellular proteins and disturbance of endogenous mitogenic signaling pathways.[81] Transformation of rodent fibroblasts by cytoplasmic (*mos*, *raf*) and membrane-associated (*ras*, *src*, *met*, *trk*), but not nuclear (*myc*, *fos*) oncoproteins may result in elevation of the cellular levels of glycerophosphoinositol (GPI).[310] This elevation may be specifically associated with the transformed state of the cells and not merely with their active state of proliferation. The GPI appears to be derived from deacylation of lysophosphatidylinositol as a result of increased phospholipase A$_2$ activity. Some aspects of the complex interactions existing between viral and cellular oncoproteins and phosphoinositide metabolism are discussed next.

A. THE *src* GENE FAMILY

Alterations in phosphoinositide metabolism may occur in cells expressing v-Src and other oncoproteins with tyrosine kinase activity, and these alterations may have a role in the mechanism of v-Src-induced transformation. The v-Src protein may enhance the changes induced by hormones and growth factors in phosphoinositide turnover. Endothelin-dependent phospholipase C activation is markedly amplified following v-Src-induced transformation, with increases in inositol 1,4,5-trisphosphate and [Ca^{2+}]$_i$.[311] The site at which v-Src acts to amplify signal transduction in response to endothelin may lie within the coupling of the endothelin receptor to phospholipase C through one or more G proteins.

Activity of phosphatidylinositol kinase activity has been detected in immunoprecipitates of the v-Src protein.[312] v-Src can stimulate phosphatidylinositol phosphorylation to form mono- and diphosphate derivatives, and diacylglycerol generated by phosphatidylinositol 4,5-bisphosphate breakdown would be phosphorylated to form phosphatidic acid.[313] v-Src would be involved in the generation of 1,2-diacylglycerol, thus contributing to the regulation of protein kinase C activity. Complexes of polyoma virus middle-T antigen and the v-Ros oncoprotein tyrosine kinase would be capable of inducing a direct alteration of phosphoinositide metabolism.[314,315] Phosphatidylinositol kinase activity is associated with the hybrid oncoproteins Gag-Ros and Gag-Fps and Gag-Crk, as well as with the complexes between polyoma middle-T antigen and the normal c-Src and c-Yes proteins.[316] However, it has been clearly demonstrated that phosphatidylinositol, phosphatidylinositol 4-phosphate, and 1,2-diacylglycerol phosphorylations are catalyzed by kinases that are distinct from v-Src and other oncoproteins with tyrosine kinase activity.[317-319] Any changes in phosphoinositide turnover or in the phosphorylation of phospholipid metabolites occurring in v-src-transformed cells must be considered as an indirect consequence of the primary action of the v-Src tyrosine kinase. The v-Ros, v-Erb-B, and v-Fps oncoproteins/tyrosine kinases are also devoid of phosphatidylinositol kinase activity.[318,320,321] Complexes between polyoma virus middle-T antigen and Src protein possess tyrosine kinase activity but no phosphatidylinositol kinase activity.[322,323] Phosphoinositides are not phosphorylated by the very active tyrosine kinase that characterizes the LSTRA lymphosarcoma cell line.[324] The basal rate of inositol phosphate is not altered solely by the presence of v-Src in mouse BALB/c 3T3 fibroblasts, but it increases when the cells are stimulated with serum.[325] It is clear that the elevated phosphatidylinositol turnover detected in cells transformed by viral oncoproteins with tyrosine kinase activity is not due to a direct effect of these oncoproteins but depends on other cellular enzymes. v-Src can activate in murine fibroblasts both protein kinase C-dependent and -independent signaling pathways.[326]

Phospholipase C-γ is constitutively phosphorylated in cells coexpressing v-Src and the EGF receptor, suggesting that the EGF receptor interacts with phospholipase C-γ.[327] Rat and mouse fibroblasts expressing either v-Src or the middle-T antigen of polyoma virus show elevated levels of inositol 1,4,5-trisphosphate due to activation of phospholipase C. However, these cellular alterations are not sufficient for transformation. Phospholipase C-γ associates in rat 3Y1 fibroblasts with both the viral and the cellular Src protein kinases.[328] However, the possible biological significance of this association is unknown.

Enzymes other than phospholipase C and metabolic changes involving pathways other than the formation of inositol trisphosphate may have a role in the phospholipid changes induced by v-Src and other oncoproteins of the Src family. These alterations may include an elevated activity of PI 3-kinase.[329,330] The specific activity of this enzyme is increased in CEF cells transformed by the v-src, v-yes, and v-fps oncogenes.[331,332] An 85-kDa subunit of the PI 3-kinase is phosphorylated by complexes between polyoma virus middle-T antigen and the c-Src protein, as well as by the activated EGF and PDGF receptors.[333] Increased accumulation of D-myo-inositol 1,4,5,6-tetrakisphosphate is observed in Rat-1 fibroblasts transformed by the v-src oncogene.[334] This lipid compound can be synthesized from D-myo-inositol 1,3,4-trisphosphate in cytosolic extracts and may represent the consequence of an aberrant inositol phosphate metabolism occurring in v-src-transformed cells. The v-Src protein induces elevated levels of diaglyceride by stimulating phosphatidylcholine hydrolysis.[335] Other oncoproteins with tyrosine kinase activity induce similar metabolic alterations but the possible biological significance of these changes is unknown.

Treatment of RSV-transformed vole cells with TPA results in increased phosphorylation of the v-Src oncoprotein on amino-terminal serine residues.[336] Such phosphorylation may depend on protein kinase C activity, and it has been shown that protein kinase C phosphorylates the Src protein at the Ser-12 residue.[337] The possible role of this phosphorylation in the function of the Src kinase is not understood. No differences were found in the K_m or V_{max} of the Ser-12 phosphorylated and unphosphorylated c-Src kinase in immunoprecipitates prepared from control and TPA-treated cells using angiotensin or casein as a substrate.[338] Untreated cells have a low but detectable level of Ser-12 phosphorylation.

Phorbol esters, serum, and the v-Src protein induce similar or identical phosphorylations of the 40S ribosomal protein S6 in serum-starved CEF cells.[339] Injection of PMA directly into ASV-induced tumors resulted in increased expression of viral antigens at the cell surface and increased production of progeny virus in comparison to untreated cells. However, the PMA-treated tumors exhibited regression and expressed diminished levels of the kinase activity associated with the v-Src oncoprotein.[340] The mechanisms involved in such opposite effects of phorbol ester are unknown.

Enzymes involved in phosphatidylinositol metabolism have been detected in the cytoskeleton of human platelets, and treatment of platelets with thrombin results not only in increased activities of the cytoskeletal enzymes but also in their association with c-Src.[341] Stimulation of platelets with thrombin may promote the translocation to the cytoskeleton of the c-Src kinase, which may be involved in the regulation of enzymes related to phospholipid phosphorylation and hydrolysis.

B. THE c-*myc* GENE

The normal c-Myc protein is highly conserved in evolution and its expression may be associated with the processes related to cell proliferation and differentiation. The c-Myc protein is localized in the nucleus and is involved in the regulation of transcription. Two other members of the Myc family, N-Myc and L-Myc, may have a similar function.

Phorbol esters may exert striking effects on cell proliferation and differentiation, and these effects are probably mediated by protein kinase C activation. Quiescent BALB/3T3 cells show a rapid and marked induction of c-*myc* mRNA upon treatment with mitogenic agents and a similar induction is produced in these cells by TPA.[342] Modulators of lymphocyte proliferation may produce changes in c-*myc* gene expression in human peripheral blood mononuclear cells,[343,344] and treatment of human T cells with phorbol ester may have a similar effect.[345] The expression of c-*myc* in normal lymphocytes is regulated at several points of the cell cycle, and protein kinase C is involved in this regulation.[343,344,346] However, c-*fos* and c-*myc* gene expression induced in human lymphocytes by phorbol ester or calcium ionophore may not be sufficient to commit the cells to DNA synthesis.[347] The precise role of c-*fos* and c-*myc* expression in mitogenically stimulated lymphocytes is not known.

Differentiation of HL-60 human myeloid leukemia cells induced by phorbol ester is probably mediated by activation of protein kinase C and is accompanied by decreased c-*myc* gene expression.[348,349] The cell-permeable diacylglycerol analog *sn*-1,2-dioctanoylglycerol causes a rapid decrease in c-*myc* gene transcription in HL-60 cells, suggesting that prolonged stimulation of protein kinase C is required for persistent inhibition of c-*myc* gene transcription in HL-60 cells.[350] All inducers of HL-60 differentiation produce a reduction of c-*myc* expression in HL-60 cells, suggesting an inverse relation between the levels of c-*myc* expression and cell differentiation. The levels of c-Myc protein remain unaltered in a mutant phorbol ester-tolerant variant line of HL-60 cells, which undergoes transient growth arrest but does not differentiate when exposed to TPA.[351] Downregulation of c-*myc* gene expression may be a prerequisite for cellular differentiation; however, increased levels of c-*myc* mRNA have been observed in the precommitment stage during HL-60 differentiation.[352]

Phorbol esters may cause altered expression of the c-*myc* gene in other cellular systems. Decreased c-*myc* mRNA levels and enhanced expression of the calcitonin gene occur in human medullary thyroid carcinoma cells treated with phorbol ester.[353] Treatment of human neuroblastoma cells with phorbol ester induces differentiation, which is associated with decreased expression of the c-*myc* and N-*myc* genes.[354-356]

Expression of genes of the *myc* family may induce alterations in phosphoinositide metabolism and protein kinase C activity in various systems. Transfection of the N-*myc* gene into neuroblastoma cells may cause changes between different protein kinase isoforms.[357] A similar effect is produced by transfection of the c-*myc* gene into cultured human small-cell lung cancer cells.[358]

C. THE c-*fos* AND c-*jun* GENES

Expression of the c-*fos* gene is frequently associated, in a manner similar to that of c-*myc*, with the processes of cell proliferation and/or differentiation.[359] Cellular Fos proteins are localized in the nucleus and are involved in the control of genomic functions through the formation of transcriptional complexes with Jun proteins. The Jun/Fos complexes recognize the AP-1 binding site at the DNA level.

Phosphoinositide hydrolysis and protein kinase C activation may have an important role in the mechanism of c-*jun* gene expression. Antigen binding to specific receptors on T cells results in a rapid and transient phosphoinositide hydrolysis and activation of protein kinase C, followed by c-*fos* and c-*jun* gene expression. Increased intracellular levels of cAMP block protein kinase C-mediated activation of T lymphocytes by inhibiting c-*jun* gene transcription.[360] Expression of c-*fos* and c-*jun* RNA in THP-1 human monocytic leukemia cells is regulated in a coupled manner by second messengers activated after membrane phospholipid turnover.[361] Protein kinase C activation by phorbol ester or other agents may result in c-Jun protein activation.[362] c-Jun is present in resting human epithelial and fibroblastic cells in a latent form in which it is phosphorylated on its basic, DNA-binding region on serine and threonine residues. Protein kinase C activation results in dephosphorylation at this site, and this coincides with an increase in the AP-1-binding activity of c-Jun, thereby modifying specific transcriptional processes.

D. THE *ras* GENES

The biological activities of Ras and related proteins may be intimately associated with phosphoinositide metabolism. Ras proteins would be controlled biologically by interactions between phospholipids and GTPase-activating protein (GAP) molecules.[363] The transforming action of *ras* oncogenes may be mediated, at least in part, by alteration of phosphoinositide turnover and activation of protein kinase C. Constitutive activation of protein kinase C and growth factor independence occurs in IL-3-dependent myeloid cells expressing a mutant c-H-*ras* gene.[364] Transformation of mouse fibroblasts by *ras* oncogenes may be associated with translocation of protein kinase C from the cytoplasm to the nucleus, which may result in increased phosphorylation of nuclear proteins.[365] Expression of a cloned N-*ras* gene in PC12 pheochromocytoma cells interferes with the ability of phorbol esters to induce downregulation of protein kinase C.[366] NIH/3T3 cells transformed with *ras* oncogenes exhibit a protein kinase C-mediated inhibition of agonist-stimulated Ca^{2+} influx.[367] Arachidonic acid and some of its metabolites released from membrane phospholipids by the action of phospholipase A_2 or other routes could act as second messengers in Ras regulation through alterations in GAP activity.[368]

The levels of inositol 4,5-bisphosphate and inositol 1,4,5-trisphosphate are elevated in *ras*-transformed NRK cells, compared to the nontransformed parental cells.[369] These changes may result in an increased production of 1,2-diacylglycerol and activation of protein kinase C.[370] A transient surge of membrane-associated protein kinase C activity would be involved in the process by which the v-Ras oncoprotein triggers G_0 to G_1 cycle transition.[371] The alterations observed in Ras-transformed cells may be the consequence of a direct activation of phospholipase C.[372,373]

Antibodies against phospholipase C can inhibit serum- and Ras-stimulated DNA synthesis.[374] Ras proteins are upstream effectors of phospholipase C activity in metabolic pathways related to phosphoinositide-specific transduction, and the activity of phospholipase C-γ may be necessary for Ras-mediated induction of DNA synthesis. Ras proteins may function as G protein-like molecules in the inositol phospholipid signal transduction. Mutant c-Ras proteins may activate the phospholipase C constitutively, thus enhancing phosphoinositide breakdown.[375] MDCK cells transformed by the v-H-*ras* oncogene show constitutive activation of phospholipid metabolism, and induction of differentiation of these cells by treatment with PGE_2 or 8-bromo-cAMP is associated with a decrease in the level of inositol trisphosphate.[376]

The precise role of Ras proteins in phosphoinositide metabolism is not clear. In C3H10T1/2 cells transfected with an activated c-H-*ras* gene, expression of the gene directly or indirectly causes an increased turnover of phosphatidylcholine.[377] Both the synthesis and degradation of phosphatidylcholine can be activated after microinjection of transforming Ras into *Xenopus laevis* oocytes, resulting in a net production of diacylglycerol.[378] Ras may not directly activate phospholipase C but may affect the levels of inositol phosphates and phosphocholine by different mechanisms.

VII. SPHINGOLIPID METABOLISM

Sphingolipids are structural and functional components of the outer membrane of animal cells.[379] Sphingosine, a natural cell component, is the backbone moiety of sphingomyelin, gangliosides, and other complex sphingolipids. These compounds are involved in the regulation of cell proliferation and cell differentiation. Alterations of sphingolipid metabolism occur during the cell cycle and may be associated with oncogenic transformation.

Sphingosine may function *in vivo* as an inhibitor of protein kinase C activity.[206,380] Independently of this effect, sphingosine may play a role as a positive regulator of cellular proliferation. Low concentrations of sphingosine stimulate DNA synthesis and act synergistically with growth factors to induce proliferation of quiescent mouse 3T3 fibroblasts.[381] The catalytic activity of protein kinase C can be inactivated by acidic phospholipids such as phosphatidylserine, and phorbol ester enhances this inactivation.[382] TPA, which activates protein kinase C, stimulates the synthesis of sphingomyelin in NIH/3T3 cells, and a diminished response is observed in NIH/3T3 cells transformed by retroviruses carrying *raf* oncogenes.[383] The mitogenic activity of phorbol esters is inhibited by a sialoglycopeptide isolated from the surface of bovine cerebral cortex cells.[384] The inhibitor counteracts TPA-induced DNA synthesis and ODC activity by postreceptor mechanisms that could involve an interference with the activation of protein kinase C and/or ion transport systems.

Exogenous sphingolipids may inhibit phorbol ester-induced differentiation of neoplastic cell lines. Treatment of HL-60 human leukemia cells with sphinganine or other sphingoid bases results in pronounced inhibition of PMA-induced differentiation.[385] This inhibition is apparently due to a direct effect of the sphingoid bases on protein kinase C. Sphingomyelin may be involved in the mechanism of action of some growth factors. The action of TNF-α appears to implicate a signaling pathway involving sphingomyelin hydrolysis to ceramide by a sphingomyelinase and stimulation of a ceramide-activated protein kinase.[386]

Glycosphingolipids are ubiquitous compounds in animal cells. They derive from glucosylceramide, a compound generated from ceramide and UDP-glucose by a glucosyl-transferase. There are more than 300 different species of glycosphingolipids, including gangliosides, most of them synthesized from glucosylceramide by the attachment of additional sugars and sulfate. Stimulation of phospholipase C activity by hormones and growth factors may depend on regulatory influences exerted by glucosphingolipids.[387]

The gangliosides, which are acidic sialic acid-containing glycosphingolipids, are constituents of the plasma membrane of mammalian cells and act as endogenous bimodal regulators of positive and negative signals for cell growth.[388-391] Interactions occurring between specific gangliosides and the receptors for growth factors such as insulin, EGF, and PDGF may contribute to the modulation of growth factor biological responses.[392-394] Gangliosides may have profound effects on the endogenous phosphorylation of cellular proteins. Ganglioside GM3 can activate protein kinase C acting as a substitute for phosphatidylserine.[395] Recent evidence indicates that at least some of the ganglioside-stimulated phosphorylation reactions may be mediated through activation of a ganglioside-stimulated protein kinase.[396] This kinase is distinct from the other protein phosphotransferases but the molecular basis for its activation by gangliosides is unknown. The physiological significance of the ganglioside-stimulated protein kinase and its relation to other kinases are also unknown. The biological importance of gangliosides is indicated by their conservation in normal and tumor tissues from phylogenetically distant vertebrate species such as humans and fishes.[397]

The B subunit of cholera toxin, which is pentavalent and binds only to gangliosides on the cell surface, stimulates DNA synthesis and cell division in quiescent, nontransformed mouse 3T3 cells in a dose-dependent manner. In addition, the B subunit of cholera toxin potentiates the response of 3T3 cells to other mitogens like EGF, PDGF, and insulin.[388] In contrast, the B subunit inhibits the growth of 3T3 cells transformed by the *ras* oncogene, which indicates that the same cells may exhibit different responses to the B subunit of cholera toxin depending on their state of growth. The mechanism by which the B subunit of cholera toxin is able to modulate the proliferative response of cells in a positive or negative manner is unknown, but there is evidence that gangliosides can directly suppress the activity of protein kinase C and its response to diacylglycerol.[398]

The v-*myc* and v-*ras* oncogenes may have synergistic effects on ganglioside synthesis by mouse fibroblasts.[399] Cells transfected with both oncogenes exhibit a transformed phenotype and have distinct changes in the acidic fraction of glycosphingolipids, which are characterized by a higher degree of sialylation. Thus, the action of particular oncogenes on the regulation of growth in particular types of cells may be mediated, at least in part, by their influence on ganglioside synthesis.

REFERENCES

1. **Michell, R. H.,** Inositol phospholipids and cell surface receptor function, *Biochim. Biophys. Acta,* 415, 81, 1975.
2. **Farese, R. V.,** Phosphoinositide metabolism and hormone action, *Endocrine Rev.,* 4, 78, 1983.
3. **Berridge, M. J.,** Inositol trisphosphate and diacylglycerol as second messengers, *Biochem. J.,* 220, 345, 1984.
4. **Catt, K. J. and Balla, T.,** Phosphoinositide metabolism and hormone action, *Annu. Rev. Med.,* 40, 487, 1989.
5. **Whitman, M. and Cantley, L.,** Phosphoinositide metabolism and the control of cell proliferation, *Biochim. Biophys. Acta,* 948, 327, 1989.
6. **Berridge, M. J. and Irvine, R. F.,** Inositol phosphates and cell signalling, *Nature,* 341, 197, 1989.
7. **Majerus, P. W., Ross, T. S., Cunningham, T. W., Caldwell, K. K., Jefferson, A. B., and Bansal, V. S.,** Recent insights in phosphatidylinositol signaling, *Cell,* 63, 459, 1990.
8. **Bansal, V. S. and Majerus, P. W.,** Phosphatidylinositol-derived precursors and signals, *Annu. Rev. Cell Biol.,* 6, 41, 1990.
9. **Majerus, P. W.,** Inositol phosphate biochemistry, *Annu. Rev. Biochem.,* 61, 251, 1992.

10. **Berridge, M. J.,** Inositol trisphosphate and calcium signalling, *Nature,* 361, 315, 1993.

11. **Paris, S., Chambard, J.-C., and Pouysségur, J.,** Tyrosine kinase-activating growth factors potentiate thrombin-and AlF$_4$-induced phosphoinositide breakdown in hamster fibroblasts. Evidence for positive cross-talk between the two mitogenic signaling pathways, *J. Biol. Chem.,* 263, 12893, 1988.

12. **Abdel-Latif, A. A.,** Biochemical and functional interactions between the inositol 1,4,5-trisphosphate-Ca^{2+} and cyclic AMP signalling systems in smooth muscle, *Cell. Signal.,* 3, 371, 1991.

13. **McIntosh, R. P. and Catt, K. J.,** Coupling of inositol phospholipid hydrolysis to peptide hormone receptors expressed from adrenal and pituitary mRNA in *Xenopus laevis* oocytes, *Proc. Natl. Acad. Sci. U.S.A.,* 84, 9045, 1987.

14. **Newell, P. C., Europe-Finner, G. N., Small, N. V., and Liu, G.,** Inositol phosphates, G-proteins and *ras* genes involved in chemotactic signal transduction of *Dictyostelium, J. Cell Sci.,* 89, 123, 1988.

15. **Boss, W. F. and Massel, M. O.,** Polyphosphoinositides are present in plant tissue culture cells, *Biochem. Biophys. Res. Commun.,* 132, 1018, 1985.

16. **Reddy, A. S. N., McFadden, J. J., Friedman, M., and Poovaiah, B. W.,** Signal transduction in plants: evidence for the involvement of calcium and turnover of inositol phospholipids, *Biochem. Biophys. Res. Commun.,* 149, 334, 1987.

17. **Ettlinger, C. and Lehle, L.,** Auxin induces rapid changes in phosphatidylinositol metabolites, *Nature,* 331, 176, 1988.

18. **Einspahr, K. J. and Thompson, G. A., Jr.,** Transmembrane signaling via phosphatidylinositol 4,5-bisphosphate hydrolysis in plants, *Plant Physiol.,* 93, 361, 1990.

19. **Drobak, B. K.,** The plant phosphoinositide system, *Biochem. J.,* 288, 697, 1992.

20. **Low, M. G.,** Biochemistry of the glycosyl-phosphatidylinositol membrane protein anchors, *Biochem. J.,* 244, 1, 1987.

21. **Kent, C., Carman, G. M., Spence, M. W., and Dowhan, W.,** Regulation of eukaryotic phospholipid metabolism, *FASEB J.,* 5, 2258, 1991.

22. **Kaiser, E., Chiba, P., and Zaky, K.,** Phospholipases in biology and medicine, *Clin. Biochem.,* 23, 349, 1990.

23. **Whitman, M., Downes, C. P., Keeler, M., Keller, T., and Cantley, L.,** Type I phosphatidylinositol kinase makes a novel inositol phospholipid, phosphatidylinositol-3-phosphate, *Nature,* 332, 644, 1988.

24. **Sylvia, V., Curtin, G., Norman, J., Stec, J., and Busbee, D.,** Activation of a low specific activity form of DNA polymerase α by inositol-1,4-bisphosphate, *Cell,* 54, 651, 1988.

25. **Fain, J. N.,** Regulation of phosphoinositide-specific phospholipase C, *Biochim. Biophys. Acta,* 1053, 81, 1990.

26. **Wilson, D. B., Bross, T. E., Hofmann, S., and Majerus, P. W.,** Hydrolysis of polyphosphoinositides by purified sheep seminal vesicle phospholipase C enzymes, *J. Biol. Chem.,* 259, 11718, 1984.

27. **Kikkawa, U. and Nishizuka, Y.,** The role of protein kinase C in transmembrane signalling, *Annu. Rev. Cell Biol.,* 2, 149, 1986.

28. **Stone, R. M., Weber, B. L., Spriggs, D. R., and Kufe, D. W.,** Phospholipase C activates protein kinase C and induces monocytic differentiation of HL-60 cells, *Blood,* 72, 739, 1988.

29. **Putney, J. W., Jr., Aub, D. L., Taylor, C. W., and Merritt, J. E.,** Formation and biological action of inositol 1,4,5-trisphosphate, *Fed. Proc.,* 45, 2634, 1986.

30. **Yoshimoto, A., Nakanishi, K., Anzai, T., and Komine, S.,** The activation of phosphatidylinositol-specific phospholipase C in mammary epithelial cells of lactating mouse, *Cell Biochem. Funct.,* 8, 163, 1990.

31. **Cockroft, S. and Thomas, G. M. H.,** Inositol-lipid-specific phospholipase C isoenzymes and their differential regulation by receptors, *Biochem. J.,* 288, 1, 1992.

32. **Manne, V.,** Identification of polyphosphoinositide-specific phospholipase C and its resolution from phosphoinositide-specific phospholipase C from human platelet extract, *Oncogene,* 2, 49, 1987.

33. **Banfic, H., Zizak, M., Divecha, N., and Irvine, R. F.,** Nuclear diacylglycerol is increased during cell proliferation *in vivo, Biochem. J.,* 290, 633, 1993.

34. **Fukui, T., Lutz, R. J., and Lowenstein, J. M.,** Purification of a phospholipase C from rat liver cytosol that acts on phosphatidylinositol 4,5-bisphosphate and phosphatidylinositol 4-phosphate, *J. Biol. Chem.,* 263, 17730, 1988.

35. **Manne, V.,** A novel candidate for receptor-coupled phospholipase C purified from human platelets, *Oncogene,* 3, 579, 1988.

36. **Nakanishi, H., Nomura, H., Kikkawa, U., Kishimoto, A., and Nishikura, Y.,** Rat brain and liver soluble phospholipase C: resolution of two forms with different requirements for calcium, *Biochem. Biophys. Res. Commun.,* 132, 582, 1985.

37. **Ryu, S. H., Suh, P.-G., Cho, K. S., Lee, K.-Y., and Rhee, S. G.,** Bovine brain cytosol contains three immunologically distinct forms of inositolphospholipid-specific phospholipase C, *Proc. Natl. Acad. Sci. U.S.A.,* 84, 6649, 1987.

38. **Suh, P.-G., Ryu, S. H., Moon, K. H., Suh, H. W., and Rhee, S. G.,** Cloning and sequence of multiple forms of phospholipase C, *Cell,* 54, 161, 1988.

39. **Suh, P.-G., Ryu, S. H., Moon, K. H., Su, H. W., and Rhee, S. G.,** Inositol phospholipid-specific phospholipase C: complete cDNA and protein sequences and sequence homology to tyrosine kinase-related oncogene products, *Proc. Natl. Acad. Sci. U.S.A.,* 85, 5419, 1988.

40. **Burgess, W. H., Dionne, C. A., Kaplow, J., Mudd, R., Friesel, R., Zilberstein, A., Schlessinger, J., and Jaye, M.,** Characterization and cDNA cloning of phospholipase C-γ, a major substrate for heparin-binding growth factor 1 (acidic fibroblast growth factor)-activated tyrosine kinase, *Mol. Cell. Biol.,* 10, 4770, 1990.

41. **Emori, Y., Homma, Y., Sorimachi, H., Kawasaki, H., Nakanishi, O., Suzuki, K., and Takenawa, T.,** A second type of rat phosphoinositide-specific phospholipase C containing a *src*-related sequence not essential for phosphoinositide-hydrolyzing activity, *J. Biol. Chem.,* 264, 21885, 1989.

42. **Stahl, M. L., Ferenz, C. R., Kelleher, K. L., Kriz, R. W., and Knopf, J. L.,** Sequence similarity of phospholipase C with the non-catalytic region of src, *Nature,* 332, 269, 1988.

43. **Mayer, B. J., Hamaguchi, M., and Hanafusa, H.,** A novel viral oncogene with structural similarity to phospholipase C, *Nature,* 332, 272, 1988.

44. **Bennett, C. F., Balcarek, J. M., Varrichio, A., and Crooke, S. T.,** Molecular cloning and complete amino-acid sequence of form-I phosphoinositide-specific phospholipase C, *Nature,* 334, 268, 1988.

45. **Smith, M. R., Ryu, S., Suh, P., Rhee, S., and Kung, H.,** S-phase induction and transformation of quiescent NIH 3T3 cells by microinjection of phospholipase C, *Proc. Natl. Acad. Sci. U.S.A.,* 86, 3659, 1989.

46. **Anthes, J. C., Billah, M. M., Cali, A., Egan, R. W., and Siegel, M. I.,** Chemotactic peptide, calcium and guanine nucleotide regulation of phospholipase C activity in membranes from DMSO-differentiated HL60 cells, *Biochem. Biophys. Res. Commun.,* 145, 825, 1987.

47. **Banno, Y., Nagao, S., Katada, T., Nagata, K., Ui, M., and Nozawa, Y.,** Stimulation by GTP-binding proteins (G_i, G_0) of partially purified phospholipase C activity from human platelet membranes, *Biochem. Biophys. Res. Commun.,* 146, 861, 1987.

48. **Ryu, S. H., Kim, U.-H., Wahl, M. I., Brown, A. B., Carpenter, G., Huang, K.-P., and Rhee, S. G.,** Feedback regulation of phospholipase C-β by protein kinase C, *J. Biol. Chem.,* 265, 17941, 1990.

49. **Divecha, N., Rhee, S. G., Letcher, A. J., and Irvine, R. F.,** Phosphoinositide signalling enzymes in rat liver nuclei — phosphoinositidase-C isoform β1 is specifically, but not predominantly, located in the nucleus, *Biochem. J.,* 289, 617, 1993.

50. **Liao, F., Shin, H. S., and Rhee, S. G.,** *In vitro* tyrosine phosphorylation of PLC-γ1 and PLC-γ2 by Src-family protein tyrosine kinases, *Biochem. Biophys. Res. Commun.,* 191, 1028, 1993.

51. **Wahl, M. I., Daniel, T. O., and Carpenter, G.,** Antiphosphotyrosine recovery of phospholipase C activity after EGF treatment of A-431 cells, *Science,* 241, 968, 1988.

52. **Nishibe, S., Wahl, M. I., Rhee, S. G., and Carpenter, G.,** Tyrosine phosphorylation of phospholipase C-II *in vitro* by the epidermal growth factor receptor, *J. Biol. Chem.,* 264, 10335, 1989.

53. **Margolis, B., Rhee, S. G., Felder, S., Mervic, M., Lyall, R., Levitzki, A., Ullrich, A., Zilberstein, A., and Schlessinger, J.,** EGF induces tyrosine phosphorylation of phospholipase C-II: a potential mechanism for EGF receptor signaling, *Cell,* 57, 1101, 1989.

54. **Wahl, M. I., Nishibe, S., Kim, J. W., Kim, H., Rhee, S. G., and Carpenter, G.,** Identification of two epidermal growth factor-sensitive tyrosine phosphorylation sites of phospholipase C-γ in intact HSC-1 cells, *J. Biol. Chem.,* 265, 3944, 1990.

55. **Wahl, M. I., Olashaw, N. E., Nishibe, S., Rhee, S. G., Pledger, W. J., and Carpenter, G.,** Platelet-derived growth factor induces rapid and sustained tyrosine phosphorylation of phospholipase C-γ in quiescent BALB/c 3T3 cells, *Mol. Cell. Biol.,* 9, 2934, 1989.

56. **Margolis, B., Zilberstein, Franks, C., Felder, S., Kremer, S., Ullrich, A., Rhee, S. G., Skorecki, K., and Schlessinger, J.,** Effect of phospholipase C-γ overexpression on PDGF-induced second messengers and mitogenesis, *Science,* 248, 607, 1990.

57. **Downing, J. R., Margolis, B. L., Zilberstein, A., Ashmun, R. A., Ullrich, A., Sherr, C. J., and Schlessinger, J.,** Phospholipase C-γ, a substrate for PDGF receptor kinase, is not phosphorylated on tyrosine during the mitogenic response to CSF-1, *EMBO J.,* 8, 3345, 1989.

58. **Nishibe, S., Wahl, M. I., Wedegaertner, P. B., Kim, J. J., Rhee, S. G., and Carpenter, G.,** Selectivity of phospholipase C phosphorylation by the epidermal growth factor receptor, the insulin receptor, and their cytoplasmic domains, *Proc. Natl. Acad. Sci. U.S.A.,* 87, 424, 1990.

59. **Wahl, M. I., Nishibe, S., Suh, P.-G., Rhee, S. G., and Carpenter, G.,** Epidermal growth factor stimulates tyrosine phosphorylation of phospholipase C-II independently of receptor internalization and extracellular calcium, *Proc. Natl. Acad. Sci. U.S.A.,* 86, 1568, 1989.

60. **Kim, U. H., Kim, H. S., and Rhee, S. G.,** Epidermal growth factor and platelet-derived growth factor promote translocation of phospholipase-C-γ from cytosol to membrane, *FEBS Lett.,* 270, 33, 1990.

61. **Vetter, M. L., Martin-Zanca, D., Parada, L. F., Bishop, J. M., and Kaplan, D. R.,** Nerve growth factor rapidly stimulates tyrosine phosphorylation of phospholipase C-γ 1 by a kinase activity associated with the product of the *trk* protooncogene, *Proc. Natl. Acad. Sci. U.S.A.,* 88, 5650, 1991.

62. **Ohmichi, M., Decker, S. J., Pang, L., and Saltiel, A. R.,** Nerve growth factor binds to the 140 kd *trk* proto-oncogene product and stimulates its association with the *src* homology domain of phospholipase Cγ1, *Biochem. Biophys. Res. Commun.,* 179, 217, 1991.

63. **Weiss, A., Koretzky, G., Schatzman, R. C., and Kadlecek, T.,** Functional activation of the T-cell antigen receptor induces tyrosine phosphorylation of phospholipase C-γ1, *Proc. Natl. Acad. Sci. U.S.A.,* 88, 5484, 1991.

64. **Secrist, J. P., Karnitz, L., and Abraham, R. T.,** T-cell antigen receptor ligation induces tyrosine phosphorylation of phospholipase C-γ1, *J. Biol. Chem.,* 266, 12135, 1991.

65. **Arteaga, C. L., Johnson, M. D., Todderud, G., Coffey, R. J., Carpenter, G., and Page, D. L.,** Elevated content of the tyrosine kinase substrate phospholipase C-γ1 in primary human breast carcinomas, *Proc. Natl. Acad. Sci. U.S.A.,* 88, 10435, 1991.

66. **Pribluda, V. S. and Metzger, H.,** Calcium-independent phosphoinositide breakdown in rat basophilic leukemia cells. Evidence for an early rise in inositol 1,4,5-trisphosphate which precedes the rise in other inositol phosphates and in cytoplasmic calcium, *J. Biol. Chem.,* 262, 11449, 1987.

67. **Guillemette, G., Balla, T., Baukal, A. J., Spät, A., and Catt, K. J.,** Intracellular receptors for inositol 1,4,5-trisphosphate in angiotensin II target tissues, *J. Biol. Chem.,* 262, 1010, 1987.

68. **Guillemette, G., Balla, T., Baukal, A. J., and Catt, K. J.,** Inositol 1,4,5-trisphosphate binds to a specific receptor and releases microsomal calcium in the anterior pituitary gland, *Proc. Natl. Acad. Sci. U.S.A.,* 84, 8195, 1987.

69. **Mignery, G. A., Newton, C. L., Archer, B. T., III, and Südhof, T. C.,** Structure and expression of the rat inositol 1,4,5-trisphosphate receptor, *J. Biol. Chem.,* 265, 12679, 1990.

70. **Pietri, F., Hilly, M., Claret, M., and Mauger, J.-M.,** Characterization of two forms of inositol 1,4,5-trisphosphate receptor in rat liver, *Cell. Signal.,* 2, 253, 1990.

71. **Nakagawa, T., Okano, H., Furuichi, T., Aruga, J., and Mikoshiba, K.,** The subtypes of the mouse inositol 1,4,5-trisphosphate receptor are expressed in a tissue-specific and developmentally specific manner, *Proc. Natl. Acad. Sci. U.S.A.,* 88, 6244, 1991.

72. **Bradford, P. G., Jin, Y. Y., Hui, P., and Wang, X. H.,** IL-3 and GM-CSF induce the expression of the inositol trisphosphate receptor in K562 myeloblast cells, *Biochem. Biophys. Res. Commun.,* 187, 438, 1992.

73. **Matter, N., Ritz, M.-F., Freyermuth, S., Rogue, P., and Malviya, A. N.,** Stimulation of nuclear protein kinase C leads to phosphorylation of nuclear inositol 1,4,5-trisphosphate receptor and accelerated calcium release by inositol 1,4,5-trisphosphate from isolated rat liver nuclei, *J. Biol. Chem.,* 268, 732, 1993.

74. **Furuichi, T., Shiota, C., and Miroshiba, K.,** Distribution of inositol 1,4,5-trisphosphate receptor messenger RNA in mouse tissues, *FEBS Lett.,* 267, 85, 1990.

75. **Furuichi, T., Yoshikawa, S., Miyawaki, A., Wada, K., Maeda, N., and Mikoshiba, K.,** Primary structure and functional expression of the inositol 1,4,5-trisphosphate-binding protein P_{400}, *Nature,* 342, 32, 1989.

76. **Mauger, J.-P., Claret, M., Pietri, F., and Hilly, M.,** Hormonal regulation of inositol 1,4,5-trisphosphate receptor in rat liver, *J. Biol. Chem.,* 264, 8821, 1989.

77. **Storey, D. J., Shears, S. B., Kirk, C. J., and Michell, R. H.,** Stepwise enzymatic dephosphorylation of inositol 1,4,5-trisphosphate to inositol in liver, *Nature,* 312, 374, 1984.

78. **Hata, Y., Ogata, E., and Kojima, I.,** Platelet-derived growth factor stimulates synthesis of 1,2-diacylglycerol from monoacylglycerol in Balb/c 3T3 cells, *Biochem. J., 262*, 947, 1989.

79. **Maroney, A. C. and Macara, I. G.,** Phorbol ester-induced translocation of diacylglycerol kinase from the cytosol to the membrane in Swiss 3T3 fibroblasts, *J. Biol. Chem., 264*, 2537, 1989.

80. **Kato, M., Kawai, S., and Takenawa, T.,** Defect in phorbol acetate-induced translocation of diacylglycerol kinase in erbB-transformed fibroblast cells, *FEBS Lett., 247*, 247, 1989.

81. **Zeisel, S. H.,** Choline phospholipids: signal transduction and carcinogenesis, *FASEB J., 7*, 551, 1993.

82. **Exton, J. H.,** Signaling through phosphatidylcholine breakdown, *J. Biol. Chem., 265*, 1, 1990.

83. **Billah, M. M. and Anthes, J. C.,** The regulation and cellular functions of phosphatidylcholine hydrolysis, *Biochem. J., 269*, 281, 1990.

84. **Stasek, J. E., Natarajan, V., and Garcia, J. G. N.,** Phosphatidic acid directly activates endothelial cell protein kinase C, *Biochem. Biophys. Res. Commun., 191*, 134, 1993.

85. **Inhorn, R. C., Bansal, V. S., and Majerus, P. W.,** Pathway for inositol 1,3,4-trisphosphate and 1,4-bisphosphate metabolism, *Proc. Natl. Acad. Sci. U.S.A., 84*, 2170, 1987.

86. **Bansal, V. S., Inhorn, R. C., and Majerus, P. W.,** The metabolism of inositol 1,3,4-trisphosphate to inositol 1,3-bisphosphate, *J. Biol. Chem., 262*, 9444, 1987.

87. **Cunha-Melo, J. R., Dean, N. M., Moyer, J. D., Maeyama, K., and Beaven, M. A.,** The kinetics of phosphoinositide hydrolysis in rat basophilic leukemia (RBL-2H3) cells varies with the type of IgE receptor cross-linking agent used, *J. Biol. Chem., 262*, 11455, 1987.

88. **Cockroft, S.,** The dependence on Ca^{2+} of the guanine-nucleotide-activated polyphosphoinositide phosphodiesterase in neutrophil plasma membranes, *Biochem. J., 240*, 503, 1986.

89. **Cockroft, S. and Gomperts, B. D.,** Role of guanine nucleotide binding protein in the activation of polyphosphoinositide phosphodiesterase, *Nature, 314*, 534, 1985.

90. **Rapuano, B. E. and Bockman, R. S.,** A23187 and protein kinase C activators stimulate phosphatidylinositol metabolism and prostaglandin synthesis in a human lung cancer cell line, *Biochem. Biophys. Res. Commun., 156*, 644, 1988.

91. **Cook, S. J. and Wakelam, M. J. O.,** Stimulated phosphatidylcholine hydrolysis as a signal transduction pathway in mitogenesis, *Cell. Signal., 3*, 273, 1991.

92. **Martin, T. W., Feldman, D. R., and Michaelis, K. C.,** Phosphatidylcholine hydrolysis stimulated by phorbol myristate acetate is mediated principally by phospholipase D in endothelial cells, *Biochim. Biophys. Acta, 1053*, 162, 1990.

93. **Cabot, M. C., Welsh, C. J., Cao, H., and Chabbott, H.,** The phosphatidylcholine pathway of diacylglycerol formation stimulated by phorbol diesters occurs via phospholipase D activation, *FEBS Lett., 233*, 153, 1988.

94. **Cabot, M. C., Welsh, C. J., Zhang, Z., and Cao, H.,** Evidence for a protein kinase C-directed mechanism in the phorbol diester-induced phospholipase D pathway of diacylglycerol generation from phosphatidylcholine, *FEBS Lett., 245*, 85, 1989.

95. **Huang, C. and Cabot, M. C.,** Phorbol diesters stimulate the accumulation of phosphatidate, phosphatidylethanol, and diacylglycerol in three cell types. Evidence for the indirect formation of phosphatidylcholine-derived diacylglycerol by a phospholipase D pathway and direct formation of diacylglycerol by a phospholipase C pathway, *J. Biol. Chem., 265*, 14858, 1990.

96. **Rosoff, P. M., Savage, N., and Dinarello, C. A.,** Interleukin-1 stimulates diacylglycerol production in T lymphocytes by a novel mechanism, *Cell, 54*, 73, 1988.

97. **MacDonald, M. L., Mack, K. F., Richardson, C. N., and Glomset, J. A.,** Regulation of diacylglycerol kinase reaction in Swiss 3T3 cells: increased phosphorylation of endogenous diacylglycerol and decreased phosphorylation of didecanoylglycerol in response to platelet-derived growth factor, *J. Biol. Chem., 263*, 1575, 1988.

98. **MacDonald, M. L., Mack, K. F., Williams B. W., King, W. C., and Glomset, J. A.,** A membrane-bound diacylglycerol kinase that selectively phosphorylates arachidonoyl-diacylglycerol: Distinction from cytosolic diacylglycerol kinase and comparison with the membrane-bound enzyme from *Escherichia coli*, *J. Biol. Chem., 263*, 1584, 1988.

99. **Fukami, K. and Takenawa, T.,** Phosphatidic acid that accumulates in platelet-derived growth factor-stimulated Balb/c 3T3 cells is a potential mitogenic signal, *J. Biol. Chem., 267*, 10988, 1992.

100. **Scallon, B. J., Fung, W.-J. C., Tsang, T. C., Li, S., Kado-Fong, H., Huang, K.-S., and Kochan, J. P.,** Primary structure and functional activity of a phosphatidylinositol-glycan-specific phospholipase D, *Science, 252*, 446, 1991.

101. **Axelrod, J.,** Receptor-mediated activation of phospholipase A_2 and arachidonic acid release in signal transduction, *Biochem. Soc. Trans.,* 18, 503, 1990.
102. **Lai, C.-Y. and Wada, K.,** Phospholipase A_2 from human synovial fluid: purification and structural homology to the placental enzyme, *Biochem. Biophys. Res. Commun.,* 157, 488, 1988.
103. **Billah, M. M. and Siegel, M. I.,** Phospholipase A_2 activation in chemotactic peptide-stimulated HL60 granulocytes: synergism between diacylglycerol and Ca^{2+} in a protein kinase C-independent mechanism, *Biochem. Biophys. Res. Commun.,* 144, 683, 1987.
104. **Parker, J., Daniel, L. W., and Waite, M.,** Evidence of protein kinase C involvement in phorbol diester-stimulated arachidonic acid release and prostaglandin synthesis, *J. Biol. Chem.,* 262, 5385, 1987.
105. **Weber, K. and Johnsson, N.,** Repeating sequence homologies in the p36 target protein of retroviral protein kinases and lipocortin, the p37 inhibitor of phospholipase A_2, *FEBS Lett.,* 203, 95, 1986.
106. **Huang, K.-S., Wallner, B. P., Mattaliano, R. J., Tizard, R., Burne, C., Frey, A., Hession, C., McGray, P., Sinclair, L. K., Chow, E. P., Browning, J. L., Ramachandran, K. L., Tang, J., Smart, J. E., and Pepinsky, R. B.,** Two human 35-kd inhibitors of phospholipase A_2 are related to substrates of pp60[v-src] and of the epidermal growth factor receptor/kinase, *Cell,* 46, 191, 1986.
107. **Brugge, J. S.,** The p35/p36 substrates of protein-tyrosine kinases as inhibitors of phospholipase A_2, *Cell,* 46, 149, 1986.
108. **William, F., Mroczkowski, B., Cohen, S., and Kraft, A. S.,** Differentiation of HL-60 cells is associated with an increase in the 35-kDa protein lipocortin I, *J. Cell. Physiol.,* 137, 402, 1988.
109. **Nakano, T., Ohara, O., Teraoka, H., and Arita, H.,** Glucocorticoids suppress group II phospholipase A_2 production by blocking mRNA synthesis and post-transcriptional expression, *J. Biol. Chem.,* 265, 12745, 1990.
110. **Goldberg, H. J., Viegas, M. M., Margolis, B. L., Schlessinger, J., and Skorecki, K. L.,** The tyrosine kinase activity of the epidermal growth factor receptor is necessary for phospholipase A_2 activation, *Biochem. J.,* 267, 461, 1990.
111. **Pike, L. J.,** Phosphatidylinositol 4-kinases and the role of polyphosphoinositides in cellular regulation, *Endocrine Rev.,* 13, 692, 1992.
112. **Auger, K. R. and Cantley, L. C.,** Novel polyphosphoinositides in cell growth and activation, *Cancer Cells,* 3, 263, 1991.
113. **Varticovski, L., Druker, B., Morrison, D., Cantley, L., and Roberts, T.,** The colony-stimulating factor-1 receptor associates with and activates phosphatidylinositol-3 kinase, *Nature,* 342, 699, 1989.
114. **Dadi, H. K., Ke, S., and Roifman, C. M.,** Interleukin 7 receptor mediates the activation of phosphatidylinositol 3-kinase in human B-cell precursors, *Biochem. Biophys. Res. Commun.,* 192, 459, 1993.
115. **Raffioni, S. and Bradshaw, R. A.,** Activation of phosphatidylinositol 3-kinase by epidermal growth factor, basic fibroblast growth factor, and nerve growth factor in PC12 pheochromocytoma cells, *Proc. Natl. Acad. Sci. U.S.A.,* 89, 9121, 1992.
116. **Hayashi, H., Kamohara, S., Nishioka, Y., Kanai, F., Miyare, N., Fukui, Y., Shibasaki, F., Takenawa, T., and Ebina, Y.,** Insulin treatment stimulates the tyrosine phosphorylation of the α-type 85-kDa subunit of phosphatidylinositol 3-kinase *in vivo,* *J. Biol. Chem.,* 267, 22575, 1992.
117. **Remillard, B., Petrillo, R., Maslinski, W., Tsudo, M., Strom, T. B., Cantley, L., and Varticovski, L.,** Interleukin-2 receptor regulates activation of phosphatidylinositol 3-kinase, *J. Biol. Chem.,* 266, 14167, 1991.
118. **Choi, K. Y., Kim, H. K., Lee, S. Y., Moon, K. H., Sim, S. S., Kim, J. W., Chung, H. K., and Rhee, S. G.,** Molecular cloning and expression of a complementary DNA for inositol 1,4,5-trisphosphate 3-kinase, *Science,* 248, 64, 1990.
119. **Cannizzaro, L. A., Skolnik, E. Y., Margolis, B., Croce, C. M., Schlesinger, J., and Huebner, K.,** The human gene encoding phosphatidylinositol-3 kinase associated p85 α is at chromosome region 5q12-13, *Cancer Res.,* 51, 3818, 1991.
120. **DelVecchio, R. L. and Pilch, P. F.,** Phosphatidylinositol 4-kinase is a component of glucose transporter (GLUT-4)-containing vesicles, *J. Biol. Chem.,* 266, 13278, 1991.
121. **Cochet, C., Filhol, O., Payrastre, B., Hunter, T., and Gill, G. N.,** Interaction between the epidermal growth factor receptor and phosphoinositide kinases, *J. Biol. Chem.,* 266, 637, 1991.
122. **Wilson, D. B., Bross, T. E., Sherman, W. R., Berger, R. A., and Majerus, P. W.,** Inositol cyclic phosphates are produced by cleavage of phosphatidylphosphoinositoids (polyphosphoinositides) with purified sheep seminal vesicle phospholipase C enzymes, *Proc. Natl. Acad. Sci. U.S.A.,* 82, 4013, 1985.

123. **Irvine, R. F., Letcher, A. J., Lander, D. J., and Berridge, M. J.,** Specificity of inositol phosphate-stimulated Ca^{2+} mobilization from Swiss-mouse 3T3 cells, *Biochem. J.,* 240, 301, 1986.

124. **Snyder, F.,** Biochemistry of platelet-activating factor: a unique class of biologically active phospholipids, *Proc. Soc. Exp. Biol. Med.,* 190, 125, 1989.

125. **Prescott, S. M., Zimmerman, G. A., and McIntyre, T. M.,** Platelet-activating factor, *J. Biol. Chem.,* 265, 17381, 1990.

126. **Chao, W. and Olson, M. S.,** Platelet-activating factor — receptors and signal transduction, *Biochem. J.,* 292, 617, 1993.

127. **Ye, R. D., Prossnitz, E. R., Zou, A. H., and Cochrane, C. G.,** Characterization of a human cDNA that encodes a functional receptor for platelet activating factor, *Biochem. Biophys. Res. Commun.,* 180, 105, 1991.

128. **Sugimoto, T., Tsuchimochi, H., McGregor, C. G. A., Mutoh, H., Shimizu, T., and Kurachi, Y.,** Molecular cloning and characterization of the platelet-activating factor receptor gene expressed in the human heart, *Biochem. Biophys. Res. Commun.,* 189, 617, 1992.

129. **Hwang, S.-B.,** Specific receptors of platelet-activating factor, receptor heterogeneity, and signal transduction mechanisms, *J. Lipid Metab.,* 2, 123, 1990.

130. **Homma, H. and Hanahan, D. J.,** Attenuation of platelet activating factor (PAF)-induced stimulation of rabbit platelet GTPase by phorbol ester, dibutyryl cAMP, and desensitization: concomitant effects on PAF receptor binding characteristics, *Arch. Biochem. Biophys.,* 262, 32, 1988.

131. **Schulam, P. G., Kuruvilla, A., Putcha, G., Mangus, L., Franklin-Johnson, J., and Shearer, W. T.,** Platelet-activating factor induces phospholipid turnover, calcium influx, arachidonic acid liberation, eicosanoid generation, and oncogene expression in a human B cell line, *J. Immunol.,* 146, 1642, 1991.

132. **Tripathi, Y. B., Lim, R. W., Fernandez-Gallardo, S., Kandala, J. C., Guntaka, R. V., and Shukla, S. D.,** Involvement of tyrosine kinase and protein kinase C in platelet-activating-factor-induced c-*fos* gene expression in A-431 cells, *Biochem. J.,* 286, 527, 1992.

133. **Gomez-Cambronero, J., Wang, E., Johnson, G., Huang, C. K., and Shaafi, R. I.,** Platelet-activating factor induces tyrosine phosphorylation in human neutrophils, *J. Biol. Chem.,* 266, 6240, 1991.

134. **Dhar, A. and Shukla, S. D.,** Involvement of pp60^{c-src} in platelet-activating factor-stimulated platelets. Evidence for translocation from cytosol to membrane, *J. Biol. Chem.,* 266, 18797, 1991.

135. **Martiny-Baron, G. and Scherer, G. F. E.,** Phospholipid-stimulated protein kinase in plants, *J. Biol. Chem.,* 264, 18052, 1989.

136. **Kuo, J. F., Schatzman, R. C., Turner, R. S., and Mazzei, G. J.,** Phospholipid-sensitive Ca^{2+}-dependent protein kinase: a major protein phosphorylation system, *Mol. Cell. Endocrinol.,* 35, 65, 1984.

137. **Ashendel, C. L.,** The phorbol ester receptor: a phospholipid-regulated protein kinase, *Biochim. Biophys. Acta,* 822, 219, 1985.

138. **Bell, R. M.,** Protein kinase C activation by diacylglycerol second messengers, *Cell,* 45, 631, 1986.

139. **Parker, P. J. and Ullrich, A.,** Protein kinase C, *J. Cell. Physiol.,* Suppl. 5, 53, 1987.

140. **Nishizuka, Y.,** Studies and perspectives of protein kinase C, *Science,* 233, 305, 1986.

141. **Blumberg, P. M.,** Protein kinase C as the receptor for the phorbol ester tumor promoters, *Cancer Res.,* 48, 1, 1988.

142. **Parker, P. J., Kour, G., Marais, R. M., Mitchell, F., Pears, C., Schaap, D., Stabel, S., and Webster, C.,** Protein kinase C — a family affair, *Mol. Cell. Endocrinol.,* 65, 1, 1989.

143. **Kikkawa, U., Kishimoto, A., and Nishizuka, Y.,** The protein kinase C family: heterogeneity and its implications, *Annu. Rev. Biochem.,* 58, 31, 1989.

144. **O'Brian, C. A. and Ward, N. E.,** Biology of the protein kinase C family, *Cancer Metast. Rev.,* 8, 199, 1989.

145. **Farago, A. and Nishizuka, Y.,** Protein kinase-C in transmembrane signalling, *FEBS Lett.,* 268, 350, 1990.

146. **Blumberg, P. M.,** Complexities of the protein kinase C pathway, *Mol. Carcinogenesis,* 4, 339, 1991.

147. **Azzi, A., Boscoboinik, D., and Hensey, C.,** The protein kinase C family, *Eur. J. Biochem.,* 208, 547, 1992.

148. **Nishizuka, Y.,** Intracellular signaling by hydrolysis of phospholipids and activation of protein kinase C, *Science,* 258, 607, 1992.

149. **Hug, H. and Sarre, T. F.,** Protein kinase C isoenzymes — divergence in signal transduction, *Biochem. J.,* 291, 329, 1993.

150. **Ogita, K., Miyamoto, S., Koide, H., Iwai, T., Oka, M., Ando, K., Kishimoto, A., Ikeda, K., Fukami, Y., and Nishizuka, Y.,** Protein kinase C in *Saccharomyces cerevisiae*: comparison with the mammalian enzyme, *Proc. Natl. Acad. Sci. U.S.A.*, 87, 5011, 1990.

151. **Schäfer, A., Bygrave, F., Matzenauer, S., and Marmé, D.,** Identification of a calcium- and phospholipid-dependent protein kinase in plant tissue, *FEBS Lett.*, 187, 25, 1985.

152. **Horn, F., Gschwendt, M., and Marks, F.,** Partial purification and characterization of the calcium-dependent and phospholipid-dependent protein kinase C from chick oviduct, *Eur. J. Biochem.*, 148, 533, 1985.

153. **Girard, P. R., Mazzei, G. J., and Kuo, J. F.,** Immunological quantitation of phospholipid/Ca^{2+}-dependent protein kinase and its fragments: tissue levels, subcellular distribution, and ontogenic changes in brain and heart, *J. Biol. Chem.*, 261, 370, 1986.

154. **Malkinson, A. M., Conway, K., Bartlett, S., Butley, M. S., and Conroy, C.,** Strain differences among inbred mice in protein kinase C activity, *Biochem. Biophys. Res. Commun.*, 122, 492, 1984.

155. **Makowske, M., Birnbaum, M. J., Ballester, R., and Rosen, O. M.,** A cDNA encoding protein kinase C identifies two species of mRNA in brain and GH$_3$ cells, *J. Biol. Chem.*, 261, 13389, 1986.

156. **Chida, K., Kato, N., Yamada, S., and Kuroki, T.,** Protein kinase C activities and bindings of a phorbol ester tumor promoter in 41 cell lines, *Biochem. Biophys. Res. Commun.*, 157, 1, 1988.

157. **Inoue, M., Kishimoto, A., Takai, Y., and Nishizuka, Y.,** Studies on a cyclin nucleotide-independent protein kinase and its proenzyme in mammalian tissues, *J. Biol. Chem.*, 252, 7610, 1977.

158. **Parker, P. J., Coussens, L., Totty, N., Rhee, L., Young, S., Chen, E., Stabel, S., Waterfield, M. D., and Ullrich, A.,** The complete primary structure of protein kinase C — the major phorbol ester receptor, *Science*, 233, 853, 1986.

159. **Mochly-Rosen, D. and Koshland, D. E., Jr.,** Domain structure and phosphorylation of protein kinase C, *J. Biol. Chem.*, 262, 2291, 1987.

160. **Nakamura, H., Kishi, Y., Pajares, M. A., and Rando, R. R.,** Structural basis of protein kinase C activation by tumor promoters, *Proc. Natl. Acad. Sci. U.S.A.*, 86, 9672, 1989.

161. **Ishikawa, F., Takaku, F., Nagao, M., and Sugimura, T.,** Cysteine-rich regions conserved in amino-terminal halves of *raf* gene family products and protein kinase C, *Jpn. J. Cancer Res.*, 77, 1183, 1986.

162. **Koenen, M., Sippel, A. E., Trachmann, C., and Bister, K.,** Primary structure of the chicken c-*mil* protein: identification of domains shared or absent from the retroviral v-*mil* protein, *Oncogene*, 2, 179, 1988.

163. **Rosenthal, A., Rhee, L., Yadegari, R., Paro, R., Ullrich, A., and Goeddel, D. V.,** Structure and nucleotide sequence of a *Drosophila melanogaster* protein kinase C gene, *EMBO J.*, 6, 433, 1987.

164. **Coussens, L., Parker, P. J., Rhee, L., Yang-Feng, T. L., Chen, E., Waterfield, M. D., Francke, U., and Ullrich, A.,** Multiple, distinct forms of bovine and human protein kinase C suggest diversity in cellular signaling pathways, *Science*, 233, 859, 1986.

165. **Housey, G. M., O'Brian, C. A., Johnson, M. D., Kirschmeier, P., and Weinstein, I. B.,** Isolation of cDNA clones encoding protein kinase C: evidence for a protein kinase C-related gene family, *Proc. Natl. Acad. Sci. U.S.A.*, 84, 1065, 1987.

166. **Nishizuka, Y.,** The molecular heterogeneity of protein kinase C and its implications for cellular regulation, *Nature*, 334, 661, 1988.

167. **Huang, K.-P., Huang, F. L., Nakabayashi, H., and Yoshida, Y.,** Expression and function of protein kinase C isozymes, *Acta Endocrinol.*, 121, 307, 1989.

168. **Woodgett, J. R. and Hunter, T.,** Isolation and characterization of two distinct forms of protein kinase C, *J. Biol. Chem.*, 262, 4836, 1987.

169. **Ono, Y., Kurokawa, T., Kawahara, K., Nishimura, O., Marumoto, R., Igarashi, K., Sugino, Y., Kikkawa, U., Ogita, K., and Nishizuka, Y.,** Cloning of rat brain protein kinase C complementary DNA, *FEBS Lett.*, 203, 111, 1986.

170. **Ono, Y., Kurokawa, T., Fujii, T., Kawahara, K., Igarashi, K., Kikkawa, U., Ogita, K., and Nishizuka, Y.,** Two types of complementary DNAs of rat brain protein kinase C, *FEBS Lett.*, 206, 347, 1986.

171. **Knopf, J. L., Lee, M.-H., Sutzman, L. A., Kriz, R. W., Loomis, C. R., Hewick, R. M., and Bell, R. M.,** Cloning and expression of multiple protein kinase C cDNAs, *Cell*, 46, 491, 1986.

172. **Huang, K.-P., Nakabayashi, H., and Huang, F. L.,** Isozymic forms of rat brain Ca^{2+}-activated and phospholipid-dependent protein kinase, *Proc. Natl Acad. Sci. U.S.A.*, 83, 8535, 1986.

173. **Pelosin, J.-M., Vilgrain, I., and Chambaz, E. M.,** A single form of protein kinase C is expressed in bovine adrenocortical tissue, as compared to four chromatographically resolved isozymes in rat brain, *Biochem. Biophys. Res. Commun.*, 147, 382, 1987.

174. **Ono, Y., Fujii, T., Ogita, K., Kikkawa, U., Igarashi, K., and Nishizuka, Y.,** The structure, expression, and properties of additional members of the protein kinase C family, *J. Biol. Chem.,* 263, 6927, 1988.

175. **Ono, Y., Kikkawa, U., Ogita, K., Fujii, T., Kurokawa, T., Asaoka, Y., Sekiguchi, K., Ase, K., Igarashi, K., and Nishizuka, Y.,** Expression and properties of two types of protein kinase C: alternative splicing from a single gene, *Science,* 236, 1116, 1987.

176. **Brandt, S. J., Niedel, J. E., Bell, R. M., and Young, W. S., III,** Distinct patterns of expression of different protein kinase C mRNAs in rat tissues, *Cell,* 49, 57, 1987.

177. **Ohno, S., Kawasaki, H., Imajoh, S., Suzuki, K., Inagaki, M., Yokokura, M., Sakoh, T., and Hidaka, H.,** Tissue-specific expression of three distinct types of rabbit protein kinase C, *Nature,* 325, 161, 1987.

178. **Ohno, S., Akita, Y., Konno, Y., Imajoh, S., and Suzuki, K.,** A novel phorbol ester receptor/protein kinase, nPKC, distantly related to the protein kinase C family, *Cell,* 53, 731, 1988.

179. **Akita, Y., Ohno, S., Konno, Y., Yano, A., and Suzuki, K.,** Expression and properties of two distinct classes of the phorbol ester receptor family, four conventional protein kinase C types, and a novel protein kinase C, *J. Biol. Chem.,* 265, 354, 1990.

180. **Bacher, N., Zisman, Y., Berent, E., and Livneh, E.,** Isolation and characterization of PKC-L, a new member of the protein kinase C-related gene family specifically expressed in lung, skin, and heart, *Mol. Cell. Biol.,* 11, 126, 1991.

181. **Coussens, L., Rhee, L., Parker, P. J., and Ullrich, A.,** Alternative splicing increases the diversity of the human protein kinase C family, *DNA,* 6, 389, 1987.

182. **Guadagno, S. N., Borner, C., and Weinstein, I. B.,** Altered regulation of a major substrate of protein kinase C in rat 6 fibroblasts overproducing PKC βI, *J. Biol. Chem.,* 267, 2697, 1992.

183. **Blackshear, P. J.,** The MARCKS family of cellular protein kinase C substrates, *J. Biol. Chem.,* 268, 1501, 1993.

184. **McCaffrey, P. G. and Rosner, M. R.,** Characterization of protein kinase C from normal and transformed cultured murine fibroblasts, *Biochem. Biophys. Res. Commun.,* 146, 140, 1987.

185. **Jaken, S. and Kiley, S. C.,** Purification and characterization of three types of protein kinase C from rabbit brain cytosol, *Proc. Natl. Acad. Sci. U.S.A.,* 84, 4418, 1987.

186. **Sawamura, S., Ase, K., Berry, N., Kikkawa, U., McCaffrey, P. G., Minowada, J., and Nishizuka, Y.,** Expression of protein kinase C subspecies in human leukemia-lymphoma cell lines, *FEBS Lett.,* 247, 353, 1989.

187. **Cochet, C., Gill, G. N., Meisenhelder, J., Cooper, J. A., and Hunter, T.,** C-kinase phosphorylates the epidermal growth factor receptor and reduces its epidermal growth factor-stimulated tyrosine protein kinase activity, *J. Biol. Chem.,* 259, 2553, 1984.

188. **Witters, L. A., Vater, C. A., and Lienhard, G. E.,** Phosphorylation of the glucose transporter *in vitro* and *in vivo* by protein kinase C, *Nature,* 315, 777, 1985.

189. **McCaffrey, P. G. and Rosner, M. R.,** Growth state-dependent regulation of protein kinase C in normal and transformed murine cells, *Cancer Res.,* 47, 1081, 1987.

190. **Gomez, M. L., Medrano, E. E., Cafferatta, E. G. A., and Tellez-Iñon, M. T.,** Protein kinase C is differentially regulated by thrombin, insulin, and epidermal growth factor in human mammary tumor cells, *Exp. Cell Res.,* 175, 74, 1988.

191. **Bazzi, M. D. and Nelsestuen, G. L.,** Properties of membrane-inserted protein kinase C, *Biochemistry,* 27, 7589, 1988.

192. **Miloszewska, J., Trawicki, W., Janik, P., Moraczewski, J., Przybyszewska, M., and Szaniawska, B.,** Protein kinase C translocation in relation to proliferative state of C3H 20T1/2 cells, *FEBS Lett.,* 206, 283, 1986.

193. **Patel, J. and Kassis, S.,** Concanavalin A prevents phorbol-mediated redistribution of protein kinase C and β-adrenergic receptors in rat glioma C6 cells, *Biochem. Biophys. Res. Commun.,* 144, 1265, 1987.

194. **White, J. R., Pluznik, D. H., Ishizaka, K., and Ishizaka, T.,** Antigen-induced increase in protein kinase C activity in plasma membrane of mast cells, *Proc. Natl. Acad. Sci. U.S.A.,* 82, 8193, 1985.

195. **Masmoudi, A., Labourdette, G., Mersel, M., Huang, F. L., Huang, K.-P., Vincendon, G., and Malviya, A. N.,** Protein kinase C located in rat liver nuclei. Partial purification and biochemical and immunochemical characterization, *J. Biol. Chem.,* 264, 1172, 1989.

196. **Buckley, A. R., Crowe, P. D., and Russell, D. H.,** Rapid activation of protein kinase C in isolated rat liver nuclei by prolactin, a known hepatic mitogen, *Proc. Natl. Acad. Sci. U.S.A.,* 85, 8649, 1988.

197. **Mochly-Rosen, D., Khaner, H., Lopez, J., and Smith, B. L.,** Intracellular receptors for activated protein kinase C. Identification of a binding site for the enzyme, *J. Biol. Chem.,* 266, 14866, 1991.

198. **Snoek, G. T., Boonstra, J., Ponek, M., and de Laat, S. W.,** Phorbol ester binding and protein kinase C activity in normal and transformed human keratinocytes, *Exp. Cell Res.,* 172, 146, 1987.

199. **Chauhan, V. P. S. and Brockerhoff, H.,** Phosphatidylinositol-4,5-bisphosphate may antecede diacylglycerol as activator of protein kinase C, *Biochem. Biophys. Res. Commun.,* 155, 18, 1988.

200. **Huang, C.-K., Devanney, J. F., and Kanaho, Y.,** Regulation of membrane associated protein kinase C activity by guanine nucleotide in rabbit peritoneal neutrophils, *Biochem. Biophys. Res. Commun.,* 142, 242, 1987.

201. **Wolf, M., Cuatrecasas, P., and Sahyoun, N.,** Interaction of protein kinase C with membranes is regulated by Ca^{2+}, phorbol esters, and ATP, *J. Biol. Chem.,* 260, 15718, 1985.

202. **Bazzi, M. D. and Nelsestuen, G. L.,** Constitutive activity of membrane-inserted protein kinase C, *Biochem. Biophys. Res. Commun.,* 152, 336, 1988.

203. **Spina, A., Chiosi, E., and Illiano, G.,** Protein kinase C activities are increased in rat thyroid epithelial cells expressing v-*ras* genes, *Biochem. Biophys. Res. Commun.,* 157, 1093, 1988.

204. **Phillips, J. L., Boldt, D. H., and Harper, J.,** Iron-transferrin-induced increase in protein kinase C activity in CCRF-CEM cells, *J. Cell. Physiol.,* 132, 349, 1987.

205. **O'Brian, C. A., Arthur, W. L., and Weinstein, I. B.,** The activation of protein kinase C by the polyphosphoinositides phosphatidylinositol 4,5-diphosphate and phosphatidylinositol 4-monophosphate, *FEBS Lett.,* 214, 339, 1987.

206. **Hannun, Y. A., Loomis, C. R., Merrill, A. H., Jr., and Bell, R. M.,** Sphingosine inhibition of protein kinase C activity and of phorbol dibutyrate binding *in vitro* and in human platelets, *J. Biol. Chem.,* 261, 12604, 1986.

207. **Murakami, K. and Routtenberg, A.,** Direct activation of purified protein kinase C by unsaturated fatty acids (oleate and arachidonate) in the absence of phospholipids and Ca^{2+}, *FEBS Lett.,* 192, 189, 1985.

208. **Eyster, K. M.,** An endogenous inhibitor of protein kinase C in the pseudopregnant rat ovary, *Biochem. Biophys. Res. Commun.,* 168, 609, 1990.

209. **McDonald, J. R., Gröschel-Stewart, U., and Walsh, M. P.,** Properties and distribution of the protein inhibitor (M_r 17,000) of protein kinase C, *Biochem. J.,* 242, 695, 1987.

210. **Pearson, J. D., DeWald, D.B., Mathews, W.R., Mozier, N.M., Zürcher-Neely, H.A., Heinrikson, R.L., Morris, M.A., McCubbin, W.D., McDonald, J.R., Fraser, E.D., Vogel, H.J., Kay, C.M., and Walsh, M.P.,** Amino acid sequence and characterization of a protein inhibitor of protein kinase C, *J. Biol. Chem.,* 265, 4583, 1990.

211. **Kobayashi, E., Nakano, H., Morimoto, M., and Tamaoki, T.,** Calphostin C (UCN-1028C), a novel microbial compound, is a highly potent and specific inhibitor of protein kinase C, *Biochem. Biophys. Res. Commun.,* 159, 548, 1989.

212. **Yamada, K., Iwahashi, K., and Kase, H.,** K252a, a new inhibitor of protein kinase C, concomitantly inhibits 40K protein phosphorylation and serotonin secretion in phorbol ester-stimulated platelets, *Biochem. Biophys. Res. Commun.,* 144, 35, 1987.

213. **Loomis, C. R. and Bell, R. M.,** Sangivamycin, a nucleoside analogue, is a potent inhibitor of protein kinase C, *J. Biol. Chem.,* 263, 1682, 1988.

214. **Mills, G. B., Girard, P., Grinstein, S., and Gelfand, E. W.,** Interleukin-2 induces proliferation of T lymphocyte mutants lacking protein kinase C, *Cell,* 55, 91, 1988.

215. **Huang, K.-P., Chan, K.-F. J., Singh, T. J., Nakabayashi, H., and Huang, F. L.,** Autophosphorylation of rat brain Ca^{2+}-activated and phospholipid-dependent protein kinase, *J. Biol. Chem.,* 261, 12134, 1986.

216. **Kase, H., Iwahashi, K., Nakanishi, S., Matsuda, Y., Yamada, K., Takahashi, M., Murakata, C., Sato, A., and Kaneko, M.,** K-252 compounds, novel and potent inhibitors of protein kinase C and cyclic nucleotide-dependent protein kinases, *Biochem. Biophys. Res. Commun.,* 142, 436, 1987.

217. **Vila, J., Walker, J. M., Itarte, E., Weber, M. J., and Sando, J. J.,** Phosphorylation of protein kinase C by casein kinase-1, *FEBS Lett.,* 255, 205, 1989.

218. **Chida, K., Kato, N., and Kuroki, T.,** Down regulation of phorbol diester receptors by proteolytic degradation of protein kinase C in a cultured cell line of fetal rat skin keratinocytes, *J. Biol. Chem.,* 261, 13013, 1986.

219. **Ohno, S., Konno, Y., Akita, Y., Yano, A., and Suzuki, K.,** A point mutation at the putative ATP-binding site of protein kinase Cα abolishes the kinase activity and renders it down-regulation-insensitive. A molecular link between autophosphorylation and down-regulation, *J. Biol. Chem.,* 265, 6296, 1990.

220. **Klemm, D. J. and Elias, L.,** A distinctive phospholipid-stimulated protein kinase of normal and malignant murine hemopoietic cells, *J. Biol. Chem.,* 262, 7580, 1987.

221. **Klemm, D. J. and Elias, L.,** Purification and assay of a phosphatidylglycerol-stimulated protein kinase from murine leukemic cells and its perturbation in response to IL-3 and PMA treatment, *Exp. Hematol.,* 16, 855, 1988.

222. **Megidish, T. and Mazurek, N.,** A mutant protein kinase C that can transform fibroblasts, *Nature,* 342, 807, 1989.

223. **Rozengurt, E., Rodriguez-Pena, A., Coombs, M., and Sinnett-Smith, J.,** Diacylglycerol stimulates DNA synthesis and cell division in mouse 3T3 cells: role of Ca^{2+}-sensitive phospholipid-dependent protein kinase, *Proc. Natl. Acad. Sci. U.S.A.,* 81, 5748, 1984.

224. **Whitfield, J. F., Durkin, J. P., Franks, D. J., Kleine, L. P., Raptis, L., Rixon, R. H., Sikorska, M., and Walker, P. R.,** Calcium, cyclic AMP and protein kinase C — partners in mitogenesis, *Cancer Metast. Rev.,* 5, 105, 1987.

225. **Berry, N., Ase, K., Kishimoto, A., and Nishizuka, Y.,** Activation of resting human T cells requires prolonged stimulation of protein kinase C, *Proc. Natl. Acad. Sci. U.S.A.,* 87, 2294, 1990.

226. **Gordon, E. A. and Carney, D. H.,** Thrombin receptor occupancy initiates cell proliferation in the presence of phorbol myristic acetate, *Biochem. Biophys. Res. Commun.,* 141, 650, 1986.

227. **Matuoka, K., Fukami, K., Nakanishi, O., Kawai, S., and Takenawa, T.,** Mitogenesis in response to PDGF and bombesin abolished by microinjection of antibody to PIP_2, *Science,* 239, 640, 1988.

228. **Moolenaar, W. H.,** G-protein-coupled receptors, phosphoinositide hydrolysis, and cell proliferation, *Cell Growth Differ.,* 2, 359, 1991.

229. **Sylvia, V. L., Norman, J. O., Curtin, G. M., and Busbee, D. L.,** Monoclonal antibody that blocks phosphoinositide-dependent activation of mouse tumor DNA polymerase alpha, *Biochem. Biophys. Res. Commun.,* 141, 60, 1986.

230. **Donnelly, T. E., Sittler, R., and Scholar, E. M.,** Relationship between membrane-bound protein kinase C activity and calcium-dependent proliferation of BALB/c 3T3 cells, *Biochem. Biophys. Res. Commun.,* 126, 741, 1985.

231. **Moody, T. W., Murphy, A., Mahmoud, S., and Fiskum, G.,** Bombesin-like peptides elevate cytosolic calcium in small cell lung cancer cells, *Biochem. Biophys. Res. Commun.,* 147, 189, 1987.

232. **Nabika, T., Velletri, P. A., Lovenberg, W., and Beaven, M. A.,** Increase in cytosolic calcium and phosphoinositide metabolism induced by angiotensin II and (Arg)vasopresin in vascular smooth muscle cells, *J. Biol. Chem.,* 260, 4661, 1985.

233. **Henne, V., Piiper, A., and Sölling, H.-D.,** Inositol 1,4,5-trisphosphate and 5′-GTP induce calcium release from different intracellular pools, *FEBS Lett.,* 218, 153, 1987.

234. **Wolf, M., Le Vine, H., III, May, W. S., Jr., Cuatrecasas, P., and Sahyoun, N.,** A model for intracellular translocation of protein kinase C involving synergism between Ca^{2+} and phorbol esters, *Nature,* 317, 546, 1985.

235. **May, W. S., Jr., Sahyoun, N., Wolf, M., and Cuatrecasas, P.,** Role of intracellular calcium mobilization in the regulation of protein kinase C-mediated membrane processes, *Nature,* 317, 549, 1985.

236. **Boynton, A. L., Kleine, L. P., Whitfield, J. F., and Bossi, D.,** Involvement of the Ca^{2+}/phospholipid-dependent protein kinase in the G_1 transit of T51B rat liver epithelial cells, *Exp. Cell Res.,* 160, 197, 1985.

237. **Harris, K. M., Kongsamut, S., and Miller, R. J.,** Protein kinase C mediated regulation of calcium channels in PC-12 pheochromocytoma cells, *Biochem. Biophys. Res. Commun.,* 134, 1298, 1986.

238. **Whitaker, M. and Irvine, R. F.,** Inositol 1,4,5-trisphosphate microinjection activates sea urchin eggs, *Nature,* 312, 636, 1984.

239. **Aberdam, E. and Dekel, N.,** Activators of protein kinase C stimulate meiotic maturation of rat oocytes, *Biochem. Biophys. Res. Commun.,* 132, 570, 1985.

240. **Kuo, J. F., Andersson, R. G. G., Wise, B. G., Mackerlova, L., Salomonsson, I., Brackett, N. L., Katoh, N., Shoji, M., and Wrenn, R. W.,** Calcium-dependent protein kinase: widespread occurrence in various tissues and phyla of the animal kingdom and comparison of the effects of phospholipid, calmodulin, and trifluoperizine, *Proc. Natl. Acad. Sci. U.S.A.,* 77, 7039, 1980.

241. **Hill, T. D., Campos-Gonzalez, R., Kindmark, H., and Boynton, A. L.,** Inhibition of inositol trisphosphate-stimulated calcium mobilization by calmodulin antagonists in rat liver epithelial cells, *J. Biol. Chem.,* 263, 16479, 1988.

242. **Leffert, H. L.,** Monovalent cations, cell proliferation and cancer: an overview, in *Ions, Cell Proliferation, and Cancer,* Boynton, A. L., McKeehan, W. L., and Whitfield, J. F., Eds., Academic Press, New York, 1982, 93.

243. **Rozengurt, E.,** Monovalent ion fluxes, cyclic nucleotides and the stimulation of DNA synthesis in quiescent cells, in *Ions, Cell Proliferation, and Cancer,* Boynton, A. L., McKeehan, W. L., and Whitfield, J. F., Eds., Academic Press, New York, 1982, 259.

244. **Rossoff, P. M. and Cantley, L. C.,** Lipopolysaccharide and phorbol esters induce differentiation but have opposite effects on phosphatidylinositol turnover and Ca^{2+} mobilization in 70Z/3 pre-B lymphocytes, *J. Biol. Chem.,* 260, 9209, 1985.

245. **Vara, F., Schneider, J. A., and Rozengurt, E.,** Ionic responses rapidly elicited by activation of protein kinase C in quiescent Swiss 3T3 cells, *Proc. Natl. Acad. Sci. U.S.A.,* 82, 2384, 1985.

246. **Vicentini, L. M. and Villereal, M. L.,** Activation of Na^+/H^+ exchange in cultured fibroblasts: synergism and antagonism between phorbol ester, Ca^{2+} ionophore, and growth factors, *Proc. Natl. Acad. Sci. U.S.A.,* 82, 8053, 1985.

247. **Macara, I. G.,** Activation of $45Ca^{2+}$ influx and $^{22}Na^+/H^+$ exchange by epidermal growth factor and vanadate in A431 cells is independent of phosphatidylinositol turnover and is inhibited by phorbol ester and diacylglycerol, *J. Biol. Chem.,* 261, 9321, 1986.

248. **Magnaldo, I., L'Allemain, G., Chambard, J. C., Moenner, M., Barritault, D., and Pouysségur, J.,** The mitogenic signaling pathway of fibroblast growth factor is not mediated through polyphosphoinositide hydrolysis and protein kinase C activation in hamster fibroblasts, *J. Biol. Chem.,* 261, 16916, 1986.

249. **Mustelin, T., Pösö, H., Lapinjoki, S. P., Gynther, J., and Andersson, L. C.,** Growth signal transduction: rapid activation of covalently bound ornithine decarboxylase during phosphatidylinositol breakdown, *Cell,* 49, 171, 1987.

250. **Kleine, L. P., Whitfield, J. F., and Boynton, A. L.,** Ca^{2+}-dependent cell surface protein phosphorylation may be involved in the initiation of DNA synthesis, *J. Cell. Physiol.,* 129, 303, 1986.

251. **Yoshimasa, T., Sibley, D. R., Bouvier, M., Lefkowitz, R. J., and Caron, M. G.,** Cross-talk between cellular signalling pathways suggested by phorbol ester-induced adenylate cyclase phosphorylation, *Nature,* 327, 67, 1987.

252. **Parker, P. J., Katan, M., Waterfield, M. D., and Leader, D. P.,** The phosphorylation of eukaryotic ribosomal protein S6 by protein kinase C, *Eur. J. Biochem.,* 148, 579, 1985.

253. **Khanna, N. C., Tokuda, M., Chong, S. M., and Waisman, D. M.,** Phosphorylation of p36 *in vitro* by protein kinase C, *Biochem. Biophys. Res. Commun.,* 137, 397, 1986.

254. **Gould, K. L., Woodgett, J. R., Isacke, C. M., and Hunter, T.,** The protein-tyrosine kinase substrate p36 is also a substrate for protein kinase C *in vitro* and *in vivo,* *Mol. Cell. Biol.,* 6, 2738, 1986.

255. **Johnsson, N., Van, P. N., Sölling, H.-D., and Weber, K.,** Functionally distinct serine phosphorylation sites of p36, the cellular substrate of retroviral protein kinase; differential inhibition of reassociation with p11, *EMBO J.,* 5, 3455, 1986.

256. **Albert, K. A., Walaas, S. I., Wang, J. K.-T., and Greengard, P.,** Widespread occurrence of "87 kDa," a major specific substrate for protein kinase C, *Proc. Natl. Acad. Sci. U.S.A.,* 83, 2822, 1986.

257. **Werth, D. K., Wiedel, J. E., and Pastan, I.,** Vinculin, a cytoskeletal substrate of protein kinase C, *J. Biol. Chem.,* 258, 11423, 1983.

258. **Werth, D. K. and Pastan, I.,** Vinculin phosphorylation in response to Ca^{2+} and phorbol esters in intact cells, *J. Biol. Chem.,* 259, 5264, 1984.

259. **Akiyama, T., Nishida, E., Ishida, J., Saji, N., Ogawara, H., Hoshi, M., Miyata, Y., and Sakai, H.,** Purified protein kinase C phosphorylates microtubule-associated protein 2, *J. Biol. Chem.,* 261, 15648, 1986.

260. **Lassing, I. and Lindberg, U.,** Evidence that the phosphatidylinositol cycle is linked to cell motility, *Exp. Cell Res.,* 174, 1, 1988.

261. **Kimura, K., Katoh, N., Sakurada, K., and Kubo, S.,** Phosphorylation of high mobility group 1 protein by phospholipid-sensitive Ca^{2+}-dependent protein kinase from pig testis, *Biochem. J.,* 227, 271, 1985.

262. **Huang, L. F., Cooper, R. H., Yau, P., Bradbury, E. M., and Chuang, R. Y.,** Protein kinase C phosphorylates leukemia RNA polymerase II, *Biochem. Biophys. Res. Commun.,* 145, 1376, 1987.

263. **Jeng, A. Y., Srivastava, S. K., Lacal, J. C., and Blumberg, P. M.,** Phosphorylation of *ras* oncogene product by protein kinase C, *Biochem. Biophys. Res. Commun.,* 145, 782, 1987.

264. **Vila, J. and Weber, M. J.,** Mitogen-stimulated tyrosine phosphorylation of a 42-kD cellular protein: evidence for a protein kinase-C requirement, *J. Cell. Physiol.,* 135, 285, 1988.

265. **Sozeri, O., Vollmer, K., Liyanage, M., Frith, D., Kour, G., Mark, G. E., and Stabel, S.,** Activation of the c-Raf protein kinase by protein kinase C phosphorylation, *Oncogene,* 7, 2259, 1992.

266. **O'Flaherty, J. T., Redman, J. F., and Jacobson, D. P.,** Protein kinase C regulates leukotriene B$_4$ receptors in human neutrophils, *FEBS Lett.,* 206, 279, 1986.

267. **Andersson, T., Schlegel, W., Monod, A., Krause, K.-H., Stendhal, O., and Lew, D. P.,** Leukotriene B$_4$ stimulation of phagocytes results in the formation of inositol 1,4,5-trisphosphate, *Biochem. J.,* 240, 333, 1986.

268. **Chambard, J. C., Paris, S., L'Allemain, G., and Pouysségur, J.,** Two growth factor signalling pathways in fibroblasts distinguished by pertussis toxin, *Nature,* 326, 800, 1987.

269. **Moenner, M., Magnaldo, I., L'Allemain, G., Barritault, D., and Pouysségur, J.,** Early and late mitogenic events induced by FGF on bovine epithelial lens cells are not triggered by hydrolysis of polyphosphoinositides, *Biochem. Biophys. Res. Commun.,* 146, 32, 1987.

270. **Poyner, D. R., Hawkins, P. T., Benton, H. P., and Hanley, M. R.,** Changes in inositol lipids and phosphates after stimulation of the MAS-transfected NG115-401L-C3 cell line by mitogenic and non-mitogenic stimuli, **Biochem. J.,** 271, 605, 1990.

271. **Carney, D. H., Scott, D. L., Gordon, E. A., and LaBelle, E. F.,** Phosphoinositides in mitogenesis: neomycin inhibits thrombin-stimulated phosphoinositide turnover and initiation of cell proliferation, *Cell,* 42, 479, 1985.

272. **Fisher, G. J., Bakshian, S., and Baldassare, J. J.,** Activation of human platelets by ADP causes a rapid rise in cytosolic free calcium without hydrolysis of phosphatidylinositol-4,5-bisphosphate, *Biochem. Biophys. Res. Commun.,* 129, 958, 1985.

273. **Sasaki, T. and Hasegawa-Sasaki, H.,** Breakdown of phosphatidylinositol 4,5-bisphosphate in a T-cell leukaemia line stimulated by phytohaemagglutinin is not dependent on Ca^{2+} mobilization, *Biochem. J.,* 227, 971, 1985.

274. **Bradford, P. G. and Autieri, M.,** Increased expression of the inositol 1,4,5-trisphosphate receptor in human leukaemia (HL-60) cells differentiated with retinoic acid or dimethylsulfoxide, *Biochem. J.,* 280, 205, 1991.

275. **Makishima, M., Honma, Y., Hozumi, M., Sampi, K., Hattori, M., Umezawa, K., and Motoyoshi, K.,** Effects of inhibitors of protein tyrosine kinase activity and/or phosphatidylinositol turnover on differentiation of some human myelomonocytic leukemia cells, *Leukemia Res.,* 15, 701, 1991.

276. **Aihara, H., Asaoka, Y., Yoshida, K., and Nishizuka, Y.,** Sustained activation of protein kinase C is essential to HL-60 cell differentiation to macrophage, *Proc. Natl. Acad. Sci. U.S.A.,* 88, 11062, 1991.

277. **Tang, W., Ziboh, V. A., Isseroff, R., and Martinez, D.,** Turnover of inositol phospholipids in cultured murine keratinocytes: possible involvement of inositol triphosphate in cellular differentiation, *J. Invest. Dermatol.,* 90, 37, 1988.

278. **Nishizawa, N., Okano, Y., Chatani, Y., Amano, F., Tanaka, E., Nomoto, H., Nozawa, Y., and Kohno, M.,** Mitogenic signaling pathways of growth factors can be distinguished by the involvement of pertussis toxin-sensitive guanosine triphosphate-binding protein and of protein kinase C, *Cell Regul.,* 1, 747, 1990.

279. **Leeb-Lundberg, L. M. F., Cotecchia, S., Lomasney, J. W., DeBernardis, J. F., Lefkowitz, R. J., and Caron, M. G.,** Phorbol esters promote alpha$_1$-adrenergic receptor phosphorylation and receptor uncoupling from inositol phospholipid metabolism, *Proc. Natl. Acad. Sci. U.S.A.,* 82, 5651, 1985.

280. **Warden, C. H. and Friedkin, M.,** Regulation of phosphatidylcholine biosynthesis by mitogenic growth factors, *Biochim. Biophys. Acta,* 792, 270, 1984.

281. **Warden, C. H. and Friedkin, M.,** Regulation of choline kinase activity and phosphatidylcholine biosynthesis by mitogenic growth factors in 3T3 fibroblasts, *J. Biol. Chem.,* 260, 6006, 1985.

282. **Besterman, J. M., Duronio, V., and Cuatrecasas, P.,** Rapid formation of diacylglycerol from phosphatidylcholine: a pathway for generation of a second messenger, *Proc. Natl. Acad. Sci. U.S.A.,* 83, 6785, 1986.

283. **Besterman, J. M., Watson, S. P., and Cuatrecasas, P.,** Lack of association of epidermal growth factor-, insulin-, and serum-induced mitogenesis with stimulation of phosphoinositide degradation in BALB/c 3T3 fibroblasts, *J. Biol. Chem.,* 261, 723, 1986.

284. **L'Allemain, G. and Pouysségur, J.,** EGF and insulin action in fibroblasts. Evidence that phosphoinositide hydrolysis is not an essential mitogenic pathway, *FEBS Lett.,* 197, 344, 1986.

285. **Taylor, D., Uhing, R. J., Blackmore, P. F., Prpic, V., and Exton, J. H.,** Insulin and epidermal growth factor do not affect phosphoinositide metabolism in rat liver plasma membranes and hepatocytes, *J. Biol. Chem.,* 260, 2011, 1985.

286. **Vaartjes, W. J., de Haas, C. G. M., and van den Bergh, S. G.,** Phorbol esters, but not epidermal growth factor or insulin, rapidly decrease soluble protein kinase C activity in rat hepatocytes, *Biochem. Biophys. Res. Commun.,* 138, 1328, 1986.

287. **Thompson, D. M., Cochet, C., Chambaz, E. M., and Gill, G. N.,** Separation and characterization of a phosphatidylinositol kinase activity that co-purifies with the epidermal growth factor receptor, *J. Biol. Chem.,* 260, 8824, 1985.

288. **Olson, J. W.,** Enhanced phosphatidylinositol kinase activity is associated with early stages of hepatocarcinogensis and hepatocellular carcinoma, *Biochem. Biophys. Res. Commun.,* 132, 969, 1985.

289. **Sharoni, Y., Teuerstein, I., and Levy, J.,** Phosphoinositide phosphorylation precedes growth in rat mammary tumors, *Biochem. Biophys. Res. Commun.,* 134, 876, 1986.

290. **Jackowski, S., Rettenmier, C. W., Sherr, C. J., and Rock, C. O.,** A guanine nucleotide-dependent phosphatidylinositol 4,5-diphosphate phospholipase C in cells transformed by the v-*fms* and v-*fes* oncogenes, *J. Biol. Chem.,* 261, 4978, 1986.

291. **Simek, S. L., Kligman, D., Patel, J., and Colburn, N. H.,** Differential expression of an 80-kDa protein kinase C substrate in preneoplastic and neoplastic mouse JB6 cells, *Proc. Natl. Acad. Sci. U.S.A.,* 86, 7410, 1989.

292. **Persons, D. A., Wilkison, W. O., Bell, R. M., and Finn, O. J.,** Altered growth regulation and enhanced tumorigenicity of NIH 3T3 fibroblasts transfected with protein kinase C-I cDNA, *Cell,* 52, 447, 1988.

293. **Miñana, M.-D., Felipo, V., and Grisolía, S.,** Inhibition of protein kinase C induces differentiation in Neuro-2a cells, *Proc. Natl. Acad. Sci. U.S.A.,* 87, 4335, 1990.

294. **Wolfson, M., Aboud, M., Ofir, R., Weinstein, Y., and Segal, S.,** Modulation of protein kinase C and Ca^{2+} lipid-independent protein kinase in lymphoma induced by Moloney murine leukemia virus in BALB/c mice, *Int. J. Cancer,* 37, 589, 1986.

295. **Borner, C., Wyss, R., Regazzi, R., Eppenberger, U., and Fabbro, D.,** Immunological quantitation of phospholipid/Ca^{2+}-dependent protein kinase of human mammary carcinoma cells: inverse relationship to estrogen receptors, *Int. J. Cancer,* 40, 344, 1987.

296. **O'Brian, C. A., Vogel, V. G., Singletary, S. E., and Ward, N. E.,** Elevated protein kinase C expression in human breast tumor biopsies relative to normal breast tissue, *Cancer Res.,* 49, 3215, 1989.

297. **Guillem, J. G., O'Brian, C. A., Fitzer, C. J., Forde, K. A., LoGerfo, P., Treat, M., and Weinstein, I. B.,** Altered levels of protein kinase C and Ca^{2+}-dependent protein kinases in human colon carcinomas, *Cancer Res.,* 47, 2036, 1987.

298. **Hashimoto, Y., Chida, K., Huang, M., Katayama, M., Nishihira, T., and Kuroki, T.,** Levels of protein kinase C activity in human gastrointestinal cancers, *Biochem. Biophys. Res. Commun.,* 163, 406, 1989.

299. **Linnenbach, A. J., Huebner, K., Reddy, E. P., Herlyn, M., Parmiter, A. H., Nowell, P. C., and Koprowski, H.,** Structural alteration in the *MYB* protooncogene and deletion within the gene encoding alpha-type protein kinase C in human melanoma cell lines, *Proc. Natl. Acad. Sci. U.S.A.,* 85, 74, 1988.

300. **Becker, D., Beebe, S. J., and Herlyn, M.,** Differential expression of protein kinase C and cAMP-dependent protein kinase in normal human melanocytes and malignant melanomas, *Oncogene,* 5, 1133, 1990.

301. **Gopalakrishna, R. and Barsky, S. H.,** Tumor promoter-induced membrane-bound protein kinase C regulates hematogenous metastasis, *Proc. Natl. Acad. Sci. U.S.A.,* 85, 612, 1988.

302. **Yamauchi, P. S. and Purich, D. L.,** Modulation of microtubule assembly and stability by phosphatidylinositol action on microtubule-associated protein-2, *J. Biol. Chem.,* 262, 3369, 1987.

303. **Takenawa, T. and Fukami, K.,** Phosphoinositide metabolism and oncogenes, *Clin. Chim. Acta,* 185, 309, 1989.

304. **Auger, K. R. and Cantley, L. C.,** Novel polyphosphoinositides in cell growth and activation, *Cancer Cells,* 3, 263, 1991.

305. **Chiarugi, V., Bruni, P., Pasquali, F., Magnelli, L., Basi, G., Ruggiero, M., and Farnararo, M.,** Synthesis of diacylglycerol *de novo* is responsible for permanent activation and down-regulation of protein kinase C in transformed cells, *Biochem. Biophys. Res. Commun.,* 164, 816, 1989.

306. **Sefton, B. M., Trowbridge, I. S., Cooper, J. A., and Scolnick, E. M.,** The transforming proteins of Rous sarcoma virus, Abelson sarcoma virus, and Harvey sarcoma virus contain tightly-bound lipid, *Cell,* 31, 465, 1982.

307. **Richert, N. D., Blithe, D. L., and Pastan, I.,** Properties of the *src* kinase purified from Rous sarcoma virus-induced rat tumors, *J. Biol. Chem.,* 257, 7143, 1982.

308. **Graziani, Y., Erikson, E., and Erikson, R. L.,** Evidence that the Rous sarcoma virus transforming gene product is associated with glycerol kinase activity, *J. Biol. Chem.,* 258, 2126, 1983.

309. **Fukami, K., Matsuoka, K., Nakanishi, O., Yamakawa, A., Kawai, S., and Takenawa, T.,** Antibody to phosphatidylinositol 4,-5-bisphosphate inhibits oncogene-induced mitogenesis, *Proc. Natl. Acad. Sci. U.S.A.,* 85, 9057, 1988.

310. **Alonso, T. and Santos, E.,** Increased intracellular glycerophosphoinositol is a biochemical marker for transformation by membrane-associated and cytoplasmic oncogenes, *Biochem. Biophys. Res. Commun.,* 171, 14, 1990.

311. **Mattingly, R. R., Wasilenko, W. J., Woddring, P. J., and Garrison, J. C.,** Selective amplification of endothelin-stimulated inositol 1,4,5-trisphosphate and calcium signaling by v-*src* transformation of Rat-1 fibroblasts, *J. Biol. Chem.,* 267, 7470, 1992.

312. **Sugimoto, Y., Whitman, M., Cantley, L. C., and Erikson, R. L.,** Evidence that the Rous sarcoma virus transforming gene product phosphorylates phosphatidylinositol and diacylglycerol, *Proc. Natl. Acad. Sci. U.S.A.,* 81, 2117, 1984.

313. **Fukui, Y. and Hanafusa, H.,** Phosphatidylinositol kinase activity associates with viral p60*src* protein, *Mol. Cell. Biol.,* 9, 1651, 1989.

314. **Whitman, M., Kaplan, D. R., Schaffhausen, B., Cantley, L., and Roberts, T. M.,** Association of phosphatidylinositol kinase activity with polyoma middle-T competent for transformation, *Nature,* 315, 239, 1985.

315. **Macara, I. G., Marinetti, G. V., and Balduzzi, P. C.,** Transforming protein of avian sarcoma virus UR2 is associated with phosphatidylinositol kinase activity: possible role in tumorigenesis, *Proc. Natl. Acad. Sci. U.S.A.,* 81, 2728, 1984.

316. **Fukui, Y., Kornbluth, S., Jong, S.-M., Wang, L.-H., and Hanafusa, H.,** Phosphatidylinositol kinase type I activity associates with various oncogene products, *Oncogene Res.,* 4, 283, 1989.

317. **MacDonald, M. L., Kuenzel, E. A., Glomset, J. A., and Krebs, E. G.,** Evidence from two transformed cell lines that the phosphorylation of peptide tyrosine and phosphatidylinositol are catalyzed by different proteins, *Proc. Natl. Acad. Sci. U.S.A.,* 82, 3993, 1985.

318. **Sugano, S. and Hanafusa, H.,** Phosphatidylinositol kinase activity in virus-transformed and nontransformed cells, *Mol. Cell. Biol.,* 5, 2399, 1985.

319. **Fukami, Y., Owada, M. K., Sumi, M., and Hayashi, F.,** A p60v-src-related tyrosine kinase in the acethylcholine receptor-rich membranes of *Narke japonica*: association of phosphatidylinositol kinase activity, *Biochem. Biophys. Res. Commun.,* 139, 473, 1986.

320. **Balduzzi, P. C., Chovav, M., Christensen, J. R., and Macara, I. G.,** Specific inhibition of tyrosine kinase activity by an antibody to the v-*ros* oncogene product, *J. Virol.,* 60, 765, 1986.

321. **Kato, M., Kawai, S., and Takenawa, T.,** Altered signal transduction in erbB-transformed cells. Implication of enhanced inositol phospholipid metabolism in erbB-induced transformation, *J. Biol. Chem.,* 262, 5696, 1987.

322. **Koch, W., Carbone, A., and Walter, G.,** Purified polyoma virus medium T antigen has tyrosine-specific protein kinase activity but no significant phosphatidylinositol kinase activity, *Mol. Cell. Biol.,* 6, 1866, 1986.

323. **Kaplan, D. R., Whitman, M., Schaffhausen, B., Raptis, L., Garcea, R. L., Pallas, D., Roberts, T. M., and Cantley, L.,** Phosphatidylinositol metabolism and polyoma-mediated transformation, *Proc. Natl. Acad. Sci. U.S.A.,* 83, 3624, 1986.

324. **Fischer, S., Fagard, R., Comoglio, P., and Gacon, G.,** Phosphoinositides are not phosphorylated by the very active tyrosine protein kinase from the murine lymphoma LSTRA, *Biochem. Biophys. Res. Commun.,* 132, 481, 1985.

325. **Gray, G. M. and Macara, I. G.,** Serum-stimulated phosphatidylinositol turnover is enhanced in 3T3 cells with active pp60v-src, *Oncogene,* 4, 1213, 1989.

326. **Qureshi, S. A., Joseph, C. K., Rim, M. H., Maroney, A., and Foster, D. A.,** v-Src activates both protein kinase C-dependent and -independent signalling pathways in murine fibroblasts, *Oncogene,* 6, 995, 1991.

327. **Wasilenko, W. J., Payne, D. M., Fitzgerald, D. L., and Weber, M. J.,** Phosphorylation and activation of epidermal growth factor receptors in cells transformed by the *src* oncogene, *Mol. Cell. Biol.,* 11, 309, 1991.

328. **Nakanishi, O., Shibasaki, F., Hidaka, M., Homma. Y., and Takenawa, T.,** Phospholipase C-τ1 associates with viral and cellular *src* kinases, *J. Biol. Chem.,* 268, 10754, 1993.

329. **Johnson, R. M., Wasilenko, W. J., Mattingly, R. R., Weber, M. J., and Garrison, J. C.,** Fibroblasts transformed with v-*src* show enhanced formation of an inositol tetrakisphosphate, *Science,* 246, 121, 1989.

330. **Gorga, F. R., Riney, C. E., and Benjamin, T. L.,** Inositol trisphosphate levels in cells expressing wild-type and mutant polyomavirus middle T antigens: evidence for activation of phospholipase C via activation of pp60$^{c\text{-}src}$, *J. Virol.,* 64, 105, 1990.

331. **Fukui, Y., Saltiel, A. R., and Hanafusa, H.,** Phosphatidylinositol-3 kinase is activated in v-*src*, v-*yes*, and v-*fps* transformed chicken embryo fibroblasts, *Oncogene,* 6, 407, 1991.

332. **Fukui, Y. and Hanafusa, H.,** Requirement of phosphatidylinositol-3 kinase modification for its association with p60^{src}, *Mol. Cell. Biol.,* 11, 1972, 1991.

333. **Otsu, M., Hiles, I., Gout, I., Fry, M. J., Ruiz-Larrea, F., Panayotou, G., Thompson, A., Dhand, R., Hsuan, J., Totty, N., Smith, A. D., Morgan, S. J., Courtneidge, S. A., Parker, P. J., and Waterfield, M. D.,** Characterization of two 85 kd proteins that associate with receptor tyrosine kinases, middle-T/pp60$^{c\text{-}src}$ complexes, and PI3-kinase, *Cell,* 65, 91, 1991.

334. **Mattingly, R. R., Stephens, L. R., Irvine, R. F., and Garrison, J. C.,** Effects of transformation with the v-*src* oncogene on inositol phosphate metabolism in Rat-1 fibroblasts. D-*myo*-inositol 1,4,5,6-tetrakisphosphate is increased in v-*src*-transformed Rat-1 fibroblasts and can be synthesized from D-*myo*-inositol 1,3,4-trisphosphate in cytosolic extracts, *J. Biol. Chem.,* 266, 15144, 1991.

335. **Wyke, A. W., Cook, S. J., MacNulty, E. E., and Wakelam, M. J. O.,** v-Src induces elevated levels of diglyceride by stimulation of phosphatidylcholine hydrolysis, *Cell. Signal.,* 4, 267, 1992.

336. **Purchio, A. F., Shoyab, M., and Gentry, L. E.,** Site-specific increased phosphorylation of pp60$^{v\text{-}src}$ after treatment of RSV-transformed cells with a tumor promoter, *Science,* 229, 1393, 1985.

337. **Gould, K. L., Woodgett, J. R., Cooper, J. A., Buss, J. E., Shalloway, D., and Hunter, T.,** Protein kinase C phosphorylates pp60^{src} at a novel site, *Cell,* 42, 849, 1985.

338. **Gentry, L. E., Chaffin, K. E., Shoyab, M., and Purchio, A. F.,** Novel serine phosphorylation of pp60$^{c\text{-}src}$ in intact cells after tumor promoter treatment, *Mol. Cell. Biol.,* 6, 735, 1986.

339. **Blenis, J., Spivak, J. G., and Erikson, R. L.,** Phorbol ester, serum, and Rous sarcoma virus transforming gene product induce similar phosphorylations of ribosomal protein S6, *Proc. Natl. Acad. Sci. U.S.A.,* 81, 6408, 1984.

340. **Wainberg, M. A., Skalski, V., and Poulin, L.,** Differential effects of phorbol ester on tumor cells induced by avian sarcoma virus, *Anticancer Res.,* 7, 81, 1987.

341. **Grondin, P., Plantavid, M., Sultan, C., Breton, M., Mauco, G., and Chap, H.,** Interaction of pp60$^{c\text{-}src}$, phospholipase C, inositol-lipid, and diacylglycerol kinases with the cytoskeletons of thrombin-stimulated platelets, *J. Biol. Chem.,* 266, 15705, 1991.

342. **Kelly, K., Cochran, B. H., Stiles, C. D., and Leder, P.,** Cell-specific regulation of the c-*myc* gene by lymphocyte mitogens and platelet-derived growth factors, *Cell,* 35, 603, 1983.

343. **Reed, J. C., Nowell, P. C., and Hoover, R. G.,** Regulation of c-*myc* mRNA levels in normal human lymphocytes by modulators of cell proliferation, *Proc. Natl. Acad. Sci. U.S.A.,* 82, 4221, 1985.

344. **Reed, J. C., Sabath, D. E., Hoover, R. G., and Prystowsky, M. B.,** Recombinant interleukin 2 regulates levels of c-*myc* mRNA in a cloned murine T lymphocyte, *Mol. Cell. Biol.,* 5, 3361, 1985.

345. **Friedrich, B., Gullberg, M., and Lundgren, E.,** Uncoupling of c-*myc* mRNA expression from G1 events in human T lymphocytes, *Anticancer Res.,* 8, 23, 1988.

346. **Coughlin, S. R., Lee, W. M. F., Williams, P. W., Giels, G. M., and Williams, L. T.,** c-*myc* gene expression is stimulated by agents that activate protein kinase C and does not account for the mitogenic effect of PDGF, *Cell,* 43, 243, 1985.

347. **Pompidou, A., Corral, M., Michel, P., Defer, N., Kruh, J., and Curran, T.,** The effects of phorbol ester and Ca ionophore on c-*fos* and c-*myc* expression and on DNA synthesis in human lymphocytes are not directly related, *Biochem. Biophys. Res. Commun.,* 148, 435, 1987.

348. **Vandenbark, G. R., Kuhn, L. J., and Niedel, J. E.,** Possible mechanism of phorbol diester-induced maturation of human promyelocytic leukemia cells, *J. Clin. Invest.,* 73, 448, 1984.

349. **Grosso, L. E. and Pitot, H. C.,** Transcriptional regulation of c-*myc* during chemically induced differentiation of HL-60 cultures, *Cancer Res.,* 45, 847, 1985.

350. **Salehi, Z., Taylor, J. D., and Niedel, J. E.,** Dioctanoylglycerol and phorbol esters regulate transcription of c-*myc* in human promyelocytic leukemia cells, *J. Biol. Chem.,* 263, 1898, 1988.

351. **Gailani, D., Cadwell, F. J., O'Donnell, P. S., Hromas, R. A., and Macfarlane, D. E.,** Absence of phorbol ester-induced down-regulation of *myc* protein in the phorbol ester-tolerant mutant of HL-60 promyelocytes, *Cancer Res.,* 49, 5329, 1989.

352. **Yen, A. and Guernsey, D. L.,** Increased c-*myc* RNA levels associated with the precommitment state during HL-60 myeloid differentiation, *Cancer Res.,* 46, 4156, 1986.

353. **de Bustros, A., Baylin, S. B., Berger, C. L., Roos, B. A., Leong, S. S., and Nelkin, B. D.,** Phorbol esters increase calcitonin gene transcription and decrease c-*myc* mRNA levels in cultured human medullary thyroid carcinoma, *J. Biol. Chem.,* 260, 98, 1985.

354. **Thiele, C. J., Reynolds, C. P., and Israel, M. A.,** Decreased expression of N-*myc* precedes retinoic acid-induced morphological differentiation of human neuroblastoma, *Nature,* 313, 404, 1985.

355. **Amatruda, T. T., III, Sidell, N., Ranyard, J., and Koeffler, H. P.,** Retinoic acid treatment of human neuroblastoma cells is associated with decreased N-*myc* expression, *Biochem. Biophys. Res. Commun.,* 126, 1189, 1985.

356. **Hammerling, U., Bjelfman, C., and Pählman, S.,** Different regulation of N- and c-*myc* expression during phorbol ester-induced maturation of human SH-SY5Y neuroblastoma cells, *Oncogene,* 2, 73, 1987.

357. **Bernards, R.,** N-*myc* disrupts protein kinase C-mediated signal transduction in neuroblastoma, *EMBO J.,* 10, 1119, 1991.

358. **Barr, L. F., Mabry, M., Nelkin, B. D., Tyler, G., May, W. S., and Baylin, S. B.,** c-*myc* gene-induced alterations in protein kinase C expression: a possible mechanism facilitating *myc-ras* gene complementation, *Cancer Res.,* 51, 5514, 1991.

359. **Müller, R.,** Cellular and viral *fos* genes: structure, regulation of expression and biological properties of their encoded products, *Biochim. Biophys. Acta,* 823, 207, 1986.

360. **Tamir, A. and Isakov, N.,** Increased intracellular cyclic AMP levels block PKC-mediated T cell activation by inhibition of c-*jun* transcription, *Immunol. Lett.,* 27, 95, 1991.

361. **Auwerx, J., Staels, B., and Sassone-Corsi, P.,** Coupled and uncoupled induction of *fos* and *jun* transcription by different second messengers in cells of hematopoietic origin, *Nucleic Acids Res.,* 18, 221, 1990.

362. **Boyle, W. J., Smeal, T., Defize, L. H. K., Angel, P., Woodgett, J. R., Karin, M., and Hunter, T.,** Activation of protein kinase C decreases phosphorylation of c-Jun at sites that negatively regulate its DNA-binding activity, *Cell,* 64, 573, 1991.

363. **Tsai, M.-H., Hall, A., and Stacey, D. W.,** Inhibition by phospholipids of the interactions between R-ras, Rho, and their GTPase-activating proteins, *Mol. Cell. Biol.,* 9, 5260, 1989.

364. **Boswell, H. S., Nahreini, T. S., Burgess, G. S., Srivastava, A., Gabig, T. G., Inhorn, L., Srour, E. F., and Harrington, M. A.,** A RAS oncogene imparts growth factor independence to myeloid cells that abnormally regulate protein kinase C: a nonautocrine transformation pathway, *Exp. Hematol.,* 18, 452, 1990.

365. **Chiarugi, V., Magnelli, L., Pasquali, F., Vannucchi, S., Bruni, P., Quattrone, A., Basi, G., Capaccioli, S., and Ruggiero, M.,** Transformation by ras oncogene induces nuclear shift of protein kinase C, *Biochem. Biophys. Res. Commun.,* 173, 528, 1990.

366. **Lacal, J. C., Cuadrado, A., Jones, J. E., Trotta, R., Burstein, D. E., Thomson, T., and Pellicer, A.,** Regulation of protein kinase C activity in neuronal differentiation induced by the N-*ras* oncogene in PC-12 cells, *Mol. Cell. Biol.,* 10, 2983, 1990.

367. **Polverino, A. J., Hughes, B. P., and Barritt, G. J.,** NIH-3T3 cells transformed with a *ras* oncogene exhibit a protein kinase C-mediated inhibition of agonist-stimulated Ca^{2+} inflow, *Biochem. J.,* 271, 309, 1990.

368. **Rozengunrt, E.,** A role for arachidonic acid and its metabolites in the regulation of p21^{ras} activity, *Cancer Cells,* 3, 397, 1991.

369. **Fleischman, L. F., Chahwala, S. B., and Cantley, L.,** *ras*-transformed cells: altered levels of phosphatidylinositol-4,5-bisphosphate and catabolites, *Science,* 231, 407, 1986.

370. **Wolfman, A. and Macara, I. G.,** Elevated levels of diacylglycerol and decreased phorbol ester sensitivity in *ras*-transformed fibroblasts, *Nature,* 325, 359, 1987.

371. **Durkin, J. P. and Whitfield, J. F.,** Characterization of the mitogenic signal from an oncogene *ras* protein, *Anticancer Res.,* 9, 1313, 1989.

372. **Olinger, P. L. and Gorman, R. R.,** NIH-3T3 cells expressing high levels of the c-*ras* proto-oncogene display reduced platelet derived growth factor-stimulated phospholipase activity, *Biochem. Biophys. Res. Commun.,* 150, 937, 1988.

373. **Cockroft, S. and Bar-Sagi, D.,** Effect of H-*ras* proteins on the activity of polyphosphoinositide phospholipase C in HL60 membranes, *Cell. Signal.,* 2, 227, 1990.

374. **Smith, M. R., Liu, Y., Kim, H., Rhee, S. G., and Kung, H.,** Inhibition of serum- and Ras-stimulated DNA synthesis by antibodies to phospholipase C, *Science,* 247, 1074, 1990.

375. **Wakelam, M. J. O., Houslay, M. D., Davies, S. A., Marshall, C. J., and Hall, A.,** The role of N-*ras* p21 in the coupling of growth factor receptors to inositol phospholipid turnover, *Biochem. Soc. Trans.,* 15, 45, 1987.

376. **Wu, Y. Y. and Lin, M. C.,** Induction of differentiation in v-Ha-*ras*-transformed MDCK cells by prostaglandin E_2 and 8-bromo-cyclic AMP is associated with a decrease in steady-state level of inositol 1,4,5-trisphosphate, *Mol. Cell. Biol.,* 10, 57, 1990.

377. **Teegarden, D., Taparowsky, E. J., and Kent, C.,** Altered phosphatidylcholine metabolism in C3H10T1/2 cells transfected with the Harvey-*ras* oncogene, *J. Biol. Chem.,* 265, 6042, 1990.

378. **Lacal, J. C.,** Diacylglycerol production in *Xenopus laevis* oocytes after microinjection of p21^{ras} proteins is a consequence of activation of phosphatidylcholine metabolism, *Mol. Cell. Biol.,* 10, 333, 1990.

379. **Kanfer, J. N. and Hakomori, S.,** Eds., *Sphingolipid Biochemistry,* Plenum Press, New York, 1983.

380. **Merrill, A. H., Jr. and Stevens, V. L.,** Modulation of protein kinase C and diverse cell functions by sphingosine — a pharmacologically interesting compound linking sphingolipids and signal transduction, *Biochim. Biophys. Acta,* 1010, 131, 1989.

381. **Zhang, H., Buckley, N. E., Gibson, K., and Spiegel, S.,** Sphingosine stimulates cellular proliferation via a protein kinase C-independent pathway, *J. Biol. Chem.,* 265, 76, 1990.

382. **Inagaki, M., Hagiwara, M., Saitoh, M., and Hidaka, H.,** Protein kinase C negatively modulated by phorbol ester, *FEBS Lett.,* 202, 277, 1986.

383. **Kiss, Z., Rapp, U. R., and Anderson, W. B.,** Phorbol ester stimulates the synthesis of sphingomyelin in NIH 3T3 cells. A diminished response in cells transformed with human A-*raf* carrying retrovirus, *FEBS Lett.,* 240, 221, 1988.

384. **Chou, H.-H. J., Sharafi, B. G., Bascom, C. C., Johnson, T. C., and Perchellet, J.-P.,** A unique sialoglycopeptide growth regulator that inhibits mitogenic activity of a phorbol ester tumor promoter, *Cancer Lett.,* 35, 119, 1987.

385. **Merrill, A. H., Jr., Sereni, A. M., Stevens, V. L., Hannun, Y. A., Bell, R. M., and Kinkade, J. M., Jr.,** Inhibition of phorbol ester-dependent differentiation of human promyelocytic leukemic (HL-60) cells by sphinganine and other long-chain bases, *J. Biol. Chem.,* 261, 12610, 1986.

386. **Dressler, K. A., Mathias, S., and Kolesnick, R. N.,** Tumor necrosis factor-alpha activates the sphingomyelin signal transduction pathway in a cell-free system, *Science,* 155, 1575, 1992.

387. **Shayman, J. A., Mahdiyoun, S., Deshmukh, G., Barcelon, F., Inokuchi, J., and Radin, N. S.,** Glucosphingolipid dependence of hormone-stimulated inositol trisphosphate formation, *J. Biol. Chem.,* 265, 12135, 1990.

388. **Spiegel, S. and Fishman, P. H.,** Gangliosides as bimodal regulators of cell growth, *Proc. Natl. Acad. Sci. U.S.A.,* 84, 141, 1987.

389. **Curatolo, W.,** Glycolipid function, *Biochim. Biophys. Acta,* 906, 137, 1987.

390. **Zeller, C. B. and Marchase, R. B.,** Gangliosides as modulators of cell function, *Am. J. Physiol.,* 262, C1341, 1992.

391. **Van Echten, G. and Sandhoff, K.,** Ganglioside metabolism: enzymology, topology, and regulation, *J. Biol. Chem.,* 268, 5341, 1993.

392. **Weis, F. M. B. and Davis, R. J.,** Regulation of epidermal growth factor receptor signal transduction. Role of gangliosides, *J. Biol. Chem.,* 268, 12059, 1990.

393. **De Cristan, G., Morbidelli, L., Alessandri, G., Ziche, M., Cappa, A. P. M., and Gullino, P. M.,** Synergism between gangliosides and basic fibroblast growth factor in favouring survival, growth, and motility of capillary endothelium, *J. Cell. Physiol.,* 144, 505, 1990.

394. **Yates, A. J., VanBrocklyn, J., Saqr, H. E., Guan, Z., Stokes, B. T., and O'Dorisio, M. S.,** Mechanisms through which gangliosides inhibit PDGF-stimulated mitogenesis in intact Swiss 3T3 cells: receptor tyrosine phosphorylation, intracellular calcium, and receptor binding, *Exp. Cell Res.,* 204, 38, 1993.

395. **Momoi, T.,** Activation of protein kinase C by ganglioside GM3 in the presence of calcium and 12-*O*-tetradecanoylphorbol-13-acetate, *Biochem. Biophys. Res. Commun.,* 138, 865, 1986.

396. **Chan, K.-F. J.,** Ganglioside-modulated protein phosphorylation. Partial purification and characterization of a ganglioside-stimulated protein kinase in brain, *J. Biol. Chem.,* 262, 5248, 1987.

397. **Felding-Habermann, B., Anders, A., Dippold, W. G., Stallcup, W. B., and Wiegandt, H.,** Melanoma-associated gangliosides in the fish genes *Xiphophorus, Cancer Res.,* 48, 3454, 1988.

398. **Kreutter, D., Kim, J. Y. H., Goldenring, J. R., Rasmussen, H., Ukomadu, C., DeLorenzo, R. U., and Yu, R. K.,** Regulation of protein kinase C activity by gangliosides, *J. Biol. Chem.,* 262, 1633, 1987.

399. **Takimoto, M., Hirakawa, T., Oikawa, T., Naiki, M., Miyoshi, I., and Kobayashi, H.,** Synergistic effects of the *myc* and *ras* oncogenes on ganglioside synthesis in BALB/c 3T3 fibroblasts, *J. Biochem.,* 100, 813, 1986.

Chapter 8

Protein Phosphorylation

I. INTRODUCTION

Covalent reversible phosphorylation of cellular proteins is an important regulatory mechanism involved in the control of metabolism as well as in the regulation of cell proliferation and differentiation.[1-3] The phosphorylations regulated by protein kinases occur most frequently on serine/threonine residues, but phosphorylation on tyrosine, although less frequent, is very important for the functional modification of cellular proteins. Some of the phosphorylated cellular proteins are located in the nucleus and are directly involved in the control of RNA synthesis and DNA replication. Nuclear proteins whose functions are regulated by phosphorylation at specific amino acid residues include proto-oncogene products such as Myc, Myb, Fos, Jun, and Ets, as well as the products of tumor suppressor genes such as the p53 and RB proteins. Transcription factors such as CREB and NF-κB are substrates for kinases involved in the control of cell growth.

Protein kinases with specificity for either serine/threonine or tyrosine residues are the enzymes responsible for the phosphorylation of different cellular proteins, including growth factor receptors.[4-11] A number of such kinases are directly involved in the control of cell division.[12-14] Several oncoproteins, as well as some receptors for hormones and growth factors, possess kinase activity with specificity for tyrosine or serine/threonine residues.[15-19] Metabolic reactions associated with the two types of protein kinases (tyrosine kinases and serine/threonine kinases) are not strictly separated but are interrelated in the form of a cascade involving the phosphorylation of multiple substrates.[20] An exceptional protein kinase, Syt, which appears to have specificity for both serine/threonine and tyrosine residues, was cloned from the P19 embryonal carcinoma cell line.[21] Expression of Syt is developmentally regulated at the transcriptional level in various adult mouse tissues.

Protein phosphorylation is implicated in the transmission of environmental signals in prokaryotes,[22] and kinases with specificity for serine/threonine or tyrosine residues are present not only in all eukaryotic cells but also in many prokaryotes.[23] The prokaryotic kinases have a wide range of substrate specificity including endogenous and heterologous prokaryotic proteins as well as eukaryotic proteins and synthetic polypeptides. The prokaryotic kinases may have an important role in the virulence properties of some microorganisms and in the development of infections.

The extent and specificity of phosphorylation of cellular proteins depends not only on the phosphorylating activities of protein kinases but also on the dephosphorylating activities of protein phosphatases with specificities for serine/threonine or tyrosine residues. Such phosphatases may show a high degree of substrate specificity.[24] The functional activities of both protein kinases and protein phosphatases are subjected to complex regulatory mechanisms that may include specific phosphorylation of the enzymes themselves.[25] Regulation of the specific activity of protein kinases and protein phosphatases depends, at least in part, on the action of hormones, growth factors, and oncoproteins.

Phosphotyrosine accounts for only a small fraction of the total cellular phosphoaminoacids, but the phosphorylation of proteins on tyrosine residues has an important role in the regulation of cell proliferation and differentiation by hormones, growth factors, regulatory peptides, and other mitogens. Although proteins containing phosphotyrosine are present in unstimulated normal cells, dramatic changes in the patterns of phosphorylation of these proteins occur upon stimulation of the cells with fresh serum or purified growth factors contained in serum.[26] Different hormones, growth factors, oncoproteins, and other mitogens can use different metabolic pathways associated with protein phosphorylation for exerting their mitogenic effects in defined biological systems.[27] Stimulation of Swiss 3T3 cells with different mitogenic agents (EGF, PDGF, FGF, insulin, fetal calf serum, trypsin, TPA) results in the increased phosphorylation on tyrosine of a number of proteins.[28] Phosphorylation of proteins with molecular weights of 220, 120, and 70 kDa, most of them diffusely distributed throughout the cell, was increased by all of the agents tested. Phosphorylation of these proteins on tyrosine increases within a few minutes after the mitogenic stimulation, reaches a peak, and returns more slowly to basal levels. Prominent among the phosphorylated proteins is a 120-kDa, membrane-associated protein that is phosphorylated on tyrosine in cells by activated growth factor receptors with tyrosine kinase activity such as the PDGF, CSF-1, and EGF receptors, as well as by a mutationally activated c-Src protein.[29] Growth factors and oncoproteins may share signal transduction mechanisms associated with the phosphorylation of cellular substrates on tyrosine residues.

II. TYROSINE-SPECIFIC PROTEIN KINASE ACTIVITY

In 1978 it was discovered that the product of the oncogene v-*src*, which is transduced by the Rous sarcoma virus (RSV), is a 60-kDa protein with kinase activity.[30] Two years later this oncoprotein, termed pp60[v-*src*] or v-Src, was shown to be a protein kinase with specificity for tyrosine residues.[31] Subsequently, a similar activity (tyrosine-specific protein kinase activity, tyrosine-protein kinase activity or, more simply, tyrosine kinase activity) was detected in other viral and cellular proteins.[32-34] In addition to v-Src, the viral oncoproteins v-Yes, v-Fes, v-Fgr, v-Fms, v-Ros, v-Erb-B, v-Abl, v-Sea, v-Kit, and v-Ryk possess tyrosine kinase activity, and the normal cellular, proto-oncogene-encoded counterparts of these proteins have the same activity. Proto-oncogenes encoding proteins with tyrosine kinase activity are present in all eukaryotes. The oncoprotein tyrosine kinases may be localized in different cellular compartments, including the plasma membrane, the cytoplasm, and the nucleus, and may be involved in important regulatory functions, including the function of hematopoietic cells.[35-37] In addition to proto-oncogenes, other normal cellular genes with no known viral counterparts may encode tyrosine kinases. Such genes, frequently considered as putative proto-oncogenes, encode a variety of proteins: Met, Fyn/Syn, Lck, Blk, Hck, Lyn, Tkl, Arg, Trk, Flg, Flt, Ret, Tec, and others. The proteins Fgr, Yes, Fyn, Hck, Lck, Lyn, Hck, and Yrk are highly homologous to Src and are classified, in addition to Src, as members of the Src family or tyrosine kinases.[38] The kinase activity of the members of the Src protein family may be negatively regulated by an enzyme, CSK, which phosphorylates a specific tyrosine residue in the Src-related proteins.[39]

According to their structure, subcellular localization, and function, the tyrosine kinases can be classified in two main groups: membrane-associated tyrosine kinases and cytoplasmic tyrosine kinases. The membrane-associated tyrosine kinases are receptor-like proteins that span the plasma membrane and are known in some instances to recognize and mediate the actions of peptide hormones and growth factors. In contrast, the cytoplasmic tyrosine kinases do not cross the membrane and are entirely intracellular. Prototypes of cytoplasmic tyrosine kinases are the Src, Fps/Fes, and Abl oncoproteins. The functions of cytoplasmic tyrosine kinases are largely unknown but are well conserved in evolution.

Tyrosine kinase activity is intrinsic to the receptors for certain growth factors.[15-19] The cell surface receptors for insulin, IGF-I, EGF, FGF, NGF, HCGF, PDGF, and CSF-1 possess this type of activity. Activation of the receptor tyrosine kinase upon growth factor binding to its surface receptor is an important signaling transduction mechanism across the cell membrane. In addition to growth factors, a number of hormones, regulatory peptides, mitogens, neurotransmitters, and other extracellular signaling agents use a similar transductional mechanism. Phosphorylation of specific tyrosine residues is necessary, for example, for T-cell receptor-mediated signal transduction.[40] The signaling mechanisms associated with activation of receptor tyrosine kinases directly result in the phosphorylation of various cellular proteins, including enzymes. These mechanisms may interact with other transductional mechanisms such as those represented by activation of phosphoinositide breakdown, intracellular mobilization of calcium, and activation of protein kinase C.[41] An important connection exists between cell surface receptor tyrosine kinases and intracellular signaling pathways through the activation of Ras proteins.

Each cellular tyrosine kinase may be independently regulated and may have unique substrate specificities.[42] The use of site-specific antibodies, prepared with synthetic peptides derived from cDNA-predicted amino acid sequences contained in growth factor receptors possessing tyrosine kinase activity, shows that regions of the receptor molecules with homology to the v-Src protein are essential for the preservation of biological activity.[43] Substrate specificity of tyrosine kinases depends, at least in some cases, on the presence of Src homology (SH) domains in the cellular substrates.[44,45] Other molecular features in addition to SH domains may also participate in substrate specificity. Studies with synthetic oligopeptides corresponding to the amino acid sequence 841 to 845 of the EGF receptor kinase, whose Tyr-845 residue is homologous to the main phosphorylation site of the v-Src oncoprotein, suggest that the site specificity of tyrosine kinases results from the balance of positive and negative determinants whose influence on the catalytic activity of the individual enzymes may differ greatly.[46]

Tyrosine kinases may depend on autophosphorylation for their catalytic activity.[47] Phosphorylation of a tyrosine residue near the carboxy-terminal region of the members of the Src family contributes to the regulation of the biological activity of these proteins. Autophosphorylation of tyrosine residues within or surrounding the catalytic active site may enhance the rate of substrate phosphorylation by tyrosine kinases and may represent a general mechanism for the activation of these enzymes.[48] Binding of substrate blocks autophosphorylation, and this in turn inhibits kinase activation. Binding of synthetic peptides that function as specific nonphosphorylatable substrate analogs can block autophosphorylation, thus preventing activation of the kinase function.

Immediately amino-terminal to the kinase domain there is a sequence of approximately 100 residues that is present in all tyrosine kinases, either localized in the cytoplasm or having a transmembrane region. This region, termed SH2, is not essential for catalytic activity but may have a role in directing the cellular actions of the kinase domain by determining specific interactions of the enzyme with cell components. The three-dimensional structure of the SH2 domain of the v-Src oncoprotein (residues 144-249) has been determined by X-ray crystallography.[49] The c-Src and c-Abl proteins share an additional noncatalytic domain of 50 residues, SH3, located amino-terminal to SH2. Elements similar to SH2 and SH3 are present in other cellular and viral proteins, including the Ras GTPase-activating protein (GAP), protein-tyrosine phosphatases, phospholipase C-γ, and the regulatory subunit of phosphatidyinositol 3-kinase (PI 3-kinase).[52] Binding of a fusion protein construct containing two SH2 domains from phospholipase C-γ to the EGF receptor is absolutely dependent on EGF-induced tyrosine phosphorylation of the receptor.[53] The v-Crk oncoprotein, which is the product of the v-*crk* oncogene contained in the avian sarcoma virus (AVS) CT10, as well as the normal cellular counterpart of this gene, the c-Crk protein, consist mainly of SH2 and SH3 regions that are involved in the recognition of phosphotyrosines contained in cellular proteins.[54,55] The 35-kDa Crk protein does not contain a catalytic domain and is expressed in CEF cells and embryonic chicken tissues.

Two proteins of 46 and 52 kDa, termed Shc proteins, possess a carboxyl-terminal SH2 domain close to those of the Fer and Fps/Fes tyrosine kinases and an amino-terminal region rich in glycine and proline residues but do not contain a catalytic domain.[56,57] Overexpression of the Shc proteins in NIH/3T3 mouse fibroblasts results in neoplastic transformation and induces the cells to the transient G_1 phase of the cycle in the absence of growth factors. The Shc proteins associate with, and become phosphorylated by, the EGF receptor in EGF-stimulated cells.

A 47-kDa protein, Nck, which consists almost exclusively of SH2 and SH3 domains (three SH3 domains followed by one SH2 domain) was cloned from a human melanoma cDNA library.[58] The Nck protein may be involved in signal transduction of growth factors whose receptors possess tyrosine kinase activity. Nck is associated with tyrosine autophosphorylated EGF and PDGF receptors via its SH2 domain and becomes phosphorylated on tyrosine, serine, and threonine residues in cells stimulated by EGF or PDGF.[59,60] Nck mRNA and protein are expressed in various murine tissues as well as in cell lines from human, murine, and rat origins. Nck is a potential oncoprotein, as demonstrated by its overexpression in NIH/3T3 cells, which leads to the formation of dense foci of transformed cells. Overexpression of Nck in rat fibroblasts also results in transformation, as judged by alteration of cell morphology, colony formation in soft agar, and tumor formation in nude BALB/c mice.[61] However, Nck overexpression does not induce elevation of the phosphotyrosine content of specific proteins such as the SH2/SH3-containing oncoprotein v-Crk.

Posttranslational modification of protein tyrosine kinases by reversible phosphorylation may be a critical component of their biochemical regulation. This modification can occur either by autophosphorylation or through phosphorylation by other kinases. Reversible autophosphorylation at a conserved tyrosine residue may constitute a kind of molecular switch by which the enzyme is activated.[62] In contrast, phosphorylation at other sites of the enzyme may act as a negative regulatory signal.[63] Therefore, phosphorylation at different sites of tyrosine kinase molecules may have either a positive or an inhibitory effect on their specific activity. The kinase activity of these enzymes may be modified by phosphorylation on serine and/or threonine residues by action of protein kinase C. Phosphorylation on tyrosine of a 42-kDa cytosolic protein which is a common substrate for phosphorylation stimulated by hormones and growth factors, depends on the serine/threonine-specific activity of protein kinase C.[64]

Interaction of tyrosine kinases with other cellular proteins or with nonprotein cellular components may greatly influence the specific enzyme activity. Oncoprotein tyrosine kinases, as well as steroid hormone receptors, can form complexes with specific heat shock proteins (HSPs).[65] The precise biological significance of this association is not understood, however. Some of the proteins associated with tyrosine kinases may be substrates for these enzymes.[66]

Cellular tyrosine kinases can be identified not only on the basis of their enzymatic activity but also by screening of cDNA libraries with degenerate oligonucleotides designed to hybridize with highly conserved regions of the molecule or by screening with antibodies to phosphotyrosine.[67,68] The PCR technique, which serves to amplify defined genetic sequences of any origin,[69] allows the amplification of cDNA sequences related to putative tyrosine kinases previously identified with degenerate oligonucleotide probes corresponding to invariant amino acid sequence motifs within the catalytic domains of these enzymes.[70]

Monoclonal or polyclonal antibodies with specificity for phosphotyrosine are useful in the identification of the substrates involved in tyrosine phosphorylation as well as in measuring the phosphotyrosine content of proteins and in the purification of phosphotyrosine-containing proteins.[71-73] However, these antibodies by themselves are of little value to the understanding of the physiological role of tyrosine phosphorylation of particular proteins in the regulation of cellular proliferation and neoplastic transformation.[74] Elucidation of the function of tyrosine kinase activities associated with growth factor receptors or oncoproteins requires the identification of their specific substrates and the demonstration of a functional role of the tyrosine phosphorylation of these substrate proteins.

The use of specific inhibitors of tyrosine kinase activity may greatly contribute to the elucidation of the role of this activity in the functional properties of growth factor receptors as well as in the mechanisms of action of oncoproteins.[75-77] Tyrosine kinase inhibitors are of varied structures and include the isoflavone compound, genistein, and synthetic 4-hydroxycinnamamide derivatives.[78,79] Genistein is able to inhibit tyrosine kinase-dependent processes without affecting responses to growth factors that are not dependent upon tyrosine kinase activation.[80] Other members of the flavone family with tyrosine kinase inhibitory activity are apigenin and kaempferol.[81] Staurosporine, a microbial alkaloid with antifungal activity, inhibits tyrosine kinases, but it also acts as an inhibitor of protein kinase C. Staurosporine is 100-times more potent an inhibitor of insulin receptor- than IGF-I receptor-associated tyrosine kinase. The anthracycline antibiotic Adriamycin inhibits the enzymatic activities of tyrosine kinases in a dose-dependent fashion.[82] It is not clear, however, if the therapeutic properties of Adriamycin are related to its effect on protein phosphorylation. Synthetic compounds have been designed in an attempt to mimic the transition state of tyrosine kinases. A synthetic multisubstrate analog (RP 53801) inhibits the v-Src kinase and is capable of reducing tyrosine kinase activity in intact cells with some selectivity at 100 μM.[83]

A. TYROSINE KINASE ONCOPROTEINS

A number of cellular oncoproteins, as well as some proteins with oncoprotein-like properties, possess tyrosine kinase activity and are discussed in the following. Many of these proteins are localized in the cytoplasm. The Erb-B1 and Erb-B2 tyrosine kinases are transmembrane proteins identical or closely related, respectively, to the EGF receptor and are discussed in Volume 2. The Trk and Met tyrosine kinases, which function as transmembrane receptors for NGF and other neurotrophic factors and for the liver growth factor, HCGF, respectively, are also discussed in Volume 2. The Kit and Fms proteins are tyrosine kinases identical to the receptors for the hematopoietic growth factors SCGF and CSF-1, respectively, and are discussed in Volume 3.

1. The Src Protein

The Rous sarcoma virus (RSV) is an acutely transforming retrovirus from the avian sarcoma virus (ASV) family and is characterized by the presence of the v-*src* oncogene.[84-86] The products of the v-*src* oncogene and its normal cellular counterpart, the c-*src* proto-oncogene, are the phosphoproteins v-Src (pp60[v-src]) and c-Src (pp60[c-src]), respectively. Both proteins possess tyrosine kinase activity. They differ from each other by the presence of scattered point mutations and the total replacement at their carboxyl termini; the last 19 residues of c-Src (amino acids 515 to 533) are replaced by 12 unrelated residues in v-Src. This replacement is sufficient to convert the normal c-Src protein into a transforming protein.[87] Deletion after amino acid 515 abolishes the specific kinase activity of the protein. Other alterations at the carboxyl terminus of the v-Src molecule may also alter the functional properties of the protein. Substitution at residue 507 has pleiotropic effects on v-Src protein function, and a second mutation, at residue 427, can restore the wild phenotype.[88] In addition to the carboxyl terminus, c-Src contains multiple tyrosine phosphorylation sites on its amino-terminal region, capable of modulating the kinase activity of the protein.[89] However, the precise regulatory role of tyrosine phosphorylation in the amino-terminal region of c-Src in intact cells remains to be determined. Purification to homogeneity of the normal c-Src protein from human platelets allows the study of its enzymatic properties.[90] Maximal activity of c-Src *in vitro* requires either Mn^{2+} or Mg^{2+}, and both ATP and GTP can be utilized as the phosphate donor. Autophosphorylation of the Src protein is a reversible reaction that occurs by an intramolecular mechanism.

The functions of the c-Src protein depend, at least in part, on its subcellular location. Cellular and viral Src contains domains in the amino-terminal region that target and attach the protein to cell membranes in distinct subcellular locations.[91] Specific post-translational modifications of the c-Src molecule may contribute to determining its predominant localization. Myristic acid, a rare fatty acid, is attached cotranslationally to an amino-terminal glycine of Src of either viral or cellular origin.[92] This modification may contribute to the membrane localization and biological activity of Src.[93] The six amino-terminal residues of the Src protein are sufficient for myristoylation of the protein.[94] Osteoclasts express high levels of c-Src protein in association with intracellular membranes.[95] In mammalian fibroblasts the c-Src protein is associated with endosomal membranes.[96] Translocation of the c-Src protein to the cytoskeleton is observed during platelet aggregation.[97] In interphase NIH/3T3 fibroblasts overexpressing c-Src, the protein shows two distinct distributions: one that appears uniform and in association with the cell surface and another that is patchy and juxtanuclear and coincides with the centrosomes.[98] The juxtanuclear aggregation of c-Src protein-containing patches depends on microtubules, and at the G_2 to M phase transition a drastic change in the localization patterns of c-Src takes place. The biological significance of the c-Src locations is unknown but an early cellular response to the v-Src oncoprotein is the increased expression of the transcription factor Egr-1, which is controlled at the transcriptional level via serum-response elements.[99] Thus, the Src protein may be involved, probably through indirect mechanisms, in the regulation of gene expression.

The physiological substrates of the v-Src and c-Src tyrosine kinases remain little characterized. A number of cellular proteins are phosphorylated on tyrosine by the v-Src kinase *in vitro* and *in vivo*. Phosphotyrosine-containing proteins of different molecular weights are immunoprecipitated by anti-tyrosine-specific antibodies in RSV-transformed cells, but the functions of many of these proteins are unknown. Some of the substrates of Src-induced tyrosine phosphorylation are located on the cell membrane. Src may be involved in the regulation of two types of glucose transporters expressed by CEF cells.[100] One of these transporters shows mRNA induction upon *src*-induced transformation and the other is regulated at a posttranslational level, which may involve phosphorylation of the transporter protein at specific sites. Constitutive expression of the v-Src oncoprotein in an antigen-specific murine T-cell hybridoma results in phosphorylation of the T-cell receptor and production of IL-2, two events associated with T-cell activation.[101]

The mechanism by which v-Src exerts its oncogenic effect is unknown, but malignant transformation of RSV-infected cells is absolutely dependent on the specific kinase activity of v-Src.[102] Genetic and biochemical analyses of *ts* mutants of the v-Src oncoprotein indicate that the specific kinase function of the protein is essential for the stimulation of cell proliferation, morphological alteration, and anchorage independence.[103-105] The v-Src kinase is inhibited by the monoclonal antibody R2D2, which also acts as an inhibitor of the tyrosine kinase codified by two other oncoproteins of the Src family, v-Fgr and v-Yes.[106] The tyrosine kinase activity of the v-Src oncoprotein can be inhibited by the isoflavone compound genistein, which does not affect the activity of protein kinases with specificity for serine residues.[78] Inhibition of the v-Src tyrosine kinase may result in reversion of RSV-transformed cells to a normal phenotype.[107] Transformation of 3Y1 rat cells by the v-*src* oncogene relieves the serum dependence of transcription from the RSV LTR.[108] This effect may be due to the constitutive activation of some intracellular signal-transduction pathways by the v-Src-associated tyrosine kinase activity. Transformation of cells induced by the v-*src* oncogene may depend on multiple alterations occurring at the level of the plasma membrane, the cytoplasm, and the nucleus. v-Src expression alters the ionic permeability of rat cell plasma membrane, inducing a permanent depolarization of the membrane.[109] Cytoskeletal components may be affected by v-Src expression. Sequence similarity exists between the noncatalytic region of v-Src and intermediate filament proteins.[110] Structural and/or functional alterations of the cell genome are probably most important for the expression of a transformed phenotype.

Different animal species may have different susceptibilities to v-*src*-induced oncogenic transformation. The v-Src protein can convert embryonic rodent fibroblasts to a transformed and tumorigenic phenotype, but it has no demonstrable effect on diploid human fibroblasts.[111] RSV-induced tumors in chickens frequently grow progressively for several weeks and then regress, and Src kinase activity is markedly reduced in the regressing tumors when compared to progressively growing tumor cells.[112] Expression of adequate levels of the kinase may be essential to progressive tumor growth.

The functional properties of Src proteins partially depend on their state of phosphorylation. In comparison to the v-Src oncoprotein, the normal c-Src protein is characterized by a lower level of tyrosine kinase activity. The major sites of phosphorylation are partially different for the viral and cellular Src proteins: in v-Src the sites are Ser-17 and Tyr-416 and in c-Src are Ser-17 and Tyr-527. The Ser-17 site is phosphorylated by cAMP-dependent kinase and Tyr-527 is phosphorylated by an independent cellular tyrosine kinase.[113] Phosphorylation at Ser-17 is a cell cycle-associated event, but it does not affect the Src functional properties in CEF cells.[114,115] Overexpressed c-Src protein is phosphorylated at Tyr-416 and exhibits increased kinase activity when isolated from cells incubated with orthovanadate, a protein tyrosine phosphatase inhibitor.[116] Analysis of c-Src protein mutants containing a phenylalanine substitution for tyrosine at position 416 indicates that phosphorylation of Tyr-416 has a positive regulatory effect on the biological activity of the protein, but this effect does not directly correlate with a general effect on the total level of tyrosine kinase activity *in vitro* or the level of tyrosine phosphorylation of cellular proteins *in vivo*.[117]

Phosphorylation of Tyr-527, a residue situated at position 6 from the carboxyl terminus, is important for maintaining the lower kinase activity of the normal c-Src product.[118-121] Thus the kinase activity of the c-*src* protein is negatively regulated by the phosphorylation of Tyr-527. Avian c-Src produced in manipulated yeast is underphosphorylated at Tyr-527, and this is accompanied by an increase in its kinase activity.[122,123] Dephosphorylation of c-Src by phosphatase treatment causes a marked increase in kinase activity of the protein, and a similar increase is obtained by the binding of specific antibody to the region of c-Src where phosphorylated Tyr-527 is contained.[118] Tyr-527 dephosphorylation requires membrane localization of the c-Src protein, and the protein phosphatase involved in this dephosphorylation is probably localized in the membrane.[124] Myristoylation is required for dephosphorylation of c-Src in Tyr-527 and, consequently, for the mitotic activation of the protein. Overexpression of c-Src kinase activity by mutation at Tyr-527 causes cellular alterations associated with changes such as reduction in junctional cell-to-cell communication.[125] Studies with site-directed mutagenesis show that carboxyl-terminal alterations that either remove or replace Tyr-527 activate the protein, resulting in increased *in vivo* kinase activity of c-Src and the capability for inducing transformation.[126] Constructed mutants in which Phe replaces for Tyr-527 strongly activate the transforming and kinase activities of c-Src, whereas the additional introduction of Phe for Tyr residue at position 416 suppresses these activities.[127] c-Src converted to a transforming protein by amino acid substitution at position 63, 95, 96, or 338 and expressed by an RSV variant was found to be phosphorylated on both Tyr-416 and Tyr-527 in CEF cells, suggesting that both positions are relevant to differences in the transforming ability of the protein.[128] Sequences homologous to Tyr-416 and its surrounding amino acids are found in almost all the tyrosine kinases. In addition to these posttranslational modifications, the c-Src protein may be phosphorylated at other amino acid residues in particular types of cells, but the biological significance of these additional phosphorylations is not understood. The Ser-97 residue of the c-Src protein is phosphorylated in the human retinoblastoma cell line Y79 but not in the human fibroblast line RT59, although the tyrosine kinase activities of the two cell lines are nearly equal.[129]

Distinct enzymes are responsible for the phosphorylation of the c-Src kinase. The Src protein is a substrate for protein kinase C-induced phosphorylation at Ser-12 both *in vitro* and *in vivo*.[130] Agents that activate phosphoinositide turnover (serum, PDGF, FGF, vasopressin, orthovanadate, $PGF_2\alpha$) can lead to phosphorylation of c-Src at Ser-12. PDGF induces additional phosphorylation of c-Src in two other serines as well as in a tyrosine residue located in the amino-terminal region of the molecule. Phosphorylation of Tyr-527 in the c-Src protein depends not on autophosphorylation but on the activity of a distinct enzyme that is present in normal cells.[131,132] A protein kinase isolated from fetal rat brain phosphorylates c-Src at Tyr-527, which results in suppression of the specific kinase activity of the protein.[133,134] This kinase, termed c-Src kinase (CSK), is specific for c-Src and may act as a physiologic regulator of its activity in intact cells. A gene encoding the CSK was cloned and sequenced.[135] The human CSK gene is located on human chromosome region 15q23-25.[136] The molecular cloning and expression of the chicken CSK gene indicates a high degree of interspecies conservation of the protein, supporting an important functional role for the CSK protein.[137] The enzyme may be involved in phosphorylating other members of the Src protein family. CSK gene transcripts are present in all adult rat tissues, and are markedly elevated in neonatal brain.

The normal physiological activities of the c-Src protein are poorly understood but at least some of them are related to developmental processes during embryogenesis and specialized functions in adult animals. The c-Src protein is expressed in many tissues but particularly high amounts are present in developing nervous tissues where its appearance coincides with the end of the proliferative stage and the beginning of differentiation and morphogenesis. In adrenal medulla and brain, c-Src is specifically located in chromaffin granule membranes in association with a 38-kDa protein.[138] The functions of c-Src in non-neural tissues are almost unknown. c-Src expression is required for mature osteoclasts to form ruffled borders and resorb bone.[139] Several independent mutations in the c-*src* gene can cause osteopetrosis in mice.[140] Osteopetrosis is characterized by impairment of the formation of normal bone marrow cavities.

Neural tissues such as brain, neural retina, and dorsal root ganglion contain high levels of c-Src tyrosine kinase, as compared to the levels observed in non-neural tissues or in CEF cells.[141-143] The c-Src protein from neuronal origin (c-Src+ protein) displays higher kinase activity than that of the protein of non-neuronal origin.[144] Increased c-Src kinase activity is found in human neuroblastoma cells, and the alteration is associated with amino-terminal phosphorylation of c-Src at tyrosine residues.[145] In contrast to the c-src product present in non-neuronal cells, the c-Src+ protein contains a serine phosphorylation site within its amino-terminal region as well as a structural change that alters its electrophoretic mobility.[146,147] The structural change of the neuronal-specific c-Src is not due to posttranslational modification of the protein but results from alternative RNA splicing events. A c-src cDNA clone isolated from a mouse brain cDNA library encoded the c-Src+ protein, which differs from chicken or human c-Src of non-neuronal origin in having an insert of six extra amino acids within the amino-terminal 16-kDa of the molecule.[148] Brain c-src mRNA contains an 18-nucleotide insertion at the position corresponding to the six extra amino acids of the protein. These differences are not species-specific but are organ-specific.[149] The c-Src+ protein is developmentally regulated in the mouse, which indicates its physiological importance.[150] The modified protein is present only in the central nervous system and is not expressed in peripheral nervous system tissues.[151] The c-Src+ protein is detected in the brain of mammals, birds, and reptiles, but not amphibians and fish, suggesting that it may play a role in events associated with higher brain functions.[152] However, the exact role of the c-Src+ protein, especially in relation to neuronal differentiation processes, is not clear. Analysis of c-Src+ expression in cell extracts from human neuronal tumor cell lines and fibroblasts does not show a correlation between the level of Src tyrosine kinase activity and the presence or absence of c-Src+.[153] The six extra amino acids present in c-Src+ could influence the phosphorylation state of the protein, which may affect its subcellular localization or the substrate specificity *in vivo*. More recently, a second neuronal-specific species of c-src mRNA was identified in human brain.[154] This mRNA is also generated by differential splicing and its expression is developmentally regulated. The general importance of c-Src proteins in the nervous system is indicated by the presence of a protein related to c-Src in the nerve cells of the early metazoan hydra (coelenterates), which is the most primitive organism possessing nerve cells.[155] However, the precise function of c-Src in the nervous tissue of different animal species is unknown.

In skeletal muscle of the chicken, c-src gene expression is different from that of other tissues, including cardiac and smooth muscle.[156,157] While other adult chicken tissues, as well as embryonic chicken skeletal muscle, are characterized by a single c-src mRNA of 4.0 kb, shortly before hatching this transcript is replaced in skeletal muscle by a smaller sized (2.8 to 3.3 kb) class of c-src transcripts that persist into adulthood. Such smaller c-src mRNAs are generated by alternative splicing and code for a membrane-associated protein of 24 kDa, which is rich in cysteine and proline residues and lacks most of the tyrosine kinase domain. The function of the skeletal muscle-specific protein is unknown but its production is inversely related to that of c-Src during skeletal muscle development.

The mechanisms through which the c-Src kinase exerts its mitogenic activity remain unknown but they could include the activation of cellular responses to growth factors. Murine fibroblasts overexpressing a transfected chicken c-src gene display an EGF-induced mitogenic response that is much higher than the response exhibited by control cells.[158] c-Src potentiates mitogenic signaling generated by EGF but not by all growth factors. The GAP protein, which enhances the rate of GTP hydrolysis by c-Ras proteins and is implicated in mitogenic signal transduction, is phosphorylated on tyrosine in cells expressing activated Src protein kinases.[159] Stable association of GAP with c-Src is dependent on activation of c-Src kinase and, in particular, on phosphorylation of the Tyr-527 residue. The SH2 region of the c-Src protein kinase contributes to stable GAP-Src association.

2. The Abl Protein

The Abelson murine leukemia virus, A-MuLV, is characterized by the presence of an oncogene, v-abl, whose protein product possesses tyrosine kinase activity.[160-163] The specific kinase activity of the v-Abl oncoprotein is crucial for its transforming ability. Antibodies to phosphotyrosine produced by immunization of rabbits with the v-Abl protein can detect tyrosine-phosphorylated proteins in normal cells, including the PDGF receptor.[164] The substrates of the v-Abl protein kinase remain little characterized, but v-abl-transformed mouse lymphocytes and fibroblasts exhibit alterations in the pattern of protein phosphorylation at tyrosine residues.[165] The v-Abl kinase can phosphorylate the catalytic subunit of type-1 protein phosphatase.[166] Protein phosphatases may be involved in signal transduction across the cell membrane.

The v-*abl* oncogene is derived from the c-*abl* proto-oncogene that is present in all tested vertebrate genomes as well as in invertebrates such as *Drosophila melanogaster*.[167,168] The normal function of the c-Abl product is unknown but may be different in different tissues and different species. During *Drosophila* embryogenesis, the c-Abl protein is expressed in the axons of the central nervous system, suggesting its participation in the formation and/or maintenance of axonal connections.[169] Mutations of the c-*abl* gene in *Drosophila* result in death of the insect during metamorphosis or in the appearance of adults exhibiting a variety of altered phenotypes. The tyrosine kinase portion of the c-Abl protein is conserved in all animal species.

The c-*abl* proto-oncogene encodes a protein with intrinsic tyrosine kinase activity and is transcribed in the cells of many tissues from at least two promoters that give rise to transcripts of two sizes, approximately 5 and 6 kb in length. In the testis, a 4-kb c-*abl* mRNA is synthesized in the haploid stage of spermatogenesis due to the use of an alternative polyadenylation site.[170,171] Apparently, the 4-kb c-*abl* mRNA has a higher stability than the usual c-*abl* mRNAs synthesized outside the testis. The two transcripts give rise in human and mouse cells to two distinct products, the type I and type IV c-Abl proteins, which differ at their amino-termini.[172] The type IV c-*abl* product is a 150-kDa protein that consists of 1142 amino acids and is myristoylated on the amino-terminal glycine. The type I c-Abl protein has 1122 amino acids and is not myristoylated. Overexpression of the myristoylated type IV form of c-Abl is not capable of inducing transformation but has a profound effect on cell growth.[173] Deletion of 53 amino acids from an amino-terminal region of the type IV c-Abl protein is sufficient to activate its transforming potential, both with respect to fibroblast and B lymphocyte transformation *in vitro* and leukemogenicity *in vivo*.[174] Type IV c-Abl is largely nuclear in location, but deletion of the 53 amino-terminal residues of the protein results in its localization in the cytoplasm and activation of its transforming potential.[175] Transcriptional expression of the c-*abl* proto-oncogene is stimulated in cultured human bone marrow stromal cells by specific cytokines such as IL-1α and TNF-α.[176] Because stromal cells are an essential component of the bone marrow microenvironment required for the normal regulation of hematopoietic processes, these results suggest that the c-Abl protein functions as an intermediate in the regulation of hematopoiesis by specific HGFs.

The normal c-Abl protein is phosphorylated on three sites during interphase and seven additional sites during mitosis.[177] The differential phosphorylation of c-Abl during the cell cycle is determined by an equilibrium between the cdc2 kinase and protein phosphatase activities. The c-Abl protein kinase is localized in both the cytoplasm and the nucleus.[178] The large carboxyl-terminal segment of c-Abl contains a DNA-binding domain that is necessary for the association of c-Abl with chromatin. Phosphorylation of the DNA-binding domain of c-Abl by the cdc2 kinase abolishes DNA binding.[179] Binding of c-Abl to DNA is regulated during the cell cycle and may be essential to its biological function.

The c-*abl* proto-oncogene is significantly implicated in human leukemogenesis. The t(9;22) translocation present in the tumor cells of human chronic myelogenous leukemia (CML) gives origin to the Philadelphia chromosome (Ph or Ph$_1$ chromosome) and results in the creation of a chimeric gene generated by fusion of sequences from the c-*abl* gene, normally located on human chromosome 9, and the *bcr* gene located on chromosome 22.[180-182] The Ph translocation leads to the production of fused *bcr-abl* transcripts and the synthesis of a hybrid Bcr-Abl protein that possesses kinase activity. The oncogenic properties of the hybrid Bcr-Abl protein are not due to an increased tyrosine kinase activity, as compared to that of the normal c-Abl protein kinase, but could be due to differences in the subcellular location of the hybrid protein or the presence of an intracellular inhibitor of c-Abl that would not inhibit Bcr-Abl, or to altered substrate specificity of the hybrid protein.[183] The human *bcr* gene encodes a 160-kDa phosphoprotein with kinase activity specific for serine and threonine residues.[184] The *bcr* gene is well conserved in evolution and is expressed in a wide variety of vertebrate tissues.[185] The normal human *bcr* gene encodes a protein of 1271 amino acids.[186] The hybrid Bcr-Abl protein of 210-kDa (P210) present in CML cells is a tumor marker not only for CML but also for a number of cases of acute lymphocytic leukemia (ALL) occurring in adults or children. P210 may display kinase activity for both serine/threonine and tyrosine residues, and these activities may be required for the origin and/or maintenance of a malignant phenotype in CML cells.

3. The Ros Protein

The v-*ros* oncogene contained in the avian sarcoma virus (AVS) UR2 encodes an oncoprotein, v-Ros, which possesses intrinsic tyrosine kinase activity. The normal cellular counterpart of the v-*ros* oncogene is the c-*ros* proto-oncogene, which is present in the vertebrate genome and is closely related to the Src protein family and the insulin receptor tyrosine kinase.[187] Sequence analysis of chicken-derived c-*ros* cDNA predicts a transmembrane tyrosine kinase molecule composed of 2311 amino acids. The carboxyl terminus of the chicken c-Ros protein contains 58 amino acids that are not present in the v-Ros oncoprotein. The c-*ros* gene may represent the vertebrate counterpart of the *sevenless* gene of *Drosophila*,[188,189] whose product functions as a tyrosine kinase receptor and is required for the induction of a predetermined precursor cell to differentiate into photoreceptor number seven in the ommatidium of the compound eye of the insect.[190]

Among chicken tissues, c-*ros* transcripts are detected only in the kidney, and the highest level of expression occurs in 7- to 14-day-old chickens.[191] An 8.3-kb mRNA of the c-*ros* gene was detected in chicken kidney but not in CEF cells.[192] Lower levels of c-*ros* gene expression are present in chicken gonad, thymus, bursa, and brain. Examination of c-*ros* mRNA in the mouse embryo indicates a stringent pattern of expression with only kidney, intestine, and lung showing *ros* mRNA expression at significant levels.[193] The temporal and spatial arrangement of c-*ros* transcripts in the mouse is coincident with the induction and proliferation of epithelium during organogenesis of kidney and intestine. Human c-Ros is a transmembrane protein with a structure similar to that of the insulin and EGF receptors.[191,194] The c-*ros* gene is located on human chromosome region 6q22, and a homologous gene (c-*ros*-2 or *flt*, also coding for a tyrosine kinase) has been assigned to human chromosome region 13q12.[195,196]

The c-Ros transmembrane protein may function as a receptor for a growth factor or regulatory peptide, but the potential physiologic ligand and normal role of the c-Ros are unknown. A constructed hybrid receptor molecule composed of the extracellular ligand-binding domain of the human insulin receptor and the transmembrane and cytoplasmic (tyrosine kinase) domains of the v-Ros oncoprotein has been expressed in CHO cells.[197] The hybrid protein was found to be expressed on the cell surface at high levels, bound insulin, was phosphorylated on tyrosine upon ligand binding, and was capable of expressing transmembrane signaling. However, the hybrid molecule was unable to elicit some insulin-specific responses, including the activation of glucose uptake and stimulation of DNA synthesis. The responses mediated by ligand-activated tyrosine kinase transmembrane receptors may utilize distinct intracellular mechanisms for postreceptor signaling.

4. The Fes/Fps Protein

Several independent isolates of the feline sarcoma virus (FeSV) and the avian Fujinami sarcoma virus (FSV) contain the v-*fes/fps* oncogene, which encodes an oncoprotein with tyrosine kinase activity that is derived from the c-*fes/fps* proto-oncogene contained in the vertebrate genome.[198] The human c-*fes/fps* gene is located on chromosome 15, at region 15q26.[199] The nucleotide sequences of the human and feline c-*fes/fps* genes have been determined.[200,201] The deduced sequence of the human c-Fes protein has a molecular weight of 93 kDa and exhibits extensive homology with the v-Abl and v-Fms oncoproteins, the c-Src protein, and the insulin and EGF receptors. These proteins have a tyrosine phosphorylation site embedded in similar surroundings and possess a lysine residue in a similar position, which may be part of the ATP binding site. Expression of c-Fes is confined mainly to hematopoietic tissue.[202] Transcripts of the c-*fes* gene are present in myeloid cells, including undifferentiated and terminally differentiated cells, the highest levels being present in granulocytes.[203] Expression of the c-Fes kinase may be an essential component of myeloid cell differentiation and responsiveness to CSF-2.[204] Mice transgenic for a 13-kb DNA fragment containing the human c-*fes* gene expressed this gene predominantly in myeloid cells.[205] Elevated levels of human Fes protein in the transgenic mice had no noxious effects on mouse development or hematopoiesis.

The transforming potential of the normal human c-Fes protein can be activated by amino-terminal linkage to viral Gag sequences.[206] A retroviral vector coding for a fused Gag-Fps protein can induce factor (IL-3)-independent growth and tumorigenicity in the factor-dependent myeloid cell line FDC-P1 through a mechanism that may involve activation of the normal hematopoietic growth factor signaling pathway.[207] On the other hand, the human c-Fes protein expressed in rat fibroblasts does not exhibit transforming properties and has a restrained kinase activity, suggesting that regulatory interactions within the host cell modify c-Fes function and restrain its oncogenic potential.[208] The transforming ability of the normal c-Fps/Fes protein in mouse NIH/3T3 cells can be increased by incubation of the cells with orthovanadate, a protein-tyrosine phosphatase inhibitor.[209] This effect is associated with an increase in the level of phosphotyrosine in the c-Fps/Fes protein and its major cellular substrates.

5. The Fgr Protein

The Gardner-Rasheed isolate of the FeSV retrovirus (GR-FeSV) contains the v-*fgr* oncogene, which is derived from the c-*fgr* proto-oncogene and encodes the 70-kDa hybrid oncoprotein v-Fgr.[210] v-*fgr* is a tripartite gene that probably arose as a result of recombinational events involving two distant cellular genes, one coding for actin and the other for tyrosine kinase. The amino acid sequence of the hybrid protein encoded GR-FeSV indicates that, in addition to the tyrosine kinase portion, the protein has extensive homology with actin, a eukaryotic cytoskeletal protein. The hybrid Fgr oncoprotein may be directed to subcellular cytoskeletal targets at the cytoplasmic face of the cell membrane.[211] The c-*fgr* gene is related to the c-*src* gene and is located on human chromosome region 1p36.1-p36.2.[212,213] Analysis of c-*fgr* cDNA clones isolated from a cDNA library derived from the human B-cell leukemia cell line IM-9 indicated that the gene encodes a polypeptide of 529 amino acids.[214] The 55-kDa c-Fgr protein functions as a tyrosine kinase not associated with the membrane.

The 5′ untranslated region of the human c-*fgr* gene has been characterized and the major myelomonocytic c-*fgr* promoter identified.[215] Expression of the c-*fgr* gene in adult human tissues is limited to peripheral blood monocytes and granulocytes as well as alveolar and splenic macrophages.[216] In the 12-week-old human fetus, the highest level of c-*fgr* gene expression occurs in the liver, which probably reflects the specific role of the c-Fgr protein in the development of normal hematopoietic cells. Smaller amounts of c-*fgr* gene transcripts were detected in the placenta as well as in fetal lung and brain. Transient expression of the c-*fgr* gene is induced in normal murine marrow cells by CSF-1 as well as by signals that activate monocytic cells such as IFN-γ, CSF-2, and LPS.[217] The c-Fgr kinase may be essential for mediating the cellular effects of monocytic stimuli.

The possible role of the c-Fgr protein in neoplasia is unknown. Human myeloid leukemia cell lines may express very low levels of c-*frg* RNA and protein, but the induction of differentiation of these cells is frequently associated with increased c-*fgr* gene expression.[218,219] Mutagenic analysis of the c-*fgr* gene suggests that the c-Fgr protein is capable of engaging two biologic pathways, one promoting transformation and the other inhibiting cell growth.[220]

6. The Yes Protein

The Yamaguchi 73 (Y73) and Esh sarcoma viruses contain the oncogene v-*yes*, which encodes a protein with tyrosine kinase activity. The amino acid sequences of the v-Yes oncoprotein exhibit homology with v-Src, especially in the carboxy-terminal region.[221] Hybrid Gag-Yes proteins concentrate in transformed cells at cellular locations (adhesion plaque structures and needle-like interdigitating cell junctions) similar to those occupied by the v-Src protein.[222] The v-*yes* oncogene contains most of the c-*yes* coding sequences, except the extreme carboxyl terminus.[223] The region missing from the v-Yes oncoprotein is the part that is highly conserved in products of the tyrosine kinase family. DNA sequences related to v-*yes* are conserved in evolution. Two of such sequences have been identified in *Drosophila*.[224] Expression of the c-*yes* proto-oncogene is subjected to developmental regulation in both vertebrates and invertebrates.

The c-Yes protein belongs to the nonreceptor types of protein kinases from the Src family. Two c-Yes proteins of 59 and 62 kDa are present in normal chicken fibroblasts.[225] These proteins are phosphorylated *in vitro* exclusively on tyrosine, whereas *in vivo* their phosphorylation occurs predominantly on serine and, to a lesser extent, on tyrosine. High levels of c-Yes protein expression occur in various chicken tissues, including brain, retina, kidney, and liver, whereas c-Yes expression in muscle, heart, marrow, and spleen is very low.[225] The highest levels of c-Yes protein are found in cerebellar Purkinje cells.[226]

The 60-kDa human c-Yes protein is composed of 543 amino acid residues. High levels of c-Yes expression occur in platelets and spermatid achrosomes, but the protein is expressed at different levels in a wide variety of organs and tissues, suggesting its involvement in various pathways.[227] c-Yes is differentially expressed in the normal, hyperplastic, and neoplastic human epidermis.[228] The precise physiological role of the c-Yes protein is unknown but autophosphorylation of c-Yes in rat liver is associated with stimulation of its tyrosine kinase activity.[229]

A Yes-related kinase, Yrk, is a protein of 536 amino acids that exhibits 72% sequence identity with the Yes protein.[230] Elevated levels of Yrk RNA and protein expression have been detected in chicken neural and hematopoietic tissues. The Yrk protein is a member of the Src protein family but its precise function is unknown.

7. The Sea Protein

The v-*sea* oncogene contained in the avian erythroblastosis retrovirus S13 encodes a protein of the tyrosine kinase family.[231] The v-*ski* oncogene cooperates with the v-*sea* oncogene in erythroid transformation by blocking erythroid cell differentiation.[232] The function of the normal c-*sea* proto-oncogene product is unknown, but there is evidence that this product is a transmembrane protein with tyrosine kinase activity, related to the product of the c-*met* proto-oncogene, which encodes the hepatocyte growth factor (HCGF) receptor.[233] The putative ligand of the c-Sea receptor protein has not been identified.

8. The Ryk Protein

The viruses RPL25, RPL28, and RPL30 were isolated from field cases of avian lymphomatosis. Injection of these viruses into newly hatched chickens induces sarcomas that contain acutely transforming retroviruses. Characterization of the viruses isolated from these tumors showed that while RPL25 and RPL28 transduce the v-*erb*B oncogene, RPL30 contains a novel oncogene, v-*ryk*, which encodes a truncated receptor-like tyrosine kinase.[234] The Ryk oncoprotein is synthesized as a 150-kDa precursor that is cleaved into the mature 69-kDa transmembrane fusion oncoprotein. v-Ryk transforms CEF cells and induces tumor in chickens. It is derived from the normal c-Ryk protein, which is a receptor tyrosine kinase-related molecule with unusual kinase domain motifs.[235] In contrast to other acute retroviruses, the RPL30 virus does not lose any of the viral sequences in transducing a proto-oncogene, although the coding region of the viral envelope protein gp37 was disrupted by insertion of the v-*ryk* oncogene. The normal physiologic ligand of the putative c-Ryk receptor tyrosine kinase is unknown.

B. OTHER PUTATIVE TYROSINE KINASE ONCOPROTEINS

In addition to the authentic proto-oncogenes that encode proteins with tyrosine kinase activity and are transduced in altered forms by acute retroviruses, other cellular genes may code for putative oncoproteins with this type of kinase activity but their acute retroviral counterparts have not been described as yet. The putative oncoproteins may display transforming potential under certain conditions. Some of these proteins are members of the Src family and are located mainly in the cytoplasm, whereas others are located on the cell surface and function as receptors for specific growth factors. The Flg and Bek proteins, which are receptors for FGFs, and the Flt protein, which may be a receptor for VEFG, are discussed in Volume 2. Examples of other tyrosine kinases that may display oncoprotein-like activities are discussed next.

1. The Fyn/Syn/Slk Protein

The *fyn* gene, also termed *syn* or *slk*, is located on human chromosome band 6q21 and encodes a 537-amino acid protein of 59 kDa, termed p59*fyn* of Fyn, which is closely related to the products of other members of the tyrosine kinase gene family, in particular to the Src, Yes, and Fgr proteins.[236-238] The human Fyn kinase is 86% identical to the chicken Src protein over a stretch of 191 amino acids at its carboxyl terminus. In contrast, only 6% amino acid homology exists within the amino-terminal 82 residues of the two proteins. A *fyn* gene of *Xenopus laevis* expresses a 537-amino acid protein with 96% amino acid identity to the product of the human *fyn* gene.[239] Transcripts of the *fyn* gene are a component of the maternal RNA pool in the frog oocyte, suggesting that Fyn may play an important role in oogenesis and early embryogenesis.

The Fyn protein kinase has covalently attached myristic acid, is phosphorylated on tyrosine, and possesses tyrosine kinase activity. This activity is negatively regulated by phosphorylation of the protein on tyrosine.[240] Human cell lines express various levels of Fyn protein. In addition to the Src protein, human platelets express relatively high levels of Fyn.[241] The normal *fyn* gene can acquire transforming activity by substituting two thirds of its coding sequence for an analogous region of the v-*fgr* oncogene.[242] The resulting hybrid protein molecule expressed in transformed cells shows tyrosine kinase activity.

The Fyn tyrosine kinase is, in concert with the Lck kinase, a signal transduction element of the T-cell antigen receptor complex.[243] The Fyn kinase is activated following T-cell antigen receptor crosslinking.[244] Both Fyn and Lck are involved in phospholipase C-γ phosphorylation and activation, as well as in protein kinase C activation and intracellular Ca^{2+} mobilization. The CDC45 phosphotyrosine phosphatase regulates Fyn, but not Lck, tyrosine kinase activity in HPB-ALL T cells.[245] In the YT human natural killer-like cell line, Fyn is physically associated with the low-affinity Fc receptor for IgE.[246] This receptor is identical to the lymphocyte differentiation antigen CD23 and is involved not only in the regulation of IgE production but also in the activation or transformation of lymphoid cells, particularly of B-cell lineage. The Fyn tyrosine kinase is also involved in the mechanism of action of IL-7.[247] The IL-7 receptor may function by recruiting the Fyn protein through a segment of its cytoplasmic tail, which may lead to the activation of PI 3-kinase, an enzyme involved in the regulation of cell proliferation.

2. The Lck/Lsk/Tck Protein

The *lck* gene, also termed *lsk* or *tck*, has been the focus of much attention for the relation of its protein product with the function of T lymphocytes.[248-250] The *lck* gene was first detected and identified in the LSTRA cell line, a line that was derived from a M-MuLV-infected mouse and contains an elevated level of tyrosine kinase activity.[251-254] The *lck* gene product is a protein of 56 kDa, termed p56^{lck} or Lck, with tyrosine kinase activity. Lck is expressed at low levels in most types of murine and human cells, with the exception of T lymphocytes, in which it is specifically associated with cell surface glycoproteins.[255-257] The *lck* gene is transcribed via two separate promoters that behave in a developmentally regulated fashion; while both promoters are active in thymocytes, only the distal promoter is active in mature T lymphocytes.[258] The chicken gene *tkl*, which encodes a 457-residue protein with tyrosine kinase activity,[259] is the avian homolog of the mammalian *lck* gene.[260] Transcripts of the *tkl/lck* gene are expressed in CEF cells as well as in chicken spleen and brain.

The functional domains of the Lck kinase can be divided into four regions on the basis of sequence comparison with the c-Src protein.[248] These regions include the sequences in the amino terminus involved in myristoylation and membrane association, the presumed substrate interactive domain, the catalytic domain, and the carboxyl-terminal domain which is involved in governing the specific activity of the enzyme. The tyrosine kinase activity of Lck correlates with its state of autophosphorylation.[261] Exogenous addition of ATP to LSTRA cells results in Lck phosphorylation.[262] The Lck protein is phosphorylated *in vivo* at the carboxyl-terminal residue Tyr-505, which is analogous to the Tyr-527 residue of c-Src.[263] Substitution of phenylalanine at this position results in increased phosphorylation of a second tyrosine residue, Tyr-394, and this change is associated with an increased tyrosine kinase activity and unmasking of the oncogenic potential of Lck in the NIH/3T3 cell transformation assay.

The Lck protein is complexed to the CD4/CD8 antigens on the surface of T lymphocytes.[264,265] CD4 is a polypeptide of 55 kDa that is expressed on the surface of lymphocytes whose receptor is specific for MHC class II molecules (generally, helper T cells). CD4 serves as a receptor for HIV, the presumptive causative agent of AIDS. CD8 is a 32-kDa polypeptide that is expressed in cytotoxic and suppressor T cells and is implicated in the recognition of MHC class I antigens.

The Lck kinase is involved in signal transduction by the T-cell antigen receptor.[266] Lck interacts with the cytoplasmic domain of CD4 through its amino-terminal region.[267] The amino-terminal region of Lck binds to the membrane-proximal 10 and 28 cytoplasmic residues of the CD8 and CD4 antigens, respectively.[268] Two cysteine residues in each of the critical sequences in CD4, CD8, and Lck are required for the association. The CD4/CD8:Lck complex has an important role in signal transduction in T cells, leading to their functional activation.[269,270] The T-cell receptor complexes CD4-Lck and CD8-Lck may include a 32-kDa phosphoprotein (p32) that can be recognized by an antiserum to a consensus GTP-binding region in G proteins.[272] The Lck protein is associated with the glycophosphatidylinositol-linked proteins CD59, CD55, and CD24 on the surface of human cells and with the Thy-1 protein in mouse cells.[272] Activation of the CD4-associated Lck protein in human T cells results in phosphorylation of the serine/threonine-specific protein kinase Raf-1.[273]

The Lck protein may have a role in the mechanism of action of IL-2. Activated Lck kinase stimulates antigen-independent IL-2 production in T cells.[274] On the other hand, stimulation of T cells results in a prompt decline in the levels of Lck mRNA and protein expression, which coincides with the maximal induction of lymphokine (IL-2 and IFN) production by the stimulated cells.[275] Addition of IL-2 to IL-2-dependent human T cells transiently stimulates Lck specific kinase activity, suggesting that Lck may participate in IL-2-mediated signal transduction in T cells.[276] The cytoplasmic domain of the β chain of the IL-2 receptor forms a complex with Lck and, as a consequence, the receptor may become phosphorylated.[277] However, the possible role of Lck in IL-2 signal transduction is not clear. The level of Lck expression may not correlate with the amount of IL-2-stimulated tyrosine phosphorylation in sensitive cells exposed to the interleukin, suggesting that other tyrosine kinases may be involved.[278] Further studies are required for a better characterization of the possible role of Lck in the transductional mechanisms of IL-2 action.

Expression of the *lck* gene is developmentally regulated during the maturation of T cells, suggesting that its product may play a role in the proliferation and differentiation of these cells.[279] High levels of *lck* transcripts are present in human leukemia T-cell lines corresponding to the more mature thymocyte. The *lck* gene, as well as the *fyn* and *c-yes* genes, is expressed at high levels in some human colon carcinoma cell lines.[280] The *lck* gene is located at the distal end of mouse chromosome 4 and on human chromosome region 1p32-35, near a site of frequent structural abnormalities in human lymphomas and neuroblastomas.[253] In the LSTRA cell line, an internally rearranged M-MuLV genome is interposed between two *lck* gene promoters that normally generate *lck* transcripts differing only in 5' untranslated regions.[281] Overexpression of the Lck protein can induce thymic tumorigenesis.[282] The normal cellular substrates of the Lck protein kinase are known only in part. In mouse fibroblastic cells artificially expressing the Lck protein from introduced DNA molecules, Lck is associated with cytoskeletal structures.[283] The Ras GTPase-activating protein, GAP, is a substrate for the Lck tyrosine kinase and a potential binding protein of Lck.[284]

3. The Blk Protein

A murine gene, *blk*, encodes a 55-kDa protein, p55[blk] or Blk, which possesses tyrosine kinase activity.[285] The *blk* gene is located on mouse chromosome 14. The Blk protein is expressed specifically in lymphoid cells of the B lineage, not in T lymphocytes or cells of the myeloid and erythroid lineages, and may function in a signal transduction pathway that is restricted to B lymphoid cells.

4. The Hck/Bmk Protein

The *hck* gene was isolated and characterized by screening a cDNA library from mitogen-stimulated human leukocytes using a murine *lck* probe.[276,287] Structural analysis of the human *hck* promoter has been reported.[288] The gene encodes a 505-residue polypeptide of 59 kDa, termed p59[hck] or Hck, which is closely related to the product of the *lck* gene and possesses tyrosine kinase activity. The *hck* gene is a member of the *src* family and its product is expressed mainly in cells of hematopoietic origin, especially in terminally differentiated granulocytes. Increased *hck* gene expression occurs in the granulocytic differentiation of human myeloid leukemia cell lines.[219] Expression of *hck* is induced by LPS in human macrophages, which depends on the presence of an LPS-responsive element located within the *hck* promoter.[289,290] In contrast, the *fgr* gene, which is structurally related to *hck*, is downregulated during macrophage activation. The Hck protein is likely to be a cell surface membrane protein and may be a receptor for an unidentified hematopoietic growth factor. Hck may assist in regulating signal transduction in myeloid cells. The *hck* gene has been mapped to human chromosome region 20q11-12, which is not far from the c-*src* locus.

The fact that the *lck* gene is expressed at maximum levels in nonproliferating lymphoid cells and the *hck* gene is expressed mainly in peripheral blood granulocytes whose lifespan is less than 12 h indicates that the expression of tyrosine kinases is not always correlated with cell proliferation. As is true for other members of the Src protein family, the biological activity of the Hck protein is regulated by the phosphorylation of a tyrosine residue (in this case, Tyr-501), located near the carboxyl terminus of the protein. Conversion of Tyr-501 to a phenylalanine residue in the Hck protein by oligonucleotide-directed mutagenesis yields a product with potent transforming activity in DNA transfection assays using NIH/3T3 mouse fibroblasts.[291] The ability of the mutated Hck protein to transform NIH/3T3 cells is abolished by a second mutation that destroys the ATP-binding domain of the protein.

A murine homolog of the human *hck* gene is *bmk*.[292] This gene is expressed in mouse organs involved in hematopoiesis, predominantly in cells of the myeloid and B-cell lineages. The B-cell line W279.1 coexpresses the *lck* and *hck*/*bmk* genes. Coexpression of the two genes was also detected in A-MuLV-transformed thymic mouse cells.

5. The Lyn Protein

The *lyn* gene was cloned from a human cDNA library and its product was characterized as a membrane-associated protein of 56-kDa and 512 amino acids, termed p56[lyn] or Lyn, whose sequences are similar to those of the Lck protein.[293,294] The *lyn* gene is expressed in various tissues of the human fetus, with particularly high levels in the liver. The Lyn protein possesses tyrosine kinase activity and is expressed mainly in macrophages/monocytes, platelets and B lymphocytes, but not in granulocytes, erythrocytes, or T lymphocytes. However, human T-cell lines infected with the HTLV-I retrovirus express the protein. Murine cells express two forms of the Lyn kinase that differ in the presence or absence of 21 amino acid residues in the amino-terminal domain.[295,296] These two forms are generated by alternative splicing and possess similar functional properties. They are associated with membrane fractions and may interact with the intracellular domain of cell surface receptors. Lyn proteins are autophosphorylated and may function in the phosphorylation of other cellular proteins in IgM-mediated signal transduction.

B lymphocytes have membrane-bound Igs on their surface that are receptors for specific antigens. Crosslinking of the membrane-bound Igs by antigens or antibodies to Ig activates B lymphocytes to enter the G_1 phase of the cycle, where they become susceptible to proliferative signals provided by helper T cells. These responses are preceded by increased phosphatidylinositol turnover, protein kinase C activation, and intracellular Ca2+ mobilization, all of which depend on G protein functions. The Lyn protein is physically associated with membrane-bound IgM in B cells and has an important function in antigen-mediated signal transduction which includes activation of PI 3-kinase.[297,298] IL-2 regulates the activity of the Lyn tyrosine kinase in a B-cell line.[299] Induction of differentiation of HL-60 leukemia cells into monocyte- or granulocyte-like cells is accompanied by Lyn expression.[300]

The high-affinity IgE receptor expressed on the surface of mast cells and basophils has a central role in immediate allergic responses. Engagement of the receptor results in activation of kinases of the Src family, in particular the Yes and Lyn tyrosine kinases.[301] These kinases may participate in the phosphorylation of cellular proteins on tyrosine observed after IgE receptor engagement. In addition to cells of hematopoietic origin, the Lyn kinase is expressed in neuroendocrine cells and may be involved in the development and function of the sympathetic system. The *lyn* gene is expressed in primary human neuroblastomas, and retinoic acid-induced glial differentiation of cell lines derived from these tumors may be associated with reduced expression of *lyn* gene transcripts.[301]

6. The Arg Protein

The product of the mammalian *arg* gene is a 145-kDa protein, Arg, with intrinsic tyrosine kinase activity.[303,304] The *arg* gene is located on human chromosome region 1q24-25 and consists of two exons. It is highly homologous to the v-*abl* and c-*abl* oncogenes and is widely expressed in normal human cells including brain tissue and fibroblasts. The *arg* gene also is expressed in some human tumor cell lines. The precise normal function of the Arg protein is unknown.

7. The Eph Protein

The *eph* gene was identified and characterized by screening of a human genomic library for gene sequences homologous to the tyrosine kinase domain of the v-*fps* oncogene.[305] The complete nucleotide sequence of the *eph* gene indicates that it encodes a primary translation product of 984 amino acids with a molecular weight of 108,801. The sequence between residues 638 and 882 includes the residues involved in ATP binding and a tyrosine residue homologous to the major autophosphorylation site (Tyr-416) of the v-Src oncoprotein. The Eph protein is a transmembrane glycoprotein with tyrosine kinase activity. Eph functions as a receptor for an unidentified growth factor. Overexpression of the *eph* gene is observed in some human tumors, including colon, lung, breast, and liver carcinomas.

8. The Ret Protein

The putative proto-oncogene *ret* was first detected by its behavior as a dominant transforming gene of NIH/3T3 cells transfected with human T-cell lymphoma DNA. The gene was activated by recombination between two unlinked human DNA segments, possibly by cointegration during transfection. The sequence of the *ret* gene indicates that the deduced product is a transmembrane glycoprotein of 802 amino acids residues with tyrosine kinase domain.[306,307] The Ret protein is phosphorylated on tyrosine residues *in vivo* and may function as a growth factor receptor. The c-*ret* proto-oncogene product is a 150-kDa cytoplasmic protein that is further processed to a 170-kDa protein that is present in the plasma membrane.[308] Transcripts of the *ret* gene are expressed at very low levels in adult rat tissues but are increased in rat conceptuses on days 9 to 11 of gestation.[309] Transcripts of the *ret* gene have been found in a minority of human tumor cell lines surveyed. Studies with transgenic mice suggest that the Ret protein may be involved in melanocyte development.[310] Ret protein expression can compensate for a defect of the c-*kit* proto-oncogene in mice carrying the W^v mutation and can induce development of melanin-producing cells *in vitro* in the absence of the specific hormone, α-MSH.

9. The Fer/Flk Protein

Two tyrosine kinases, Elk and Flk, were identified by screening a rat brain cDNA library using anti-tyrosine antibodies.[311] Elk shows a high degree of homology with Eph (a presumed transmembrane receptor-like tyrosine kinase), and its expression is most prominent in the brain. Flk is a member of the Fes/Fps subfamily of cytoplasmic tyrosine kinases and is widely expressed in rat tissues, the higher levels occurring in the testis. A tyrosine kinase related to, but distinct from, the c-Fes/Fps protein is widely distributed among normal chicken tissues.[312] This kinase may be similar or identical to rat Flk. Moreover, the rat *flk* gene may be identical to the human *fer* gene, which was identified by using v-*abl* probes.[313] The *fer* gene is located on human chromosome region 5q21-q22 and may be deleted in myeloid leukemias.[314] The gene encodes a 94-kDa protein that exhibits homology to the tyrosine kinase domain of certain oncoproteins and growth factor receptors but lacks a transmembrane region. Thus, the *fer* gene encodes a tyrosine kinase of the nonreceptor type. The Fer kinase was found to be present in both the cytoplasm and the nucleus, where a substantial amount is associated with chromatin.[315] However, there are two types of Fer protein produced by alternative transcript splicing mechanisms, a 94-kDa evolutionarily conserved Fer tyrosine kinase and a 51-kDa tyrosine kinase termed Fer-T, whose expression is restricted to meiotically dividing spermatocytes in the testis.[316] The Fer T protein is unique in being a meiosis-specific nuclear tyrosine kinase.

The amino acid sequence of the 94-kDa Fer protein resembles that of c-Fes/Fps. In contrast to c-Fes/Fps, which is biosynthesized only in some hematopoietic cells, Fer is found in a wide range of cell types.[317] The Fer protein is identical to NCP92, a tyrosine kinase detected in mammalian hematopoietic cells by antibodies directed against an amino acid sequence of the conserved domain of the v-Fps oncoprotein.[318]

10. The Tec Protein

The *tec* gene, which encodes a putative tyrosine kinase, was detected by screening of a murine liver cDNA library with the tyrosine kinase domain of the v-*fps* oncogene as a probe.[319] The *tec* cDNA clone contains an ORF encoding a polypeptide of 527 amino acids with a calculated molecular weight of 61,556 Da. The carboxyl-terminal portion of the Tec protein exhibits a high degree of homology to the catalytic domain of other tyrosine kinases, especially with the v-Abl and v-Src oncoproteins, and contains a single putative autophosphorylation site. The Tec protein is likely to be a nonreceptor tyrosine kinase. The *tec* gene is expressed mainly in liver. High levels of *tec* expression were found in two human hepatocellular carcinoma cell lines.

11. The Ltk Protein

The transmembrane receptor tyrosine kinase Ltk is devoid of an extracellular domain and was discovered in murine leukocytes.[320] However, Ltk uses an upstream non-AUG translational initiator and may be a typical, albeit small, glycoprotein receptor.[321] Ltk may be the receptor for a pre-B lymphocyte growth or differentiation factor. In addition to leukocytes, Ltk is expressed in adult, but not embryonic, mouse brain (cerebral cortex and hippocampus).

12. The Nyk Protein

A cDNA encoding a receptor-like tyrosine kinase, termed Nyk or Nyk-r, has been cloned from murine plasmacytoma cells.[322] The extracellular domain of Nyk shows no homology with other receptor families, but its intracellular kinase-related region exhibits some similarity with members of the insulin receptor family, in particular, to the c-Met and Trk-B receptor proteins. The *nyk* gene is expressed in a wide range of rodent tissues and cell lines. The putative ligand of the Nyk receptor-like kinase is unknown.

13. The Hyk Protein

The receptor-like tyrosine kinase Hyk, which was identified in mouse embryonic stem cells, is a transmembrane protein of 1125 amino acids.[323] Hyk is characterized by the coexistence of three EGF-like repeats, immunoglobulin-like domains, and fibronectin type III (FNIII) repeats in its extracellular region. Hyk is expressed at high levels in mouse embryonic tissues. In the adult mouse, Hyk mRNA is found at relatively high levels in lung, liver, kidney, and ovary, but is also expressed in other tissues, including brain. The precise functions of Hyk and its putative physiologic ligand remain to be determined.

14. The Syk Protein

By using antibodies against a portion of amino-terminus of a 40-kDa kinase, a tyrosine kinase of 72 kDa was detected in porcine spleen homogenates.[324] The gene encoding this enzyme, *syk*, was cloned from a porcine spleen library, and its sequence showed the existence of an ORF capable of coding a 628-amino acid polypeptide with a calculated molecular weight of 71,618. The 40-kDa protein kinase is a proteolytic fragment of the 72-kDa holoprotein p72^{syk} or Syk. This protein is a tyrosine kinase of the nonreceptor type. Syk contains two SH2 domains that may play a role in intracellular signaling through binding to phosphotyrosine-containing proteins. Syk is activated by thrombin in porcine platelets and is negatively regulated through Ca^{2+} signals in the stimulated platelets.[325] The major activating pathways evoked by thrombin in platelets are represented by stimulation of phosphatidylinositide metabolism, intracellular mobilization of Ca^{2+}, and activation of protein kinase C. In addition of these pathways, the phosphorylation of platelet proteins on tyrosine by Syk and other kinases may result in the activation of platelet aggregation and secretion.

15. The Axl Protein

A gene, *axl*, isolated from the DNA of CML patients, encodes a receptor-like protein of 140 kDa composed of 894 amino acids which probably possesses tyrosine kinase activity.[326] The *axl* gene shows significant homology to genes encoding other tyrosine kinases, in particular *eph*, *eck*, *elk*, *ros*, and *trk*, as well as the genes encoding the insulin receptor and IGF-I. The extracellular domain of the Axl protein contains two fibronectin type III (FNIII) repeats juxtaposed to two Ig-like repeats. The *axl* gene is widely expressed in normal and neoplastic cells of various origins, suggesting an important normal function for the Axl receptor kinase. The normal *axl* gene is located on human chromosome region 19q13.2. Overexpression of the normal *axl* gene is sufficient to induce transformation in the NIH/3T3 assay system.

16. The Tie Protein

The cloning and characterization of a gene, *tie*, encoding an endothelial cell surface receptor tyrosine kinase, has been reported.[327] The *tie* gene is located on human chromosome region 1p33-p34 and its product appears to be a glycoprotein of 117 kDa. Large amounts of *tie* mRNA are expressed in some myeloid cell lines as well as in endothelial cell lines. The Tie tyrosine kinase may have evolved for multiple protein-protein interactions, possibly including cell adhesion to the vascular endothelium. The Tie protein contains EGF-like, Ig-like, fibronectin-like, and tyrosine kinase domains, thus belonging to four different gene families.

17. The Ror Proteins

Analysis of human cDNA clones encoding two proteins, Ror-1 and Ror-2, indicated that these proteins contain a region with strong homology to the tyrosine kinase domain of several growth factor receptors.[328] In the Ror proteins, a secretion signal sequence and a transmembrane domain delimit the extracellular portion, which contains Ig-like, cysteine-rich, and kringle domains. The cytoplasmic portion contains a tyrosine kinase-like domain which is followed by proline- and serine/threonine-rich motifs. The unique structural features of the extracellular and cytoplasmic domains of the Ror proteins suggest that they are members of a new family of receptors for unidentified growth factor(s).

18. The Tek Protein

Analysis of a cDNA clone derived from murine embryonic heart encoding a protein, termed Tek, showed that this protein may define a new subfamily of receptor tyrosine kinases.[329] The deduced 1122-residue polypeptide contains an intracellular kinase region interrupted by a 21-amino acid insert linked via a transmembrane region to a complex extracellular domain which comprises three FNIII repeats fused to two Ig-like loops that are themselves separated by three tandem EGF-like repeats. The 140-kDa Tek protein may be involved in the regulation of endothelial cell proliferation and differentiation. Transcripts of the *tek* gene are found in highly vascularized embryonic tissues. The putative ligand of the Tek tyrosine kinase receptor has not been identified.

C. TYROSINE PHOSPHORYLATION AND THE MECHANISM OF ACTION OF GROWTH FACTORS

Serum and certain growth factors purified from serum can produce dramatic changes in the patterns of phosphorylation of cellular proteins on tyrosine residues. Stimulation of quiescent BALB/c 3T3 mouse cells with serum causes a major, yet transient, increase in protein tyrosine phosphorylation, which occurs predominantly at the G_0/G_1 transition of the cell cycle.[26] Phosphorylation of tyrosine residues occurs in different subcellular fractions, including the membrane, the cytoplasm, the mitochondria, and the nucleus.

Tyrosine kinase activity is associated with ligand binding-induced activation of the receptors for a number of growth factors,[17-19] including the receptors for insulin,[329,330] IGF-I,[331] EGF/TGF-α[332-334] PDGF,[335] FGF,[336,337] NGF,[338] and CSF-1.[339] In addition to these receptors, other cell surface receptors may possess the same type of activity. A family of embryonic tyrosine kinase receptors was identified by screening a chicken embryo cDNA library with anti-phosphotyrosine antibodies.[340] Two such receptors, Cek-2 and Cek-3, are structurally related to Cek-1, a chicken receptor for basic FGF. The mechanisms of action of many hematopoietic growth factors and interleukins whose receptors are members of the hematopoietin receptor superfamily and do not possess tyrosine kinase activity, may be associated with increased phosphorylation of cellular proteins on tyrosine residues through the secondary activation of specific tyrosine kinases.[341]

In addition to growth factors, other extracellular signaling agents, including mitogens and nonmitogens, can induce changes in the phosphorylation of cellular proteins on tyrosine. Platelets are no longer capable of developing a proliferative response but contain unusually high levels of phosphotyrosine, and the lectin wheat germ agglutinin (WGA) can rapidly induce protein-tyrosine phosphorylation in human platelets.[342] Phosphorylation of tyrosine in several platelet proteins is stimulated by thrombin, which is a physiological platelet activator.[343] Crosslinking of membrane Ig promotes activation of mature B lymphocytes for clonal expansion and antibody production against foreign antigens, and this process is associated with the phosphorylation of proteins on tyrosine.[344] Stimulation of neutrophils with certain chemotactic agents may result in rapid regulation of tyrosine phosphorylation.[345] Productive infection of cells by viruses may be associated with alteration in the phosphorylation of various cellular proteins.[346] Various chemicals stimulate the formation of phosphotyrosine in cellular proteins. The insulinomimetic agents H_2O_2 and vanadate stimulate tyrosine phosphorylation in intact cells.[347] Phosphorylation of proteins on tyrosine is stimulated in human fibroblasts by bacterial components such as lipoteichoic acid.[348] Protein phosphorylation highly specific for tyrosine residues is stimulated in isolated rat liver plasma membranes by oxygen free radicals generated through redox cycling naphthoquinones, including menadione.[349] DMSO stimulates the activity of a tyrosine kinase present in rat lung.[350] Phosphorylation of a 66-kDa soluble cytoplasmic protein on tyrosine is induced by DMSO in human peripheral blood T lymphocytes.[351] The function of this protein is unknown but its phosphorylation is induced by EGF and PDGF. Calmodulin stimulates the phosphorylation of proteins on tyrosine in rat brain membranes.[352] The tyrosine kinases involved in many of these phosphorylations have not been identified.

The intracellular transductional pathways between increased phosphotyrosine formation and propagation of a mitogenic signal elicited by the interaction of a growth factor with its tyrosine kinase receptor are little known but there is evidence that they include the activation of Ras proteins. Growth factor receptors with tyrosine kinase activity, as well as oncoproteins with this activity, may increase the amount of Ras-GTP complexes in the stimulated cells, and Ras proteins may have an important role in the transduction of signals for both normal cell proliferation and malignant transformation.[353] A widely expressed protein, called growth factor receptor-bound protein 2 (Grb2), associates with activated tyrosine kinase receptors via its carboxy-terminal SH2 domain and binds through its two SH3 domains to the Ras exchange factor Sos, which leads to an increase in the active, GTP-bound form of Ras.[353-356] Another protein, Shc, encoded by the mammalian *shc* gene, also contains a carboxy-terminal SH2 domain and can bind Grb2, thereby regulating the Ras signaling pathway through the formation of a Shc-Grb2 complex.[357] In some manner, the GTP-activated Ras protein transmits signals required for cell proliferation or differentiation, according to the specific type of cell and its physiological condition. In hematopoietic cells, the protein Vav may represent a link between receptor tyrosine kinases and Ras-like GTPases by functioning in a manner similar to that of Grb2.[358] Vav is a 95-kDa protein that contains a cysteine-rich region with similarity to protein kinase C, a region similar to that of proteins with guanine nucleotide exchange activity, and a carboxy-terminal SH2 domain flanked by two SH3 domains. The Vav protein is expressed exclusively in hematopoietic cells. Further studies are required for the characterization of the mechanisms involved in the transduction of mitogenic signals by proteins with tyrosine kinase activity in different types of cells.

D. SUBSTRATES OF TYROSINE-SPECIFIC PROTEIN KINASES

Substrate specificity is important to preserve the specificity of the response to growth factor signal after the formation of the activated growth factor-receptor complex. Such specificities can be examined by comparison of the abilities of different protein kinases to phosphorylate exogenous substrates.[359] Sharing of common substrates suggests the existence of common mechanisms by which growth factor signals are processed. The activated receptors for PDGF, CSF-1, and EGF share with a mutationally activated c-Src protein the phosphorylation on tyrosine of a membrane-associated protein substrate of 120-kDa in NIH/3T3 cells,[29] suggesting that some intracellular signaling pathways are shared by oncoproteins and growth factor receptors with tyrosine kinase activity. However, substrate specificity is very important for the specific actions of different kinases in different types of cells. The insulin and EGF receptor tyrosine kinases have similar but not identical substrate specificities.[360] Proteins phosphorylated on tyrosine by the action of growth factors can be identified and purified by using antiphosphotyrosine-specific polyclonal or monoclonal antibodies.[71-74] Proteins located at different cellular sites may become phosphorylated on tyrosine as a result of the activity of tyrosine kinases. Cell surface receptors for hormones, growth factors, regulatory peptides, and neurotransmitters, may be substrates for tyrosine kinases. Several cytoplasmic proteins are phosphorylated on tyrosine by the action of specific kinases. The ribosomal protein S6, which is involved in the mitogen-activated kinase cascade, is a protein kinase that can be activated by phosphorylation. S6 is a substrate for several tyrosine kinases, including growth factor receptors and oncoproteins. Similar activation of the S6-associated protein kinase is found in insulin-treated 3T3-L1 cells and RSV-transformed CEF cells.[361] However, the mechanism of S6 activation is not totally clear. Only phosphoserine and phosphothreonine are found in the activated S6 kinase purified from serum-stimulated Swiss mouse 3T3 cells.[362]

Enzymes involved in phospholipid metabolism may be substrates for tyrosine phosphorylation induced by receptor tyrosine kinases or cytoplasmic tyrosine kinases. Phospholipase C-γ isozymes, which catalyze the hydrolysis of phosphatidylinositol 4,5-bisphosphate for the generation of the second messengers, inositol 1,4,5-trisphosphate and 1,2-diacylglycerol, are substrates of tyrosine kinase receptors such as those for EGF, PDGF, FGF, and NGF. The phospholipase C-γ subtypes, PLC-γ1 and PLC-γ2, are also phosphorylated on tyrosine by several kinases of the Src family, including the Lck, Lyn, Hck, Fyn, and Src protein kinases.[363]

Factors involved in RNA translation may be substrates of the Src kinase. The eukaryotic initiation factor 4F (IF-4F) is a three-subunit complex that binds the 5' cap structure of eukaryotic mRNA. The IF-4F facilitates ribosome binding by unwinding the secondary structure in the mRNA 5' noncoding region. The limiting component of the IF-4F complex is a 24-kDa cap-binding phosphoprotein, IF-4E, and the Src kinase is involved in the phosphorylation of this protein.[364] Src variants with transforming potential, including the v-Src oncoprotein, could release cells from the requirement for extracellular signals to modulate 1F-4E phosphorylation, allowing the cells to proliferate in the absence of serum or growth factors. Phosphorylation of RNA translation factors may be a common pathway for the signaling mechanisms of oncoproteins and growth factors.

An important substrate of viral oncoproteins tyrosine kinases is the Ca^{2+}- and lipid-binding 36- to 38-kDa protein p36/p38 (calpactin I), which is also a substrate for phosphorylation on serine and threonine residues by protein kinase C activity *in vitro* and *in vivo*.[365,366] A method for calcium-dependent isolation of p36 has been proposed, and the fractionation of the phosphorylated and unphosphorylated p36 species has been reported.[367] Modulation of p36 phosphorylation on tyrosine and serine residues induced in human cells by growth factors such as EGF and PDGF can be studied by using monoclonal antibodies to human p36.[368] The p36 substrate is a member of the annexin family of Ca^{2+}/phospholipid-binding proteins, but its normal function is not understood.[369] Expression of p36 is developmentally regulated in the avian embryonic limb, and it has been suggested that p36 may have a structural or mechanical function.[370] Two related but distinct forms of calpactins may interact with phospholipid and actin in a Ca^{2+}-dependent manner.[371] In the absence of Ca^{2+} chelators, three cellular proteins copurify with p36 extracted from a cytosolic or membrane fraction of CEF cells.[372] Protein p36 exhibits structural homology with a 67-kDa protein isolated from the bovine aorta.[373] It also exhibits a high structural homology with the glia-specific protein S-100 as well as with phospholipase A_2, an enzyme involved in the generation of leukotrienes and prostaglandins from phospholipid metabolite precursors.[374-376] Calpactin I (p36) and calpactin II (p35) are identical with lipocortin II and lipocortin I, respectively.[377,378] The lipocortins are phosphoproteins with an inhibiting effect on the activity of phospholipase A2. Lipocortin is a misnomer for calpactins because glucocorticoids have no effect on the synthesis or secretion of these proteins *in vivo*.[379] Calpactin II is not a true member of the lipocortin family of glucocorticoid-inducible anti-inflammatory proteins, despite its ability to inhibit phospholipase A_2 *in vitro*.[380]

Cytoskeletal proteins are substrates for tyrosine kinases *in vitro* and probably also *in vivo*.[381] The different kinases may exhibit different specificities for microfilament and microtubule protein components. A low-abundance protein of 42 kDa (pp42) is transiently phosphorylated on tyrosine after stimulation of fibroblasts by a variety of mitogens, including EGF, PDGF, IGF-II, and PMA.[382] The induction of pp42 phosphorylation by a diversity of mitogens suggests an important role for pp42 in the cascade of events related to the cell cycle. The pp42 protein has been identified as a serine/threonine-specific protein kinase and is identical with the microtubule-associated protein 2 (MAP-2).[383] These results suggest that MAP kinase is a tyrosine-phosphorylated form of pp42 involved in the control of the cell cycle.

Protein kinase activity specific for tyrosine residues has been detected in the mitochondrial fraction purified from human fibroblasts and sarcoma 180 ascites tumor cells.[384,385] The activity may be associated with the outer mitochondrial membrane. An inverse relationship may exist between the rate of cell proliferation and phosphorylation of mitochondrial proteins on tyrosine. A marked increase in mitochondrial tyrosine kinase activity occurs in serum-deprived cells. In mitochondria from resting cells, a protein band with an apparent molecular weight of 50 kDa is phosphorylated on tyrosine.

Phosphorylation of nuclear proteins on tyrosine, as well as on serine and threonine residues, is of utmost importance for the regulation of genomic functions, including DNA replication and transcription. The phosphorylated nuclear proteins may include chromatin components, oncoproteins, and transcription factors.[2,3]

E. TYROSINE KINASES IN NORMAL CELLS

The tyrosine kinases play essential regulatory functions in normal cells from most, if not all, tissues. They are particularly active during developmental processes and are present at high amounts in fetal tissues. A relatively large fraction (about 0.1%) of the total mRNA from chicken embryos encodes proteins with tyrosine kinase activity, suggesting that phosphorylation of proteins on tyrosine is most important for the embryonic developmental processes.[68] Protein tyrosine phosphorylation is a very active process during embryogenesis, and proteins phosphorylated on tyrosine are found in all embryonic chicken tissues.[386] The overall level of protein tyrosine phosphorylation in most tissues decreases during embryonic development, falling to very low or undetectable levels in tissues of the adult chicken. The substrates of tyrosine kinases in different chicken tissues are similar or identical proteins with molecular weights between 35 and 220 kDa. The levels of phosphotyrosine-containing proteins in embryonic tissues do not simply correlate with actively growing tissues, and the observed changes suggest that the major phosphotyrosine-containing proteins may play a role in the control or expression of cell differentiation.

Variable amounts of tyrosine kinases are present in normal adult animal tissues including liver,[387-389] placenta,[390] and testis, especially in the Leydig cells.[391] The normal rat spleen expresses tyrosine kinase activity at levels comparable to those of RSV-transformed avian cells.[392] Tyrosine kinase activity is present in bone marrow,[393] lymphoid cells,[394-396] and platelets. Relatively high levels of tyrosine kinase activity have been detected in nonproliferating, terminally differentiated cells such as human peripheral blood cells.[397]

Tyrosine kinases present in normal cells have been only partially characterized but, as discussed earlier, many of them correspond to the products of proto-oncogenes or putative proto-oncogenes. Other tyrosine kinases have been detected in various types of cells. A tyrosine kinase of 56 kDa was found in lymphocytes and a tyrosine kinase of 40 kDa was purified from bovine thymocytes.[398] These enzymes would be distinct from oncoproteins with tyrosine kinase activity. Two tyrosine kinases, TPK-I (35 kDa) and TPK-II (40 kDa), were partially purified form the particulate fraction of rat spleen.[399] These kinases were found to be distinct from the EGF receptor, the insulin receptor, and the c-Src kinase. A tyrosine kinase of 56 kDa was purified from rat spleen,[400] and an enzyme with similar characteristics was purified from bovine spleen.[401] This kinase shares some common properties with the v-Abl oncoprotein. Many other tyrosine kinases have been found in animal tissues, but their possible role in the mechanisms of growth factor action remains to be elucidated. Tyrosine kinases are implicated in the regulation of proliferative processes occurring during injury, fibrosis, and compensatory growth. Activation of tyrosine kinases was found, for example, in two experimental models of lung repair, one associated with lung fibrosis and sustained cell proliferation and the other with compensatory growth in which the cellular proliferation was only transient.[402] In general, activation of tyrosine kinases may be associated with the increase in cellular turnover that occurs during periods of tissue injury and repair.

F. TYROSINE KINASES IN MALIGNANT CELLS

Proteins with tyrosine kinase activity have an important role in the complex oncogenic processes. Some tyrosine kinases are encoded by proto-oncogenes. However, the precise role of these kinases in oncogenesis is not clear. High levels of tyrosine kinase activity and increased amounts of phosphotyrosine in proteins have been detected in some, but not all, tumor cells.[403-406] No significant difference in total tyrosine kinase activity was detected in freshly isolated cells from various types of human leukemia and normal human blood cells.[407] However, many primary human solid tumors and tumor cell lines may express relatively high levels of tyrosine kinase activity. A number of human colon and breast cancers exhibit high levels of this activity, suggesting that tyrosine phosphorylation may play an important role in the growth of these tumors.[408,409] There may be a correlation between enhanced expression of tyrosine kinase in cytosol of primary breast carcinomas and early systemic relapse of the disease. Benign breast tumors show moderately elevated levels of tyrosine kinase activity in comparison to normal breast tissues.

Variable levels of tyrosine kinase activity have been found in different natural and experimental tumors as well as in tumor cell lines. Tyrosine kinase activity has been detected in the hormone-responsive murine Leydig tumor cell line M5480A; however, the level of this activity is similar to that of normal mouse Leydig cells.[410] The specific enzymatic activity was not altered when the cells were incubated with either human chorionic gonadotropin (hCG) or prolactin, in spite of the fact that these hormones display mitogenic effects on Leydig cells of either normal or tumor origin. The major substrate for tyrosine kinase activity in M5480A cells is a protein of 50 to 54 kDa. Tyrosine kinases may participate in autocrine loops that may serve to maintain tumor growth. Tyrosine phosphorylation of a 145-kDa protein (p145) detected in human gastric carcinoma cells is sustained by an extracellular factor produced by the tumor cells themselves.[411] p145 is a membrane-associated glycoprotein that exhibits protein kinase activity *in vitro*, but its precise structure and function were not determined.

Protein phosphorylation on tyrosine in response to growth factor stimulation may be different in malignant cells and normal cells. However, there is little evidence in favor of this possibility. A growth factor derived from retinoblastoma cells stimulated more tyrosine phosphorylation in the human retinoblastoma Y-79 cell line than in normal retina.[412] Amplification and overexpression of the N-*myc* proto-oncogene has been detected in Y-79 cells,[413] but the possible relation, if any, between this alteration and the increased growth factor-dependent phosphorylation is unknown.

Phosphorylation of specific cellular proteins on tyrosine may have a crucial role in the induction of tumor cell differentiation. Oncoproteins with tyrosine kinase activity are capable of inducing differentiation in certain types of neoplastic cells maintained in culture. For example, the v-Src oncoprotein is able to induce PC12 rat pheochromocytoma cells to differentiate into a neuron-like phenotype in a manner similar to the differentiation induced by the specific growth factor, NGF.[414] However, there are some differences in the phenotypic characteristics of NGF- and v-*src*-induced PC12 cell differentiation. While neurite extension in v-*src*-transformed cells, like NGF-induced differentiation, is accompanied by an increase in an NGF-inducible large external protein, neurite extension in v-*src*-transformed cells is not blocked by the protein kinase inhibitor K-252a, which completely blocks NGF-induced neurite extension. The different phenotypic characteristics of PC12 cells induced to differentiation by NGF and v-Src protein indicate that these inducers may use different pathways or, alternatively, the pathway used by the oncoprotein may be a distal component of a final common pathway of the NGF-induced differentiation of PC12 cells.

Phorbol ester tumor promoters may paradoxically act as inducers of tumor cell differentiation. These compounds are able to induce phosphorylation on tyrosine in specific cellular proteins as well as in some tyrosine-containing artificial substrates and exogenous proteins such as casein.[415,416] Protein p36, which is a substrate for tyrosine kinases, is also a substrate for serine/threonine phosphorylation by protein kinase C and other serine/threonine kinases *in vitro* and *in vivo*.[417-419] In the T-cell lymphoma cell line LSTRA, PMA causes changes in the state of phosphorylation of the Lck tyrosine kinase.[420] Tyrosine kinase activity is stimulated in a specific manner by inducers of differentiation other than phorbol esters. Induction of differentiation of human colon carcinoma cell lines with sodium butyrate results in diminished tyrosine kinase activities and abundance of the c-Src and Lck protein kinases.[421] Removal of the butyrate results in reversion of the decreased expression of both proto-oncogene products. The HL-60 human leukemia cell line does not possess detectable membrane-bound tyrosine kinase activity, but induction of HL-60 cell differentiation by IFN-γ or TNF-α, as well as by the synergistic action of retinoic acid and the calcium ionophore A23187, results in a marked increase in this activity.[422,423] Differentiation of HL-60 cells is accompanied by a marked reduction in the expression of the c-*myc* proto-oncogene.

III. PHOSPHOTYROSINE-SPECIFIC PROTEIN PHOSPHATASES

The amount of phosphotyrosine in cellular proteins depends not only on the phosphorylating activity of tyrosine-specific protein kinases but also on the dephosphorylating activity of a family of enzymes that specifically dephosphorylates phosphotyrosine residues, the phosphotyrosine-specific protein phosphatases (phosphotyrosyl protein phosphatases, protein-tyrosine-phosphate phosphohydrolases, or protein-tyrosine phosphatases).[424-427] Characterization of these enzymes showed that they can be divided into two types: low molecular weight (<80 kDa) cytosolic protein-tyrosine phosphatases and high molecular weight (>80 kDa) membrane-associated (transmembrane) protein-tyrosine phosphatases. The enzymes of the cytosolic type have a single phosphatase domain, whereas those of the transmembrane type have two phosphatase domains. Such domains are about 260 amino acids long and exhibit various degrees of sequence similarity. Alternative RNA splicing mechanisms may generate different tyrosine phosphatase domains.[428]

Protein-tyrosine phosphatases are evolutionarily conserved, are widely distributed in nature, and are present in both normal and neoplastic cells.[429-436] These enzymes are involved in determining the levels of protein tyrosine phosphorylation in cells involved in important functions, such as the T lymphocytes.[437]

They may participate in the regulatory mechanisms of cell proliferation and differentiation. The transmembrane protein-tyrosine phosphatases may be involved in signal transduction. Some of the immediate-early genes induced by growth factors in their target cells may encode protein-tyrosine phosphatases. The growth factor-inducible immediate-early gene *3CH134* encodes a protein-tyrosine phosphatase capable of dephosphorylating the p42 MAP kinase.[438] The substrates of protein-tyrosine phosphatases may include tyrosine kinases, which may be inactivated as a consequence of the protein-tyrosine phosphatase-induced dephosphorylation of tyrosine residues.[439] The activities of protein-tyrosine phosphatases may have an important role in regulating the signal transduction cascade associated with cell surface receptors with tyrosine kinase activity.

A. SUBCELLULAR LOCALIZATION AND FUNCTION OF PROTEIN-TYROSINE PHOSPHATASES

Protein-tyrosine phosphatases are associated with various cellular structures, including the plasma membrane and the cytoplasm. Membrane-associated protein-tyrosine phosphatases may play an important role in signal transduction across the membrane through the dephosphorylation of specific targets and may contribute to the regulation of cell proliferation. Density-dependent inhibition of cell proliferation *in vitro* involves the regulated elevation of a specific protein-tyrosine phosphatase activity.[440] Screening of human cDNA libraries allowed the identification of a multigene family of membrane-associated, receptor-linked protein-tyrosine phosphatases, each containing a distinct extracellular domain, a single hydrophobic transmembrane region, and two tandemly repeated conserved cytoplasmic domains.[441] At least some members of this family are present in *Drosophila*.

The leukocyte common antigen (LCA) CD45, also called Ly-5, possesses tyrosine-specific protein phosphatase activity. CD45 is represented by a family of structurally related transmembrane proteins of 180 to 240 kDa, which are present in all hematopoietic cells except mature erythrocytes.[442] The fact that CD45 molecules have large cytoplasmic segments suggests that they may interact with and dephosphorylate other structural components of the membrane-cytoplasm interface, including hormone and growth factor receptors. Surface expression of CD45 is required for normal signaling events to occur in T lymphocytes, B lymphocytes, and NK cells.[443] CD45 is involved in the modulation of B and T lymphocyte responses and could alter the expression of the IL-2 receptor. CD45 is a component of a complex of proteins associated with the antigen receptor on the lymphocyte surface and may regulate signal transduction by modulating the phosphorylation state of the antigen receptor subunits.[444] There are multiple molecular CD45 isoforms differing in the extracellular domain, but studies with chimeric transmembrane EGF receptor/CD45 enzymes suggest that the extracellular and transmembrane domains of the molecule are not required for CD45 signaling through the T-cell receptor.[445] CD45 antigens are linked to stimulation of early human myeloid progenitor cells by IL-3, CSF-2, and the Kit ligand, SCGF/MGF.[446]

Enzymes with protein-tyrosine phosphatase activity have been detected in cytosolic fractions but they would be primarily associated with the plasma membrane, and the cytosolic forms may be derived by proteolysis from the enzymes located on the membrane.[447] In HL-60 leukemia cells, tyrosine kinases are located mainly on the plasma membrane and protein-tyrosine phosphatases are concentrated predominantly on internal membranes.[448] A protein-tyrosine phosphatase purified from the cytosol and particulate fractions of human placenta catalyzes the dephosphorylation of receptors for insulin, IGF-I, and EGF.[449] Microinjection into *Xenopus* oocytes of a 35-kDa protein-tyrosine phosphatase (PTPase 1B), which dephosphorylates the insulin receptor *in vitro*, can block the phosphorylation of the insulin and IGF-I receptor β subunit and antagonizes early events triggered by insulin, including the activation of ribosomal protein S6 kinase and the phosphorylation of ribosomal protein S6 *in vivo*.[450] The demonstration that protein-tyrosine phosphatases are activated by proteolysis implies that there are protein factors blocking the specific activity of these enzymes.

B. MOLECULAR FORMS OF PROTEIN-TYROSINE PHOSPHATASES

Two major protein-tyrosine phosphatases have been purified from the human placenta and are characterized as proteins of 30 to 40 kDa.[451,452] These placental enzymes are absolutely specific for tyrosine residues. The complete amino acid sequence of a cytosolic human protein-tyrosine phosphatase has been determined.[453] The enzyme consists of 321 residues and contains an unusually proline-rich carboxyl-terminal region. This human placental protein-tyrosine phosphatase is related to a protein, termed LAR, which resembles the human leukocyte antigen CD45 in its cytoplasmic domains but has an external segment displaying structural characteristics of a neural cell adhesion molecule (N-CAM). Mutational analysis of the cytoplasmic domains of LCA and LAR indicated that a cysteine to serine missense mutation in the first of the two homologous tyrosine phosphatase domains of both LCA and LAR destroys the enzyme activity, and that the second domain does not possess intrinsic protein-tyrosine phosphatase activity but may be involved in the regulation of substrate specificity.[454]

Synthetic oligonucleotides encoding segments conserved among protein-tyrosine phosphatase domains have been used for the identification of a cDNA clone present in cDNA isolated from a human T-cell cDNA library which encodes a protein with homology to the human placental protein-tyrosine phosphatase.[455,456] The 48.4-kDa protein encoded by the human T-cell cDNA clone contains 415 amino acid residues. A rat homolog of the human T-cell protein-tyrosine phosphatase was isolated from a rat spleen cDNA library.[457] The cloned rat gene encodes a protein of 363 amino acids. The noncatalytic region of this protein-tyrosine phosphatase is located at the carboxyl terminus and shows homology with the basic domains of the transcription factors encoded by the c-*fos* and c-*jun* proto-oncogenes. The highest levels of the rat protein-tyrosine phosphatase are found in macrophages.

Screening of a rat brain cDNA library with oligonucleotides whose sequences were deduced from the amino acid sequence of human placental protein-tyrosine phosphatase has led to the isolation of a cDNA clone encoding a protein of 432 amino acids that exhibited 97% sequence identity with the corresponding 321 residues of the placental enzyme.[458] Expression of the coding sequence in *Escherichia coli* yielded a 50-kDa protein which showed absolute specificity for phosphotyrosine-containing protein substrates. Sequences found in the carboxyl-terminal portion of the protein suggested mechanisms by which the protein may be attached to cell membranes. Transcripts corresponding to this enzyme were detected in a number of rat tissues.

A major protein-tyrosine phosphatase present in the bovine brain and the human placenta is calcineurin, a calmodulin-binding protein.[459,460] The major secreted isoenzyme of human prostatic acid phosphatase is also a protein-tyrosine phosphatase.[461] Phosphotyrosine, but not phosphoserine or phosphothreonine, residues are specifically amplified in prostate tumor cells, and this modification is associated with repressed levels of acid phosphatase in the gland, suggesting a possible role of the enzyme in the growth of human prostatic tumors.[462] There is an inverse relationship between prostatic acid phosphatase and protein-tyrosine kinase activity.[463] The addition of testosterone to prostatic carcinoma cell lines, which results in decreased acid phosphatase activity, is accompanied by an increase in tyrosine kinase activity. Several different isoenzymes of acid phosphatase are present in prostatic cells, and it is not clear which form of them may interact with the cellular tyrosine kinases.

The protein-tyrosine phosphatases are widely distributed in nature, suggesting a central role in important functions. Two receptor-linked protein-tyrosine phosphatase genes, DLAR and DPTP, were detected in *Drosophila* by using degenerate oligonucleotide probes containing consensus sequences present in the human genes encoding the HD45 and LAR protein-tyrosine phosphatases.[464] The extracellular segments of DLAR and DPTP are composed of multiple immunoglobulin-like domains and FNIII-like domains. The cytoplasmic region of DLAR and DPTP consists of two tandemly repeated tyrosine phosphatase domains. Site-directed mutagenesis indicated that a conserved cysteine residue is essential for protein-tyrosine phosphatase activity.

Additional molecular forms of protein-tyrosine phosphatases have been identified by modern methodology such as PCR amplification of cDNAs with oligonucleotide primers to the conserved regions of the known enzymes. One of these novel phosphatases, PTPty42 or HCP, is expressed mainly in human and murine hematopoietic cells.[465] The HCP amino-terminal region contains two SH2 domains, but the levels of HCP mRNA are not altered by growth factors in hematopoietic cells. The HCP gene is located on human chromosome 12p11-p13, a region that is frequently involved in translocations or deletions in acute lymphocytic leukemias. Screening of human breast cancer and umbilical cord cDNA libraries allowed the isolation of cDNA clones encoding two protein-tyrosine phosphatases, PTP1C and PTP2C, which are expressed in many tissues and are characterized by the presence of a single phosphatase domain and two adjacent copies of SH2 domains at their amino-terminal region.[466,467] The SH2 domains of PTP1C form high-affinity complexes with the EGF receptor and other phosphotyrosine-containing proteins, which suggests that the PTP1C SH2 sequences may interact with other cellular components to modulate their own phosphatase activity against interacting substrates. Whereas the expression of PTP1C is restricted to hematopoietic and epithelial cells, PTP2C is ubiquitously expressed in human tissues, with relatively high levels of expression in heart, brain, and skeletal muscle.

A subfamily of SH2 domains is found specifically in protein-tyrosine phosphatases.[468] Activation of these enzymes depends on tyrosine phosphorylation,[469] and tyrosine phosphatases containing SH2 domains may be a target of tyrosine kinases. One of these enzymes, Syp, contains two SH2 domains and is rapidly phosphorylated on tyrosine in EGF- and PDGF-stimulated cells.[470] Syp exhibits homology to the product of the *corkscrew* (*csw*) gene of *Drosophila*, which is required for signal transduction downstream of the Torso receptor tyrosine kinase of the insect.[471] The Syp gene is widely expressed throughout mouse embryonic development as well as in adult mouse tissues. The activity of some tyrosine phosphatases may link growth factor receptors and other signaling cellular proteins with tyrosine kinase activity for the regulation of signal transduction and other related functions through phosphorylation/dephosphorylation of tyrosine.

A transmembrane receptor-like protein-tyrosine phosphatase of 793 amino acids, LRP or PTPα, is expressed in various tissues.[468] The enzyme contains an extracellular domain of 123 amino acids that is predicted to be highly glycosylated, a 24-amino acid membrane-spanning region, and a 627-amino acid cytoplasmic domain. Two regions contained in the cytoplasmic domain of PTPα show homology to the tyrosine phosphatase protein family. Overexpression of PTPα in REF cells may result in persistent activation of the c-Src kinase, which may lead to neoplastic transformation and tumorigenesis.[472] Thus, PTPα exhibits oncoprotein-like activity. Activation of the c-Src kinase induced by PTPα is accompanied by dephosphorylation at the Tyr-527 residue.

C. REGULATION OF PROTEIN-TYROSINE PHOSPHATASE ACTIVITY

Multiple factors of cellular and extracellular origin are involved in regulating the activities of protein-tyrosine phosphatases. Serine/threonine-specific protein kinase may have an important role in the regulation of protein-tyrosine phosphatase activity through phosphorylation of serine and threonine residues in a regulatory component complexed with the 55-kDa catalytic subunit of the enzyme.[473] The PTP1B isoenzyme undergoes mitosis-specific phosphorylation on serine.[474]

Specific protein inhibitors may have an important role in the regulation of protein-tyrosine phosphatases. Seven distinct protein-tyrosine phosphatases with molecular weights ranging from 24,000 to 104,000 were isolated from bovine brain, and their activities were found to be regulated by two specific inhibitors, termed H (500 kDa) and L (38 kDa).[475,476]

Polycation-stimulated protein phosphatases with specificity for serine/threonine residues can be converted *in vitro* into protein-tyrosine phosphatases by incubation with ATP.[477] This conversion may involve a change in the catalytic subunit of the phosphatase, but its possible role in the physiological regulation of the enzymes remains to be elucidated.

D. ROLE OF PROTEIN-TYROSINE PHOSPHATASES IN TRANSFORMATION

The role of tyrosine phosphorylation/dephosphorylation processes occurring in cellular proteins can be evaluated by using sodium orthovanadate, which is a potent inhibitor of protein-tyrosine phosphatase activity.[431,478,479] Treatment of cells with orthovanadate may allow the detection of tyrosine phosphorylation of viral and cellular proteins that are not detected in untreated cells. Vanadate reversibly induces transformation of cultured cells in a dose-dependent manner, so the degree of transformation can be controlled by the concentration of vanadate added to the medium.[480] The vanadate-induced transformation may not depend on changes in phospholipid metabolism, which remains almost unaltered, but on the phosphorylation of one or more protein targets. The effects of vanadate on the transformation of NRK cells are dependent on growth factor conditions in a bimodal way, being stimulatory in the presence of suboptimal concentrations of growth factors and stimulatory when the cells are stimulated by TGF-β added alone or in combination with other growth factors.[481]

The possibility that altered dephosphorylation of proteins on tyrosine, dependent on an uncontrolled activity of protein-tyrosine phosphatases, may be a mechanism capable of leading to neoplastic transformation has been considered.[458] It was suggested that members of the protein-tyrosine phosphatase family may be the products of tumor suppressor genes, whose absence or inactivation would lead to unrestrained growth control and the expression of a transformed phenotype. A specific protein-tyrosine phosphatase, PTPase-γ, may have tumor suppressor properties.[482] The gene of PTPase-γ is located on human chromosome region 3p21, and this region is frequently affected by nonrandom chromosome deletions that occur in both renal cell carcinoma and lung carcinoma. Hemizygous deletions involving the PTPase-γ gene have been observed in a subset of clinical samples and cell lines from these tumors. Thus PTPase-γ would be a tumor suppressor gene whose functional loss is involved in the pathogenesis of human kidney and lung tumors. However, as is true for many proto-oncogenes, the classification of a given gene as a tumor suppressor gene may be a matter of controversy because these genes are essentially normal genes that may display tumor suppressor activity only under certain conditions.

IV. SERINE/THREONINE-SPECIFIC PROTEIN KINASES

Phosphorylation of cellular proteins on serine and threonine is more common than the phosphorylation on tyrosine residues. Hormones and growth factors, as well as oncoproteins and tumor promoters, can stimulate the phosphorylation of cellular proteins not only on tyrosine but also on serine and threonine residues, and this modification may be important for regulating the functional activity of the substrates. Distinct types of serine/threonine-specific protein kinases (serine/threonine kinases) are involved in such phosphorylations.[483] These enzymes depend on either activation of the adenylate cyclase system (cAMP-dependent protein kinases) or activation of other types of protein kinases (cAMP-independent protein kinases), the latter including Ca^{2+}-phospholipid-dependent protein kinase (protein kinase C).[484] A unique type of serine/threonine kinase is the DNA-activated protein kinase (DNA-PK), which requires double-stranded DNA for activity *in vitro* and may have a role in DNA transcription as well as in DNA replication, recombination, and/or repair.[485] Many DNA binding proteins, including a number of transcription factors, are substrates for DNA-PK *in vitro*. Phosphorylation of serines by DNA-PK in the amino-terminal transactivation domain of DNA-bound p53 protein may alter p53 function, thus contributing to the regulation of cellular proliferation.[486]

Serine/threonine kinases capable of transmitting growth-regulating information include growth factor receptors such as the type III TGF-β receptor.[487] They also include hormone- and growth factor-regulated enzymes such as cAMP- and cGMP-dependent protein kinases, members of the protein kinase C family of enzymes, and Ca^{2+}/calmodulin-dependent protein kinases. In addition, there are several growth-regulated serine/threonine kinases that do not require known second messengers for activity, including the 65- to 70-kDa S6 protein kinases, MAP-2 kinase, casein kinase II, c-Raf kinase, and a 100-kDa EGF-regulated kinase. A phospholipid-stimulated kinase, protein kinase P, is also involved in the phosphorylation of proteins on serine and threonine residues.[488]

The genes of two serine/threonine kinases, Rac-α and Rac-β, have been cloned from human cell lines.[489,490] The Rac-α protein is encoded by the *AKT1* gene, which is the human homolog of the viral oncogene v-*akt*. The β form of Rac is a 60-kDa protein that has a carboxyl-terminal extension of 40 amino acids in comparison to the α form. A third human gene, *AKT2*, closely related to those encoding Rac-α and Rac-β, codes for a 56-kDa protein, Akt-2, which has serine/threonine kinase activity and contains an SH2 domain.[491] The *AKT2* gene is localized on human chromosome 19, at region 19q.13.1-q13.2 and was found to be amplified in some human ovarian carcinoma cell lines and primary ovarian carcinomas. The Rac/Akt kinases show a high degree of homology to both the protein kinase C and cAMP-dependent families, and may be universally expressed in human cells. These kinases are most probably involved in intracellular signal transduction. The gene encoding another serine/threonine kinase, Mak, is expressed almost exclusively in testicular meiotic cells. The *mak* gene is not expressed in ovarian cells, including oocytes after the dictyotene stage.[492] DNA sequences homologous to the *mak* gene are highly conserved in mammals and may play an important role in spermatogenesis.

Serine/threonine kinases involved in transductional responses to hormones, growth factors, and other mitogens are the ribosomal S6 protein kinases (RSK kinases) and the mitogen-activated protein kinases (MAP kinases).[493-495] Members of the RSK family of S6 protein kinases are enzymes of 85 to 92 kDa (pp90^{RSK}), as well as S6 kinases of 70 to 85 kDa (pp70^{S6K}).[496-500] These enzymes are activated in the early G_1 phase of the cycle and their activities are regulated by phosphorylation. The macrolide rapamycin blocks the phosphorylation and activation of pp70^{S6K} in a variety of animal cells and delays entry of serum-stimulated 3T3 cells into S phase of the cycle. Rapamycin blocks the activation of pp70^{S6K} in early G_1 but has no effect on the serum-stimulated activation of MAP kinases or pp90^{RSK}, suggesting a role for p70^{S6K} in the regulation of cellular proliferation.[501]

The MAP kinases were originally isolated from insulin-treated 3T3-L1 adipocytes, but these enzymes are activated by an array of hormones and growth factors in different cell types. Growth hormone stimulates MAP kinase activity in target cells such as 3T3-F441A fibroblasts.[502] The activity of MAP kinases is regulated by phosphorylation on serine/threonine and tyrosine residues, which depends, in turn, on the activity of kinases of the MAP kinases (MAP kinase kinases), whose activity is regulated by hormone and growth factor receptors associated with stimulation of tyrosine phosphorylation of cellular proteins.[503-505] The enzymatic activity of two MAP-2 kinases of 42 and 44 kDa (also called myelin basic protein [MBP] kinases) is regulated at different stages of the cycle of mammalian cells by tyrosine phosphorylation in murine cells.[506] A 44-kDa MAP-related serine kinase, p44^{mpk} or Mpk, is present in sea star oocytes and *Xenopus* ovary.[507] While the oocyte Mpk kinase is activated at the time of entry into metaphase of meiosis I, the MAP kinases present in somatic vertebrate cells are activated during the G_0

to G_1 phase transition. The MAP kinases and the Mpk kinase share many similarities with respect to size, physical properties, substrate specificity, and regulation. Both types of kinases may be activated by phosphorylation on tyrosine residues. Progesterone induces oocyte maturation and may be involved in activation of the MAP/Mpk kinases in the ovary.

MAP kinases may represent a point of convergence of different receptor-mediated signaling pathways and may play a role in integrating those signals. The activity and phosphorylation on tyrosine of the 42-kDa MAP kinase are stimulated by α-thrombin and basic FGF.[508] A MAP-2-related protein kinase, termed ERT kinase, is involved on phosphorylation of the EGF receptor on the Tyr-669 residue. The ERT kinase recognizes the consensus primary sequence Pro-Leu-Ser/Thr-Pro and is also involved in phosphorylating the human c-Myc protein at Ser-62 and the rat c-Jun protein at Ser-246.[509] Another kinase, TIK, is recognized by antibodies to phosphotyrosine but its activity is specifically directed to serine and threonine residues.[510] Transcripts of the TIK gene are expressed in all murine tissues. A serine/threonine kinase encoded by the mouse gene *Erk-1* has the ability to autophosphorylate on tyrosine.[511] Myelin basic protein and the MAP-2 kinase are substrates of the Erk-1 kinase. Members of the Erk and RSK family of protein kinases are present in both the cytoplasm and the nucleus.[512] Addition of growth factors to serum-deprived cells results in increased tyrosine and threonine phosphorylation and activation of cytosolic and nuclear MAP kinases. It is thus clear that within the cell there is a cascade of phosphorylations of discrete proteins on serine, threonine, and tyrosine residues, implicating the existence of complex interactions between serine/threonine-specific and tyrosine-specific kinases involved in the regulation of specific cell functions.

The casein kinases I and II are serine/threonine kinases with an important role in signal transduction mechanisms.[513] The casein kinases can recognize specific sequences in endogenous and exogenous substrates, including *src*-related sequences present in some cellular polypeptides.[514] The casein kinase I recognizes residues located on the carboxyl-terminal edge of acidic stretches. The enzyme phosphorylates the 25-kDa mRNA cap-binding protein.[515] Phosphotyrosyl-containing side chains can act as specificity determinants for casein kinase II. Some oncoproteins are phosphorylated on serine/threonine by casein kinase II, including the c-Erb-A/thyroid hormone receptor.[516] Both casein kinase II and the ribosomal protein S6 kinase are two important elements in the kinase cascade that leads to the initiation of cell proliferation stimulated by mitogens, hormones, and growth factors. Casein kinase activity may be regulated by phosphorylation. The cell cycle-associated protein kinase cdc2 catalyzes the phosphorylation of casein kinase II *in vitro* and at mitosis.[517]

Serine/threonine kinases involved in the phosphorylation of plasma membrane proteins when cells are stimulated to proliferate may be independent of Ca^{2+}, cAMP, and cGMP, and may require Mn^{2+} as an optimum cofactor, although Mg^{2+} can replace Mn^{2+} in one of these enzymes.[343] A ganglioside-stimulated kinase that is distinct from other phosphotransferases was purified from particulate fractions of guinea pig brain.[518]

Two types of cAMP-dependent protein kinases, type I and type II, are involved in phosphorylation of proteins on serine and threonine residues. Differences between the type I and type II isoenzymes reside not in the catalytic subunits, which are identical, but in the regulatory components of the enzymes.[519,520] The type I and type II isoenzymes may have different substrates and different functions and may respond to cAMP in a hormone- and tissue-specific manner. For example, in the human lung cancer cell line BEN, calcitonin selectively stimulates the type I isoenzyme, while in the human breast cancer cell lines T47D and MCF7 calcitonin exclusively activates the type II isoenzyme.[521] A cAMP-independent, Ca^{2+}-independent protein kinase, termed protease-activated kinase II (PAK II), may be involved in the phosphorylation of the ribosomal protein S6 by EGF and other mitogenic agents.[522] S6 is an important component of a mitogen-activated kinase cascade, and its activation in serum-stimulated Swiss 3T3 cells would be associated with its phosphorylation on serine and threonine residues.[362] Nuclear and cytosolic cAMP-independent protein kinases have been purified from the rat ventral prostate, but their precise structural and functional relationships are poorly understood.[523,524] Substances capable of inducing differentiation of neoplastic cells may act through changes in protein phosphorylations mediated by the activities of cAMP-dependent and -independent protein kinases.[525]

The role of serine/threonine-specific protein kinases in the malignant transformation of cells is less well understood than that of tyrosine-specific kinases. No differences in total serine/threonine-specific protein kinase activity were found between normal human gastric mucosa and samples from human gastric carcinomas, but a selective elevation of type I cAMP-dependent isoenzyme was detected in the carcinomas independent of the histological type.[526] The type I isoenzyme was elevated in xenotransplantable human gastric carcinomas in nude mice. These results suggest that type I cAMP-dependent protein kinase activity may be a marker for malignant transformation and transplantability of gastric tumors. Further studies are required for a proper evaluation of the role of serine/threonine kinases in different types of natural and experimental tumors.

Several oncoproteins possess serine/threonine protein kinase activity. This type of activity is intrinsically present in the v-Mos and v-Raf oncoproteins.[527-531] A similar activity is present in the products of the normal proto-oncogene counterparts of these viral oncogenes. The putative cellular oncoprotein Pim-1, as well as the Akt and Cot proteins, also possesses kinase activity specific for serine and threonine residues.

A. THE Mos PROTEIN

The product of the v-mos oncogene contained in the Moloney murine sarcoma virus (M-MuSV) is a soluble cytoplasmic protein, termed p37^{v-mos} or v-Mos.[532] This protein exhibits homology to the EGF precursor polypeptide.[533] The v-Mos oncoprotein is also related to both v-Src and the catalytic subunit of mammalian cAMP-dependent protein kinase.[534] A sequence located at amino acids 115 to 128 of the predicted v-Mos sequence shows homology with the ATP-binding site of bovine cAMP-dependent protein kinase as well as with the v-Abl, v-Yes, and v-Fms oncoproteins, which are members of the src gene superfamily and possess tyrosine kinase activity.[349] In all of these oncoproteins including v-Mos, a group of glycines with the sequence Gly-x-Gly-x-x-Gly lies 16 to 28 residues to the amino-terminal side of a lysine residue implicated in binding ATP. The v-Mos protein possesses cAMP-independent protein kinase activity with specificity for serine and threonine residues.[535,536] This activity is required for the neoplastic transformation of cells induced by the v-Mos oncoprotein.[537] The protein kinase C pathway, but not the protein kinase A pathway, modulates v-Mos protein kinase activity.[538] Histidine-to-tyrosine substitution at position 221 of v-Mos abolishes both the associated protein kinase activity of the protein and its transforming activity.[539] The mechanism of v-mos-induced neoplastic transformation remains unknown.

The normal c-Mos protein has an important physiologic role in oocyte maturation and metaphase arrest.[540] The c-mos proto-oncogene was initially considered to be silent in all normal tissues. However, by using a sensitive S$_1$ nuclease assay to screen RNA preparations, c-mos-related transcripts of different sizes have been detected in normal mouse embryos as well as in mouse testis and ovaries.[541] Transcription of the c-mos gene in the rat testis depends on binding of a testis-specific nuclear factor to a promoter.[542] Transcripts of the c-mos gene are present in human testis as well as in monkey ovaries, testis, epididymis, and seminal vesicle.[543] The transcripts are of different sizes in the testis and the ovary, and two major c-mos transcripts are present in the mouse embryo. Accumulation of the c-Mos protein is correlated with postnatal development of skeletal muscle.[544] A negative regulatory element located in the 5'-flanking side of the c-mos gene inhibits transcription of c-mos in somatic cells.[545]

The gonads are organs where c-mos gene expression plays an important role. A c-Mos protein of 43 kDa is expressed before meiosis during spermatogenesis in rodents.[546,547] Testicular c-mos mRNA expression correlates with the presence of haploid germ cells in several sterile mouse mutants.[548] The round spermatid, a haploid postmitotic germ cell, is the major source of c-mos mRNA in mouse testis, but grown oocytes contain higher concentrations of c-mos mRNA than round spermatids.[549] The c-mos mRNA accumulated in mature oocytes may be utilized during fertilization, the first cleavage division, and the early stages of embryogenesis. Transcripts of c-mos are not found in resting oocytes but accumulate soon after the oocyte enters the growth phase.[550] High levels of c-mos gene transcripts are present throughout oocyte growth and maturation as well as in ovulated eggs prior to fertilization and decline abruptly thereafter. Mouse oocyte maturation can be blocked by the introduction of anti-Mos antibodies into immature functional oocytes.[551,552] The c-Mos and cdc2 proteins form a complex associated with tubulin and localized on microtubules.[553,554] The c-Mos protein may have a role in regulating the assembly and/or function of the spindle during meiotic division. Transcripts of c-mos are found in undifferentiated embryonal carcinoma cells, suggesting that the c-mos product may play a role in early stages of development.[555] The c-Mos kinase can activate cyclin B, which together with cdc2 is an essential component of the maturation promoting factor (MPF), a universal regulator of mitotic and meiotic cell cycles.[556] However, the c-Mos kinase is not directly involved in the phosphorylation of cyclin B.[557]

The mouse c-mos proto-oncogene, which resides on chromosome 4,[558] is expressed not only in gonadal tissues and late-term embryos but also, at much lower levels, in adult tissues including brain, kidney, mammary gland, and epididymis.[559] Marked differences are observed in the size of c-mos gene transcripts detected in different mouse tissues. Mouse NIH/3T3 fibroblasts in culture express the c-mos gene in a cell cycle-dependent manner, with highest levels of c-*mos* transcripts (approximately five copies per cell) present in the G_2 phase of the cycle.[560] These results suggest that the mammalian c-Mos protein may have functions in addition to those related to gametogenesis.

The product of the c-*mos* gene of *Xenopus laevis* oocytes is a 39-kDa protein that is necessary for meiotic maturation.[561] Expression of the c-Mos protein can induce germinal vesicle breakdown and MPF activation in *Xenopus* oocytes, even in the absence of protein synthesis.[562] However, the oocytes do not proceed into meiosis II, indicating that the synthesis of additional proteins is required to complete their maturation. Microinjection of *mos*-specific antisense oligonucleotides into *Xenopus* oocytes inhibits c-Mos protein expression, resulting in prevention of germinal vesicle breakdown and inhibition of meiotic maturation.[563] Mos expression can be induced by treatment of *Xenopus* oocytes with progesterone and its synthesis precedes the activation of MPF and germinal vesicle breakdown in the oocytes.[564] The c-Mos protein may interact, directly or indirectly, with MPF, an M-phase-specific protein kinase complex that is composed of cdc2 and cyclin. The *Xenopus* c-Mos protein functions downstream of c-Ras but upstream of cdc2 and S6 kinase activation.[565] The c-Mos protein is the active component of the cytostatic factor, which is responsible for the arrest of unfertilized eggs at meiotic metaphase II.[566] The c-Mos protein is a tubulin-associated kinase that may participate in the modification of microtubules and contribute to the formation of the spindle.[567] The use of a highly sensitive method combining reverse transcription and PCR showed the ubiquitous, low-level expression of the c-*mos* gene in human tissues.[568] The c-Mos protein is probably required for the mechanism of division of both germ cells and somatic cells in vertebrates.

B. THE Raf/Mil PROTEINS

The v-*raf*/*mil* oncogene is present in several different acute retroviruses, including the avian carcinoma virus MH2 and the murine sarcoma virus 3611. The translation product of the MH2 retrovirus is a hybrid oncoprotein of 100 kDa whose biological effects, including its ability to induce production of a myelomonocytic growth factor in chicken macrophages, are mediated by the phosphorylation of cellular substrates on serine/threonine residues.[569]

DNA sequences with homology to v-*raf*/*mil* are present in all vertebrate species examined thus far, including man. At least five v-*raf*-related genes are contained in the human genome: three active genes, c-*raf*/c-*raf*-1, A-*raf*-1/*pks*, and B-*raf*, and two inactive genes, c-*raf*-2 and A-*raf*-2.[570-573] The proteins encoded by the three active *raf* genes contained in the human genome are termed Raf-1, A-Raf, and B-Raf. These proteins are kinases with specificity for serine and threonine residues that are involved in signaling pathways originated from surface receptors which may or may not possess tyrosine kinase activity. Studies with c-*raf*-1-specific antisense RNA show that the mammalian Raf-1 kinase functions as an essential signal transducer downstream of growth factor receptors, protein kinase C, and Ras proteins.[574] Both the normal and oncogenic Ras proteins bind to the amino-terminal domain of the c-Raf-1 protein kinase.[575] The normal Raf-1 kinase activates the MAP kinase pathway.[576,577] In turn, the Raf-1 protein is a substrate for MAP serine/threonine kinase, which may be involved in the regulation of the Raf-1 kinase activity *in vitro* and *in vivo*.[578] The sites phosphorylated on Raf-1 by MAP kinase undergo increased phosphorylation following stimulation of intact HIRC-B rat fibroblasts with insulin. CSF-2 and IL-3 phosphorylate the Raf-1 protein kinase in human myeloid cells, suggesting that Raf-1 may be involved in the signal transduction cascade of growth factors.[579] Studies using an inhibitory mutant of c-Raf-1 indicate that the c-Raf-1 kinase is also a component of an intracellular signaling pathway initiated by the v-Src kinase that leads to the induction of the mitogen-responsive transcription factor Egr-1.[580]

Genes homologous to mammalian *raf* sequences are present in insects and amphibians. Two c-*raf* have been detected in the *Drosophila* genome, and one of them, D-*raf*-1, has an important role in the early embryogenesis of the insect.[581] The D-*raf*-1 gene encodes a protein that exhibits homology to mammalian serine/threonine kinases and may have a function similar or identical to that of these enzymes. The D-Raf-1 protein is required maternally for the normal establishment of the anterior-posterior patterning during embryogenesis in *Drosophila*. D-Raf may function as a signal transducer in the action of the *torso* gene product, which is a tyrosine kinase that exhibits homology to the mammalian PDGF receptor.[582,583] c-*raf* gene transcripts of *Xenopus laevis* are members of the class of maternal RNAs and are expressed at relatively high levels during embryogenesis.[584] Several adult tissues of the amphibian also express c-*raf* gene transcripts. The Raf-1 protein kinase is important for progesterone-induced *Xenopus* oocyte maturation and acts downstream of c-Mos.[585]

The chicken c-Raf/Mil protein, as deduced from a cDNA clone, is composed of 647 amino acids and has a calculated molecular weight of 73,132.[586] Two domains are recognized in the protein, one of 250 amino acids on the carboxyl-terminal half exhibiting homology to the Src oncoprotein and protein kinase

C, and the other on the amino-terminal half containing a cysteine-rich segment, which shares homology with two similar repetitive domains of protein kinase C.

The rat c-Raf polypeptide, as deduced from a cDNA clone, is composed of 648-amino acids and has a molecular weight of 72,927, whereas the A-Raf product is a 604-amino acid polypeptide of 67,551 mol wt.[587] The carboxyl-terminal regions of these polypeptides show high homology with conserved ATP-binding sites and protein kinase catalytic domains, whereas the cysteine-rich amino-terminal regions of the same polypeptides exhibit striking homologies with the cysteine-rich regions present in protein kinase C. The c-Raf protein is activated by phosphorylation dependent on protein kinase C activity.[588] Activation of the T-cell receptor stimulates phosphorylation of c-Raf on serine residues and induces c-Raf kinase activity via a protein kinase C-dependent pathway.[589] Treatment of PC12 rat pheochromocytoma cells with NGF does not alter the levels of B-Raf protein but rapidly induces B-Raf phosphorylation on serine residues.[590] A similar effect is induced by treatment of the cells with either EGF or TPA.

Three raf-related genes are active in the mouse genome: c-raf-1, A-raf, and B-raf.[591,592] The c-raf-1 gene is located on human chromosome 3 at region 3p25, whereas the A-raf gene is located on the X chromosome in mouse and human.[593,594] Transcripts of c-raf-1 are universally present in mouse tissues with highest levels in striated muscle, cerebellum, and fetal brain. Apparently, c-raf-1 is a "housekeeping gene" whose level of expression in each tissue may be influenced, either positively or negatively, by exogenous signals such as hormones and growth factors. In contrast to c-raf-1, the expression of A-raf appears to be subjected to a much tighter control and shows great variation among different tissues, with higher levels in epididymis and prostate, where its expression would be regulated by hormones such as prolactin, aldosterone, and calcitriol. The most stringently regulated member of the raf family is B-raf, whose expression is restricted mainly to the fetal and adult brain, suggesting a specialized role of its protein product in neural tissue.

The structure of the human c-raf-1 gene has been determined and the complete amino acid sequence of the protein product was deduced from the cDNA sequence.[595] The Raf-1 protein is phosphorylated on serine in vivo by the ligand-activated EGF receptor tyrosine kinase.[596] This phosphorylation occurs by a mechanism that is independent of EGF receptor phosphorylation and internalization. Insulin and other growth factors may also induce Raf-1 protein phosphorylation, which results in enhanced kinase activity of the protein. The Raf-1 protein may be involved in the intracellular transduction mechanisms of hormones and growth factors. Studies with antisense RNA indicate that the Raf-1 kinase is an essential component of the oncogenic signal cascade shared by EGF and PDGF in NRK cells.[597] Other types of growth stimulation may also involve the Raf-1 protein. Proliferative stimulation of human B cells, either normal or neoplastic, with anti-Ig antibody results in rapid phosphorylation of Raf-1 on serine residues, which may be due to the activation of protein kinase C.[598]

All the human raf genes encode serine/threonine-specific protein kinases. A full-length cDNA clone derived from the human A-raf gene encodes a 606-amino acid protein of 67,530 mol wt which shows 75% amino acid identities with the human Raf-1 protein.[599] The A-Raf protein is expressed in human hematopoietic cell lines of the myeloid and T-cell lineages. Analysis of a cDNA of the B-raf gene isolated from a human testis cDNA library showed the presence of three conserved regions previously identified in Raf kinases.[600] Human B-raf cDNA encodes a protein of 651 amino acids. This gene is expressed in testis, suggesting that its product may have a role in germ cell maturation. The B-raf gene is also expressed in neural tissues. B-Raf is the major isotype of Raf protein present in hippocampal neurons. Redistribution of kinase from the cytoplasm into the nucleus was observed in injured hippocampal neurons following an ischemic insult.[601] B-Raf expressed in PC12 rat neuroblastoma cells and mouse brain tissues is a protein of 95 kDa.[602] Treatment of PC12 cells with NGF induces autophosphorylation of B-Raf on the Thr-372 residue.

Expression of the c-raf gene in human hematopoietic cells may not be directly related to either cell proliferation or cell differentiation.[603] However, the human and murine 74-kDa Raf protein may be a second messenger that mediates the signal transduction of CSF-1 in its target cells.[604,605] Although the Raf-1 protein is probably not a substrate for the CSF-1 receptor tyrosine kinase, its serine/threonine kinase activity may be stimulated in a particular step of the phosphorylation cascade initiated by the binding of CSF to its receptor at the cell surface, which may eventually lead to the stimulation of cell proliferation. The c-Raf kinase may also be involved in the transduction of IL-2-elicited signals. IL-2 induces rapid phosphorylation of the c-Raf kinase in the IL-2 dependent murine T-cell line CTLL-2, and this phosphorylation is associated with an increased kinase activity of the proto-oncogene product.[606] Raf-1 is required for serum-, TPA-, and Ras-induced expression of target genes in mouse NIH/3T3 cells.[607] The Raf-1 kinase may function as a switch that connects growth factor receptor activation at the membrane with transcriptional processes occurring in the nucleus. According to a model, in quiescent cells the Raf-1 protein is held in an inactive conformation by the action of a repressor which is relieved upon the action of growth factors by a Ras-induced factor, which binds to a cysteine finger region of Raf-1, resulting in activation of the Raf-1 kinase. This activation may lead to the transcriptional activation of specific target genes and, ultimately, to the stimulation of cell growth.[607]

C. THE Pim-1 PROTEIN

A putative proto-oncogene, *pim*-1, was found to be activated in a high proportion of MuLV-induced murine T-cell lymphomas due to proviral insertion either 5′ or 3′ to the gene.[608,609] The *pim*-1 gene was mapped within the *t* complex on mouse chromosome 17.[610] The homologous human gene is located on chromosome 6 at region 6p21. High levels of *pim*-1 gene transcripts are expressed in the erythroleukemia cell line K562, which contains a rearrangement involving the 6p21 region.[611] However, enhanced expression of the *pim*-1 gene occurs in many human leukemias by mechanisms other than translocation or amplification.[612]

The murine and human *pim*-1 gene code for a 313-amino acid Pim-1 protein of 34 kDa.[613-615] In addition, the murine *pim*-1 gene encodes an amino-terminal extension protein of 44 kDa, which is synthesized by alternative translation initiation of an upstream CUG codon.[616] The human *pim*-1 gene product was initially believed to be represented by a tyrosine kinase,[617] but recent evidence unequivocally indicates that the kinase activity intrinsic to the murine and human Pim-1 protein is specific for serine/threonine residues.[618-620] The *pim*-1 gene is expressed in hematopoietic tissues as well as in testis and ovaries, but the precise role of the Pim-1 protein kinase is unknown.[621] Consecutive inactivation of both *pim*-1 proto-oncogene alleles by homologous recombination in mouse embryonic stem cells shows that the Pim-1 protein is not required for the normal growth and differentiation of these cells.

D. THE Akt/Rac PROTEIN

The AKT8 virus, the only acute retrovirus isolated from a rodent T-cell lymphoma (AKR), can transform mink lung cells in culture. A defective clone of the AKT8 virus contained an oncogene, v-*akt*, derived from a mouse cellular gene.[622] The AKT8 retrovirus is tumorigenic and is capable of inducing thymic lymphoma upon inoculation into newborn mice. Murine v-Akt is a 105-kDa fusion phosphoprotein containing viral Gag sequences at its amino-terminus.[623] The oncogenic activation of the v-*akt* gene may be due to its fusion to a viral *gag* gene. The v-Akt oncoprotein, but not the normal c-Akt protein, is myristoylated and the two proteins exhibit different intracellular distribution.[624] While the c-Akt protein is almost exclusively cytoplasmic, v-Akt is associated with the nucleus. The normal murine c-Akt phosphoprotein is a serine/threonine kinase related to protein kinase C. The c-Akt noncatalytic domain contains an SH2-like region, suggesting that c-Akt may represent a functional link between serine/threonine and tyrosine protein phosphorylation pathways. The c-*akt* gene is expressed in all mouse tissues, with highest levels of expression in thymus. A human homolog of the murine c-*akt* gene, termed *rac*, encodes a serine/threonine kinase related to protein kinase C and cAMP-dependent protein kinase.[625] The human c-*akt*/*rac* gene is localized on chromosome region 14p32, proximal to the Ig heavy chain locus.

E. THE Cot PROTEINS

A putative proto-oncogene, *cot*, identified by transfection assay of the hamster embryo-derived cell line SHOK with DNA from the human cell line TCO-4, derived from an anaplastic thyroid cancer, encodes two serine kinases of 46 and 52 kDa.[626,627] The structure and tissue-specific expression of the murine *cot* gene have been determined.[628] Cot proteins are localized predominantly in the cytosol. Oncogenic activation of the *cot* gene probably occurred during the DNA transfection experiment and was not responsible for the original thyroid cancer.

F. SUBSTRATES FOR SERINE/THREONINE-SPECIFIC PROTEIN KINASES

Protein phosphorylation on serine and threonine may be associated with the action of hormones, growth factors, and oncoproteins. The phosphorylated cellular proteins include enzymes as well as growth factor receptors, which may result in changes in the functional properties of the receptors.

The basic 39-kDa protein S6, contained in the 40S ribosomal subunit, is phosphorylated on serine and threonine residues when cultured cells are stimulated with serum, mitogens, or growth factors.[362] This phosphorylation may occur, at least in part, through the activation of a specific protein kinase, the mitogen-activated S6 kinase.[629,630] In turn, the activation of this kinase by growth factors would depend on signals involving the membrane potential and/or K+ fluxes as well as on redistribution of intracellular Ca^{2+}.[631] S6 phosphorylation may play an important role in the regulation of cell proliferation. Phosphorylation of the S6 protein is constitutively stimulated, in the presence or absence of serum, by the v-Abl oncoprotein.[632] Because v-Abl is a tyrosine kinase, the phosphorylation of S6 protein on serine may be an indirect consequence of the primary action of the v-Abl oncoprotein. S6 protein phosphorylation is stimulated by the PMA, serum, and the v-Src oncoprotein, probably through common pathways.[633,634] The same serine sites of S6 can be phosphorylated under the influence of the different agents. The precise role of S6 phosphorylation in cell proliferation is unknown. Estrogen stimulates the growth of target cells (human mammary tumor cells ZR-75) without activation of S6 kinase or S6 phosphorylation.[635] The yeast *Saccharomyces cerevisiae* has a ribosomal protein, S10, which is equivalent to the S6 protein of animal cells, but the results obtained with oligonucleotide mutagenesis studies indicate that S10 phosphorylation is not essential for growth.[636]

Cellular proteins of high molecular weight may be substrates for serine/threonine protein kinases. The 300-kDa microtubule-associated proteins 1 and 2 (MAP-1 and MAP-2) are substrates for serine phosphorylation stimulated by various growth factors (insulin, EGF, FGF, and transferrin) as well as by diacylglycerol and phorbol ester in the presence of Ca^{2+}.[637] Phosphorylation of MAP-1 and MAP-2 is associated with their translocation from the cytoplasm to the nucleus. MAP-1 and MAP-2 are associated with the mitotic apparatus. Many other proteins, including membrane and nuclear proteins, are substrates for serine/threonine-specific protein kinases present in normal or transformed cells, but the precise role of the protein modifications induced by these kinases in the respective protein functions are largely unknown.

G. SERINE/THREONINE KINASES AND NEOPLASTIC TRANSFORMATION

Both serine/threonine- and tyrosine-specific protein kinases have an important role in oncogenesis. Several oncoproteins, including Mos and Raf, have serine/threonine kinase activity. The activity of various serine/threonine kinases is altered in malignant neoplasms. Elevated levels of cAMP- and cGMP-dependent protein kinases and casein kinase II activity, as well as protein kinase C activity, have been found, for example, in human squamous cell carcinomas of the upper aerodigestive tract, as compared with normal local mucosa.[638] The enhanced serine/threonine kinase activities present in malignant tissues may result in increased levels of phosphorylation of several cellular proteins, including histones. The possible role of hormones and growth factors in the altered serine/threonine kinase activity detected in tumor tissues is almost unknown.

V. SERINE/THREONINE-SPECIFIC PROTEIN PHOSPHATASES

The levels of phosphorylation of cellular proteins on serine and threonine residues are determined by an equilibrium between the activity of serine/threonine-specific protein kinases and serine/threonine-specific protein phosphatases (protein-serine/threonine phosphatases).[639,640] Four major types of serine/threonine phosphatases identified in eukaryotic cells have been termed PP1 (PPH-1), PP2A (PPH-2A), PP2B (PPH-2B), and PP2C (PPH-2C). These isoenzymes can be distinguished by their action on phosphorylase kinase and by their sensitivity to certain activators and inhibitors. More recently, several serine/threonine phosphatases (PPV, PP2B$_w$, PPX, PPY, and PPZ) have been identified by cloning cDNA from mammalian and *Drosophila* libraries. The complete amino acid sequences of some protein-serine/threonine phosphatases have been deduced from the analysis of cloned cDNA, but the precise function and regulation of these enzymes remains to be elucidated. PP1, the most abundant serine/threonine kinase in many cells and tissues, is activated by insulin and EGF during the cell cycle. The v-Abl oncoprotein phosphorylates the catalytic subunit of PP1.[641] The activity of PP2A is regulated by phosphorylation on tyrosine residues.[642] The general biological importance of serine/threonine-specific protein phosphatases is indicated by their extreme conservation in evolution. The regulatory action of PP1 and PP2A on the progression of the cell cycle may be associated with regulation of the phosphorylation state of the RB tumor suppressor protein.[643]

Okadaic acid, a complex fatty acid polyketal produced by marine dinoflagellates such as black sponge, is an inhibitor of serine/threonine phosphatase activity and may thus serve as a probe for identifying biological processes that are controlled by protein dephosphorylation. In primary human fibroblast cultures, okadaic acid induces a composite pattern of early phosphorylation changes and gene expression that is strikingly similar to those induced by TNF-α and IL-1.[644] Like TNF-α, okadaic acid induces c-*jun*, *egr*-1, and IL-6 gene transcription in human fibroblasts. This mimicry suggests that regulation of protein phosphatase activity may play a role in the transductional mechanisms of action of growth factors.

Treatment of human leukemic cell lines with okadaic acid may result in cell differentiation, and this effect may be accompanied by enhanced expression of the c-*jun* proto-oncogene.[645] Okadaic acid has a potential to revert the transformed phenotype induced by oncogenes such as *raf* and *ret*.[646] Differentiation of human breast cancer cell lines (MCF-7, Au-565, and MB-231) by okadaic acid correlates with its ability to inhibit PP1 and PP2A activities.[647] In certain systems, okadaic acid acts as a tumor promoter, but it is not oncogenic and is not capable of inducing morphological transformation of hamster embryo cells.[648]

Protein phosphatases may have an important role in mitogenic signaling pathways. Studies performed on early melanocyte passages suggest that protein phosphatase inhibitors behave as growth suppressors and that both serine/threonine-specific and tyrosine-specific protein phosphatase activities are required early during mitogenesis in cells of melanocytic origin and probably also in other types of cells.[649] Homologs of the fission yeast mitotic inducer cdc25 function as vertebrate protein phosphatases which activate complexes formed by the cdc2 kinase and cyclin B.[650] This complex is an essential regulator of the onset of mitosis and is inhibited by phosphorylation of cdc2 on Thr-14 and Tyr-15. On the other hand, complexes between cdk2 and cyclin A or cyclin E have an important role in entry of cells into S phase of the cycle.

There are three human homologs of yeast cdc25, termed CDC25A, -B, and -C. While CDC25C functions as a positive regulator of entry into mitosis in vertebrate cells, CDC25A appears to be required for progression through M phase of the cycle and cell division. It is thus clear that both protein kinases and protein phosphatases are centrally involved in the regulation of the cell cycle.

REFERENCES

1. **Cohen, P.,** Protein phosphorylation and hormone action, *Proc. R. Soc. London,* B234, 115, 1988.
2. **Meek, D. W. and Street, A. J.,** Nuclear protein phosphorylation and growth control, *Biochem. J.,* 287, 1, 1992.
3. **Wang, J. Y. J.,** Oncoprotein phosphorylation and cell cycle control, *Biochim. Biophys. Acta,* 1114, 179, 1992.
4. **Cohen, P.,** The role of protein phosphorylation in neural and hormonal control of cellular activity, *Nature,* 296, 613, 1982.
5. **Fischer, E. H.,** Cellular regulation by protein phosphorylation, *Bull. Inst. Pasteur (Paris),* 81, 7, 1983.
6. **Krebs, E. G.,** The phosphorylation of proteins: a major mechanism for biological regulation, *Biochem. Soc. Trans.,* 13, 813, 1985.
7. **Cooper, E. and Spaulding, S. W.,** Hormonal control of the phosphorylation of histones, HMG proteins and other nuclear proteins, *Mol. Cell. Endocrinol.,* 39, 1, 1985.
8. **Sibley, D. R., Benovic, J. L., Caron, M. G., and Lefkowitz, R. J.,** Phosphorylation of cell surface receptors: a mechanism for regulating signal transduction pathways, *Endocrine Rev.,* 9, 38, 1988.
9. **Hanks, S. K., Quinn, A. M., and Hunter, T.,** The protein kinase family: conserved features and deduced phylogeny of the catalytic domains, *Science,* 241, 42, 1988.
10. **Ralph, R. K., Darkin-Rattray, S., and Schofield, P.,** Growth-related protein kinases, *BioEssays,* 12, 121, 1990.
11. **Panayotou, G. and Waterfield, M. D.,** The assembly of signalling complexes by receptor tyrosine kinases, *BioEssays,* 15, 171, 1993.
12. **Pelech, S. L., Sanghera, J. S., and Daya-Makin, M.,** Protein kinase cascades in meiotic and mitotic cell cycle control, *Biochem. Cell Biol.,* 68, 1297, 1990.
13. **Kidd, V. J.,** Cell division control-related protein kinases: putative origins and functions, *Mol. Carcinogen.,* 5, 95, 1992.
14. **Nurse, P.,** Eukaryotic cell cycle control, *Biochem. Soc. Trans.,* 20, 239, 1992.
15. **Hunter, T. and Cooper, J. A.,** Protein tyrosine kinases, *Annu. Rev. Biochem.,* 54, 897, 1985.
16. **Mäkelä, T. P. and Alitalo, K.,** Tyrosine kinases in control of cell growth and transformation, *Med. Biol.,* 64, 325, 1986.
17. **Feige, J.-J. and Chambaz, E. M.,** Membrane receptors with protein-tyrosine kinase activity, *Biochimie,* 69, 379, 1987.
18. **Yarden, Y. and Ullrich, A.,** Growth factor receptor tyrosine kinases, *Annu. Rev. Biochem.,* 57, 443, 1988.
19. **Cadena, D. L. and Gill, G. N.,** Receptor tyrosine kinases, *FASEB J.,* 6, 2332, 1992.
20. **Ahn, N. G. and Krebs, E. G.,** Evidence for an epidermal growth factor-stimulated protein kinase cascade in Swiss 3T3 cells. Activation of serine peptide kinase activity by myelin basic protein kinases *in vitro, J. Biol. Chem.,* 265, 11495, 1990.

21. **Howell, B. W., Afar, D. E. H., Lew, J., Douville, E. M. J., Icely, P. L. E., Gray, D. A., and Bell, J. C.,** SYT, a tyrosine-phosphorylating enzyme with sequence homology to serine/threonine kinases, *Mol. Cell. Biol.,* 11, 568, 1991.

22. **Bourret, R. B., Borkovich, K. A., and Simon, M. I.,** Signal transduction involving protein phosphorylation in prokaryotes, *Annu. Rev. Biochem.,* 60, 401, 1991.

23. **Chiang, T. M., Reizer, J., and Beachey, E. H.,** Serine and tyrosine protein kinase activities in *Streptococcus pyogenes.* Phosphorylation of native and synthetic peptides of streptococcal M proteins, *J. Biol. Chem.,* 264, 2957, 1989.

24. **Blumenthal, D. K., Takio, K., Hansen, R. S., and Krebs, E. G.,** Dephosphorylation of cAMP-dependent protein kinase regulatory subunit (type II) by calmodulin-dependent protein phosphatase: determinants of substrate specificity, *J. Biol. Chem.,* 261, 8140, 1986.

25. **Tung, H. Y. L.,** Phosphorylation of the calmodulin-dependent protein phosphatase by protein kinase C, *Biochem. Biophys. Res. Commun.,* 138, 783, 1986.

26. **Morla, A. O. and Wang, J. Y. J.,** Protein tyrosine phosphorylation in the cell cycle of BALB/c 3T3 fibroblasts, *Proc. Natl. Acad. Sci. U.S.A.,* 83, 8191, 1986.

27. **Contor, L., Lamy, F., Lecocq, R., Roger, P. P., and Dumont, J. E.,** Differential protein phosphorylation in induction of thyroid cell proliferation by thyrotropin, epidermal growth factor, or phorbol ester, *Mol. Cell. Biol.,* 8, 2494, 1988.

28. **Pasquale, E. B., Maher, P. A., and Singer, S. J.,** Comparative study of the tyrosine phosphorylation of proteins in Swiss 3T3 fibroblasts stimulated by a variety of mitogenic agents, *J. Cell. Physiol.,* 137, 146, 1988.

29. **Downing, J. R. and Reynolds, A. B.,** PDGF, CSF-1, and EGF induce tyrosine phosphorylation of p120, a pp60*src* transformation-associated substrate, *Oncogene,* 6, 607, 1991.

30. **Collett, M. S. and Erikson, R. L.,** Protein kinase activity associated with the avian sarcoma virus *src* gene product, *Proc. Natl. Acad. Sci. U.S.A.,* 75, 2021, 1978.

31. **Hunter, T. and Sefton, B. M.,** Transforming gene product of Rous sarcoma virus phosphorylates tyrosine, *Proc. Natl. Acad. Sci. U.S.A.,* 77, 1311, 1980.

32. **Erikson, E., Cook, R., Miller, G. J., and Erikson, R. L.,** The same normal cell protein is phosphorylated after transformation by avian sarcoma viruses with unrelated transforming genes, *Mol. Cell. Biol.,* 1, 43, 1981.

33. **Cooper, J. A. and Hunter, T.,** Four different classes of retroviruses induce phosphorylation of tyrosines present in similar cellular proteins, *Mol. Cell. Biol.,* 1, 394, 1981.

34. **Patschinsky, T. and Sefton, B. M.,** Evidence that there exist four classes of RNA tumor viruses which encode proteins associated with tyrosine protein kinase activity, *J. Virol.,* 39, 104, 1981.

35. **Eiseman, E. and Bolen, J. B.,** *src*-related tyrosine protein kinases as signaling components in hematopoietic cells, *Cancer Cells,* 2, 303, 1990.

36. **Punt, C. J. A.,** Regulation of hematopoietic cell function by protein tyrosine kinase-encoding oncogenes, a review, *Leukemia Res.,* 16, 551, 1992.

37. **Bolen, J. B., Rowley, R. B., Spana, C., and Tsygankov, A. Y.,** The Src family of tyrosine protein kinases in hematopoietic signal transduction, *FASEB J.,* 6, 3403, 1992.

38. **Brickell, P. M.,** The c-*src* family of protein-tyrosine kinases, *Int. J. Exp. Pathol.,* 72, 97, 1991.

39. **Okada, M., Nada, S., Yamanashi, Y., Yamamoto, T., and Nakagawa, H.,** CSK: a protein-tyrosine kinase involved in regulation of *src* family kinases, *J. Biol. Chem.,* 266, 24249, 1991.

40. **June, C. H., Fletcher, M. C., Ledbetter, J. A., Schieven, G. L., Siegel, J. N., Phillips, A. F., and Samelson, L. E.,** Inhibition of tyrosine phosphorylation prevents T-cell receptor-mediated signal transduction, *Proc. Natl. Acad. Sci. U.S.A.,* 87, 772, 1990.

41. **Paris, S., Chambard, J.-C., and Pouysségur, J.,** Tyrosine kinase-activating growth factors potentiate thrombin-and AlF$_4$-induced phosphoinositide breakdown in hamster fibroblasts. Evidence for positive cross-talk between the two mitogenic signaling pathways, *J. Biol. Chem.,* 263, 12893, 1988.

42. **Willman, C. L., Stewart, C. C., Griffith, J. K., Stewart, S. J., and Tomasi, T. B.,** Differential expression and regulation of the c-*src* and c-*fgr* protooncogenes in myelomonocytic cells, *Proc. Natl. Acad. Sci. U.S.A.,* 84, 4480, 1987.

43. **Izumi, T., Saeki, Y., Akanuma, Y., Takaku, F., and Kasuga, M.,** Requirement for receptor-intrinsic tyrosine kinase activities during ligand-induced membrane ruffling of KB cells. Essential sites of *src*-related growth factor receptor kinases, *J. Biol. Chem.,* 263, 10386, 1988.

44. **Carpenter, G.,** Receptor tyrosine kinase substrates: *src* homology domains and signal transduction, *FASEB J.,* 6, 3283, 1992.

45. **Pawson, T. and Gish, G. D.,** SH2 and SH3 domains: from structure to function, *Cell,* 71, 359, 1992.
46. **Cola, C., Brunati, A. M., Borin, G., Ruzza, P., Calderan, A., De Castiglione, R., and Pinna, L. A.,** Synthetic peptides reproducing the EGF-receptor segment homologous to the pp60^{v-src} phosphoacceptor site. Phosphorylation by tyrosine protein kinases, *Biochim. Biophys. Acta,* 1012, 191, 1989.
47. **Kong, S.-K., Chohan, I. S., and Wang, J. H.,** Autophosphorylation and autoactivation of spleen protein tyrosine kinase, *J. Biol. Chem.,* 263, 14523, 1988.
48. **Shoelson, S. E., White, M. F., and Kahn, C. R.,** Nonphosphorylatable substrate analogs selectively block autophosphorylation and activation of the insulin receptor, epidermal growth factor receptor, and pp60^{v-src} kinases, *J. Biol. Chem.,* 264, 7831, 1989.
49. **Waksman, G., Kominos, D., Robertson, S. C., Pant, N., Baltimore, D., Birge, R. B., Cowburn, D., Hanafusa, H., Mayer, B. J., Overduin, M., Resh, M. D., Rios, C. B., Silverman, L., and Kuriyan, J.,** Crystal structure of the phosphotyrosine recognition domain SH2 of v-*src* complexed with tyrosine-phosphorylated peptides, *Nature,* 358, 646, 1992.
50. **Pawson, T.,** Non-catalytic domains of cytoplasmic protein-tyrosine kinases: regulatory elements in signal transduction, *Oncogene,* 3, 491, 1988.
51. **Hanks, S. K., Quinn, A. M., and Hunter, T.,** The protein kinase family: conserved features and deduced phylogeny of the catalytic domains, *Science,* 241, 42, 1988.
52. **Koch, C. A., Anderson, D., Moran, M. F., Ellis, C., and Pawson, T.,** SH2 and SH3 domains: elements that control interactions of cytoplasmic signaling proteins, *Science,* 252, 668, 1991.
53. **Zhu, G., Decker, S. J., and Saltiel, A. R.,** Direct analysis of the binding of Src-homology 2 domains of phospholipase C to the activated epidermal growth factor receptor, *Proc. Natl. Acad. Sci. U.S.A.,* 89, 9559, 1992.
54. **Matsuda, M., Mayer, B. J., and Hanafusa, H.,** Identification of domains of the v-*crk* oncogene product sufficient for association with phosphotyrosine-containing proteins, *Mol. Cell. Biol.,* 11, 1607, 1991.
55. **Reichman, C. T., Mayer, B. J., Keshav, S., and Hanafusa, H.,** The product of the cellular *crk* gene consists primarily of SH2 and SH3 regions, *Cell Growth Differ.,* 3, 451, 1992.
56. **Pelicci, G., Lanfrancone, L., Grignani, F., McGlade, J., Cavallo, F., Forni, G., Nicoletti, I., Grignani, F., Pawson, T., and Pelicci, P. G.,** A novel transforming protein (SHC) with an SH2 domain is implicated in signal transduction, *Cell,* 70, 93, 1992.
57. **McGlade, J., Cheng, A., Pelicci, G., Pelicci, P. G., and Pawson, T.,** Shc proteins are phosphorylated and regulated by the v-Src and v-Fps protein-tyrosine kinases, *Proc. Natl. Acad. Sci. U.S.A.,* 89, 8869, 1992.
58. **Lehmann, J. M., Riethmüller, G., and Johnson, J. P.,** Nck, a melanoma cDNA encoding a cytoplasmic protein consisting of the src homology units SH2 and SH3, *Nucleic Acids Res.,* 18, 1048, 1990.
59. **Park, D. and Rhee, S. G.,** Phosphorylation of Nck in response to a variety of receptors, phorbol myristate acetate, and cyclic AMP, *Mol. Cell. Biol.,* 12, 5816, 1992.
60. **Wi, W., Hu, P., Skolnik, E. Y., Ullrich, A., and Schlessinger, J.,** The SH2 and SH3 domain-containing Nck protein is oncogenic and a common target for phosphorylation by different surface receptors, *Mol. Cell. Biol.,* 12, 5824, 1992.
61. **Chou, M. M., Fajardo, J. E., and Hafusa, H.,** The SH2-and SH3-containing Nck protein transforms mammalian fibroblasts in the absence of elevated phosphotyrosine levels, *Mol. Cell. Biol.,* 12, 5834, 1992.
62. **Meckling-Hansen, K., Nelson, R., Branton, P., and Pawson, T.,** Enzymatic activation of Fujinami sarcoma virus *gag-fps* transforming proteins by autophosphorylation at tyrosine, *EMBO J.,* 6, 659, 1987.
63. **Courtneidge, S. A.,** Activation of the pp60^{c-src} kinase by middle T antigen binding or by dephosphorylation, *EMBO J.,* 4, 1471, 1985.
64. **Vila, J. and Weber, M. J.,** Mitogen-stimulated tyrosine phosphorylation of a 42-kD cellular protein: evidence for a protein kinase-C requirement, *J. Cell. Physiol.,* 135, 285, 1988.
65. **Ziemicki, A., Catelli, M.-G., Joab, I., and Moncharmont, B.,** Association of the heat shock protein HSP90 with steroid hormone receptors and tyrosine kinase oncogene products, *Biochem. Biophys. Res. Commun.,* 138, 1298, 1986.
66. **Sartor, O., Sameshima, J. H., and Robbins, K. C.,** Differential association of cellular proteins with family protein-tyrosine kinases, *J. Biol. Chem.,* 266, 6462, 1991.
67. **Lindberg, R. A., Thompson, D. P., and Hunter, T.,** Identification of cDNA clones that code for protein-tyrosine kinases by screening expression libraries with antibodies against phosphotyrosine, *Oncogene,* 3, 629, 1988.

68. **Pasquale, E. B. and Singer, S. J.,** Identification of a developmentally regulated protein-tyrosine kinase by using anti-phosphotyrosine antibodies to screen a cDNA expression library, *Proc. Natl. Acad. Sci. U.S.A.,* 86, 5449, 1989.

69. **Peake, I.,** The polymerase chain reaction, *J. Clin. Pathol.,* 42, 673, 1989.

70. **Wilks, A. F.,** Two putative protein-tyrosine kinases identified by application of the polymerase chain reaction, *Proc. Natl. Acad. Sci. U.S.A.,* 86, 1603, 1989.

71. **Frackelton, A. R., Jr., Tremble, P. M., and Williams, L. T.,** Evidence for the platelet-derived growth factor-stimulated tyrosine phosphorylation of the platelet-derived growth factor receptor *in vivo*: immunopurification using a monoclonal antibody to phosphotyrosine, *J. Biol. Chem.,* 259, 7909, 1984.

72. **Seki, J., Owada, M. K., Sakato, N., and Fujio, H.,** Direct identification of phosphotyrosine-containing proteins in some retrovirus-transformed cells by use of anti-phosphotyrosine antibody, *Cancer Res.,* 46, 907, 1986.

73. **Di Renzo, M. F., Ferracini, R., Naldini, L., Giordano, S., and Comoglio, P. M.,** Immunological detection of proteins phosphorylated at tyrosine in cells stimulated by growth factors or transformed by retroviral-oncogene-coded tyrosine kinases, *Eur. J. Biochem.,* 158, 383, 1986.

74. **Wang, J. Y. J.,** Antibodies for phosphotyrosine: analytical and preparative tool for tyrosyl-phosphorylated proteins, *Anal. Biochem.,* 172, 1, 1988.

75. **Kruse, C. H., Holden, K. G., Pritchard, M. L., Feild, J. A., Rieman, D. J., Greig, R. G., and Poste, G.,** Synthesis and evaluation of multisubstrate inhibitors of an oncogene-encoded tyrosine-specific protein kinase. 1, *J. Med. Chem.,* 31, 1762, 1988.

76. **Kruse, C. H., Holden, K. G., Offen, P. H., Pritchard, M. L., Feild, J. A., Rieman, D. J., Bender, P. E., Ferguson, B., Greig, R. G., and Poste, G.,** Synthesis and evaluation of multisubstrate inhibitors of an oncogene-encoded tyrosine-specific protein kinase. 2, *J. Med. Chem.,* 31, 1768, 1988.

77. **Shiraishi, T., Owada, M. K., Tatsuka, M., Fuse, Y., Watanabe, K., and Kakunaga, T.,** A tyrosine-specific protein kinase inhibitor, alpha-cyano-3-ethoxy-4-hydroxy-5-phenylthiomethylcinnamide, blocks the phosphorylation of tyrosine substrate in intact cells, *Jpn. J. Cancer Res.,* 81, 645, 1990.

78. **Akiyama, T., Ishida, J., Nakagawa, S., Ogawara, H., Watanabe, S., Itoh, N., Shibuya, M., and Fukami, Y.,** Genistein, a specific inhibitor of tyrosine-specific protein kinases, *J. Biol. Chem.,* 262, 5592, 1987.

79. **Shiraishi, T., Domoto, T., Imai, N., Shimada, Y., and Watanabe, K.,** Specific inhibitors of tyrosine-specific protein kinase, synthetic 4-hydroxycinnamamide derivatives, *Biochem. Biophys. Res. Commun.,* 147, 322, 1987.

80. **Dean, N. M., Kanemitsu, M., and Boynton, A. L.,** Effects of the tyrosine-kinase inhibitor genistein on DNA synthesis and phospholipid-derived second messenger generation in mouse 10T1/2 fibroblasts and rat liver T51B cells, *Biochem. Biophys. Res. Commun.,* 165, 795, 1989.

81. **Fujita-Yamaguchi, Y. and Kathuria, S.,** Characterization of receptor tyrosine-specific protein kinases by the use of inhibitors. Staurosporine is a 100-times more potent inhibitor of insulin receptor than IGF-I receptor, *Biochem. Biophys. Res. Commun.,* 157, 955, 1988.

82. **Donella-Deana, A., Monti, E., and Pinna, L. A.,** Inhibition of tyrosine protein kinases by the antineoplastic agent adriamycin, *Biochem. Biophys. Res. Commun.,* 160, 1309, 1989.

83. **Baginski, I., Commerçon, A., Tocqué, B., Colson, G., and Zerial, A.,** Selective inhibition of tyrosine protein kinase by a synthetic multisubstrate analog, *Biochem. Biophys. Res. Commun.,* 165, 1324, 1989.

84. **Svoboda, J.,** Rous sarcoma virus, *Intervirology,* 26, 1, 1986.

85. **Wyke, J. A. and Stoker, A. W.,** Genetic analysis of the form and function of the viral *src* oncogene product, *Biochim. Biophys. Acta,* 907, 47, 1987.

86. **Wang, L.-H.,** The mechanism of transduction of proto-oncogene c-*src* by avian retroviruses, *Mutat. Res.,* 186, 135, 1987.

87. **Kato, J., Hirota, Y., Nakamura, N., Nakamura, N., and Takeya, T.,** Structural features of the carboxy terminus of p60[c-*src*] that are required for the regulation of its intrinsic kinase activity, *Jpn. J. Cancer Res.,* 78, 1354, 1987.

88. **Welham, M. J. and Wyke, J. A.,** A single point mutation has pleiotropic effects on pp60[v-*src*] function, *J. Virol.,* 62, 1898, 1988.

89. **Espino, P. C., Harvey, R., Schweickhardt, R. L., White, G. A., Smith, A. E., and Cheng, S. H.,** The amino-terminal region of pp60[c-*src*] has a modulatory role and contains multiple sites of tyrosine phosphorylation, *Oncogene,* 5, 283, 1990.

90. **Feder, D. and Bishop, J. M.,** Purification and enzymatic characterization of pp60$^{c\text{-}src}$ from human platelets, *J. Biol. Chem.,* 265, 8205, 1990.

91. **Kaplan, J. M., Varmus, H. E., and Bishop, J. M.,** The *src* protein contains multiple domains for specific attachment to membranes, *Mol. Cell. Biol.,* 10, 1000, 1990.

92. **Buss, J. E. and Sefton, B. M.,** Myristic acid, a rare fatty acid, is the lipid attached to the transforming protein of Rous sarcoma virus and its cellular homolog, *J. Virol.,* 53, 7, 1985.

93. **Goddard, C., Arnold, S. T., and Felsted, R. L.,** High affinity binding of an N-terminal myristoylated p60^{src} peptide, *J. Biol. Chem.,* 264, 15173, 1989.

94. **Buss, J. E., Der, C. J., and Solski, P. A.,** The six amino-terminal amino acids of p60^{src} are sufficient to cause myristylation of p21$^{v\text{-}ras}$, *Mol. Cell. Biol.,* 8, 3960, 1988.

95. **Horne, W. C., Neff, L., Chatterjee, D., Lomri, A., Levy, J. B., and Baron, R.,** Osteoclasts express high levels of pp60 (c-*src*) in association with intracellular membranes, *J. Cell Biol.,* 119, 1003, 1992.

96. **Kaplan, K. B., Swedlow, J. R., Varmus, H. E., and Morgan, D. O.,** Association of p60 (c-*src*) with endosomal membranes in mammalian fibroblasts, *J. Cell Biol.,* 118, 321, 1992.

97. **Horvath, A. R., Musbek, L., and Kellie, S.,** Translocation of pp60$^{c\text{-}src}$ to the cytoskeleton during platelet aggregation, *EMBO J.,* 11, 855, 1992.

98. **David-Pfeuty, T. and Nouvian-Dooghe, Y.,** Immunolocalization of the cellular *src* protein in interphase and mitotic NIH c-*src* overexpresser cells, *J. Cell Biol.,* 111, 3097, 1990.

99. **Qureshi, S. A., Cao, X., Sukhatme, V. P., and Foster, D. A.,** v-Src activates mitogen-responsive transcription factor Egr-1 via serum response elements, *J. Biol. Chem.,* 266, 10802, 1991.

100. **White, M. K., Rall, T. B., and Weber, M. J.,** Differential regulation of glucose transporter isoforms by the *src* oncogene in chicken embryo fibroblasts, *Mol. Cell. Biol.,* 11, 44448, 1991.

101. **O'Shea, J. J., Ashwell, J. D., Bailey, T. L., Cross, S. L., Samelson, L. E., and Klausner, R. D.,** Expression of v-*src* in a murine T-cell hybridoma results in constitutive T-cell receptor phosphorylation and interleukin 2 production, *Proc. Natl. Acad. Sci. U.S.A.,* 88, 1741, 1991.

102. **Snyder, M. A., Bishop, J. M., McGrath, J. P., and Levinson, A. D.,** A mutation at the ATP-binding site of pp60$^{v\text{-}src}$ abolishes kinase activity, transformation, and tumorigenicity, *Mol. Cell. Biol.,* 5, 1772, 1985.

103. **Jove, R., Mayer, B. J., Iba, H., Laugier, D., Poirier, F., Calothy, G., Hanafusa, T., and Hanafusa, H.,** Genetic analysis of p60$^{v\text{-}src}$ domains involved in the induction of different cell transformation parameters, *J. Virol.,* 60, 840, 1986.

104. **Jove, R., Garber, E. A., Iba, H., and Hanafusa, H.,** Biochemical properties of p60$^{v\text{-}src}$ mutants that induce different cell transformation parameters, *J. Virol.,* 60, 849, 1986.

105. **Mayer, B. J., Jove, R., Krane, J. F., Poirier, F., Calothy, G., and Hanafusa, H.,** Genetic lesions involved in temperature sensitivity of the *src* gene products of four Rous sarcoma virus mutants, *J. Virol.,* 60, 858, 1986.

106. **McCarley, D. J., Parsons, T. J., Benjamin, D. C., and Parsons, S. J.,** Inhibition of the tyrosine kinase activity of v-*src*, v-*fgr*, and v-*yes* gene products by a monoclonal antibody which binds both amino and carboxy peptide fragments of pp60$^{v\text{-}src}$, *J. Virol.,* 61, 1927, 1987.

107. **Uehara, Y., Hori, M., Takeuchi, T., and Umezawa, H.,** Screening of agents which convert "transformed morphology" of Rous sarcoma virus-infected rat kidney cells to "normal morphology": identification of an active agent as herbimycin and its inhibition of intracellular *src* kinase, *Jpn. J. Cancer Res.,* 76, 675, 1985.

108. **Dutta, A., Hamaguchi, M., and Hanafusa, H.,** Serum independence of transcription from the promoter of an avian retrovirus in v-*src*-transformed cells is a primary, intracellular effect of increased tyrosine phosphorylation, *Proc. Natl. Acad. Sci. U.S.A.,* 87, 608, 1990.

109. **van der Valk, J., Verlaan, I., de Laat, S. W., and Moolenaar, W. H.,** Expression of pp60$^{v\text{-}src}$ alters the ionic permeability of the plasma membrane in rat cells, *J. Biol. Chem.,* 262, 2431, 1987.

110. **Tachibana, H., Inoue, Y., Kanehisu, T., and Fukami, Y.,** Local similarity in the amino acid sequence between the non-catalytic region of Rous sarcoma virus oncogene product p60$^{v\text{-}src}$ and intermediate filament proteins, *J. Biochem.,* 104, 869, 1988.

111. **Hjelle, B., Liu, E., and Bishop, J. M.,** Oncogene v-*src* transforms and establishes embryonic rodent fibroblasts but not diploid human fibroblasts, *Proc. Natl. Acad. Sci. U.S.A.,* 85, 4355, 1988.

112. **Poulin, L., Grisé-Miron, L., and Wainberg, M. A.,** Immunological responsiveness against tumors induced by avian sarcoma virus: reduced expression of pp60^{src} kinase activity in regressing tumors, *J. Virol.,* 53, 800, 1985.

113. **Thomas, J. E., Soriano, P., and Brugge, J. S.,** Phosphorylation of c-Src on tyrosine 527 by another protein tyrosine kinase, *Science,* 254, 568, 1991.

114. **Hirota, Y., Kato, J., and Takeya, T.,** Substitution of Ser-17 of pp60^{c-src}: biological and biochemical characterization in chicken embryo fibroblasts, *Mol. Cell. Biol.,* 8, 1826, 1988.

115. **Chackalaparampil, I. and Shalloway, D.,** Altered phosphorylation and activation of pp60^{c-src} during fibroblast mitosis, *Cell,* 52, 801, 1988.

116. **Kmiecik, T. E., Johnson, P. J., and Shalloway, D.,** Regulation by the autophosphorylation site in overexpressed pp60^{c-src}, *Mol. Cell. Biol.,* 8, 4541, 1988.

117. **Ferracini, R. and Brugge, J.,** Analysis of mutant forms of the c-*src* gene product containing a phenylalanine substitution for tyrosine 416, *Oncogene Res.,* 5, 205, 1990.

118. **Cooper, J. A. and King, C. S.,** Dephosphorylation or antibody binding to the carboxy terminus stimulates pp60^{c-src}, *Mol. Cell. Biol.,* 6, 4467, 1986.

119. **Cheng, S. H., Piwnica-Worms, H., Harvey, R. W., Roberts, T. M., and Smith, A. E.,** The carboxy terminus of pp60^{c-src} is a regulatory domain and is involved in complex formation with the middle-T antigen of polyomavirus, *Mol. Cell. Biol.,* 8, 1736, 1988.

120. **MacAuley, A. and Cooper, J. A.,** Structural differences between repressed and derepressed forms of p60^{c-src}, *Mol. Cell. Biol.,* 9, 2648, 1989.

121. **Harvey, R., Hehir, K. M., Smith, A. E., and Cheng, S. H.,** pp60^{c-src} variants containing lesions that affect phosphorylation at tyrosines 416 and 527, *Mol. Cell. Biol.,* 9, 3647, 1989.

122. **Kornbluth, S., Jove, R., and Hanafusa, H.,** Characterization of avian and viral p60^{src} proteins expressed in yeast, *Proc. Natl. Acad. Sci. U.S.A.,* 84, 4455, 1987.

123. **Cooper, J. A. and Runge, K.,** Avian pp60^{c-src} is more active when expressed in yeast than in vertebrate fibroblasts, *Oncogene Res.,* 1, 297, 1987.

124. **Bagrodia, S., Taylor, S. J., and Shalloway, D.,** Myristoylation is required for Tyr-527 dephosphorylation and activation of pp60^{c-src} in mitosis, *Mol. Cell. Biol.,* 13, 1464, 1993.

125. **Azarnia, R., Reddy, S., Kmiecik, T. E., Shalloway, D., and Loewenstein, W. R.,** The cellular *src* gene product regulates junctional cell-to-cell communication, *Science,* 239, 398, 1988.

126. **Reynolds, A. B., Vila, J., Lansing, T. J., Potts, W. M., Weber, M. J., and Parsons, J. T.,** Activation of the oncogenic potential of the avian cellular *src* protein by specific structural alteration of the carboxy terminus, *EMBO J.,* 6, 2359, 1987.

127. **Kmiecik, T. E. and Shalloway, D.,** Activation and suppression of pp60^{c-src} transforming ability by mutation of its primary sites of tyrosine phosphorylation, *Cell,* 49, 65, 1987.

128. **Sato, M., Kato, J., and Takeya, T.,** Characterization of partially activated p60^{c-src} in chicken embryo fibroblasts, *J. Virol.,* 63, 683, 1989.

129. **Kato, G. and Wakabayashi, K.,** Novel serine phosphorylation occurs in the fibroblast form of pp60^{c-src} from Y79 retinoblastoma cells, *Biochem. Biophys. Res. Commun.,* 178, 764, 1991.

130. **Gould, K. L. and Hunter, T.,** Platelet-derived growth factor induces multisite phosphorylation of pp60^{c-src} and increases its protein-tyrosine kinase activity, *Mol. Cell. Biol.,* 8, 3345, 1988.

131. **Jove, R., Kornbluth, S., and Hanafusa, H.,** Enzymatically inactive p60^{c-src} mutant with altered ATP-binding site is fully phosphorylated in its carboxy-terminal regulatory region, *Cell,* 50, 937, 1987.

132. **Schuh, S. M. and Brugge, J. S.,** Investigation of factors that influence phosphorylation of pp60^{c-src} on tyrosine 527, *Mol. Cell. Biol.,* 8, 2465, 1988.

133. **Okada, M. and Nakagawa, H.,** Identification of a novel protein tyrosine kinase that phosphorylates pp60^{c-src} and regulates its activity in neonatal rat brain, *Biochem. Biophys. Res. Commun.,* 154, 796, 1988.

134. **Okada, M. and Nakagawa, H.,** A protein tyrosine kinase involved in regulation of pp60^{c-src} function, *J. Biol. Chem.,* 264, 20886, 1989.

135. **Nada, S., Okada, M., MacAuley, A., Cooper, J. A., and Nakagawa, H.,** Cloning of a complementary DNA for a protein-tyrosine kinase that specifically phosphorylates a negative regulatory site of p60^{c-src}, *Nature,* 351, 69, 1991.

136. **Armstrong, E., Cannizzaro, L., Bergman, M., Huebner, K., and Alitalo, K.,** The c-*src* tyrosine kinase (CSK) gene, a potential antioncogene, localizes to human chromosome region 15q23-q25, *Cytogenet. Cell Genet.,* 60, 119, 1992.

137. **Sabe, H., Knudsen, B., Okada, M., Nada, S., Nakagawa, H., and Hanafusa, H.,** Molecular cloning and expression of chicken C-terminal Src kinase: lack of stable association with c-Src protein, *Proc. Natl. Acad. Sci. U.S.A.,* 89, 2190, 1992.

138. **Grandori, C. and Hanafusa, H.,** p60$^{c\text{-}src}$ is complexed with a cellular protein in subcellular compartments involved in exocytosis, *J. Cell. Biol.,* 107, 2125, 1988.

139. **Boyce, B. F., Yoneda, T., Lowe, C., Soriano, P., and Mundy, G. R.,** Requirement of pp60$^{c\text{-}src}$ expression for osteoclasts to form ruffled borders and resorb bone in mice, *J. Clin. Invest.,* 90, 1622, 1992.

140. **Soriano, P., Montgomery, C., Geske, R., and Bradley, A.,** Targeted disruption of the c-*src* proto-oncogene leads to osteopetrosis in mice, *Cell,* 64, 693, 1991.

141. **Cotton, P. C. and Brugge, J. S.,** Neural tissues express high levels of the cellular *src* gene product pp60$^{c\text{-}src}$, Mol. Cell. Biol., 3, 1157, 1983.

142. **Levy, B. T., Sorge, L. K., Meymandi, A., and Maness, P. F.,** pp60$^{c\text{-}src}$ kinase is in chick and human embryonic tissues, *Dev. Biol.,* 104, 9, 1984.

143. **Schartl, M. and Barnekow, A.,** Differential expression of the cellular *src* gene during vertebrate development, *Dev. Biol.,* 105, 415, 1984.

144. **Levy, J. B. and Brugge, J. S.,** Biological and biochemical properties of the c-*src*$^+$ gene product overexpressed in chicken embryo fibroblasts, *Mol. Cell. Biol.,* 9, 3332, 1989.

145. **Bolen, J. B., Rosen, N., and Israel, M. A.,** Increased pp60$^{c\text{-}src}$ tyrosyl kinase activity in human neuroblastomas is associated with amino-terminal tyrosine phosphorylation of the *src* gene product, *Proc. Natl. Acad. Sci. U.S.A.,* 82, 7275, 1985.

146. **Brugge, J. S., Cotton, P. C., Queral, A. E., Barrett, J. N., Nonner, D., and Keane, R. W.,** Neurones express high levels of a structurally modified, activated form of pp60$^{c\text{-}src}$, *Nature,* 316, 554, 1985.

147. **Brugge, J., Cotton, P., Lusting, A., Yonemoto, W., Lipsich, L., Coussens, P., Barrett, J. N., Nonner, D., and Keane, R. W.,** Characterization of the altered form of the c-*src* gene product in neuronal cells, *Genes Dev.,* 1, 287, 1987.

148. **Martinez, R., Mathey-Prevot, B., Bernards, A., and Baltimore, D.,** Neuronal pp60$^{c\text{-}src}$ contains a six-amino acid insertion relative to its non-neuronal counterpart, *Science,* 237, 411, 1987.

149. **Pyper, J. M. and Bolen, J. B.,** Neuron-specific splicing of C-SRC RNA in human brain, *J. Neurosci. Res.,* 24, 89, 1989.

150. **Wiestler, O. D. and Walter, G.,** Developmental expression of two forms of pp60$^{c\text{-}src}$ in mouse brain, *Mol. Cell. Biol.,* 8, 502, 1988.

151. **Le Beau, J. M., Wiestler, O. D., and Walter, G.,** An altered form of pp60$^{v\text{-}src}$ is expressed primarily in the central nervous system, *Mol. Cell. Biol.,* 7, 4115, 1987.

152. **Yang, X., Martinez, R., Le Beau, J., Wiestler, O., and Walter, G.,** Evolutionary expression of the neuronal form of the src protein in the brain, *Proc. Natl. Acad. Sci. U.S.A.,* 86, 4751, 1989.

153. **Yang, X. and Walter, G.,** Specific kinase activity and phosphorylation state of pp60$^{c\text{-}src}$ from neuroblastomas and fibroblasts, *Oncogene,* 3, 237, 1988.

154. **Pyper, J. M. and Bolen, J. B.,** Identification of a novel neuronal C-SRC exon expressed in human brain, *Mol. Cell. Biol.,* 10, 2035, 1990.

155. **Schartl, M., Holstein, T., Robertson, S. M., and Barnekow, A.,** Preferential expression of a pp60$^{c\text{-}src}$ related protein tyrosine kinase activity in nerve cells of the early metazoan hydra (coelenterates), *Oncogene,* 4, 1185, 1989.

156. **Wang, L.-H., Iijima, S., Dorai, T., and Lin, B.,** Regulation of the expression of proto-oncogene c-*src* by alternative RNA splicing in chicken skeletal muscle, *Oncogene Res.,* 1, 43, 1987.

157. **Dorai, T. and Wang, L.-H.,** An alternative non-tyrosine protein kinase product of the c-*src* gene in chicken skeletal muscle, *Mol. Cell. Biol.,* 10, 4068, 1990.

158. **Wilson, L. K., Luttrell, D. K., Parsons, J. T., and Parsons, S. J.,** pp60$^{c\text{-}src}$ tyrosine kinase, myristoylation, and modulatory domains are required for enhanced mitogenic responsiveness to epidermal growth factor seen in cells overexpressing c-*src*, *Mol. Cell. Biol.,* 9, 1536, 1989.

159. **Brott, B. K., Decker, S., O'Brien, M. C., and Jove, R.,** Molecular features of the viral and cellular Src kinases involved in interactions with the GTPase-activating protein, *Mol. Cell. Biol.,* 11, 5059, 1991.

160. **Goff, S. P.,** The Abelson murine leukemia virus oncogene, *Proc. Soc. Exp. Biol. Med.,* 179, 403, 1985.

161. **Risser, R. and Green, P. L.,** Abelson virus: current status of a viral oncogene, *Proc. Soc. Exp. Biol. Med.,* 188, 235, 1988.

162. **Rosenberg, N. and Witte, O. N.,** The viral and cellular forms of the Abelson (*abl*) oncogene, *Adv. Virus Res.,* 35, 39, 1988.

163. **Ramakrishnan, L. and Rosenberg, N.,** *abl* genes, *Biochim. Biophys. Acta,* 989, 209, 1989.

164. **Wang, J. Y. J.,** Isolation of antibodies for phosphotyrosine by immunization with a v-*abl* oncogene-encoded protein, *Mol. Cell. Biol.,* 5, 3640, 1985.

165. **Saggioro, D., Ferracini, R., DiRenzo, M. F., Naldini, L., Chieco-Bianchi, L., and Comoglio, P. M.,** Protein phosphorylation at tyrosine residues in v-*abl* transformed mouse lymphocytes and fibroblasts, *Int. J. Cancer,* 37, 623, 1986.

166. **Villa-Moruzzi, E., Zonca, P. D., and Crabb, J. W.,** Phosphorylation of the catalytic subunit of type-1 protein phosphatase by the v-abl tyrosine kinase, *FEBS Lett.,* 293, 67, 1991.

167. **Goff, S. P., Gilboa, E., Witte, O. N., and Baltimore, D.,** Structure of the Abelson murine leukemia virus genome and the homologous cellular gene: studies with cloned viral DNA, *Cell,* 22, 777, 1980.

168. **Henkemeyer, M. J., Gertler, F. B., Goodman, W., and Hoffmann, F. M.,** The Drosophila Abelson proto-oncogene homolog: identification of mutant alleles that have pleiotropic effects late in development, *Cell,* 51, 821, 1987.

169. **Gertler, F. B., Bennett, R. L., Clark, M. J., and Hoffmann, F. M.,** Drosophila *abl* tyrosine kinase in embryonic CNS axons: a role in axonogenesis is revealed through dosage-sensitive interactions with disabled, *Cell,* 58, 103, 1989.

170. **Oppi, C., Shore, S. K., and Reddy, E. P.,** Nucleotide sequence of testis-derived c-*abl* cDNAs: implications for testis-specific transcription and *abl* oncogene activation, *Proc. Natl. Acad. Sci. U.S.A.,* 84, 8200, 1987.

171. **Meijer, D., Hermans, A., von Lindern, M., van Agthoven, T., de Klein, A., Mackenbach, P., Grootegoed, A., Talarico, D., Della Valle, G., and Grosveld, G.,** Molecular characterization of the testis specific c-*abl* mRNA in mouse, *EMBO J.,* 6, 4041, 1987.

172. **Shtivelman, E., Lifshitz, B., Gale, R. P., Roe, B. A., and Canaani, E.,** Alternative splicing of RNAs transcribed from the human *abl* gene and from the *bcr-abl* fused gene, *Cell,* 47, 277, 1986.

173. **Jackson, P. and Baltimore, D.,** N-terminal mutations activate the leukemogenic potential of the myrystoylated form of c-*abl*, *EMBO J.,* 8, 449, 1989.

174. **Jackson, P. and Baltimore, D.,** N-terminal mutations activate the leukemogenic potential of the myristoylated form of c-*abl*, *EMBO J.,* 8, 449, 1989.

175. **Van Etten, R. A., Jackson, P., and Baltimore, D.,** The mouse type IV c-*abl* gene product is a nuclear protein, and activation of transforming ability is associated with cytoplasmic localization, *Cell,* 58, 669, 1989.

176. **Andrews, D. F., III, Nemunaitis, J. J., and Singer, J. W.,** Recombinant tumor necrosis factor alpha and interleukin 1 alpha increase expression of c-*abl* protooncogene mRNA in cultured human marrow stromal cells, *Proc. Natl. Acad. Sci. U.S.A.,* 86, 6788, 1989.

177. **Kipreos, E. T. and Wang, J. Y. J.,** Differential phosphorylation of c-Abl in cell cycle determined by *cdc2* kinase and phosphatase activity, *Science,* 248, 217, 1990.

178. **Dhut, S., Chaplin, T., and Young, B. D.,** Normal c-*abl* gene protein — a nuclear component, *Oncogene,* 6, 1459, 1991.

179. **Kipreos, E. T. and Wang, J. Y. J.,** Cell cycle-regulated binding of c-Abl tyrosine kinase to DNA, *Science,* 256, 382, 1992.

180. **Groffen, J. and Heisterkamp, N.,** Philadelphia chromosome translocation, *Crit. Rev. Oncogenesis,* 1, 53, 1989.

181. **Campbell, M. L. and Arlinghaus, R. B.,** Current status of the *BCR* gene and its involvement with human leukemia, *Adv. Cancer Res.,* 57, 227, 1991.

182. **Sawyers, C. L.,** The *bcr-abl* gene in chronic myelogenous leukaemia, *Cancer Surv.,* 15, 37, 1992.

183. **Reynolds, C. H., Willson, M. G., Groffen, J., Heisterkamp, N., Peakman, T. C., and Page, M. J.,** Comparison of baculovirus-expressed c-Abl and BCR/ABL protein tyrosine kinases, *Biochim. Biophys. Acta,* 1181, 122, 1993.

184. **Stam, K., Heisterkamp, N., Reynolds, F. H., Jr., and Groffen, J.,** Evidence that the *phl* gene encodes a 160,000-dalton phosphoprotein with associated kinase activity, *Mol. Cell. Biol.,* 7, 1955, 1987.

185. **Maru, Y. and Witte, O. N.,** The *BCR* gene encodes a novel serine threonine kinase activity within a single exon, *Cell,* 67, 459, 1991.

186. **Collins, S., Coleman, H., and Groudine, M.,** Expression of *bcr* and *bcr-abl* fusion transcripts in normal and leukemic cells, *Mol. Cell. Biol.,* 7, 2870, 1987.

187. **Ullrich, A., Bell, J. R., Chen, E. Y., Herrera, R., Petruzzelli, L. M., Dull, T. J., Gray, A., Coussens, L., Liao, Y.-C., Tsubokawa, M., Mason, A., Seeburg, P. H., Grunfeld, C., Rosen, O. M., and Ramachandran, J.,** Human insulin receptor and its relationship to tyrosine kinase family of oncogenes, *Nature,* 313, 756, 1985.

188. **Birchmeier, C., O'Neill, K., Riggs, M., and Wigler, M.,** Characterization of *ROS1* cDNA from a human glioblastoma cell line, *Proc. Natl. Acad. Sci. U.S.A.,* 87, 4799, 1990.

189. **Chen, J., Heller, D., Poon, B., Kang, L., and Wang, L.-H.,** The proto-oncogene c-*ros* codes for a transmembrane tyrosine protein kinase sharing sequence and structural homology with *sevenless* protein of *Drosophila melanogaster, Oncogene,* 6, 257, 1991.

190. **Basler, K. and Hafen, E.,** Control of photoreceptor cell fate by the *sevenless* protein requires a functional tyrosine kinase domain, *Cell,* 54, 299, 1988.

191. **Neckameyer, W. S., Shibuya, M., Hsu, M.-T., and Wang, L.-H.,** Proto-oncogene c-*ros* codes for a molecule with structural features common to those of growth factor receptors and display tissue-specific and developmentally regulated expression, *Mol. Cell. Biol.,* 6, 1478, 1986.

192. **Podell, S. B. and Sefton, B. M.,** Chicken proto-oncogene c-*ros* cDNA clones: identification of a c-*ros* RNA transcript and deduction of the amino acid sequence of the carboxyl terminus of the c-*ros* product, *Oncogene,* 2, 9, 1987.

193. **Tessarollo, L., Nagarajan, L., and Parada, L. F.,** c-*ros*: the vertebrate homolog in the *sevenless* tyrosine kinase is tightly regulated during organogenesis in mouse embryonic development, *Development,* 115, 11, 1992.

194. **Matsushime, H., Wang, L.-H., and Shibuya, M.,** Human c-*ros*-1 gene homologous to the v-*ros* sequence of UR2 sarcoma virus encodes for a transmembrane receptorlike molecule, *Mol. Cell. Biol.,* 6, 3000, 1986.

195. **Matsushime, H., Yoshida, M. C., Sasaki, M., and Shibuya, M.,** A possible new member of tyrosine kinase family, human *frt* sequence, is highly conserved in vertebrates and located on human chromosome 13, *Jpn. J. Cancer Res.,* 68, 655, 1987.

196. **Satoh, H., Yoshida, M. C., Matsushime, H., Shibuya, M., and Sasaki, M.,** Regional localization of the human c-*ros*-1 on 6q22 and *flt* on 13q12, *Jpn. J. Cancer Res.,* 78, 772, 1987.

197. **Ellis, L., Morgan, D. O., Jong, S.-M., Wang, L.-H., Roth, R. A., and Rutter, W. J.,** Heterologous transmembrane signaling by a human insulin receptor-v-*ros* hybrid in Chinese hamster ovary cells, *Proc. Natl. Acad. Sci. U.S.A.,* 84, 5101, 1987.

198. **Hanafusa, H.,** The *fps/fes* oncogene, in *The Oncogene Handbook,* Reddy, E. P., Skalka, A. M., and Curran, T., Eds., Elsevier, Amsterdam, 1988, 39.

199. **Mathew, S., Murty, V. V. S., German, J., and Chaganti, R. S. K.,** Confirmation of 15q26. 1 as the site of the FES protooncogene by fluorescence *in situ* hybridization, *Cell Growth Differ.,* 4, 33, 1993.

200. **Roebroek, A. J. M., Schalken, J. A., Verbeek, J. S., Van den Ouweland, A. M. W., Onnekink, C., Bloemers, H. P. J., and Van de Ven, W. J. M.,** The structure of the human c-*fes/fps* proto-oncogene, *EMBO J.,* 4, 2897 1985.

201. **Roebroek, A. J. M., Schalken, J. A., Onnekink, C., Bloemers, H. P. J., and Van de Ven, W. J. M.,** Structure of the feline c-*fes/fps* proto-oncogene: genesis of a retroviral genome, *J. Virol.,* 61, 2009, 1987.

202. **Mathey-Prevot, B., Hanafusa, H., and Kawai, S.,** A cellular protein is immunologically crossreactive with and functionally homologous to the Fujinami sarcoma virus transforming protein, *Cell,* 28, 897, 1982.

203. **Lanfrancone, L., Mannoni, P., Pebusque, M. F., Carè, A., Peschle, C., Grignani, F., and Pelicci, P. G.,** Expression pattern of c-*fes* oncogene mRNA in human myeloid cells, *Int. J. Cancer,* Suppl. 4, 35, 1989.

204. **Smithgall, T. E., Yu, G., and Glazer, R. I.,** Identification of the differentiation-associated p93 tyrosine protein kinase of HL-60 leukemia cells as the product of the human c-*fes* locus and its expression in myelomonocytic cells, *J. Biol. Chem.,* 263, 15050, 1988.

205. **Greer, P., Maltby, V., Rossant, J., Bernstein, A., and Pawson, T.,** Myeloid expression of the human c-*fps/fes* proto-oncogene in transgenic mice, *Mol. Cell. Biol.,* 10, 2521, 1990.

206. **Feldman, R. A., Vass, W. C., and Tambourin, P. E.,** Human cellular *fps/fes* cDNA rescued via retroviral shuttle vector encodes myeloid cell NCP92 and has transforming potential, *Oncogene Res.,* 1, 441, 1987.

207. **Meckling-Gill, K. A., Yee, S.-P., Schrader, J. W., and Pawson, T.,** A retrovirus encoding the v-*fps* protein-tyrosine kinase induces factor-independent growth and tumorigenicity in FDC-P1 cells, *Biochim. Biophys. Acta,* 1137, 65, 1992.

208. **Greer, P. A., Meckling-Hansen, K., and Pawson, T.,** The human c-*fps/fes* gene product expressed ectopically in rat fibroblasts is nontransforming and has restrained protein-tyrosine kinase activity, *Mol. Cell. Biol.,* 8, 578, 1988.

209. **Feldman, R. A., Lowy, D. R., and Vass, W. C.,** Selective potentiation of c-*fps/fes* transforming activity by a phosphatase inhibitor, *Oncogene Res.,* 5, 187, 1990.

210. **Naharro, G., Robbins, K. C., and Reddy, E. P.,** Gene product of v-*fgr onc*: hybrid protein containing a portion of actin and a tyrosine-specific protein kinase, *Science,* 223, 63, 1984.

211. **Manger, R., Rasheed, S., and Rohrschneider, L.,** Localization of the feline sarcoma virus *fgr* gene product (P70$^{gag-actin-fgr}$): association with the plasma membrane and detergent-insoluble matrix, *J. Virol.,* 59, 66, 1986.

212. **Tronick, S. R., Popescu, N. C., Cheah, M. S. C., Swan, D. C., Amsbaugh, S. C., Lengel, C. R., DiPaolo, J. A., and Robbins, K. C.,** Isolation and chromosomal localization of the human *fgr* protooncogene, a distinct member of the tyrosine kinase gene family, *Proc. Natl. Acad. Sci. U.S.A.,* 82, 6595, 1985.

213. **Nishizawa, M., Semba, K., Yoshida, M. C., Yamamoto, T., Sasaki, M., and Toyoshima, K.,** Structure, expression, and chromosomal localization of the human c-*fgr* gene, *Mol. Cell. Biol.,* 6, 511, 1986.

214. **Inoue, K., Ikawa, S., Semba, K., Sukegawa, J., Yamamoto, T., and Toyoshima, K.,** Isolation and sequencing of cDNA clones homologous to the v-*fgr* oncogene from a human B lymphocyte cell line, IM-9, *Oncogene,* 1, 301, 1987.

215. **Link, D. C., Gutkind, S. J., Robbins, K. C., and Ley, T. J.,** Characterization of the 5' untranslated region of the human c-*fgr* gene and identification of the major myelomonocytic c-*fgr* promoter, *Oncogene,* 7, 877, 1992.

216. **Ley, T. J., Connolly, N. L., Katamine, S., Cheah, M. S. C., Senior, R. M., and Robbins, K. C.,** Tissue-specific expression and developmental regulation of the human *fgr* proto-oncogene, *Mol. Cell. Biol.,* 9, 92, 1989.

217. **Yi, T.-L. and Willman, C. L.,** Cloning of the murine c-*fgr* proto-oncogene cDNA and induction of c-*fgr* expression by proliferation and activation factors in normal bone marrow-derived monocytic cells, *Oncogene,* 4, 1081, 1989.

218. **Notario, V., Gutkind, J. S., Imaizumi, M., Katamine, S., and Robbins, K. C.,** Expression of the *fgr* protooncogene product as a function of myelomonocytic cell maturation, *J. Cell Biol.,* 109, 3129, 1989.

219. **Willman, C. L., Stewart, C. C., Longacre, T. L., Head, D. R., Habbersett, R., Ziegler, S. F., and Perlmutter, R. M.,** Expression of the c-*fgr* and *hck* protein-tyrosine kinase in acute myeloid leukemic blasts is associated with early commitment and differentiation events in the monocytic and granulo-cytic lineages, *Blood,* 77, 726, 1991.

220. **Sartor, O., Moriuchi, R., Sameshima, J. H., Severino, M., Gutkind, J. S., and Robbins, K. C.,** Diverse biologic properties imparted by the c-*fgr* proto-oncogene, *J. Biol. Chem.,* 267, 3460, 1992.

221. **Kitamura, N., Kitamura, A., Toyoshima, K., Hirayama, Y., and Yoshida, M.,** Avian sarcoma virus Y73 genome sequence and structural similarity of its transforming gene product to that of Rous sarcoma virus, *Nature,* 297, 205, 1982.

222. **Gentry, L. E. and Rohrschneider, L. R.,** Common features of the *yes* and *src* gene products defined by peptide-specific antibodies, *J. Virol.,* 51, 539, 1984.

223. **Sukegawa, J., Semba, K., Yamanishi, Y., Nishizawa, M., Miyajima, N., Yamamoto, T., and Toyoshima, K.,** Characterization of cDNA clones for the human c-*yes* gene, *Mol. Cell. Biol.,* 7, 41, 1987.

224. **Bhargava, A., Deobagkar, D. N., and Deobagkar, D. D.,** Identification and characterization of oncogene *yes*-homologous genomic clones from *Drosophila melanogaster, J. Genet.,* 70, 181, 1991.

225. **Sudol, M. and Hanafusa, H.,** Cellular proteins homologous to the viral *yes* gene product, *Mol. Cell. Biol.,* 6, 2839, 1986.

226. **Sudol, M., Kuo, C. F., Shigemitsu, L., and Alvarez-Buylla, A.,** Expression of the *yes* proto-oncogene in cerebellar Purkinje cells, *Mol. Cell. Biol.,* 9, 4545, 1989.

227. **Zhao, Y.-H., Krueger, J. G., and Sudol, M.,** Expression of cellular-*yes* protein in mammalian tissues, *Oncogene,* 5, 1629, 1990.

228. **Krueger, J., Zhao, Y. H., Murphy, D., and Sudol, M.,** Differential expression of p62^{c-yes} in normal, hyperplastic and neoplastic human epidermis, *Oncogene,* 6, 933, 1991.

229. **Azuma, K., Ariki, M., Miyauchi, T., Usui, H., Takeda, M., Semba, K., Matsuzawa, Y., Yamamoto, T., and Toyoshima, K.,** Purification and characterization of a rat liver membrane tyrosine-protein kinase, the possible protooncogene c-*yes* product, p60^{c-yes}, *J. Biol. Chem.,* 266, 4831, 1991.

230. **Sudol, M., Greulich, H., Newman, L., Sarkar, A., Sukegawa, J., and Yamamoto, T.,** A novel Yes-related kinase, Yrk, is expressed at elevated levels in neural and hematopoietic tissues, *Oncogene,* 8, 823, 1993.

231. **Smith, D. R., Vogt, P. K., and Hayman, M. J.,** The v-*sea* oncogene of avian erythroblastosis retrovirus S13: another member of the protein-tyrosine kinase gene family, *Proc. Natl. Acad. Sci. U.S.A.,* 86, 5291, 1989.

232. **Larsen, J., Beug, H., and Hayman, M. J.,** The v-*ski* oncogene cooperates with the v-*sea* oncogene in erythroid transformation by blocking erythroid differentiation, *Oncogene,* 7, 10, 1992.

233. **Huff, J. L., Jelinek, M. A., Borgman, C. A., Lansing, T. J., and Parsons, J. T.,** The protooncogene c-*sea* encodes a transmembrane protein-tyrosine kinase related to the Met/hepatocyte growth factor/ scatter factor receptor, *Proc. Natl. Acad. Sci. U.S.A.,* 90, 6140, 1993.

234. **Jia, R., Mayer, B. J., Hanafusa, T., and Hanafusa, H.,** A novel oncogene, v-*ryk*, encoding a truncated receptor tyrosine kinase is transduced into the RPL30 virus without loss of viral sequences, *J. Virol.,* 66, 5975, 1992.

235. **Hovens, C. M., Stacker, S. A., Andres, A. C., Harpur, A. G., Ziemicki, A., and Wilks, A. F.,** RYK, a receptor tyrosine kinase-related molecule with unusual kinase domain motifs, *Proc. Natl. Acad. Sci. U.S.A.,* 89, 11818, 1992.

236. **Semba, K., Nishizawa, M., Miyajima, N., Yoshida, M. C., Sukegawa, J., Yamanishi, Y., Sasaki, M., Yamamoto, T., and Toyoshima, K.,** *yes*-related protooncogene, *syn*, belongs to the protein-tyrosine kinase family, *Proc. Natl. Acad. Sci. U.S.A.,* 83, 5459, 1986.

237. **Yoshida, M. C., Satoh, H., Sasaki, M., Semba, K., Yamamoto, T., and Toyoshima, K.,** Regional location of a novel *yes*-related proto-oncogene, *syn*, on human chromosome 6 at band q21, *Jpn. J. Cancer Res.,* 77, 1059, 1986.

238. **Popescu, N. C., Kawakami, T., Matsui, T., and Robbins, K. C.,** Chromosomal localization of the human *fyn* gene, *Oncogene,* 1, 449, 1987.

239. **Steele, R. E., Deng, J. C., Ghosn, C. R., and Fero, J. B.,** Structure and expression of *fyn* genes in *Xenopus laevis, Oncogene,* 5, 369, 1990.

240. **Kypta, R. M., Hemming, A., and Courtneidge, S. A.,** Identification and characterization of p59^{fyn} (a *src*-like protein tyrosine kinase) in normal and polyoma virus transformed cells, *EMBO J.,* 7, 3837, 1988.

241. **Horak, I. D., Corcoran, M. L., Thompson, P. A., Wahl, L. M., and Bolen, J. B.,** Expression of p60^{fyn} in human platelets, *Oncogene,* 5, 597, 1990.

242. **Kawakami, T., Pennington, C. Y., and Robbins, K. C.,** Isolation and oncogenic potential of a novel human *src*-like gene, *Mol. Cell. Biol.,* 6, 4195, 1986.

243. **Cooke, M. P., Abraham, K. M., Forbush, K. A., and Perlmutter, R. M.,** Regulation of T cell receptor signaling by a *src* family protein-tyrosine kinase (p59^{fyn}), *Cell,* 65, 281, 1991.

244. **Tsygankov, A. Y., Broker, B. M., Fargnoli, J., Ledbetter, J. A., and Bolen, J. B.,** Activation of tyrosine kinase p60^{fyn} following T-cell antigen receptor cross-linking, *J. Biol. Chem.,* 267, 18259, 1992.

245. **Shiroo, M., Goff, L., Biffen, M., Shivnan, E., and Alexander, D.,** CD45 tyrosine phosphatase-activated p59^{fyn} couples the T cell antigen receptor to pathways of diacylglycerol production, protein kinase C activation and calcium, *EMBO J.,* 11, 4887, 1992.

246. **Sugie, K., Kawakami, T., Maeda, Y., Kawabe, T., Uchida, A., and Yodoi, J.,** Fyn tyrosine kinase associated with Fc epsilon RII/CD23: possible multiple roles in lymphocyte activation, *Proc. Natl. Acad. Sci. U.S.A.,* 88, 9132, 1991.

247. **Venkitaraman, A. R. and Cowling, R. J.,** Interleukin-7 receptor functions by recruiting the tyrosine kinase p59^{fyn} through a segment of its cytoplasmic tail, *Proc. Natl. Acad. Sci. U.S.A.,* 89, 12083, 1992.

248. **Bolen, J. B. and Veillette, A.,** A function for the *lck* proto-oncogene, *Trends Biochem. Sci.,* 14, 404, 1989.

249. **Rudd, C. E.,** CD4, CD8 and the TCR-CD3 complex: a novel class of protein-tyrosine kinase receptor, *Immunol. Today,* 11, 400, 1990.

250. **Sefton, B. M.,** The *lck* tyrosine protein kinase, *Oncogene,* 6, 683, 1991.

251. **Marth, J. D., Peet, R., Krebs, E. G., and Perlmutter, R. M.,** A lymphocyte-specific protein-tyrosine kinase gene is rearranged and overexpressed in the murine T cell lymphoma LSTRA, *Cell,* 43, 393, 1985.

252. **Voronova, A. F. and Sefton, B. M.,** Expression of a new tyrosine protein kinase is stimulated by retrovirus promoter insertion, *Nature,* 319, 682, 1986.

253. **Marth, J. D., Disteche, C. , Pravtcheva, D., Ruddle, F., Krebs, E. G., and Perlmutter, R. M.,** Localization of a lymphocyte-specific protein tyrosine kinase gene (*lck*) at a site of frequent chromosomal abnormalities in human lymphomas, *Proc. Natl. Acad. Sci. U.S.A.,* 83, 7400, 1986.

254. **Thom, R. E. and Casnellie, J. E.,** Demonstration that LSTRA cells have an elevated level of proteins phosphorylated on tyrosine residues, *FEBS Lett.,* 222, 104, 1987.

255. **Voronova, A. F., Buss, J. E., Patschinsky, T., Hunter, T., and Sefton, B. M.,** Characterization of the protein apparently responsible for the elevated tyrosine protein kinase activity in LSTRA cells, *Mol. Cell. Biol.,* 4, 2705, 1984.

256. **Trevillyan, J. M., Lin, Y., Chen, S. J., Phillips, C. A., Canna, C., and Linna, T. J.,** Human T lymphocytes express a protein-tyrosine kinase homologous to p56[LSTRA], *Biochim. Biophys. Acta,* 888, 286, 1986.

257. **Trevillyan, J. M., Canna, C., Maley, D., Linna, T. J., and Phillips, C. A.,** Identification of the human T-lymphocyte protein-tyrosine kinase by peptide-specific antibodies, *Biochem. Biophys. Res. Commun.,* 140, 392, 1986.

258. **Wildin, R. S., Garvin, A. M., Pawar, S., Lewis, D. B., Abraham, K. M., Forbush, K. A., Ziegler, S. F., Allen, J. M., and Perlmutter, R. M.,** Developmental regulation of *lck* gene expression in T lymphocytes, *J. Exp. Med.,* 173, 383, 1991.

259. **Strebhardt, K., Mullins, J. I., Bruck, C., and Rübsamen-Waigmann, H.,** Additional member of the protein-tyrosine kinase family: the *src-* and *lck*-related protooncogene c-*tkl*, *Proc. Natl. Acad. Sci. U.S.A.,* 84, 8778, 1987.

260. **Chow, L. M. L., Ratcliffe, M. J. H., and Veillette, A.,** *tkl* is the avian homolog of the mammalian *lck* tyrosine protein kinase gene, *Mol. Cell. Biol.,* 12, 1226, 1992.

261. **Boulet, I., Fagard, R., and Fischer, S.,** Correlation between phosphorylation and kinase activity of a tyrosine protein kinase: p56 *lck*, *Biochem. Biophys. Res. Commun.,* 149, 56, 1987.

262. **Allée, G., Fagard, R., Danielian, S., Boulet, I., Soula, M., and Fischer, S.,** Phosphorylation of p56[lck] by external ATP in intact cells, *Biochem. Biophys. Res. Commun.,* 162, 51, 1989.

263. **Marth, J. D., Cooper, J. A., King, C. S., Ziegler, S. F., Tinker, D. A., Overell, R. W., Krebs, E. G., and Perlmutter, R. M.,** Neoplastic transformation induced by an activated lymphocyte-specific protein tyrosine kinase (pp56[lck]), *Mol. Cell. Biol.,* 8, 540, 1988.

264. **Rudd, C. E., Trevillyan, J. M., Dasgupta, J. D., Wong, L. L., and Schlossman, S. F.,** The CD4 receptor is complexed in detergent lysates to a protein-tyrosine kinase (pp58) from human T lympho-cytes, *Proc. Natl. Acad. Sci. U.S.A.,* 85, 5190, 1988.

265. **Rudd, C., Helms, S., Barber, E. K., and Schlossman, S. F.,** The CD4/CD8:p56[lck] complex in T lymphocytes: a potential mechanism to regulate T-cell growth, *Biochem. Cell Biol.,* 67, 581, 1989.

266. **Straus, D. B. and Weiss, A.,** Genetic evidence for the involvement of the *lck* tyrosine kinase in signal transduction through the T-cell antigen receptor, *Cell,* 70, 585, 1992.

267. **Shaw, A. S., Amrein, K. E., Hammond, C., Stern, D. F., Sefton, B. M., and Rose, J. K.,** The *lck* tyrosine protein kinase interacts with the cytoplasmic tail of the CD4 glycoprotein through its unique amino-terminal domain, *Cell,* 59, 627, 1989.

268. **Turner, J. M., Brodsky, M. H., Irving, B. A., Levin, S. D., Perlmutter, R. M., and Littman, D. R.,** Interaction of the unique N-terminal region of tyrosine kinase p56[lck] with cytoplasmic domains of CD4 and CD8 is mediated by cysteine motifs, *Cell,* 60, 755, 1990.

269. **Glaichenhaus, N., Shastri, N., Littman, D. R., and Turner, J. M.,** Requirement of association of p56[lck] with CD4 in antigen-specific signal transduction in T cells, *Cell,* 64, 511, 1991.

270. **Chalupny, N. J., Ledbetter, J. A., and Kavathas, P.,** Association of CD8 with p56[lck] is required for early T-cell signalling events, *EMBO J.,* 10, 1201, 1991.

271. **Telfer, J. C. and Rudd, C. E.,** A 32-kD GTP-binding protein associated with the CD4-p56[lck] and CD8-p56[lck] T cell receptor complexes, *Science,* 254, 439, 1991.

272. **Stefanova, I., Horejsi, V., Ansotegui, I. J., Knapp, W., and Stockinger, H.,** GPI-anchored cell-surface molecules complexed to protein tyrosine kinases, *Science,* 254, 1016, 1991.

273. **Thompson, P. A., Ledbetter, J. A., Rapp, U. R., and Bolen, J. B.,** The Raf-1 serine-threonine kinase is a substrate of the p56[lck] protein tyrosine kinase in human T-cells, *Cell Growth Differ.,* 2, 609, 1991.

274. **Luo, K. X. and Sefton, B. M.,** Activated *lck* tyrosine protein kinase stimulates antigen-independent interleukin-2 production in T cells, *Mol. Cell. Biol.,* 12, 4724, 1992.

275. **Marth, J. D., Lewis, D. B., Wilson, C. B., Gearn, M. E., Krebs, E. G., and Perlmutter, R. M.,** Regulation of pp56[lck] during T-cell activation: functional implications for the *src*-like protein tyrosine kinases, *EMBO J.,* 2727, 1987.

276. **Horak, I. D., Gress, R. E., Lucas, P. J., Horak, E. M., Waldmann, T. A., and Bolen, J. B.,** T-lymphocyte interleukin 2-dependent tyrosine protein kinase signal transduction involves the activation of p56[lck], *Proc. Natl. Acad. Sci. U.S.A.,* 88, 1996, 1991.

277. **Hatakeyama, M., Kono, T., Kobayashi, N., Kawahara, A., Levin, S. D., Perlmutter, R. M., and Taniguchi, T.,** Interaction of the IL-2 receptor with the *src*-family kinase p56*lck*: identification of novel intermolecular association, *Science*, 252, 1523, 1991.

278. **Shackelford, D. A. and Trowbridge, I. S.,** Ligand-stimulated tyrosine phosphorylation of the IL-2 receptor beta chain and receptor-associated proteins, *Cell Regul.*, 2, 73, 1991.

279. **Koga, Y., Kimura, N., Minowada, J., and Mak, T. K.,** Expression of the human T-cell-specific tyrosine kinase YT16 (*lck*) message in leukemic T-cell lines, *Cancer Res.*, 48, 856, 1988.

280. **Veillette, A., Foss, F. M., Sausville, E. A., Bolen, J. B., and Rosen, N.,** Expression of the *lck* tyrosine kinase gene in human colon carcinoma and other non-lymphoid human tumor cell lines, *Oncogene Res.*, 1, 357, 1987.

281. **Garvin, A. M., Pawar, S., Marth, J. D., and Perlmutter, R. M.,** Structure of the murine *lck* gene and its rearrangement in a murine lymphoma cell line, *Mol. Cell. Biol.*, 8, 3058, 1988.

282. **Abraham, K. M., Levin, S. D., Marth, J. D., Forbush, K. A., and Perlmutter, R. M.,** Thymic tumorigenesis induced by overexpression of p56*lck*, *Proc. Natl. Acad. Sci. U.S.A.*, 88, 3977, 1991.

283. **Louie, R. R., King, C. S., MacAuley, A., Marth, J. D., Perlmutter, R. M., Eckhart, W., and Cooper, J. A.,** p56*lck* protein-tyrosine kinase is cytoskeletal and does not bind to polyomavirus middle T antigen, *J. Virol.*, 62, 4673, 1988.

284. **Amrein, K. E., Flint, N., Panholzer, B., and Burn, P.,** Ras GTPase-activating protein — a substrate and a potential binding protein of the protein-tyrosine kinase p56*lck*, *Proc. Natl. Acad. Sci. U.S.A.*, 89, 3343, 1992.

285. **Dymecki, S. M., Niederhuber, J. E., and Desiderio, S. V.,** Specific expression of a tyrosine kinase gene, *blk*, in B lymphoid cells, *Science*, 247, 332, 1990.

286. **Quintrell, N., Lebo, R., Varmus, H., Bishop, J. M., Pettenati, M. J., Le Beau, M. M., Diaz, M. O., and Rowley, J. D.,** Identification of a human gene (*HCK*) that encodes a protein-tyrosine kinase and is expressed in hemopoietic cells, *Mol. Cell. Biol.*, 7, 2267, 1987.

287. **Ziegler, S. F., Marth, J. D., Lewis, D. B., and Perlmutter, R. M.,** Novel protein-tyrosine kinase gene (*hck*) preferentially expressed in cells of hematopoietic origin, *Mol. Cell. Biol.*, 7, 2276, 1987.

288. **Lichtenberg, U., Quintrell, N., and Bishop, J. M.,** Human protein-tyrosine kinase gene *HCK*: expression and structural analysis of the promoter region, *Oncogene*, 7, 859, 1992.

289. **Ziegler, S. F., Wilson, C. B., and Perlmutter, R. M.,** Augmented expression of a myeloid-specific protein tyrosine kinase gene (*hck*) after macrophage activation, *J. Exp. Med.*, 168, 1801, 1988.

290. **Ziegler, S. F., Pleiman, C. M., and Perlmutter, R. M.,** Structure and expression of the murine *hck* gene, *Oncogene*, 6, 283, 1991.

291. **Ziegler, S. F., Levin, S. D., and Perlmutter, R. M.,** Transformation of NIH 3T3 fibroblasts by an activated form of p59*hck*, *Mol. Cell. Biol.*, 9, 2724, 1989.

292. **Holtzman, D. A., Cook, W. D., and Dunn, A. R.,** Isolation and sequence of a cDNA corresponding to a *src*-related gene expressed in murine hemopoietic cells, *Proc. Natl. Acad. Sci. U.S.A.*, 84, 8325, 1987.

293. **Yamanashi, Y., Fukushige, S.-I., Semba, K., Sukegawa, J., Miyajima, N., Matsubara, K.-I., Yamamoto, T., and Toyoshima, K.,** The *yes*-related cellular gene *lyn* encodes a possible tyrosine kinase similar to p56*lck*, *Mol. Cell. Biol.*, 7, 237, 1987.

294. **Yamanashi, Y., Mori, S., Yoshida, M., Kishimoto, T., Inoue, K., Yamamoto, T., and Toyoshima, K.,** Selective expression of a protein-tyrosine kinase, p56*lyn*, in hematopoietic cells and association with production of human T-cell lymphotropic virus type I, *Proc. Natl. Acad. Sci. U.S.A.*, 86, 6538, 1989.

295. **Li, T., Bolen, J. B., and Ihle, J. N.,** Hematopoietic cells express two forms of *lyn* kinase differing by 21 amino acids in the amino terminus, *Mol. Cell. Biol.*, 11, 2391, 1991.

296. **Stanley, E., Ralph, S., McEwen, S., Boulet, I., Holtzman, D. A., Lock, P., and Dunn, A. R.,** Alternatively spliced murine *lyn* mRNAs encode distinct proteins, *Mol. Cell. Biol.*, 11, 3399, 1991.

297. **Yamanashi, Y., Kakiuchi, T., Mizuguchi, J., Yamamoto, T., and Toyoshima, K.,** Association of B cell antigen receptor with protein tyrosine kinase Lyn, *Science*, 251, 192, 1991.

298. **Yamananshi, Y., Fukui, Y., Wongsasant, B., Kinoshita, Y., Ichimori, Y., Toyoshima, K., and Yamamoto, T.,** Activation of Src-like protein-tyrosine kinase Lyn and its association with phosphatidylinositol 3-kinase upon B-cell antigen receptor-mediated signaling, *Proc. Natl. Acad. Sci. U.S.A.*, 89, 1118, 1992.

299. **Torigoe, T., Saragovi, H. U., and Reed, J. C.,** Interleukin-2 regulates the activity of the *lyn* protein-tyrosine kinase in a B-cell line, *Proc. Natl. Acad. Sci. U.S.A.*, 89, 2674, 1992.

300. **Meier, R. W., Bielke, W., Chen, T., Niklaus, G., Friis, R. R., and Tobler, A.,** *lyn,* a *src*-like tyrosine-specific protein kinase, is expressed in HL60 cells induced to monocyte-like or granulocyte-like cells, *Biochem. Biophys. Res. Commun.,* 185, 91, 1992.

301. **Eiseman, E. and Bolen, J. B.,** Engagement of the high-affinity IgE receptor activates *src* protein-related tyrosine kinases, *Nature,* 355, 78, 1992.

302. **Bielke, W., Ziemieki, A., Kappos, L., and Miescher, G. C.,** Expression of the B cell-associated tyrosine kinase gene *lyn* in primary neuroblastoma tumours and its modulation during the differentiation of neuroblastoma cell lines, *Biochem. Biophys. Res. Commun.,* 186, 1403, 1992.

303. **Kruh, G. D., King, C. R., Kraus, M. H., Popescu, N. C., Amsbaugh, S. C., McBride, W. O., and Aaronson, S. A.,** A novel gene closely related to the *abl* proto-oncogene, *Science,* 234, 1545, 1986.

304. **Perego, R., Ron, D., and Kruh, G. D.,** *arg* encodes a widely expressed 145-kDa protein-tyrosine kinase, *Oncogene,* 6, 1899, 1991.

305. **Hirai, H., Maru, Y., Hagiwara, K., Nishida, J., and Takaku, F.,** A novel putative tyrosine kinase receptor encoded by the *eph* gene, *Science,* 238, 1717, 1990.

306. **Takahashi, M. and Cooper, G. M.,** *ret* transforming gene encodes a fusion protein homologous to tyrosine kinases, *Mol. Cell. Biol.,* 7, 1378, 1987.

307. **Taniguchi, M., Iwamoto, T., Hamaguchi, M., Matsuyama, M., and Takahashi, M.,** The *ret* oncogene products are membrane-bound glycoproteins phosphorylated on tyrosine residues *in vivo, Biochem. Biophys. Res. Commun.,* 181, 416, 1991.

308. **Miyazaki, K., Asai, N., Iwashita, T., Taniguchi, M., Isomura, T., Funahashi, H., Takagi, H., Matsuyama, M., and Takahashi, M.,** Tyrosine kinase activity of the ret proto-oncogene products *in vitro, Biochem. Biophys. Res. Commun.,* 193, 565, 1993.

309. **Tahira, T., Ishizaka, Y., Sugimura, T., and Nagan, M.,** Expression of proto-ret mRNA in embryonic and adult rat tissues, *Biochem. Biophys. Res. Commun.,* 153, 1290, 1988.

310. **Iwamoto, T., Takahashi, M., Ohbayashi, M., and Nakashima, I.,** The *ret* oncogene can induce melanogenesis and melanocyte development in W^v/W^v mice, *Exp. Cell Res.,* 200, 410, 1992.

311. **Letwin, K., Yee, S.-P., and Pawson, T.,** Novel protein-tyrosine kinase cDNA related to *fps/fes* and *eph* cloned using anti-phosphotyrosine antibody, *Oncogene,* 3, 621, 1988.

312. **Feldman, R. A., Tam, J. P., and Hanafusa, H.,** Antipeptide antiserum identifies a widely distributed cellular tyrosine kinase related to but distinct from the c-*fps/fes*-encoded protein, *Mol. Cell. Biol.,* 6, 1065, 1986.

313. **Hao, Q.-L., Heisterkamp, N., and Groffen, J.,** Isolation and sequence analysis of a novel human tyrosine kinase gene, *Mol. Cell. Biol.,* 9, 1587, 1989.

314. **Morris, C., Heisterkamp, N., Hao, Q. L., Testa, J. R., and Groffen, J.,** The human tyrosine kinase gene (*FER*) maps to chromosome 5 and is deleted in myeloid leukemias with a del(5q), *Cytogenet. Cell Genet.,* 53, 196, 1990.

315. **Hao, Q.-L., Ferris, D. K., White, G., Heisterkamp, N., and Groffen, J.,** Nuclear and cytoplasmic location of the FER tyrosine kinase, *Mol. Cell Biol.,* 11, 1180, 1991.

316. **Hazan, B., Bern, O., Carmel, M., Lejbkowizcz, F., Goldstein, R. S., and Nir, U.,** ferT encodes a meiosis-specific nuclear tyrosine kinase, *Cell Growth Differ.,* 4, 443, 1993.

317. **Pawson, T., Letwin, K., Lee, T., Hao, Q.-L., Heisterkamp, N., and Groffen, J.,** The *FER* gene is evolutionarily conserved and encodes a widely expressed member of the *FPS/FES* protein-tyrosine kinase family, *Mol. Cell. Biol.,* 9, 5722, 1989.

318. **Feldman, R. A., Grabrilove, J. L., Tam, J. P., Moore, M. A. S., and Hanafusa, H.,** Specific expression of the human cellular *fps/fes*-encoded protein NCP92 in normal and leukemic myeloid cells, *Proc. Natl. Acad. Sci. U.S.A.,* 82, 2379, 1985.

319. **Mano, H., Ishikawa, F., Nishida, J., Hirai, H., and Takaku, F.,** A novel protein-tyrosine kinase, *tec* is preferentially expressed in liver, *Oncogene,* 5, 1781, 1990.

320. **Ben-Neriah, Y. and Bauskin, A. R.,** Leukocytes express a novel gene encoding a putative transmembrane protein-tyrosine kinase devoid of an extracellular domain, *Nature,* 333, 672, 1988.

321. **Bernards, A. and de la Monte, S. M.,** The *ltk* receptor tyrosine kinase is expressed in pre-B lymphocytes and cerebral neurons and uses a non-AUG translational initiator, *EMBO J.,* 9, 2279, 1990.

322. **Paul, S. R., Merberg, D., Finnerty, H., Morris, G. E., Morris, J. C., Jones, S. S., Turner, K. J., and Wood, C. R.,** Molecular cloning of the cDNA encoding a receptor tyrosine kinase-related molecule with a catalytic region homologous to c-*met, Int. J. Cell Cloning,* 10, 309, 1992.

323. **Horita, K., Yagi, T., Kohmura, N., Tomooka, Y., Ikawa, Y., and Aizawa, S.,** A novel tyrosine kinase, *hyk,* expressed in murine embryonic stem cells, *Biochem. Biophys. Res. Commun.,* 189, 1747, 1992.

324. **Taniguchi, T., Kobayashi, T., Kondon, J., Takahashi, K., Nakamura, H., Suzuki, J., Nagai, K., Yamada, T., Nakamura, S., and Yamamura, H.,** Molecular cloning of a porcine gene *syk* that encodes a 72-kDa protein-tyrosine kinase showing high susceptibility to proteolysis, *J. Biol. Chem.,* 266, 15790, 1991.

325. **Taniguchi, T., Kitagawa, H., Yasue, S., Yanagi, S., Sakai, K., Asahi, M., Ohta, S., Takeuchi, F., Nakamura, S., and Yamamura, H.,** Protein-tyrosine kinase p72^{syk} is activated by thrombin and is negatively regulated through Ca^{2+} mobilization in platelets, *J. Biol. Chem.,* 268, 2277, 1993.

326. **O'Bryan, J. P., Frye, R. A., Cogswell, P. C., Neubauer, A., Kitch, B., Prokop, C., Espinosa, R., III, Le Beau, M. M., Earp, H. S., and Liu, E. T.,** *axl,* a transforming gene isolated from primary human myeloid leukemia cells, encodes a novel receptor tyrosine kinase, *Mol. Cell. Biol.,* 11, 5016, 1991.

327. **Partanen, J., Armstrong, E., Mäkelä, T. P., Korhonen, J., Sandberg, M., Renkonen, R., Knuutila, S., Huebner, K., and Alitalo, K.,** A novel endothelial cell surface receptor tyrosine kinase with extracellular epidermal growth factor homology domains, *Mol. Cell. Biol.,* 12, 1698, 1992.

328. **Masiakowski, P. and Carroll, R. D.,** A novel family of cell surface receptors with tyrosine kinase-like domain, *J. Biol. Chem.,* 267, 26181, 1992.

329. **Dumont, D. J., Gradwohl, G. J., Fong, G.-H., Auerbach, R., and Breitman, M. L.,** The endothelial-specific receptor tyrosine kinae, *tek* is a member of a new subfamily of receptors, *Oncogene,* 8, 1293, 1993.

330. **Kasuga, M., Fujita-Yamaguchi, T., Blithe, D. L., and Kahn, C. R.,** Tyrosine-specific protein kinase activity is associated with the purified insulin receptor, *Proc. Natl. Acad. Sci. U.S.A.,* 80, 2137, 1983.

331. **Rubin, J. B., Shia, M. A., and Pilch, P. F.,** Stimulation of tyrosine-specific phosphorylation *in vitro* by insulin-like growth factor I, *Nature,* 305, 438, 1983

332. **Hunter, T. and Cooper, J. A.,** Epidermal growth factor induces rapid tyrosine phosphorylation of proteins in A431 tumor cells, *Cell,* 24, 741, 1981.

333. **Reynolds, F. H., Todaro, G. J., Fryling, C., and Stephenson, J. R.,** Human transforming growth factors induce tyrosine phosphorylation of EGF receptors, *Nature,* 292, 259, 1981.

334. **Pike, L. J., Marquardt, H., Todaro, G. J., Gallis, B., Casnellie, J. E., Bornstein, P., and Krebs, E. G.,** Transforming growth factor and epidermal growth factor stimulate the phosphorylation of a synthetic, tyrosine-containing peptide in a similar manner, *J. Biol. Chem.,* 257, 14628, 1982.

335. **Ek, B. and Heldin, C.-H.,** Characterization of a tyrosine-specific kinase activity in human fibroblast membranes stimulated by platelet-derived growth factor, *J. Biol. Chem.,* 257, 10486, 1982.

336. **Huang, S. S. and Huang, J. S.,** Association of bovine brain-derived growth factor receptor with protein tyrosine kinase activity, *J. Biol. Chem.,* 261, 9568, 1986.

337. **Kuo, M.-D., Huang, S. S., and Huang, J. S.,** Acidic fibroblast growth factor receptor purified from bovine liver is a novel protein tyrosine kinase, *J. Biol. Chem.,* 265, 16455, 1990.

338. **Meakin, S. O. and Shooter, E. M.,** Tyrosine kinase activity coupled to the high-affinity nerve growth factor receptor complex, *Proc. Natl. Acad. Sci. U.S.A.,* 88, 5862, 1991.

339. **Downing, J. R., Rettenmier, C. W., and Sherr, C. J.,** Ligand-induced tyrosine kinase activity of the colony-stimulating factor 1 receptor in a murine macrophage cell line, *Mol. Cell. Biol.,* 8, 1795, 1988.

340. **Pasquale, E. B.,** A distinctive family of embryonic protein-tyrosine kinase receptors, *Proc. Natl. Acad. Sci. U.S.A.,* 87, 5812, 1990.

341. **Kanakura, Y., Druker, B., Cannistra, S. A., Furukawa, Y., Torimoto, Y., and Griffin, J. D.,** Signal transduction of the human granulocyte-macrophage colony-stimulating factor and interleukin-3 receptors involves tyrosine phosphorylation of a common set of cytoplasmic proteins, *Blood,* 76, 706, 1990.

342. **Inazu, T., Taniguchi, T., Ohta, S., Miyabo, S., and Yanamura, H.,** The lectin wheat germ agglutinin induces rapid protein-tyrosine phosphorylation in human platelets, *Biochem. Biophys. Res. Commun.,* 174, 1154, 1991.

343. **Gold, M. R., Law, D. A., and DeFranco, A. L.,** Stimulation of protein tyrosine phosphorylation by the B-lymphocyte antigen receptor, *Nature,* 345, 810, 1990.

344. **Huang, C.-K., Bonak, V., Laramee, G. R., and Casnellie, J. E.,** Protein tyrosine phosphorylation in rabbit peritoneal neutrophils, *Biochem. J.,* 269, 431, 1990.

345. **Leader, D. P. and Katan, M.,** Viral aspects of protein phosphorylation, *J. Gen. Virol.,* 69, 1441, 1988.

346. **Heffetz, D., Bushkin, I., Dror, R., and Zick, Y.,** The insulinomimetic agents H_2O_2 and vanadate stimulate protein tyrosine phosphorylation in intact cells, *J. Biol. Chem.,* 265, 2896, 1990.

347. **Ferrell, J. E., Jr. and Martin, G. S.,** Platelet tyrosine- specific protein phosphorylation is regulated by thrombin, *Mol. Cell. Biol.,* 8, 3603, 1988.

348. **Ganguly, C. L., Dale, J. B., Courtney, H. S., and Beachey, E. H.,** Tyrosine phosphorylation of a 94-kDa protein of human fibroblasts stimulated by streptococcal lipoteichoic acid, *J. Biol. Chem.,* 260, 13342, 1985.

349. **Chan, T. M., Chen, E., Tatoyan, A., Shargill, N. S., Pleta, M., and Hochstein, P.,** Stimulation of tyrosine-specific protein phosphorylation in the rat liver plasma membrane by oxygen radicals, *Biochem. Biophys. Res. Commun.,* 139, 439, 1986.

350. **Srivastava, A. K.,** Stimulation of tyrosine protein kinase activity by dimethyl sulfoxide, *Biochem. Biophys. Res. Commun.,* 126, 1042, 1985.

351. **Wedner, H. J. and Bass, G.,** Induction of the tyrosine phosphorylation of a 66 KD soluble protein by DMSO in human peripheral blood T lymphocytes, *Biochem. Biophys. Res. Commun.,* 140, 743, 1986.

352. **Michiel, D. F. and Wang, J. H.,** Stimulation of tyrosine phosphorylation of rat brain membrane proteins by calmodulin, *FEBS Lett.,* 190, 11, 1985.

353. **Satoh, T., Endo, M., Nakafuku, M., Akiyama, T., Yamamoto, T., and Kaziro, Y.,** Accumulation of p21^{ras}-GTP in response to stimulation with epidermal growth factor and oncogene products with tyrosine kinase activity, *Proc. Natl. Acad. Sci. U.S.A.,* 87, 7926, 1990.

354. **Lowenstein, E. J., Daly, R. J., Batzer, A. G., Li, W., Margolis, B., Lammers, R., Ullrich, A., Skolnik, E. Y., Bar-Sagi, D., and Schlessinger, J.,** The SH2 and SH3 domain-containing protein GRB2 links receptor tyrosine kinases to *ras* signaling, *Cell,* 70, 431, 1992.

355. **Egan, S. E., Giddings, B. W., Brooks, M. W., Buday, L., Sizeland, A. M., and Weinberg, R. A.,** Association of Sos Ras exchange protein with Grb2 is implicated in tyrosine kinase signal transduction and transformation, *Nature,* 363, 45, 1993.

356. **Buday, L. and Downward, J.,** Epidermal growth factor regulates p21^{ras} through the formation of a complex of receptor, Grb2 adapter protein, and Sos nucleotide exchange factor, *Cell,* 73, 611, 1993.

357. **Rozakis-Adcock, M., McGlade, J. M., Mbamalu, G., Pelicci, G., Daly, R., Li, W., Batzer, A., Thomas, S., Brugge, J., Pelicci, P. G., Schlessinger, J., and Pawson, T.,** Association of the Shc and Grb2/Sem5 SH2-containing proteins is implicated in activation of the Ras pathway by tyrosine kinases, *Nature,* 360, 689, 1992.

358. **Hu, P., Margolis, B., and Schlessinger, J.,** *vav:* a potential link between tyrosine kinases and *ras*-like GTPases in hematopoietic cell signaling, *BioEssays,* 15, 179, 1993.

359. **Klein, H. H., Freidenberg, G. R., Cordera, R., and Olefsky, J. M.,** Substrate specificities of insulin and epidermal growth factor receptor kinases, *Biochem. Biophys. Res. Commun.,* 127, 254, 1985.

360. **Pike, L. J., Kuenzel, E. A., Casnellie, J. E., and Krebs, E. G.,** A comparison of the insulin- and epidermal growth factor-stimulated protein kinases from human placenta, *J. Biol. Chem.,* 259, 9913, 1984.

361. **Cobb, M. H., Burr, J. G., Linder, M. E., Gray, T. B., and Gregory, J. S.,** Similar ribosomal protein S6 kinase activity is found in insulin-treated 3T3-L1 cells and chick embryo fibroblasts transformed by Rous sarcoma virus, *Biochem. Biophys. Res. Commun.,* 137, 702, 1986.

362. **Ballou, L. M., Siegmann, M., and Thomas, G.,** S6 kinase in quiescent Swiss mouse 3T3 cells is activated by phosphorylation in response to serum treatment, *Proc. Natl. Acad. Sci. U.S.A.,* 85, 7154, 1988.

363. **Liao, F., Shin, H. S., and Rhee, S. G.,** *In vitro* tyrosine phosphorylation of PLC-γ1 and PLC-γ2 by Src-family protein tyrosine kinases, *Biochem. Biophys. Res. Commun.,* 191, 1028, 1993.

364. **Frederickson, R. M., Montine, K. S., and Sonenberg, N.,** Phosphorylation of eukaryotic translation initiation factor 4E is increased in Src-transformed cell lines, *Mol. Cell. Biol.,* 11, 2896, 1991.

365. **Khanna, N. C., Tokuda, M., Chong, S. M., and Waisman, D. M.,** Phosphorylation of p36 *in vitro* by protein kinase C, *Biochem. Biophys. Res. Commun.,* 137, 397, 1986.

366. **Gould, K. L., Woodgett, J. R., Isacke, C. M., and Hunter, T.,** The protein-tyrosine kinase substrate p36 is also a substrate for protein kinase C *in vitro* and *in vivo, Mol. Cell. Biol.,* 6, 2738, 1986.

367. **Soric, J. and Gordon, J. A.,** Calcium-dependent isolation of the 36-kilodalton substrate of pp60^{src}-kinase: fractionation of the phosphorylated and unphosphorylated species, *J. Biol. Chem.,* 261, 14490, 1986.

368. **Isacke, C. M., Trowbridge, I. S., and Hunter, T.,** Modulation of p36 phosphorylation in human cells: studies using anti-p36 monoclonal antibodies, *Mol. Cell. Biol.,* 6, 2745, 1986.

369. **Gerke, V.,** Tyrosine protein kinase substrate p36: a member of the annexin family of Ca^{2+}/phospholipid-binding proteins, *Cell Motil. Cytoskel.,* 14, 449, 1989.

370. **Carter, V. C., Howlett, A. R., Martin, G. S., and Bissell, M. J.,** The tyrosine phosphorylation substrate p36 is developmentally regulated in embryonic avian limb and is induced in cell culture, *J. Cell Biol.,* 103, 2017, 1986.

371. **Glenney, J.,** Two related but distinct forms of the M_r 36,000 tyrosine kinase substrate (calpactin) that interact with phospholipid and actin in a Ca^{2+}-dependent manner, *Proc. Natl. Acad. Sci. U.S.A.,* 83, 4258, 1986.

372. **Simon, M., Arrigo, A.-P., and Spahr, P.-F.,** Association of three chicken proteins with the 34-kD target of Rous sarcoma virus tyrosine kinase, *Exp. Cell Res.,* 169, 419, 1987.

373. **Martin, F., Derancourt, J., Capony, J.-P., Colote, S., and Cavadore, J.-C.,** Sequence homologies between p36, the substrate of pp60^{src} tyrosine kinase and a 67 kDa protein isolated from bovine aorta, *Biochem. Biophys. Res. Commun.,* 145, 961, 1987.

374. **Weber, K. and Johnsson, N.,** Repeating sequence homologies in the p36 target protein of retroviral protein kinases and lipocortin, the p37 inhibitor of phospholipase A_2, *FEBS Lett.,* 203, 95, 1986.

375. **Gerke, V. and Weber, K.,** The regulatory chain in the p36-kd substrate complex of viral tyrosine-specific protein kinases is related to the S-100 protein of glial cells, *EMBO J.,* 4, 2917, 1985.

376. **Johnsson, N., Vandekerckhove, J., Van Damme, J., and Weber, K.,** Binding sites for calcium, lipid and p11 on p36, the substrate of retroviral tyrosine-specific protein kinases, *FEBS Lett.,* 198, 361, 1986.

377. **Brugge, J. S.,** The p35/p36 substrates of protein-tyrosine kinases as inhibitors of phospholipase A_2, *Cell,* 46, 149, 1986.

378. **William, F., Mroczkowski, B., Cohen, S., and Kraft, A. S.,** Differentiation of HL-60 cells is associated with an increase in the 35-kDa protein lipocortin I, *J. Cell. Physiol.,* 137, 402, 1988.

379. **Isacke, C. M., Lindberg, R. A., and Hunter, T.,** Synthesis of p36 and p35 is increased when U-937 cells differentiate in culture but expression is not inducible by glucocorticoids, *Mol. Cell. Biol.,* 9, 232, 1989.

380. **Northup, J. K., Valentine-Braun, K. A., Johnson, L. K., Severson, D. L., and Hollenberg, M. D.,** Evaluation of the anti-inflammatory and phospholipase-inhibitory activity of calpactin II/lipocortin I, *J. Clin. Invest.,* 82, 1347, 1988.

381. **Akiyama, T., Kadowaki, T., Nishida, E., Kadooka, T., Ogawara, H., Fukami, Y., Sakai, H., Takaku, F., and Kasuga, M.,** Substrate specificities of tyrosine-specific protein kinases toward cytoskeletal proteins *in vitro, J. Biol. Chem.,* 261, 14797, 1986.

382. **Cooper, J. A. and Hunter, T.,** Major substrate for growth factor-activated protein-tyrosine kinases is a low-abundance protein, *Mol. Cell. Biol.,* 5, 3304, 1985.

383. **Rossomando, A. J., Payne, D. M., Weber, M. J., and Sturgill, T. W.,** Evidence that pp42, a major tyrosine kinase target protein, is a mitogen-activated serine/threonine protein kinase, *Proc. Natl. Acad. Sci. U.S.A.,* 86, 6940, 1989.

384. **Piedimonte, G., Silvotti, L., Chamaret, S., Borghetti, A. F., and Montagnier, L.,** Association of tyrosine protein kinase activity with mitochondria in human fibroblasts, *J. Cell. Biochem.,* 32, 113, 1986.

385. **Piedimonte, G., Chamaret, S., Dauget, C., Borghetti, A. F., and Montagnier, L.,** Identification and characterization of tyrosine kinase activity associated with mitochondrial outer membrane in sarcoma 180 cells, *J. Cell. Biochem.,* 36, 91, 1988.

386. **Maher, P. A. and Pasquale, E. B.,** Tyrosine phosphorylated proteins in different tissues during chick embryo development, *J. Cell Biol.,* 106, 1747, 1988.

387. **Wong, T. W. and Goldberg, A. R.,** Tyrosyl protein kinases in normal rat liver: identification and partial characterization, *Proc. Natl. Acad. Sci. U.S.A.,* 80, 2529, 1983.

388. **Wong, T. W. and Goldberg, A. R.,** Purification and characterization of the major species of tyrosine protein kinase in rat liver, *J. Biol. Chem.,* 259, 8505, 1984.

389. **Yoshikawa, K., Usui, H., Imazu, M., Tsukamoto, H., and Takeda, M.,** Comparison of tyrosine protein kinases in membrane fractions from mouse liver and Ehrlich ascites tumor, *J. Biol. Chem.,* 260, 15091, 1985.

390. **Galski, H., De Groot, N., and Hochberg, A. A.,** Phosphorylation of tyrosine in cultured human placenta, *Biochim. Biophys. Acta,* 761, 284, 1983.

391. **Dangott, L. J., Puett, D., Garbers, D. L., and Melner, M. H.,** Tyrosine protein kinase activity in purified rat Leydig cells, *Biochem. Biophys. Res. Commun.,* 116, 400, 1983.

392. **Swarup, G., Dasgupta, J. D., and Garbers, D. L.,** Tyrosine protein kinase activity of rat spleen and other tissues, *J. Biol. Chem.,* 258, 10341, 1983.

393. **Nakamura, S., Takeuchi, F., Kondo, H., and Yamamura, H.,** High tyrosine protein kinase activities in soluble and particulate fractions in bone marrow cells, *FEBS Lett.,* 170, 139, 1984.

394. **Gacon, G., Piau, J.-P., Blaineau, C., Fagard, R., Genetet, N., and Fischer, S.,** Tyrosine phosphorylation in human T lymphoma cells, *Biochem. Biophys. Res. Commun.,* 117, 843, 1983.

395. **Harrison, M. L., Low, P. S., and Geahlen, R. L.,** T and B lymphocytes express distinct tyrosine protein kinases, *J. Biol. Chem.,* 259, 9348, 1984.

396. **Earp, H. S., Austin, K. S., Gillespie, G. Y., Buesson, S. C., Davies, A. A., and Parker, P. J.,** Characterization of distinct tyrosine-specific protein kinases in B and T lymphocytes, *J. Biol. Chem.,* 260, 4351, 1985.

397. **Tuy, F. P. D., Henry, J., Rosenfeld, C., and Kahn, A.,** High tyrosine kinase activity in normal nonproliferating cells, *Nature,* 305, 435, 1983.

398. **Zioncheck, T. F., Harrison, M. L., and Geahlen, R. L.,** Purification and characterization of a protein-tyrosine kinase from bovine thymus, *J. Biol. Chem.,* 261, 15637, 1986.

399. **Tokuda, M., Khanna, N. C., Arora, A. K., and Waisman, D. M.,** Identification and characterization of the tyrosine protein kinases of rat spleen, *Biochem. Biophys. Res. Commun.,* 139, 910, 1986.

400. **Swarup, G. and Subrahmanyanm, G.,** Activation of a cellular tyrosine-specific protein kinase by phosphorylation, *FEBS Lett.,* 188, 131, 1985.

401. **Kong, S.-K. and Wang, J. H.,** Purification and characterization of a protein tyrosine kinase from bovine spleen, *J. Biol. Chem.,* 262, 2597, 1987.

402. **Dubaybo, B. A., Marwah, G. S., Fligiel, S. E. G., Hatfield, J. S., and Majumdar, A. P. N.,** Tyrosine kinase activation during lung injury, fibrosis, and compensatory lung growth, *Exp. Lung Res.,* 16, 257, 1990.

403. **Montagnier, L., Chamaret, S., and Dauguet, C.,** Augmentation, dans des extraits de cellules cancéreuses ou transformées, de l'activité d'une protéine-kinase phosphorylant la tyrosine, *Compt. Rend. Acad. Sci. Paris,* 295, 375, 1982.

404. **Wickremasinghe, R. G., Piga, A., Mire, A. R., Reza Taheri, M., Yaxley, J. C., and Hoffbrand, A. V.,** Tyrosine protein kinases and their substrates in human leukemia cells, *Leukemia Res.,* 9, 1443, 1985.

405. **Fischer, S., Fagard, R., Piau, J.-P., Genetet, N., Blaineau, C., Reibel, L., Le Prise, Y., and Gacon, G.,** Acute myeloblastic leukemia with active tyrosine protein kinase, *Leukemia Res.,* 9, 1345, 1985.

406. **Ogawa, R., Ohtsuda, M., Sasadaira, H., Hirasa, H., Yabe, H., Uchida, H., and Watanabe, Y.,** Increase of phosphotyrosine-containing proteins in human carcinomas, *Jpn. J. Cancer Res.,* 76, 1049, 1985.

407. **Punt, C. J. A., Rijksen, G., Vlug, A. M. C., Dekker, A. W., and Staal, G. E. J.,** Tyrosine protein kinase activity in normal and leukaemic human blood cells, *Br. J. Haematol.,* 73, 51, 1989.

408. **Kéri, G., Balogh, A., Teplán, I., and Csuka, O.,** Comparison of the tyrosine kinase activity with the proliferation rate in human colon solid tumors and tumor cell lines, *Tumor Biol.,* 2, 315, 1988.

409. **Hennipman, A., van Oirschot, B. A., Smits, J., Rijksen, G., and Staal, G. E. J.,** Tyrosine kinase activity in breast cancer, benign breast disease, and normal breast tissue, *Cancer Res.,* 49, 516, 1989.

410. **Dangott, L. J., Puett, D., and Melner, M. H.,** Characterization of protein tyrosine kinase activity in murine Leydig tumor cells, *Biochim. Biophys. Acta,* 886, 187, 1986.

411. **Giordano, S., Di Renzo, M. F., Narsimhan, R. P., Tamagnone, L., Gerbaudo, E. V., Chiadó-Piat, L., and Comoglio, P. M.,** Evidence for autocrine activation of a tyrosine kinase in a human gastric carcinoma cell line, *J. Cell. Biochem.,* 38, 229, 1988.

412. **Gentleman, S., Russell, P., Martensen, T. M., and Chader, G. J.,** Characteristics of protein tyrosine kinase activities of Y-79 retinoblastoma cells and retina, *Arch. Biochem. Biophys.,* 239, 130, 1985.

413. **Lee, W.-H., Murphree, A. L., and Benedict, W. F.,** Expression and amplification of the N-*myc* gene in primary retinoblastoma, *Nature,* 309, 458, 1984.

414. **Rausch, D. M., Dickens, G., Doll, S., Fujita, K., Koizumi, S., Rudkin, B. B., Tocco, M., Eiden, L. E., and Guroff, G.,** Differentiation of PC12 cells with v-*src*: comparison with nerve growth factor, *J. Neurosci. Res.,* 24, 49, 1989.

415. **Bishop, R., Martinez, R., Nakamura, K. D., and Weber, M. J.,** A tumor promoter stimulates phosphorylation on tyrosine, *Biochem. Biophys. Res. Commun.,* 115, 536, 1983.

416. **Grunberger, G., Zick, Y., Taylor, S. I., and Gorden, P.,** Tumor-promoting phorbol ester stimulates tyrosine phosphorylation in U-937 monocytes, *Proc. Natl. Acad. Sci. U.S.A.,* 81, 2762, 1984.

417. **Khanna, N. C., Tokuda, M., Chong, S. M., and Waisman, D. M.,** Phosphorylation of p36 *in vitro* by protein kinase C, *Biochem. Biophys. Res. Commun.,* 137, 397, 1986.

418. **Gould, K. L., Woodgett, J. R., Isacke, C. M., and Hunter, T.,** The protein-tyrosine kinase substrate p36 is also a substrate for protein kinase C *in vitro* and *in vivo*, *Mol. Cell. Biol.*, 6, 2738, 1986.

419. **Johnsson, N., Van, P. N., Sölling, H.-D., and Weber, K.,** Functionally distinct serine phosphorylation sites of p36, the cellular substrate of retroviral protein kinase; differential inhibition of reassociation with p11, *EMBO J.*, 5, 3455, 1986.

420. **Casnellie, J. E. and Lamberts, R. J.,** Tumor promoters cause changes in the state of phosphorylation and apparent molecular weight of a tyrosine protein kinase in T lymphocytes, *J. Biol. Chem.*, 261, 4921, 1986.

421. **Foss, F. M., Veillette, A., Sartor, O., Rosen, N., and Bolen, J. B.,** Alterations in the expression of pp60$^{c\text{-}src}$ and p56^{lck} associated with butyrate-induced differentiation of human colon carcinoma cells, *Oncogene Res.*, 5, 13, 1989.

422. **Chapekar, M. S., Hartman, K. D., Knode, M. C., and Glazer, R. I.,** Synergistic effect of retinoic acid and calcium ionophore A23187 on differentiation, c-*myc* expression, and membrane tyrosine kinase activity in human promyelocytic leukemia cell line HL-60, *Mol. Pharmacol.*, 31, 140, 1987.

423. **Glazer, R. I., Chapekar, M. S., Hartman, K. D., and Knode, M. C.,** Appearance of membrane-bound tyrosine kinase during differentiation of HL-60 leukemia cells by immune interferon and tumor necrosis factor, **Biochem. Biophys. Res. Commun.**, 140, 908, 1986.

424. **Fischer, E. H., Charbonneau, H., and Tonks, N. K.,** Protein tyrosine phosphatases: a diverse family of intracellular and transmembrane enzymes, *Science,* 253, 401, 1991.

425. **Pot, D. A. and Dixon, J. E.,** A thousand and two protein tyrosine phosphatases, *Biochim. Biophys. Acta,* 1136, 35, 1992.

426. **Walton, K. M. and Dixon, J. E.,** Protein tyrosine phosphatases, *Annu. Rev. Biochem.*, 62, 101, 1993.

427. **Lau, K.-H. W. and Baylink, D. J.,** Phosphotyrosyl protein phosphatases: potential regulators of cell proliferation and differentiation, *Crit. Rev. Oncogenesis,* 4, 451, 1993.

428. **Matthews, R. J., Cahir, E. D., and Thomas, M. L.,** Identification of an additional member of the protein-tyrosine-phosphatase family: evidence for alternative splicing in the tyrosine phosphatase domain, *Proc. Natl. Acad. Sci. U.S.A.*, 87, 4444, 1990.

429. **Swarup, G., Cohen, S., and Garbers, D. L.,** Selective dephosphorylation of proteins containing phosphotyrosine by alkaline phosphatases, *J. Biol. Chem.*, 256, 8197, 1981.

430. **Swarup, G., Speeg, K. V., Jr., Cohen, S., and Garbers, D. L.,** Phosphotyrosyl-protein phosphatase of TCRC-2 cells, *J. Biol. Chem.,* 257, 7298, 1982.

431. **Leis, J. F. and Kaplan, N. O.,** An acid phosphatase in the plasma membrane of human astrocytoma showing marked specificity toward phosphotyrosine protein, *Proc. Natl. Acad. Sci. U.S.A.,* 79, 6507, 1982.

432. **Foulkes, J. G., Erikson, E., and Erikson, R. L.,** Separation of multiple phosphotyrosyl- and phosphoseryl-protein phosphatases from chicken brain, *J. Biol. Chem.*, 258, 431, 1983.

433. **Shriner, C. L. and Brautigan, D. L.,** Cytosolic protein phosphotyrosine phosphatases from rabbit kidney: purification of two distinct enzymes that bind to Zn^{2+}-iminodiacetate agarose, *J. Biol. Chem.,* 259, 11383, 1984.

434. **Boivin, P. and Galand, C.,** The human red cells acid phosphatase is a phosphotyrosine protein phosphatase which dephosphorylates the membrane protein band 3, *Biochem. Biophys. Res. Commun.,* 134, 557, 1986.

435. **Clari, G., Brunati, A. M., and Moret, V.,** Partial purification and characterization of phosphotyrosyl-protein phosphatase(s) from human erythroid cytosol, *Biochem. Biophys. Res. Commun.,* 137, 566, 1986.

436. **Frank, D. A. and Sartorelli, A. C.,** Regulation of protein phosphotyrosine content by changes in tyrosine kinase and protein phosphotyrosine phosphatase activities during induced granulocytic and monocytic differentiation of HL-60 leukemia cells, *Biochem. Biophys. Res. Commun.,* 140, 440, 1986.

437. **Garcia-Morales, P., Minami, Y., Luong, E., Klausner, R. D., and Samelson, L. E.,** Tyrosine phosphorylation in T-cells is regulated by phosphatase activity — studies with phenylarsine oxide, *Proc. Natl. Acad. Sci. U.S.A.,* 87, 9255, 1990.

438. **Charles, C. H., Sun, H., Lau, L. F., and Tonks, N. K.,** The growth factor-inducible immediate-early gene 3CH134 encodes a protein-tyrosine phosphatase, *Proc. Natl. Acad. Sci. U.S.A.,* 90, 5292, 1993.

439. **Swarup, G. and Subrahmanyam, G.,** Purification and characterization of a protein-phosphotyrosine phosphatase from rat spleen which dephosphorylates and inactivates a tyrosine-specific protein kinase, *J. Biol. Chem.,* 264, 7801, 1989.

440. **Pallen, C. J. and Tong, P. H.,** Elevation of membrane tyrosine phosphatase activity in density-dependent growth-arrested fibroblasts, *Proc. Natl. Acad. Sci. U.S.A.,* 88, 6996, 1991.

441. **Kaplan, R., Morse, B., Huebner, K., Croce, C., Howk, R., Ravera, M., Ricca, G., Jaye, M., and Schlessinger, J.,** Cloning of three human tyrosine phosphatases reveals a multigene family of receptor-linked protein-tyrosine-phosphatases expressed in brain, *Proc. Natl. Acad. Sci. U.S.A.,* 87, 7000, 1990.

442. **Tonks, N. K., Charbonneau, H., Diltz, C. D., Fischer, E. H., and Walsh, K. A.,** Demonstration that the leukocyte common antigen CD45 is a protein tyrosine phosphatase, *Biochemistry,* 27, 8695, 1988.

443. **Koretzky, G. A.,** Role of the CD45 tyrosine phosphatase in signal transduction in the immune system, *FASEB J.,* 7, 420, 1993.

444. **Justement, L. B., Campbell, K. S., Chien, N. C., and Cambier, J. C.,** Regulation of B cell antigen receptor signal transduction and phosphorylation by CD45, *Science,* 252, 1839, 1991.

445. **Desai, D. M., Sap, J., Schlessinger, J., and Weiss, A.,** Ligand-mediated negative regulation of a chimeric transmembrane receptor tyrosine phosphatase, *Cell,* 73, 541, 1993.

446. **Broxmeyer, H. E., Lu, L., Hagoc, G., Cooper, S., Hendrie, P. C., Ledbetter, J. A., Xiao, M., Williams, D. E., and Shen, F.-W.,** CD45 cell surface antigens are linked to stimulation of early human myeloid progenitor cells by interleukin 3 (IL-3), granulocyte/macrophage colony-stimulating factor (GM-CSF), a GM-CSF/IL-3 fusion protein, and mast cell growth factor (a c-*kit* ligand), *J. Exp. Med.,* 174, 447, 1991.

447. **Rottenberg, S. A. and Brautigan, D. L.,** Membrane protein phophotyrosine phosphatase in rabbit kidney: proteolysis activates the enzyme and generates soluble catalytic fragments, *Biochem. J.,* 243, 747, 1987.

448. **Frank, D. A. and Sartorelli, A. C.,** Biochemical characterization of tyrosine kinase and phosphotyrosine phosphatase activities of HL-60 leukemia cells, *Cancer Res.,* 48, 4299, 1988.

449. **Roome, J., O'Hare, T., Pilch, P. F., and Brautigan, D. L.,** Protein phosphotyrosine phosphatase purified from the particulate fraction of human placenta dephosphorylates insulin and growth-factor receptors, *Biochem. J.,* 256, 493, 1988.

450. **Cicirelli, M. F., Tonks, N. K., Diltz, C. D., Weiel, J. E., Fischer, E. H., and Krebs, E. G.,** Microinjection of a protein-tyrosine-phosphatase inhibits insulin action in *Xenopus* oocytes, *Proc. Natl. Acad. Sci. U.S.A.,* 87, 5514, 1990.

451. **Nelson, R. L. and Branton, P. E.,** Identification, purification, and characterization of phosphotyrosine-specific protein phosphatases from cultured chicken embryo fibroblasts, *Mol. Cell. Biol.,* 4, 1003, 1984.

452. **Tonks, N. K, Diltz, C. D., and Fischer, E. H.,** Purification of the major protein-tyrosine-phosphatases of human placenta, *J. Biol. Chem.,* 263, 6722, 1988.

453. **Tonks, N. K., Diltz, C. D., and Fischer, E. H.,** Characterization of the major protein-tyrosine-phosphatases of human placenta, *J. Biol. Chem.,* 263, 6731, 1988.

454. **Streuli, M., Krueger, N. X., Thai, T., Tang, M., and Saito, H.,** Distinct functional role of the two intracellular phosphatase like domains of the receptor-linked protein tyrosine phosphatases LCA and LAR, *EMBO J.,* 9, 2399, 1990.

455. **Charbonneau, H., Tonks, N. K., Kumar, S., Diltz, C. D., Harrylock, M., Cool, D. E., Krebs, E. G., Fischer, E. H., and Walsh, K. A.,** Human placenta protein-tyrosine-phosphatase: amino acid sequence and relationship to a family of receptor-like proteins, *Proc. Natl. Acad. Sci. U.S.A.,* 86, 5252, 1989.

456. **Cool, D. E., Tonks, N. K., Charbonneau, H., Walsh, K. A., Fischer, E. H., and Krebs, E. G.,** cDNA isolated from a human T-cell library encodes a member of the protein-tyrosine-phosphatase family, *Proc. Natl. Acad. Sci. U.S.A.,* 86, 5257, 1989.

457. **Swarup, G., Kamatkar, S., Radha, V., and Rema, V.,** Molecular cloning and expression of a protein-tyrosine phosphatase showing homology with transcription factors Fos and Jun, *FEBS Lett.,* 280, 65, 1991.

458. **Guan, K., Haun, R. S., Watson, S. J., Geahlen, R. L., and Dixon, J. E.,** Cloning and expression of a protein-tyrosine-phosphatase, *Proc. Natl. Acad. Sci. U.S.A.,* 87, 1501, 1990.

459. **Pallen, C. J., Valentine, K. A., Wang, J. H., and Hollenberg, M. D.,** Calcineurin-mediated dephosphorylation of the human placental membrane receptor for epidermal growth factor urogastrone, *Biochemistry,* 24, 4727, 1985.

460. **Chan, C. P., Gallis, B., Blumenthal, D. K., Pallen, C. J., Wang, J. H., and Krebs, E. G.,** Characterization of the phosphotyrosyl protein phosphatase activity of calmodulin-dependent protein phosphatase, *J. Biol. Chem.,* 261, 9890, 1986.

461. **Kincaid, R. L., Martensen, T. M., and Vaughan, M.,** Modulation of calcineurin phosphotyrosyl protein phosphatase activity by calmodulin and protease treatment, *Biochem. Biophys. Res. Commun.,* 140, 320, 1986.

462. **Lin, M.-F. and Clinton, G. M.,** Human prostatic acid phosphatase has phosphotyrosyl protein phosphatase activity, *Biochem. J.,* 235, 351, 1986.

463. **Lin, M.-F., Lee, C.-L., and Clinton, G. M.,** Tyrosyl kinase activity is inversely related to prostatic acid phosphatase activity in two human prostate carcinoma cell lines, *Mol. Cell. Biol.,* 6, 4753, 1986.

464. **Streuli, M., Krueger, N. X., Tsai, A. Y. M., and Saito, H.,** A family of receptor-linked tyrosine phosphatases in humans and *Drosophila, Proc. Natl. Acad. Sci. U.S.A.,* 86, 8698, 1989.

465. **Yi, T., Cleveland, J. L., and Ihle, J. N.,** Protein tyrosine phosphatase containing SH2 domains: characterization, preferential expression in hematopoietic cells, and localization to human chromosome 12p12-p13, *Mol. Cell. Biol.,* 12, 836, 1992.

466. **Shen, S.-H., Bastien, L., Posner, B. I., and Chrétien, P.,** A protein-tyrosine phosphatase with sequence similarity to the SH2 domain of the protein-tyrosine kinases, *Nature,* 352, 736, 1991.

467. **Ahmad, S., Banville, D., Zhao, Z., Fischer, E. H., and Shen, S.-H.,** A widely expressed human protein-tyrosine phosphatase containing *src* homology 2 domains, *Proc. Natl. Acad. Sci. U.S.A.,* 90, 2197, 1993.

468. **Krueger, N. X., Streuli, M., and Saito, H.,** Structural diversity and evolution of human receptor-like protein tyrosine phosphatases, *EMBO J.,* 9, 3241, 1990.

469. **Vogel, W., Lammers, R., Huang, J. T., and Ullrich, A.,** Activation of phosphotyrosine phosphatase by tyrosine phosphorylation, *Science,* 259, 1611, 1993.

470. **Feng, G. S., Hui, C. C., and Pawson, T.,** SH2-containing phosphotyrosine phosphatase as a target of protein-tyrosine kinases, *Science,* 259, 1607, 1993.

471. **Freeman, R. M., Jr., Plutzky, J., and Neel, B. G.,** Identification of a human src homology 2-containing protein-tyrosine-phosphatase: a putative homolog of *Drosophila* corkscrew, *Proc. Natl. Acad. Sci. U.S.A.,* 89, 11239, 1992.

472. **Zheng, X. M., Wang, Y., and Pallen, C. J.,** Cell transformation and activation of pp60^{c-src} by overexpression of a protein tyrosine phosphatase, *Nature,* 359, 336, 1992.

473. **Brautigan, D. L. and Pinault, F. M.,** Activation of membrane protein-tyrosine phosphatase involving cAMP- and Ca^{2+}/phospholipid-dependent protein kinases, *Proc. Natl. Acad. Sci. U.S.A.,* 88, 6696, 1991.

474. **Schievella, A. R., Paige, L. A., Johnson, K. A., Hill, D. E., and Erikson, R. L.,** Protein tyrosine phosphatase-1B undergoes mitosis-specific phosphorylation on serine, *Cell Growth Differ.,* 4, 239, 1993.

475. **Jones, S. W., Erikson, R. L., Ingebritsen, V. M., and Ingebritsen, T. S.,** Phosphotyrosyl-protein phosphatases. I. Separation of multiple forms from bovine brain and purification of the major form to near homogeneity, *J. Biol. Chem.,* 264, 7747, 1989.

476. **Ingebritsen, T. S.,** Phosphotyrosyl-protein phosphatases. II. Identification and characterization of two heat-stable protein inhibitors, *J. Biol. Chem.,* 264, 7754, 1989.

477. **Goris, J., Pallen, C. J., Parker, P. J., Hermann, J., Waterfield, M. D., and Merlevede, W.,** Conversion of a phosphoseryl/threonyl phosphatase into a phosphotyrosyl phosphatase, *Biochem. J.,* 256, 1029, 1988.

478. **Okada, M., Owada, K., and Nakagawa, H.,** Phosphotyrosine/protein phosphatase in rat brain. A major phosphotyrosine/protein phosphatase is a 23 kDa protein distinct from acid phosphatase, *Biochem. J.,* 239, 155, 1986.

479. **Yonemoto, W., Filson, A. J., Queral-Lustig, A. E., Wang, J. Y. J., and Brugge, J. S.,** Detection of phosphotyrosine-containing proteins in polyomavirus middle tumor antigen-transformed cells after treatment with a phosphotyrosine phosphatase inhibitor, *Mol. Cell. Biol.,* 7, 905, 1987.

480. **Klarlund, J. K.,** Transformation of cells by an inhibitor of phosphatases acting on phosphotyrosine in proteins, *Cell,* 41, 707, 1985.

481. **Rijksen, G., Völler, M. C. W., and Van Zoelen, E. J. J.,** Orthovanadate both mimics and antagonizes the transforming growth factor β action on normal rat kidney cells, *J. Cell. Physiol.,* 154, 393, 1993.

482. **LaForgia, S., Morse, B., Levy, J., Barnea, G., Cannizzaro, L. A., Li, F., Nowell, P. C., Boghosian-Sell, L., Glick, J., Weston, A., Harris, C. C., Drabkin, H., Patterson, D., Croce, C. M., Schlessinger, J., and Huebner, K.,** Receptor protein-tyrosine phosphatase gamma is a candidate tumor suppressor gene at human chromosome region 3p21, *Proc. Natl. Acad. Sci. U.S.A.,* 88, 5036, 1991.

483. **Edelman, A. M., Blumenthal, D. K., and Krebs, E. G.,** Protein serine/threonine kinases, *Annu. Rev. Biochem.,* 56, 567, 1987.

484. **Mackie, K., Lai, Y., Nairn, A. C., Greengard, P., Pitt, B. R., and Lazo, J. S.,** Protein phosphorylation in cultured endothelial cells, *J. Cell. Physiol.,* 128, 367, 1986.

485. **Lees-Miller, S. P. and Anderson, C. W.,** The DNA-activated protein kinase, DNA-PK: a potential coordinator of nuclear events, *Cancer Cells,* 3, 341, 1991.

486. **Lees-Miller, S. P., Sakaguchi, K., Ullrich, S. J., Appella, E., and Anderson, C. W.,** Human DNA-activated protein kinase phosphorylates serines 15 and 37 in the amino-terminal transactivation domain of human p53, *Mol. Cell. Biol.,* 12, 5041, 1992.

487. **Lin, H. Y., Wang, X.-F., Ng-Eaton, E., Weinberg, R. A., and Lodish, H. F.,** Expression of the TGF-beta type II receptor, a functional transmembrane serine/threonine kinase, *Cell,* 68, 775, 1992.

488. **Klemm, D. J. and Elias, L.,** A distinctive phospholipid-stimulated protein kinase of normal and malignant murine hemopoietic cells, *J. Biol. Chem.,* 262, 7580, 1987.

489. **Jones, P. F., Jakubowicz, T., and Hemmings, B. A.,** Molecular cloning of a second form of *rac* protein kinase, *Cell Regul.,* 2, 1001, 1991.

490. **Jones, P. F., Jakubowicz, T., and Hemmings, B. A.,** Molecular cloning of a second form of *rac* protein kinase, *Cell Regul.,* 2, 1001, 1991.

491. **Cheng, J. Q., Godwin, A. K., Bellacosa, A., Taguchi, T., Franke, T. F., Hamilton, T. C., Tsichlis, P. N., and Testa, J. R.,** *AKT2,* a putative oncogene encoding a member of a subfamily of protein-serine/threonine kinases, is amplified in human ovarian carcinomas, *Proc. Natl. Acad. Sci. U.S.A.,* 89, 9267, 1992.

492. **Matsushime, H., Jinno, A., Takagi, N., and Shibuya, M.,** A novel mammalian protein kinase gene (*mak*) is highly expressed in testicular germ cells at and after meiosis, *Mol. Cell. Biol.,* 10, 2261, 1990.

493. **Kozma, S. C., Ferrari, S., and Thomas, G.,** Unmasking a growth factor/oncogene-activated S6 phosphorylation cascade, *Cell. Signal.,* 1, 219, 1989.

494. **Blenis, J.,** Growth-regulated signal transduction by the MAP kinases and RSKs, *Cancer Cells,* 3, 445, 1991.

495. **Anderson, N. G.,** MAP kinases — ubiquitous signal transducers and potentially important components of the cell cycling machinery in eukaryotes, *Cell. Signal.,* 4, 239, 1992.

496. **Chen, R.-H. and Blenis, J.,** Identification of *Xenopus* S6 protein kinase homologs (pp90^{rsk}) in somatic cells: phosphorylation and activation during initiation of cell proliferation, *Mol. Cell. Biol.,* 10, 3204, 1990.

497. **Chen, R.-H., Chung, J., and Blenis, J.,** Regulation of pp60^{rsk} phosphorylation and S6 phosphotransferase activity in Swiss 3T3 cells by growth factor-, phorbol ester-, and cyclic AMP-mediated signal transduction, *Mol. Cell Biol.,* 11, 1861, 1991.

498. **Chung, J., Chen, R.-H., and Blenis, J.,** Coordinate regulation of pp90^{rsk} and a distinct protein-serine/threonine kinase activity that phosphorylates recombinant pp90^{rsk} *in vitro, Mol. Cell. Biol.,* 11, 1868, 1991.

499. **Chung, J., Pelech, S. L., and Blenis, J.,** Mitogen-activated Swiss mouse 3T3 RSK kinases I and II are related to pp44^{mpk} from sea star oocytes and participate in the regulation of pp90^{rsk} activity, *Proc. Natl. Acad. Sci. U.S.A.,* 88, 4981, 1991.

500. **Blenis, J., Chung, J., Erikson, E., Alcorta, D. A., and Erikson, R. L.,** Distinct mechanisms for the activation of the RSK kinases/MAP2 kinase/pp90^{rsk} and pp70-S6 kinase signaling systems are indicated by inhibition of protein synthesis, *Cell Growth Regul.,* 2, 279, 1991.

501. **Chung, J., Kuo, C. J., Crabtree, G. R., and Blenis, J.,** Rapamycin-FKBP specifically blocks growth-dependent activation of and signaling by the 70 kd S6 protein kinases, *Cell,* 69, 1227, 1992.

502. **Campbell, G. S., Pang, L., Miyasaka, T., Saltiel, A. R., and Carter-Su, C.,** Stimulation by growth hormone of MAP kinase activity in 3T3-F442A fibroblasts, *J. Biol. Chem.,* 267, 6074, 1992.

503. **Gómez, N. and Cohen, P.,** Dissection of the protein kinase cascade by which nerve growth factor activates MAP kinases, *Nature,* 353, 170, 1991.

504. **Posada, J. and Cooper, J. A.,** Requirements for phosphorylation of MAP kinase during meiosis in *Xenopus* oocytes, *Science,* 255, 212, 1992.

505. **Adams, P. D. and Parker, P. J.,** Activation of mitogen-activated protein (MAP) kinase by a MAP kinase-kinase, *J. Biol. Chem.,* 267, 13135, 1992.

506. **Rossomando, A. J., Sanghera, J. S., Marsden, L. A., Weber, M. J., Pelech, S. L., and Sturgill, T. W.,** Biochemical characterization of a family of serine/threonine protein kinases regulated by tyrosine and serine/threonine phosphorylation, *J. Biol. Chem.,* 266, 20270, 1991.

507. **Posada, J., Sanghera, J., Pelech, S., Aebersold, R., and Cooper, J. A.,** Tyrosine phosphorylation and activation of homologous protein kinases during oocyte maturation and mitogenic activation of fibroblasts, *Mol. Cell. Biol.,* 11, 2517, 1991.

508. **L'Allemain, G., Pouysségur, J., and Weber, M. J.,** p42/ mitogen-activated protein kinase as a converging target for different growth factor signaling pathways: use of pertussis toxin as a discrimination factor, *Cell Regul.,* 2, 675, 1991.

509. **Alvarez, E., Northwood, I. C., Gonzalez, F. A., Latour, D. A., Seth, A., Abate, C., Curran, T., and Davis, R. G.,** Pro-Leu-Ser/Thr-Pro is a consensus primary sequence for substrate protein phosphorylation. Characterization of the phosphorylation of c-*myc* and c-*jun* proteins by an epidermal growth factor receptor threonine 669 protein kinase, *J. Biol. Chem.,* 266, 15277, 1991.

510. **Icely, P. L., Gros, P., Bergeron, J. J. M., Devault, A., Afar, D. E. H., and Bell, J. C.,** TIK, a novel serine/threonine kinase, is recognized by antibodies against phosphotyrosine, *J. Biol. Chem.,* 266, 16073, 1991.

511. **Crews, C. M., Alessandri, A. A., and Erikson, R. L.,** Mouse *erk-1* gene product is a serine/threonine protein kinase that has the potential to phosphorylate tyrosine, *Proc. Natl. Acad. Sci. U.S.A.,* 88, 8845, 1991.

512. **Chen, R.-H., Sarnecki, C., and Blenis, J.,** Nuclear localization and regulation of *erk-* and *rsk*-encoded protein kinases, *Mol. Cell. Biol.,* 12, 915, 1992.

513. **Krebs, E. G., Eisenman, R. N., Kuenzel, E. A., Lichtfield, D. W., Lozeman, F. J., Lüscher, B., and Sommercorn, J.,** Casein kinase II is a potentially important enzyme concerned with signal transduction, *Cold Spring Harbor Symp. Quant. Biol.,* 53, 77, 1988.

514. **Kitas, E. A., Meggio, F., Valerio, R. M., Perich, J. W., Johns, R. B., and Pinna, L. A.,** Phosphorylation of *src*-phosphopeptides by casein kinases-1 and -2: favourable effect of phosphotyrosine, *Biochem. Biophys. Res. Commun.,* 170, 635, 1990.

515. **Haas, D. W. and Hagedorn, C. H.,** Casein kinase-I phosphorylates the 25-kDa messenger RNA cap-binding protein, *Arch. Biochem. Biophys.,* 284, 84, 1991.

516. **Glineur, C., Bailly, M., and Ghysdael, J.,** The c-*erbA* alpha-encoded thyroid hormone receptor is phosphorylated in its amino terminal domain by casein kinase II, *Oncogene,* 4, 1247, 1989.

517. **Litchfield, D. W., Luscher, B., Lozeman, F. J., Eisenman, R. N., and Krebs, E. G.,** Phosphorylation of casein kinase II by p34 (cdc2) *in vitro* and at mitosis, *J. Biol. Chem.,* 267, 13943, 1992.

518. **Chan, K.-F. J.,** Ganglioside-modulated protein phosphorylation. Partial purification and characterization of a ganglioside-stimulated protein kinase in brain, *J. Biol. Chem.,* 262, 5248, 1987.

519. **Corbin, J. D., Keely, S. L., and Park, C. R.,** The distribution and dissociation of cyclic adenosine 3':5'-monophosphate-dependent protein kinases in adipose, cardiac, and other tissues, *J. Biol. Chem.,* 250, 218, 1975.

520. **Hofmann, F., Beavo, J. A., Bechtel, P., and Krebs, E. G.,** Comparison of adenosine 3':5'-monophosphate-dependent protein kinases from rabbit skeletal and bovine heart muscle, *J. Biol. Chem.,* 250, 7795, 1975.

521. **Zajac, J. D., Livesey, S. A., and Martin, T. J.,** Selective activation of cyclic AMP dependent protein kinase by calcitonin in a calcitonin secreting lung cancer cell line, *Biochem. Biophys. Res. Commun.,* 122, 1040, 1984.

522. **Perisic, O. and Traugh, J. A.,** Protease-activated kinase II as the mediator of epidermal growth factor-stimulated phosphorylation of ribosomal protein S6, *FEBS Lett.,* 183, 215, 1985.

523. **Goueli, S. A., Davis, A. T., and Ahmed, K.,** Purification of nuclear cAMP-independent protein kinases from rat ventral prostate, *Int. J. Biochem.,* 18, 861, 1986.

524. **Goueli, S. A., Ferkul, K. M., and Ahmed, K.,** Purification of cytosolic cAMP-independent protein kinases from rat ventral prostate, *Int. J. Biochem.,* 18, 875, 1986.

525. **Fontana, J. A., Emler, C., Ku, K., McClung, J. K., Butcher, F. R., and Durham, J. P.,** Cyclic AMP-dependent and -independent protein kinases and protein phosphorylation in human promyelocytic leukemia (HL60) cells induced to differentiate by retinoic acid, *J. Cell. Physiol.,* 120, 49, 1984.

526. **Yasui, W., Sumiyoshi, H., Ochiai, A., Yamahara, M., and Tahara, E.,** Type I and II cyclic adenosine 3':5'-monophosphate-dependent protein kinase in human gastric mucosa and carcinomas, *Cancer Res.,* 45, 1565, 1985.

527. **Kloetzer, W. S., Maxwell, S. A., and Arlinghaus, R. B.,** Further characterization of the p85*gag-mos*-associated protein kinase activity, *Virology,* 138, 143, 1984.

528. **Moelling, K., Heimann, B., Beimling, P., Rapp, U. R., and Sander, T.,** Serine- and threonine-specific protein kinase activities of purified *gag-mil* and *gag-raf* proteins, *Nature,* 312, 558, 1984.

529. **Maxwell, S. A. and Arlinghaus, R. B.,** Serine kinase activity associated with Moloney murine sarcoma virus-124-encoded p37^{mos}, *Virology,* 143, 321, 1985.

530. **Maxwell, S. A. and Arlinghaus, R. B.,** cAMP-independent serine threonine kinase activity is associated with the *mos* sequences of ts 110 Moloney murine sarcoma virus-encoded p85 *gag-mos, J. Gen. Virol.,* 66, 2135, 1985.

531. **Rice, N. R., Copeland, T. D., Simek, S., Oroszlan, S., and Gilden, R. V.,** Detection and characterization of the protein encoded by the v-*rel* oncogene, *Virology,* 149, 217, 1986.

532. **Papkoff, J., Nigg, E. A., and Hunter, T.,** The transforming protein of Moloney murine sarcoma virus is a soluble cytoplasmic protein, *Cell,* 33, 161, 1983.

533. **Baldwin, G. S.,** Epidermal growth factor precursor is related to the translation product of the Moloney sarcoma virus oncogene *mos, Proc. Natl. Acad. Sci. U.S.A.,* 82, 1921, 1985.

534. **Barker, W. C. and Dayhoff, M. O.,** Viral *src* gene products are related to the catalytic chain of mammalian cAMP-dependent protein kinase, *Proc. Natl. Acad. Sci. U.S.A.,* 79, 2836, 1982.

535. **Singh, B., Wittenberg, C., Reed, S. I., and Arlinghaus, R. B.,** Moloney murine sarcoma virus encoded p37^{mos} expressed in yeast has protein kinase activity, *Virology,* 152, 502, 1986.

536. **Herzog, N. K., Nash, M., Ramagli, L. S., and Arlinghaus, R. B.,** v-*mos* protein produced by *in vitro* translation has protein kinase activity, *J. Virol.,* 64, 3093, 1990.

537. **Singh, B., Hannink, M., Donoghue, D. J., and Arlinghaus, R. B.,** p37^{mos}-associated serine/threonine protein kinase activity correlates with the cellular transformation function of v-*mos, J. Virol.,* 60, 1148, 1986.

538. **Al-Bagdadi, F., Singh, B., and Arlinghaus, R. B.,** Evidence for involvement of the protein kinase C pathway in the activation of p37$^{v\text{-}mos}$ protein kinase, *Oncogene,* 5, 1251, 1990.

539. **Singh, B., Wittenberg, C., Hannink, M., Reed, S. I., Donoghue, D. J., and Arlinghaus, R. B.,** The histidine-221 to tyrosine substitution in v-*mos* abolishes its biological function and its protein kinase activity, *Virology,* 164, 114, 1988.

540. **Singh, B. and Arlinghaus, R. B.,** The *mos* proto-oncogene product: its role in oocyte maturation, metaphase arrest, and neoplastic transformation, *Mol. Carcinogenesis,* 6, 182, 1992.

541. **Propst, F. and Vande Woude, G. F.,** Expression of c-*mos* proto-oncogene transcripts in mouse tissues, *Nature,* 315, 516, 1985.

542. **van der Hoorn, F. A.,** Identification of the testis c-*mos* promoter: specific activity in a seminiferous tubule-derived extract and binding of a testis-specific nuclear factor, *Oncogene,* 7, 1093, 1992.

543. **Paules, R. S., Propst, F., Dunn, K. J., Blair, D. G., Kaul, K., Palmer, A. E., and Vande Woude, G. F.,** Primate c-*mos* proto-oncogene structure and expression: transcription initiation both upstream and within the gene in a tissue-specific manner, *Oncogene,* 3, 59, 1988.

544. **Leibovitch, S. A., Guillier, M., Lenormand, J. L., and Leibovitch, M. P.,** Accumulation of the c-*mos* protein is correlated with post-natal development of skeletal muscle, *Oncogene,* 6, 1617, 1991.

545. **Zinkel, S. S., Pal, S. K., Szeberenyi, J., and Cooper, G. M.,** Identification of a negative regulatory element that inhibits c-*mos* transcription in somatic cells, *Mol. Cell. Biol.,* 12, 2029, 1992.

546. **Herzog, N. K., Singh, B., Elder, J., Lipkin, I., Trauger, R. J., Millette, C. F., Goldman, D. S., Wolfes, H., Cooper, G. M., and Arlinghaus, R. B.,** Identification of the protein product of the c-*mos* proto-oncogene in mouse testes, *Oncogene,* 3, 225, 1988.

547. **Van der Hoorn, F. A., Spiegel, J. E., Maylie-Pfenninger, M. F. and Nordeen, S. K.,** A 43 kD c-*mos* protein is only expressed before meiosis during rat spermatogenesis, *Oncogene,* 6, 929, 1991.

548. **Propst, F., Rosenberg, M. P., Oskarsson, M. K., Russell, L. B., Nguyen-Huu, M. C., Nadeau, J., Jenkins, N. A., Copeland, N. G., and Vande Woude, G. F.,** Genetic analysis and developmental regulation of testis-specific RNA expression of *Mos, Abl,* actin and *Hox*-1.4, *Oncogene,* 2, 227, 1988.

549. **Goldman, D. S., Kiessling, A. A., Millette, C. F., and Cooper, G. M.,** Expression of c-*mos* RNA in germ cells of male and female mice, *Proc. Natl. Acad. Sci. U.S.A.,* 84, 4509, 1987.

550. **Keshet, E., Rosenberg, M. P., Mercer, J. A., Propst, F., Vande Woude, G. F., Jenkins, N. A., and Copeland, N. G.,** Developmental regulation of ovarian-specific *Mos* expression, *Oncogene,* 2, 235, 1988.

551. **Zhao, X., Batten, B., Singh, B., and Arlinghaus, R. B.,** Requirement of the c-*mos* protein kinase for murine meiotic maturation, *Oncogene,* 5, 1727, 1990.

552. **Zhao, X., Singh, B., and Batten, B. E.,** The role of c-*mos* proto-oncoprotein in mammalian meiotic maturation, *Oncogene,* 6, 43, 1991.

553. **Zhou, R., Rulong, S., Pinto da Siva, P., and Vande Woude, G. F.,** *In vitro* and *in vivo* characterization of pp39^{mos} association with tubulin, *Cell Growth Differ.,* 2, 257, 1991.

554. **Zhou, R., Daar, I., Ferris, D. K., White, G., Paules, R., and Vande Woude, G.,** p39^{mos} is associated with p34^{cdc2} kinase in c-*mos*^{xe}-transformed NIH 3T3 cells, *Mol. Cell. Biol.*, 12, 3583, 1992.

555. **Ogiso, Y., Matsumoto, M., Morita, T., Nishino, H., Iwashima, A., and Matsushiro, A.,** Expression of c-*mos* proto-oncogene in differentiated teratocarcinoma cells, *Biochem. Biophys. Res. Commun.*, 140, 477, 1986.

556. **O'Keefe, S. J., Kiessling, A. A., and Cooper, G. M.,** The c-*mos* gene product is required for cyclin B accumulation during meiosis of mouse eggs, *Proc. Natl. Acad. Sci. U.S.A.*, 88, 7869, 1991.

557. **Xu, W., Ladner, K. J., and Smith, L. D.,** Evidence that Mos protein may not act directly on cyclin, *Proc. Natl. Acad. Sci. U.S.A.*, 89, 4573, 1992.

558. **Dandoy, F., De Maeyer-Guignard, J., and De Maeyer, E.,** Linkage analysis of the murine *mos* proto-oncogene on chromosome 4, *Genomics*, 4, 546, 1989.

559. **Propst, F., Rosenberg, M. P., Iyer, A., Kaul, K., and Vande Woude, G. F.,** c-*mos* proto-oncogene RNA transcripts in mouse tissues: structural features, developmental regulation, and localization in specific cell types, *Mol. Cell. Biol.*, 7, 1629, 1987.

560. **Tsui, L. V., Ramagli, L. S., Singh, B., Nash, M., and Arlinghaus, R. B.,** Somatic cell expression of the *mos* proto-oncogene is cell cycle regulated: highest RNA expression in the G$_2$ phase, *Int J. Oncol.*, 2, 493, 1993.

561. **Paules, R. S., Buccione, R., Moschel, R. C., Vande Woude, G. F., and Eppig, J. J.,** Mouse *Mos* protooncogene product is present and functions during oogenesis, *Proc. Natl. Acad. Sci. U.S.A.*, 86, 5395, 1989.

562. **Yew, N., Mellini, M. L., and Vande Woude, G. F.,** Meiotic initiation by the Mos protein in *Xenopus*, *Nature*, 355, 649, 1992.

563. **Sagata, N., Oskarsson, M., Copeland, T., Brumbaugh, J., and Vande Woude, G. F.,** Function of c-*mos* proto-oncogene in meiotic maturation in *Xenopus* oocytes, *Nature*, 335, 519, 1988.

564. **Sagata, N., Daar, I., Oskarsson, M., Showalter, S. D., and Vande Woude, G. F.,** The product of the *mos* proto-oncogene as a candidate "initiator" for oocyte maturation, *Science*, 245, 643, 1989.

565. **Barrett, C. B., Schroetke, R. M., van der Hoorn, F. A., Nordeen, S. K., and Maller, J. L.,** Ha-*ras*$^{Val-12,Thr-59}$ activates S6 kinase and p34^{cdc2} kinase in *Xenopus* oocytes: evidence for c-*mos*^{xe}-dependent and -independent pathways, *Mol. Cell. Biol.*, 10, 310, 1990.

566. **Sagata, N., Watanabe, N., Vande Woude, G. F., and Ikawa, Y.,** The c-*mos* proto-oncogene product is a cytostatic factor responsible for meiotic arrest in vertebrate eggs, *Nature*, 342, 512, 1989.

567. **Zhou, R., Oskarsson, M., Paules, R. S., Schulz, N., Cleveland, D., and Vande Woude, G. F.,** Ability of the c-*mos* product to associate with and phosphorylate tubulin, *Science*, 251, 671, 1991.

568. **Li, C.-C. H., Chen, E., O'Connell, C. D., and Longo, D. L.,** Detection of c-*mos* proto-oncogene expression in human cells, *Oncogene*, 8, 1685, 1993.

569. **Denhez, F., Heimann, B., d'Auriol, L., Graf, T., Coquillaud, M. Coll, J., Galibert, F., Moelling, K., Stehelin, D., and Ghysdael, J.,** Replacement of lys 622 in the ATP binding domain of P100$^{gag-mil}$ abolishes the *in vitro* autophosphorylation of the protein and the biological properties of the v-*mil* oncogene of MH2 virus, *EMBO J.*, 7, 541, 1988.

570. **Storm, S. M., Brennscheidt, U., Sithanandam, G., and Rapp, U. R.,** *raf* oncogenes in carcinogenesis, *Crit. Rev. Oncogenesis*, 2, 1, 1990.

571. **Morrison, D. K.,** The Raf-1 kinase as a transducer of mitogenic signals, *Cancer Cells*, 2, 377, 1990.

572. **Li, P., Wood, K., Mamon, H., Haser, W., and Roberts, T.,** Raf-1 — a kinase currently without a cause but not lacking in effects, *Cell*, 64, 479, 1991.

573. **Rapp, U. R.,** Role of the Raf-1 serine/threonine protein kinase in growth factor signal transduction, *Oncogene*, 6, 495, 1991.

574. **Kolch, W., Heidecker, G., Lloyd, P., and Rapp, U. R.,** Raf-1 protein kinase is required for growth of induced NIH/3T3 cells, *Nature*, 349, 426, 1991.

575. **Zhang, X. F., Settleman, J., Kyriakis, J. M., Takeuchi-Suzuki, E., Elledge, S. J., Marshall, M. S., Bruder, J. T., Rapp, U. R., and Avruch, J.,** Normal and oncogenic p21^{ras} proteins bind to the amino-terminal domain of c-Raf-1, *Nature*, 364, 308, 1993.

576. **Kyriakis, J. M., App, H., Zhang, X. F., Banerjee, P., Brautigan, D. L., Rapp, U. R., and Avruch, J.,** Raf-1 activates MAP kinase-kinase, *Nature*, 358, 417, 1992.

577. **Howe, L. R., Leevers, S. J., Gómez, N., Nakielny, S., Cohen, P., and Marshall, C. J.,** Activation of the MAP kinase pathway by the protein kinase raf, *Cell*, 71, 335, 1992.

578. **Anderson, N. G., Li, P., Marsden, L. A., Williams, N., Roberts, T. M., and Sturgill, T. W.,** Raf-1 is a potential substrate for mitogen-activated protein kinase *in vivo*, *Biochem. J.*, 277, 573, 1991.

579. **Kanakura, Y., Druker, B., Wood, K. W., Mamon, H. J., Okuda, K., Roberts, T. M., and Griffin, J. D.,** Granulocyte-macrophage colony-stimulating factor and interleukin-3 induce rapid phosphorylation and activation of the proto-oncogene Raf-1 in a human factor-dependent myeloid cell line, *Blood,* 77, 243, 1991.

580. **Qureshi, S. A., Rim, M., Bruder, J., Kolch, W., Rapp, U., Sukhatme, V. P., and Foster, D. A.,** An inhibitory mutant of c-Raf-1 blocks v-Src-induced activation of the Egr-1 promoter, *J. Biol. Chem.,* 266, 20594, 1991.

581. **Mark, G. E., MacIntyre, R. J., Digan, M. E., Ambrosio, L., and Perrimon, H.,** *Drosophila melanogaster* homologs of the *raf* oncogene, *Mol. Cell. Biol.,* 7, 2134, 1987.

582. **Sprenger, F., Stevens, L. M., and Nüsslein-Volhard, C.,** The *Drosophila* gene *torso* encodes a putative receptor tyrosine kinase, *Nature,* 338, 478, 1989.

583. **Ambrosio, L., Mahowald, A. P., and Perrimon, N.,** Requirement of the *Drosophila raf* homologue for torso function, *Nature,* 342, 288, 1989.

284. **Le Guellec, R., Couturier, A., Le Guellec, K., Paris, J., Le Fur, N., and Philippe, M.,** *Xenopus* c-*raf* proto-oncogene: cloning and expression during oogenesis and early development, *Biol. Cell,* 72, 39, 1991.

585. **Muslin, A. J., MacNicol, A. M., and Williams, L. T.,** Raf-1 protein kinase is important for progesterone-induced *Xenopus* oocyte maturation and acts downstream of *mos, Mol. Cell. Biol.,* 13, 4197, 1993.

586. **Koenen, M., Sippel, A. E., Trachmann, C., and Bister, K.,** Primary structure of the chicken c-*mil* protein: identification of domains shared or absent from the retroviral v-*mil* protein, *Oncogene,* 2, 179, 1988.

587. **Ishikawa, F., Takaku, F., Nagao, M., and Sugimura, T.,** The complete primary structure of the rat A-*raf* cDNA coding region: conservation of the putative regulatory regions present in rat c-*raf, Oncogene Res.,* 1, 243, 1987.

588. **Sozeri, O., Vollmer, K., Liyanage, M., Frith, D., Kour, G., Mark, G. E., and Stabel, S.,** Activation of the c-Raf protein kinase by protein kinase C phosphorylation, *Oncogene,* 7, 2259, 1992.

589. **Siegel, J. N., Klausner, R. D., Rapp, U. R., and Samelson, L. E.,** T cell antigen receptor engagement stimulates c-*raf* phosphorylation and induces c-*raf*-associated kinase activity via a protein kinase C-dependent pathway, *J. Biol. Chem.,* 265, 18472, 1990.

590. **Oshima, M., Sithanandam, G., Rapp, U. R., and Guroff, G.,** The phosphorylation and activation of B-*raf* in PC12 cells stimulated by nerve growth factor, *J. Biol. Chem.,* 266, 23753, 1991.

591. **Huleihel, M., Goldsborough, M., Cleveland, J., Gunnell, M., Bonner, T., and Rapp, U. R.,** Characterization of murine A-*raf*, a new oncogene related to the v-*raf* oncogene, *Mol. Cell. Biol.,* 6, 2655, 1986.

592. **Storm, S. M., Cleveland, J. L., and Rapp, U. R.,** Expression of *raf* family proto-oncogenes in normal mouse tissues, *Oncogene,* 5, 345, 1990.

593. **Huebner, K., ar-Rushdi, A., Griffin, C. A., Isobe, M., Kozak, C., Emanuel, B. S., Nagarajan, L., Cleveland, J. L., Bonner, T. I., Goldsborough, M. D., Croce, C. M., and Rapp, U.,** Actively transcribed genes in the *raf* oncogene group, located on the X chromosome in mouse and human, *Proc. Natl. Acad. Sci. U.S.A.,* 83, 3934, 1986.

594. **Mark, G. E., Seeley, T. W., Shows, T. B., and Mountz, J. D.,** *pks*, a *raf*-related sequence in humans, *Proc. Natl. Acad. Sci. U.S.A.,* 83, 6312, 1986.

595. **Bonner, T. I., Oppermann, H., Seeburg, P., Kerby, S. B., Gunnell, M. A., Young, A. C., and Rapp, U. R.,** The complete coding sequence of the human *raf* oncogene and the corresponding structure of the c-*raf*-1 gene, *Nucleic Acids Res.,* 14, 1009, 1986.

596. **Baccarini, M., Gill, G. N., and Stanley, E. R.,** Epidermal growth factor stimulates phosphorylation of RAF-1 independently of receptor autophosphorylation and internalization, *J. Biol. Chem.,* 266, 10941, 1991.

597. **Kizaka-Kondoh, S., Sato, K., Tamura, K., Nojima, H., and Okayama, H.,** Raf-1 protein kinase is an integral component of the oncogenic signal cascade shared by epidermal growth factor and platelet-derived growth factor, *Mol. Cell. Biol.,* 12, 5087, 1992.

598. **Tamaki, T., Kanakura, Y., Kuriu, A., Ikeda, H., Mitsui, H., Yagura, H., Matsumura, I., Druker, B., Griffin, J. D., Kanayama, Y., and Yonezawa, T.,** Surface immunoglobulin-mediated signal transduction involves rapid phosphorylation and activation of the protooncogene product *Raf*-1 in human B-cells, *Cancer Res.,* 52, 566, 1992.

599. **Beck, T. W., Huleihel, M., Gunnell, M., Bonner, T. I., and Rapp, U. R.,** The complete coding sequence of the human A-*raf*-1 oncogene and transforming activity of a human A-*raf* carrying retrovirus, *Nucleic Acids Res.,* 15, 595, 1987.

600. **Sithanandam, G., Kolch, W., Duh, F.-M., and Rapp, U. R.,** Complete coding sequence of a human B-*raf* cDNA and detection of B-*raf* protein kinase with isozyme specific antibodies, *Oncogene,* 5, 1755, 1990.

601. **Oláh, Z., Komoly, S., Nagashima, N., Joó, F., Rapp, U. R., and Anderson, W. B.,** Cerebral ischemia induces transient intracellular redistribution and intranuclear translocation of the *raf* proto-oncogene product in hippocampal pyramidal cells, *Exp. Brain Res.,* 84, 403, 1991.

602. **Stephens, R. M., Sithanandam, G., Copeland, T. D., Kaplan, D. R., Rapp, U. R., and Morrison, D. K.,** 95-kilodalton B-Raf serine/threonine kinase: identification of the protein and its major autophosphorylation site, *Mol. Cell. Biol.,* 12, 3733, 1992.

603. **Sariban, E., Mitchell, T., and Kufe, D.,** Expression of the c-*raf* protooncogene in human hematopoietic cells and cell lines, *Blood,* 69, 1437, 1987.

604. **Choudhury, G. G., Sylvia, V. L., Pfeifer, A., Wang, L.-M., Smith, E. A., and Sakaguchi, A. Y.,** Human colony stimulating factor-1 receptor activates the c-*raf*-1 proto-oncogene kinase, *Biochem. Biophys. Res. Commun.,* 172, 154, 1990.

605. **Baccarini, M., Sabatini, D. M., App, H., Rapp, U. R., and Stanley, E. R.,** Colony stimulating factor-1 (CSF-1) stimulates temperature dependent phosphorylation and activation of the RAF-1 proto-oncogene product, *EMBO J.,* 9, 3649, 1990.

606. **Turner, B., Rapp, U., App, H., Greene, M., Dobashi, K., and Reed, J.,** Interleukin 2 induces tyrosine phosphorylation and activation of p72-74 Raf-1 kinase in a T-cell line, *Proc. Natl. Acad. Sci. U.S.A.,* 88, 1227, 1991.

607. **Bruder, J. T., Heidecker, G., and Rapp, U. R.,** Serum-, TPA-, and Ras-induced expression from Ap-1/Ets-driven promoters requires Raf-1 kinase, *Genes Dev.,* 6, 545, 1992.

608. **Cuypers, H. T., Selten, G., Quint, W., Zijlstra, M., Maandag, E. R., Boelens, W., van Wezenbeek, P., Melief, C., and Berns, A.,** Murine leukemia virus-induced T-cell lymphomagenesis: integration of proviruses in a distinct chromosomal region, *Cell,* 37, 141, 1984.

609. **Selten, G., Cuypers, T., and Berns, A.,** Proviral activation of the putative oncogene *Pim*-1 in MuLV induced T-cell lymphomas, *EMBO J.,* 4, 1793, 1985.

610. **Nadeau, J. H. and Phillips, S. J.,** The putative oncogene *Pim*-1 in the mouse: its linkage and variation among *t* haplotypes, *Genetics,* 117, 533, 1987.

611. **Nagarajan, L., Louie, E., Tsujimoto, Y., ar-Rushdi, A., Huebner, K., and Croce, C. M.,** Localization of the human *pim* oncogene (*PIM*) to a region of chromosome 6 involved in translocations in acute leukemias, *Proc. Natl. Acad. Sci. U.S.A.,* 83, 2556, 1986.

612. **Amson, R., Sigaux, F., Przedborski, S., Flandrin, G., Givol, D., and Telerman, A.,** The human protooncogene product p33^{pim} is expressed during fetal hematopoiesis and in diverse leukemias, *Proc. Natl. Acad. Sci. U.S.A.,* 86, 8857, 1989.

613. **Selten, G., Cuypers, H. T., Boelens, W., Robanus-Maandag, E., Verbeek, J., Domen, J., van Beveren, C., and Berns, A.,** The primary structure of the putative oncogene *pim*-1 shows extensive homology with protein kinases, *Cell,* 46, 603, 1986.

614. **Zakut-Houri, R., Hazum, S., Givol, D., and Telerman, A.,** The cDNA sequence and gene analysis of the human *pim* oncogene, *Gene,* 54, 105, 1987.

615. **Domen, J., von Lindern, M., Hermans, A., Breuer, M., Grosveld, G., and Berns, A.,** Comparison of the human and mouse *PIM*-1 cDNAs: nucleotide sequence and immunological identification of the *in vitro* synthesized *PIM*-1 protein, *Oncogene Res.,* 1, 103, 1987.

616. **Saris, C. J. M., Domen, J., and Berns, A.,** The *pim*-1 oncogene encodes two related protein-serine/threonine kinases by alternative initiation at AUG and CUG, *EMBO J.,* 10, 655, 1991.

617. **Telerman, A., Amson, R., Zakut-Houri, R., and Givol, D.,** Identification of the human *pim*-1 gene product as a 33-kilodalton cytoplasmic protein with tyrosine kinase activity, *Mol. Cell. Biol.,* 8, 1498, 1988.

618. **Padma, R. and Nagarajan, L.,** The human *PIM*-1 gene product is a protein serine kinase, *Cancer Res.,* 51, 2486, 1991.

619. **Hoover, D., Friedmann, M., Reeves, R., and Magnuson, N. S.,** Recombinant human Pim-1 protein exhibits serine threonine kinase activity, *J. Biol. Chem.,* 266, 14018, 1991.

620. **Friedmann, M., Nissen, M. S., Hoover, D. S., Reeves, R., and Magnuson, N. S.,** Characterization of the proto-oncogene Pim-1: kinase activity and substrate recognition sequence, *Arch. Biochem. Biophys.,* 298, 594, 1992.

621. **te Riele, H., Maandag, E. R., Clarke, A., Hooper, M., and Berns, A.,** Consecutive inactivation of both alleles of the *pim-*1 proto-oncogene by homologous recombination in embryonic stem cells, *Nature,* 348, 649, 1990.

622. **Staal, S. P.,** Molecular cloning of the *akt* oncogene and its human homologues *AKT1* and *AKT2*: amplification and *AKT1* in a primary human gastric adenocarcinoma, *Proc. Natl. Acad. Sci. U.S.A.,* 84, 5034, 1987.

623. **Bellacosa, A., Testa, J. R., Staal, S. P., and Tsichlis, P. N.,** A retroviral oncogene, *akt,* encoding a serine-threonine kinase containing an SH2-like region, *Science,* 254, 274, 1991.

624. **Ahmed, N. N., Franke, T. F., Bellacosa, A., Datta, K., Gonzalez-Portal, M.-E., Taguchi, T., Testa, J. R., and Tsichlis, P. N.,** The proteins encoded by c-*akt* and v-*akt* differ in post-translational modification, subcellular localization and oncogenic potential, *Oncogene,* 8, 1957, 1993.

625. **Jones, P. F., Jakubowicz, T., Pitossi, F. J., Maurer, F., and Hemmings, B. A.,** Molecular cloning and identification of a serine threonine protein kinase of the 2nd messenger subfamily, *Proc. Natl. Acad. Sci. U.S.A.,* 88, 4171, 1991.

626. **Miyoshi, J., Higashi, T., Mukai, H., Ohuchi, T., and Kakunaga, T.,** Structure and transforming potential of the human *cot* oncogene encoding a putative protein kinase, *Mol. Cell. Biol.,* 11, 4088, 1991.

627. **Aoki, M., Akiyama, T., Miyoshi, J., and Toyoshima, K.,** Identification and characterization of protein products of the *cot* oncogene with serine kinase activity, *Oncogene,* 6, 1515, 1991.

628. **Ohara, R., Miyoshi, J., Aoki, M., and Toyoshima, K.,** The murine *cot* proto-oncogene — genome structure and tissue-specific expression, *Jpn. J. Cancer Res.,* 84, 518, 1993.

629. **Pelech, S. L. and Krebs, E. G.,** Mitogen-activated S6 kinase is stimulated via protein kinase C-dependent and independent pathways in Swiss 3T3 cells, *J. Biol. Chem.,* 262, 11598, 1987.

630. **Jenö, P., Ballou, L. M., Novak-Hofer, I., and Thomas, G.,** Identification and characterization of a mitogen-activated S6 kinase, *Proc. Natl. Acad. Sci. U.S.A.,* 85, 406, 1988.

631. **Novak-Hofer, I., Küng, W., and Eppenberger, U.,** Role of extracellular electrolytes in the activation of ribosomal protein S6 kinase by epidermal growth factor, insulin-like growth factor 1, and insulin in ZR-75-1 cells, *J. Cell Biol.,* 106, 395, 1988.

632. **Maller, J. L., Foulkes, J. G., Erikson, E., and Baltimore, D.,** Phosphorylation of ribosomal protein S6 on serine after microinjection of the Abelson murine leukemia virus tyrosine-specific protein kinase into *Xenopus* oocytes, *Proc. Natl. Acad. Sci. U.S.A.,* 82, 272, 1985.

633. **Blenis, J., Spivack, J. G., and Erikson, R. L.,** Phorbol ester, serum, and Rous sarcoma virus transforming gene product induce similar phosphorylations of ribosomal protein S6, *Proc. Natl. Acad. Sci. U.S.A.,* 81, 6408, 1984.

634. **Blenis, J. and Erikson, R. L.,** Regulation of a ribosomal protein S6 kinase activity by the Rous sarcoma virus transforming protein, serum, or phorbol ester, *Proc. Natl. Acad. Sci. U.S.A.,* 82, 7621, 1985.

635. **Novak-Hofer, I., Küng, W., Fabbro, D., and Eppenberger, U.,** Estrogen stimulates growth of mammary tumor cells ZR-75 without activation of S6 kinase and S6 phosphorylation: difference from epidermal growth factor and alpha-transforming growth-factor-induced proliferation, *Eur. J. Biochem.,* 164, 445, 1987.

636. **Kruse, C., Johnson, S. P., and Warner, J. R.,** Phosphorylation of the yeast equivalent of ribosomal protein S6 is not essential for growth, *Proc. Natl. Acad. Sci. U.S.A.,* 82, 7515, 1985.

637. **Sato, C., Nishikawa, K., Nakayama, T., Ohtsuka, K., Nakamura, H., Kobayashi, T., and Inagaki, M.,** Rapid phosphorylation of MAP-2-related cytoplasmic and nuclear M_r 300,000 protein by serine kinases after growth stimulation in quiescent cells, *Exp. Cell Res.,* 175, 136, 1988.

638. **Rydell, E. L., Axelsson, K. L., Olofsson, J., and Hellem, S.,** Protein kinase activities in neoplastic squamous epithelia and normal epithelia from the upper aero-digestive tract, *Cancer Biochem. Biophys.,* 11, 187, 1990.

639. **Cohen, P. and Cohen, P. T. W.,** Protein phosphatases come of age, *J. Biol. Chem.,* 264, 21435, 1989.

640. **Cohen, P. T. W., Brewis, N. D., Hughes, V., and Mann, D. J.,** Protein serine/threonine phosphatases; an expanding family, *FEBS Lett.,* 268, 355, 1990.

641. **Villa-Moruzzi, E., Dalla Zonca, P., and Crabb, J. W.,** Phosphorylation of the catalytic subunit of type-1 protein phosphatase by the v-*abl* tyrosine kinase, *FEBS Lett.,* 293, 67, 1991.

642. **Chen, J., Martin, B. L., and Brautigan, D. L.,** Regulation of protein serine-threonine phosphatase type 2A by tyrosine phosphorylation, *Science,* 257, 1261, 1992.

643. **Alberts, A. S., Thorburn, A. M., Shenolikar, S., Mumby, M. C., and Feramisco, J. R.,** Regulation of cell cycle progression and nuclear affinity of the retinoblastoma protein by protein phosphatases, *Proc. Natl. Acad. Sci. U.S.A.,* 90, 388, 1993.

644. **Guy, G. R., Cao, X., Chua, S. P., and Tan, Y. H.,** Okadaic acid mimics multiple changes in early protein phosphorylation and gene expression induced by tumor necrosis factor or interleukin-1, *J. Biol. Chem.,* 267, 1846, 1992.

645. **Adunyah, S. E., Unlap, T. M., Franklin, C. C., and Kraft, A. S.,** Induction of differentiation and c-*jun* expression in human leukemic cells by okadaic acid, an inhibitor of protein phosphatases, *J. Cell. Physiol.,* 151, 415, 1992.

646. **Sakai, R., Ikeda, I., Kitani, H., Fujiki, H., Takaku, F., Rapp, U., Sugimura, T., and Nagao, M.,** Flat reversion by okadaic acid of *raf* and *ret*-II transformants, *Proc. Natl. Acad. Sci. U.S.A.,* 86, 9946, 1989.

647. **Kiguchi, K., Giometti, C., Chubb, C. H., Fujiki, H., and Huberman, E.,** Differentiation induction in human breast tumor cells by okadaic acid and related inhibitors of protein phosphatases 1 and 2A, *Biochem. Biophys. Res. Commun.,* 189, 1261, 1992.

648. **Rivedal, E., Mikalsen, S.-O., and Sanner, T.,** The nonphorbol ester tumor promoter okadaic acid does not promote morphological transformation or inhibit junctional communication of hamster embryo cells, *Biochem. Biophys. Res. Commun.,* 167, 1302, 1990.

649. **Rieber, M. and Rieber, M. S.,** Early inhibition of protein phosphatases preferentially blocks phorbol ester-stimulated mitogenic signalling in melanocytes: increase in specific tyrosine phosphoproteins, *Biochem. Biophys. Res. Commun.,* 192, 483, 1993.

650. **Sebastian, B., Kakizuka, A., and Hunter, T.,** Cdc25M2 activation of cyclin-dependent kinases by dephosphorylation of threonine-14 and tyrosine-15, *Proc. Natl. Acad. Sci. U.S.A.,* 90, 3521, 1993.

Chapter 9

Proto-Oncogene and Onco-Suppressor Gene Expression

I. INTRODUCTION

Evidence derived from a multitude of studies clearly indicates the existence of important structural and functional relationships between growth factors and oncoproteins.[1-5] The oncoproteins are the products of oncogenes and proto-oncogenes, which are defined as genes with potential properties for the induction of neoplastic transformation in either natural or experimental conditions.[6] Classic oncogenes were isolated from acute retroviruses, which are defective, nonreplicating viruses that act as oncogene transducers (Table 9.1). Acute retroviruses are not infectious under natural conditions but have been isolated from different types of animal tumors and are originated from recombination events that may occur between cellular sequences, called proto-oncogenes, and sequences derived from various infectious chronic retroviruses (Figure 9.1). No transforming sequences (oncogenes) are contained in chronic retroviruses. The viral oncogenes (v-*onc* genes) contained in acute retroviruses are responsible for the high oncogenic potential exhibited by these viruses under experimental conditions. The v-*onc* genes are derived from sequences of cellular origin, the proto-oncogenes or c-*onc* genes. Some putative proto-oncogenes, represented by cellular sequences with oncogenic potential, have not been detected in acute retroviruses but have been isolated directly from cellular genomes.

Viral oncogenes code for oncoproteins that are responsible for the initiation and maintenance of transformation induced by the transducing virus in susceptible cells. The cellular and viral oncogenes are indicated in the literature in the form of symbols of three letters in italic. The letters "v" and "c" preceding these symbols indicate the viral or cellular origin of the respective oncogenes. The proteins encoded by oncogenes are indicated by the same symbols but in letters of the usual type, and the symbols are initiated with a capital letter. Again, their viral or cellular origin can be indicated by a preceding "v" or "c" letter, respectively. Although the term oncoprotein should be properly reserved for the viral oncogene products, it is frequently used for proto-oncogene products as well.

Proto-oncogenes are present in all vertebrates as well as invertebrates. They are normal constituents of the genome of all multicellular organisms, and genes homologous to proto-oncogenes are present in unicellular eukaryotes. According to the current definition criteria, the total number of proto-oncogenes in any biological species is limited to about 50. A list of the human proto-oncogenes, with their respective chromosome localization, appears in Table 9.2 and is represented in the human idiogram of Figure 9.2.

The normal functions of cellular oncoproteins are understood only in part, but most of these products may be involved in the control of specific metabolic processes as well as in processes related to cell proliferation and differentiation. Proto-oncogene expression is subjected to developmental and tissue-specific regulation, and hormones and growth factors have an important role in regulating this expression.[7] There are intimate structural and functional relationships between hormones and growth factors, the cellular receptors for hormones and growth factors, and oncoproteins (Table 9.3). Moreover, certain proto-oncogenes encode for growth factors or for the cellular receptors of hormones or growth factors.[8]

Structural and/or functional alterations of proto-oncogenes may be associated with neoplastic diseases, including human cancer. Four basic types of mechanisms may be responsible for the association between proto-oncogenes and carcinogenic processes: (1) increased expression of proto-oncogenes at unscheduled sites or times; (2) amplification of proto-oncogene sequences; (3) translocation or rearrangement of proto-oncogene sequences; and (4) point mutation of proto-oncogenes.[9] The frequency and relative importance of each one of these proto-oncogene abnormalities may vary according to the type of tumor and its grade of differentiation. The study of abnormalities occurring in proto-oncogenes and their products may be clinically useful for the diagnosis and prognosis of cancer.[10] Studies with antisense inhibition of proto-oncogene expression may contribute to a better definition of the physiologic role of the different proto-oncogene products as well as to understanding the role of these products in malignant diseases.[11]

In addition to the alterations of proto-oncogenes, recent evidence indicates the existence of tumor suppressor genes (onco-suppressor genes or antioncogenes) whose deletion or inactivation may result in or contribute to the expression of a transformed phenotype. The retinoblastoma gene (*RB1*), the gene encoding the nuclear protein p53, and the Wilms' tumor gene (*WT1*) are reliable examples of tumor

Table 1 **Acute transforming retroviruses and their respective oncogenes**

Oncogene	Isolate Origin	Prototype Virus Strain
abl	Rodent	Abelson murine leukemia virus (A-MuLV)
akt	Rodent	Rodent virus AKT8
crk	Avian	CT10 avian sarcoma virus (CT10-ASV)
erb-A	Avian	Avian erythroblastosis virus (AEV)
erb-B	Avian	Avian erythroblastosis virus (AEV)
ets	Avian	E26 avian leukemia virus (E26-ALV)
fes	Feline	Snyder-Theilen feline sarcoma virus (ST-FeSV)
fgr	Feline	Gardner-Rasheed feline sarcoma virus (GR-FeSV)
fms	Feline	McDonough feline sarcoma virus (MD-FeSV)
fos	Rodent	FBJ murine osteosarcoma virus (FBJ-MOV)
jun	Avian	ASV17 avian sarcoma virus (ASV17-ASV)
maf	Avian	AS42 avian fibrosarcoma virus (AS42-AFV)
mht	Avian	Mill Hill 2 avian carcinoma virus (MH2-ACV)
mos	Rodent	Moloney murine sarcoma virus (M-MuSV)
mpl	Rodent	Myeloproliferative leukemia virus (MPLV)
myb	Avian	Avian myeloblastosis virus (AMV)
myc	Avian	MC29 avian myelocytomatosis virus (MC29-AMCV)
raf	Rodent	3611 murine sarcoma virus (3611-MuSV)
H-*ras*	Rodent	Harvey murine sarcoma virus (H-MuSV)
K-*ras*	Rodent	Kirsten murine sarcoma virus (K-MuSV)
rel	Avian	Avian reticuloendotheliosis virus (ARV)
ros	Avian	Rochester URII avian sarcoma virus (URII-ASV)
ryk	Avian	RPL30 avian sarcoma virus (RPL-30 ASV)
sea	Avian	S13 avian erythroblastosis virus (S13 AEV)
sis	Primate	Simian sarcoma virus (SSV)
ski	Avian	Sloan-Kettering avian virus (SKV)
src	Avian	Rous sarcoma virus (RSV)
yes	Avian	Yamaguchi avian sarcoma virus (Y-ASV)

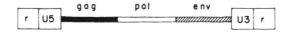

GENOME OF A CHRONIC RETROVIRUS

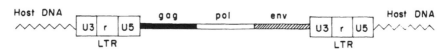

PROVIRUS OF A CHRONIC RETROVIRUS

ACUTE TRANSFORMING RETROVIRAL PROVIRUS

Figure 9.1 Schematic representation of the general structures of chronic and acute retroviral genomes.

Table 2 **Chromosomal localization of human proto-oncogenes**

Chromosome	Localization	Proto-Oncogene	Chromosome	Localization	Proto-Oncogene
1	1p11-p13	N-*ras*	8	8q24.12-q24.13	*myc*
1	1p31-p32	*jun*	9	9q34.1	*abl*
1	1p32	L-*myc*	10	10q11.2	*ret*
1	1p32	B-*lym*-1	11	11p14.1	H-*ras*-1
1	1p32-p35	*lck/lsk/tck*	11	11q13	*int*-2
1	1p34	*mpl*	11	11q13	*hst*
1	1p36.1-p36.2	*fgr*	11	11q13.3	*bcl*-1
1	1q22-q24	*ski*	11	11q23-q24	*ets*-1
1	1q24-q25	*arg*	12	12p11.1-p12.1	K-*ras*-2
1	1q23-q24	*trk*	12	12q13	*int*-1
2	2p12-p13	*rel*	13	13q12	*flt*
2	2p23-p24	N-*myc*	14	14q24.3-q31	*fos*
3	3p21-p25	*erb*-A2	14	14q32	*akt*-1
3	3p25	*raf*-1/*mil*	15	15q26.1	*fes*
4	4q11-q12	*kit*	17	17p13	p53
5	5q21-q22	*fer*	17	17q11.2	*erb*-A1
5	5q34	*fms*	17	17q21	*neu/erb*-B2
6	6p11-p12	K-*ras*-1	17	17q23-qter	*bek*
6	6p23-q12	*yes*-2	18	18q21	*bcl*-2
6	6p21	*pim*-1	18	18q21.3	*yes*-1
6	6q21	*fyn/syn/slk*	19	19p12-p13.2	*vav*
6	6q22	*ros*	19	19p13.2-q13.2	*mel*
6	6q22-q23	*myb*	19	19q13.1-q13.2	*akt*-2
7	7p12-p14	*erb*-B1	19	19q13.2	*axl*
7	7p14-q21	A-*raf*-2	20	20q11-q12	*hck*
7	7q21-q31	*met*	20	20q11.2	*src*
7	7q33-q36	B-*raf*	21	21q22.1-q22.3	*ets*-2
8	8p12	*flg*	22	22q13.1	*sis*
8	8q11 or q22	*mos*	X	Xp21-q11	A-*raf*-1
8	8q13-qter	*lyn*	X	Xq27	*mcf*-2

suppressor genes. These genes may have an important role as negative regulators of the growth of both normal and transformed cells. Available evidence indicates that structural and functional alterations of both proto-oncogenes and onco-suppressor genes may contribute, in concert with alterations of other cellular genes, to multistage oncogenic processes.

II. EXPRESSION AND FUNCTION OF NUCLEAR ONCOPROTEINS

Growth factors and other extracellular signaling agents are involved in regulating the expression of cellular genes, including the expression of genes associated with the control mechanisms of cell differentiation and cell proliferation. The levels of expression of certain genes, called "immediate-early" genes or early-response genes, are rapidly altered by the exposure of sensitive cells to the signaling agents. Other genes may be regulated secondarily to the early cellular response and are considered as late-response genes. Frequently, the genes regulated by growth factors and other extracellular signaling molecules are proto-oncogenes or tumor suppressor genes whose products are localized mainly in the nucleus. The nuclear localization of these products may depend on the presence of specific amino acid sequences that are required for their interaction with the nuclear pore complex.[12] The proto-oncogene products Myc, Myb, Fos, Jun, Ets, and Rel function as transcription factors and their expression may be regulated at the transcriptional and/or posttranscriptional level by extracellular signaling agents. The regulation may include reversible phosphorylation/dephosphorylation of the protein.[13] The p53 and RB tumor suppressor proteins are localized mainly in the nucleus, and their functions are also regulated by phosphorylation at specific amino acid residues. All these proto-oncogene and tumor suppressor gene products have an important role in growth control.

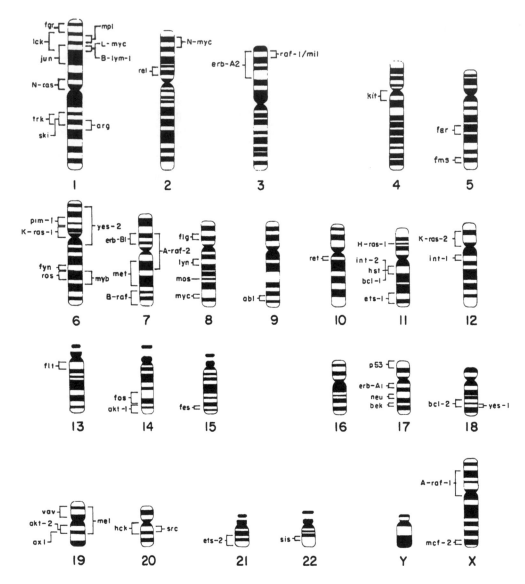

Figure 9.2 Chromosomal localization of proto-oncogenes.

The levels of expression of certain cellular genes, called cell cycle-dependent genes, show important fluctuation in association with the different stages of the cell cycle. It has been suggested that some proto-oncogenes, including c-*myc*, c-*fos*, and c-*jun*, may be considered as examples of cell cycle-dependent genes.[14-19] The products of these genes are nuclear proteins involved in the control of gene expression, and their levels of expression can be readily altered upon stimulation of proliferation induced in susceptible cells by mitogens, hormones, and growth factors.[20] The products of c-*myb* and c-*ets*-2, and c-*rel* are also nuclear proteins that may show an early response to the action of mitogens and that may be related in some way to the cell cycle.[21,22] The proteins Myc, Fos, Jun, Myb, Ets, and Rel, as well as the Erb-B-A protein, which is a receptor for thyroid hormone, have an important role in the control of transcription. However, multiple evidence indicates that some or all of the abovementioned proto-oncogenes cannot be strictly considered as cell cycle-dependent genes, i.e., as genes exhibiting marked fluctuations in their levels of transcriptional expression during the different phases of the cell cycle. A continuum model has been proposed in which the regulation of cell cycle events occurs in all phases of the cycle and not only in the form of a control mechanism specifically associated with the G_1 phase of the cycle.[23] The variations observed in c-*fos* and c-*myc* gene expression in serum-stimulated rat 3Y1 fibroblasts, either normal or transformed, are in accordance with a continuum model of regulation of the cell cycle.[24] However, there is no doubt that several oncoproteins are involved in some manner in the

Table 3 **Homologies between oncoproteins and hormones and growth factors, their receptors, or other cellular proteins**

Oncoprotein	Cellular Protein
Src	Insulin receptor
Erb-B1	EGF receptor*
Mos	EGF precursor
Sis	PDGF-B*
Erb-A	Thyroid hormone receptor*
Fms	CSF-1 receptor*
Kit	SCGF receptor*
Int-2	Basic FGF
Flg	FGF receptor*
Flt	ECGF receptor*
Met	HCGF receptor*
Trk-A	NGF receptor*
Trk-B	BDNF receptor*
Trk-C	NT-3 receptor*
Ras	α subunit of G proteins
Jun	Transcription factor AP-1*
Ets	Transcription factor PU.1
Crk	Phospholipase C

Note: * Indicates identity.

regulation of the cycle, and it is clear that proto-oncogenes and some other genes that code for nuclear proteins can be considered as primary response genes involved in generating cellular responses to extracellular agents such as hormones and growth factors.[25] The functional activity of nuclear oncoproteins is modulated by phosphorylation/dephosphorylation processes that may depend on the action of growth factors.[26] Phosphorylation of transcription factors may represent a link between the transduction of signals generated by extracellular agents that interact with the cell membrane and the regulation of gene expression in the nucleus. Complementary patterns of expression of nuclear proteins encoded by proto-oncogenes are observed during developmental processes in vertebrate embryogenesis.[27]

A. THE Myc PROTEIN FAMILY

The c-*myc* proto-oncogene was identified through the transduction of its sequences by the acute avian retrovirus MC29, which carries the v-*myc* oncogene. The normal c-*myc* gene is highly conserved in evolution. Sequences related to c-*myc* are present in the mammalian genome in the form of genes encoding a *myc* family of structurally related proteins. The members of this family include the c-*myc*, N-*myc*, L-*myc*, B-*myc*, and S-*myc* genes. Myc are short-lived proteins involved in the control of transcriptional processes within the nucleus.[28-33] The Myc proteins contain at their carboxyl-terminal end a tripartite segment comprising a basic region, a helix-loop-helix (HLH) motif, and a leucine zipper sequence that are involved in DNA binding and protein-protein interaction.

The Myc proteins may have a diversity of functional roles but, in particular, they participate in processes associated with cell proliferation and/or differentiation. In many cell types, c-Myc expression correlates with cell proliferation.[34,35] Myc proteins may have a role in facilitating the G_0 to G_1 transition of the cycle and/or in DNA replication, and there is evidence that c-Myc may be a component of the complex of proteins involved in DNA replication.[36] However, the different types of Myc proteins may have distinct functions, according to tissue and stage of development. Expression of cellular *myc* genes is subjected to differential regulation at both the transcriptional and posttranscriptional levels during growth and development.[37] The functional activity of c-Myc would depend, at least in part, on phosphorylation at specific residues. Transactivation of gene expression by Myc is inhibited by mutation at the phosphorylation sites Thr-58 and Ser-62.[38] c-Myc is a substrate for casein kinase II at a region near its carboxyl terminus.[39] Myc proteins are phosphorylated in their amino-terminal domain by growth factor-regulated MAP kinases, and this modification is associated with an increase in the level of transactivation of gene expression.[40] Myc proteins may be substrates for the cell cycle-dependent kinase cdc2.

Human c-Myc is a short-lived nuclear phosphoprotein of 439-amino acids that binds DNA and is involved in the control of DNA transcription. As other Myc proteins, human c-Myc has a tripartite carboxyl-terminal domain that contains three sequence motifs: a basic region, an HLH motif, and a leucine heptad repeat.[41] Similar motifs are present in other eukaryotic transcription factors and have been shown to facilitate protein-DNA and protein-protein interactions. Myc specifically binds to double-stranded DNA containing the palindromic hexanucleotide core sequence CACGTG, which is part of the binding sites for human transcription factors USF and TFE3. However, the hexanucleotide is necessary but not sufficient to define c-Myc DNA-binding specificity. The specific DNA sequence recognized by the c-Myc protein may be the 12-nucleotide-long palindrome GACCACGTGGTC.[42] However, at present it is controversial whether intact Myc proteins can recognize this sequence. Myc may instead recognize E-boxes containing the central CACGTG sequence. Point mutations within the c-Myc HLH motif can abolish both the DNA-binding and transforming activities of c-Myc. The DNA-binding domain of c-Myc may be structurally similar to that of restriction endonuclease EcoRI.[43]

The DNA-binding activity of c-Myc requires the formation of homo- or heterodimers. c-Myc may form oligomers with other proteins through its carboxyl-terminal leucine zipper motif,[44] a structure that is common to a specific class of DNA-binding proteins.[45] A protein called Max interacts with c-Myc in a manner dependent on the integrity of HLH and leucine zipper motifs.[46,47] Max readily homodimerizes and binds DNA with high affinity through its basic HLH domain.[48] The Max protein is associated with Myc *in vivo* and enables Myc to bind DNA under physiological conditions. A dominant negative mutant of Max was found to inhibit sequence-specific DNA binding of Myc.[49]

Max-Myc heterodimers have an important role in the regulation of gene expression and cell proliferation by growth factors. Max and Myc may exert opposite regulatory effects on gene transcription and cell proliferation.[50] In quiescent cells, there are little amounts of Myc protein but high levels of Max protein. Max binds to DNA E-boxes containing CACGTG sequences, but may not activate transcription and may therefore act as a suppressor of the expression of genes recognized by the Myc protein. In growth factor-stimulated cells, the predominant species is the Myc/Max heterodimer complex, which carries the Myc transcriptional activation domain and activates E-box-linked genes. Oncogenic activity of c-Myc requires its dimerization with Max.[51]

The c-*myc* gene is universally expressed in dividing cells. The activity of the c-*myc* promoter is negatively regulated by the c-Myc protein itself.[52] c-*myc* has been considered to be an example of a cell cycle-regulated gene, but in a wide variety of cell types the levels of c-*myc* mRNA expression remain constant throughout the cell cycle.[53-55] Analysis of c-*myc* expression in individual cells using flow cytometry showed that the c-Myc protein is expressed at similar levels in all phases of the cycle.[56] However, in murine M1 myeloid leukemia cells, c-*myc* is expressed at high levels in early G_1 phase and at low levels in late G_1 phase.[57] Thus, it appears that c-Myc protein expression is expressed at constant levels in some types of cells, whereas in other cell types the levels of c-Myc expression may show variation according to the phase of the cycle. In any case, there may be little doubt that the Myc protein is functionally related in some manner to cell cycle events. The level of c-*myc* expression correlates with the rate of growth in different types of cells, including both normal and transformed cells.[58] The transition of cells from G_1 to S phase of the cycle may depend on a threshold level of c-*myc* gene expression.[59] The cloned c-*myc* gene is capable of inducing the expression of transition genes associated with serum-inducible G_0/G_1 transition.[60]

Expression of c-*myc* may be a necessary component of activation-induced programmed cell death (apoptosis).[61] c-*myc* gene expression may determine either continuous cell proliferation or apoptosis, depending on the availability of some critical growth factors.[62] Cultured immortalized murine fibroblasts expressing the c-*myc* gene constitutively may be able to exhibit DNA replication and cell division in the absence of growth factors, but many of these cells die by apoptosis after withdrawal of serum growth factors. c-*myc*-mediated apoptosis can be blocked by ectopic expression of the Bcl-2 protein.[63]

Expression of c-*myc* in different cells and tissues depends on the complex regulatory actions of endogenous and exogenous stimuli, including hormones and growth factors. The concentration of c-*myc* mRNA may be regulated at both the transcriptional and post-transcriptional levels.[64] Regulation of c-*myc* expression at the transcriptional level by serum requires DNA sequences located at both the 5' and 3' regions of the first exon.[65]

The c-Myc protein may be involved in the control of DNA synthesis and cell proliferation. Early increase of c-Myc is observed in synchronized mouse 3T3 cells upon stimulation by growth factors contained in serum.[66] In exponentially growing BALB/c 3T3 mouse fibroblasts, withdrawal of serum causes a marked decline in c-*myc* mRNA, whereas cells made quiescent at subconfluence or confluence contain low levels of c-*myc* mRNA which rise after stimulation of growth by fresh serum.[67] Quiescent human diploid fibroblasts WI-38 are capable of entering DNA synthesis and proliferating in response to the addition of platelet-poor plasma to the medium as long as they express the c-*myc* gene, but not after this expression is ceased.[68] In stimulated B lymphocytes, expression of c-*myc* may be necessary not only for leading to a competence state, rendering cells able to enter G_1 phase, but also for the progression of the activated cells to initiation of DNA synthesis.[69,70] Addition of polyclonal or monoclonal antibodies against human c-Myc protein to nuclei isolated from different types of human cells resulted in inhibition of DNA synthesis and DNA polymerase activity, but not DNA transcription.[71] However, the reported inhibition of DNA synthesis by Myc-directed antibodies was later shown to be due to the presence of an inhibitor of DNA polymerases α and β, and not to the action of Myc protein.[72] The use of oligonucleotides containing antisense sequences specific to those of c-*myc* DNA or RNA can result in suppression of c-Myc protein synthesis and may allow a better definition of the precise role of c-Myc in cell proliferation and differentiation. An antisense oligonucleotide complementary to c-*myc* sequences and capable of inhibiting Myc protein expression prevented entry of mitogen-stimulated lymphocytes into S phase of the cycle but, interestingly, it did not inhibit G_0 to G_1 transversally.[73] Myc proteins are associated in the nucleus with small ribonucleoprotein particles.[74] A peptide sequence contained in the Myc product, as well as in the products of the related genes, N-*myc* and L-*myc*, serves as a nuclear transport signal peptide, inducing a rapid translocation of the protein to the nucleus.[75] Although the level of Myc protein is invariant during the cell cycle, the nuclear matrix-bound portion of Myc is higher during the S phase of the cycle than in G_2.[76] In addition to its role in transcription, Myc may act at specific posttranscriptional levels in the regulation of RNA export, splicing, or nuclear RNA turnover. Two genes (*mr1* and *mr2*) that are induced in quiescent rat fibroblast cultures in response to growth factors are constitutively expressed in *myc*-immortalized cells, but not in growth-arrested normal cells.[77] Induction of *mr1* and *mr2* by c-*myc* does not alter the transcription or mRNA turnover of these two genes, suggesting that some posttranscriptional mechanism operating in the nucleus is responsible for the altered regulation.

The mechanism of action of Myc in the processes associated with cell proliferation and differentiation are poorly understood. Myc may have a role in the regulation of transcription through its affinity to promoter sequences and may be involved in both the positive and negative regulation of gene expression.[78,79] The Myc protein would act as an intracellular competence factor, rendering cells competent to enter the S phase of the cycle. Myc protein microinjected into the nucleus of quiescent mouse fibroblasts in culture cooperates with platelet-poor plasma in the stimulation of DNA synthesis.[80] Constitutive expression of a cloned c-*myc* gene increases the response of rodent cells to PDGF and other growth factors of both the competence and progression classes.[81] Amplification and overexpression of c-*myc* gene sequences in some immortalized cell lines is correlated with a decreased requirement for growth factors.[82]

Other members of the Myc protein family may have functions similar to those of c-Myc. The N-Myc protein binds DNA and functions as a transcription factor.[83] N-Myc is associated with two nuclear proteins, Max p20 and p22, which are phosphorylated by casein kinase II.[84] The HLH region and leucine zipper motif of the N-Myc protein are essential to the formation of functional N-Myc/Max complexes. Expression of the N-*myc* gene depends on regulatory elements contained in its flanking sequences.[85] The N-Myc protein may have a role in developmental processes. Regulation of N-*myc* gene expression by growth factors or during development is partially different from that of c-*myc*, suggesting different functions of the respective products.[86] The c-*myc* and N-*myc* genes play distinct but overlapping roles in the regulation of the growth and differentiation of various types of cells.[87] Expression of the N-*myc* gene is a feature of certain differentiation processes associated with embryogenesis.[88,89] In contrast to the ubiquitous expression of the c-*myc*, expression of N-*myc* in the mouse embryo is restricted mainly to epithelial cells. High levels of N-*myc* mRNA are present in the brain, kidney, thymus, spleen, and liver of newborn mice, but these levels drop to very low or undetectable by the time of weaning. Studies with an antisense oligodeoxynucleotide specific to N-*myc* DNA sequences suggest that N-Myc may be involved in proliferative and differentiative processes.[90] There are contrasting patterns of c-*myc* and N-*myc* gene expression in proliferating, quiescent, and differentiating cells of the embryonic chicken lens.[91] Studies with a targeted leaky mutation showed that the N-*myc* gene product has a role in branching morphogenesis of the mouse lung.[92] Animals homozygous for the mutation die immediately after birth due to an inability to oxygenate their blood, and histological examination reveals marked underdevelopment in the lung airway epithelium. The mutant mice were slightly smaller than normal and had a marked reduction in spleen size. Mice embryos homozygous for null N-*myc* gene mutations develop normally until the onset of organogenesis but thereafter begin to display a wide variety of pathologic features in the proliferation of partially differentiated structures.[93] Normal human lymphocyte precursors do not express the N-*myc* gene at significant levels.[94] The possible role of the N-Myc protein in the mechanism of action of growth factors is not clear. Expression of the N-*myc* gene in human fibroblasts potentiates the mitogenic responsiveness to PDGF-BB and basic FGF by an unknown mechanism.[95] The N-myc protein could act by increasing the concentration of a limiting factor in the signal transduction pathways to growth factors, thereby amplifying the mitogenic signal induced by them.

Other members of the *myc* gene family are L-*myc* and B-*myc*. Human L-*myc* encodes multiple nuclear proteins from alternatively processed mRNAs.[96] Expression of the L-*myc* gene is controlled by short-lived proteins.[97,98] The B-*myc* gene has been cloned from rat genomic and cDNA libraries.[99] In contrast to the stage-specific expression of c-*myc*, B-*myc* is expressed at high levels in all fetal tissues and during subsequent postnatal development in the rat. B-Myc is a 168-amino acid protein with extensive homology to the c-Myc amino-terminal region. The B-Myc protein is involved in the regulation of transcription, but its role may be unique because it lacks the DNA-binding motif found in other Myc proteins. B-Myc can inhibit the transcriptional activation and neoplastic transformation induced by c-Myc in cellular systems such as rat embryo cells, possibly by competing for factors necessary for c-Myc in the regulation of genes implicated in cellular growth.[100] In general, the normal functions of the c-Myc-related proteins are little known. Analyses of the expression of different Myc proteins in the developing and adult mouse kidney suggest that they are involved in quite disparate differentiation processes, even within one tissue and that they may participate in processes not necessarily related to either cell proliferation or cell differentiation.[101] The c-*myc*, L-*myc*, and N-*myc* genes are expressed in the developing human brain, and this expression is uncoupled from cell division.[102] The L-*myc* gene is also expressed in other human fetal tissues, particularly in fetal skin.

B. THE Myb PROTEIN FAMILY

The v-*myb* oncogene is transduced by two acute retroviruses, AMV and E26, which cause myeloblastic leukemia in birds. The normal counterpart of the v-*myb* oncogene is the c-*myb* proto-oncogene, which encodes a nuclear protein of 75 kDa that is structurally related to c-Myc and is functionally involved in the control of cell proliferation. The c-*myb* gene is expressed in mainly immature hematopoietic cells. The levels of c-*myb* gene expression decrease during the differentiation of these cells.[103] The c-Myb protein is located mainly in the nucleus, although in proliferating human T lymphocytes it was found to be almost exclusively distributed in the cytoplasm.[104] In addition to c-*myb* two additional genes, A-*myb* and B-*myb*, have been isolated in the form of cDNA clones from the human genome.[105]

The abundance of c-*myb* mRNA is regulated by transcriptional and posttranscriptional mechanisms.[106] Binding of nuclear factors to a specific site within the first intron of the c-*myb* gene may be responsible for the regulation of c-*myb* expression in hematopoietic cells.[107] A block to elongation of transcription in the first exon of the c-*myb* gene may be responsible for the downregulation of c-*myb* gene expression that occurs during the process of maturation of hematopoietic cells.[108] This mechanism may also operate in other types of cells,[109] but regulatory mechanisms in addition to transcription arrest, including transcription initiation and RNA turnover, may be active in c-*myb*-expressing cells. Alternative internal splicing in c-*myb* mRNAs occurs commonly in normal and tumor cells.[110] The 5'-terminal region of chicken c-*myb* mRNA shows hematopoietic lineage-specific heterogeneity, probably related to differential splicing processes.[111]

The c-Myb protein is involved in DNA transcription. The protein is associated with the nuclear matrix in human leukemic cells.[112] The purified 75-kDa human c-Myb protein binds double-stranded DNA *in vitro*.[113] The c-Myb protein recognizes a hexanucleotide consensus DNA sequence as well as other flanking nucleotides.[114] Deletion analysis of the c-*myb* gene expressed in chimeric vectors allowed the identification of a transcriptional activation domain that corresponds to amino acid residues 275 to 327.[115] Three imperfect repeats containing regularly spaced tryptophan residues form the DNA binding domain of the c-Myb protein.[116] This binding involves the presence in c-Myb of two consecutive HTH motifs with unconventional turns. The c-Myb protein may be involved in the control of the cell cycle. It mediates an increase in $[Ca^{2+}]_i$ during the late G_1 phase of the cycle.[117]

The c-Myb protein plays a critical role in the normal regulation of hematopoiesis. Although most human bone marrow cells express some level of Myb, the highest amount of Myb is present in the progenitor cell population.[118] c-Myb is required for proliferation of intermediate-late myeloid and erythroid progenitors, but is less important for lineage commitment and early progenitor cell amplification.[119] Constitutive expression of an exogenous c-*myb* gene may cause maturation block in monocyte-macrophage differentiation, demonstrating that downregulation of the c-Myb protein is required during monocyte-macrophage differentiation for the maturation of promonocytes to mature monocytes.[120] Exposure of normal human bone marrow mononuclear cells to c-*myb*-specific sense and antisense oligodeoxynucleotides may result in a decrease in colony size and number, without apparent effect on the maturation of residual colony cells.[121] G_1-S transition of mitogen- or antigen-stimulated normal human T lymphocytes requires both c-*myb* mRNA and protein expression.[122] IL-2-induced progression of T cells through G_1 to S phase of the cell cycle is associated with a severalfold increase in c-*myb* gene expression.[123]

The c-*myb* gene is expressed by certain nonhematopoietic cells maintained in culture. It is expressed at low levels by quiescent bovine vascular smooth muscle (VSM) cells, and addition of serum to these cells results in increased c-*myb* mRNA expression.[124] The levels of c-*myb* mRNA increase as cells enter G_1 phase and reach maximum levels in S phase, suggesting that the c-Myb protein is required for cell cycle progression and DNA synthesis. In different types of chicken cells, including fibroblasts (CEF cells), a T-cell line, and B cells from the bursa of Fabricius, the levels of c-*myb* mRNA vary as a function of cell proliferation.

The *myb* gene family is composed, in addition to c-*myb*, of the genes A-*myb* and B-*myb*. The functions of the products of these other genes are little understood. As c-*myb*, the A-*myb* gene is expressed predominantly in hematopoietic cells. In contrast, B-*myb* is expressed in various types of cells in a growth-regulated fashion.[125] The product of B-*myb* is a 100-kDa nuclear protein that appears to be involved in inhibition of specific transcriptional events.[126] Moreover, B-Myb can antagonize *trans*-activation of some promoters by c-Myb.[127] Stimulation of the proliferation of human B and T lympho-cytes or quiescent mouse 3T3 fibroblasts results in the induction of B-*myb* gene expression in late G_1 and S phases of the cycle.[128] The c-Myb and B-Myb proteins may both play a role in G1 to S phase transition, but their functions in relation to the regulation of gene expression may be partially distinct. An E2F-binding site mediates cell cycle-regulated repression of mouse B-*myb* gene transcription.[129]

C. THE Fos PROTEIN FAMILY

The c-*fos* proto-oncogene is the cellular homolog of the v-*fos* oncogene transduced by the murine osteogenic sarcoma viruses FBJ-MSV and FBR-MSV.[130] The murine and human c-Fos proteins consist of 380 amino acids and their expression is stimulated by serum or purified growth factors. Two other cellular genes closely related to c-*fos* in their structure and function, r-*fos* and *fos*-B, were discovered through their expression in growth factor-stimulated mouse cells. DNA sequences related to c-*fos* are highly conserved in evolution and are involved in cellular functions related to both cell proliferation and differentiation. The Fos protein is associated with chromatin and binds DNA *in vitro*, suggesting that it may be involved in the regulation of gene expression.[131]

The susceptibility of cells to the mitogenic action of growth factors could be predicted as a function of the level of c-*fos* gene expression elicited by the factor.[132] Complex regulatory mechanisms and a wide diversity of endogenous and exogenous factors may participate in the regulation of c-*fos* expression in different cell types and under various physiologic conditions.[133] A variety of hormones and growth factors can induce c-*fos* gene expression in susceptible cells. Serum-induced expression of the c-*fos* gene is under negative regulation mediated by the c-Fos protein.[134] Such a feedback mechanism (*trans*-repression) depends on the promoter region of c-*fos* that is required for serum inducibility and that binds the AP-1 transcription factor. The AP-1 factor itself is a homo- or heterodimer composed of the products of genes from the *fos* and *jun* families. Fos proteins are able to modulate gene expression through their interaction with Jun proteins in the form of the sequence-specific factor AP-1.[135-137] Complexes containing Fos protein are able to recognize a *cis*-acting heptanucleotide consensus sequence (TGACTCA) that is also recognized by AP-1 factors in mammalian cells and the GCN4 factor in yeast.[138,139]

The role of c-*fos* gene expression in cell cycle-associated events is not totally clear. The levels of c-*fos* mRNA are extremely low, or even undetectable, throughout the cycle of cells such as mouse fibroblasts, suggesting that c-*fos* may not always contribute to regulation of the normal cell cycle and may not be required for the continuous cycling of certain types of cells.[140] However, c-*fos* transcripts are rapidly expressed when quiescent cultured cells are stimulated with serum or specific purified components contained in serum. When rat 3Y1-B diploid fibroblasts resting at a saturation density are mitotically stimulated with serum, the c-*fos* mRNA level increases markedly and rapidly and decreases thereafter to undetectable levels.[141] However, this increase is not necessarily parallel with the transition from the resting state to the proliferating state, and expression of the c-*fos* gene after serum stimulation may not represent a cell cycle-specific event. Notwithstanding, it is clear that continuous expression of the c-*fos* gene at a minimum level is required for the eventual initiation of S phase. Studies with a *fos*-specific antisense RNA suggest that c-*fos* expression may be necessary for a mitogenic response to growth factors to occur in certain types of cells.[142]

Multiple regulatory elements located upstream of the c-*fos* gene coding sequences gene can modulate transcriptional expression of the gene by eliciting independent responses to different external stimuli.[143,144] Some of these elements are involved in regulatory actions exerted by growth factors at the transcriptional level. cAMP response elements (CREs) located inside and outside of the c-*fos* transcribed sequences are involved in the regulation of c-*fos* expression.[145] The CREs are recognized by CRE-binding proteins (CREBs). A serum responsive element (SRE) contained within the c-*fos* gene promoter is required for serum-induced stimulation of c-*fos* expression. The inner core of the SRE mediates both the rapid induction and subsequent repression of c-*fos* gene transcription following serum stimulation.[146] Stable DNA bending of the SRE occurs upon transcription factor binding to the c-*fos* promoter.[147] The c-*fos* gene SRE contains a dyad symmetry element (DSE) that interacts with a protein — the serum response factor (SRF) —which is represented by a 63- to 67-kDa protein whose synthesis is activated in mouse fibroblasts within minutes of serum stimulation.[148] Newly synthesized SRF is transported to the nucleus, where it may undergo several posttranslational alterations, including phosphorylations, which can modify its affinity for SRE binding. Negative regulation of c-*fos* transcription observed in serum-deprived cells is mediated through SRF binding to the DSE.[149,150] The SRE is specifically recognized by a 67-kDa protein (p67) which may contribute to the regulation of c-*fos* gene expression by extracellular stimuli, including factors contained in serum.[151] Another protein of 62 kDa (p62) may form a ternary complex with p67 and the SRE.[152,153] The p62 alone may be unable to bind to SRE. Mutations that abolish formation of the ternary complex reduce the serum responsiveness of the c-*fos* promoter. However, ternary complex formation at the SRE is not a prerequisite for the transient expression of c-*fos* expression induced by serum or phorbol ester in vivo.[154] A specific structure is maintained at the c-*fos* SRE, regardless of changes in the transcriptional state of the gene.[155] The SRE is necessary and sufficient for response to protein kinase C-dependent and -independent intracellular signaling pathways but not for response to the cAMP pathway. Other DNA sequences are also involved in the regulation of c-*fos* gene expression. At least 9 interdependent regulatory sites, located between the TATA box and position –610 are contained in the c-*fos* promoter-enhancer region of the mouse c-*fos* gene, and they may bind different proteins.[156,157] Immediately 5′ to the SRE there is a protein-binding site that resembles a TPA-response element (TRE) or a CRE. Some of these multiple sites exhibit cell type-specific effects and their functions are dependent on the context of the promoter, i.e., the presence or absence of other sites.

In addition to regulation at the transcriptional level, c-*fos* gene expression is regulated by mechanisms acting at the posttranscriptional level, as shown with cell lines stimulated by different types of growth factors (EGF, PDGF, FGF, TGF-β).[158] Downregulation of c-*fos* gene transcripts in growth-stimulated murine cells may partially depend on sequences located on the 3′ untranslated region of the mRNA, which destabilize the c-*fos* transcript.[159]

The function of c-Fos protein may depend, at least in part, on phosphorylation at specific serine residues.[160] Multiple sites of the Fos protein, clustered within its regulatory regions, are phosphorylated by the cdc2 kinase, the cAMP-dependent protein kinase, and protein kinase C.[161] Phosphorylation of c-Fos on serine residues, in particular Ser-362, by cAMP-dependent protein kinase (protein kinase A) may result in important alterations of its functional properties and may also allow c-Fos to act as a regulator of its own synthesis by downregulating c-*fos* expression at a transcriptional level.[162] Mutation of the Ser-362 residue enhances c-Fos transforming potential to a level comparable to that of the v-Fos oncoprotein. It is clear that the regulation of c-*fos* gene expression depends on multiple molecular mechanisms.

The c-Fos protein may have an important role in the control of cell proliferation and differentiation. Stimulation of quiescent cultured mouse fibroblasts with serum results in rapid induction of c-*fos* expression even in the presence of protein synthesis inhibitors.[163] Studies with either *fos*-directed antisense RNA or microinjection of *fos*-specific antibodies indicate that the c-Fos protein is required for the progression of cells through the cycle.[142,164,165] Mutational analysis shows that only a central region of the Fos protein which is highly conserved among human, mouse, and chicken Fos proteins, is required for stimulating the proliferation of chicken neuroretina cells.[166] Fos shares some homology with the DNA-binding domain of the GCN4 and Jun proteins. However, there is no firm evidence that expression of the c-*fos* gene is directly responsible for the induction of DNA synthesis and cell proliferation. A direct role for the Fos protein in DNA synthesis is unlikely.

The c-*fos* gene is expressed at high levels in prenatal mice, in particular in the placenta and during later stages of development (after day 16 of gestation).[167-169] During embryonic development, c-*fos* expression is associated with differentiation-dependent growth processes of fetal bone and mesodermal web tissue.[170] However, the precise role of the c-Fos protein in developmental processes is not understood. Transcripts of the c-*fos* gene are expressed at very low levels in the rat brain during the period of development of this organ that is characterized by rapid mitosis, whereas much higher concentrations of c-*fos* mRNA are found in the brain of older neonatal animals and adults.[171] The c-Fos protein may be involved in the regulation of genomic functions in highly differentiated, noncycling cells such as nervous cells. Disruption of both copies of the c-*fos* gene in mouse embryonic stem cell (SC) lines by homologous recombination has no detectable effect on the viability, proliferation, or differentiation potential of the SC cells.[172] Although this may reflect an ability of related genes to compensate for c-*fos* functions, the evolutionary conservation of c-*fos* and other proto-oncogenes encoding nuclear proteins suggests that these genes are involved in unique functions. Mice carrying nonfunctional (null) c-*fos* gene mutations exhibit pleiotropic phenotypic alterations.[173] Bone and hematopoietic defects occur in mice lacking a functional c-*fos* gene.[174] Heterozygous *fos* +/– mice appear normal; homozygous *fos* –/– mice are growth retarded, develop osteopetrosis, and have hematopoietic alterations. Differential c-*fos* expression occurs in hematopoietic cells, correlating with differentiation of myelomonocytic cells.[175]

Growth factors regulate gene expression by pathways that may be either dependent or independent of c-*fos* gene expression. The products of the other members of the *fos* gene family may exert functions partially similar to those of the c-Fos protein or the growth factor responses may be effected by pathways independent of Fos-related proteins. In addition to the c-*fos*, r-*fos*, and *fos*-B genes, the *fos* gene family is composed of two other *fos*-related genes, *fra*-1 and *fra*-2. These genes can be induced by serum stimulation in susceptible cells.[177-179] Moreover, the induction follows an orderly mechanism and the Fra proteins participate, in a manner similar to the Fos proteins, in the formation of the AP-1 transcription factor. Fra proteins are continuously synthesized in cultured mouse fibroblasts during exponential growth.[180] Fra-1 and Fra-2 are the main proteins complexed with the Jun proteins in asynchronously growing Swiss 3T3 cells. While the activities of c-Fos and Fos-B are required mostly during the G_0 to G_1 transition in serum-stimulated cells, Fra-1 and Fra-2 are involved both in this transition and in asynchronous growth. Stimulation of growth-arrested CEF cells with serum results in expression of *fra*-2 mRNA and phosphorylation of the Fra-2 protein on serine residues.[181] Constitutive expression of *fra*-1 in PC12 rat pheochromocytoma cells can inhibit the differentiation of these cells and the transcriptional activation of c-*fos*, c-*jun*, *jun*-B, and *egr*-1, but not *jun*-D, induced by NGF, EGF, and dibutyryl cAMP.[182]

D. THE Jun PROTEIN FAMILY

The c-*jun* proto-oncogene is the normal cellular counterpart of the v-*jun* oncogene transduced by the avian sarcoma virus ASV-17. Expression of c-*jun*, alone or in coordination with c-*fos* and other genes, may occur as a consequence of a variety of physiological stimuli acting on different types of cells. The c-*jun* gene promoter contains sequence elements responsive to serum, EGF, and phorbol ester.[183] Induction of cell proliferation in fibroblast cultures by the addition of serum or purified hormones or growth factors is accompanied by a rapid change in the expression of immediate-early genes, which include members of both the *fos* and *jun* gene families. The members of the Jun and Fos protein families are both required for cell cycle progression and for entry into the S phase in serum-stimulated cultured fibroblasts.[184] The c-*jun* gene is transcriptionally activated during G_0 to G_1 phase transition of serum-stimulated mouse fibroblasts.[185,186]

The *jun* gene family is composed of the c-*jun* proto-oncogene and the related genes, *jun*-B and *jun*-D. The three *jun* genes (c-*jun*, *jun*-B, and *jun*-D) are expressed in different types of cells, including human peripheral blood granulocytes, and functional activation of these cells may result in elevated levels of *jun* mRNAs.[187] A specific transcription factor, NF-jun, may function as a signal transduction molecule to mediate the rapid induction of c-*jun* in a cell type- and stimulus-specific manner.[188] Binding of NF-jun to its DNA recognition site is enhanced by growth factors such as TNF-α and phorbol esters such as TPA.

The *jun* genes are early response genes. The synthesis of Jun proteins rapidly increases following serum stimulation of quiescent Swiss 3T3 cells, with Jun-B showing the highest levels of expression.[189] In addition to the 338-amino acid Jun-B protein, a shorter but functional form of the protein, Jun-B/SF, is generated in about equal amounts by alternative RNA splicing in serum-stimulated mouse fibroblasts.[190] The products of the different *jun* genes may have subtly differing functions. The pattern of expression of *jun*-D RNA differs from that of c-*jun* and *jun*-B RNA during male germinal cell differentiation in the mouse, with higher levels of *jun*-D transcripts occurring in adult postmeiotic cells.[191] In human placenta, c-*jun* mRNA expression occurs predominantly during early gestation and is associated with cytotrophoblastic proliferation, whereas maximal expression of *jun*-B RNA occurs in the late stages of placental development and may be related to the terminal differentiation of trophoblastic cells.[192] In the golden hamster, expression of *jun*-B in the suprachiasmatic nucleus of the hypothalamus is selectively induced by photic stimuli related to circadian rhythms.[193] Differential expression of members of the *jun* family is also observed in the rat brain.[194] In general, expression of Fos and Jun proteins is by induced growth factors as well as by a number of mitogenic, differentiation-inducing, and neuronal-specific stimuli. The c-*jun* gene displays limited oncogenic potential when overexpressed in some types of cells (CEF cells), lowering the requirement of these cells for the exogenous supply of growth factors.[195]

Multiple molecular interactions between the different Jun and Fos proteins, involving the formation of leucine zipper structures, occur *in vivo* and may have an important role in the regulation of gene expression by growth factors and other extracellular stimuli. Jun proteins function as transcription factors involved in the regulation of gene expression.[196] A protein of 39 kDa associated in the form of a complex with the Fos protein within the nucleus of fibroblastic cells was identified as the product of the c-*jun* proto-oncogene.[197] Fos and Jun proteins form a complex that binds to specific DNA sequences in the form of the transcription factor activator protein AP-1.[198,199] Fos and Jun interact through respective motifs that constitute a heptad repeat of leucine residues, called the "leucine zipper".[200-203] Homodimers or heterodimers containing the Jun protein function as sequence-specific transcriptional activators.[204-209] In addition to their positive action on gene transcription, the Jun and Fos proteins can act as negative regulators of the transcription of certain genes, such as the atrial natriuretic factor (ANF) gene.[210] The effects of Fos and Jun on the expression of steroid hormone receptor genes can be either inhibitory or stimulatory and are receptor-, promoter-, and cell type-specific.[211]

Regulation of transcription by the Fos-Jun heterodimer involves a complex integration of multiple activator and regulatory domains.[212] By associating with Jun, Fos is responsible for the formation of a protein complex that has greater affinity for the DNA target sequence than does the Jun protein alone.[213] Site-specific mutagenesis analysis indicates that both the Fos and Jun proteins contribute directly to the DNA binding of the heterodimer through their leucine zipper domains.[214,215] Reciprocal mutations in the basic region of Fos and Jun can influence the binding of the heterodimer to DNA, implying a symmetrical binding site.[216] Association of Fos and Jun with the AP-1 site results in a conformational change in the basic amino acid regions that constitute the DNA-binding domain, and Fos-Jun dimers induce an alteration in the conformation of the DNA helix.[217,218] Fos-Jun heterodimers are capable of stimulating promoters containing AP-1 recognition sites even in plant cells.[219]

The AP-1 complex is involved in different cellular functions. AP-1 may play a role in the regulation of DNA replication.[220] AP1 is involved in the control of cell proliferation and differentiation, as well as in neoplastic transformation.[221,222] Loss of AP-1 activity may occur in senescent human cells in culture and may contribute to the inability of these cells to proliferate in response to serum or purified mitogens.[223] Growth factors and other extracellular stimuli may contribute to regulating c-*jun* gene expression. Enhanced expression of c-*jun* is an early genomic response to TGF-β stimulation.[224] An AP-1-binding site in the c-*fos* gene mediates its induction by EGF and TPA.[225] EGF-induced expression of the *jun*-B gene in transfected murine embryonal carcinoma cells expressing EGF receptors depends on the Jun D protein.[226] Jun proteins are themselves involved in controlling the expression of *jun* genes. In addition to regulation at the transcriptional level, expression of Jun proteins depends on posttranslational mechanisms. Multiple signal transduction pathways mediate c-Jun protein phosphorylation.[227] Transcriptional activity of c-Jun is modified by phosphorylation of its amino-terminal region (Ser-63 and Ser-73) by a c-Jun amino-terminal protein kinase (cJAT-PK), whose activity depends, in turn, on serine and threonine phosphorylation by protein kinase C.[228-230] Growth factors such as IGF-I may be able to regulate c-Jun protein activity through its phosphorylation on tyrosine residues.[231] The transcriptional activity of the c-Jun protein is also regulated through phosphorylation of serine residues located near the carboxyl terminus induced by MAP kinases.[232,233] Activation of c-Jun by MAP kinases may underlie the common stimulation of AP-1 by growth factors and oncoproteins.

The c-*jun*-like genes *jun*-B and *jun*-D are functionally important members of the *jun* gene family.[234-236] These two genes were first identified by screening of cDNA libraries prepared from serum-stimulated NIH/3T3 cells. The tissue distribution and levels of expression of the members of the *jun* family may be different. Whereas *jun*-B behaves like c-*jun* in relation to its high inducibility by serum and may be considered as an immediate-early gene in the mitogenic response, *jun*-D is already expressed in serum-starved mouse fibroblasts and is not stimulated by the addition of serum. Both c-*jun* and *jun*-B are developmentally regulated in the immature and mature mouse testis.[237] Expression of the c-*jun* and *jun*-B genes is increased after dissociation of spermatogenic cells, with maximal induction in prepuberal animals. Differential expression of the genes c-*jun* and *jun*-B is observed in some types of cells, for example, in TGF-α-stimulated BC₃H1 muscle cells.[238] The physiological activities of c-Jun and Jun-B have been found to differ in several aspects, such as in their ability to activate AP-1 responsive genes.[239] In general, Jun-B is much less active than c-Jun for its gene *trans*-activation and cellular transformation abilities, and these differences depend on only a few amino acid differences within the conserved DNA-binding domain of the two proteins.[240] In some systems, the Jun-B protein may function as a negative regulator of c-*jun* gene expression.

The different homo- and heterodimeric Jun/Fos proteins may have similar but distinctive cellular functions.[241] A wide diversity of homo- and heterodimeric nuclear proteins may participate in the transcriptional regulation of DNA elements that respond to extracellular signals such as hormones and growth factors. Two major classes of regulatory elements that contribute to this regulation by extracellular signaling agents are the DNA sequence motifs AP-1/TRE and ATF/CRE. The AP-1/TRE element (TGACTCA) is the AP-1 binding site or phorbol ester (TPA) responsive element (TRE). The ATF/CRE element (TGACGTCA) is the activating transcription factor (ATF) binding site or cAMP responsive element (CRE). Several proteins bind to such *cis*-acting elements.[242] The AP-1/TRE site is recognized by a group of proteins that includes those of the Fos and Jun families. The ATF/CRE site is recognized by a family of proteins called ATF or CRE-binding proteins (CREBs). The ATF/CREB family is implicated in cAMP- and calcium-induced alterations in transcription.

The members of the Fos/Jun family of transcription factors (i.e., the Fos, Fos-B, Fra-1, Fra-2, Jun, Jun-B, and Jun-D proteins) can form functional heterodimers not only between them but also in association with members of the ATF/CREB family (ATF-1 to ATF-4). The array of homo- and heterodimers that is formed by members of the Fos/Jun family and the ATF/CREB family provides an enormous diversity of mechanisms for transcriptional regulation by growth factors and other exogenous stimuli, and also opens possibilities for compensatory effects in cases of alterations in one or more of these members. Disruption of both copies of the c-*jun* gene has no apparent effect on mouse embryonic stem cell viability, growth rate, and differentiation potential,[243] which could be due to the ability of other proteins to compensate for the c-Jun protein function.

E. THE Ets PROTEIN FAMILY

The c-*ets*-1 gene is the normal cellular homolog of the v-*ets* oncogene contained in the avian leukemia virus E26. It is a member of a family of genes, the *ets* family, which includes, in addition to *ets*-1 and *ets*-2, the genes *erg*, *elk*-1, *elk*-2, *elf*-1, *SAP1*, *PEA3*, *erg*-B/*Fli*-1, and *PU.1*/*Spi*-1.[244-247] The products of these genes are nuclear phosphoproteins that bind DNA through a conserved carboxyl-terminal basic domain and regulate gene expression in a specific manner due to their distinct amino-terminal regions.[248] The carboxyl-terminal portion of Ets proteins exhibits a high level of sequence identity with the transcription factor PU.1.[24] This factor binds to a purine-rich sequence, the PU box, and acts as a transcriptional activator that is preferentially expressed in macrophages and B lymphocytes. The Ets proteins show sequence-specific binding to the PEA-3 (CAGGAAGT) motif. The Ets proteins do not display, as the Fos and Jun proteins, typical leucine zipper or zinc finger structures but have, as the Myb protein, a conserved triplet of tryptophan residues. However, the Ets proteins can cooperate with Fos and Jun proteins for transcriptional activation.[250] The c-*ets*-1 and c-*ets*-2 proto-oncogenes are actively expressed in lymphoid cells and macrophages, respectively. The Ets-1 protein binds to the T-cell receptor α gene enhancer and regulates the maturation and differentiation of T lymphocytes, while expression of the Ets-2 protein may correlate with cell proliferation.

Nucleotide sequences exhibiting a high degree of homology to the avian and mammalian c-*ets* genes are highly conserved in evolution.[251,252] Sequences related to *ets* are present in amphibians such as *Xenopus laevis*, in the sea urchin, and in insects such as *Drosophila melanogaster*. Two distinct c-*ets*-2 genes present in *Xenopus laevis* are actively expressed during oogenesis and early embryogenesis as well as in some adult tissues.[253] Expression of Ets-2 protein is required for *Xenopus* oocyte maturation.[254] Transcripts of the c-*ets*-1 gene are also part of the maternal pool of mRNAs in the *Xenopus* oocyte, but at the adult stage c-*ets*-1 is expressed only in ovary and spleen.[255] Expression of the *ets*-related gene E74 is induced in *Drosophila* by the insect steroid hormone ecdysone.

Molecular heterogeneity of Ets proteins is due to alternative RNA splicing or posttranslational modifications including phosphorylation. The chicken c-*ets*-1 gene encodes two proteins of 54 and 68 kDa, which are generated by alternative splicing.[256] The chicken 54-kDa Ets-1 product is a nuclear DNA-binding protein that is expressed in lymphoid organs such as thymus, spleen, and bursa. The 68-kDa Ets-1 product is also a nuclear protein expressed only in spleen cells. Posttranslational modification of Ets proteins is related to phosphorylation on serine and, to a lesser extent, on threonine.[257] These phosphorylations are stimulated by calcium ionophore and are abolished by lowering the extracellular concentration of calcium. Mitogenic doses of concanavalin A stimulate Ets-1 protein phosphorylation in thymocytes, suggesting that Ets-1 mediates early events linked to cell activation. Ligation of membrane Ig in B lymphocytes leads to calcium mobilization from intracellular stores, which stimulates Ets-1 phosphorylation, thus contributing to the alteration of gene expression during B-cell activation.[258] Phorbol ester treatment of the CEM human lymphocytic cell line leads to activation of protein kinase C, which results in Ets-1 phosphorylation and activation.[259] At least six different phosphorylated isoforms of Ets-1 are expressed in mitogen-stimulated human lymphocytes.[260] The Ets-1 protein is hyperphosphorylated during the mitotic phase of the cycle, which may be due to changes in cellular phosphatase activity during this phase.[261]

The products of the *ets* gene family are expressed in different types of cells at distinct stages of development. The activity of the human c-*ets*-1 promoter is inducible by serum and the Fos-Jun/AP-1 protein as well as by the Ets-1 protein itself.[262] The c-*ets*-1 gene is expressed mainly in lymphoid B and T cells, but it is also expressed in other cell types such as growing endothelial cells of human embryos, as well as in granulation tissue and during tumor angiogenesis.[263] The theca cells of adult mouse ovary express both the *ets*-1 and *ets*-2 genes.[264] The two genes are also expressed in astrocytes, but not neurons, of the adult human brain.[265] Because astrocytes (unlike neurons) are glial cells able to divide, this expression suggests that c-Ets proteins may be implicated in the control of cell proliferation. Expression of Ets-1 protein can render REF-1 fibroblasts capable of growing in the absence of growth factors.[266] Overexpression of Ets-2 by introduction of an expression vector containing the *ets*-2 gene into NIH/3T3 mouse cells results in stimulation of cell proliferation and abolishment of serum requirement.[267] The c-*ets*-2-transfected cells form colonies in semisolid medium and can induce tumors in nude mice, indicating that the c-*ets*-2 gene may display, under certain specific conditions, oncogenic activity.

The Ets-related protein Elk-1 is expressed in adult lung and testis as well as in lymphoid and myeloid cells. Elk-1 forms a ternary complex with the SRF and SRE, suggesting that Elk-1 could be homologous to the p62 c-Fos regulatory factor.[268] Elk-1, like the c-Ets-1 protein, binds in a sequence-specific manner to the PEA-3 DNA motif.

The *erg* gene, identified in the human colon carcinoma cell line COLO 320, is structurally related to v-*ets* oncogene and encodes two proteins, Erg-1 and Erg-2, by alternative RNA splicing and usage of the initiation codon.[269,270] The Erg proteins are sequence-specific transcriptional activators.[271] The *erg*/*erg*-A gene resides on human chromosome 21, which also contains the c-*ets*-2 gene. A second *erg* gene, *erg*-B, resides on human chromosome 11 and is transposed to chromosome 4 as a result of the t(14;11) translocation associated with acute leukemia.[272] The *erg*-B gene is the human homolog of the murine *fli*-1 gene and is expressed in a subset of human erythroleukemia cell lines, suggesting a role for the Fli-1/Erg-B protein in hematopoietic lineage commitment.[273]

The *fli*-1 gene is expressed at highest levels in proliferating, unstimulated macrophages, and inflammatory mediators LPS and IFN-γ lower the expression of *fli*-1 mRNA. The mouse genes *fli*-1 and *PU.1*/*Spi*-1 may be involved in the induction of erythroleukemia by some strains of the chronic retrovirus F-MuLV.[274] While *erg*-A transcripts are detected mainly in the mouse thymus, *fli*-1/*erg*-B exhibits a broader distribution of expression that includes, in addition to thymus and lymphocytes, ovary, bone marrow, spleen, and heart, as well as some established cell lines. The Fli-1/Erg-B protein may function as an activator of gene transcription.

Another member of the *ets* gene family, *PEA3*, encodes in the mouse a 61-kDa protein composed of 555 amino acids that shares extensive sequence similarity with the ETS domain, a stretch of 102 amino acids that is common to all members of the family and is involved in DNA binding.[275] Expression of *PEA3* in mouse tissues is highly restricted to brain and epididymis. The *PEA3* gene is expressed to a lesser degree in the mammary gland, but is not expressed in hematopoietic cells. Expression of *PEA3* is downregulated during retinoic acid-induced differentiation of mouse embryonic cell lines, suggesting that the PEA3 protein may have a regulatory role during embryogenesis.[276]

F. THE Rel PROTEIN FAMILY

The v-*rel* oncogene is contained in the reticuloendotheliosis virus strain T (REV-T), which was isolated from a turkey and transduces c-*rel* proto-oncogene sequences.[277-279] REV-T is able to transform hematopoietic cells.[280] The v-Rel oncoprotein has been detected in the cytoplasm of REV-T-infected cells, but it contains sequences that determine its partial localization in the nucleus.[281] v-Rel functions as a promoter-specific transcriptional activator, capable of modulating host cell genome expression. There are multiple structural and functional differences between the viral v-Rel oncoprotein and the normal c-Rel protein. The cloning and expression of chicken c-*rel* cDNA sequences have been reported.[282] In contrast to the 57-kDa v-Rel protein, c-Rel is a 68-kDa protein that is localized in the cytoplasm of CEF cells. However, the normal c-Rel protein is also localized, at least in part, in the nucleus and may be involved in the regulation of transcription.

The Rel proteins are members of a family of transcription factors that include the transcription factor NF-κB and the *Drosophila* maternal morphogen *dorsal*.[283,284] The NF-κB factor is ubiquitously present in an inactive form in the cytosol but is a constitutively active nuclear protein in mature B lymphocytes. Exposure of cells to mitogens and growth factors results in stimulation of NF-κB DNA-binding activity and a rapid transport of the protein from the cytoplasm to the nucleus. A 40-kDa Rel-associated protein (pp40) may inhibit this transport, thus retaining NF-κB in the cytoplasm and inhibiting the transcription of genes recognized by NK-KB.[285]

The turkey c-*rel* locus is expressed at low levels in many tissues and at higher levels in hematopoietic tissues.[286] In the chicken, c-*rel* gene transcripts are expressed in hematopoietic organs.[282] The bursa of Fabricius contains the highest levels of c-*rel* mRNA, followed by liver, marrow, spleen, and thymus. Nonhematopoietic organs, such as muscle and brain, contain very low levels of c-*rel* mRNA. Relatively high levels of c-*rel* transcripts are observed in mouse B and T lymphocytes, whereas lower levels have been found in functionally immature thymocytes.[287] In contrast to c-Myb and c-Ets, which may be involved in the earlier stages of hematopoietic differentiation, c-Rel may have a role in the later stages of lymphocyte differentiation.[288] Transcription of the c-*rel* gene is induced transiently when resting NIH/3T3 mouse fibroblasts are stimulated with serum or phorbol ester.[289] c-*rel* may be a member of the early-response gene family.

The human c-*rel* gene is located on chromosome region 2p12-p13. Sequence analysis of human c-*rel* shows the presence of a portion that corresponds to the chicken and turkey c-*rel* exons 4 and 5 and contains a 12-amino acid sequence that may define a site of serine phosphorylation conserved in the c-Rel sequences of man, chicken, and turkey.[290,291] Human c-*rel* is expressed most abundantly in differentiated lymphoid cells.[292] A c-*rel*-related gene, *rel*-B, which encodes the transcription activator protein Rel-B, is also a serum-inducible gene.[293]

G. THE Egr/Krox PROTEINS

The genes *egr*-1/*Krox*-24 and *egr*-2/*Krox*-20 code for nuclear phosphoproteins that possess three zinc finger motifs and function as sequence-specific transcriptional activators. Zinc fingers are DNA-binding peptide domains that were first identified in transcription factor IIIA, a protein required for initiation of the transcription of 5S RNA in *Xenopus laevis*.[294] The zinc fingers consist of tandemly repeated units of 28 to 30 amino acids that require Zn^{2+} for DNA-binding activity and contain two cysteines and two histidines at invariant positions. The *erg*/*Krox* genes are immediate-early genes that are rapidly induced by serum and growth factors in different cell types and are involved in the modulation of cell proliferation and differentiation.[295-298] They were termed *Krox* in the mouse for the Krüppel box (*Kr*), which was previously identified in *Drosophila*.[299]

The *egr*-1/*Krox*-24 gene, also called *zif*-268 or NGF1-A, codes for a nuclear phosphoprotein of 75 to 80 kDa which is involved in the regulation of gene expression.[300] The Egr-1 protein is able to regulate transcription of the rat cardiac α-myosin heavy chain gene.[301] The product of the tumor suppressor gene *WT1* is a zinc finger-containing protein that binds to the same DNA sequences as the Egr-1 protein.[302] The WT1 protein functions as a repressor of transcription when bound to the DNA site recognized by Egr-1.

The *egr*-1 gene encodes two distinct proteins of 82 and 88 kDa, depending on the site of initiation at an AUG or non-AUG codon.[303] The two Egr-1 proteins bind to DNA in a sequence-specific manner, and this binding results in the activation of transcription. The single human *egr*-1 gene is located on chromosome region 5q23-q31.

Expression of the *egr*-1 gene is rapidly induced by several growth factors in various types of cells and is frequently, but not invariably, coregulated with c-*fos* gene expression. Normal development of the mouse skeleton depends, in part, on a coregulated expression of the c-*fos* and *egr*-1 genes.[304] The *egr*-1 gene is induced by NGF in neural cells and its expression increases markedly during cardiac and neural cell differentiation. In mouse fibroblasts stimulated with serum or growth factors, the Egr-1 protein is expressed mainly during G_0/G_1 transition.[305] Human fibroblasts stimulated with serum or EGF express high levels of Egr-1 protein through mechanisms that do not include activation of the Na^+/H^+ exchanger.[306]

The *egr*-2/*Krox*-20 gene also encodes a protein containing a DNA-binding domain with three zinc fingers.[307] Expression of this gene is increased in quiescent mouse cells that are stimulated to re-enter the proliferative cell cycle.[308] However, expression of *egr*-2 mRNA is not restricted to the G_0 or G_1 phase of the cycle, but its levels are similar in all stages of the cycle of growing mouse fibroblasts.[309] The possible role of *egr*-2 gene expression in relation to the proliferation of growth factor-stimulated cells is unknown. Adherence of circulating monocytes to endothelium, which is a prerequisite for their migration into injured tissues and their conversion into tissue macrophages, may depend on the expression of the PDGF-B gene, and this is associated with increased expression of c-*jun*, c-*fos*, and *egr*-2 mRNA.[310]

H. THE Vav PROTEIN

The *vav* gene was identified through its oncogenic activation in the course of transfer assays involving DNA extracted from human esophageal carcinoma cells. The normal mouse 95-kDa Vav protein contains, from the amino-terminal to the carboxyl-terminal end, a region rich in acidic amino acid residues, a guanine nucleotide exchange motif, a cysteine-rich domain similar to the phorbol ester-binding domain of protein kinase C, and a carboxyl-terminal SH2 motif flanked by SH3 motifs.[311,312] The presence in Vav of a proposed amino-terminal HLH/leucine-zipper-like motif sharing sequence similarity with motifs contained in the Myc and Max proteins, as well as two potential nuclear localization signals, was later called into question. Deletion of the amino-terminal region results in activation of the oncogenic potential of the Vav protein.[313,314]

The *vav* proto-oncogene is primarily expressed in mammalian tissues engaged in active hematopoiesis, and its product may be involved in regulating the development and maintenance of the hematopoietic system.[315,316] The Vav protein is phosphorylated on tyrosine by growth factor receptors with tyrosine kinase activity. Vav may play an important role by coupling tyrosine kinase pathways to Ras-like GTPases through the regulation of guanine nucleotide exchange. Stimulation of the T-cell antigen receptor complex on normal or leukemic human lymphocytes promotes the phosphorylation of Vav on tyrosine, and activation of B cells by engagement of their IgM antigen receptors has a similar effect.[317-320] Vav may be associated with the EGF and PDGF receptor kinases through its SH2 domain, and activation of these receptors by ligand binding results in the phosphorylation of Vav on tyrosine residues. The 25-kDa molecule, Grb2, which is present in nonhematopoietic cells also contains an SH2 domain flanked by two SH3 domains. Grb2 associates with the activated EGF and PDGF receptor kinases via its SH2 domain and may function as an adapter protein that couples tyrosine kinases to Ras.[321] Vav and Grb2 may play similar roles in signal transduction in different types of cells, linking surface receptor tyrosine kinases to Ras signaling.

III. EXPRESSION AND FUNCTION OF TUMOR SUPPRESSOR PROTEINS

The tumor suppressor genes (onco-suppressor genes) function as negative regulators of the proliferation of both normal and neoplastic cells.[322-330] The products of these genes include the tumor suppressor proteins p53, RB, WT1, NF1, and BTG1. These proteins have important roles in the control of the cell cycle and are involved in normal physiological processes related to cell proliferation and differentiation as well as in cellular senescence.[331] The effects of antisense oligodeoxynucleotides specific to RB and p53 RNA in TIG-1 human diploid fibroblasts at different population doubling levels exposed to serum growth factors indicate a cooperative action of RB and p53 on the responsiveness of the cells to serum and suggest that RB and p53 may function in separate yet complementary pathways in the transduction of signals related to senescence.[332] RB and p53 would control cellular senescence at the G_1/S boundary of the cell cycle, and p53 would modulate the function of RB to suppress cell proliferation. The tumor suppressor proteins may exert a negative effect on the proliferation of tumor cells. Little is known about the regulation of tumor suppressor gene expression by hormones, growth factors, regulatory peptides, and other extracellular signaling agents. Wild-type RB and p53 proteins, but not their transforming mutants, may act as transcriptional repressors of the genes for specific growth factors such as IL-6.[333] Modulation of the expression or function of growth factors by tumor suppressor proteins may have regulatory effects on the proliferation and differentiation of normal cells as well as in oncogenesis.

A. THE p53 GENE

The p53 antigen is a 53-kDa nuclear phosphoprotein that is involved in the regulation of events related to the cell cycle in both normal and transformed cells.[334-340] Growth arrest induced by p53 occurs prior to or near a restriction point, the R point, in late G_1 phase of the cycle and may be affected by the repression of the expression of a specific subset of genes, in particular the c-*myb* proto-oncogene and the DNA polymerase-α gene.[341] The wild-type p53 protein is an unusually shaped tetramer that binds directly to DNA.[342] The p53 protein contains a domain that is involved in the regulation of transcription through the recognition of a consensus palindrome sequence in the DNA.[343] Wild-type p53 protein binds to the TATA-binding protein and represses transcription.[344] Upon binding DNA, wild-type p53 changes conformation at both its amino- and carboxyl-terminal regions.[345] Transcription from the human *HSP70* promoter is repressed by p53, and this effect is mediated by a CCAAT binding factor (CBF). Thus, protein-protein interaction between p53 and specific transcription factors may be a mechanism by which p53 regulates gene expression.[346]

Expression of the p53 protein in different types of cells is variable, according to the type of cells and its predominant physiological condition. High levels of p53 mRNA expression are observed in early stages of mouse embryogenesis, but, as development proceeds, p53 expression becomes more pronounced during the differentiation of tissues, and during terminal differentiation of specific tissues the level of p53 mRNA declines markedly.[347] The p53 gene is upregulated during skeletal muscle cell differentiation.[348] A normal p53 gene is dispensable for embryonic mouse development, but its absence predisposes the animal to neoplastic disease.[349] p53 expression increases during maturation of human hematopoietic cells but becomes barely detectable in mature, myeloid, lymphoid, and monocytic cell populations.[350] The p53 gene product may play an important role in spermatogenesis.[351] It is expressed in the testis in a cyclical and spatial-specific manner, suggesting that it is involved in the cessation of DNA replication that is necessary for the transition from tetraploid primary spermatocytes to haploid spermatides that differentiate to mature sperm cells. Wild-type p53 is capable of inducing programmed cell death (apoptosis) in certain types of tumor cells (for example, in murine erythroleukemia cells) and this effect occurs mainly in the G_1 phase of the cycle.[352] On the other hand, the p53 gene is overexpressed in a wide variety of human tumors,[353] suggesting that it may act in some cases as a dominant oncogene.

The cellular mechanism of action of the p53 protein is not understood. It has been suggested that a mechanism involved in the negative regulation of cell proliferation by p53 is associated with the regulation of GTP synthesis.[354] Growth suppression induced by wild-type p53 is accompanied by selective downregulation of cyclin/PCNA expression.[355] Nuclear localization is essential for the activity of the p53 protein.[356] The p53 protein is probably involved in the regulation of gene expression. DNA-binding activity is intrinsic to the p53 protein, and p53 purified from both normal and transformed cells is able to bind to double- and single-stranded DNA *in vitro*.[357] The p53 protein is composed of separate functional domains. The DNA-binding activity of p53 is cryptic and requires activation by specific cellular factors that act on a carboxyl-terminal regulatory domain of the protein.[358] Activation of p53 DNA-binding activity may be critical to regulation of its ability to arrest cell proliferation. A DNA-binding domain is contained in the carboxyl-terminal region of wild-type p53 protein, and wild-type p53 has a higher DNA-binding affinity than mutant p53 forms.[359] A human DNA sequence of as few as 33 bp is sufficient to confer specific binding to the p53 protein.[360] Certain guanines within this 33-bp region are critical, because their methylation or substitution by thymine can abrogate binding. A transcription-activating sequence is contained in the p53 protein.[361]

The notion that p53 is essential for cell proliferation is supported by the observation that microinjection of a monoclonal antibody to p53 into the nuclei of quiescent Swiss 3T3 cells inhibits serum-induced entry of the cells into the S phase of the cycle.[362,363] Introduction of DNA constructs encoding p53-specific antisense RNA into transformed and nontransformed mouse cells results in the complete cessation of the proliferation of these cells.[364] The results from different studies support the idea that one function of the p53 protein is to activate the transcription of genes involved in the suppression of cell proliferation. The antiproliferative effects of p53 are related to the acquisition of a unique conformational state associated with an increased level of phosphorylation of the protein.[365] Phosphorylation of wild-type and mutant p53 proteins at specific amino acid residues depends on associated protein kinases.[366] Protein kinase C phosphorylates the p53 protein both *in vivo* and *in vitro*, and interaction of p53 with S100b (a member of the S100 protein family involved in cell cycle progression and cell differentiation) is inhibited by p53 phosphorylation by protein kinase C.[367]

Expression of p53 in different types of cells is regulated by complex endogenous and exogenous factors. The effects of mitogens on human peripheral blood lymphocytes suggest that p53 may be considered as a cell cycle-dependent gene.[368] Expression of the p53 protein is downregulated by posttranscriptional mechanisms during embryonal development in mouse and chicken.[369] Human p53 protein is phosphorylated by the cdc2 protein kinase *in vitro* and coprecipitates with cdc2 *in vivo*, suggesting that phosphorylation of p53 by the cdc2 kinase may regulate the activity of p53 in the initiation of DNA replication in mammalian cells.[370] Human p53 contains a sequence that may be involved in its transport to the nucleus.[371] Ser-315, contained within this sequence, is phosphorylated *in vitro* by cdc2. Binding of mutant p53 proteins to the cdc2 kinase is greatly reduced relative to wild-type.[372] Casein kinase II, a ubiquitous enzyme that is present in both the cytoplasm and the nucleus, forms a complex with the p53 protein and phosphorylates p53 at Ser-389.[373,374] Casein kinase II is stimulated in response to mitogens and growth factors, suggesting that p53 may play a role in the transduction of extracellular signals to the nucleus. The p53 protein may be phosphorylated at the amino-terminal end (on residues Ser-4, Ser-6, and Ser-9) by a casein kinase I-like enzyme.[375] The precise role of p53 phosphorylation in the specific functions of the protein is not understood. The p53 protein is covalently linked to 5.8S ribosomal RNA, suggesting that it may be involved in the expression or function of this RNA.[376]

The wild-type p53 gene behaves as a tumor suppressor gene or antioncogene. Mutations of evolutionary conserved codons of the p53 gene occur in a diversity of human tumors.[377,378] Individuals with germline p53 mutations may have a high risk of developing a wide spectrum of malignancies.[379] The tumor suppressing activity of the p53 gene is demonstrated by the results of experiments in which expression of an introduced wild-type p53 gene was shown to be capable of suppressing the growth and/or the tumorigenic phenotype of human cancer cells, including breast cancer cells and acute lymphoblastic leukemia cells.[380,381] Expression of p53 suppresses the tumorigenicity but not the growth of human A637 cells.[382] Thus, the normal p53 gene product may be considered as a tumor suppressor protein involved in the negative regulation of cell proliferation. Oncogenic forms of p53 may interfere with the activation of DNA sequences that are recognized by the normal p53 protein.[383] Thus, the mutant p53 protein may inhibit p53-regulated gene expression.

The mechanism of the growth inhibitory effect of p53 could involve the action of negative growth factors, in particular TGF-β. There is evidence that the growth inhibitory action of p53 in thyroid follicular cells may be mediated by TGF-β. While normal thyroid epithelial cells show marked growth inhibition in response to TGF-β, malignant thyroid cells do not respond to TGF-β; this lack of response appears to be mediated by some structural and/or functional alteration of the p53 protein.[384] In turn, p53 may be involved in mediating the growth inhibitory action of TGF-β. In the MCF-7 human breast cancer cell line, TGF-β-induced inhibition of cell proliferation is associated with alterations in the phosphorylation and subcellular localization of the p53 protein.[385] Human bronchial epithelial cell clones expressing mutant p53 are more resistant to TGF-β than clones expressing wild-type p53.[386] These results suggest that p53 may function in multistage oncogenesis by reducing responsiveness of cells to negative growth factors.

B. THE *RB1* GENE

Retinoblastoma is a rare malignant tumor of the retina in infants. Both heritable and nonheritable forms of the tumor are observed, and almost all patients with bilateral tumors have an inherited disease.[387] Chromosomal fragility has been found in patients with sporadic unilateral retinoblastoma.[388] The *RB1* tumor suppressor gene is an autosomal recessive gene whose physical or functional loss results in increased susceptibility to retinoblastoma and other tumors including osteosarcomas and soft-tissue sarcomas, as well as carcinomas of lung, breast, prostate, and bladder. The human *RB1* gene encodes a 110-kDa protein, termed RB, Rb, or RB1, which is composed of 928 amino acids. The RB protein is localized in the nucleus and contains multiple phosphorylation sites.[389,390]

Expression of the *RB1* gene is regulated by both positive and negative factors. A common set of nuclear factors — the RB control proteins (RCPs) — bind to a 30-bp element, the *RB1* control element (RCE), within the *RB1* promoter and regulate *RB1* gene expression in a positive manner.[391] The RCE is also contained in the c-*fos* and c-*myc* proto-oncogenes as well as in the gene for TGF-β, and these genes are also regulated by the RCPs. On the other hand, *RB1* gene expression is negatively regulated by the p53 protein.[392] Retinoic acid and calcitriol downregulate *RB1* and c-*myc* gene expression during the myeloid or monocytic differentiation of HL-60.[393]

The RB protein has an important role in the control of the cell cycle.[394] Accumulation of RB may be required for growth arrest during differentiation of certain types of cells.[395] RB protein dissociates from the condensing chromosomes during mitosis, localizes in the cytoplasm during metaphase and anaphase, and reassociates with the chromatin of the daughter nuclei. Interphase nuclei show fine, speckled, granular distribution of RB, excepting the nucleoli. RB may regulate cell proliferation by restricting cycle progression at a specific point in G_1.[396] The c-Myc protein can abrogate G_1 phase arrest induced by RB but not by p53.[397] A marked increase in the content of nuclear RB is observed during G_0 and G_1, in comparison with cells in G_2/M phase of the cycle.[398] A high rate of mitosis and dense focus formation occurs in cultured human embryonic lung fibroblasts by treatment with antisense oligonucleotides capable of inhibiting *RB1* gene expression.[399] Wild-type p53 protein can suppress *RB1* gene transcription.[400]

The state of RB protein phosphorylation varies according to the different phases of the cell cycle.[401] RB is underphosphorylated and tightly bound to the nucleus only in the G_0 and G_1 phases of the cycle.[402] Newly synthesized RB is phosphorylated at multiple sites only at the G_1/S boundary and S phase of the cycle.[403,404] However, the patterns of RB phosphorylation may be different in different cell types; in

BV173 cells RB protein phosphorylation occurs relatively early in the G_1 phase.[405] The state of RB protein phosphorylation is critical for the regulation of the cell cycle. RB is underphosphorylated in G_0 and early G_1 and becomes hyperphosphorylated just before the onset of S phase, suggesting that RB phosphorylation results in inactivation of its suppressor function and is necessary for cells to enter the S phase of the cycle. Both protein kinases and protein phosphatases may be involved in regulating the state of RB protein phosphorylation. In nontransformed human fibroblasts (IMR-90 and WI-38 cells), the extracellular concentrations of Ca^{2+} play a critical role in the control of RB phosphorylation, whereas in transformed cells RB phosphorylation may be little affected by calcium.[406] The effects of calcium on RB protein phosphorylation are probably mediated by changes in the activity of the cdc2 kinase. The protein phosphatases PP1 and PP2A may have an important role in regulating the state of RB protein phosphorylation and, consequently, the action of the protein in the progression of the cell cycle.[407]

Hyperphosphorylation of RB protein can regulate RB interaction with the nucleus.[408] The RB protein may be a substrate of the cdc2 kinase.[409,410] This kinase is associated with the RB protein and is involved in RB protein phosphorylation *in vivo*.[411,412] However, in human T cells cdc2 kinase activation may not be required for RB phosphorylation in early phases of the cell cycle,[413] suggesting that other kinases may be involved in this phosphorylation. The cdk2 kinase, in the form of a complex with cyclin E or cyclin A, may have a role in RB protein phosphorylation.[414-416] The G_1 to S transition in higher eukaryotes may be regulated by the cdk2 kinase, whereas the G_2 to M transition is controlled by the cdc2 kinase. The cdk2 protein kinase is involved in the activation of DNA synthesis in human cells.[417] Another cyclin-dependent kinase, cdk4, and the cyclin subtypes D2 and D3 also may form stable complexes with the RB protein and may function as regulators of RB protein function in mammalian cells.[418] It is thus clear that RB functional activity depends on its phosphorylation by multiple cellular protein kinases.[419] The state of RB phosphorylation also depends on the activity of specific protein phosphatases. Further studies are required for a proper evaluation of the functional roles of RB protein complexes containing different cellular protein kinases in the regulation of the proliferation of different cell types under different physiological conditions.

The biological activity of RB depends on its interaction with distinct cellular factors. RB may act as a transducer of afferent signals (via the kinase that phosphorylates it) and efferent signals (through transcription factor binding).[420] RB protein dephosphorylation may allow the cells to enter a quiescent state.[421] Growth arrest caused by deprivation of growth factors or by induction of cell differentiation is associated with the disappearance of RB hyperphosphorylated forms. RB phosphorylation and RB mRNA and protein levels may be regulated in some types of cells by TGF-β.[422] Inhibition of cell proliferation induced by TGF-β is associated with inhibition of RB phosphorylation.[423,424] However, the accumulation of hypophosphorylated RB associated with the negative action of TGF-β on human epithelial cell growth is not due to a direct modulation of the RB phosphorylation process but is instead a consequence of the TGF-β-induced G_1 growth arrest.[425] RB activates expression of the human TGF-β$_2$ gene through the transcription factor ATF-2.[426] The possible role of TGF-β in the growth inhibitory effect of the RB protein, as well as the possible role of RB in the growth inhibitory effect of TGF-β, is unknown.[427] Interferons and IL-6 suppress RB protein phosphorylation in growth-sensitive hematopoietic cells, and this effect is independent of the reduction of c-*myc* gene expression observed in these cells.[428] Senescent cells in culture, including senescent human fibroblasts, exhibit decreased ability to phosphorylate the RB protein.[429,430]

The RB protein is involved in transcriptional regulation, and part of this action is exerted through its association with certain transcription factors. Cyclin A and the RB protein form a complex with a transcription factor.[431] RB can negatively regulate the activity of E2F, a factor involved in the transcription of several cellular and viral genes.[432,433] E2F represents a molecular link between the RB protein and viral oncoproteins.[434] A 60-kDa nuclear protein allows the RB-E2F complex to interact with DNA and to inhibit the expression of genes associated with the regulation of cell proliferation, which may result in growth arrest.[435,436] E2F can also form complexes with other proteins involved in the proliferative response, including the RB-related protein p107.[437] The p107 protein is able to form complexes with cyclin A and the cdk2 kinase, and p107 can inhibit E2F-dependent gene transcription during the G_1 phase of the cycle. Overexpression of cyclin A and/or E can override an RB-induced block in cell cycling and induce cells to enter S phase.[438] Interactions between the positive action of various cyclins and the negative action of the RB or p107 proteins may be most important for the normal regulation of the cell cycle.

The genes regulated by E2F-RB protein complexes remain little characterized, but the c-*myc* proto-oncogene may be one of these genes. Expression of the TGF-β$_1$ gene and the c-*myc* and c-*fos* proto-oncogenes may be regulated, either negatively or positively, by the RB protein, depending on the cell type.[439] An amino-terminal domain present in both the c-Myc and N-Myc proteins mediates their binding to the RB protein, suggesting that c-Myc and RB may cooperate through direct binding to control cell proliferation.[440] Genes negatively regulated by RB may include those encoding dihydrofolate reductase and DNA polymerase-α.

Transcription factors other than E2F may be regulated by the RB protein. In contrast to the negative regulation of E2F, RB protein functions as a positive regulator of IGF-II gene transcription through its interaction with the transcription factor Sp1, which recognizes the RCE motif represented by the sequence CCACCC.[441]

The RB protein may have an important role during vertebrate development. Homozygous *RB1* mouse mutants may die before the 16th day of embryonic development with multiple defects, including abnormalities in the hematopoietic and nervous systems.[442,443] Young mice heterozygous for *RB1* mutation are not abnormal and do not develop retinoblastoma. In adult mouse fetal tissues, the RB protein is not ubiquitous and is not restricted to proliferating and postmitotic cells.[444] According to the results of RB-specific immunofluorescence staining techniques, RB protein is expressed in both actively dividing cells, such as hematopoietic cells, as well as in nonproliferating cells, such as megakaryocytes and skeletal muscle cells. Terminally differentiated cells such as neurons and glial cells in the central nervous system exhibit RB protein immunostaining. RB is not expressed in the mitotically most active compartments of squamous epithelia, mucous membranes of the gastrointestinal tract, and retina, but it can be detected in the more differentiated layers of these tissues, suggesting that RB may play a role in cells committed to a certain differentiation program. RB is involved in the antiproliferative action of certain cytokines on normal and neoplastic cells. IFN-γ directly modulates RB mRNA expression in human monocytoid cells.[445]

Although it is generally believed that the RB protein functions as a tumor suppressor, the precise role of RB in tumorigenesis is not understood. At least in some cases, RB may not behave as a tumor suppressor protein. In normal human peripheral blood lymphocytes, stimulation with pokeweed mitogen (PWM) results in an early reduction in the amount of RB protein per cell, and this reduction precedes the entry of the stimulated cells into S phase.[446] EBV-transformed lymphocytes and HL-60 human leukemia cells contain elevated amounts of RB protein per cell, compared to normal peripheral blood lymphocytes. Early downregulation of RB protein expression occurs in HL-60 cells induced to differentiate along either the myeloid or the monocytic pathway. These results may be difficult to reconcile with the concept of *RB1* as a tumor suppressor gene, i.e., a gene whose loss or inactivation allows an increased proliferative activity and relatively dedifferentiated state of the cell. Stable transfection of an *RB1* gene into the human bladder carcinoma cell line HTB9, which lacks the *RB1* gene, results in growth retardation and partial loss of tumorigenic potential, but the expression of *RB1* is not sufficient to induce a complete reversion of the malignant phenotype of HTB9 cells.[447] *RB1* gene expression may fail to reverse the malignant phenotype of different types of human tumor cell lines.[448] Further studies are required for a better understanding of the role of the RB protein in normal cell physiology and neoplasia.

C. THE *WT1* GENE

Wilms' tumor, a malignant kidney tumor (nephroblastoma), is one of the most common solid tumors of childhood. The tumor is usually unilateral and can be associated with congenital abnormalities such as aniridia, genitourinary anomalies, and hemihypertrophy. Wilms' tumor is usually nonhereditary, but in a small minority of cases an autosomal dominant mode of inheritance has been observed. Two regions on the short arm of human chromosome 11 (11p13 and 11p15) may be implicated in Wilms' tumor origin and/or development. In particular, the putative tumor suppressor gene, *WT1*, located at region 11p13, may be associated with Wilms' tumor.[449,450] *WT1* is deleted or mutated in some patients with Wilms' tumor. A genomic map or the human *WT1* gene with restriction sites and positions of exons has been elaborated.[451]

The product of the *WT1* gene is a 52-kDa nuclear protein that contains four zinc fingers and a glutamine- and proline-rich amino terminus.[452-457] The WT1 protein is involved in the regulation of transcription. WT1 binds the same DNA sequence as the Egr-1 protein, which is the product of a mitogen-inducible immediate-early gene. In contrast to Egr-1, the WT1 protein is not induced by serum. The WT1 protein may function as a repressor of transcription when bound to the DNA sequences recognized by the Egr-1 protein. WT1 may be a tissue-specific antagonist of transcriptional activation of Egr-1 by hormones, growth factors, and mitogens. WT1 can act as a repressor for the expression of genes encoding growth factors such as IGF-II and PDGF-A.[458-460] These results suggest that the antiproliferative effects of tumor suppressor gene products may be exerted, at least in part, through the inhibition of growth factor gene expression. The WT1 protein is expressed at high levels in the metanephric mesenchyme.[461] WT1 may have an important role in mediating developmental processes in the kidney and other organs.

D. THE *NF1* GENE

Neurofibromatosis type 1 (von Recklinghausen disease) is a human genetic disease of the nervous system characterized by a variable phenotype that may include neurofibromas, pigmented skin lesions, and other abnormalities. In addition to benign neurofibromas, some patients develop tumors derived from the neural crest such as pheochromocytomas and malignant schwannomas. The gene *NF1*, coding for the neurofibromatosis-associated protein, has been cloned, sequenced, and assigned to human chromosome region 17q.11.2.[462] The coding sequence of *NF1* exhibits a large region of homology with the products of the yeast *IRA1* and *IRA2* genes and a smaller region of homology with the mammalian GTPase activating protein (GAP). The GAP-related domain of the NF1 protein interacts with Ras protein and displays GAP-like activity.[463] The 280-kDa NF1 protein, also called neurofibromin, may be the product of a tumor suppressor gene involved in the negative regulation of Ras proteins, which appears to have a role in neurofibromatosis-associated malignant diseases.[464] Cell lines derived from malignant schwannomas in neurofibromatosis patients exhibit high concentrations of GTP-bound Ras and low levels of neurofibromin.

E. THE *BTG1* GENE

Molecular cloning of a t(8;12)(q24;q22) translocation that occurred in a case of B-cell chronic lymphocytic leukemia (CLL) enabled the isolation of a gene, *BTG1*, which exhibits antiproliferative properties.[465] Transfection experiments showed that *BTG1* gene expression negatively regulates NIH/3T3 cell proliferation. Expression of *BTG1* is a marker of quiescent cells in the G_0-G_1 phases of the cycle in the presence of serum, and this expression is downregulated in response to optimal doses of mitogenic factors or after the suppression of cell contact inhibition, allowing the cells to progress further through the cycle. Treatment of PC12 rat pheochromocytoma cells in culture with NGF results in cessation of cell proliferation and induction of their differentiation into neuron-like cells, and this is associated with induction of *BTG1* gene expression. Sequence analyses show that *BTG1* is homologous, but not cognate, to *PC3*, an immediate-early gene induced by NGF in PC12 cells, and that it encodes a polypeptide of 171 amino acids. The precise function of the BTG1 protein is unknown.

IV. REGULATION OF ONCOPROTEIN AND TUMOR SUPPRESSOR PROTEIN EXPRESSION BY EXTRACELLULAR AGENTS

A wide diversity of endogenous and exogenous extracellular agents is involved in the regulation of proto-oncogene and tumor suppressor gene expression. The levels of expression of several proto-oncogenes, in particular those of the *myc*, *fos*, and *jun* gene families, may exhibit transient variation when resting cells are stimulated by exogenous signaling agents such as serum, mitogens, hormones, growth factors, regulatory peptides, neurotransmitters, and tumor promoters, as well as by some specific endogenous signals that may depend on cell-to-cell contact. The precise role of altered proto-oncogene expression in the mechanisms of action of signaling agents is not clear, however.

A. MITOGENIC AGENTS

In addition to hormones and growth factors, a wide variety of other extracellular agents with mitogenic potential are able to regulate the levels of expression of genes such as the proto-oncogenes and tumor suppressor genes, whose products are involved in the positive or negative control of cellular proliferation, respectively.[466] Interesting results have been obtained with mitogenically stimulated human peripheral blood cells *in vitro*. From unstimulated populations of blood cells, only granulocytes express c-*fos* gene transcripts; no such transcripts are found in monocytes and lymphocytes or in alveolar macrophages.[467] Induction of c-*myc* and c-*myb* expression occurs very rapidly following the mitogenic activation of T lymphocytes with concanavalin A and B lymphocytes with LPS.[468-471] Stimulation of T lymphocytes with the lectin mitogen PHA, the phorbol ester PMA, the calcium ionophore ionomycin, or the monoclonal antibody OKT3 (directed against antigen receptor complex) produces marked increases in the levels of c-*myc* mRNA.[472] T lymphocytes of the CD4/T4+ phenotype (helper/inducer T cells), but not the CD8/T8+ phenotype (cytotoxic/suppressor T cells) exhibit increased levels of c-*myc* transcripts after stimulation with PHA.[473] Combined stimulation of murine B lymphocytes with anti-Ig plus the cytoskeleton-perturbing agent, cytochalasin, may result in intense and prolonged c-*myc* expression.[474] The product of the putative proto-oncogene *bcl-2*, the 26-kDa Bcl-2 protein, is expressed at low levels in resting peripheral blood cells but undergoes marked increases after stimulation of the cells with mitogenic lectins and cytokines.[475] The Bcl-2 protein may be able to enhance hematolymphoid cell survival by interfering with apoptosis.

The precise role of proto-oncogene expression in relation to the onset of DNA synthesis is not understood. Expression of c-*myc* transcripts is markedly increased after stimulation of lymphocytes with concanavalin A and is downregulated at the onset of DNA synthesis.[476] A clear correlation between onset and strength of c-*myc* expression or steady-state level of c-*myc* mRNA and commitment of the cells to grow is not obvious, however. Concanavalin A-induced expression of c-*myc* in T lymphocytes occurs at the transcriptional level and requires at least two signals, one from protein kinase C and the other involving the Ca^{2+}/calmodulin system.[477] Expression of c-Myb protein occurs in PHA-stimulated T cells only after the initiation of DNA synthesis.[478] Studies on T lymphocytes stimulated with monoclonal antibodies to different epitopes of T-cell surface antigens suggest that whereas c-*fos* and early c-*myc* gene induction are responses to initial membrane signaling events, c-*myb* and late c-*myc* induction may be linked in some way to cell proliferation.[479]

PHA stimulation of normal human T cells is associated with the early accumulation of c-*fos* and c-*myc* mRNA as well as IL-2 and the IL-2 receptor mRNA.[480] Stimulation of thymocytes with concanavalin A causes a rapid increase in c-*fos* and c-*myc* mRNAs.[481] Both transcriptional and posttranscriptional mechanisms contribute to the regulation of c-*fos* and c-*myc* expression in mouse lymphocytes under basal conditions and after mitogenic stimulation with concanavalin A.[482] However, neither early c-*fos* nor c-*myc* expression is sufficient to commit the cells to DNA synthesis, which requires additional changes in gene expression. In addition to c-*myc* and c-*fos* gene expression, mitogens can also induce an increase in the levels of c-*jun* gene expression, as well as in AP-1 binding and AP-1 transcriptional activity.[483] PHA-induced expression of the c-*jun* gene, but not the c-*fos* and c-*jun* genes, is decreased in lymphocytes from elderly individuals.[484] In contrast to early-response genes such as c-*fos*, c-*myc*, and c-*jun*, other genes such as c-*myb*, N-*ras*, *dbl*-2, and the transferrin receptor genes behave as late-response genes during T-lymphocyte mitogenesis.[480,485] Transcripts of other proto-oncogenes such as c-*abl*, c-*ets*, c-*fgr*, and c-*yes* are expressed in unstimulated peripheral blood mononuclear cells.

Stimulation of nonhematic cells by mitogenic agents may also be associated with alterations in proto-oncogene expression. The deazaguanine-derivative queuine behaves as a growth factor for HeLa cells and its action is accompanied by increased expression of the c-*fos* and c-*myc* genes.[486] Stimulation of monkey kidney epithelial cells (BSC-1 cell line) with ADP, which is the most potent mitogen described for these cells, results in activation of c-H-*ras* and c-*myc* gene expression before the initiation of DNA synthesis.[487] Proto-oncogene responses may exhibit variation among different types of mitogen-stimulated cells, as demonstrated with the use of specific inhibitors. Inhibitors of the nuclear enzyme ADP-ribosyltransferase (ADPRT) block differentiation of several eukaryotic cells, including an early event in lymphocyte activation. In PHA-stimulated human lymphocytes, ADPRT inhibition completely blocks the proliferative response and the increase in c-*myc* gene expression without affecting c-*fos* significantly.[488] Conversely, in fibroblasts the serum-induced growth is not affected by ADPRT inhibitor, and both genes, c-*myc* and c-*fos*, are superinduced. Hence, there are differences between the responses of lymphocytes and fibroblasts to mitogenic stimulation, and also between the modes of regulation of c-*fos* and c-*myc* expression. A given proto-oncogene may serve different functions in different types of cells.

The action of anti-mitogenic agents in certain types of cells may be associated with altered expression of specific proto-oncogenes. Heparin, a glycosaminoglycan composed of repeating glucosamine and uronic acid sugar residues, is a potent inhibitor of serum-induced growth of VSM cells in culture, and this effect is associated with suppression of c-*myc* and c-*fos* expression.[489] The effect of heparin on proto-oncogene expression is exerted at the transcriptional level, is independent of ongoing protein synthesis, and specifically affects gene induction dependent on activation of protein kinase C.

B. TISSUE REGENERATION AND COMPENSATORY CELL GROWTH

Extracellular signaling agents such as hormones and growth factors, and intracellular agents such as oncoproteins and tumor suppressor proteins, are important for the regulation of tissue regeneration and compensatory cell growth that occurs in organs such as the liver after injury by chemical or infectious agents or after partial resection of the organ.[490] Sequential changes in the patterns of proto-oncogene expression occur in the liver following injury, partial hepatectomy, or the administration of hepatotoxic agents such as lead nitrate or carbon tetrachloride.[491-496] After partial hepatectomy in the adult rat, the steady-state levels of mRNAs for the c-*fos*, c-*myc*, N-*myc*, c-*erb*-B, c-*ets*-2, and c-*ras*, as well as those of the p53 protein, increase sequentially in the liver during the prereplicative phase that precedes DNA synthesis. The c-*fos*, c-*myc*, and N-*myc* proto-oncogenes exhibit, in a manner similar to the ODC gene, a biphasic pattern of expression during liver regeneration. Expression of the c-*jun* and *jun*-B genes is also transiently induced in both parenchymal and nonparenchymal liver cells during liver regeneration.[497] The complex interactions that occur among the products of c-*myc*, c-*fos*, c-*jun*, *jun*-B, and other early response genes may be functionally important for the regulation of delayed response genes during extended times, which is required for regeneration of the organ.[498]

The mechanisms involved in genetic regulatory alterations that occur during liver regeneration are poorly understood. Endogenous humoral factors synthesized in the liver after local injury or hepatectomy, such as the hepatocyte growth factor (HCGF), may be involved in regulating the expression of proto-oncogenes and other cellular genes at both the transcriptional and posttranscriptional levels. Circulating endogenous estrogens may have a role in liver regeneration. Administration of estradiol to ovariectomized rats induces a weak but significant increase in the proliferative activity of hepatocytes, and this effect is associated with changes in the levels of c-*myc* and c-H-*ras* gene transcripts in the liver cells that are similar to those observed after partial hepatectomy.[499] However, the physiological significance of these changes in relation to the regeneration of the liver is unknown. The transient and sequential expression of proto-oncogenes may exhibit variations in response to different proliferative stimuli, and increased expression of c-*fos* and c-*myc* may not be a necessary prerequisite for the entry of cells into the cycle.[500] Expression of c-*fos* gene is rapidly increased in rat livers reperfused after non-necrogenic ischemia, which is not followed by DNA synthesis.[501] The role of proto-oncogene expression in regenerating human liver is also not understood. In a clinical study on the levels of expression of 9 proto-oncogenes in human liver tissue from 70 patients with various non-neoplastic diseases of the liver associated with regeneration of the organ, no particular proto-oncogene was found to be expressed at high levels in any specific type or stage of the disease.[502]

Proto-oncogene expression during compensatory growth of the kidney after unilateral nephrectomy in rat and mouse differs in some aspects from the changes that occur after hepatectomy.[503,504] Enhanced expression of c-*myc* and c-*fos* may have an important role in normal and malignant renal growth.[505] Induction of c-*fos* and c-*myc* mRNAs occurs in the contralateral kidney 15 min after nephrectomy in the rat, suggesting the influence of humoral factors.[506] Expression of c-H-*ras* and c-K-*ras* increases in the contralateral kidney after nephrectomy in the mouse which suggests their participation in the process of compensatory renal growth.[507] The effects of kidney regeneration on proto-oncogene expression have been examined after folic acid administration to adult mice, which causes renal injury.[508] The folic acid-induced regenerative process is also associated with a marked increase in the levels of c-*myc* and c-*fos* gene transcripts in the kidney. It should be noted that a difference between the regenerative processes in kidney and liver is the large increase in DNA synthesis associated with liver regeneration but not with compensatory renal hypertrophy.

In general, the biological significance of altered proto-oncogene expression in liver and kidney regeneration is not understood. This expression is not tissue-specific, because c-*myc* and c-H-*ras* gene expression is also increased in the kidneys and livers of rats crosscirculated (parabiotized) with either partially or totally hepatectomized animals.[509] Thus, the early increase in proto-oncogene expression is a nonorgan-specific response to hepatectomy that does not ensure subsequent cell proliferation. The results from several studies suggest that the production and secretion of an extrahepatic serum factor, or the loss of a normally present growth inhibitory factor, may be responsible for the early increase in c-*myc* expression in regenerating rat liver. Similar results were obtained in nephrectomized rats,[510] indicating that increased proto-oncogene expression may be necessary, but not sufficient to stimulate cellular proliferation in regenerating organs.

Subtotal pancreatectomy is a stimulus capable of repressing certain genes expressed in differentiated pancreatic tissue and of derepressing other genes probably necessary for starting and/or maintaining the process of pancreatic regeneration.[511] While the concentrations of transcripts encoding amylase, chymotrypsinogen B, and trypsinogen I decrease during the pancreatic regeneration time, the proto-oncogenes c-*myc* and c-H-*ras* are overexpressed 12 to 24 and 48 h, respectively, and then return to basal levels. The concentrations of mRNAs for proinsulin and actin may also be increased after subtotal pancreatectomy.

Increased expression of specific proto-oncogenes, possibly mediated by growth factors, may have a role in the myocardial hypertrophy that occurs in response to severe hemodynamic loading.[512] Pressure-load myocardial hypertrophy may increase the expression of *myc*, *fos*, *jun*, and *ras* genes, which would be mediated by stimulation of α_1-adrenergic receptors.

Limb regeneration in Amphibia, especially in Urodeles, is a well documented biological phenomenon. Forelimb regenerative outgrowth in the young *Xenopus laevis* is strongly correlated with enhanced c-*myc* gene expression.[513] The expression of c-*myc* may play a role in regulating the continued growth of specific cells involved in the regenerative process, whereas repression of the c-*myc* gene in the epidermis correlates with terminal differentiation of keratinocytes.

C. DIETARY INFLUENCES

The levels of expression of particular proto-oncogenes may exhibit changes related to the quantity and quality of the diet. Normal components of the diet such as glucose and amino acids may influence proto-oncogene expression in the liver and other organs. Glucose may contribute to regulate the expression of c-*myc* in human cells.[514] Administration of glycine to rats results in rapid and transient induction of c-*fos*, c-*myc*, and c-H-*ras* expression in the liver.[515] Short-time fasting in the rat is associated with a marked decrease in the level of c-*myc* mRNA in the liver.[516] This effect would not depend on glucagon, because administration of glucagon leads to an increase in c-*myc* mRNA in the liver. Energy restriction in female rats results in decreased levels of c-*fos* and c-H-*ras* gene expression in the mammary gland.[517] Fasting in broiler chickens results in increased c-*myc* mRNA levels in the adipose tissue, but not in muscle and liver.[518] The levels of c-*myc* mRNA in the chicken adipose tissue are negatively correlated with serum concentrations of glucose, triglycerides, and IGF. These results suggest that proto-oncogene products may be involved in the metabolic changes that occur during energy restriction.

Amino acid starvation in both unicellular and multicellular organisms may induce a protein synthesis-dependent increase in the expression of growth-associated genes. Deprivation of a single essential amino acid in CHO cells may induce an increased expression of several genes, including the proto-oncogenes c-*myc*, c-*fos*, c-*jun*, and *jun*-B and the ODC gene.[519] Such gene inductions may represent a compensatory reaction by which cells try to continue the synthesis of certain vital proteins in unfavorable growth conditions. Deprivation of dietary protein in rats for several days followed by a meal containing protein results in an increased rate of DNA synthesis in the liver. A gradual increase in c-*myc* expression is observed in the liver of rats submitted to protein-deficient diets, and the levels of c-*myc* transcripts are abruptly decreased when the animals are fed a meal containing an appropriate amount of protein.[520]

D. GROWTH FACTORS

The action of growth factors frequently includes alterations in the levels of expression of proto-oncogenes, in particular those associated with immediate-early growth responses.[521] However, the precise role of the observed alterations in proto-oncogene expression in relation to the growth response is not apparent. Senescent cells, which are usually unable to proliferate, may also show alterations in the expression of several proto-oncogenes when they are stimulated with growth factors in culture. Both young and senescent rodent fibroblasts may respond to growth factor-induced mitogenic stimulation with a sharp increase in the levels of c-*myc* mRNA, although the senescent cells are not able to initiate DNA synthesis after the stimulation.[522] Expression of the c-*fos* gene is altered in senescent human fibroblasts.[523] Fibroblasts from progeria patients, which are characterized by diminished capacity for growth *in vitro* culture, have a decreased growth response to potent mitogens such as PDGF, and this alteration is associated with reduced levels of c-*fos* mRNA expression.[524] The c-*jun* and *jun*-B genes are expressed at normal levels in senescent WI-38 human cells.[525]

Serum is composed of a complex mixture of hormones, growth factors, mitogens, and other extracellular signaling agents, and it exerts a potent mitogenic action on certain types of cells. Characterization of the factors contained in serum that are capable of inducing a mitogenic stimulation of specific cell types can be achieved by testing the purified substances for their capacity to replace serum in the medium of cultured cells and to mimic its action. In the osteoblast-like cell line MC3T3-E1, derived from newborn mouse calvaria, serum caused a rapid and transient induction of c-*fos* and c-*jun* gene expression, and the effects of serum could be mimicked by PDGF, EGF, and basic FGF, but not IGF-I or TGF-β.[526] The ability of serum and purified growth factors to induce the transcriptional expression of proto-oncogenes in MC3T3-E1 cells correlates well with their ability to stimulate the proliferation of the cells.

Purified growth factors may have effects on the levels of expression of proto-oncogenes that are similar to those of serum. Many studies related to this aspect have been performed with cultured mouse fibroblasts. Exposure of Swiss 3T3 fibroblasts to the potent mitogen bombesin results in a rapid and concentration-dependent increase in c-*fos* and c-*myc* mRNA levels, followed by stimulation of DNA synthesis.[527-529] PDGF is a strong mitogenic agent for cultured fibroblastic cells, but the effects of PDGF depend on the conditions of the culture system and the state of the cells. In contrast to the expression of c-*fos* which occurs almost exclusively when mouse fibroblasts are in dense cultures, the PDGF-induced expression of c-*myc* occurs preferentially in sparse cell cultures.[530] Spontaneously immortalized (established) rodent fibroblasts express higher levels of c-*myc* mRNA than their mortal counterparts, but both types of cells respond equally with enhanced c-*myc* expression to growth factor stimulation. Exposure of quiescent diploid human FS-4 fibroblasts to IL-1 and/or TNF-α results in enhanced expression of c-*fos* and c-*myc* mRNA.[531] However, the biological significance of this enhancement is not understood, because a marked increase in c-*fos* and c-*myc* mRNA levels occurs when FS-4 cells are exposed to the protein synthesis inhibitor cycloheximide alone.

The kinetics and extent of proto-oncogene expression may be related to the cell type, degree of quiescence, and response to serum deprivation. Starved NIH/3T3 and HeLa S3 cells show c-*fos* mRNA induction 20 to 30 min after addition of serum but, in contrast, Swiss/3T3 cells express c-*fos* constitutively following serum starvation.[532] Primary cultures of guinea pig glandular epithelial cells respond with a transient increase in c-*fos* mRNA expression to fetal calf serum (FCS) or to estradiol plus insulin or EGF, but do not respond with this increase to estradiol, EGF, or insulin alone.[533] In primary cultures of mature rat hepatocytes, expression of the c-*myc* gene is biphasic, and the second phase of this expression depends on cell density.[534]

Stimulation of serum-deprived, quiescent NIH/3T3 cells with FCS can lead to synchronous cell cycle initiation.[535] The stimulated cells exhibit the following characteristics: (1) expression of the c-*fos* gene increases a few minutes after serum stimulation and reaches at 1 h levels that are two orders of magnitude higher than those existing before stimulation, decreasing thereafter to basal levels; (2) transcripts of the c-*myc* gene increase after 30 min of stimulation and are induced more than 20-fold 1 h after the addition of serum, followed thereafter by a slow decrease; (3) the levels of c-H-*ras* and c-K-*ras* mRNA increase at any time point no more than threefold after stimulation; (4) transcripts of c-*abl* and c-*raf* do not show significant changes in the level of expression; and (5) transcripts of c-*sis* are not detected in this cell system. When the action of purified growth factors was studied in the same system, the following results were obtained:[535] (1) Expression of c-*fos* increased markedly after stimulation with FGF or PDGF and was then abolished rapidly, whereas in cells treated with EGF the induction of c-*fos* was much less pronounced. (2) The strong initial induction of c-*myc* expression by FCS was reproduced using either FGF, PDGF, or EGF alone, but it was much more stable in FCS- or FGF-treated cells than in PDGF- or EGF-stimulated cells. (3) No changes in c-H-*ras* proto-oncogene expression were observed after stimulation with FGF, PDGF, or EGF.

The kinetics of c-*fos* and c-*myc* proto-oncogene expression induced by specific growth factors in other systems may be different from that observed in cultured mouse fibroblasts. Exposure of quiescent diploid human FS-4 fibroblasts to IL-1 or TNF-α results in increased levels of c-*fos* and c-*myc* mRNA levels within 20 min, with a peak at 30 min, followed by a decline of c-*fos* mRNA to undetectable levels and c-*myc* to basal levels by 60 or 90 min.[531] In the presence of the protein synthesis inhibitor cycloheximide, the levels of both proto-oncogene mRNAs continue to rise throughout the period of observation (90 min). The fact that the patterns of c-*fos* and c-*myc* induction by IL-1 and TNF-α are similar in FS-4 cells suggests that a common intracellular signaling pathway, shared by the two cytokines, may be responsible for the stimulation. This may not be necessarily true for other types of cells, however.

Expression of some proto-oncogenes may not be enhanced but may be reduced when target cells are exposed to the specific factor. For example, the transcriptional expression of the c-*myc* gene is reduced after exposure of HeLa cells to either IFN-γ or TNF-α, and this effect is associated with a specific arrest in the G_0/G_1 phase of the cell cycle.[536] Addition of both cytokines together enhances the inhibition of c-*myc* expression. IFN-γ and TNF-α seem to modify c-*myc* gene expression by different mechanisms. While the reduction of c-*myc* mRNA by IFN-γ is dependent on protein synthesis, the inhibitory effect of TNF-α on the c-*myc* mRNA level is direct and is not abrogated by cycloheximide.

V. MECHANISMS OF PROTO-ONCOGENE REGULATION BY GROWTH FACTORS

The transductional mechanisms involved in the regulation of proto-oncogene expression by hormones, growth factors, and other extracellular signaling agents can include activation of the adenylyl cyclase system, changes in Ras guanine nucleotide-binding activity, increased phosphoinositide turnover, changes in ion fluxes and distribution, and activation of protein kinases and protein phosphatases.[537] These mechanisms may exert either positive or negative effects on the expression of specific proto-oncogenes, and the final result may depend on a balance between the opposite influences.

The precise nature of the signals transmitted from ligand-activated surface receptors to the nucleus for the regulation of gene expression is poorly understood but the signals may result in an altered activity of specific nuclear proteins, in particular chromatin components and transcription factors. Several types of molecules may be involved in intracellular transductional mechanisms in different types of cells. One of these molecules could be represented by the intermediate filament protein vimentin. The vimentin gene belongs to the early response gene family, and its product exhibits affinity to nucleic acids and core histones.[538] Mouse vimentin is composed of 466 amino acids and comprises a non-α helical domain at both its carboxyl-and amino-terminal ends and a central rod-like domain that arises from three α-helices.[539] There are domains of homology between α helical regions of vimentin and DNA binding-leucine zipper domains of the Fos, Jun, and Raf proteins, as well as with the transcription regulator CREB protein. The possible role of vimentin in the transduction of signals between the membrane and the nucleus requires further evaluation.

The use of specific inhibitors may contribute to elucidation of the mechanisms of regulation of gene expression by growth factors and other extracellular stimuli. The application of the protein synthesis inhibitor cycloheximide in serum-stimulated human and murine cells suggests that an unstable protein is involved in modulating the expression of cell cycle-associated genes,[540] including the expression of c-*fos* and c-*myc*.[535] The effects of the protein kinase inhibitor, 2-aminopurine, suggest that a protein kinase may be involved in serum-stimulated c-*fos* and c-*myc* expression as well as in the induction of the IFN-β gene by viruses or synthetic double-stranded RNAs.[541] The use of the protein synthesis inhibitor anisomycin indicates that the regulatory mechanisms of c-*fos* and c-*myc* gene expression induced by growth factors may be different in different types of cells.[163] Thus, there are multiple levels and mechanisms through which c-*fos* and c-*myc* gene expression is regulated.

Exchange of specific types of molecules between the cytoplasm and the nucleus may play an important role in the regulation of genomic functions and the behavior of cells during the different phases of the cycle. In sparse cultures of NIH/3T3 mouse cells, two replicative proteins (DNA polymerase-α and cyclin/PCNA) and two oncoproteins (c-Myc and c-Fos) were found to be localized mainly in the nucleus.[542] When cellular proliferation was stopped by the high density of culture cells or serum starvation, these proteins left the nucleus for the cytoplasm and during serum stimulation they relocalized into the nucleus.

Regulation of transcriptional proto-oncogene expression may be mediated by transcription factors capable of recognizing specific DNA sequences.[543] Inducible DNA-binding factors may act as interme-diates in the signaling pathway, which couples surface receptors to specific regulation of gene expression within the nucleus. Transcription of the c-*fos* gene may be mediated by both negatively and positively acting cellular factors. An intracellular protein may act as a repressor of c-*fos* gene transcription throughout the cell cycle.[544,545] Transient inactivation of this repressor by a mechanism independent of protein synthesis would occur during the stimulation of cell proliferation by growth factors and other mitogens, allowing the action of a preformed positive factor which stimulates the expression of the gene.

Proteins involved in regulation of proto-oncogene expression may include oncoproteins, and some oncoproteins may function as transcription factors. Analysis of exponentially growing human tumor cell lines has shown that at mitosis there may be an asymmetrical distribution of oncoproteins such as Ras, Myc, Fos, and Erb-B2, as well as tumor suppressor proteins such as p53.[546] This uneven distribution may provide an explanation for the high degree of heterogeneity exhibited by postmitotic cells and the asynchrony in cell cycle traverse of cultured cells.

Isoprenylated proteins may have a role in the stimulation of cell proliferation induced by growth factors. Pretreatement of quiescent human fibroblasts with mevinolin, an inhibitor of hydroxymethylglutaryl coenzyme A reductase (an enzyme that catalyzes the synthesis of mevalonate, a precursor of cholesterol and several isoprenoid compounds), partially suppresses accumulation of c-*fos* and c-*myc* mRNA in response to serum stimulation.[547] Depletion of isoprenylated proteins may block cellular growth either because of their essential role in nuclear architecture (lamins) or because they are necessary for cell cycle progression (Ras).

A. MECHANISMS OF REGULATION OF c-*myc* GENE EXPRESSION

Regulation of c-*myc* gene expression in cells such as mitogen-stimulated human lymphocytes occurs at both transcriptional and posttranscriptional levels.[548] Complex mechanisms may be responsible for the regulation of c-*myc* expression at each of these two levels. Moreover, different types of cells may preferentially employ different modes of regulation of c-*myc* gene expression under different physiologic conditions.[549]

Regulation of the c-*myc* gene expression at the level of transcription depends, at least in part, on sites sensitive to S1 nuclease located upstream of the gene coding sequences. Both positive and negative mechanisms are involved in the regulation of c-*myc* expression. A protein that recognizes a specific site 290 bp 5′ to the transcription start site within the murine c-*myc* promoter may be responsible for suppressing the expression of c-*myc* in certain cell types or in particular physiological conditions.[550] Short RNA molecules may bind directly to DNA sites located within the 5′ end of the human c-*myc* gene sequences (from −270 to −1400 bp), thereby shifting the underlying conformational equilibrium and changing the level of c-*myc* gene expression.[551] In addition to transcription initiation, the transcriptional expression of c-*myc* may be regulated by an RNA elongation block within the first exon of the gene.[552]

The nuclear phosphoprotein Max has an important role in the regulation of Myc activity. The *max* gene resides on human chromosome region 14q23. Max shares with Myc the presence of binding region, HLH, and leucine zipper motif in its sequence and is able to form specific dimers with Myc. The levels of Max RNA are relatively stable and Max protein expression is modestly induced after serum stimulation of resting mouse fibroblasts and are not altered during the process of cell differentiation.[553] The inverse levels of Max and Myc protein expression in growth-arrested cells vs. proliferating cells suggest that Max may modulate Myc function in an antagonistic fashion.[50] In quiescent cells, Max may exist predominantly in the form of a homodimer, whereas in growth factor-stimulated cells it forms a heterodimer with Myc. Phosphorylation of Max may also regulate its DNA-binding ability and its dimerization with Myc. Max-Myc heterodimers may activate gene transcription in a specific manner through the recognition of DNA elements (E boxes) containing the central CACGTG sequence. In quiescent cells there are little amounts of Myc protein and high amounts of Max protein, and Max would be bound to E boxes but not capable of activating transcription. The Max protein would thus act as a repressor of the expression of genes containing E box sequences in their promoter. In growth factor-stimulated cells, the concentration of Myc-Max heterodimers increases, which leads to activation of the transcriptional expression of genes containing E boxes.

An important level of regulation of c-*myc* gene expression is represented by posttranscriptional mechanisms.[554] c-*myc* gene transcripts are extremely unstable in various types of normal and transformed cells. The c-*myc* gene is transcribed at a high rate in G_0-arrested hamster lung fibroblasts, but mature c-*myc* mRNA is present only at barely detectable levels in these cells. Stimulation of cells by growth factors is not always accompanied by any appreciable change in c-*myc* gene transcription rate, suggesting that the expression of c-*myc* is frequently regulated at the level of mRNA degradation.[555] Regulation of c-*myc* mRNA stability may depend on the action of labile, nucleic acid-containing cytosolic factors that are not associated with polysomes.[556] In addition, the expression of c-*myc* may be regulated at the translational level. The 5' noncoding region of mouse c-*myc* mRNA has a negative effect on its translational efficiency, which is related to a restrictive element confined to a 240-nucleotide sequence that acts by a *cis* mechanism.[557] The regulation of c-*myc* gene expression in a transfection system in which introduced c-*myc* genes exhibit serum-responsive activity indicates that the responsiveness is not mediated by increased initiation of transcription but is rather effected at a point between transcription and stabilization of the c-*myc* RNA.[558] It is thus evident that mechanisms acting at multiple levels are responsible for the regulation of c-*myc* expression.

B. MECHANISMS OF REGULATION OF c-*fos* GENE EXPRESSION

Expression of c-*fos* is regulated by positive and negative factors at both the transcriptional and posttranscriptional levels, depending on the type of cell and the predominant physiologic conditions.[559] Protein kinase activation and phosphorylation of specific cellular proteins at tyrosine and/or serine/threnonine residues may be necessary for the induction of c-*fos* gene expression by growth factors.

Regulation of c-*fos* expression at the level of transcription is due to the specific interaction of nuclear proteins with three nucleotide sequences located at the 5'-flanking region of the c-*fos* gene.[560,561] One of these regions is a site exhibiting hypersensitivity to DNase I in both human and mouse genes and its deletion eliminates induction of c-*fos* expression by serum. This sequence may be necessary for the induction of c-*fos* gene transcription by growth factors. The other two DNA sequences would contribute to basal c-*fos* gene promoter activity. An enhancer element composed of a 20-bp sequence with dyad symmetry, located from positions –317 to –298 of the c-*fos* gene, relative to the mRNA cap site in the 5'-flanking region, is recognized with high affinity by a 68-kDa protein, the c-*fos* enhancer-binding protein, which has been purified from HeLa cell nuclear extracts.[562]

Induction of c-*fos* expression by EGF requires the presence of the c-*fos* gene enhancer as well as other controlling elements contained in the gene sequences.[561] CREs are contained in the transcribed and nontranscribed regions of the c-*fos* gene. Three elements contained in the 5'-flanking side of the gene (CRE, AP-1, and SRE) are constitutively occupied *in vivo*.[563] Transient accumulation of c-*fos* mRNA after serum stimulation requires not only conserved 5' sequences of the c-*fos* gene but also c-*fos* 3' sequences.[564] Induction of c-*fos* by serum or growth factors may include reversible changes in c-*fos* chromatin structure, with increased sensitivity to DNase I.[565]

The SRE contained within the c-*fos* promoter is recognized by transcription factors.[566] A 63- to 67-kDa protein, the serum response factor (SRF), specifically recognizes the SRE and mediates the regulation of c-*fos* expression exerted by serum factors.[148] In serum-stimulated mouse fibroblasts, newly synthesized SRF is translocated to the nucleus, where it undergoes posttranslational modifications, including phosphorylations, capable of modifying the affinity of SRF for the specific DNA sequences represented by the SRE. There is a synergism between the SRF protein and a p62 factor, the ternary complex factor (TCF), in their interaction with the SRE.[567] The TCF does not bind directly to SRE sequences but interacts with the SRF protein. Activation of the SRF depends on its phosphorylation by protein kinase C, which in turn is associated with phospholipase C-dependent generation of diacylglycerol.[568] Casein kinase II can also activate c-*fos* gene expression through the phosphorylation of the SRF protein.[569]

In addition to the SRF, other nuclear proteins of molecular weights between 36 and 112 kDa can interact with the SRE DNA sequence.[570] Transcriptional expression of the c-*fos* gene may depend on the negative effect exerted by a short-lived repressor protein. Treatment of mouse peritoneal macrophages with the protein synthesis inhibitor cycloheximide results in a very rapid and marked increase in the levels of c-*fos* mRNA.[571] This type of control may be specific for c-*fos* and does not operate with other proto-oncogenes expressed in macrophages, including c-*myc*, c-*sis*, and c-*fms*. It is thus clear that multiple factors with either positive or negative effects are specifically involved in the regulation of c-*fos* gene expression at the transcriptional level.

In addition to regulation at the level of transcription, c-*fos* expression may also be regulated at posttranscriptional levels. The growth-promoting effect of PDGF on 3T3 mouse fibroblasts depends not only on its effect at the level of c-*fos* gene transcription, but also on an increased posttranslational stability of the nascent c-Fos protein.[572] Growth factors such as NGF and EGF may regulate the function of c-Fos through phosphorylation of the protein at the Ser-362 residue, which is located on a site at the carboxyl terminus of the molecule that is critically implicated in the capacity of c-Fos to exhibit *trans*-repressive activity in the regulation of gene expression.[573] Phosphorylation of the c-Fos protein at Ser-362 is mediated by a particular type of protein kinase.

C. INTRACELLULAR MEDIATORS INVOLVED IN THE REGULATION OF c-*fos* AND c-*myc* GENE EXPRESSION

Regulation of proto-oncogene expression by growth factors and other extracellular signaling agents can occur through different intracellular signaling pathways. Expression of the c-*fos* gene in macrophages and fibroblasts may depend on increased production of cAMP. The direct activator of adenylyl cyclase, forskolin, and the cAMP analog, 8-bromo-cAMP, enhance c-*fos* gene expression in 3T3 fibroblasts.[574] A similar activation is obtained with arachidonic acid and PGE_2, as well as with PDGF.

Monovalent ions may have a role as mediators in the regulation of proto-oncogene expression by growth factors. Activation of the Na^+/H^+ antiporter may result in c-*fos* expression in U937 cells.[575] However, activation of the Na^+/H^+ antiporter and intracellular alkalinization may not be required for the induction of c-*fos* expression with other types of serum-stimulated cells, for example, in VSM cells.[576]

In some cell types, the regulation of proto-oncogene expression by growth factors and other extracellular agents includes activation of phosphoinositide turnover, generation of 1,2-diacylglycerol, and subcellular Ca^{2+} mobilization. In 3T3 cells, PDGF induces Ca^{2+} mobilization and c-*myc* gene expression, and a similar stimulation of c-*myc* expression is induced by the calcium ionophores A23187 and ionomycin.[577] Ca^{2+} may serve as a messenger for c-myc gene expression induced by PDGF and other growth factors in certain cell types, which may involve stimulation of phosphoinositide turnover to activate protein kinase C and to generate inositol trisphosphate, the latter stimulating intracellular mobilization of Ca^{2+}.[578] Expression of c-*fos* in fibroblasts and macrophages is activated by growth factor-induced breakdown of cellular phospholipids with ensuing activation of protein kinase C.[579,580] Treatment of U-937 human leukemia cells with the protein kinase C agonist, 1,2-diacylglycerol, results in transient expression to high levels of the c-*fos* gene.[581] Induction of c-*fos* gene expression in serum-deprived 3T3-L1 fibroblasts and adipocytes stimulated by serum, hormones, growth factors, or phorbol ester may occur by two separate pathways, one involving protein kinase C activation and the other independent of protein kinase C.[582] In certain cell types, growth factor-induced expression of proto-oncogenes may occur independently of phosphoinositide turnover.[583]

An intact cytoskeleton as well as membrane ruffling may be required for the stimulation of proto-oncogene expression by growth factors. Treatment of cultured mouse fibroblast cells with a mixture of EGF and insulin results in accumulation of c-*fos* mRNA, and this accumulation is partially inhibited by cytochalasin D, an inhibitor of cell motility.[584] It thus may be concluded that different growth factors can activate the levels of expression of particular proto-oncogenes through different molecular mechanisms.

VI. PHYSIOLOGICAL SIGNIFICANCE OF GROWTH FACTOR-REGULATED PROTO-ONCOGENE EXPRESSION

A rapid induction of the *fos*, *myc*, and *jun* genes, which code for nuclear proteins, occurs in certain cells stimulated by growth factors,[585] suggesting that the products of these genes may be involved in the control of cell proliferation and/or cell differentiation.[586] In some cases, the expression of these proto-oncogenes may be associated with both cell proliferation and cell differentiation. Expression of the c-*myc* gene in rat and chicken growth plate chondrocytes occurs during both the proliferation and differentiation of these cells.[587] However, in many cases there is a clear dissociation between cell proliferation and/or cell differentiation and the expression of proto-oncogenes encoding nuclear proteins. Expression of the c-Myc protein is not always correlated with either cell proliferation or cell differentiation. Myc protein expression in the nuclei of hematopoietic cells is usually not correlated with cellular proliferation.[588] In human peripheral blood lymphocytes stimulated with BCGF, Myc expression is not, by itself, sufficient for B-cell progression into the S phase.[589] Exposure of MG-63 human osteosarcoma cells to PDGF results in accumulation of c-*myc* mRNA, but this is not followed by DNA synthesis and cell proliferation.[590] Inhibition of the proliferation of MEL cells by the inhibitor ODC activity, difluoromethylornithine (DFMO), is not associated with detectable changes in c-*myc* expression.[591] The depletion of spermidine by DFMO also has no clear effects on the biphasic changes of c-*myc* mRNA that are observed during differentiation of MEL cells. It is thus clear that Myc protein expression is frequently, but not always, correlated with either cell differentiation or cell proliferation and that the physiologic role of Myc as well as its mechanism of action remain to be elucidated. However, there may be little doubt that the mechanism of action of growth factors is in some way associated with Myc expression.

The importance of proto-oncogene protein products localized in the nucleus (nuclear oncoproteins) in cell proliferation and differentiation, in particular during embryogenesis, is indicated by the results of experiments with nonfunctional (null) alleles of the genes encoding these proteins. Null c-*myc* gene alleles obtained by homologous recombination in embryonic stem cells (ES cells) have been used to generate heterozygous and homozygous c-*myc* mutant ES cell lines.[592] The null alleles are not able to encode functional c-Myc protein. Mouse chimeras from heterozygous ES cell lines are able to transmit the mutant allele to their offspring, and the mutation is lethal in homozygotes between 6.5 and 10.5 days postconception. Female heterozygotes have reduced fertility owing to embryonic resorption before 9.5 days postconception. Mice carrying null mutations of the c-*fos* gene are viable, although the genetic defect is associated with pleiotropic phenotypic effects.[172-174]

A. NUCLEAR ONCOPROTEINS AND CELLULAR PROLIFERATION

The levels of expression of proto-oncogenes encoding nuclear proteins may be higher in rapidly proliferating cells than in quiescent cells. In many biological systems, both *in vivo* and *in vitro*, there is a correlation between the expression of *myc*, *fos*, *jun*, and/or *ets* genes and cellular proliferative activity. However, cellular proliferation is not universally associated with altered expression of these genes, and induction of their expression does not necessarily lead to cell proliferation.

The c-*myc* gene may have an important function in relation to cell proliferation. Expression of c-*myc* may correlate with the rate of proliferative activity in normal and transformed cells in culture. An essential role for c-*myc* expression in the processes leading to cell proliferation is suggested by the growth suppressive effects of c-*myc*-specific antisense oligonucleotides in normal and neoplastic cells.[593-596] The inhibitory effects of antisense oligodeoxynucleotides targeting c-*myc* mRNA suggest that the c-*myc* gene product is important for the signal transduction pathways mediating smooth muscle cell proliferation and migration.[597] Even small perturbations of c-*myc* expression caused by disruption of one copy of the c-*myc* gene and introduction of a c-*myc* transgene into Rat-1 cells results in changes in the growing status of the cells.[598] The sharp induction of c-*myc* expression following growth factor stimulation may be a prerequisite for a normal transition to the S phase of the cycle. The mechanism by which the c-Myc protein may influence cell proliferation is not understood, but it could be related to its association with the RB tumor suppressor protein. The amino-terminal domain present in both the c-Myc and N-Myc proteins mediates RB protein binding,[599] suggesting that Myc proteins may stimulate cell proliferation, at least in part, by inactivating the growth inhibitory action of RB protein. However, the precise relationship between cellular proliferation and c-*myc* expression *in vivo* is not understood. In certain tissues there is no correlation between cell proliferation and c-*myc* expression. The study of c-*myc* expression in normal human intestinal epithelium by *in situ* molecular hybridization and immunohistochemistry showed that c-*myc* is expressed uniformly throughout the entire thickness of the intestine epithelium and is not confined to proliferating cells.[600] Expression of c-*myc* mRNA and protein in the adult mouse testis is related to the cell cycle only in spermatogonia and not in spermatocytes and spermatids, suggesting that c-*myc* may play a role in somatic and germ cell division but not in meiotic cell division.[601]

In many different cellular systems, inhibition of DNA synthesis and proliferation and induction of differentiation is associated with decreased c-*myc* gene expression. TNF-α can inhibit the proliferation and induce the differentiation of HL-60 human leukemia cells, and this effect is associated with decreased expression of c-*myc* at the transcriptional level.[602] A cause-effect relationship between the decreased expression of c-*myc* and the inhibition of DNA synthesis and the induction of cell differentiation could not exist, however. In HL-60 cells induced to differentiation by calcitriol, inhibition of DNA synthesis precedes downregulation of c-*myc* expression.[603] Moreover, calcitriol directly inhibits DNA synthesis in isolated nuclei of HL-60 cells independently of changes in c-*myc* expression. CSF-1-induced proliferation of the murine macrophage cell line BAC1.2F5 is mediated by cAMP and is accompanied by reduced expression of c-*myc* mRNA, but this reduction is neither necessary nor sufficient for the inhibition of BAC1.2F5 cell growth.[604] Overexpression of an exogenous c-*myc* gene is not sufficient to initiate DNA synthesis in primary cultures of rat hepatocytes.[605] The mechanism by which IFN-β slows the growth rate of human lung cancer cell lines does not involve downregulation of the c-*myc* gene, which is amplified and overexpressed in these cells.[606] In different human B-cell lines, downregulation of c-*myc* gene expression was not found to be a prerequisite for the reduction of cell proliferation induced by TNF-α or a monoclonal antibody to an MHC antigen.[607] Expression of c-myc is induced in human thymocytes by agents that act as mitogens (IL-2, TPA, and concanavalin A), but the activation of c-*myc* gene expression is elicited by the binding of the mitogen or growth factor to its receptor, rather than by the growth response itself.[608] Expression of Myc protein in the nuclei of human hematopoietic cells is unrelated to cell proliferation.[75] PCC7 embryonal carcinoma cells do not express c-*myc* transcripts even at an exponential rate of growth.[609] It is thus clear that a correlation between c-*myc* expression and cell proliferation is not observed in many different cellular systems *in vitro* and *in vivo*.

The expression of other proto-oncogenes encoding nuclear proteins may also not be required for, or may not correlate with, cellular proliferation, as shown by the following examples. Insulin-dependent re-entry of myoblasts into the cycle is not preceded by accumulation of c-*fos* mRNA.[610] Human MCF-7 breast cancer cells show a proliferative response to IGF-I and EGF, but the levels of expression of the c-*myc*, c-*fos*, and c-*jun* do not correlate with the increase of growth stimulation, and the expression of the proto-oncogenes is not growth rate limiting for the stimulated MCF-7 cells.[611] Inhibition of the growth of human leukemic cell lines by treatment with IFN-α is not necessarily associated with alteration in the levels of expression of proto-oncogenes encoding nuclear proteins.[612] Interferon-induced inhibition of the growth of normal human fibroblasts occurs without a reduction in the transcriptional expression of the c-*fos* and c-*myc* proto-oncogenes.[613] Treatment of human lymphocytes with phorbol ester or calcium ionophore results in induction of c-*fos* and c-*myc* gene expression, but this induction is not sufficient by itself to commit the cells to DNA synthesis.[614] Stimulation of the P388D1 murine macrophage cell line with TPA, calcium ionophore, or CSF-1 results in induction of c-*fos* and c-*myc* gene expression, but no direct correlation is observed between this expression and the entry of cells into the cycle and the synthesis of DNA.[615]

It has been postulated that transient induction of c-*fos* and c-*myc* gene expression in quiescent stimulated cells would make the cells competent for entering the cycle. At least a part of the population of WI-38 cells is rendered capable of initiating DNA synthesis when they express an inducible construct containing a c-*fos* gene.[616] However, persistence of the competent state may be independent of c-*fos* and c-*myc* gene expression. Murine fibroblasts can remain competent in spite of expressing c-*fos* and c-*myc* at very low levels (i.e., at levels similar to those of quiescent cells), suggesting that the products of these genes are not directly involved in the maintenance of competence.[617] Cells rendered competent by PDGF may not require high levels of c-*fos* and c-*myc* gene expression to progress through G_1.[618] Anti-Ig-induced crosslinking of surface Ig on murine B cells causes a rapid and transient increase in the level of c-*fos* mRNA, which is associated not with stimulation but with inhibition of cell proliferation.[619] Rat epithelial liver cells overexpressing the c-*fos* gene exhibit increased sensitivity to the growth-promoting effect of EGF and to the growth-inhibiting effect of TGF-β.[620] Thus, the effects of the c-Fos protein on cell proliferation would depend on the type of cell, the level of c-Fos expression, and the environmental conditions created by specific growth factors.

Expression of the genes of the Jun family is frequently, but not always, correlated with cellular proliferation. Studies on primary cultures of dog thyrocytes stimulated by different pathways (TSH stimulating the cAMP pathway, diacylglycerol and TPA stimulating the protein kinase C pathway, and EGF stimulating the tyrosine kinase pathway) indicate that the c-*jun* and *jun*-D genes can be differentially regulated and that the expression of c-*jun* is not universally correlated with the stimulation of cell proliferation. Downregulation of c-*jun* expression is observed in dog thyrocytes stimulated to proliferate with TSH.[621]

The action of different mitogenic stimuli in different types of cells may result in the expression of different genes, including some proto-oncogenes. Comparison of the effects of concanavalin A and IL-2 on the expression of c-*myc*, c-*fos*, and c-*myb* in cloned T lymphocytes indicates that these mitogens induce expression of overlapping, but not identical sets of proto-oncogenes.[622] Accumulation of c-*myc* mRNA is induced by both concanavalin A and IL-2, expression of c-*fos* mRNA is stimulated by concanavalin A, and increase of c-*myb* mRNA is induced predominantly by IL-2. Although the precise role of proto-oncogenes in T-cell functions remains to be defined, they apparently are not directly involved in the commitment of T cells to proliferation: expression of c-*myb* does not occur in IL-2-stimulated T cells, IL-2 does not induce accumulation of c-*fos* mRNA in the same cells, and concanavalin A-induced c-*myc* accumulation in the cells is insufficient to induce cellular growth. Even the combined expression of c-*fos* and c-*myc* genes may be insufficient to induce proliferation in T cells stimulated by mitogens. On the other hand, expression of the c-*ras* proto-oncogene is required for G_1 phase progression as well as for G_2 phase transit and proliferation of certain types of cells.[623] It is thus clear that an effective mitogenic response may require the cooperative expression of different types of cellular genes, including some proto-oncogenes. However, these genes may not necessarily be the same in all types of cells.

B. NUCLEAR ONCOPROTEINS AND CELLULAR DIFFERENTIATION

Multiple lines of evidence indicate that the expression of proto-oncogenes encoding nuclear oncoproteins is associated in some systems not with cell proliferation but with cell differentiation. Extracellular agents capable of inducing or modulating the processes of cell differentiation frequently induce changes in the levels of expression of these genes. Sodium butyrate, an agent capable of modifying the program of cell differentiation, alters the expression of c-*myc* and c-*fos* in HTC rat hepatoma cells.[624] Okadaic acid, an inhibitor of protein phosphatases, is capable of inducing the differentiation of certain human leukemic cell lines, and this effect is accompanied by enhanced c-*jun* expression.[625] The effects of growth factors on cell differentiation may depend, at least in part, on alterations of proto-oncogene expression.

The c-*myc* proto-oncogene may have an important role in some developmental processes. Two distinct c-*myc* genes (c-*myc*-I and c-*myc*-II) are contained in the genome of *Xenopus laevis*, and they are differentially regulated during developmental processes occurring in this amphibian species, with one c-*myc* gene active in oocytes and the other active in both oocytes and postgastrula embryos.[626] Expression of the c-Myc protein may play an important role in the differentiation of specific organs and tissues in other vertebrates. Selected examples are mentioned in the following. Explants of the central region of embryonic chicken lens epithelia can be induced to differentiate into lens fiber cells when cultured in the presence of fetal calf serum, insulin, IGF, or vitreous humor (which contains IGF). The levels of c-*myc* transcripts are transiently elevated in the differentiating lens epithelial explants in the absence of mitogenic stimulation, as the cells withdraw from the cell cycle.[627] Differentiation of SHE myoblasts *in vitro* is associated with enhanced expression of the c-*myc* gene.[628] Myc proteins can induce the expression of a differentiated phenotype in quail neural crest cells, with the appearance of catecholaminergic traits.[629] The Myc protein has been detected in mature mammalian sperm cells, including human sperm cells, and it may have a role in sperm cell function, especially in capacitation and/or acrosome reaction.[630]

The precise relationship between the expression of nuclear oncoproteins and the processes of cell differentiation is unknown. In some cellular systems c-*myc* gene expression is not associated with cell differentiation but with dedifferentiation. Establishment of mammalian thyroid follicular cells in monolayer culture is associated with an acute period of dedifferentiation in which an increased expression of c-*myc* is observed.[631] Expression of c-*myc* is dissociated from cell differentiation in certain systems. The mechanism controlling differentiation operates independently of the level of c-*myc* expression in F9 teratocarcinoma cells.[632] Antisense oligodeoxynucleotides specific to c-*myc* or N-*myc* sequences cause growth inhibition and limited differentiation of NIE-115 mouse neuroblastoma cells, but their effects are not coupled to neurite formation.[633] Antisense c-*myc*-specific oligomer inhibits the proliferation but does not induce the differentiation of cultured human keratinocytes,[634] indicating that cell proliferation and differentiation are not necessarily coupled in some systems. Persistent c-*myc* expression is not required for growth factor-mediated inhibition of differentiation of the nonfusing mouse myogenic cell line BC_3H_1.[635] Exposure of differentiated BC_3H_1 cells to either FGF or TGF-β causes the disappearance of muscle-specific gene products, and this effect is accompanied by only transient, low-level induction of c-*myc* mRNA.

In many systems, especially in the differentiation of tumor cells induced by specific agents *in vitro*, cell differentiation is not associated with upregulation but with downregulation of c-*myc* gene expression. Experiments using c-*myc*-specific antisense RNA suggest that downregulation of c-*myc* expression may be sufficient for inducing differentiation in F9 mouse teratocarcinoma cells.[636] However, the possible role of downregulation of c-*myc* expression in the induced differentiation of neoplastic cells is not clear. Although most frequently the induction of HL-60 human leukemia cell differentiation is associated with downregulation of c-*myc* expression,[637] the differentiation of HL-60 cells induced by IFN-γ is not associated with altered levels of c-*myc* mRNA.[638]

Nuclear oncoproteins other than c-Myc, or other proteins, may have a role in the processes of differentiation of different types of cells. Late downregulation of the c-*myb* gene precedes monocytic differentiation.[639] Induction of HL-60 cell differentiation into monocytoid cells by the glutamine antagonist acivicin is associated with decreased expression of both c-*myc* and c-*myb* genes, as well as with induction of TNF-α and IL-1β, suggesting that these two cytokines may have an autocrine role in the process of differentiation.[640]

Expression of the c-*myc* gene in different types of cells may depend on the state of cell differentiation. Growth factor-induced expression of the c-*myc* gene in human colon carcinoma cell lines depends on their state of differentiation.[641] In quiescent, well-differentiated cell lines, stimulation of the cells with a combination of EGF, insulin, and transferrin results in upregulation of c-*myc* expression. In contrast, quiescent poorly differentiated cells show altered regulation of c-*myc* expression and only minimal differences in c-*myc* regulation by nutrients and exogenous growth factors. Because the proliferation of well-differentiated, but not poorly differentiated, colon carcinoma cell lines can be regulated by growth stimulatory factors, these results indicate that differences in the mitogenic responses and regulation of c-*myc* expression by exogenous hormones and growth factors are related to the differentiation status of the cells.

The c-*fos* gene may participate, in association with other genes, in regulatory events related to the expression of cellular differentiation.[642-644] Studies on different types of systems (fetal membranes and placenta, bone marrow, fetal liver, differentiated macrophages) suggest a correlation between c-*fos* expression and cell differentiation.[645-647] Expression of c-*fos* may be related to epithelial cell differentiation, specifically in relation to the process of cornification and cell death during keratinization.[648]

As it is true for c-*myc*, the precise relationship between c-*fos* gene expression and cell differentiation remains elusive. At least in some cases, expression of c-*fos* is associated more with the performance of a differentiated function than with an active process of cell proliferation or differentiation. Expression of c-*fos* in amnion cells may be regulated by factors derived from the placenta and the embryo, but proliferation of the amnion cells is not dependent on high levels of c-*fos* gene expression, suggesting that the function of the c-*fos* gene product is more likely to be associated with cellular functions other than proliferation in the differentiated amnion cells.[649]

The differentiation of tumor cells induced by growth factors and other agents may be associated with enhanced expression of c-*fos* and other proto-oncogenes.[650] Rapid increase in c-*fos* expression occurs when PC12 rat pheochromocytoma cells are induced to differentiate by treatment with NGF.[651] This induction is enhanced in the presence of peripherally active benzodiazepines. Constitutive expression of the c-*fos* gene may inhibit NGF-induced differentiation of PC12 cells, indicating that deregulated expression of c-*fos* may interfere with the normal role of NGF in neuronal differentiation.[652] Expression of c-*fos* is induced when quiescent terminally differentiated macrophages are treated with CSF-1, although the kinetics of c-*fos* induction in this system are different from those in growth factor-stimulated fibroblasts, supporting the view that the c-Fos protein may serve different functions in different cell types. Induction of c-*fos* is not observed when HL-60 cells are induced to differentiation by either calcitriol or retinoic acid,[653] nor is it observed in F9 mouse teratocarcinoma cells induced to differentiate by treatment with retinoic acid and dibutyryl cAMP,[171] which indicates that c-*fos* gene induction is not universally related to cell differentiation processes.

Expression of c-*fos* may be neither sufficient nor obligatory for the differentiation of some cell types, for example, for the differentiation of monomyelocytes to macrophages.[581] Histamine-induced differentiation of bovine chromaffin cells to a neuronal phenotype is repressed by protein kinase C but does not depend on increased levels of expression of the c-Fos protein.[654] Expression of c-*fos* induced by different stimuli (phorbol ester, calcium ionophore, diacylglycerol, cholera toxin, LPS, dexamethasone) in peritoneal macrophages is independent of both cell proliferation and expression of differentiated functions.[572,575]

Treatment of peritoneal macrophages with bacterial LPS alters the expression of both c-*fos* and c-*myc*.[655] On the other hand, IFN-β and IFN-γ, which are potent macrophage activators eliciting tumoricidal activity, may not alter the levels of c-*fos* mRNA.[575]

The c-*myb* proto-oncogene may have an important role in the differentiation of certain types of cells, in particular hematopoietic cells. Studies with *lpr/lpr* homozygous mice, which spontaneously develop massive lymphoproliferation and an associated lupus-like autoimmune disease, have shown an inverse relationship between c-*myb* gene expression and cell proliferation or differentiation.[656] The abnormal lymphocytes from *lpr* mice express increased levels of c-*myb* mRNA, and these levels are decreased after treatment *in vitro* with phorbol ester or calcium ionophore, which is concomitantly associated with activation of IL-2 receptor gene expression and progression of the T cells through the cell cycle. Expression of c-*myb* is regulated mainly at the level of transcription in untreated T cells, but the gene is negatively regulated via posttranscriptional mechanisms in the cells treated with phorbol ester or calcium ionophore.

C. UNSPECIFIC STIMULATION OF PROTO-ONCOGENE EXPRESSION

The physiological significance of the expression of proto-oncogenes encoding nuclear proteins may be obscured by the fact that a wide diversity of apparently unspecific stimuli can induce large changes in these genes. In particular, the results of experiments performed in cells cultured *in vitro* should be interpreted with caution. Wounding of fibroblast monolayer cultures may result in a rapid induction of c-*fos* expression in the absence of any growth factor.[657-659] Short-term primary cultures of rat hepatocytes express high levels of c-*myc* transcripts either in the absence or presence of a growth factor such as EGF, which is capable of inducing DNA synthesis in these cells.[660] A greater level of c-*myc* gene expression in cultured cells may be due to changes occurring during cell isolation or may represent an adaptive response to culture condition.

Exposure of murine and human cells in culture to stressors such as heat shock or sodium arsenite results in transient increase of c-*fos* mRNA.[661] Human lymphoid cells exposed to heat shock *in vitro* express increased levels of c-*myc*, c-*fos*, and c-*jun* mRNA, which may be involved in the cellular response to acute stress.[662] Exposure of HeLa cells to hypergravity results in high levels of expression of c-*myc* mRNA, followed by reduction of G_1 duration, DNA synthesis, and increased rate of cell proliferation.[663] Hypogravity decreases EGF-induced c-*fos* expression in A431 cells, while hypergravity increases this expression.[664,665] Exposure of HL-60 cells to an electromagnetic field of 45 Hz for 20 min results in a fourfold increase in the levels of c-*myc* and H2B histone transcripts.[666] Expression of the c-*fos* gene is induced by exposure to a static magnetic field in HeLa S3 cells.[667] Similar changes may occur by the action of mechanical or chemical stimuli *in vivo*.

Altered levels of c-*fos* gene expression occur in various types of cells subjected to apparently unspecific stimuli. The mere dissociation of rat hepatocytes may result in increased c-*fos* gene expression.[668] Activation of c-*fos* expression occurs when tissue slices or isolated lobes of murine submandibular glands are incubated at 37°C in physiologic solutions or in tissue culture media.[669] Dissociation of the cells with collagenase-hyaluronidase treatment also results in increased steady-state levels of c-*fos* mRNA without any additional stimulus. The synthetic chemotactic peptide, fMet-Leu-Phe, induces expression of c-*fos* mRNA in purified preparations of human peripheral blood granulocytes.[670]

The levels of proto-oncogene expression may be altered in isolated organs or intact animals subjected to different types of noxious and non-noxious stimuli. The mere increase in portal flow can induce c-*myc* gene expression in isolated perfused rat liver.[671] Nonmitogenic signals such as insulin-induced hypoglycemia *in vivo* and exposure to nicotine or angiotensin *in vitro* induce rapid and transient expression of c-*fos* mRNA in adrenal medullary cells.[672] The proto-oncogenes c-*fos*, c-*jun*, *jun*-B, and *egr*-1, but not c-*myc* and *fos*-B, are induced in porcine myocardium subjected to ischemia.[673] Considerable expression of the same genes also occurs in the nonischemic area of the heart, suggesting that it may be nonspecific or unrelated to ischemia itself. Induction of c-Fos protein is observed in the spinal cord of the rat at various periods of time following section of the sciatic nerve.[674] Intraperitoneal injection of hypertonic saline solutions to rats can induce c-*fos* expression in regions of the forebrain that are not necessarily involved in osmotic regulation. Moreover, injection of isotonic saline injections may have similar effects, suggesting that c-*fos* expression may result from the stress of handling and injecting the animals. Circadian rhythmicity in c-*fos* gene expression is observed in cortical adrenal cells of intact, nonstressed adult rats.[675] Thus, different types of exogenous and endogenous stimuli that do not appear to have a common characteristic are capable of inducing c-*fos* gene expression in different types of cells. It is conceivable, however, that such dissimilar stimuli may act through some final common pathway at the intracellular level, resulting in altered gene expression, including c-*fos* induction.

In conclusion, expression of proto-oncogenes encoding nuclear proteins is frequently associated with, and probably required for, the proliferation or differentiation of certain types of cells. However, in some cases, this expression may be dissociated from both cell proliferation and cell differentiation. Although it is clear that nuclear oncoproteins are involved in the control of genomic functions, their precise roles in the regulation of cell proliferation and cell differentiation processes by growth factors and other extracellular signaling agents is unknown. The role of growth factors in the regulation of tumor suppressor protein expression, and the role of tumor suppressor proteins in the complex processes associated with cell differentiation and cell proliferation is only beginning to be characterized.

REFERENCES

1. **Pimentel, E.,** *Hormones, Growth Factors, and Oncogenes,* CRC Press, Boca Raton, FL, 1987.
2. **Sinkovits, J. G.,** Oncogenes and growth factors, *CRC Crit. Rev. Immunol.,* 8, 217, 1988.
3. **Pierce, J. H.,** Oncogenes, growth factors and hematopoietic cell transformation, *Biochim. Biophys. Acta,* 989, 179, 1989.
4. **Travali, S., Koniecki, J., Petralia, S., and Baserga, R.,** Oncogenes in growth and development, *FASEB J.,* 4, 3209, 1990.
5. **Lemoine, N. R.,** Growth factors and oncogenes, *J. Pathol.,* 168, 419, 1992.
6. **Pimentel, E.,** *Oncogenes,* 2nd ed., Volumes I and II, CRC Press, Boca Raton, FL, 1989.
7. **Rollins, B. J. and Stiles, C. D.,** Regulation of c-*myc* and c-*fos* proto-oncogene expression by animal cell growth factors, *In Vitro Cell. Dev. Biol.,* 24, 81, 1988.
8. **Sherr, C. J.,** Growth factor receptors and cell transformation, *Mol. Biol. Med.,* 4, 1, 1987.
9. **Pimentel, E.,** Oncogenes and human cancer, *Cancer Genet. Cytogenet.,* 14, 347, 1985.
10. **McKenzie, S. J.,** Diagnostic utility of oncogenes and their product in human cancer, *Biochim. Biophys. Acta,* 1072, 193, 1991.
11. **Neckers, L., Whitesell, L., Rosolen, A., and Geselowitz, D. A.,** Antisense inhibition of oncogene expression, *Crit. Rev. Oncogenesis,* 3, 175, 1992.
12. **Dang, C. V. and Lee, W. M. F.,** Nuclear and nucleolar targeting sequences of c-*erb*-A, c-*myb*, N-*myc*, p53, HSP70, and HIV *tat* proteins, *J. Biol. Chem.,* 264, 18019, 1989.
13. **Meek, D. W. and Street, A. J.,** Nuclear protein phosphorylation and growth control, *Biochem. J.,* 287, 1, 1992.
14. **Kaczmarek, L.,** Protooncogene expression during the cell cycle, *Lab. Invest.,* 54, 365, 1986.
15. **Denhardt, D. T., Edwards, D. R., and Parfett, C. L. J.,** Gene expression during the mammalian cell cycle, *Biochim. Biophys. Acta,* 865, 83, 1986.
16. **Ferrari, S. and Baserga, R.,** Oncogenes and cell cycle genes, *BioEssays,* 7, 9, 1987.
17. **Calabretta, B., Kaczmarek, L., Mars, W., Ochoa, D., Gibson, C. W., Hirschhorn, R. R., and Baserga, R.,** Cell-cycle-specific genes differentially expressed in human leukemias, *Proc. Natl. Acad. Sci. U.S.A.,* 82, 4463, 1985.
18. **Kaczmarek, L., Calabretta, B., and Baserga, R.,** Expression of cell-cycle-dependent genes in phytohemagglutinin-stimulated human lymphocytes, *Proc. Natl. Acad. Sci. U.S.A.,* 82, 5375, 1985.
19. **Rittling, S. R. and Baserga, R.,** Regulatory mechanisms in the expression of cell cycle dependent genes, *Anticancer Res.,* 7, 541, 1987.
20. **Schönthal, A.,** Nuclear protooncogene products: fine-tuned components of signal transduction pathways, *Cell. Signal.,* 2, 215, 1990.
21. **Bhat, N. K., Fisher, R. J., Fujiwara, S., Ascione, R., and Papas, T. S.,** Temporal and tissue-specific expression of mouse *ets* genes, *Proc. Natl. Acad. Sci. U.S.A.,* 84, 3161, 1987.
22. **Bull, P., Hunter, T., and Verma, I. M.,** Transcriptional induction of the murine c-*rel* gene with serum and phorbol-12-myristate-13-acetate in fibroblasts, *Mol. Cell. Biol.,* 9, 5239, 1989.
23. **Cooper, S.,** The continuum model and c-*myc* synthesis during the division cycle, *J. Theor. Biol.,* 135, 393, 1988.
24. **Okuda, A., Masuzaki, A., and Kimura, G.,** Increase in c-*fos* and c-*myc* mRNA levels in untransformed and SV40-transformed 3Y1 fibroblasts after addition of serum: its relationship to the control of initiation of S phase, *Exp. Cell Res.,* 185, 258, 1989.
25. **McMahon, S. B. and Monroe, J. G.,** Role of primary response genes in generating cellular responses to growth factors, *FASEB J.,* 6, 2707, 1992.
26. **Bohmann, D.,** Transcription factor phosphorylation: a link between signal transduction and the regulation of gene expression, *Cancer Cell,* 2, 337, 1990.

27. **Vandenbunder, B., Pardanaud, L., Jaffredo, T., Mirabel, M. A., and Stéhelin, D.,** Complementary patterns of expression of c-*ets* 1, c-*myb*, and c-*myc* in the blood forming system of the chick embryo, *Development,* 107, 265, 1989.

28. **Zimmerman, K. and Alt, F. W.,** Expression and function of *myc* family genes, *Crit. Rev. Oncogenesis,* 2, 75, 1990.

29. **Dang, C. V.,** c-Myc oncoprotein function, *Biochim. Biophys. Acta,* 1072, 103, 1991.

30. **De Pinho, R. A., Schreiber-Agus, N., and Alt, F. W.,** *myc* family of oncogenes in the development of normal and neoplastic cells, *Adv. Cancer Res.,* 57, 1, 1991.

31. **Kato, G. J. and Dang, C. V.,** Function of the c-Myc oncoprotein, *FASEB J.,* 6, 3065, 1992.

32. **Marcu, K. B., Bossone, S. A., and Patel, A. J.,** Myc function and regulation, *Annu. Rev. Biochem.,* 61, 809, 1992.

33. **Meichle, A., Philipp, A., and Eilers, M.,** The functions of Myc proteins, *Biochim. Biophys. Acta,* 1114, 129, 1992.

34. **Kelly, K. and Siebenlist, U.,** The regulation and expression of c-*myc* in normal and malignant cells, *Annu. Rev. Immunol.,* 4, 317, 1986.

35. **Cole, M. D.,** The *myc* oncogene: its role in transformation and differentiation, *Annu. Rev. Genet.,* 20, 361, 1986.

36. **Studzinski, G. P., Shankavaram, U. T., Moore, D. C., and Reddy, P. V.,** Association of c-Myc protein with enzymes of DNA replication in high molecular weight fractions from mammalian cells, *J. Cell. Physiol.,* 147, 412, 1991.

37. **Xu, L., Morgenbesser, S. D., and DePinho, R. A.,** Complex transcriptional regulation of *myc* family gene expression in the developing mouse brain and liver, *Mol. Cell. Biol.,* 11, 6007, 1991.

38. **Gupta, S., Seth, A., and Davis, R. J.,** Transactivation of gene expression by Myc is inhibited by mutation at the phosphorylation sites Thr-58 and Ser-62, *Proc. Natl. Acad. Sci. U.S.A.,* 90, 3216, 1993.

39. **Lüscher, B., Kuenzel, E. A., Krebs, E. G., and Eisenman, R. N.,** Myc oncoproteins are phosphorylated by casein kinase II, *EMBO J.,* 8, 1111, 1989.

40. **Seth, A., Alvarez, E., Gupta, S., and Davis, R. J.,** A phosphorylation site located in the NH$_2$-terminal domain of c-Myc increases transactivation of gene expression, *J. Biol. Chem.,* 266, 23521, 1991.

41. **Kerkhoff, E., Bister, K., and Klempnauer, K.-H.,** Sequence-specific DNA binding by Myc proteins, *Proc. Natl. Acad. Sci. U.S.A.,* 88, 4323, 1991.

42. **Halazonetis, T. D. and Kandil, A. N.,** Determination of the c-Myc DNA-binding site, *Proc. Natl. Acad. Sci. U.S.A.,* 88, 6162, 1991.

43. **Halazonetis, T. D. and Kandil, A. N.,** Predicted structural similarities of the DNA binding domains of c-Myc and endonuclease Eco RI, *Science,* 255, 464, 1992.

44. **Dang, C. V., McGuire, M., Buckmire, M., and Lee, W. M. F.,** Involvement of the "leucine zipper" region in the oligomerization and transforming activity of human c-Myc protein, *Nature,* 337, 664, 1989.

45. **Landschulz, W. H., Johnson, P. F., and McKnight, S. L.,** The leucine zipper: a hypothetical structure common to a new class of DNA binding proteins, *Science,* 240, 1759, 1988.

46. **Kato, G. J., Lee, W. M. F., Chen, L., and Dang, C.,** Max: functional domains and interaction with c-Myc, *Genes Dev.,* 6, 81, 1992.

47. **Davis, L. J. and Halazonetis, T. D.,** Both the helix-loop-helix and the leucine zipper motifs of c-Myc contribute to its dimerization specificity with Max, *Oncogene,* 8, 125, 1993.

48. **Ferré-D'Amaré, A. R., Prendergast, G. C., Ziff, E. B., and Burley, S. K.,** Recognition by Max of its cognate DNA through a dimeric b/HLH/Z domain, *Nature,* 363, 38, 1993.

49. **Billaud, M., Isselbacher, K. J., and Bernards, R.,** A dominant-negative mutant of Max that inhibits sequence-specific DNA binding of Myc, *Proc. Natl. Acad. Sci. U.S.A.,* 90, 2739, 1993.

50. **Gu, W., Cechova, K., Tassi, V., and Dalla Favera, R.,** Opposite regulation of gene transcription and cell proliferation by c-Myc and Max, *Proc. Natl. Acad. Sci. U.S.A.,* 90, 2935, 1993.

51. **Amati, B., Brooks, M. W., Levy, N., Littlewood, T. D., Evan, G. I., and Land, H.,** Oncogenic activity of the c-Myc protein requires dimerization with Max, *Cell,* 72, 233, 1993.

52. **Kitaura, H., Galli, I., Taira, T., Iguchi-Ariga, S. M. M., and Ariga, H.,** Activation of c-*myc* promoter by c-*myc* protein in serum-starved cells, *FEBS Lett.,* 290, 247, 1991.

53. **Thompson, C. B., Challoner, P. B., Neiman, P. E., and Groudine, M.,** Levels of c-*myc* oncogene mRNA are invariant throughout the cell cycle, *Nature,* 314, 363, 1985.

54. **Hann, S. R., Thompson, C. B., and Eisenman, R. N.,** c-*myc* oncogene protein synthesis is independent of the cell cycle in human and avian cells, *Nature,* 314, 366, 1985.

55. **Barker, K. A. and Newburger, P. E.,** Relationships between the cell cycle and the expression of c-*myc* and transferrin receptor genes during induced myeloid differentiation, *Exp. Cell Res.,* 186, 1, 1990.

56. **Mohamed, A. N., Nakeff, A., Mohammad, R. M., KuKuruga, M., and Al-Katib, A.,** Modulation of c-*myc* oncogene expression by phorbol ester and interferon gamma: appraisal by flow cytometry, *Oncogene,* 3, 429, 1988.

57. **Neckers, L. M., Tsuda, H., Weiss, E., and Pluznik, D. H.,** Differential expression of c-*myc* and the transferrin receptor in G_1 synchronized M1 myeloid leukemia cells, *J. Cell. Physiol.,* 135, 339, 1988.

58. **Billings, P. C., Shuin, T., Lillehaug, J., Miura, T., Roy-Burman, P., and Landolph, J. R.,** Enhanced expression and state of the c-*myc* oncogene in chemically and X-ray-transformed C3H/10T1/2 Cl 8 mouse embryo fibroblasts, *Cancer Res.,* 47, 3643, 1987.

59. **Phillips, N. E. and Parker, D. C.,** Fc gamma receptor effects on induction of c-*myc* mRNA expression in mouse B lymphocytes by anti-immunoglobulin, *Mol. Immunol.,* 24, 1199, 1987.

60. **Schweinfest, C. W., Fujiwara, S., Lau, L. F., and Papas, T. S.,** c-*myc* can induce expression of G_0/G_1 transition genes, *Mol. Cell. Biol.,* 8, 3080, 1988.

61. **Shi, Y., Glynn, J. M., Guilbert, L. J., Cotter, T. G., Bissonnette, R. P., and Green, D. R.,** Role for c-*myc* in activation-induced apoptotic cell death in T cell hybridomas, *Science,* 257, 212, 1992.

62. **Evan, G. I., Wyllie, A. H., Gilbert, C. S., Littlewood, T. D., Land, H., Brooks, M., Waters, C. M., Penn, L. Z., and Hancock, D. C.,** Induction of apoptosis in fibroblasts by c-Myc protein, *Cell,* 69, 119, 1992.

63. **Wagner, A. J., Small, M. B., and Hay, N.,** Myc-mediated apoptosis is blocked by ectopic expression of Bcl-2, *Mol. Cell. Biol.,* 13, 2432, 1993.

64. **Rao, G. N. and Church, R. L.,** Regulation of expression of c-*myc* protooncogene in a clonal line of mouse lens epithelial cells by serum growth factors, *Exp. Cell Res.,* 183, 140, 1989.

65. **Götschl, M. and Eick, D.,** A stably transfected c-*myc* cat hybrid gene is not regulated by serum in NIH3T3 cells, *Biochem. Biophys. Res. Commun.,* 157, 379, 1988.

66. **Persson, H., Gray, H. E., and Godeau, F.,** Growth-dependent synthesis of c-*myc*-encoded proteins: early stimulation by serum factors in synchronized mouse 3T3 cells, *Mol. Cell. Biol.,* 5, 2903, 1985.

67. **Dean, M., Levine, R. A., Ran, W., Kindy, M. S., Sonenshein, G. E., and Campisi, J.,** Regulation of c-*myc* transcription and mRNA abundance by serum growth factors and cell contact, *J. Biol. Chem.,* 261, 9161, 1986.

68. **Ferrari, S., Calabretta, B., Battini, R., Cosenza, S. C., Owen, T. A., Soprano, K. J., and Baserga, R.,** Expression of c-*myc* and induction of DNA synthesis by platelet-poor plasma in human diploid fibroblasts, *Exp. Cell Res.,* 174, 25, 1988.

69. **Friedrich, B., Gullberg, M., and Lundgren, E.,** Uncoupling of c-*myc* mRNA expression from G_1 events in human T lymphocytes, *Anticancer Res.,* 8, 23, 1988.

70. **Buckler, A. J., Rothstein, T. L., and Sonenshein, G. E.,** Two-step stimulation of B lymphocytes to enter DNA synthesis: synergy between anti-immunoglobulin antibody and cytochalasin on expression of c-*myc* and a G_1-specific gene, *Mol. Cell. Biol.,* 8, 1371, 1988.

71. **Studzinski, G. P., Brelvi, Z. S., Feldman, S. C., and Watt, R. A.,** Participation of c-Myc protein in DNA synthesis of human cells, *Science,* 234, 467, 1986.

72. **Gutierrez, C., Guo, Z.-S., Farrell-Towt, J., Ju, G., and DePamphilis, M. L.,** c-*myc* protein and DNA replication: separation of c-*myc* antibodies from an inhibitor of DNA synthesis, *Mol. Cell. Biol.,* 7, 4594, 1987.

73. **Heikkila, R., Schwab, G., Wickstrom, E., Loke, S. L., Pluznik, D. H., Watt, R., and Neckers, L. M.,** A c-*myc* antisense oligodeoxynucleotide inhibits entry into S phase but not progress from G_0 to G_1, *Nature,* 328, 445, 1987.

74. **Spector, D. L., Watt, R. A., and Sullivan, N. F.,** The v- and c-*myc* oncogene proteins colocalize *in situ* with small nuclear ribonucleoprotein particles, *Oncogene,* 1, 5, 1987.

75. **Dang, C. V. and Lee, W. M. F.,** Identification of the human c-Myc protein nuclear translocation signal, *Mol. Cell. Biol.,* 8, 4048, 1988.

76. **Waitz, W. and Loidl, P.,** Cell cycle dependent association of c-Myc protein with the nuclear matrix, *Oncogene,* 6, 29, 1991.

77. **Prendergast, G. C. and Cole, M. D.,** Posttranscriptional regulation of cellular gene expression by the c-*myc* oncogene, *Mol. Cell. Biol.,* 9, 124, 1989.

78. **Lech, K., Anderson, K., and Brent, R.,** DNA-bound Fos proteins activate transcription in yeast, *Cell,* 52, 179, 1988.

79. **Kaddurah-Daouk, R., Greene, J. M., Baldwin, A. S., Jr., and Kingston, R. E.,** Activation and repression of mammalian gene expression by the c-Myc protein, *Genes Dev.,* 1, 347, 1987.

80. **Kaczmarek, L., Hyland, J., Watt, R., Rosenberg, M., and Baserga, R.,** Microinjected c-*myc* as a competence factor, *Science,* 228, 1313, 1985.

81. **Sorrentino, V., Drozdoff, V., McKinney, M. D., Zeitz, L., and Fleissner, E.,** Potentiation of growth factor activity by exogenous c-*myc* expression, *Proc. Natl. Acad. Sci. U.S.A.,* 83, 8167, 1986.

82. **Madsen, M. W., Lykkesfeldt, A. E., Laursen, I., Nielsen, K. V., and Briand, P.,** Altered gene expression of c-*myc*, epidermal growth factor receptor, transforming growth factor-α, and c-*erb*-B2 in an immortalized human breast epithelial cell line, HMT-3522, is associated with decreased growth factor requirements, *Cancer Res.,* 52, 1210, 1992.

83. **Ramsay, G., Stanto, L., Schwab, M., and Bishop, J. M.,** Human proto-oncogene N-*myc* encodes nuclear proteins that bind DNA, *Mol. Cell. Biol.,* 6, 4450, 1986.

84. **Wenzel, A., Cziepluch, C., Hamann, U., Schürmann, J., and Schawab, M.,** The N-Myc oncoprotein is associated *in vivo* with the phosphoprotein Max(p20/22) in human neuroblastoma cells, *EMBO J.,* 10, 3703, 1991.

85. **Hiller, S., Breit, S., Wang, Z. Q., Wagner, E. F., and Schwab, M.,** Localization of regulatory elements controlling human *MYCN* expression, *Oncogene,* 6, 969, 1991.

86. **Sejersen, T., Rahm, M., Szabo, G., Ingvarsson, S., and Sümegi, J.,** Similarities and differences in the regulation of N-*myc* genes in murine embryonal carcinoma cells, *Exp. Cell Res.,* 172, 304, 1987.

87. **Stanton, B. R. and Parada, L. F.,** The N-*myc* proto-oncogene: developmental expression and *in vivo* site-directed mutagenesis, *Brain Pathol.,* 2, 71, 1992.

88. **Mugrauer, G., Alt, F. W., and Ekblom, P.,** N-*myc* proto-oncogene expression during organogenesis in the developing mouse as revealed by *in situ* hybridization, *J. Cell Biol.,* 107, 1325, 1988.

89. **Hirning, U., Schmid, P., Schulz, W. A., Rettenberger, G., and Hameister, H.,** A comparative analysis of N-*myc* and c-*myc* expression and cellular proliferation in mouse organogenesis, *Mechan. Dev.,* 33, 119, 1991.

90. **Negroni, A., Scarpa, S., Romeo, A., Ferrari, S., Modesti, A., and Raschellà, G.,** Decrease of proliferation rate and induction of differentiation by a N-*myc* antisense DNA oligomer in a human neuroblastoma cell line, *Cell Growth Differ.,* 2, 511, 1991.

91. **Harris, L. L., Talian, J. C., and Zelenka, P. S.,** Contrasting patterns of c-*myc* and N-*myc* expression in proliferating, quiescent, and differentiating cells of the embryonic chicken lens, *Development,* 115, 813, 1992.

92. **Moens, C. B., Auerbach, A. B., Conlon, R. A., Joyner, A. L., and Rossant, J.,** A targeted mutation reveals a role for N-*myc* in branching morphogenesis in the embryonic mouse lung, *Genes Dev.,* 6, 691, 1992.

93. **Stanton, B. R., Perkins, A. S., Tessarollo, L., Sassoon, D. A., and Parada, L. F.,** Loss of N-*myc* function results in embryonic lethality and failure of the epithelial component of the embryo to develop, *Genes Dev.,* 6, 2235, 1992.

94. **Wetherall, N. T. and Vogler, L. B.,** Human lymphocyte precursors do not express the N-*myc* gene, *Pathobiology,* 60, 87, 1992.

95. **Brondyk, W. H., Boeckman, F. A., and Fahl, W. E.,** N-*myc* oncogene enhances mitogenic responsiveness of diploid human fibroblasts to growth factors but fails to immortalize, *Oncogene,* 6, 1269, 1991.

96. **De Greve, J., Battey, J., Fedorko, J., Birrer, M., Evan, G., Kaye, F., Sausville, E., and Minna, J.,** The human L-*myc* gene encodes multiple nuclear phosphoproteins from alternatively processed mRNAs, *Mol. Cell. Biol.,* 8, 4381, 1988.

97. **Saksela, K.,** Expression of the L-*myc* gene is under positive control by short-lived proteins, *Oncogene,* 1, 291, 1987.

98. **DePinho, R. A., Hatton, K. S., Tesfaye, A., Yancopoulos, G. D., and Alt, F. W.,** The human *myc* gene family: structure and activity of L-*myc* and an L-*myc* pseudogene, *Genes Dev.,* 1, 1311, 1987.

99. **Ingvarsson, S., Asker, C., Axelson, H., Klein, G., and Sümegi, J.,** Structure and expression of B-*myc*, a new member of the *myc* gene family, *Mol. Cell. Biol.,* 8, 3168, 1988.

100. **Resar, L. M. S., Dolde, C., Barrett, J. F., and Dang, C. V.,** B-Myc inhibits neoplastic transformation and transcriptional activation by c-Myc, *Mol. Cell. Biol.,* 13, 1130, 1993.

101. **Mugrauer, G. and Ekblom, P.,** Contrasting expression patterns of three members of the *myc* family of protooncogenes in the developing and adult mouse kidney, *J. Cell Biol.,* 112, 13, 1991.

102. **Hirvonen, H., Mäkelä, T. P., Sandberg, M., Kalimo, H., Vuorio, E., and Alitalo, K.,** Expression of the *myc* proto-oncogenes in developing human fetal brain, *Oncogene,* 5, 1787, 1990.

103. **Shen-Ong, G. L. C.,** The *myb* oncogene, *Biochim. Biophys. Acta,* 1032, 39, 1990.

104. **Bading, H., Gerdes, J., Schwarting, R., Stein, H., and Moelling, K.,** Nuclear and cytoplasmic distribution of cellular Myb protein in human haematopoietic cells evidenced by monoclonal antibody, *Oncogene,* 3, 257, 1988.

105. **Nomura, N., Takahashi, M., Matsui, M., Ishii, S., Date, T., Sasamoto, S., and Ishizaki, R.,** Isolation of human cDNA clones of *myb*-related genes, A-*myb* and B-*myb*, *Nucleic Acids Res.,* 16, 11075, 1988.

106. **Thompson, C. B., Challoner, P. B., Neiman, P. E., and Groudine, M.,** Expression of the c-*myb* proto-oncogene during cellular proliferation, *Nature,* 319, 374, 1986.

107. **Reddy, C. D. and Reddy, E. P.,** Differential binding of nuclear factors to the intron 1 sequences containing the transcriptional pause site correlates with c-*myb* expression, *Proc. Natl. Acad. Sci. U.S.A.,* 86, 7326, 1989.

108. **Bender, T. P., Thompson, C. B., and Kuehl, W. M.,** Differential expression of c-*myb* mRNA in murine B lymphomas by a block to transcription elongation, *Science,* 237, 1473, 1987.

109. **Watson, R. J.,** A transcriptional arrest mechanism involved in controlling constitutive levels of mouse c-*myb* mRNA, *Oncogene,* 2, 267, 1988.

110. **Shen-Ong, G. L. C.,** Alternative internal splicing in c-*myb* RNAs occurs in normal and tumor cells, *EMBO J.,* 6, 4035, 1987.

111. **Kim, W.-K. and Baluda, M. A.,** Hematopoietic lineage-specific heterogeneity in the 5′-terminal region of the chicken proto-*myb* transcript, *Mol. Cell. Biol.,* 9, 3771, 1989.

112. **Slamon, D. J., Boone, T. C., Murdock, D. C., Keith, D. E., Press, M. F., Larson, R. A., and Souza, L. M.,** Studies of the human c-*myb* gene and its product in human acute leukemias, *Science,* 233, 347, 1986.

113. **Bading, H., Hansen, J., and Moelling, K.,** Selective DNA binding of the human cellular Myb protein isolated by immunoaffinity chromatography using a monoclonal antibody, *Oncogene,* 1, 395, 1987.

114. **Howe, K. M. and Watson, R. J.,** Nucleotide preferences in sequence-specific recognition of DNA by c-Myb protein, *Nucleic Acids Res.,* 19, 3913, 1991.

115. **Kalkbrenner, F., Guehmann, S., and Moelling, K.,** Transcriptional activation by human c-*myb* and v-*myb* genes, *Oncogene,* 5, 657, 1990.

116. **Gabrielsen, O. S., Sentenac, A., and Fromageot, P.,** Specific DNA binding by c-Myb: evidence for a double helix-turn-helix-related motif, *Science,* 253, 1140, 1991.

117. **Simons, M., Morgan, K. G., Parker, C., Collins, E., and Rosenberg, R. D.,** The proto-oncogene c-*myb* mediates an intracellular calcium rise during the late G_1 phase of the cell cycle, *J. Biol. Chem.,* 268, 627, 1993.

118. **Kastan, M. B., Slamon, D. J., and Civin, C. I.,** Expression of protooncogene c-*myb* in normal human hematopoietic cells, *Blood,* 73, 1444, 1989.

119. **Caracciolo, D., Venturelli, D., Valtieri, M., Peschle, C., Gewirtz, A. M., and Calabretta, B.,** Stage-related proliferative activity determines c-*myb* functional requirements during normal human hematopoiesis, *J. Clin. Invest.,* 85, 55, 1990.

120. **Yanagisawa, H., Nagasawa, T., Kuramochi, S., Abe, T., Ikawa, Y., and Todokoro, K.,** Constitutive expression of exogenous c-*myb* gene causes maturation block in monocyte-macrophage differentiation, *Biochim. Biophys. Acta,* 1088, 380, 1991.

121. **Gewirtz, A. M. and Calabretta, B.,** A c-*myb* antisense oligodeoxynucleotide inhibits normal human hematopoiesis *in vitro*, *Science,* 242, 1303, 1988.

122. **Gewirtz, A. M., Anfossi, G., Venturelli, D., Valpreda, S., Sims, R., and Calabretta, B.,** G_1/S transition in normal human T-lymphocytes requires the nuclear protein encoded by c-*myb*, *Science,* 245, 181, 1989.

123. **Stern, J. B. and Smith, K. A.,** Interleukin-2 induction of T-cell G_1 progression and c-*myb* expression, *Science,* 233, 203, 1986.

124. **Brown, K. E., Kindy, M. S., and Sonenshein, G. E.,** Expression of the c-*myb* proto-oncogene in bovine vascular smooth muscle cells, *J. Biol. Chem.,* 267, 4625, 1992.

125. **Reiss, K., Travali, S., Calabretta, B., and Baserga, R.,** Growth regulated expression of B-*myb* in fibroblasts and hematopoietic cells, *J. Cell. Physiol.,* 148, 338, 1991.

126. **Foos, G., Grimm, S., and Klempnauer, K.-H.,** Functional antagonism between members of the *myb* family: B-*myb* inhibits v-*myb*-induced gene activation, *EMBO J.,* 11, 4619, 1992.

127. **Watson, R. J., Robinson, C., and Lam, E. W.-F.,** Transcription by murine B-*myb* is distinct from that by c-*myb*, *Nucleic Acids Res.,* 21, 267, 1993.

128. **Golay, J., Capucci, A., Arsura, M., Castellano, M., Rizzo, V., and Introna, M.,** Expression of c-*myb* and B-*myb*, but not A-*myb*, correlates with proliferation in human hematopoietic cells, *Blood,* 77, 149, 1991.

129. **Lam, E. W.-F., and Watson, R. J.,** An E2F-binding site mediates cell cycle regulated repression of mouse B-*myb* transcription, *EMBO J.,* 12, 2705, 1993.

130. **Cohen, D. R. and Curran, T.,** The structure and function of the *fos* proto-oncogene, *Crit. Rev. Oncogenesis,* 1, 65, 1989.

131. **Sambucetti, L. C. and Curran, T.,** The Fos protein complex is associated with DNA in isolated nuclei and binds to DNA cellulose, *Science,* 234, 1417, 1986.

132. **Di Francesco, P. and Liboi, E.,** Role of the c-*fos* gene expression on the mitogenic response in EL2 rat fibroblasts, *Int. J. Tissue React.,* 10, 311, 1988.

133. **Verma, I. M. and Sassone-Corsi, P.,** Proto-oncogene *fos*: complex but versatile regulation, *Cell,* 51, 513, 1987.

134. **Sassone-Corsi, P., Sisson, J. C., and Verma, I. M.,** Transcriptional autoregulation of the proto-oncogene *fos*, *Nature,* 334, 314, 1988.

135. **Lucibello, F. C., Neuberg, M., Hunter, J. B., Jenuwein, T., Schuermann, M., Wallich, R., Stein, B., Schönthal, A., Herrlich, P., and Müller, R.,** Transactivation of gene expression by Fos protein: involvement of a binding site for the transcription factor AP-1, *Oncogene,* 3, 43, 1988.

136. **Chiu, R., Boyle, W. J., Meek, J., Smeal, T., Hunter, T., and Karin, M.,** The c-Fos protein interacts with c-*jun*/AP-1 to stimulate transcription of AP-1 responsive genes, *Cell,* 54, 541, 1988.

137. **Zerial, M., Toschi, L., Ryseck, R.-P., Schuermann, M., Müller, R., and Bravo, R.,** The product of a novel growth factor activated gene, *fos* B, interacts with JUN proteins enhancing their DNA binding activity, *EMBO J.,* 8, 805, 1989.

138. **Rauscher, F. J., III, Sambucetti, L. C., Curran, T., Distel, R. J., and Spiegelman, B. M.,** Common DNA binding site for Fos protein complexes and transcription factor AP-1, *Cell,* 52, 471, 1988.

139. **Franza, B. R., Jr., Rauscher, F. J., III, Josephs, S. F., and Curran, T.,** The Fos complex and Fos-related antigens recognize sequence elements that contain AP-1 binding sites, *Science,* 239, 1150, 1988.

140. **Bravo, R., Burckhardt, J., Curran, T., and Müller, R.,** Expression of c-*fos* in NIH3T3 cells is very low but inducible throughout the cell cycle, *EMBO J.,* 5, 695, 1986.

141. **Okuda, A., Matsuzaki, A., and Kimura, G.,** Transient increase in the c-*fos* mRNA level after change in culture condition from serum absence to serum presence and after cycloheximide addition in rat 3Y1 fibroblasts, *Biochem. Biophys. Res. Commun.,* 159, 501, 1989.

142. **Nishikura, K. and Murray, J. M.,** Antisense RNA of proto-oncogene c-*fos* blocks renewed growth of quiescent 3T3 cells, *Mol. Cell. Biol.,* 7, 639, 1987.

143. **Deschamps, J., Meijlink, F., and Verma, I. M.,** Identification of a transcriptional enhancer element upstream from proto-oncogene *fos*, *Science,* 230, 1174, 1985.

144. **Hayes, T. E., Kitchen, A. M., and Cochran, B. H.,** A rapidly inducible DNA-binding activity which binds upstream of the c-*fos* proto-oncogene, *J. Cell. Physiol.,* Suppl. 5, 63, 1987.

145. **Härtig, E., Loncarevic, I. F., Büscher, M., Herrlich, P., and Rahmsdorf, H. J.,** A new cAMP response element in the transcribed region of the human c-*fos* gene, *Nucleic Acids Res.,* 19, 4153, 1991.

146. **Rivera, R. M., Sheng, M., and Greenberg, M. E.,** The inner core of the serum response element mediates both the rapid induction and subsequent repression of c-*fos* transcription following serum stimulation, *Genes Dev.,* 4, 255, 1990.

147. **Gustafson, T. A., Taylor, A., and Kedes, L.,** DNA bending is induced by a transcription factor that interacts with the human c-*fos* and α-actin promoters, *Proc. Natl. Acad. Sci. U.S.A.,* 86, 2162, 1989.

148. **Misra, R. P., Rivera, V. M., Wang, J. M., Fan, P.-D., and Greenberg, M. E.,** The serum response factor is extensively modified by phosphorylation following its synthesis in serum-stimulated fibroblasts, *Mol. Cell. Biol.,* 11, 4545, 1991.

149. **Leung, S. and Miyamoto, N. G.,** Point mutational analysis of the human c-*fos* serum response factor binding site, *Nucleic Acids Res.,* 17, 1177, 1989.

150. **König, H., Ponta, H., Rahmsdorf, U., Büscher, M., Schönthal, A., Rahmsdorf, H. J., and Herrlich, P.,** Autoregulation of *fos*: the dyad symmetry element as the major target of repression, *EMBO J.,* 8, 2559, 1989.

151. **Schröter, H., Shaw, P. E., and Nordheim, A.,** Purification of intercalator-released p67, a polypeptide that interacts specifically with the c-*fos* serum response element, *Nucleic Acids Res.,* 15, 10145, 1987.

152. **Shaw, P. E., Schröter, H., and Nordheim, A.,** The ability of a ternary complex to form over the serum response element correlates with serum inducibility of the human c-*fos* promoter, *Cell,* 56, 563, 1989.

153. **Herrera, R. E., Shaw, P. E., and Nordheim, A.,** Occupation of the c-*fos* serum response element *in vivo* by a multi-protein complex is unaltered by growth factor induction, *Nature,* 340, 68, 1989.

154. **König, H.,** Cell-type specific multiprotein complex formation over the c-*fos* serum response element *in vivo*: ternary complex formation is not required for the induction of c-*fos, Nucleic Acids Res.,* 19, 3607, 1991.

155. **Gilman, M. Z.,** The c-*fos* serum response element responds to protein kinase C-dependent and -independent signals but not to cyclic AMP, *Gene Dev.,* 2, 394, 1988.

156. **Berkowitz, L. A., Riabowol, K. T., and Gilman, M. Z.,** Multiple sequence elements of a single functional class are required for cyclic AMP responsiveness of the mouse c-*fos* promoter, *Mol. Cell. Biol.,* 9, 4272, 1989.

157. **Lucibello, F. C., Ehlert, F., and Müller, R.,** Multiple interdependent regulatory sites in the mouse c-*fos* promoter determine basal level transcription: cell type-specific effects, *Nucleic Acids Res.,* 19, 3583, 1991.

158. **Liboi, E., Di Francesco, P., Gallinari, P., Testa, U., Rossi, G. B., and Peschle, C.,** TGFβ induces a sustained c-*fos* expression associated with stimulation or inhibition of cell growth in EL2 or NIH 3T3 fibroblasts, *Biochem. Biophys. Res. Commun.,* 151, 298, 1988.

159. **Bonnieu, A., Rech, J., Jeanteur, P., and Fort, P.,** Requirements for c-*fos* mRNA down regulation in growth stimulated murine cells, *Oncogene,* 4, 881, 1989.

160. **Ofir, R., Dwarki, V. J., Rashid, D., and Verma, I. M.,** Phosphorylation of the C terminus of Fos protein is required for transcriptional transrepression of the c-*fos* promoter, *Nature,* 348, 80, 1990.

161. **Abate, C., Marshak, D. R., and Curran, T.,** Fos is phosphorylated by p34^{cdc2}, cAMP-dependent protein kinase and protein kinase C at multiple sites clustered within regulatory regions, *Oncogene,* 6, 2179, 1991.

162. **Tratner, I., Ofir, R., and Verma, I. M.,** Alteration of a cyclic AMP-dependent protein kinase phosphorylation site in the c-Fos protein augments its transforming potential, *Mol. Cell. Biol.,* 12, 998, 1992.

163. **Greenberg, M. E., Hermanowski, A. L., and Ziff, E. B.,** Effect of protein synthesis inhibitors on growth factor activation of c-*fos*, c-*myc*, and actin gene transcription, *Mol. Cell. Biol.,* 6, 1050, 1986.

164. **Holt, J. T., Gopal, T. V., Moulton, A. D., and Nienhuis, A. W.,** Inducible production of c-*fos* antisense RNA inhibits 3T3 cell proliferation, *Proc. Natl. Acad. Sci. U.S.A.,* 83, 4794, 1986.

165. **Riabowol, K. T., Vosatka, R. J., Ziff, E. B., Lamb, N. J., and Feramisco, J. R.,** Microinjection of *fos*-specific antibodies blocks DNA synthesis in fibroblast cells, *Mol. Cell. Biol.,* 8, 1670, 1988.

166. **Yoshida, T., Shindo, Y., Ohta, K., and Iba, H.,** Identification of a small region of the v-*fos* gene product that is sufficient for transforming potential and growth-stimulating activity, *Oncogene,* 5, 79, 1989.

167. **Müller, R., Slamon, D. J., Tremblay, J. M., Cline, M. J., and Verma, I. M.,** Differential expression of cellular oncogenes during pre- and postnatal development of the mouse, *Nature,* 299, 640, 1982.

168. **Müller, R., Verma, I. M., and Adamson, E. D.,** Expression of c-*onc* genes: c-*fos* transcripts accumulate to high levels during development of mouse placenta, yolk sac and amnion, *EMBO J.,* 2, 679, 1983.

169. **Mason, I., Murphy, D., and Hogan, B. L. M.,** Expression of c-*fos* in parietal endoderm, amnion and differentiating F9 teratocarcinoma cells, *Differentiation,* 30, 76, 1985.

170. **Dony, C. and Gruss, P.,** Proto-oncogene c-*fos* expression in growth regions of fetal bone and mesodermal web tissue, *Nature,* 328, 711, 1987.

171. **Gubits, R. M., Hazelton, J. L., and Simantov, R.,** Variations in c-*fos* gene expression during rat brain development, *Mol. Brain Res.,* 3, 197, 1988.

172. **Field, S. J., Johnson, R. S., Mortensen, R. M., Papaioannou, V. E., Spiegelman, B. M., and Greenberg, M. E.,** Growth and differentiation of embryonic stem cells that lack an intact c-*fos* gene, *Proc. Natl. Acad. Sci. U.S.A.,* 89, 9306, 1992.

173. **Johnson, R. S., Spiegelman, B. M., and Papaioannou, V.,** Pleiotropic effects of a null mutation in the c-*fos* proto-oncogene, *Cell,* 71, 577, 1992.

174. **Wang, Z. Q., Ovitt, C., Grigoriadis, A. E., Möhle-Steinlein, U., Rüther, U., and Wagner, E. F.,** Bone and haematopoietic defects in mice lacking c-*fos, Nature,* 360, 741, 1992.

175. **Müller, R., Müller, D., and Guilbert, L.,** Differential expression of c-*fos* in hematopoietic cells: correlation with differentiation of monomyelocytic cells *in vitro, EMBO J.,* 3, 1887, 1984.

176. **Kerr, L. D., Holt, J. T., and Matrisian, L. M.,** Growth factors regulate transin gene expression by c-*fos*-dependent and c-*fos*-independent pathways, *Science,* 242, 1424, 1988.

177. **Cohen, D. R. and Curran, T.,** *fra*-1: a serum-inducible, cellular immediate-early gene that encodes a *fos*-related antigen, *Mol. Cell. Biol.,* 8, 2063, 1988.

178. **Nishina, H., Sato, H., Suzuki, T., Sato, M., and Iba, H.,** Isolation and characterization of *fra*-2, an additional member of the *fos* gene family, *Proc. Natl. Acad. Sci. U.S.A.,* 87, 3619, 1990.

179. **Matsui, M., Tokuhara, M., Konuma, Y., Nomura, N., and Ishizaki, R.,** Isolation of human *fos*-related genes and their expression during monocyte-macrophage differentiation, *Oncogene,* 5, 249, 1990.

180. **Kovary, K. and Bravo, R.,** Existence of different Fos/Jun complexes during the G_0-to-G_1 transition and during exponential growth in mouse fibroblasts: differential role of Fos proteins, *Mol. Cell. Biol.,* 12, 5015, 1992.

181. **Yoshida, T., Sato, H., and Iba, H.,** Transcription of *fra*-2 mRNA and phosphorylation of Fra-2 protein are stimulated by serum, *Biochem. Biophys. Res. Commun.,* 174, 934, 1991.

182. **Ito, E., Sweterlitsch, L. A., Tran, P. B.-V., Rauscher, F. J., III, and Narayanan, R.,** Inhibition of PC-12 cell differentiation by the immediate early gene *fra*-1, *Oncogene,* 5, 1755, 1990.

183. **Han, T. H., Lamph, W. W., and Prywes, R.,** Mapping of epidermal growth factor-responsive, serum-responsive, and phorbol ester-responsive sequence elements in the c-*jun* promoter, *Mol. Cell. Biol.,* 12, 4472, 1992.

184. **Kovary, K. and Bravo, R.,** The Jun and Fos protein families are both required for cell cycle progression in fibroblasts, *Mol. Cell. Biol.,* 11, 4466, 1991.

185. **Ryseck, R.-P., Hirai, S. I., Yaniv, M., and Bravo, R.,** Transcriptional activation of c-*jun* during the G_0/G_1 transition in mouse fibroblasts, *Nature,* 334, 535, 1988.

186. **Lamph, W. W., Wamsley, P., Sassone-Corsi, P., and Verma, I. M.,** Induction of proto-oncogene *JUN*/AP-1 by serum and TPA, *Nature,* 334, 629, 1988.

187. **Mollinedo, F, Vaquerizo, M. J., and Naranjo, J. R.,** Expression of c-*jun, jun* B and *jun* D proto-oncogenes in human peripheral blood granulocytes, *Biochem. J.,* 273, 477, 1991.

188. **Brach, M. A., Herrmann, F., Yamada, H., Bäuerle, P. A., and Kufe, D. W.,** Identification of NF-*jun*, a novel inducible transcription factor that regulates c-*jun* gene transcription, *EMBO J.,* 11, 1479, 1992.

189. **Kovary, K. and Bravo, R.,** Expression of different Jun and Fos proteins during the G_0-to-G_1 transition in mouse fibroblasts: *in vitro* and *in vivo* associations, *Mol. Cell. Biol.,* 11, 2451, 1991.

190. **Dobrzanski, P., Noguchi, T., Kovary, K., Rizzo, C. A., Lazo, P. S., and Bravo, R.,** Both products of the *fos*B gene, Fos B and its shorter form, FosB/SF, are transcriptional activators in fibroblasts, *Mol. Cell. Biol.,* 11, 5470, 1991.

191. **Alcivar, A. A., Hake, L. E., Kwon, Y. K., and Hecht, N. B.,** junD mRNA expression differs from c-*jun* and *jun*-B mRNA expression during male germinal cell differentiation, *Mol. Reprod. Dev.,* 30, 187, 1991.

192. **Dungy, L. J., Siddiqi, T. A., and Khan, S.,** c-*jun* and *jun*-B oncogene expression during placental development, *Am. J. Obstet. Gynecol.,* 165, 1853, 1991.

193. **Kornhauser, J. M., Nelson, D. E., Mayo, K. E., and Takahashi, J. S.,** Regulation of *jun*-B messenger RNA and AP-1 activity by light and a circadian clock, *Science,* 255, 1581, 1992.

194. **Mellström, B., Achaval, M., Montero, D., Naranjo, J. R., and Sassone-Corsi, P.,** Differential expression of the *jun* family members in rat brain, *Oncogene,* 6, 1959, 1991.

195. **Castellazzi, M., Dangy, J.-P., Mechta, F., Hirai, S.-I., Yaniv, M., Samarut, J., Lassailly, A., and Brun, G.,** Overexpression of avian or mouse c-*jun* in primary chick embryo fibroblasts confers a partially transformed phenotype, *Oncogene,* 5, 1541, 1990.

196. **Ransone, L. J. and Verma, I. M.,** Nuclear proto-oncogenes *fos* and *jun, Annu. Rev. Cell Biol.,* 6, 539, 1990.

197. **Rauscher, F. J., III, Cohen, D. R., Curran, T., Bos, T. J., Vogt, P. K., Bohmann, D., Tjian, R., and Franza, B. R., Jr.,** Fos-associated protein p39 is the product of the *jun* proto-oncogene, *Science,* 240, 1010, 1988.

198. **Rauscher, F. J., III, Voulalas, P. J., Franza, B. R., Jr., and Curran, T.,** *fos* and *jun* bind cooperatively to the AP-1 site: reconstitution *in vitro, Genes Dev.,* 2, 1687, 1988.

199. **Risse, G., Neuberg, M., Hunter, J. B., Verrier, B., and Müller, R.,** Products of the *fos* and *jun* proto-oncogenes bind cooperatively to the APÍ DNA recognition sequence, *Environ. Health Perspect.,* 88, 133, 1990.

200. **Sassone-Corsi, P., Ransone, L. J., Lamph, W. W., and Verma, I. M.,** Direct interaction between Fos and Jun nuclear oncoproteins: role of the "leucine zipper" domain, *Nature,* 336, 692, 1988.

201. **Gentz, R., Rauscher, F. J., III, Abate, C., and Curran, T.,** Parallel association of Fos and Jun leucine zippers juxtaposes DNA binding domains, *Science,* 243, 1695, 1989.

202. **Neuberg, M., Schuermann, M., Hunter, J. B., and Müller, R.,** Two functionally different regions in Fos are required for the sequence-specific DNA interaction of the Fos/Jun protein complex, *Nature,* 338, 589, 1989.

203. **Hirai, S.-I., Bourachot, B., and Yaniv, M.,** Both Jun and Fos contribute to transcription activation by the heterodimer, *Oncogene,* 5, 39, 1990.

204. **Bohmann, D., Bos, T. J., Admon, A., Nishimura, T., Vogt, P. K., and Tjian, R.,** Human proto-oncogene c-*jun* encodes a DNA binding protein with structural and functional properties of transcription factor AP-1, *Science,* 238, 1386, 1987.

205. **Angel, P., Allegretto, E. A., Okino, S. T., Hattori, K., Boyle, W. J., Hunter, T., and Karin, M.,** Oncogene *jun* encodes a sequence-specific *trans*-activator similar to AP-1, *Nature,* 332, 166, 1988.

206. **Bos, T. J., Bohmann, D., Tsuchie, H., Tjian, R., and Vogt, P. K.,** v-*jun* encodes a nuclear protein with enhancer binding properties of AP-1, *Cell,* 52, 705, 1988.

207. **Sassone-Corsi, P., Lamph, W. W., Kamps, M., and Verma, I. M.,** *fos*-associated cellular protein p39 is related to nuclear transcription factor AP-1, *Cell,* 54, 553, 1988.

208. **Ball, A. R., Jr., Bos, T. J., Löliger, C., Nagata, L. P., Nishimura, T., Su, H., Tsuchie, H., and Vogt, P. K.,** *jun*: oncogene and transcriptional regulator, *Cold Spring Harbor Symp. Quant. Biol.,* 53, 687, 1988.

209. **Vogt, P. K. and Bos, T. J.,** The oncogene *jun* and nuclear signalling, *Trends Biochem. Sci.,* 14, 172, 1989.

210. **McBride, K., Robitaille, L., Tremblay, S., Argentin, S., and Nemer, M.,** *fos/jun* repression of cardiac-specific transcription in quiescent and growth-stimulated myocytes is targeted at a tissue-specific *cis* element, *Mol. Cell. Biol.,* 13, 600, 1993.

211. **Shemshedini, L., Knauthe, R., Sassone-Corsi, P., Pormon, A., and Gronemeyer, H.,** Cell-specific inhibitory and stimulatory effects of Fos and Jun on transcription activation by nuclear receptors, *EMBO J.,* 10, 3839, 1991.

212. **Abate, C., Luk, D., and Curran, T.,** Transcriptional regulation by Fos and Jun *in vitro*: interaction among multiple activator and regulatory domains, *Mol. Cell. Biol.,* 11, 3624, 1991.

213. **Allegretto, E. A., Smeal, T., Angel, P., Spiegelman, B. M., and Karin, M.,** DNA-binding activity of Jun is increased through its interaction with Fos, *J. Cell. Biochem.,* 42, 193, 1990.

214. **Scheuermann, M., Neuberg, M., Hunter, J. B., Jenuwein, T., Ryseck, R.-P., Bravo, R., and Müller, R.,** The leucine repeat motif in Fos protein mediates complex formation with Jun/AP-1 and is required for transformation, *Cell,* 56, 507, 1989.

215. **Ransone, L. J., Visvader, J., Sassone-Corsi, P., and Verma, I. M.,** Fos-Jun interaction: mutational analysis of the leucine zipper domain of both proteins, *Genes Dev.,* 3, 770, 1989.

216. **Ransone, L. J., Visvader, J., Wamsley, P., and Verma, I. M.,** Trans-dominant mutants of Fos and Jun, *Proc. Natl. Acad. Sci. U.S.A.,* 87, 3806, 1990.

217. **Patel, L., Abate, C., and Curran, T.,** Altered protein conformation on DNA binding by Fos and Jun, *Nature,* 347, 572, 1990.

218. **Kerppola, T. K. and Curran, T.,** Fos-Jun heterodimers and Jun homodimers bend DNA in opposite orientations: implications for transcription factor cooperativity, *Cell,* 66, 317, 1991.

219. **Hilson, P., de Froidmont, D., Lejour, C., Hirai, S.-H., Jacquemin, J.-M., and Yaniv, M.,** Fos and Jun oncogenes transactivate chimeric or native promoters containing AP1/GCN4 binding sites in plant cells, *Plant Cell,* 1, 651, 1990.

220. **Murakami, Y., Satake, M., Yamaguchi-Iwai, Y., Sakai, M., Muramatusu, M., and Ito, Y.,** The nuclear protooncogenes c-*jun* and c-*fos* as regulators of DNA replication, *Proc. Natl. Acad. Sci. U.S.A.,* 88, 3947, 1991.

221. **Angel, P. and Karin, M.,** The role of Jun, Fos and the AP-1 complex in cell proliferation and transformation, *Biochim. Biophys. Acta,* 1072, 129, 1991.

222. **Lucibello, F. C. and Müller, R.,** Proto-oncogenes encoding transcriptional regulators: unravelling the mechanisms of oncogenic conversion, *Crit. Rev. Oncogenesis,* 2, 259, 1991.

223. **Riabowol, K., Schiff, J., and Gilman, M. Z.,** Transcription factor AP-1 activity is required for initiation of DNA synthesis and is lost during cellular aging, *Proc. Natl. Acad. Sci. U.S.A.,* 89, 157, 1992.

224. **Pertovaara, L., Sistonen, L., Bos, T. J., Vogt, P. K., Keski-Oja, J., and Alitalo, K.,** Enhanced *jun* gene expression is an early genomic response to transforming growth factor β stimulation, *Mol. Cell. Biol.,* 9, 1255, 1989.

225. **Fisch, T. M., Prywes, R., and Roeder, R. G.,** An AP1-binding site in the c-*fos* gene can mediate induction by epidermal growth factor and 12-*O*-tetradecanoyl phorbol-13-acetate, *Mol. Cell. Biol.,* 9, 1327, 1989.

226. **den Hertog, J., de Groot, R. P., de Laat, S. W., and Kruijer, W.,** EGF-induced *jun-B*-expression in transfected P19 embryonal carcinoma cells expressing EGF-receptors is dependent on Jun D, *Nucleic Acids Res.,* 20, 125, 1992.

227. **Franklin, C. C., Unlap, T., Adler, V., and Kraft, A. S.,** Multiple signal transduction pathways mediate c-Jun protein phosphorylation, *Cell Growth Differ.,* 4, 377, 1993.

228. **Adler, V., Franklin, C. C., and Kraft, A. S.,** Phorbol esters stimulate the phosphorylation of c-Jun but not v-Jun: regulation by the N-terminal delta domain, *Proc. Natl. Acad. Sci. U.S.A.,* 89, 5341, 1992.

229. **Franklin, C. C., Sanchez, V., Wagner, F., Woodgett, J. R., and Kraft, A. S.,** Phorbol ester-induced amino-terminal phosphorylation of human JUN but not JUNB regulates transcriptional activation, *Proc. Natl. Acad. Sci. U.S.A.,* 89, 7247, 1992.

230. **Adler, V., Polotskaya, A., Wagner, F., and Kraft, A. S.,** Affinity-purified c-Jun amino-terminal protein kinase requires serine/threonine phosphorylation for activity, *J. Biol. Chem.,* 267, 17001, 1992.

231. **Oemar, B. S., Law, N. M., and Rosenzweig, S. A.,** Insulin-like growth factor-1 induces tyrosyl phosphorylation of nuclear proteins, *J. Biol. Chem.,* 266, 24241, 1991.

232. **Pulverer, B. J., Kyriakis, J. M., Avruch, J., Nikolakaki, E., and Woodgett, J. R.,** Phosphorylation of c-*jun* mediated by MAP kinases, *Nature,* 353, 670, 1991.

233. **Chou, S., Baichwal, V., and Ferrell, J. E., Jr.,** Inhibition of c-Jun DNA binding by mitogen-activated protein kinase, *Mol. Biol. Cell,* 3, 1117, 1992.

234. **Ryder, K., Lau, L. F., and Nathans, D.,** A gene activated by growth factors is related to the oncogene v-*jun, Proc. Natl. Acad. Sci. U.S.A.,* 85, 1487, 1988.

235. **Ryder, K., Lanahan, A., Perez-Albuerne, E., and Nathans, D.,** *jun*-D: a third member of the jun gene family, *Proc. Natl. Acad. Sci. U.S.A.,* 86, 1500, 1989.

236. **Hirai, S.-I., Ryseck, R.-P., Mechta, F., Bravo, R., and Yaniv, M.,** Characterization of *jun*-D: a new member of the *jun* proto-oncogene family, *EMBO J.,* 8, 1433, 1989.

237. **Alcivar, A. A., Hake, L. E., Hardy, M. P., and Hecht, N. B.,** Increased levels of *jun*-B and c-*jun* mRNAs in male germ cells following testicular cell dissociation. Maximal stimulation in prepuberal animals, *J. Biol. Chem.,* 265, 20160, 1990.

238. **Li, L., Hu, J.-S., and Olson, E. N.,** Different members of the *jun* proto-oncogene family exhibit distinct patterns of expression in response to type β transforming growth factor, *J. Biol. Chem.,* 265, 1556, 1990.

239. **Chiu, R., Angel, P., and Karin, M.,** Jun-B differs in its biological properties from, and is a negative regulator of, c-Jun, *Cell,* 59, 979, 1989.

240. **Deng, T. and Karin, M.,** Jun-B differs from c-Jun in its DNA-binding and dimerization domains, and represses c-Jun by formation of inactive heterodimers, *Genes Dev.,* 7, 479, 1993.

241. **Ryseck, R.-P. and Bravo, R.,** c-JUN, JUN-B, and JUN-D differ in their binding affinities to AP-1 and CRE consensus sequences: effects of FOS proteins, *Oncogene,* 6, 533, 1991.

242. **Hai, T. and Curran, T.,** Cross-family dimerization of transcription factors Fos/Jun and ATF/CREB alters DNA binding specicity, *Proc. Natl. Acad. Sci. U.S.A.,* 88, 3720, 1991.

243. **Hilberg, F. and Wagner, E. F.,** Embryonic stem (ES) cells lacking functional c-*jun*: consequences for growth and differentiation, AP-1 activity and tumorigenicity, *Oncogene,* 7, 2371, 1992.

244. **Watson, D. K., Ascione, R., and Papas, T. S.,** Molecular analysis of the *ets* genes and their products, *Crit. Rev. Oncogenesis,* 1, 409, 1990.

245. **Seth, A., Ascione, R., Fisher, R. J., Mavro-Thalassitis, G. J., Bhat, N. K., and Papas, T. S.,** The *ets* gene family, *Cell Growth Differ.,* 3, 327, 1992.

246. **Macleod, K., Leprince, D., and Stehelin, D.,** The *ets* gene family, *Trends Biochem. Sci.,* 17, 251, 1992.

247. **Wasylyk, B., Hahn, S. J. L., and Giovanne, A.,** The *ets* family of transcription factors, *Eur. J. Biochem.,* 211, 7, 1993.

248. **Ho, I.-C., Bhat, N. K., Gottschalk, L. R., Lindsten, T., Thompson, C. B., Papas, T. S., and Leiden, J. M.,** Sequence-specific binding of human Ets-1 to the T cell receptor alpha gene enhancer, *Science,* 250, 814, 1990.

249. **Klemsz, M. J., McKercher, S. R., Celada, A., Van Beveren, C., and Maki, R. A.,** The macrophage and B cell-specific transcriptional factor PU.1 is related to the *ets* oncogene, *Cell,* 61, 113, 1990.

250. **Wasylyk, B., Wasylyk, C., Flores, P., Begue, A., Leprince, D., and Stehelin, D.,** The c-*ets* proto-oncogenes encode transcription factors that cooperate with c-Fos and c-Jun for transcriptional activation, *Nature,* 346, 191, 1990.

251. **Watson, D. K., McWilliams, M. J., Lapis, P., Lautenberger, J. A., Schweinfest, C. W., and Papas, T. S.,** Mammalian *ets*-1 and *ets*-2 genes encode conserved proteins, *Proc. Natl. Acad. Sci. U.S.A.,* 85, 7862, 1988.

252. **Laudet, V., Niel, C., Duterque-Coquillaud, M., Leprince, D., and Stehelin, D.,** Evolution of the *ets* gene family, *Biochem. Biophys. Res. Commun.,* 190, 8, 1993.

253. **Wolff, C.-M., Stiegler, P., Baltzinger, M., Meyer, D., Ghysdael, J., Stéhelin, D., Befort, N., and Remy, P.,** Cloning, sequencing, and expression of two *Xenopus laevis* c-*ets*-2 protooncogenes, *Cell Growth Differ.,* 2, 447, 1991.

254. **Chen, Z., Burdett, L. A., Seth, A. K., Lautenberger, J. A., and Papas, T. S.,** Requirement of *ets*-2 expression for *Xenopus* oocyte maturation, *Science,* 250, 1416, 1990.

255. **Stiegler, P., Wolff, C.-M., Meyer, D., Sénan, F., Hourdry, J., Befort, N., and Remy, P.,** The c-*ets*-1 proto-oncogenes in *Xenopus laevis*: expression during oogenesis and embryogenesis, *Mechan. Dev.,* 41, 163, 1993.

256. **Leprince, D., Gesquière, J. C., and Stéhelin, D.,** The chicken cellular progenitor of the v-*ets* oncogene, p68^c-ets-1, is a nuclear DNA-binding protein not expressed in lymphoid cells of the spleen, *Oncogene Res.,* 5, 255, 1990.

257. **Pognonec, P., Boulukos, K. E., Gesquière, J. C., Stéhelin, D., and Ghysdael, J.,** Mitogenic stimulation of thymocytes results in the calcium-dependent phosphorylation of c-*ets*-1 proteins, *EMBO J.,* 7, 977, 1988.

258. **Fisher, C. L., Ghysdael, J., and Cambier, J. C.,** Ligation of membrane Ig leads to calcium-mediated phosphorylation of the proto-oncogene product, Ets-1, *J. Immunol.,* 146, 1743, 1991.

259. **Fujiwara, S., Fisher, R. J., Bhat, N. K., Moreno Diaz de la Espina, S., and Papas, T. S.,** A short-lived nuclear phosphoprotein encoded by the human *ets*-2 proto-oncogene is stabilized by activation of protein kinase C, *Mol. Cell. Biol.,* 8, 4700, 1988.

260. **Fisher, R. J., Koizumi, S., Kondoh, A., Mariano, J. M., Mavrothalassitis, G., Bhat, N. K., and Papas, T. S.,** Human ETS1 oncoprotein. Purification, isoforms, -SH modification, and DNA sequence-specific binding, *J. Biol. Chem.,* 267, 17957, 1992.

261. **Fleischman, L. F., Pilaro, A. M., Murakami, K., Kondoh, A., Fisher, R. J., and Papas, T. S.,** c-Ets-1 protein is hyperphosphorylated during mitosis, *Oncogene,* 8, 771, 1993.

262. **Majérus, M.-A., Bibollet-Ruche, F., Telliez, J.-B., Wasylyk, B., and Bailleul, B.,** Serum, AP-1 and Ets-1 stimulate the human *ets*-1 promoter, *Nucleic Acids Res.,* 20, 2699, 1992.

263. **Wernert, N., Raes, M.-B., Lassalle, P., Dehouck, M.-P., Gosselin, B., Vandenbunder, B., and Stehelin, D.,** c-*ets*-1 proto-oncogene is a transcription factor expressed in endothelial cells during tumor vascularization and other forms of angiogenesis in humans, *Am. J. Pathol.,* 140, 119, 1992.

264. **Rowe, A. and Propst, F.,** *ets*-1 and *ets*-2 protooncogene expression in theca cells of the adult mouse ovary, *Exp. Cell Res.,* 202, 199, 1992.

265. **Amouyel, P., Gégonne, A., Delacourte, A., Défossez, A., and Stéhelin, D.,** Expression of Ets proto-oncogenes in astrocytes in human cortex, *Brain Res.,* 447, 149, 1988.

266. **Topol, L.Z., Tatosyan, A.G., Ascione, R., Thompson, D.M., Blair, D.G., Kola, I., and Seth, A.,** C-*ets*-1 protooncogene expression alters growth properties of immortalized rat fibroblasts, *Cancer Lett.,* 67, 71, 1992.

267. **Seth, A., Watson, D. K., Blair, D. G., and Papas, T. S.,** c-*ets*-2 protooncogene has mitogenic and oncogenic activity, *Proc. Natl. Acad. Sci. U.S.A.,* 86, 7833, 1989.

268. **Rao, V. N. and Reddy, E. S. P.,** A divergent *ets*-related protein, Elk-1, recognizes similar c-*ets*-1 proto-oncogene target sequences and acts as a transcriptional activator, *Oncogene,* 7, 65, 1992.

269. **Rao, V. N., Papas, T. S., and Reddy, E. S. P.,** *erg,* a human *ets*-related gene on chromosome 21: alternative splicing, polyadenylation, and translation, *Science,* 237, 635, 1987.

270. **Rao, V. N., Zweig, M., Papas, T. S., and Reddy, E. S. P.,** Expression in *E. coli* of *erg*: a novel gene in humans related to the *ets* oncogene, *Oncogene Res.,* 2, 95, 1987.

271. **Reddy, E. S. P. and Rao, V. N.,** *erg,* an *ets*-related gene, codes for sequence-specific transcriptional activators, *Oncogene,* 6, 2285, 1991.

272. **Watson, D. K., Smyth, F. E., Thompson, D. M., Cheng, J. Q., Testa, J. R., Papas, T. S., and Seth, A.,** The *Erb*-B/*Fli*-1 gene: isolation and characterization of a new member of the family of human *ETS* transcription factors, *Cell Growth Differ.,* 3, 705, 1992.

273. **Klemsz, M. J., Maki, R. A., Papayannopoulou, T., Moore, J., and Hromas, R.,** Characterization of the *ets* oncogene family member, *fli*-1, *J. Biol. Chem.,* 268, 5769, 1993.

274. **Ben-David, Y., Giddens, E. B., Letwein, K., and Bernstein, A.,** Erythroleukemia induction by Friend murine leukemia virus — insertional activation of a new member of the *ets* gene family, *Genes Dev.,* 5, 908, 1991.

275. **Xin, J.-H., Cowie, A., Lachance, P., and Hassell, J. A.,** Molecular cloning and characterization of PEA3, a new member of the *ets* oncogene family that is differentially expressed in mouse embryonic cells, *Genes Dev.,* 6, 481, 1992.

276. **Xin, J.-H., Cowie, A., Lachance, P., and Hassell, J. A.,** Molecular cloning and characterization of PEA3, a new member of the *ets* oncogene family that is differentially expressed in mouse embryonic cells, *Genes Dev.,* 6, 481, 1992.

277. **Theilen, G. H., Zeigel, R. T., and Twiehaus, M. J.,** Biological studies with RE virus (strain T) that induces reticuloendotheliosis in turkeys, chickens, and Japanese quails, *J. Natl. Cancer Inst.,* 37, 731, 1966.

278. **Gélinas, C. and Temin, H. M.,** The v-*rel* oncogene encodes a cell-specific transcriptional activator of certain promoters, *Oncogene,* 3, 349, 1988.

279. **Moore, B. E. and Bose, H. R., Jr.,** Transformation of avian lymphoid cells by reticuloendotheliosis virus, *Mutat. Res.,* 195, 79, 1988.

280. **Beug, H., Müller, H., Grieser, S., Doederlein, G., and Graf, T.,** Hematopoietic cells transformed *in vitro* by REV-T avian reticuloendotheliosis virus express characteristics of very immature lymphoid cells, *Virology,* 115, 295, 1981.

281. **Gilmore, T. D. and Temin, H. M.,** v-Rel oncoproteins in the nucleus and in the cytoplasm transform chicken spleen cells, *J. Virol.,* 62, 703, 1988.

282. **Capobianco, A. J., Simmons, D. L., and Gilmore, T. D.,** Cloning and expression of a chicken c-*rel* cDNA: unlike p59$^{v\text{-}rel}$, p68$^{v\text{-}rel}$ is a cytoplasmic protein in chicken embryo fibroblasts, *Oncogene,* 5, 257, 1990.

283. **Bose, H. R.,** The Rel family: models for transcriptional regulation and oncogenic transformation, *Biochim. Biophys. Acta,* 1114, 1, 1992.

284. **Gilmore, T. D.,** Role of *rel* family genes in normal and malignant lymphoid cell growth, *Cancer Surv.,* 15, 69, 1992.

285. **Davis, N., Ghosh, S., Simmons, D. L., Tempst, P., Liou, H., Baltimore, D., and Bose, H. R., Jr.,** Rel-associated pp40: an inhibitor of the Rel family of transcription factors, *Science,* 253, 1268, 1991.

286. **Chen, I. S. Y., Wilhelmsen, K. C., and Temin, H. M.,** Structure and expression of c-*rel*, the cellular homolog to the oncogene of reticuloendotheliosis virus strain T, *J. Virol.,* 45, 104, 1983.

287. **Herzog, N. K., Bargmann, W. J., and Bose, H. R., Jr.,** Oncogene expression in reticuloendotheliosis virus-transformed lymphoid cell lines and avian tissues, *J. Virol.,* 57, 371, 1986.

288. **Brownell, E., Mathieson, B., Young, H. A., Keller, J., Ihle, J. N., and Rice, N. R.,** Detection of c-*rel*-related transcripts in mouse hematopoietic tissues, fractionated lymphocyte populations, and cell lines, *Mol. Cell. Biol.,* 7, 1304, 1987.

289. **Bull, P., Hunter, T., and Verma, I. M.,** Transcriptional induction of the murine c-*rel* gene with serum and phorbol-12-myristate-13-acetate in fibroblasts, *Mol. Cell. Biol.,* 9, 5239, 1989.

290. **Brownell, E., O'Brien, S. J., Nash, W. G., and Rice, N.,** Genetic characterization of human c-*rel* sequences, *Mol. Cell. Biol.,* 5, 2826, 1985.

291. **Brownell, E., Kozak, C. A., Fowlen, J. R., III, Modi, W. S., Rice, N. R., and O'Brien, S. J.,** Comparative genetic mapping of cellular *rel* sequences in man, mouse, and the domestic cat, *Am. J. Hum. Genet.,* 39, 194, 1986.

292. **Brownell, E., Ruscetti, F. W., Smith, R. G., and Rice, N. R.,** Detection of *rel*-related RNA and protein in human lymphoid cells, *Oncogene,* 3, 93, 1988.

293. **Ryseck, R.-P., Bull, P., Takamiya, M., Bours, V., Siebenlist, U., Dobrzanski, P., and Bravo, R.,** RelB, a new Rel family transcription activator that can interact with p50-NF-κB, *Mol. Cell. Biol.,* 12, 674, 1992.

294. **Brown, R. S., Sander, C., and Argos, P.,** The primary structure of transcription factor TFIIIA has 12 consecutive repeats, *FEBS Lett.,* 186, 271, 1985.

295. **Sukhatme, V. P., Kartha, S., Toback, F. G., Taub, R., Hoover, R. G., and Tsai-Morris, C.-H.,** A novel early growth response gene rapidly induced by fibroblast, epithelial cell and lymphocyte mitogens, *Oncogene Res.,* 1, 343, 1987.

296. **Chavrier, P., Zerial, M., Lemaire, P., Almendral, J., Bravo, R., and Charnay, P.,** A gene encoding a protein with zinc fingers is activated during G_0/G_1 transition in cultured cells, *EMBO J.,* 7, 29, 1988.

297. **Sukhatme, V. P., Cao, X., Chang, L. C., Tsai-Morris, C.-H., Stamenkovich, D., Ferreira, P. C. P., Cohen, D. R., Edwards, S. A., Shows, T. B., Curran, T., Le Beau, M. M., and Adamson, E. D.,** A zinc finger-encoding gene coregulated with c-*fos* during growth and differentiation, and after cellular depolarization, *Cell,* 53, 37, 1988.

298. **Cortner, J. and Farnham, P. J.,** Identification of the serum-responsive transcription initiation site of the zinc finger gene *Krox*-20, *Mol. Cell. Biol.,* 10, 3788, 1990.

299. **Chavrier, P., Lemaire, P., Revelant, O., Bravo, R., and Charnay, P.,** Characterization of a mouse multigene family that encodes zinc finger structures, *Mol. Cell. Biol.,* 8, 1319, 1988.

300. **Waters, C. M., Hancock, D. C., and Evan, G. I.,** Identification and characterization of the *egr*-1 gene product as an inducible, short-lived, nuclear phosphoprotein, *Oncogene,* 5, 669, 1990.

301. **Gupta, M. P., Gupta, M., Zak, K., and Sukhatme, V. P.,** Egr-1, a serum-inducible zinc finger protein, regulates transcription of the rat cardiac α-myosin heavy chain gene, *J. Biol. Chem.,* 266, 12813, 1991.

302. **Madden, S. L., Cook, D. M., Morris, J. F., Gashler, A., Sukhatme, V. P., and Rauscher, F. J., III,** Transcriptional repression mediated by the WT1 Wilms tumor gene product, *Science,* 253, 1550, 1991.

303. **Lemaire, P., Vesque, C., Schmitt, J., Stunnenberg, H., Frank, R., and Charnay, P.,** The serum-inducible mouse gene *Krox*-24 encodes a sequence-specific transcriptional activator, *Mol. Cell. Biol.,* 10, 3456, 1990.

304. **McMahon, A. P., Champion, J. E., McMahon, J. A., and Sukhatme, V. P.,** Developmental expression of the putative transcription factor *egr*-1 suggests that *egr*-1 and c-*fos* are coregulated in some tissues, *Development,* 108, 281, 1990.

305. **Lemaire, P., Revelant, O., Bravo, R., and Charnay, P.,** Two mouse genes encoding potential transcription factors with identical DNA-binding domains are activated by growth factors in cultured cells, *Proc. Natl. Acad. Sci. U.S.A.,* 85, 4691, 1988.

306. **Jamieson, G. A., Jr., Mayforth, R. D., Villereal, M. L., and Sukhatme, V. P.,** Multiple intracellular pathways induce expression of a zinc-finger encoding gene (*egr*-1): relationship to activation of the Na/H exchanger, *J. Cell. Physiol.,* 139 262, 1989.

307. **Josephs, L. J., LeBeau, M. M., Jamieson, G. A., Jr., Acharya, S., Shows, T. B., Rowley, J. D., and Sukhatme, V. P.,** Molecular cloning, sequencing, and mapping of *egr*-2, a human early growth response gene encoding a protein with "zinc-binding finger" structure, *Proc. Natl. Acad. Sci. U.S.A.,* 85, 7164, 1988.

308. **Cortner, J. and Farnham, P. J.,** Identification of the serum-responsive transcription initiation site of the zinc finger gene Krox-20, *Mol. Cell. Biol.,* 10, 3788, 1990.

309. **Cortner, J. and Farnham, P. J.,** Cell cycle analysis of Krox-20, c-*fos*, and JE expression in proliferating NIH3T3 fibroblasts, *Cell Growth Differ.,* 2, 465, 1991.

310. **Shaw, R. J., Doherty, D. E., Ritter, A. G., Benedict, S. H., and Clark, R. A. F.,** Adherence-dependent increase in human monocyte PDGF(B) mRNA is associated with increases in c-*fos*, c-*jun*, and EGR2 mRNA, *J. Cell Biol.,* 111, 2139, 1990.

311. **Katzav, S.,** *vav*: a molecule for all haemopoiesis, *Br. J. Haematol.,* 81, 141, 1992.

312. **Hu, P., Margolis, B., and Schlessinger, J.,** *vav*: a potential link between tyrosine kinases and *ras*-like GTPases in hematopoietic cell signaling, *BioEssays,* 15, 179, 1993.

313. **Coppola, J., Bryant, S., Koda, T., Conway, D., and Barbacid, M.,** Mechanism of activation of the *vav* protooncogene, *Cell Growth Differ.,* 2, 95, 1991.

314. **Katzav, S., Cleveland, J. L., Heslop, H. E., and Pulido, D.,** Loss of the amino-terminal helix-loop-helix domain of the *vav* proto-oncogene activates its transforming potential, *Mol. Cell. Biol.,* 11, 1912, 1991.

315. **Katzav, S., Martin-Zanca, D., and Barbacid, M.,** *vav*, a novel human oncogene derived from a locus ubiquitously expressed in hematopoietic cells, *EMBO J.,* 8, 2283, 1989.

316. **Bustelo, X. R., Rubin, S. D., Suen, K.-L., Carrasco, D., and Barbacid, M.,** Developmental expression of the *vav* protooncogene, *Cell Growth Differ.,* 4, 297, 1993.

317. **Bustelo, X. R., Ledbetter, J. A., and Barbacid, M.,** Product of *vav* proto-oncogene defines a new class of tyrosine protein kinase substrates, *Nature,* 356, 68, 1992.

318. **Margolis, B., Hu, P., Katzav, S., Li, W., Oliver, J. M., Ullrich, A., Weiss, A., and Schlessinger, J.,** Tyrosine phosphorylation of *vav* proto-oncogene product containing SH2 domain and transcription factor motifs, *Nature,* 356, 71, 1992.

319. **Bustelo, X. R. and Barbacid, M.,** Tyrosine phosphorylation of the *vav* proto-oncogene product in activated B cells, *Science,* 256, 1196, 1992.

320. **Gulbins, E., Coggeshall, K. M., Baier, G., Katzav, S., Burn, P., and Altman, A.,** Tyrosine kinase-stimulated guanine nucleotide exchange activity of *vav* in T-cell activation, *Science,* 260, 822, 1993.

321. **Lowenstein, E. J., Daly, R. J., Batzer, A. G., Li, W., Margolis, B., Lammers, R., Ullrich, A., Skolnik, E. Y., Bar-Sagi, D., and Schlessinger, J.,** The SH2 and SH3 domain-containing protein GRB2 links receptor tyrosine kinases to *ras* signaling, *Cell,* 70, 431, 1992.

322. **Weinberg, R. A.,** Positive and negative controls on cell growth, *Biochemistry,* 28, 8263, 1989.

323. **Geiser, A. G. and Stanbridge, E. J.,** A review of the evidence for tumor suppressor genes, *Crit. Rev. Oncogenesis,* 1, 261, 1989.

324. **Stanbridge, E. J.,** Human tumor suppressor genes, *Annu. Rev. Genet.,* 24, 615, 1990.

325. **Hollingsworth, R. E. and Lee, W.-H.,** Tumor suppressor genes: new prospects for cancer research, *J. Natl. Cancer Inst.,* 83, 91, 1991.

326. **Cavenee, W. K.,** Recessive mutations in the causation of human cancer, *Cancer,* 67, 243, 1991.

327. **Lee, E. Y.-H. P.,** Tumor suppressor genes: a new era for the molecular genetic studies of cancer, *Breast Cancer Res. Treat.,* 19, 3, 1991.

328. **Stanbridge, E. J.,** Functional evidence for human tumour suppressor genes — chromosome and molecular genetic studies, *Cancer Surv.,* 12, 5, 1992.

329. **Evans, H. J. and Prosser, J.,** Tumor suppressor genes — cardinal factors in inherited predisposition to human cancers, *Environ. Health Perspect.,* 98, 25, 1992.

330. **Levine, A. J.,** The tumor suppressor genes, *Annu. Rev. Biochem.,* 62, 623, 1993.

331. **Shay, J. W., Pereira-Smith, O. M., and Wright, W. E.,** A role for both RB and p53 in the regulation of human cellular senescence, *Exp. Cell Res.,* 196, 33, 1991.

332. **Hara, E., Tsurui, H., Shinozaki, A., Nakada, S., and Oda, K.,** Cooperative effect of antisense-Rb and antisense-p53 oligomers on the extension of life span in human diploid fibroblasts, TIG-1, *Biochem. Biochem. Res. Commun.,* 179, 528, 1991.

333. **Santhanam, U., Ray, A., and Sehgal, P. B.,** Repression of the interleukin 6 gene promoter by p53 and the retinoblastoma susceptibility gene product, *Proc. Natl. Acad. Sci. U.S.A.,* 88, 7605, 1991.

334. **Levine, A. J.,** The p53 tumour suppressor gene and product, *Cancer Surv.,* 12, 59, 1992.

335. **Ullrich, S. J., Anderson, C. W., Mercer, W. E., and Appella, E.,** The p53 tumor suppressor protein, a modulator of cell proliferation, *J. Biol. Chem.,* 267, 15259, 1992.

336. **Vogelstein, B. and Kinzler, K. W.,** p53 function and dysfunction, *Cell,* 70, 523, 1992.

337. **Oren, M.,** p53: The ultimate tumor suppressor gene, *FASEB J.,* 6, 3169, 1992.

338. **Montenarh, M.,** Biochemical, immunological, and functional aspects of growth suppressor/oncoprotein p53, *Crit. Rev. Oncogenesis,* 3, 233, 1992.

339. **Tominaga, O., Hamelin, R., Remvikos, Y., Salmon, R. J., and Thomas, G.,** p53 from basic research to clinical applications, *Crit. Rev. Oncogenesis,* 3, 257, 1992.

340. **Levine, A. J.,** The p53 tumor suppressor gene and product, *Biol. Chem. Hoppe-Seyler,* 374, 227, 1993.

341. **Lin, D., Shields, M. T., Ullrich, S. J., Appella, E., and Mercer, W. E.,** Growth arrest induced by wild-type p53 protein blocks cells prior or near the restriction point in late G_1 phase, *Proc. Natl. Acad. Sci. U.S.A.,* 89, 9210, 1992.

342. **Friedman, P. N., Chen, X. B., Bargonetti, J., and Prives, C.,** The p53 protein is an unusually shaped tetramer that binds directly to DNA, *Proc. Natl. Acad. Sci. U.S.A.,* 90, 3319, 1993.

343. **Funk, W. D., Pak, D. T., Karas, R. H., Wright, W. E., and Shay, J. W.,** A transcriptionally active DNA-binding site for human p53 protein complexes, *Mol. Cell. Biol.,* 12, 2866, 1992.

344. **Seto, E., Usheva, A., Zambetti, G. P., Momand, J., Horikosyi, N., Weinmann, R., Levine, A. J., and Shenk, T.,** Wild-type p53 binds to the TATA-binding protein and represses transcription, *Proc. Natl. Acad. Sci. U.S.A.,* 89, 12028, 1992.

345. **Halazonetis, T. D., Davis, L. J., and Kandil, A. N.,** Wild-type p53 adopts a "mutant"-like conformation when bound to DNA, *EMBO J.,* 12, 1021, 1993.

346. **Agoff, S. N., Hou, J., Linzer, D. I. H., and Wu, B.,** Regulation of the human hsp70 promoter by p53, *Science,* 259, 84, 1993.

347. **Schmid, P., Lorenz, A., Hameister, H., and Montenarh, M.,** Expression of p53 during mouse embryogenesis, *Development,* 113, 857, 1991.

348. **Halevy, O.,** p53 gene is up-regulated during skeletal muscle cell differentiation, *Biochem. Biophys. Res. Commun.,* 192, 714, 1993.

349. **Donehower, L. A., Harvey, M., Slagle, B. L., McArthur, M. J., Montgomery, C. A., Jr., Butel, J. S., and Bradley, A.,** Mice deficient for p53 are developmentally normal but susceptible to spontaneous tumours, *Nature,* 356, 215, 1992.

350. **Kastan, M. B., Radin, A. I., Kuerbitz, S. J., Onyekwere, O., Wolkow, C. A., Civin, C. I., Stone, K. D., Woo, T., Ravindranath, Y., and Craig, R. W.,** Levels of p53 protein increase with maturation in human hematopoietic cells, *Cancer Res.,* 51, 4279, 1991.

351. **Almon, E., Goldfinger, N., Kapon, A., Schwartz, D., Levine, A. J., and Rotter, V.,** Testicular tissue-specific expression of the p53 suppressor gene, *Dev. Biol.,* 156, 107, 1993.

352. **Ryan, J. J., Danish, R., Gottlieb, C. A., and Clarke, M. F.,** Cell cycle analysis of p53-induced cell death in murine erythroleukemia cells, *Mol. Cell. Biol.,* 13, 711, 1993.

353. **Porter, P. L., Gown, A. M., Kramp, S. G., and Coltrera, M. D.,** Widespread p53 overexpression in human malignant tumors. An immunohistochemical study using methacarn-fixed, embedded tissue, *Am. J. Pathol.,* 140, 145, 1992.

354. **Sherley, J. L.,** Guanine nucleotide biosynthesis is regulated by the cellular p53 concentration, *J. Biol. Chem.,* 266, 24815, 1991.

355. **Mercer, W. E., Shields, M. T., Lin, D., Appella, E., and Ullrich, S. J.,** Growth suppression induced by wild-type p53 protein is accompanied by selective down-regulation of proliferating-cell nuclear antigen expression, *Proc. Natl. Acad. Sci. U.S.A.,* 88, 1958, 1991.

356. **Shaulsky, G., Goldfinger, N., Tosky, M. S., Levine, A. J., and Rotter, V.,** Nuclear localization is essential for the activity of p53 protein, *Oncogene,* 6, 2055, 1991.

357. **Steinmeyer, K. and Deppert, W.,** DNA binding properties of murine p53, *Oncogene,* 3, 501, 1988.

358. **Hupp, T. R., Meek, D. W., Midgley, C. A., and Lane, D. P.,** Regulation of the specific DNA binding function of p53, *Cell,* 71, 875, 1992.

359. **Foord, O. S., Bhattacharya, P., Reich, Z., and Rotter, V.,** A DNA binding domain is contained in the C-terminus of wild type p53 protein, *Nucleic Acids Res.,* 19, 5191, 1991.

360. **Kern, S. E., Kinzler, K. W., Bruskin, A., Jarosz, D., Friedman, P., Prives, C., and Vogelstein, B.,** Identification of p53 as a sequence-specific DNA-binding protein, *Science,* 252, 1708, 1991.

361. **Fields, S. and Jang, S. K.,** Presence of a potent transcription activating sequence in the p53 protein, *Science,* 249, 1046, 1990.

362. **Mercer, W. E., Nelson, D., DeLeo, A., Old, L. J., and Baserga, R.,** Microinjection of monoclonal antibody to protein p53 inhibits serum-induced DNA synthesis in 3T3 cells, *Proc. Natl. Acad. Sci. U.S.A.,* 79, 6309, 1982.

363. **Mercer, W. E., Avignolo, C., and Baserga, A.,** Role of the p53 protein in cell proliferation as studied by microinjection of monoclonal antibodies, *Mol. Cell. Biol.,* 4, 276, 1984.

364. **Shohat, O., Greenberg, M., Reisman, D., Oren, M., and Rotter, V.,** Inhibition of cell growth mediated by plasmids encoding p53 anti-sense, *Oncogene,* 1, 277, 1987.

365. **Ullrich, S. J., Mercer, W. E., and Appella, E.,** Human wild-type p53 adopts a unique conformational and phosphorylation state *in vivo* during growth arrest of glioblastoma cells, *Oncogene,* 7, 1635, 1992.

366. **Lorenz, A., Herrmann, C. P. E., Issinger, O.-G., and Montenarh, M.,** Phosphorylation of wild-type and mutant phenotypes of p53 by an associated protein kinase, *Int. J. Oncol.,* 1, 571, 1992.

367. **Baudier, J., Delphin, C., Grunwald, D., Khochbin, S., and Lawrence, J. J.,** Characterization of the tumor suppressor protein p53 as a protein kinase C substrate and a S100b-binding protein, *Proc. Natl. Acad. Sci. U.S.A.,* 89, 11627, 1992.

368. **Mercer, W. E. and Baserga, R.,** Expression of the p53 protein during the cell cycle of human peripheral blood lymphocytes, *Exp. Cell Res.,* 160, 31, 1985.

369. **Louis, J. M., McFarland, V. W., May, P., and Mora, P. T.,** The phosphoprotein p53 is down-regulated posttranscriptionally during embryogenesis in vertebrates, *Biochim. Biophys. Acta,* 950, 395, 1988.

370. **Stürzbecher, H.-W., Maimets, T., Chumakov, P., Brain, R., Addison, C., Simanis, V., Rudge, K., Philip, R., Grimaldi, M., Court, W., and Jenkins, J. R.,** p53 interacts with p34cdc2 in mammalian cells: implications for cell cycle control and oncogenesis, *Oncogene,* 5, 795, 1990.

371. **Addison, C., Jenkins, J. R., and Stürzbecher, H.-W.,** The p53 nuclear localisation signal is structurally linked to a p34cdc2 kinase motif, *Oncogene,* 5, 423, 1990.

372. **Milner, J., Cook, A., and Mason, J.,** p53 is associated with p34cdc2 in transformed cells, *EMBO J.,* 9, 2885, 1990.

373. **Meek, D. W., Simon, S., Kikkawa, U., and Eckhart, W.,** The p53 tumour suppressor protein is phosphorylated at serine 389 by casein kinase II, *EMBO J.,* 9, 3253, 1990.

374. **Filhol, O., Baudier, J., Delphin, C., Loue-Mackenbach, P., Chambaz, E. M., and Cochet, C.,** Casein kinase II and the tumor suppressor protein p53 associate in a molecular complex that is negatively regulated upon p53 phosphorylation, *J. Biol. Chem.,* 267, 20577, 1992.

375. **Milne, D. M., Palmer, R. H., Campbell, D. G., and Meek, D. W.,** Phosphorylation of the p53 tumour-suppressor protein at three N-terminal sites by a novel casein kinase I-like enzyme, *Oncogene,* 7, 1361, 1992.

376. **Fontoura, B. M. A., Sorokina, E. A., David, E., and Carroll, R. B.,** p53 is covalently linked to 5.8S rRNA, *Mol. Cell. Biol.,* 12, 5145, 1992.

377. **Hollstein, M., Sidransky, D., Vogelstein, B., and Harris, C. C.,** p53 mutations in human cancers, *Science,* 253, 49, 1991.

378. **Rotter, V. and Prokocimer, M.,** p53 and human malignancies, *Adv. Cancer Res.,* 57, 257, 1991.

379. **Frebourg, T. and Friend, S. H.,** Cancer risks from germline p53 mutations, *J. Clin. Invest.,* 90, 1637, 1992.

380. **Casey, G., Lohsueh, M., Lopez, M. E., Vogelstein, B., and Stanbridge, E. J.,** Growth suppression of human breast cancer cells by the introduction of a wild-type p53, *Oncogene,* 6, 1791, 1991.

381. **Cheng, J., Yee, J.-K., Yeargin, J., Friedmann, T., and Haas, M.,** Suppression of acute lymphoblastic leukemia by the human wild-type p53 gene, *Cancer Res.,* 52, 222, 1992.

382. **Chen, Y. M., Chen, P. L., Arnaiz, N., Goodrich¡, D., and Lee, W. H.,** Expression of wild-type p53 in human A673 cells suppresses tumorigenicity but not growth rate, *Oncogene,* 6, 1799, 1991.

383. **Kern, S. E., Pietenpol, J. A., Thiagalingam, S., Seymour, A., Kinzler, K. W., and Vogelstein, B.,** Oncogenic forms of p53 inhibit p53-regulated gene expression, *Science,* 256, 827, 1992.

384. **Wyllie, F. S., Dawson, T., Bond, J. A., Goretzki, P., Game, S., Prime, S., and Wynford-Thomas, D.,** Correlated abnormalities of transforming growth factor-β1 response and p53 expression in thyroid epithelial cell transformation, *Mol. Cell. Endocrinol.,* 76, 13, 1991.

385. **Suzuki, K., Ono, T., and Takahashi, K.,** Inhibition of DNA synthesis by TGF-β1 coincides with inhibition of phosphorylation and cytoplasmic translocation of p53 protein, *Biochem. Biophys. Res. Commun.,* 183, 1175, 1992.

386. **Gerwin, B. I., Spillare, E., Forrester, K., Lehman, T. A., Kispert, J., Welsh, J. A., Pfeifer, A. M. A., Lechner, J. F., Baker, S. J., Vogelstein, B., and Harris, C. C.,** Mutant p53 can induce tumorigenic conversion of human bronchial epithelial cells and reduce their responsiveness to a negative growth factor, transforming growth factor β_1, *Proc. Natl. Acad. Sci. U.S.A.,* 89, 2759, 1992.

387. **Vogel, F.,** Genetics of retinoblastoma, *Hum. Genet.,* 52, 1, 1979.

388. **de Nuñez, M., Penchaszadeh, V., and Pimentel, E.,** Chromosomal fragility in patients with sporadic unilateral retinoblastoma, *Cancer Genet. Cytogenet.,* 11, 139, 1984.

389. **Bookstein, R. and Lee, W.-H.,** Molecular genetics of the retinoblastoma suppressor gene, *Crit. Rev. Oncogenesis,* 2, 211, 1991.

390. **Weinberg, R.,** The retinoblastoma gene and gene product, *Cancer Surv.,* 12, 43, 1992.

391. **Udavadia, A. J., Rogers, K. T., and Horowitz, J. M.,** A common set of nuclear factors bind to promoter elements regulated by the retinoblastoma protein, *Cell Growth Differ.,* 3, 597, 1992.

392. **Shiio, Y., Yamamoto, T., and Yamaguchi, N.,** Negative regulation of Rb expression by the p53 gene product, *Proc. Natl. Acad. Sci. U.S.A.,* 89, 5206, 1992.

393. **Yen, A., Chandler, S., Forbes, M. E., Fung, Y.-K., T'Ang, A., and Pearson, R.,** Coupled down-regulation of the RB retinoblastoma and c-myc genes antecedes cell differentiation: possible role of RB as a "status quo" gene, *Eur. J. Cell Biol.,* 57, 210, 1992.

394. **Hamel, P. A., Gallie, B. L., and Phillips, R. A.,** The retinoblastoma protein and cell cycle regulation, *Trends Genet.,* 8, 180, 1992.

395. **Endo, T. and Goto, S.,** Retinoblastoma gene product Rb accumulates during myogenic differentiation and is deinduced by the expression of SV40 large T antigen, *J. Biochem.,* 112, 427, 1992.

396. **Goodrich, D. W., Wang, N. P., Qian, Y.-W., Lee, E. Y.-H. P., and Lee, W.-H.,** The retinoblastoma gene product regulates progression through the G_1 phase of the cell cycle, *Cell,* 67, 293, 1991.

397. **Goodrich, D. W. and Lee, W.-H.,** Abrogation by c-*myc* of G_1 phase arrest induced by RB protein but not by p53, *Nature,* 360, 177, 1992.

398. **Xu, H.-J., Hu, S.-X., and Benedict, W. F.,** Lack of nuclear RB protein staining in G_0/middle G_1 cells: correlation to changes in total RB protein level, *Oncogene,* 6, 1139, 1991.

399. **Strauss, M., Hering, S., Lieber, A., Herrmann, G., Griffin, B. E., and Arnold, W.,** Stimulation of cell division and fibroblast focus formation by antisense repression of retinoblastoma protein synthesis, *Oncogene,* 7, 769, 1992.

400. **Shiio, Y., Yamamoto, T., and Yamaguchi, N.,** Negative regulation of Rb expression by the p53 gene product, *Proc. Natl. Acad. Sci. U.S.A.,* 89, 5206, 1992.

401. **Szekely, L., Uzvolgyi, E., Jiang, W.-Q., Durko, M., Wiman, K. G., Klein, G., and Sümegi, J.,** Subcellular localization of the retinoblastoma protein, *Cell Growth Differ.,* 2, 287, 1991.

402. **Stokke, T., Erikstein, B., Smedshammer, L., Boye, E., and Steen, H. B.,** The retinoblastoma gene product is bound in the nucleus in early G_1 phase, *Exp. Cell Res.,* 204, 147, 1993.

403. **Mihara, K., Cao, X.-R., Yen, A., Chandler, S., Driscoll, B., Murphree, A. L., T'Ang, A., and Fung, Y.-K. T.,** Cell cycle-dependent regulation of phosphorylation of the human retinoblastoma gene product, *Science,* 246, 1300, 1989.

404. **DeCaprio, J. A., Furukawa, Y., Aschenbaum, F., Griffin, J. D., and Livingston, D. M.,** The retinoblastoma-susceptibility gene product becomes phosphorylated in multiple stages during cell cycle entry and progression, *Proc. Natl. Acad. Sci. U.S.A.,* 89, 1795, 1992.

405. **Zhang, W., Hittelman, W., Van, N., Andreeff, M., and Deisseroth, A.,** The phosphorylation of retinoblastoma gene product in human myeloid leukemia cells during the cell cycle, *Biochem. Biophys. Res. Commun.,* 184, 212, 1992.

406. **Takuwa, N., Zhou, W., Kumada, M., and Takuwa, Y.,** Ca^{2+}-dependent stimulation of retinoblastoma gene product phosphorylation and p34cdc2 kinase activation in serum-stimulated human fibroblasts, *J. Biol. Chem.,* 268, 138, 1993.

407. **Alberts, A. S., Thorburn, A. M., Shenolikar, S., Mumby, M. C., and Feramisco, J. R.,** Regulation of cell cycle progression and nuclear affinity of the retinoblastoma protein by protein phosphatases, *Proc. Natl. Acad. Sci. U.S.A.,* 90, 388, 1993.

408. **Templeton, D. J.,** Nuclear binding of purified retinoblastoma gene product is determined by cell cycle-regulated phosphorylation, *Mol. Cell. Biol.,* 12, 435, 1992.

409. **Taya, Y., Yasuda, H., Kamijo, M., Nakaya, K., Nakamura, Y., Ohba, Y., and Nishimura, S.,** *In vitro* phosphorylation of the tumor suppressor gene RB protein by mitosis-specific histone H1 kinase, *Biochem. Biophys. Res. Commun.,* 164, 580, 1989.

410. **Lin, B. T. Y., Gruenwald, S., Morla, A. O., Lee, W. H., and Wang, J. Y. J.,** Retinoblastoma cancer suppressor gene product is a substrate of the cell cycle regulator cdc2 kinase, *EMBO J.,* 10, 857, 1991.

411. **Kitagawa, M., Saitoh, S., Ogino, H., Okabe, T., Matsumoto, H., Okuyama, A., Tamai, K., Ohba, Y., Yasuda, H., Nishimura, S., and Taya, Y.,** cdc2-like kinase is associated with the retinoblastoma protein, *Oncogene,* 7, 1067, 1992.

412. **Hu, Q., Lees, J. A., Buchkovich, K. J., and Harlow, E.,** The retinoblastoma protein physically associates with the human cdc2 kinase, *Mol. Cell. Biol.,* 12, 971, 1992.

413. **Evans, G. A. and Farrar, W. L.,** Retinoblastoma protein phosphorylation does not require activation of p34 cdc2 protein kinase, *Biochem. J.,* 287, 965, 1992.

414. **Akiyama, T., Ohuchi, T., Sumida, S., Matsumoto, K., and Toyoshima, K.,** Phosphorylation of the retinoblastoma protein by cdk2, *Proc. Natl. Acad. Sci. U.S.A.,* 89, 7900, 1992.

415. **Dulic, V., Lees, E., and Reed, S. I.,** Association of human cyclin E with a periodic G_1-S phase protein kinase, *Science,* 257, 1958, 1992.

416. **Hinds, P. W., Mittnacht, S., Dulic, V., Arnold, A., Reed, S. I., and Weinberg, R. A.,** Regulation of retinoblastoma protein functions by ectopic expression of human cyclins, *Cell,* 70, 993, 1992.

417. **Pagano, M., Pepperkok, R., Lukas, J., Baldin, V., Ansorge, W., Bartek, J., and Draetta, G.,** Regulation of the cell cycle by the cdk2 protein kinase in cultured human fibroblasts, *J. Cell Biol.,* 121, 101, 1993.

418. **Kato, J., Matsushime, H., Hiebert, S. W., Ewen, M. E., and Sherr, C. J.,** Direct binding of cyclin D to the retinoblastoma gene product (pRb) and pRb phosphorylation by the cyclin D-dependent kinase CDK4, *Genes Dev.,* 7, 331, 1993.

419. **Hall, F. L., Williams, R. T., Wu, L., Wu, F., Carbonaro-Hall, D. A., Harper, J. W., and Warburton, D.,** Two potentially oncogenic cyclins, cyclin A and cyclin D1, share common properties of subunit configuration, tyrosine phosphorylation and physical association with the Rb protein, *Oncogene,* 8, 1377, 1993.

420. **Qian, Y., Luckey, C., Horton, L., Esser, M., and Templeton, D. J.,** Biological functions of the retinoblastoma protein requires distinct domains for hyperphosphorylation and transcription factor binding, *Mol. Cell. Biol.,* 12, 5563, 1992.

421. **Thomas, N. S. B., Burke, L. C., Bybee, A., and Linch, D. C.,** The phosphorylation state of the retinoblastoma (RB) protein in G_0/G_1 is dependent on growth status, *Oncogene,* 6, 317, 1991.

422. **Yan, Z., Hsu, S., Winawer, S., and Friedman, E.,** Transforming growth factor β1 (TGF-β1) inhibits retinoblastoma gene expression but not pRB phosphorylation in TGF-β1-growth stimulated colon carcinoma cells, *Oncogene,* 7, 801, 1992.

423. **Laiho, M., DeCaprio, J. A., Ludlow, J. W., Livingston, D. M., and Massagué, J.,** Growth inhibition by TGF-β linked to suppression of retinoblastoma protein phosphorylation, *Cell,* 62, 175, 1990.

424. **Whitson, R. H., Jr. and Itakura, K.,** TGF-β$_1$ inhibits DNA synthesis and phosphorylation of the retinoblastoma gene product in a rat liver epithelial cell line, *J. Cell. Biochem.,* 48, 305, 1992.

425. **Münger, K., Pietenpol, J. A., Pittelkow, M. R., Holt, J. T., and Moses, H. L.,** Transforming growth factor β$_1$ regulation of c-*myc* expression, pRB phosphorylation, and cell cycle progression in keratinocytes, *Cell Growth Differ.,* 3, 291, 1992.

426. **Kim, S. J., Wagner, S., Liu, F., O'Reilly, M. A., Robbins, P. D., and Green, M. R.,** Retinoblastoma gene product activates expression of the human TGF-β2 gene through transcription factor ATF-2, *Nature,* 358, 331, 1992.

427. **Roberts, A. B., Kim, S.-J., and Sporn, M. B.,** Is there a common pathway mediating growth inhibition by TGF-β and the retinoblastoma gene product?, *Cancer Cells,* 3, 19, 1991.

428. **Resnitzky, D., Tiefenbrun, N., Berissi, H., and Kimchi, A.,** Interferons and interleukin 6 suppress phosphorylation of the retinoblastoma protein in growth-sensitive hematopoietic cells, *Proc. Natl. Acad. Sci. U.S.A.,* 89, 402, 1992.

429. **Stein, G. H., Beeson, M., and Gordon, L.,** Failure to phosphorylate the retinoblastoma gene product in senescent human fibroblasts, *Science,* 249, 666, 1990.

430. **Futreal, P. A. and Barrett, J. C.,** Failure of senescent cells to phosphorylate the Rb protein, *Oncogene,* 6, 1109, 1991.

431. **Bandara, L. R., Adamczewski, J. P., Hunt, T., and La Thangue, N. B.,** Cyclin-A and the retinoblastoma gene product complex with a common transcription factor, *Nature,* 352, 249, 1991.

432. **Chellappan, S. P., Hiebert, S., Mudryj, M., Horowitz, J. M., and Nevins, J. R.,** The E2F transcription factor is a cellular target for the RB protein, *Cell,* 65, 1053, 1991.

433. **Bagchi, S., Weinmann, R., and Raychaudhuri, P.,** The retinoblastoma protein copurifies with E2F-I, an E1A-regulated inhibitor of the transcription factor E2F, *Cell,* 65, 1063, 1991.

434. **Nevins, J. R.,** E2F: a link between the Rb tumor suppressor protein and viral oncoproteins, *Science,* 258, 424, 1992.

435. **Ray, S. K., Arroyo, M., Bagchi, S., and Raychaudhuri, P.,** Identification of a 60-kilodalton Rb-binding protein, Rb-P60, that allows the Rb-E2F complex to bind DNA, *Mol. Cell. Biol.,* 12, 4327, 1992.

436. **Arroyo, M. and Raychaudhuri, P.,** Retinoblastoma-repression E2F-dependent transcription depends on the ability of the retinoblastoma protein to interact with E2F and is abrogated by the adenovirus E1A oncoprotein, *Nucleic Acids Res.,* 20, 5947, 1992.

437. **Schwarz, J. K., Devoto, S. H., Smith, E. J., Chellappan, S. P., Jakoi, L., and Nevins, J. R.,** Interactions of the p107 and Rb proteins with E2F during the cell proliferation response, *EMBO J.,* 12, 1013, 1993.

438. **Hinds, P. W., Mittnacht, S., Dulic, V., Arnold, A., Reed, S. I., and Weinberg, R. A.,** Regulation of retinoblastoma protein functions by ectopic expression of human cyclins, *Cell,* 70, 993, 1992.

439. **Kim, S.-J., Lee, H.-D., Robbins, P. D., Busam, K., Sporn, M. B., and Roberts, A. B.,** Regulation of transforming growth factor β1 gene expression by the product of the retinoblastoma-susceptibility gene, *Proc. Natl. Acad. Sci. U.S.A.,* 88, 3052, 1991.

440. **Rustgi, A. K., Dyson, N., and Bernards, R.,** Amino-terminal domains of c-*myc* and N-Myc proteins mediate binding to the retinoblastoma gene product, *Nature,* 352, 541, 1991.

441. **Kim, S.-J., Onwuta, U. S., Lee, Y. I., Li, R., Botchan, M. R., and Robbins, P. D.,** The retinoblastoma gene product regulates Sp1-mediated transcription, *Mol. Cell. Biol.,* 12, 2455, 1992.

442. **Lee, E. Y.-H. P., Chang, C.-Y., Hu, N., Wang, Y.-C. J., Lai, C.-C., Herrup, K., Lee, W.-H., and Bradley, A.,** Mice deficient for Rb are nonviable and show defects in neurogenesis and haematopoiesis, *Nature,* 359, 288, 1992.

443. **Clarke, A. R., Maandag, E. R., van Roon, M., van der Lugt, N. M. T., van der Valk, M., Hooper, M. L., Berns, A., and te Riele, H.,** Requirement for a functional *Rb-1* gene in murine development, *Nature,* 359, 328, 1992.

444. **Szekely, L., Jian, W.-Q., Bulic-Jakus, F., Rosen, A., Ringertz, N., Klein, G., and Wiman, K. G.,** Cell type and differentiation dependent heterogeneity in retinoblastoma protein expression in SCID mouse fetuses, *Cell Growth Differ.,* 3, 149, 1992.

445. **Mistchenko, A. S., Diez, R. A., Romquin, N., Sanceau, J., and Wietzerbin, J.,** Interferon-γ modulates retinoblastoma gene mRNA in monocytoid cells, *Int. J. Cancer,* 53, 87, 1993.

446. **Yen, A., Chandler, S., and Sturzenegger-Varvayanis, S.,** Regulated expression of the RB "tumor suppressor gene" in normal lymphocyte mitogenesis: elevated expression in transformed leukocytes and role as a "status quo" gene, *Exp. Cell Res.,* 192, 289, 1991.

447. **Takahashi, R., Hashimoto, T., Xu, H.-J., Hu, S.-X., Matsui, T., Miki, T., Bigo-Marshall, H., Aaronson, S. A., and Benedict, W. F.,** The retinoblastoma gene functions as a growth and tumor suppressor in human bladder carcinoma cells, *Proc. Natl. Acad. Sci. U.S.A.,* 88, 5257, 1991.

448. **Muncaster, M. M., Cohen, B. L., Phillips, R. A., and Gallie, B. L.,** Failure of *RB1* to reverse the malignant phenotype of human tumor cell lines, *Cancer Res.,* 52, 654, 1992.

449. **Slater, R. M. and Mannens, M. M. A. M.,** Cytogenetics and molecular genetics of Wilms' tumor of childhood, *Cancer Genet. Cytogenet.,* 61, 111, 1992.

450. **Haber, D. A. and Buckler, A. J.,** *WT1* — a novel tumor suppressor gene inactivated in Wilms' tumor, *New Biologist,* 4, 97, 1992.

451. **Tadokoro, K., Oki, N., Fujii, H., Ohshima, A., Inoue, T., and Yamada, M.,** Genomic organization of the human WT1 gene, *Jpn. J. Cancer Res.,* 83, 1198, 1992.

452. **Rose, E. A., Glaser, T., Jones, C., Smith, C. L., Lewis, W. H., Call, K. M., Minden, M., Champagne, E., Bonetta, L., Yeger, H., and Housman, D. E.,** Complete physical map of the WAGR region of 11p13 localizes a candidate Wilms' tumor gene, *Cell,* 60, 495, 1990.

453. **Call, K. M., Glaser, T., Ito, C. Y., Buckler, A. J., Pelletier, J., Haber, D. A., Rose, E. A., Kral, A., Yeger, H., Lewis, W. H., Jones, C., and Housman, D. E.,** Isolation and characterization of a zinc finger polypeptide gene at the human chromosome 11 Wilms' tumor locus, *Cell,* 60, 509, 1990.

454. **Gessler, M., Poutska, A., Cavenee, W., Neve, R. L., Orkin, S. H., and Bruns, G. A. P.,** Homozygous deletion in Wilms tumours of a zinc-finger gene identified by chromosome jumping, *Nature,* 343, 774, 1990.

455. **Rauscher, F. J., III, Morris, J. F., Tournay, O. E., Cook, D. M., and Curran, T.,** Binding of the Wilms' tumor locus zinc finger protein to the EGR-1 consensus sequence, *Science,* 250, 1259, 1990.

456. **Madden, S. L., Cook, D. M., Morris, J. F., Gashler, A., Sukhatme, V. P., and Rauscher, F. J., III,** Transcriptional repression mediated by the WT1 Wilms' tumor gene product, *Science,* 253, 1550, 1991.

457. **Morris, J. F., Madden, S. L., Tournay, O. E., Cook, D. M., Sukhatme, V. P., and Rauscher, F. J., III,** Characterization of the zinc finger protein encoded by the WT1 Wilms' tumor locus, *Oncogene,* 6, 2339, 1991.

458. **Drummond, I. A., Madden, S. L., Rohvernutter, P., Bell, G. I., Sukhatme, V. P., and Rauscher, F. J.,** Repression of the insulin-like growth factor II gene by the Wilms tumor suppressor WT1, *Science,* 257, 674, 1992.

459. **Wang, Z. Y., Madden, S. L., Deuel, T. F., and Rauscher, F. J., III,** The Wilms' tumor gene product, WT1, represses transcription of the platelet-derived growth factor A-chain gene, *J. Biol. Chem.,* 267, 21999, 1992.

460. **Gashler, A. L., Bonthron, D. T., Madden, S. L., Rauscher, F. J., Collins, T., and Sukhatme, V. P.,** Human platelet-derived growth factor-A chain is transcriptionally repressed by the Wilms' tumor suppressor WT1, *Proc. Natl. Acad. Sci. U.S.A.,* 89, 10984, 1992.

461. **Armstrong, J. F., Pritchard-Jones, K., Bickmore, W. A., Hastie, N. D., and Bard, J. B. L.,** The expression of the Wilms' tumour gene, WT1, in the developing mammalian embryo, *Mechan. Dev.,* 40, 85, 1992.

462. **Marchuk, D. A., Saulino, A. M., Tavakkol, R., Swaroop, M., Wallace, M. R., Andersen, L. B., Mitchell, A. L., Gutmann, D. H., Boguski, M., and Collins, F. S.,** cDNA cloning of the type 1 neurofibromatosis gene: complete sequence of the NF1 gene product, *Genomics,* 11, 931, 1991.

463. **Martin, G. A., Viskochil, D., Bollag, D., McCabe, P. C., Crosier, W. J., Haubruck, H., Conroy, L., Clark, R., O'Connell, P., Cawthon, R. M., Innis, M. A., and McCormick, F.,** The GAP-related domain of the neurofibromatosis type 1 gene product interacts with *ras* p21, *Cell,* 63, 843, 1990.

464. **DeClue, J. E., Papageorge, A. G., Fletcher, J. A., Diehl, S. R., Ratner, N., Vass, W. C., and Lowy, D. R.,** Abnormal regulation of mammalian p21^{ras} contributes to malignant tumor growth in von Recklinghausen (type 1) neurofibromatosis, *Cell,* 69, 265, 1992.

465. **Rouault, J.-P., Rimokh, R., Tessa, C., Paranhos, G., French, M., Duret, L., Garoccio, M., Germain, D., Samarut, J., and Magaud, J.-P.,** *BTG1*, a member of a new family of antiproliferative genes, *EMBO J.,* 11, 1663, 1992.

466. **Kujubu, D. A., Lim, R. W., Varnum, B. C., and Herschman, H. R.,** Induction of transiently expressed genes in PC-12 pheochromocytoma cells, *Oncogene,* 1, 257, 1987.

467. **Heidorn, K., Kreipe, H., Radzun, H. J., Müller, R., and Parwaresch, M. R.,** The protooncogene c-*fos* is transcriptionally active in normal human granulocytes, *Blood,* 70, 456, 1987.

468. **Kelly, K., Cochran, B. H., Stiles, C. D., and Leder, P.,** Cell-specific regulation of the c-*myc* gene by lymphocyte mitogens and platelet-derived growth factor, *Cell,* 35, 603, 1983.

469. **Persson, H., Hennighausen, L., Taub, R., De Grado, W., and Leder, P.,** Antibodies to human c-*myc* oncogene product: evidence of an evolutionarily conserved protein induced during cell proliferation, *Science,* 225, 687, 1984.

470. **Klinman, D. M., Mushinski, J. F., Honda, M., Ishigatsubo, Y., Mountz, J. D., Raveche, E. S., and Steinberg, A. D.,** Oncogene expression in autoimmune and normal peripheral blood mononuclear cells, *J. Exp. Med.,* 163, 1292, 1986.

471. **Gravekamp, C., van den Bulck, L. P., Vijg, J., van de Griend, R. J., and Bolhuis, R. L. H.,** c-*myc* gene expression and interleukin-2 receptor levels in cloned human CD^{2+}, CD^{3+}, CD_{3-} lymphocytes, *Natl. Immun. Cell Growth Regul.,* 6, 28, 1987.

472. **Reed, J. C., Nowell, P. C., and Hoover, R. G.,** Regulation of c-*myc* mRNA levels in normal human lymphocytes by modulators of cell proliferation, *Proc. Natl. Acad. Sci. U.S.A.,* 82, 4221, 1985.

473. **Löhr, H., Löhr, G. W., Kanz, L., and Fauser, A. A.,** Expression of c-*myc* in stimulated T lymphocytes of the helper/inducer phenotype producing lymphokine(s) supporting multilineage colony formation, *Acta Haematol.,* 76, 192, 1986.

474. **Buckler, A. J., Rothstein, T. L., and Sonenshein, G. E.,** Transcriptional control of c-*myc* gene expression during stimulation of murine B lymphocytes, *J. Immunol.,* 145, 732, 1990.

475. **Reed, J. C., Miyashita, T., Cuddy, M., and Cho, D.,** Regulation of p26-Bcl-2 protein levels in human peripheral blood lymphocytes, *Lab. Invest.,* 67, 443, 1992.

476. **Schneider-Schaulies, J., Hünig, T., Schimpl, A., and Wecker, E.,** Kinetics of cellular oncogene expression in mouse lymphocytes. I. Expression of c-*myc* and c-*ras*[Ha] in T lymphocytes induced by various mitogens, *Eur. J. Immunol.,* 16, 312, 1986.

477. **Morris, D. R., Allen, M. L., Rabinovitch, P. S., Kuepfer, C. A., and White, M. W.,** Mitogenic signaling pathways regulating expression of c-*myc* and ornithine decarboxylase genes in bovine T-lymphocytes, *Biochemistry,* 27, 8689, 1988.

478. **Lipsick, J. S. and Boyle, W. J.,** c-*myb* protein expression is a late event during T-lymphocyte activation, *Mol. Cell. Biol.,* 7, 3358, 1987.

479. **Shipp, M. A. and Reinherz, E. L.,** Differential expression of nuclear proto-oncogenes in T cells triggered with mitogenic and nonmitogenic T3 and T11 activation signals, *J. Immunol.,* 139, 2143, 1987.

480. **Reed, J. C., Alpers, J. D., Nowell, P. C., and Hoover, R. G.,** Sequential expression of protooncogenes during lecti-stimulated mitogenesis of normal human lymphocytes, *Proc. Natl. Acad. Sci. U.S.A.,* 83, 3982, 1986.

481. **Moore, J. P., Todd, J. A., Hesketh, T. R., and Metcalfe, J. C.,** c-*fos* and c-*myc* gene activation, ionic signals, and DNA synthesis in thymocytes, *J. Biol. Chem.,* 261, 8158, 1986.

482. **Schneider-Schaulies, J., Schimpl, A., and Wecker, E.,** Kinetics of cellular oncogene expression in mouse lymphocytes. II. Regulation of c-*fos* and c-*myc* gene expression, *Eur. J. Immunol.,* 17, 713, 1987.

483. **Kvanta, A., Kontny, E., Jondal, M., Okret, S., and Fredholm, B. B.,** Mitogen stimulation of T cells increases c-Fos and c-Jun protein levels, AP-1 binding and AP-1 transcriptional activity, *Cell. Signal.,* 4, 275, 1992.

484. **Song, L., Stephens, J. M., Kittur, S., Collins, G. D., Nagel, J. E., Pekala, P. H., and Adler, W. H.,** Expression of c-*fos*, c-*jun* and *jun* B in peripheral blood lymphocytes from young and elderly adults, *Mech. Ageing Dev.,* 65, 149, 1992.

485. **Reed, J. C., Tsujimoto, Y., Alpers, J. D., Croce, C. M., and Nowell, P. C.,** Regulation of *bcl*-2 proto-oncogene expression during normal human lymphocyte proliferation, *Science,* 236, 1295, 1987.

486. **Langgut, W. and Kersten, H.,** The deazaguanine-derivative, queuine, affects cell proliferation, protein phosphorylation and the expression of the protooncogenes c-*fos* and c-*myc* in HeLa cells, *FEBS Lett.,* 265, 33, 1990.

487. **Kartha, S., Sukhatme, V. P., and Toback, F. G.,** ADP activates protooncogene expression in renal epithelial cells, *Am. J. Physiol.,* 252, F1175, 1987.

488. **McNerney, R., Darling, D., and Johnstone, A.,** Differential control of proto-oncogene c-*myc* and c-*fos* expression in lymphocytes and fibroblasts, *Biochem. J.,* 245, 605, 1987.

489. **Pukac, L. A., Castellot, J. J., Jr., Wright, T. C., Jr., Caleb, B. L., and Karnovsky, M. J.,** Heparin inhibits c-*fos* and c-*myc* mRNA expression in vascular smooth muscle cells, *Cell Regul.,* 1, 435, 1990.

490. **Fausto, N. and Mead, J. E.,** Regulation of liver growth: protooncogenes and transforming growth factors, *Lab. Invest.,* 60, 4, 1989.

491. **Thompson, N. L., Mead, J. E., Braun, L., Goyette, M., Shank, P. R., and Fausto, N.,** Sequential protooncogene expression during rat liver regeneration, *Cancer Res.,* 46, 3111, 1986.

492. **Kruijer, W., Shelly, H., Botteri, F., van der Putten, H., Barber, J. R., Verma, I. M., and Leffert, H. L.,** Proto-oncogene expression in regenerating liver is stimulated in cultures of primary adult rat hepatocytes, *J. Biol. Chem.,* 261, 7929, 1986.

493. **Corral, M., Paris, B., Guguen-Guillouzo, C., Corcos, D., Kruh, J., and Defer, N.,** Increased expression of N-*myc* gene during normal and neoplastic rat liver growth, *Exp. Cell Res.,* 174, 107, 1988.

494. **Johnson, A. C., Garfield, S. H., Merlino, G. T., and Pastan, I.,** Expression of epidermal growth factor receptor proto-oncogene mRNA in regenerating rat liver, *Biochem. Biophys. Res. Commun.,* 150, 412, 1988.

495. **Coni, P., Bignone, F. A., Pichiri, G., Ledda-Columbano, G. M., Columbano, A., Rao, P. M., Rajalakshmi, S., and Sarma, D. S. R.,** Studies on the kinetics of expression of cell cycle dependent proto-oncogenes during mitogen-induced liver cell proliferation, *Cancer Lett.,* 47, 115, 1989.

496. **Kurokawa, N., Hirotani, Y., Arakawa, Y., Iguchi, K., Urabe, N., Yanaihara, N., Abe, K., and Yanaihara, C.,** Increased production of c-*myc* and c-*fos*-related proteins during late stage of regenerative process in rat liver after partial hepatectomy, *Biomed. Res.,* 11, 165, 1990.

497. **Alcorn, J. A., Feitelberg, S. P., and Brenner, D. A.,** Transient induction of c-*jun* during hepatic regeneration, *Hepatology,* 11, 909, 1990.

498. **Hsu, J.-C., Bravo, R., and Taub, R.,** Interactions among LRF-1, JunB, c-Jun, and c-Fos define a regulatory program in the G_1 phase of liver regeneration, *Mol. Cell. Biol.,* 12, 4654, 1992.

499. **Servais, P. and Galand, P.,** Effects of 17 β-estradiol on c-*myc* and c-Ha-*ras* expression in the liver of ovariectomized rats, *Cell Biol. Int. Rep.,* 14, 927, 1990.

500. **Coni, P., Pichiri-Coni, G., Ledda-Columbano, G. M., Rao, P. M., Rajalakshmi, S., Sarma, D. S. R., and Columbano, A.,** Liver hyperplasia is not necessarily associated with increased expression of c-*fos* and c-*myc* mRNA, *Carcinogenesis,* 11, 835, 1990.

501. **Schiaffonati, L., Rappocciolo, E., Tacchini, L., Cairo, G., and Bernelli-Zazzera, A.,** Reprogramming of gene expression in postischemic rat liver: induction of proto-oncogenes and hsp 70 gene family, *J. Cell. Physiol.,* 143, 79, 1990.

502. **Haritani, H., Esumi, M., Uchida, T., and Shikata, T.,** Oncogene expression in the liver tissue of patients with nonneoplastic liver disease, *Cancer,* 67, 2594, 1991.

503. **Beer, D,G., Zweifel, K. A., Simpson, D. P., and Pitot, H. C.,** Specific gene expression during compensatory renal hypertrophy in the rat, *J. Cell. Physiol.,* 131, 29, 1987.

504. **Cowley, B. D., Jr., Smardo, F. L., Jr., Grantham, J. J., and Calvet, J. P.,** Elevated c-*myc* protooncogene expression in autosomal recessive polycystic kidney disease, *Proc. Natl. Acad. Sci. U.S.A.,* 84, 8394, 1987.

505. **Vamvakas, S., Bittner, D., and Köster, U.,** Enhanced expression of the protooncogenes c-*myc* and c-*fos* in normal and malignant renal growth, *Toxicol. Lett.,* 67, 161, 1993.

506. **Sawczuk, I. S., Olsson, C. A., Hoke, G., and Buttyan, R.,** Immediate induction of c-*fos* and c-*myc* transcripts following unilateral nephrectomy, *Nephron,* 55, 193, 1990.

507. **Nomata, K., Igarashi, H., Kanetake, H., Miyamoto, T., and Saito, Y.,** Expression of *ras* gene family as result of compensatory renal growth in mice, *Urol. Res.,* 18, 251, 1990.

508. **Asselin, C. and Marcu, K. B.,** Mode of c-*myc* gene regulation in folic acid-induced kidney regeneration, *Oncogene Res.,* 5, 67, 1989.

509. **Roesel, J., Rigsby, D., Bailey, A., Alvarez, R., Sanchez, J. D., Campbell, V., Shrestha, K., and Miller, D. M.,** Stimulation of protooncogene expression by partial hepatectomy is not tissue-specific, *Oncogene Res.,* 5, 129, 1989.

510. **Bailey, A., Sanchez, J. D., Rigsby, D., Roesel, J., Alvarez, R., Rodu, B., and Miller, D. M.,** Stimulation of renal and hepatic c-*myc* and c-Ha-*ras* expression by unilateral nephrectomy, *Oncogene Res.,* 5, 287, 1990.

511. **Calvo, E. L., Dusetti, N. J., Cadenas, M. B., Dagorn, J.-C., and Iovanna, J. L.,** Changes in gene expression during pancreatic regeneration: activation of c-*myc* and H-*ras* oncogenes in the rat pancreas, *Pancreas,* 6, 150, 1991.

512. **Simpson, P. C.,** Role of proto-oncogenes in myocardial hypertrophy, *Am. J. Cardiol.,* 63, 13G, 1988.

513. **Géraudie, J., Hourdry, J., Vriz, S., Singer, M., and Méchali, M.,** Enhanced c-*myc* gene expression during forelimb regenerative outgrowth in the young *Xenopus laevis, Proc. Natl. Acad. Sci. U.S.A.,* 87, 3797, 1990.

514. **Briata, P., Laurino, C., and Gherzi, R.,** c-*myc* gene expression in human cells is controlled by glucose, *Biochem. Biophys. Res. Commun.,* 165, 1123, 1989.

515. **Vasudevan, S., Lee, G., Rao, P. M., Rajalakshmi, S., and Sarma, D. S. R.,** Rapid and transient induction of c-*fos,* c-*myc* and c-Ha-*ras* in rat liver following glycine administration, *Biochem. Biophys. Res. Commun.,* 152, 252, 1988.

516. **Corcos, D., Vaulont, S., Denis, N., Lyonnet, S., Simon, M.-P., Kitzis, A., Kahn, A., and Kruh, J.,** Expression of c-*myc* is under dietary control in rat liver, *Oncogene Res.,* 1, 193, 1987.

517. **Baik, M., Choi, C. B., Keller, W. L., and Park, C. S.,** Developmental stages and energy restriction affect cellular oncogene expression in tissues of female rats, *J. Nutr.,* 122, 1614, 1992.

518. **Kim, J.-W., Fletcher, D. L., Campion, D. R., Gaskins, H. R., and Dean, R.,** Effect of dietary manipulation on c-*myc* RNA expression in adipose tissue, muscle and liver of broiler chickens, *Biochem. Biophys. Res. Commun.,* 180, 1, 1991.

519. **Pohjanpelto, P. and Höltta, E.,** Deprivation of a single amino acid induces protein synthesis-dependent increases in c-*jun,* c-*myc,* and ornithine decarboxylase mRNAs ion Chinese hamster ovary cells, *Mol. Cell. Biol.,* 10, 5814, 1990.

520. **Horikawa, S., Sakata, K., Hatanaka, M., and Tsukada, K.,** Expression of c-*myc* oncogene in rat liver by a dietary manipulation, *Biochem. Biophys. Res. Commun.,* 140, 574, 1986.

521. **Curran, T., Bravo, R., and Müller, R.,** Transient induction of c-*fos* and c-*myc* is an immediate consequence of growth factor stimulation, *Cancer Surv.,* 4, 655, 1985.

522. **Tavassoli, M. and Shall, S.,** Transcription of the c-*myc* oncogene is altered in spontaneously immortalized rodent fibroblasts, *Oncogene,* 2, 337, 1988.

523. **Seshadri, T. and Campisi, J.,** Repression of c-*fos* transcription and an altered genetic program in senescent human fibroblasts, *Science,* 247, 205, 1990.

524. **Winkles, J. A., O'Connor, M. L., and Friesel, R.,** Altered regulation of platelet-derived growth factor A-chain and c-*fos* gene expression in senescent progeria fibroblasts, *J. Cell. Physiol.,* 144, 313, 1990.

525. **Phillips, P. D. and Pignolo, R. J.,** Altered expression of cell cycle dependent genes in senescent WI-38 cells, *Exp. Gerontol.,* 27, 403, 1992.

526. **Okazaki, R., Ikeda, K., Sakamoto, A., Nakano, T., Morimoto, K., Kikuchi, T., Urakawa, K., Ogata, E., and Matsumoto, T.,** Transcriptional activation of c-*fos* and c-*jun* protooncogenes by serum growth factors in osteoblast-like MC3T3-E1 cells, *J. Bone Mineral Res.,* 7, 1149, 1992.

527. **Palumbo, A. P., Rossino, P., and Comoglio, P. M.,** Bombesin stimulation of c-*fos* and c-*myc* gene expression in cultures of Swiss 3T3 cells, *Exp. Cell Res.,* 167, 276, 1986.

528. **Letterio, J. J., Coughlin, S. R., and Williams, L. T.,** Pertussis toxin-sensitive pathway in the stimulation of c-*myc* expression and DNA synthesis by bombesin, *Science,* 234, 1117, 1986.

529. **Bravo, R., Macdonald-Bravo, H., Müller, R., Hübsch, D., and Almendral, J. M.,** Bombesin induces c-*fos* and c-*myc* expression in quiescent Swiss 3T3 cells: comparative study with other mitogens, *Exp. Cell Res.,* 170, 103, 1987.

530. **Rollins, B. J., Morrison, E. D., and Stiles, C. D.,** A cell-cycle constraint on the regulation of gene expression by platelet-derived growth factor, *Science,* 238, 1269, 1987.

531. **Lin, J.-X. and Vilcek, J.,** Tumor necrosis factor and interleukin-1 cause a rapid and transient stimulation of c-*fos* and c-*myc* mRNA levels in human fibroblasts, *J. Biol. Chem.,* 262, 11908, 1987.

532. **Bird, R. C., Kung, T.-Y. T., Wu, G., and Young-White, R. R.,** Variations in c-*fos* mRNA expression during serum induction and the synchronous cell cycle, *Biochem. Cell Biol.,* 68, 858, 1990.

533. **Jouvenot, M., Pellerin, I., Maréchal, G., Royez, M., Ordener, C., Alkhalaf, M., and Adessi, G. L.,** Regulation of c-*fos* expression in primary culture of guinea pig glandular epithelial cells stimulated by growth factors and estradiol, *Reprod. Nutr. Dev.,* 30, 455, 1990.

534. **Kumatori, A., Nakamura, T., and Ichihara, A.,** Cell-density dependent expression of the c-*myc* gene in primary cultured rat hepatocytes, *Biochem. Biophys. Res. Commun.,* 178, 480, 1991.

535. **Müller, R., Bravo, R., Burckhardt, J., and Curran, T.,** Induction of c-*fos* gene and protein by growth factors precedes activation of c-*myc,* *Nature,* 312, 716, 1984.

536. **Yarden, A. and Kimchi, A.,** Tumor necrosis factor reduces c-*myc* expression and cooperates with interferon-gamma in HeLa cells, *Science,* 234, 1419, 1986.

537. **Reuse, S., Maenhaut, C., and Dumont, J. E.,** Regulation of proto-oncogenes c-*fos* and c-*myc* expressions by protein tyrosine kinase, protein kinase C, and cyclic AMP mitogenic pathways in dog primary thyrocytes: a positive and negative control by cyclic AMP on c-*myc* expression, *Exp. Cell Res.,* 189, 33, 1990.

538. **Steinert, P. M. and Roop, D. R.,** Molecular and cellular biology of intermediate filaments, *Annu. Rev. Biochem.,* 57, 593, 1988.

539. **Capetanaki, Y., Kuisk, I., Rothblum, K., and Starnes, S.,** Mouse vimentin: structural relationship to *fos, jun,* CREB and *tpr, Oncogene,* 5, 645, 1990.

540. **Rittling, S. R., Gibson, C. W., Ferrari, S., and Baserga, R.,** The effect of cycloheximide on the expression of cell cycle dependent genes, *Biochem. Biophys. Res. Commun.,* 132, 327, 1985.

541. **Zinn, K., Keller, A., Whittemore, L.-A., and Maniatis, T.,** 2-aminopurine selectively inhibits the induction of β-interferon, c-*fos*, and c-*myc* gene expression, *Science,* 240, 210, 1988.

542. **Vriz, S., Lemaitre, J.-M., Leibovici, M., Thierry, N., and Méchali, M.,** Comparative analysis of the intracellular localization of c-Myc, c-Fos, and replicative proteins during cell cycle progression, *Mol. Cell. Biol.,* 12, 3548, 1992.

543. **Hayes, T. E., Kitchen, A. M., and Cochran, B. H.,** Inducible binding of a factor to the c-*fos* regulatory region, *Proc. Natl. Acad. Sci. U.S.A.,* 84, 1272, 1987.

544. **Fort, P., Rech, J., Vie, A., Piechaczyk, M., Bonnieu, P. E., Jeanteur, P., and Blanchard, J.-M.,** Regulation of c-*fos* gene expression in hamster fibroblasts: initiation and elongation of transcription and mRNA degradation, *Nucleic Acids Res.,* 15, 5657, 1987.

545. **Sassone-Corsi, P. and Verma, I. M.,** Modulation of c-*fos* gene transcription by negative and positive cellular factors, *Nature,* 326, 507, 1987.

546. **Czerniak, B., Herz, F., Wersto, R. P., and Koss, L. G.,** Asymmetric distribution of oncogene products at mitosis, *Proc. Natl. Acad. Sci. U.S.A.,* 89, 4860, 1992.

547. **Barbu, V. and Dautry, F.,** Mevalonate deprivation alters the induction of *fos* and *myc* by growth factors, *Oncogene,* 5, 1077, 1990.

548. **Reed, J. C., Alpers, J. D., and Nowell, P. C.,** Expression of c-*myc* proto-oncogene in normal human lymphocytes: regulation by transcriptional and posttranscriptional mechanisms, *J. Clin. Invest.,* 80, 101, 1987.

549. **Nepveu, A., Levine, R. A., Campisi, J., Greenberg, M. E., Ziff, E. B., and Marcu, K. B.,** Alternative modes of c-*myc* regulation in growth factor-stimulated and differentiating cells, *Oncogene,* 1, 243, 1987.

550. **Kakkis, E. and Calame, K.,** A plasmacytoma-specific factor binds the c-*myc* promoter region, *Proc. Natl. Acad. Sci. U.S.A.,* 84, 7031, 1987.

551. **Boles, T. C. and Hogan, M. E.,** DNA structure equilibria in the human c-*myc* gene, *Biochemistry,* 26, 367, 1987.

552. **Eick, D., Berger, R., Polack, A., and Bornkamm, G. W.,** Transcription of c-*myc* in human mononuclear cells is regulated by an elongation block, *Oncogene,* 2, 61, 1987.

553. **Wagner, A. J., Le Beau, M. M., Diaz, M. O., and Hay, N.,** Expression, regulation, and chromosomal localization of the *max* gene, *Proc. Natl. Acad. Sci. U.S.A.,* 89, 3111, 1992.

554. **Dani, C., Blanchard, J. M., Piechaczyk, M., El Sabouty, S., Marty, L., and Jeanteur, P.,** Extreme instability of *myc* mRNA in normal and transformed human cells, *Proc. Natl. Acad. Sci. U.S.A.,* 81, 7046, 1984.

555. **Blanchard, J.-M., Piechaczyk, M., Dani, C., Chambard, J.-C., Franchi, A., Pouysségur, J., and Jeanteur, P.,** c-*myc* gene is transcribed at high rate in G_0-arrested fibroblasts and is post-transcriptionally regulated in response to growth factors, *Nature,* 317, 443, 1985.

556. **Brewer, G. and Ross, J.,** Regulation of c-*myc* mRNA stability *in vitro* by a labile destabilizer with an essential nucleic acid component, *Mol. Cell. Biol.,* 9, 1996, 1989.

557. **Parkin, N., Darveau, A., Nicholson, R., and Sonenberg, N.,** *cis*-Acting translational effects of the 5′ noncoding region of c-*myc* mRNA, *Mol. Cell. Biol.,* 8, 2875, 1988.

558. **Richman, A. and Hayday, A.,** Serum-inducible expression of transfected human c-*myc* genes, *Mol. Cell. Biol.,* 9, 4962, 1989.

559. **Sariban, E., Luebbers, R., and Kufe, D.,** Transcriptional and posttranscriptional control of c-*fos* gene expression in human monocytes, *Mol. Cell. Biol.,* 8, 340, 1988.

560. **Gilman, M. Z., Wilson, R. N., and Weinberg, R. A.,** Multiple protein-binding sites in the 5′-flanking region regulate c-*fos* expression, *Mol. Cell. Biol.,* 6, 4305, 1986.

561. **Fisch, T. M., Prywes, R., and Roeder, R. G.,** c-*fos* sequences necessary for basal expression and induction by epidermal growth factor, 12-*O*-tetradecanoyl phorbol-13-acetate, and the calcium iono-phore, *Mol. Cell. Biol.,* 7, 3490, 1987.

562. **Prywes, R. and Roeder, R. G.,** Purification of the c-*fos* enhancer-binding protein, *Mol. Cell. Biol.,* 7, 3482, 1987.

563. **Dey, A., Nebert, D. W., and Ozato, K.,** The AP-1 site and the cAMP- and serum response elements of the c-*fos* gene are constitutively occupied *in vivo, DNA Cell Biol.,* 10, 537, 1991.

564. **Treisman, R.,** Transient accumulation of c-*fos* RNA following serum stimulation requires a conserved 5′ element and c-*fos* 3′ sequences, *Cell,* 42, 889, 1985.

565. **Feng, J. and Villeponteau, B.,** Serum stimulation of the c-*fos* enhancer induces reversible changes in c-*fos* chromatin structure, *Mol. Cell. Biol.,* 10, 1126, 1990.

566. **Metz, R. and Ziff, E.,** cAMP stimulates the C/EBP-related transcription factor rNFIL-6 to *trans*-locate to the nucleus and induce c-*fos* transcription, *Genes Dev.,* 5, 1754, 1991.

567. **Schröter, H., Mueller, C. G. F., Meese, K., and Nordheim, A.,** Synergism in ternary complex formation between the dimeric glycoprotein p67^{SRF}, polypeptide p62^{TCF} and the c-*fos* serum response element, *EMBO J.,* 9, 1123, 1990.

568. **Schalasta, G. and Doppler, C.,** Inhibition of c-*fos* transcription and phosphorylation in the serum response factor by an inhibitor of phospholipase C-type reactions, *Mol. Cell. Biol.,* 10, 5558, 1990.

569. **Guathier-Rouvière, C., Basset, M., Blanchard, J.-M., Cavadore, J.-C., Fernandez, A., and Lamb, N. J. C.,** Casein kinase II induces c-*fos* expression via the serum response element pathway and p67^{SRF} phosphorylation in living fibroblasts, *EMBO J.,* 10, 2930, 1991.

570. **de Belle, I., Walker, P. R., Smith, I. C. P., and Sikorska, M.,** Identification of a multiprotein complex interacting with the c-*fos* serum response element, *Mol. Cell. Biol.,* 11, 2752, 1991.

571. **Collart, M. A., Belin, D., Vassalli, J.-D., and Vassalli, P.,** Modulations of functional activity in differentiated macrophages are accompanied by early and transient increase or decrease in c-*fos* gene transcription, *J. Immunol.,* 139, 949, 1987.

572. **Jackson, J. A., Holt, J. T., and Pledger, W. J.,** Platelet-derived growth factor regulation of Fos stability correlates with growth induction, *J. Biol. Chem.,* 267, 17444, 1992.

573. **Taylor, L. K., Marshak, D. R., and Landreth, G. E.,** Identification of a nerve growth factor- and epidermal growth factor-regulated protein kinase that phosphorylates the protooncogene product c-Fos, *Proc. Natl. Acad. Sci. U.S.A.,* 90, 368, 1993.

574. **Kacich, R. L., Williams, L. T., and Coughlin, S. R.,** Arachidonic acid and cyclic adenosine monophosphate stimulation of c-*fos* expression by a pathway independent of phorbol ester-sensitive protein kinase C, *Mol. Endocrinol.,* 2, 73, 1988.

575. **Shibanuma, M., Kuroki, T., and Nose, K.,** Inhibition of proto-oncogene c-*fos* transcription by inhibitors of protein kinase C and ion transport, *Eur. J. Biochem.,* 164, 15, 1987.

576. **Nabika, T., Kobayashi, A., Nara, Y., Endo, J., and Yamori, Y.,** Activation of Na$^+$/H$^+$ exchange is unnecessary in the induction of c-*fos* mRNA in serum-stimulated vascular smooth muscle cells, *Clin. Exp. Pharmacol. Physiol.,* 18, 543, 1991.

577. **Tsuda, T., Kaibuchi, K., West, B., and Takai, Y.,** Involvement of Ca^{2+} in platelet-derived growth factor-induced expression of c-*myc* oncogene in Swiss 3T3 fibroblasts, *FEBS Lett.,* 187, 43, 1985.

578. **Kaibuchi, K., Tsuda, T., Kikuchi, A., Tanimoto, T., Yamashita, T., and Takai, Y.,** Possible involvement of protein kinase C and calcium ion in growth factor-induced expression of c-*myc* oncogene in Swiss 3T3 fibroblasts, *J. Biol. Chem.,* 261, 1187, 1986.

579. **Bravo, R., Neuberg, M., Burckhardt, J., Almendral, J., Wallich, R., and Müller, R.,** Involvement of common and cell type-specific pathways in c-*fos* gene control: stable induction by cAMP in macrophages, *Cell,* 48, 251, 1987.

580. **Radzioch, D., Bottazzi, B., and Varesio, L.,** Augmentation of c-*fos* mRNA expression by activators of protein kinase C in fresh, terminally differentiated resting macrophages, *Mol. Cell. Biol.,* 7, 595, 1987.

581. **Mitchell, R. L., Henning-Chubb, C., Huberman, E., and Verma, I. M.,** c-*fos* expression is neither sufficient nor obligatory for differentiation of monomyelocytes to macrophages, *Cell,* 45, 497, 1986.

582. **Stumpo, D. J. and Blackshear, P. J.,** Insulin growth factor effects on c-*fos* expression in normal and protein kinase C-deficient 3T3-L1 fibroblasts and adipocytes, *Proc. Natl. Acad. Sci. U.S.A.,* 83, 9453, 1986.

583. **McCaffrey, P., Ran, W., Campisi, J., and Rosner, M. R.,** Two independent growth factor-generated signals regulate c-*fos* and c-*myc* mRNA levels in Swiss 3T3 cells, *J. Biol. Chem.,* 262, 1442, 1987.

584. **Rebillard, M., Leibovitch, S., Jullien, M., Talha, S., and Harel, L.,** Early stimulation by EGF plus insulin of rRNA, c-*fos*, and actin mRNA expression: inhibition by cytochalasin D, *Exp. Cell Res.,* 172, 432, 1987.

585. **Greenberg, M. E. and Ziff, E. B.,** Stimulation of 3T3 cells induces transcription of the c-*fos* proto-oncogene, *Nature,* 311, 433, 1984.

586. **Eisenman, R. N. and Thompson, C. B.,** Oncogenes with potential nuclear function: *myc, myb,* and *fos, Cancer Surv.,* 5, 309, 1986.

587. **Farquharson, C., Hesketh, J. E., and Loveridge, N.,** The proto-oncogene c-myc is involved in cell differentiation as well as cell proliferation: studies on growth plate chondrocytes *in situ, J. Cell. Physiol.,* 152, 135, 1992.

588. **Bains, M. A., Hoy, T. G., Baines, P., and Jacobs, A.,** Nuclear c-myc protein, maturation, and cell-cycle status of human haematopoietic cells, *Br. J. Haematol.,* 67, 293, 1987.

589. **Smeland, E. B., Beiske, K., Ek, B., Watt, R., Pfeiffer-Ohlsson, S., Blomhoff, H. K., Godal, T., and Ohlsson, R.,** Regulation of c-*myc* transcription and protein expression during activation of normal human B cells, *Exp. Cell Res.,* 172, 101, 1987.

590. **Womer, R. B., Frick, K., Mitchell, C. D., Ross, A. H., Bishayee, S., and Scher, C. D.,** PDGF induces c-*myc* mRNA expression in MG-63 human osteosarcoma cells but does not stimulate cell replication, *J. Cell. Physiol.,* 132, 65, 1987.

591. **Watanabe, T., Sherman, M., Shafman, T., Iwata, T., and Kufe, D.,** Effects of ornithine decarboxylase inhibition on c-myc expression during murine erythroleukemia cell proliferation and differentiation, *J. Cell. Physiol.,* 127, 480, 1986.

592. **Davis, A. C., Wims, M., Spotts, G. D., Hann, S. R., and Bradley, A.,** A null c-*myc* mutation causes lethality before 10.5 days of gestation in homozygotes and reduced fertility in heterozygous female mice, *Genes Dev.,* 7, 671, 1993.

593. **Ebbecke, M., Unterberg, C., Buchwald, A., Stohr, S., and Wiegand, V.,** Antiproliferative effects of a c-*myc* antisense oligonucleotide on human arterial smooth muscle cells, *Bas. Res. Cardiol.,* 87, 585, 1992.

594. **Yokoyama, K. and Imamoto, F.,** Transcriptional control of the endogenous *myc* protooncogene by antisense RNA, *Proc. Natl. Acad. Sci. U.S.A.,* 84, 7363, 1987.

595. **Wickstrom, E. L., Bacon, T. A., Gonzalez, A., Freeman, D. L., Lyman, G. H., and Wickstrom, E.,** Human promyelocytic leukemia HL-60 cell proliferation and c-*myc* protein expression are inhibited by an antisense pentadeoxynucleotide targeted against c-*myc* mRNA, *Proc. Natl. Acad. Sci. U.S.A.,* 85, 1028, 1988.

596. **Watson, P. H., Pon, R. T., and Shiu, R. P. C.,** Inhibition of c-*myc* expression by phosphorothioate antisense oligonucleotide identifies a critical role for c-*myc* in the growth of human breast cancer cells, *Cancer Res.,* 51, 3996, 1991.

597. **Biro, S., Fu, Y.-M., Yu, Z.-X., and Epstein, S. E.,** Inhibitory effects of antisense oligodeoxynucleotides targeting c-*myc* mRNA on smooth muscle cell proliferation and migration, *Proc. Natl. Acad. Sci. U.S.A.,* 90, 654, 1993.

598. **Shichiri, M., Hanson, K. D., and Sedivy, J. M.,** Effects of c-*myc* expression on proliferation, quiescence, and the G_0 to G_1 transition in nontransforming cells, *Cell Growth Differ.,* 4, 93, 1993.

599. **Rustgi, A. K., Dyson, N., and Bernards, R.,** Amino-terminal domains of c-Myc and N-Myc proteins mediate binding to the retinoblastoma gene product, *Nature,* 352, 541, 1991.

600. **ten Kate, J., Eidelman, S., Bosman, F. T., and Damjanov, I.,** Expression of c-*myc* proto-oncogene in normal human intestinal epithelium, *J. Histochem. Cytochem.,* 37, 541, 1989.

601. **Koji, T., Izumi, S., Tanno, M., Moriuchi, T., and Nakane, P. K.,** Localization *in situ* of c-myc mRNA and c-myc protein in adult mouse testes, *Histochem. J.,* 20, 551, 1988.

602. **Krönke, M., Schlüter. C., and Pfizenmaier, K.,** Tumor necrosis factor inhibits *MYC* expression in HL-60 cells at the level of mRNA transcription, *Proc. Natl. Acad. Sci. U.S.A.,* 84, 469, 1987.

603. **Brelvi, Z. S. and Studzinski, G. P.,** Inhibition of DNA synthesis by an inducer of differentiation of leukemic cells, $1\alpha,25$ dihydroxy vitamin D_3, precedes down regulation of the c-*myc* gene, *J. Cell. Physiol.,* 128, 171, 1986.

604. **Rock, R. O., Cleveland, J. L., and Jackowski, S.,** Macrophage growth arrest by cyclic AMP defines a distinct checkpoint in the mid-G_1 stage of the cell cycle and overrides constitutive c-*myc* expression, *Mol. Cell. Biol.,* 12, 2351, 1992.

605. **Skouteris, G. G. and Kaser, M. R.,** Expression of exogenous c-*myc* oncogene does not initiate DNA synthesis in primary rat hepatocyte cultures, *J. Cell. Physiol.,* 150, 353, 1992.

606. **Pape, K. A. and Floyd-Smith, G.,** Effects of interferon-β on Daudi cells and on small cell lung carcinoma cells which over-express the c-*myc* oncogene, *Anticancer Res.,* 9, 1737, 1989.

607. **Lomo, J., Holte, H., de Lange Davies, C., Ruud, E., Laukas, M., Smeland, E. B., Godal, T., and Blomhoff, H. K.,** Downregulation of c-*myc* RNA is not a prerequisite for reduced cell proliferation, but is associated with G₁ arrest in B-lymphoid cell lines, *Exp. Cell Res.,* 172, 84, 1987.

608. **Carding, S. and Reem, G. H.,** c-*myc* gene expression and activation of human thymocytes, *Thymus,* 10, 219, 1987.

609. **Sejersen, T., Björklund, H., Sümegi, J., and Ringertz, N. R.,** N-*myc* and c-*src* genes are differentially regulated in PCC7 embryonal carcinoma cells undergoing neuronal differentiation, *J. Cell. Physiol.,* 127, 274, 1986.

610. **Montarras, D., Pinset, C., Dubois, C., Chenevert, J., and Gros, F.,** High level of c-*fos* mRNA accumulation is not obligatory for renewed cell proliferation, *Biochem. Biophys. Res. Commun.,* 153, 1090, 1988.

611. **Wosikowski, K., Eppenberger, U., Küng, W., Nagamine, Y., and Mueller, H.,** c-*fos*, c-*jun* and c-*myc* expressions are not growth rate limiting for the human MCF-7 breast cancer cells, *Biochem. Biophys. Res. Commun.,* 188, 1067, 1992.

612. **Marshall, A. H., Alper, D., and Hiscott, J.,** Modulation of nuclear proto-oncogene expression and cellular growth in myeloid leukemic cells by human interferon α, *J. Cell. Physiol.,* 135, 324, 1988.

613. **Yaar, M., Peacocke, M., Cohen, M. S., and Gilchrest, B. A.,** Dissociation of proto-oncogene induction from growth response in normal human fibroblasts, *J. Cell. Physiol.,* 145, 39, 1990.

614. **Pompidou, A., Corral, M., Michel, P., Defer, N., Kruh, J., and Curran, T.,** The effects of phorbol ester and Ca ionophore on c-*fos* and c-*myc* expression and on DNA synthesis in human lymphocytes are not directly related, *Biochem. Biophys. Res. Commun.,* 148, 435, 1987.

615. **Teyssier, M. M., Le Garrec, Y. C., Jullian, E. H., Michel, P. J., and Pompidou, A. J.,** Expression of c-*fos* and c-*myc* oncogenes in the P388D1 murine macrophage line treated by immunomodulators: absence of direct correlation with DNA synthesis, *Immunopharmacol. Immunotoxicol.,* 14, 637, 1992.

616. **Phillips, P. D., Pignolo, R. J., Nishikura, K., and Cristofalo, V. J.,** Renewed DNA synthesis in senescent WI-38 cells by expression of an inducible chimeric c-*fos* construct, *J. Cell. Physiol.,* 151, 206, 1992.

617. **Bravo, R., Burckhardt, J., and Müller, R.,** Persistence of the competent state in mouse fibroblasts is independent of c-*fos* and c-*myc* expression, *Exp. Cell Res.,* 160, 540, 1985.

618. **Andrews, G. K., Varma, S., and Ebner, K. E.,** Regulation of expression of c-*fos* and c-*myc* in rat lymphoma Nb-2 cells, *Biochim. Biophys. Acta,* 909, 231, 1987.

619. **Monroe, J. G.,** Up-regulation of c-*fos* expression is a component of the mIg signal transduction mechanism but is not indicative of competence for proliferation, *J. Immunol.,* 140, 1454, 1988.

620. **Lagarrigue, S., Seillan-Heberden, C., Martel, P., and Gaillard-Sanchez, I.,** Altered response to growth factors in rat epithelial liver cells overexpressing human c-Fos protein, *FEBS Lett.,* 314, 399, 1992.

621. **Reuse, S., Pirson, I., and Dumont, J. E.,** Differential regulation of protooncogenes c-jun and jun D expressions by protein tyrosine kinase, protein kinase c, and cyclic-AMP mitogenic pathways in dog primary thyrocytes: TSH and cyclic-AMP induce proliferation but downregulate c-*jun* expression, *Exp. Cell Res.,* 196, 210, 1991.

622. **Reed, J. C., Alpers, J. D., Scherle, P. A., Hoover, R. G., Nowell, P. C., and Prystowsky, M. B.,** Proto-oncogene expression in cloned T lymphocytes: mitogens and growth factors induce different patterns of expression, *Oncogene,* 1, 223, 1987.

623. **Durkin, J. P. and Whitfield, J. F.,** The viral Ki-*ras* gene must be expressed in the G₂ phase if *ts* Kirsten sarcoma virus-infected NRK cells are to proliferate in serum-free medium, *Mol. Cell. Biol.,* 7, 444, 1987.

624. **Tichonicky, L., Kruh, J., and Defer, N.,** Sodium butyrate inhibits c-*myc* and stimulates c-*fos* expression in all the steps of the cell-cycle in hepatoma tissue cultured cells, *Biol. Cell,* 69, 65, 1990.

625. **Adunyah, S. E., Unlap, T. M., Franklin, C. C., and Kraft, A. S.,** Induction of differentiation and c-*jun* expression in human leukemic cells by okadaic acid, an inhibitor of protein phosphatases, *J. Cell. Physiol.,* 151, 415, 1992.

626. **Vriz, S., Taylor, M., and Méchali, M.,** Differential expression of two *Xenopus* c-*myc* proto-oncogenes during development, *EMBO J.,* 8, 4091, 1989.

627. **Nath, P., Getzenberg, R., Beebe, D., Pallansch, L., and Zelenka, P.,** c-*myc* mRNA is elevated as differentiating lens cells withdraw from the cell cycle, *Exp. Cell Res.,* 169, 215, 1987.

628. **Suzuki, K., Suzuki, F., Nikaido, O., and Watanabe, M.,** Suppression of differentiation phenotypes in myogenic cells: association of aneuploidy and altered regulation of c-*myc* gene expression, *Exp. Cell Res.,* 195, 416, 1991.

629. **Fauquet, M., Stehelin, D., and Saule, S.,** Myc products induce the expression of catecholaminergic traits in quail neural crest-derived cells, *Proc. Natl. Acad. Sci. U.S.A.,* 87, 1546, 1990.

630. **Naz, R. K., Ahmad, K., and Kumar, G.,** Presence and role of c-*myc* proto-oncogene product in mammalian sperm cell function, *Biol. Reprod.,* 14, 842, 1991.

631. **Hirayu, H., Dere, W. H., and Rapoport, B.,** Isolation of normal thyroid cells in primary culture is associated with enhanced c-*myc* messenger ribonucleic acid levels, *Endocrinology,* 120, 924, 1987.

632. **Nishikura, K., Kim, U., and Murray, J. M.,** Differentiation of F9 cells is independent of c-*myc* expression, *Oncogene,* 5, 981, 1990.

633. **Larcher, J. C., Basseville, M., Vayssiere, J. L., Cordeau-Lossouarn, L., Croizat, B., and Gros, F.,** Growth inhibition of NIE-115 mouse neuroblastoma cells by c-*myc* or N-*myc* antisense oligodeoxynucleotides causes limited differentiation but is not coupled to neurite formation, *Biochem. Biophys. Res. Commun.,* 185, 915, 1992.

634. **Hashiro, M., Matsumoto, K., Okumura, H., Hashimoto, K., and Yoshikawa, K.,** Growth inhibition of human keratinocytes by antisense c-*myc* oligomer is not coupled to induction of differentiation, *Biochem. Biophys. Res. Commun.,* 174, 287, 1991.

635. **Spizz, G., Hu, J.-S., and Olson, E. N.,** Inhibition of myogenic differentiation by fibroblast growth factor or type β transforming growth factor does not require persistent c-*myc* expression, *Dev. Biol.,* 123, 500, 1987.

636. **Griep, A. E. and Westphal, H.,** Antisense *myc* sequences induce differentiation of F9 cells, *Proc. Natl. Acad. Sci. U.S.A.,* 85, 6806, 1988.

637. **Mitchell, L. S., Neill, R. A., and Birnie, G. D.,** Temporal relationships between induced changes in c-*myc* mRNA abundance, proliferation, and differentiation in HL60 cells, *Differentiation,* 49, 119, 1992.

638. **McAchren, S. S., Jr., Salehi, Z., Weinberg, J. B., and Niedel, J. E.,** Transcription interruption may be a common mechanism of c-*myc* regulation during HL-60 differentiation, *Biochem. Biophys. Res. Commun.,* 151, 574, 1988.

639. **Yen, A., Samuel, V., and Forbes, M.,** Regulation of cell proliferation: late down-regulation of c-*myb* preceding myelomonocytic cell differentiation, *J. Cell. Physiol.,* 153, 147, 1992.

640. **Weinberg, J. B. and Mason, S. N.,** Relationship of acivicin-induced monocytoid differentiation of human myeloid leukemia cells to acivicin-induced modulation of growth factor, cytokine, and protooncogene mRNA expression, *Cancer Res.,* 51, 1202, 1991.

641. **Mulder, K. M. and Brattain, M. G.,** Effects of growth stimulatory factors on mitogenicity and c-*myc* expression in poorly differentiated and well differentiated human colon carcinoma cells, *Mol. Endocrinol.,* 3, 1215, 1989.

642. **Deschamps, J., Mitchell, R. L., Meijlink, F., Kruijer, W., Schubert, D., and Verma, I. M.,** Proto-oncogene *fos* is expressed during development, differentiation and growth, *Cold Spring Harbor Symp. Quant. Biol.,* 50, 733, 1985.

643. **Verma, I. M.,** Proto-oncogene *fos*: a multifaceted gene, *Trends Genet.,* 2, 93, 1986.

644. **Müller, R.,** Cellular and viral *fos* gene: structure, regulation of expression and biological properties of their encoded products, *Biochim. Biophys. Acta,* 823, 207, 1986.

645. **Müller, R., Tremblay, J. M., Adamson, E. D., and Verma, I. M.,** Tissue and cell type-specific expression of two human c-*onc* genes, *Nature,* 304, 454, 1983.

646. **Müller, R., Müller, D., and Guilbert, L.,** Differential expression of c-*fos* in hematopoietic cells: correlation with differentiation of monomyelocytic cells *in vitro, EMBO J.,* 3, 1887, 1984.

647. **Caubet, J.-F., Mitjavila, M.-T., Dubart, A., Roten, D., Weil, S. C., and Vanchenker, W.,** Expression of the c-*fos* protooncogene by human and murine erythroblasts, *Blood,* 74, 947, 1989.

648. **Fisher, C., Byers, M. R., Iadarola, M. J., and Powers, E. A.,** Patterns of epithelial expression of Fos protein suggest important role in the transition from viable to cornified cell during keratinization, *Development,* 111, 253, 1991.

649. **Müller, R., Müller, D., Verrier, B., Bravo, R., and Herbst, H.,** Evidence that expression of c-Fos protein in amnion cells is regulated by external signals, *EMBO J.,* 5, 311, 1986.

650. **Gonda, T. J. and Metcalf, D.,** Expression of *myb, myc* and *fos* proto-oncogenes during the differentiation of a murine myeloid leukaemia, *Nature,* 310, 249, 1984.

651. **Curran, T. and Morgan, J. I.,** Superinduction of c-*fos* by nerve growth factor in the presence of peripherally active benzodiazepines, *Science,* 229, 1265, 1985.

652. **Ito, E., Sonnenberg, J. L., and Narayanan, R.,** Nerve growth factor-induced differentiation in PC-12 cells is blocked by *fos* oncogene, *Oncogene,* 4, 1193, 1989.

653. **Müller, R., Curran, T., Müller, D., and Guilbert, L.,** Induction of c-*fos* during myelomonocytic differentiation and macrophage proliferation, *Nature,* 314, 546, 1985.

654. **Demeneix, B. A., Kley, N., and Loeffler, J.-P.,** Differentiation to a neuronal phenotype in bovine chromaffin cells is repressed by protein kinase C and is not dependent on c-*fos* oncoproteins, *DNA Cell Biol.,* 9, 335, 1990.

655. **Introna, M., Hamilton, T. A., Kaufman, E., Adams, D. O., and Bast, R. C., Jr.,** Treatment of murine peritoneal macrophages with bacterial lipopolysaccharide alters expression of c-*fos* and c-*myc* oncogenes, *J. Immunol.,* 137, 2703, 1986.

656. **Yokota, S., Yuan, D., Katagiri, T., Eisenberg, R. A., Cohen, P. L., and Ting, J. P.-Y.,** The expression and regulation of c-*myb* transcription in B6/*lpr* Lyt-2-, L3T4- T lymphocytes, *J. Immunol.,* 139, 2810, 1987.

657. **Verrier, B., Müller, D., Bravo, R., and Müller, R.,** Wounding a fibroblast monolayer results in the rapid induction of the c-*fos* proto-oncogene, *EMBO J.,* 5, 913, 1986.

658. **Wichelhaus, O., Olek, K., Wappenschmidt, C., and Wagener, C.,** Rapid expression of c-*fos* specific messenger RNA after wounding of a BALB/c-3T3 fibroblast monolayer, *J. Clin. Chem. Clin. Biochem.,* 25, 419, 1987.

659. **Womer, R. B., Frick, K., Mitchell, C. D., Ross, A. H., Bishayee, S., and Scher, C. D.,** PDGF induces c-*myc* mRNA expression in MG-63 human osteosarcoma cells but does not stimulate cell replication, *J. Cell Physiol.,* 132, 65, 1987.

660. **Kost, D. P. and Michalopoulos, G. K.,** Effect of epidermal growth factor on the expression of protooncogenes c-*myc* and c-Ha-*ras* in short-term primary hepatocyte culture, *J. Cell. Physiol.,* 144, 122, 1990.

661. **Gubits, R. M. and Fairhurst, J. L.,** c-*fos* mRNA levels are increased by the cellular stressors, heat shock and sodium arsenite, *Oncogene,* 3, 163, 1988.

662. **Bukh, A., Martinez-Valdez, H., Freedman, S. J., Freedman, M. H., and Cohen, A.,** The expression of c-*fos*, c-*jun*, and c-*myc* genes is regulated by heat shock in human lymphoid cells, *J. Immunol.,* 144, 4835, 1990.

663. **Kumei, Y., Nakajima, T., Sato, A., Kamata, N., and Enomoto, S.,** Reduction of G_1 phase duration and enhancement of c-*myc* gene expression in HeLa cells at hypergravity, *J. Cell Sci.,* 93, 221, 1989.

664. **de Groot, R. P., Rijken, P. J., den Hertoeg, J., Boonstra, J., Verkleij, A. J., de Laat, S. W., and Kruijer, W.,** Microgravity decreases c-*fos* induction and serum response element activity, *J. Cell Sci.,* 97, 33, 1990.

665. **de Groot, R. P., Rijken, P. J., Boonstra, J., Verkeij, A. J., de Laat, S. W., and Kruijer, W.,** Epidermal growth factor-induced expression of c-*fos* is influenced by altered gravity conditions, *Aviat. Space Environ. Med.,* 62, 37, 1991.

666. **Wei, L.-X., Goodman, R., Henderson, A.,** Changes in levels of c-*myc* and histone H2B following exposure of cells to low-frequency sinusoidal electromagnetic fields: evidence for a window effect, *Bioelectromagnetics,* 11, 269, 1990.

667. **Hiraoka, M., Miyakoshi, J., Li, Y.-P., Shung, B., Takebe, H., and Abe, M.,** Induction of c-*fos* gene expression by exposure to a static magnetic field in HeLa S3 cells, *Cancer Res.,* 52, 6522, 1992.

668. **Gonzúlez-Espinosa, C. and García-Sáinz, J. A.,** Angiotensin II and active phorbol esters induce proto-oncogene expression in isolated rat hepatocytes, *Biochim. Biophys. Acta,* 1136, 309, 1992.

669. **Stachowiak, M. K., Sar, M., Tuominen, R. K., Jiang, H.-K., An, S., Iadarola, M. J., Poisner, A. M., and Hong, J. S.,** Stimulation of adrenal medullary cells *in vivo* and *in vitro* induces expression of c-*fos* proto-oncogene, *Oncogene,* 5, 69, 1990.

670. **Itami, M., Kuroki, T., and Nose, K.,** Induction of c-*fos* proto-oncogene by a chemotactic peptide in human peripheral granulocytes, *FEBS Lett.,* 222, 289, 1987.

671. **Isomura, H., Sawada, N., Nakajima, Y., Sakamoto, H., Ikeda, T., Kojima, T., Enomoto, K., and Mori, M.,** Increase in portal flow induces c-*myc* expression in isolated perfused rat liver, *J. Cell. Physiol.,* 154, 329, 1993.

672. **Sharp, F. R., Sagar, S. M., Hicks, K., Lowenstein, D., and Hisanaga, K.,** c-*fos* mRNA, Fos, and Fos-related antigen induction by hypertonic saline and stress, *J. Neurosci.,* 11, 2121, 1991.

673. **Brand, T., Sharma, H. S., Fleischmann, K. E., Duncker, D. J., McFalls, E. O., Verdouw, P. D., and Schaper, W.,** Proto-oncogene expression in porcine myocardium subjected to ischemia and reperfusion, *Circulation Res.,* 71, 1351, 1992.

674. **Williams, S., Evan, G., and Hunt, S. P.,** c-*fos* induction in the spinal cord after peripheral nerve lesion, *Eur. J. Neurosci.,* 3, 887, 1991.

675. **Koistinaho, J., Roivainen, R., and Yang, G.,** Circadian rhythm in c-*fos* protein expression in the rat adrenal cortex, *Mol. Cell. Endocrinol.,* 71, R1, 1990.

Role of Growth Factors in Neoplastic Processes

I. INTRODUCTION

Growth factors and other extracellular signaling agents such as hormones and regulatory peptides have an important role in the regulation of cell proliferation and differentiation, two processes that are critically altered in neoplasia. Growth factors may participate in oncogenic processes *in vivo*, contributing to the proliferation of neoplastically transformed cells, the vascularization of the mass of tumor cells through the formation of new vessels (angiogenesis), and the dissemination of the tumor cells through the body by the invasion and metastatization of normal body tissues. The interactions that exist between growth factors and oncoproteins are of great importance in relation to the development of tumors in both human and nonhuman species.[1] Clinical studies on different human cancers indicate that these interactions may contribute to determining the presentation and pathological characteristics of human tumors and may define, at least in part, the prognosis of different tumors.[2-6] Inhibition of specific signaling pathways involving growth factors and oncoproteins may be useful for cancer therapy.[7]

Alterations in the regulation of pericellular proteolysis by growth factors locally produced by the tumor cells may allow the invasion of metastasization of normal tissues.[8] Both the levels of hormones and growth factors and the expression of cognate receptors for these signaling agents, as well as the expression of specific oncoproteins, may be important for determining the behavior and fate of the cell during the multistage processes of oncogenic transformation and tumor development. Tumor growth and metastasization may depend on endocrine, paracrine, and autocrine influences that involve multiple hormones and growth factors and local interactions between the tumor cells and stromal tissues, as well as mononuclear cells and endothelial cells.[9] The influence of multiple hormones and growth factors in the growth regulation of malignant diseases such as breast cancer may serve as a rational basis for the therapy of these diseases.[10] Not only tumors of classically hormone-responsive organs such as the mammary gland, the uterus, and the prostate, but also tumors of other organs and tissues may be influenced in their growth by the hormonal milieu. Hormones and growth factors may have direct and indirect effects in the growth of melanomas.[11] Similar influences may operate in tumors of the liver, lung, skin, and other organs and tissues.

Alterations in the synthesis and/or secretion of specific hormones may occur in certain types of tumors, especially in those affecting the endocrine glands, allowing the use of hormones as potential tumor markers for the diagnosis and follow-up of patients with these tumors.[12] Not only the mature hormone molecules but also their precursors, subunits, and fragments can find clinical application as tumor markers.[13] Growth factors may also be useful as tumor markers in clinical practice.[14] Ectopic production of hormones and growth factors occurs in a wide diversity of tumors and may represent a major useful marker for human tumors such as small cell undifferentiated carcinoma of the lung.[15] However, the use of certain growth factors for tumor marker purposes is restricted by the fact that they may be produced in high amounts during acute inflammatory diseases.

II. GROWTH FACTORS AND ONCOGENESIS

Animal tumors are composed of a population of transformed cells that show continuous, apparently unrestricted growth. Neoplastically transformed cells exhibit a relatively high degree of autonomy and are usually less dependent on growth factors and other serum components that are normal cells. However, most tumor cells are not completely autonomous but require for their survival and growth *in vitro*, and probably also *in vivo*, the exogenous supply of variable amounts of specific hormones and growth factors. Human tumor cell lines may be continuously grown in a serum-free medium, but this medium must be supplemented with some purified hormones and growth factors (transferrin, insulin, T3, and EGF) as well as with selenium.[16] The human megakaryoblastic cell line M-07, even in the presence of serum, is absolutely dependent on either IL-3 or CSF-2 for its proliferation.[17] Serum-free media can support the growth of some tumor cell lines but are unable to support the long-term growth of other lines, indicating that different types of tumor cells may vary in their degree of autonomy as well as in their nutritional requirements.

A. GROWTH FACTOR-INDUCED NEOPLASTIC TRANSFORMATION

Unregulated expression of growth factors may result in the expression of a transformed phenotype by susceptible cells. Overexpression of growth factors such as PDGF, FGF, and TGF-β can induce and modulate the *in vitro* anchorage-independent growth of certain cell types, a phenomenon that correlates with the expression of a transformed phenotype.[18,19] Normal rat kidney (NRK) cells grown in serum-free conditions express a normal phenotype in the presence of a single growth factor such as EGF or PDGF but undergo anchorage-independent growth in the presence of a combination of EGF and PDGF.[20] PDGF alone can induce the growth in soft agar of several nontransformed cell lines, and TGF-β is able to inhibit the growth response of these cells to PDGF. Complex interactions between PDGF and other platelet-derived factors, including EGF and TGF-β, are essential, however, for the appearance of a transformed phenotype in carcinogen-treated cells.[21] Expression of a vector coding for the human EGF precursor protein is capable of inducing transformation of cultured mouse fibroblasts.[22] Constitutive expression of IL-3 by murine myeloid cells transfected with a plasmid vector containing the murine IL-3 gene ligated to an active promoter may result in autonomous growth associated with production of biologically active IL-3.[23] The transfected IL-3-secreting cells display tumorigenic properties in syngeneic mice.

The oncogenic potential of FGFs has been demonstrated in a number of cellular systems. Basic FGF is itself unable to transform cells when expressed at high levels from a recombinant plasmid, but can acquire this ability after fusion with a secretory signal sequence, indicating the requirement for an interaction between the growth factor and its receptor on the cell surface.[24] The tumorigenic potential of FGFs is limited, however. Acidic FGF may enhance the tumorigenic potential of Rat-1 fibroblasts rather than act as a factor with primary transforming properties.[25] Thus, acidic FGF behaves more as a progression factor than as an oncogenic factor. Expression at high levels of growth factor receptors may result in cell transformation when the cognate ligand is present. Overexpression of the IGF-I receptor, for example, may lead to the induction of a transformed phenotype in a ligand-dependent manner.[26]

Combinations of growth factors and other growth modulating agents may be required for the phenotypic transformation of particular types of cells, for example, NRK cells.[27] However, transformation of NRK cells does not require the presence of specific growth factors, but may rather reflect a general response of these cells to the action of unphysiological concentrations of multiple growth factors.[28] Combinations of EGF, PDGF, TGF-β, and retinoic acid are required to induce phenotypic transformation of NRK cells, but none of these factors is absolutely essential for the process of transformation.

The species or origin and the type of cell are critical for the results of experiments related to growth factor-induced neoplastic transformation. Some species and some types of cells from a given species may be resistant to transformation. Infection with a retroviral vector coding for TGF-α results in tumorigenic transformation of normal mouse mammary epithelial cells but not normal rat fibroblasts.[29] This difference may be explained, at least in part, by the higher levels of EGF receptors expressed by mouse mammary cells. In general, human cells are resistant to transformation induced by either growth factors or proto-oncogene products,[30] but some types of human cells can be transformed *in vitro* by the artificial expression of high levels of certain growth factors.

The relevance of the results obtained with *in vitro* systems to the complex situations occurring during tumorigenesis *in vivo* is not readily apparent. Hormones and growth factors interacting in a complex manner may be instrumental for the progression of tumors toward disseminated disease. At least six growth factors are important for the development of melanomas: EGF, NGF, FGF, PDGF, TGF-β, and insulin.[31] Monocytes and other blood cells that frequently infiltrate tumors can secrete diffusible factors capable of modulating the growth of tumor cells.[32] Thus, it is clear that the hormonal environment of the whole organism, as well as the local action of hormones and growth factors, are of great importance for the development of neoplastic diseases, not only under experimental conditions,[33-37] but also in human cancer.[37-40] The local microenvironment may have the most important role in the normal or abnormal behavior of cells. Interactions between the mesenchyme and epithelial cells may determine, through the action of locally produced growth factors or other local signaling mechanisms, the expression or lack of expression of a transformed phenotype in a given epithelial cell.[41]

B. INFLUENCE OF THE TRANSFORMING AGENT

The particular type of transforming agent may contribute to determination of the specific requirement for the exogenous supply of hormones and growth factors. Carcinogen-transformed mouse cells are less dependent on the exogenous supply of growth factors such as EGF.[42] Reversion of transformed mouse embryo fibroblasts to a normal phenotype induced by treatment with retinoic acid is associated with a restored ability to respond to EGF.[43] The requirements of mouse 3T3 cells for exogenous insulin and PDGF are diminished or abolished after SV40-induced transformation.[44]

Transformation of different types of cells by different viruses may result in varied changes in the requirement for the exogenous supply of growth factors. Transformation of the rat fibroblast cell line NRK-49F by polyoma virus abrogates the requirement for insulin and retinoic acid but not for EGF and fibronectin. In contrast, transformation induced by human adenovirus type 5 (Ad5) does not result in elimination of the requirement for exogenous supply of these factors.[45] The biological basis of this difference is unknown. Rat T-cell lymphoma cells inoculated with a chronic retrovirus, the Moloney murine leukemia virus (M-MuLV), become independent of the exogenous supply of IL-2, and this effect depends on the integration of the virus within a specific locus, the growth factor-independence-1 (*Gfi*-1) gene.[46] The *Gfi*-1 gene encodes a protein containing six zinc finger domains in its carboxyl-terminal region, suggesting that it is probably involved in transcriptional regulation. In the adult animal, expression of the Gfi-1 protein is restricted to thymus, spleen, and testis. Gfi-1 may be involved in events occurring after the interaction of the growth factor with its receptor, probably during the transition from G_1 to S phase of the cell cycle.

Cells infected with acute retroviruses or transfected with viral oncogenes or activated proto-oncogenes may become relatively independent of the presence in the medium of serum or specific factors contained in serum. Rat adrenal cells infected with the Kirsten murine sarcoma virus (K-MuSV), which transduces the v-K-*ras* oncogene, lose the differential response to growth factors contained in serum.[47] Viral oncoproteins may act as potent mitogens, inducing a relative autonomy of the transformed cells, which is reflected in their capacity to proliferate in the absence of exogenous growth factors. The v-Src protein is a potent and complete mitogen that is able to stimulate host cell proliferation without the help of exogenous growth factors.[48,49] However, not all oncogenes are equally efficient in inducing abrogation of the requirement of cells for specific growth factors. Transfection of expression vector carrying the oncogenes v-*mos*, v-*src*, and v-*sis*, as well as a mutant c-H-*ras* proto-oncogene, eliminates the requirement of mouse fibroblasts (NIH/3T3 and BALB/c 3T3 cells) for PDGF or FGF, but transfection with the v-*fos* oncogene does not abrogate the requirement for the two growth factors.[50] Infection of rat thyroid cell lines with retroviruses carrying different oncogenes (v-H-*ras*, v-K-*ras*, v-*mos*, v-*raf*, and v-*src*) may result in abolishment of the requirement for the exogenous supply of several growth factors.[51] After infection, the thyroid cells become independent of TSH for their growth.

Activated proto-oncogenes may have effects similar to those of viral oncogenes. Introduction of a mutated c-H-*ras* gene into a murine bone marrow-derived IL-3-dependent cell line (FDC-P2 cells) results in the acquisition of growth factor independence and tumorigenic conversion of the cells.[52] An autocrine mechanism would not be responsible, however, for the alterations observed in the FDC-P2 cells expressing the activated proto-oncogene.

Cooperation between growth factors and oncoproteins may be important for the expression of a transformed phenotype in cells infected by chronic retroviruses or DNA viruses with oncogenic potential. Mitogenic growth factors may cooperate with polyoma virus middle-T antigen in the transformation of cultured rat cells.[53] Infection of mouse fibroblasts with SV40 may result in the production and release of growth factors, including PDGF-like activity as a consequence of increased transcription of the c-*sis* proto-oncogene.[54]

C. INFLUENCE OF THE TYPE OF CELL

The sensitivity of different tumors to hormones and growth factors may show great variation, according to the type of tumor. Even a given type of tumor may respond in different ways to the signaling agents according to the histological subtype and the biological behavior of the tumor. Human non-Hodgkin lymphomas, for example, may derive from both T- and B-cell precursors, and the neoplastic cells of these tumors may exhibit a normal response (increased DNA synthesis) to the respective specific interleukin growth factor, TCGF or BCGF, when the tumors are histologically classified as small T- and small B-cell non-Hodgkin lymphomas, respectively. On the other hand, the T- and B-cell types of large-cell non-Hodgkin lymphomas tend to be aggressive tumors and the cells from these tumors do not show a normal DNA synthesis response when they are exposed to the specific interleukin growth factor.[55]

The cells from both malignant and benign tumors of a given type may exhibit marked differences in their response to specific growth factors, and even within the cell population of a given tumor the response may be heterogeneous from one cell to another. Epithelial cells from benign mastopathies, for example, show in some cases responses that are similar to those of normal breast cells, whereas the responses of other cells are comparable to those of mammary cancer cells.[56] The sensitivity of cells to growth factors may show variation even among different subclones of the same cultured cells.[57] Moreover, clones responsive to growth factors may be more susceptible to transformation induced by retroviruses transducing oncogenes such as v-K-*ras* and v-*abl*. Cells transfected with *myc* and *ras* oncogenes may show differential responsiveness to particular growth factors, and *myc*-transfected cells may exhibit selective stimulation by EGF.[58]

D. REQUIREMENT FOR MULTIPLE HORMONES AND GROWTH FACTORS

Tumor growth *in vivo* depends on complex interactions between multiple hormones, growth factors, regulatory peptides, and other extracellular signaling agents. The growth of various tumor cell lines *in vitro* is stimulated more efficiently by serum than by selected purified growth factors, which is due to the fact that serum contains a highly effective mixture of growth-stimulating factors involved in both autocrine and paracrine mechanisms.[59] Serially transplantable rat mammary tumor cells consist of cell populations that have acquired autonomy of single or multiple factors for growth in serum-free culture.[60] Lines derived from these cells may exhibit neoplastic potential *in vivo*, and independence of multiple growth factors in these cells can be mediated by distinct alterations in the complex mechanisms associated with growth control, including the operation of one or more autocrine loops.

Studies performed *in vitro* may contribute to the characterization of growth factors specifically required for tumor cell growth. Comparison of human metastatic melanoma cell lines to normal human foreskin fibroblasts for the production of transcripts specific for 11 different growth factors suggests that TGF-β_2, TGF-α, and basic FGF are important for melanomagenesis, while acidic FGF and FGF-5 may play a more important role in tumor progression than in oncogenesis.[61] Analysis of the medium conditioned by incubation of the human lung carcinoma cell line A549-1 indicated that two components of the medium, TGF-α and IGF-I, are required for growth response in clonal assays of newly cultured cells derived from primary solid lung tumors.[62] Among different hormones and growth factors supplemented to a human tumor stem cell assay, cortisol was found to be the most effective single supplement for the stimulation of colony growth.[63] Supplementation with insulin, estradiol, or both had some growth-promoting effects but not as great as hydrocortisone. Moreover, the addition of insulin, estradiol, or both often resulted in a negative interaction with hydrocortisone.

In addition to hormones and growth factors, other components present in serum may influence the growth of transformed cells *in vitro* and *in vivo*. Addition of thrombin to the chemically defined medium where transformed cells (T-47D human metastatic breast carcinoma cells) are being cultured may result in further stimulation of growth in comparison to that due to the presence of typical growth factors (EGF, insulin, and transferrin).[64] The mechanism of growth stimulation by thrombin is not clear but it is known that thrombin action involves its binding to high-affinity cell surface receptors. Moreover, thrombin may be important in the metastatic spread of tumor cells *in vivo*.

E. ANGIOGENIC FACTORS AND STROMAL FACTORS

Neovascularization is tightly associated with the processes of regeneration of normal organs and tissues as well as with the growth of tumors ever since their initial stages. The angiogenic factors are factors capable of stimulating the formation and growth of blood vessels. They may be locally produced by stromal or parenchymal cells and may act directly on vascular endothelial cells to stimulate their locomotion and/or mitosis or may have an indirect action by mobilizing host cells such as macrophages to release growth factors with specificity for endothelial cells.[65,66]

Growth factors may support tumor growth not only by a direct stimulation of the proliferation of transformed cells but also by promoting angiogenic processes that are required for nutrient supply. Growth factors with angiogenic properties include EGF, acidic FGF, basic FGF, TGF-α, TGF-β, TNF-α, IL-1α, IL-11, platelet-derived ECGF, and angiotensin II. Other factors with angiogenic properties detected in different systems remain less well characterized, and some of them are represented by lipids. The action of angiogenic factors may be opposed by that of factors with anti-angiogenic activity, which may display antitumor activity.[67] Oncoproteins and tumor suppressor proteins may act as mediators of the action of extracellular factors with angiogenic or anti-angiogenic properties, respectively, thus favoring tumor progression or regression. The angiogenic factors are discussed in Volume 2.

The growth of parenchymal cells from both normal and tumor tissues partially depends on the microenvironment created by cells from the vicinity, especially by stromal cells.[68] Carcinomas often contain massive amounts of stromal cells that are probably required for tumor growth *in vivo*. In addition to the production of growth factors that stimulate tumor cell proliferation and angiogenesis, tumor cells such as mammary carcinoma cell lines may produce factors with mesenchymal cell-stimulating activity.[69] In this manner, by producing factors that stimulate their own proliferation and the growth of endothelial cells and mesenchymal cells, the tumor cells may be able to create optimal conditions for tumor growth *in vivo*.

III. AUTOCRINE MECHANISMS IN ONCOGENESIS

Neoplastic cells frequently exhibit an autonomous behavior that is characterized by a partial or total independence of the exogenous supply of hormones and growth factors. This autonomy may be associated with the operation of autocrine and/or paracrine mechanisms of growth control occurring in tumor cells.[70-74] Functional autocrine loops require the production of a given hormone or growth factor by the cell and the presence of specific receptors for the ligand in the same cell as well as the development of a cellular response to the ligand (Figure 10.1). Regulators of cell proliferation and cell differentiation such as the cytokines may also act as autostimulatory factors in various types of malignancies.[75] In order to prove that a factor secreted by a normal cell or a tumor cell acts via an autocrine pathway, there must be demonstration that both the ligand and its receptor are expressed by the cell, that interference with the ligand-receptor interaction inhibits cell growth, and that the growth inhibition caused by this interference is specific.[76] Selective antagonists capable of blocking the activity of growth factors acting through autocrine loops could be applied, in principle, to the treatment of cancer patients, particularly in cases of metastatic disease where the dissemination of tumor cells may be influenced by paracrine mediators.[77]

The proliferation of tumor cells frequently depends on the operation of autocrine mechanisms. Two or more independent autocrine loops may arise in clones derived from a single tumor cell.[78] The human pancreatic carcinoma cell line MIA-PaCa 2, for example, expresses both IGF-I and TGF-α and their receptors, which would constitute a double autocrine loop in these cells.[79] Coexpression of two or more growth factors and their receptors may occur frequently in malignant tumors,[80] but a similar coexpression is observed in various benign tumors.[81,82] Higher levels of growth factor expression may be observed in malignant cells as compared with cells from benign tumors, but there are no definite correlations between growth factor overexpression and tumor progression. Moreover, normal cells such as placental cytotrophoblasts, may coexpress growth factors and their receptors.[83]

Formal evidence for autocrine growth stimulation is obtained only when it is proved that growth is retarded by specific inhibitors of ligand-receptor interaction, such as blocking antibodies.[84] This effect would be possible in "public" autocrine loops in which the growth factor must be secreted and may act on the cell producing the factor as well as on surrounding cells of the same type. However, there is some evidence suggesting the operation, at least in some cases, of intracellular or "private" loops in which secretion of the growth factor is not required for autonomous growth.[85] In these cases, cell proliferation is not density-dependent, even in the absence of exogenous factors, and antibodies that neutralize the factor are not capable of preventing continued growth. Binding of the growth factor to its receptor would occur in these cases within intracellular compartments. Such a mechanism has been suggested to operate in the case of the v-Sis oncoprotein, which is closely related in its structure and function to the B chain of PDGF.[86] However, the validity of this type of mechanism is controversial. The autocrine mechanism of v-*sis* oncogene-induced transformation requires cell surface localization of internally activated receptors in order to functionally couple with intracellular mitogenic signaling pathways.[87] Prolactin functions as an autocrine or paracrine factor for the growth of the Nb2 rat lymphoma cell line, and secretion of prolactin into the medium is required for the hormone to stimulate the growth of Nb2 cells; when these cells are transfected with a vector directing the synthesis of a nonsecreted form of prolactin no cell growth is observed.[88] Constitutive production of basic FGF by transfection of a vector encoding the K-*fgf* oncogene in the adrenal carcinoma cell line SW-13 (which does not clone in soft agar even in the presence of fetal calf serum unless supplemented with basic FGF) requires that basic FGF is secreted in order to function as a growth stimulator *in vitro* as well as after implantation into athymic nude mice.[89] Thus, interaction of growth factors or hormones with intracellular receptors would not be a biologically significant mechanism for the autocrine growth of normal or neoplastic cells, and secretion of the growth factor or hormone is most probably required for the operation of autocrine loops.

Acquisition of the capacity to produce and utilize growth factors in an autocrine fashion may occur progressively in certain neoplastic cells and may be associated with higher degrees of malignant behavior. After prolonged cultivation, chemically transformed cells may acquire step-wisely the capacity of anchorage-independent growth in synthetic liquid or semisolid medium without the supplementation of insulin or transferrin, which may be due to an increased capacity of these cells to produce and utilize autologous growth factors.[90] Carcinogen-transformed rat mammary epithelial cells display a markedly

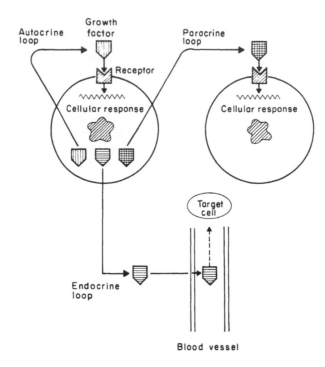

Figure 10.1 Autocrine, paracrine, and endocrine loops.

enhanced growth potential in long-term culture, as compared to that of the respective normal cells, and expression of growth factor independence in a subset from these cells is associated with high neoplastic potential *in vivo*.[91]

Autocrine factors capable of stimulating DNA synthesis and cell proliferation specifically in the tumor cells of origin *in vivo* have been isolated from some animal tumors.[92] Some autocrine factors may stimulate cell motility and may have a role in the processes of tumor invasion and metastasization. An autocrine motility factor (AMF) was identified in the human melanoma cell line A2058.[93] The AMF elicits increases in tumor cell motility and phosphoinositide metabolism via a pertussis toxin-sensitive G protein signal transduction pathway.[94] AMF and IGF-I would act in a synergistic manner through independent cell surface receptors to stimulate the motility of tumor cells and may be involved in the process of metastasization.[95] The possibility that such factors are involved in the expression of a transformed phenotype is reinforced by the observation that the expression of a cDNA encoding a growth factor in a factor-dependent cell line may result in autonomous growth and tumorigenicity.[96] A proposed model for autocrine-regulated proliferation of tumor cell populations would accommodate any growth curve that fits the data.[97]

Autocrine mechanisms are generally associated with secretion of the growth factor and interaction of the secreted factor with a functional receptor on the surface of the same cell. In some systems *in vitro*, this interaction may result in the expression of a transformed phenotype. Murine BALB/c 3T3 fibroblasts transformed with a human cDNA encoding basic FGF produce the growth factor, and binding of the synthesized basic FGF to the receptor on the cell surface may result in neoplastic transformation.[98] An insulin-related factor produced by a mouse teratoma-derived insulin-dependent cell line binds to insulin receptors on the same tumor cells and acts in an autostimulatory manner.[99] Factors associated with the autocrine growth of human tumor cell populations have been identified.[100-102] Human hepatoma cell lines may express PDGF and IGF-I as well as their receptors, suggesting that autocrine regulation may be an important mechanism for the maintenance of a transformed state in these cells.[103] IGF-II and its surface receptor are frequently expressed in human rhabdomyosarcoma tumors, which may result in stimulation of the growth and motility of the tumor cells by an autocrine mechanism.[104] Endogenous secretion of IL-1 by human AML cells stimulates secretion of CSF-2 and induces the function of an autocrine loop involved in the autonomous growth of the leukemic blast cell population.[105]

Not only classic hormones or growth factors, but also some regulatory peptides may have a role in autocrine mechanisms associated with the growth of normal or malignant cells. The gastrointestinal peptide gastrin, well known as a stimulator of gastric acid secretion, may display growth factor-like activities such as stimulation of the proliferation of colonic mucosa cells. An endogenous gastrin/cholecystokinin-like peptide displays growth stimulatory activity in certain human carcinoma cell lines, and this effect is exerted by an autocrine mechanism.[106] Gastrin receptors are frequently present in surgical specimens of human colon cancers.[107] Gastrin stimulates the clonal proliferation of small-cell lung cancer (SCLC) cells *in vitro*.[108] Human SCLC cell lines produce and secrete, in addition, bombesin-like peptides such as the gastrin-releasing peptide (GRP) and express receptors for these peptides.[109-113] A monoclonal antibody that binds to the carboxyl-terminal region of bombesin-like peptides inhibits the clonal growth of SCLC cells *in vitro* and the growth of xenografts of these cells *in vivo*.[114] An antagonist to the bombesin receptor blocks, in the absence of exogenous bombesin, the growth of a human SCLC cell line (NCI-H345) that constitutively produces GRP.[115] The autocrine growth mechanism operating in human SCLC cells appears to involve phosphorylation of proteins on tyrosine residues by the activated bombesin receptor kinase.[116]

The factors involved in putative autocrine loops operating in normal and tumor cells remain largely uncharacterized. An autostimulatory activity detected in a spontaneous murine pre-B-cell transformant, as well as in two independently derived pre-B-cell lines transformed by the A-MuLV chronic retrovirus, is apparently different from known hematopoietic growth factors.[117] Growth factor activities detected in human renal carcinoma cells corresponded, in large part, to unidentified heparin-binding growth factors.[118] Substances immunologically crossreactive with insulin are secreted in an autocrine manner by murine melanoma B16 cells.[119] These substances may be responsible, at least in part, for the hypoglycemia associated with certain nonpancreatic β-cell tumors. The characterization of factors involved in the autonomous growth of cell populations may be achieved by the analysis of factor-independent mutant cells.[120] This analysis may allow the identification of genes directly involved in the acquisition of growth autonomy by tumor cells.

The general validity of autocrine mechanisms for oncogenesis is questionable. Many different types of cells may express a transformed phenotype through mechanisms not necessarily involving the operation of autocrine loops. An unscheduled production of a growth factor by a cell that carries the respective receptor may not be expected to cause anything but a hyperplastic response of otherwise normal cells. Activation of the mitogenic pathway of a growth factor is not sufficient in itself for tumorigenesis, and the autocrine stimulation of cells needs to be associated with other genetic alterations, including a block in cell differentiation, in order to cause malignant transformation.

Both hematopoietic neoplasms and solid tumors may require the supply of exogenous growth factors for optimal growth *in vitro* and *in vivo*. Human leukemic cells usually fail to proliferate autonomously *in vitro* and may require the supply of specific growth factors.[121,122] These cells probably require the availability of exogenous growth factors for their *in vivo* proliferation as well. Autocrine growth has not been proven as a major mechanism for the growth of human tumor cells *in vivo* in myeloid and lymphocytic leukemias.[123] In general, the role of autocrine mechanisms in relation to the growth of human tumors *in vivo* remains unclear. Nonhematopoietic primary human tumors maintained in culture may frequently respond to the exogenous supply of particular growth factors with increased growth rate. A majority of the specimens of human tumors maintained *in vitro* in a medium supplemented with transferrin, insulin, cortisol, and estradiol respond to the addition of EGF with accelerated cellular division.[124] This response may be independent of tumor type, but most breast carcinomas have little growth without exogenous EGF.

The production of growth factors by malignant cells is not necessarily associated with autocrine growth. For example, P19 EPI-7 embryonal carcinoma cells produce growth factors such as TGF-β and IGF-II, as well as a PDGF-like factor, but they do not contain receptors for some of the produced factors, and the biological significance of growth factor secretion by P19 EPI-7 cells is unknown.[125] None of the growth factors secreted by P19 EPI-7 cells is capable of stimulating their proliferation. Human colon adenocarcinoma cell lines secrete several growth factors (TGF-α, TGF-β, and a PDGF-like material), but there is no detectable correlation between the levels of these factors and soft agar growth or tumorigenesis.[126] Thus, while secretion of growth factors is an invariant property of colorectal carcinoma cell lines, other regulatory mechanisms would be involved in determining the phenotypes characteristic of malignant behavior.

Autocrine growth is not unique to tumor cells but may also operate in a diversity of normal cells. Clonal expansion of T lymphocytes is regulated in an autocrine fashion by IL-2 through induction of the IL-2 receptor in the same cells.[127] The lymphokine BCGF may function in an autocrine fashion in both normal and transformed cells.[128] The developing human placenta represents a case of autocrine growth regulation, because it displays abundant high-affinity PDGF receptors and responds to PDGF by expression of c-*myc* gene transcripts and stimulation of DNA synthesis.[83] IGFs may have an important autocrine/paracrine role in the regulation of human placental growth.[129] The growth of keratinocytes *in vitro* may depend on autocrine and paracrine factors that act through activation of both EGF receptor-dependent and -independent mitogenic pathways.[130] Human and murine keratinocytes produce TGF-α and TGF-β, which act as autocrine factors in controlling the proliferation of these cells.[131,132] Autocrine growth of keratinocytes involves the opposing effects of endogenously produced TGF-α, which acts as a growth stimulator, and TGF-β, which acts as a growth inhibitor.[133] TGF-α expression in keratinocytes may be modulated by EGF or TGF-α itself. TGF-β stimulates primary human skin fibroblast DNA synthesis via an autocrine production of PDGF-related peptides.[134] Adult arterial smooth muscle cells express PDGF A-chain mRNA, secrete a PDGF-like mitogen, and utilize this mitogen in an autocrine or paracrine manner *in vitro*.[135] The same cells may use autocrine mechanisms for their growth *in vivo*, as shown by the results obtained in experiments with balloon catheter aortic injury in rats, which induces immediate-early gene expression followed by elevated levels of transcripts coding for PDGF-A, TGF-β, and basic FGF.[136] Autocrine growth factors are produced and secreted by both normal and immortalized Schwann cells.[137] These examples clearly show that autocrine stimulatory and inhibitory phenomena are not an exclusive property of tumor cells and that similar or identical mechanisms may operate in both normal and neoplastic cells.

IV. GROWTH FACTORS AND VIRAL ONCOPROTEINS

Acute retroviruses carrying viral oncogenes are capable of inducing malignant transformation in a diversity of cells both *in vivo* and *in vitro*. In some cases this effect may be mediated by the endogenous production of growth factors that are capable of producing an autostimulatory effect in the cells expressing the oncogene. However, the production of both a viral oncoprotein and an endogenous growth factor may not suffice in all cases for maintaining the expression of a transformed phenotype.

Cells transformed by acute retroviruses, or by oncogenes cloned from these viruses, may become independent of the exogenous supply of specific growth factors and may produce different types of growth factors, especially TGFs.[138-149] Unlike PDGF, the v-Src oncoprotein is able to stimulate the proliferation of NRK cells without the help of serum growth factors.[150] Thus, for some types of cells the v-Src protein behaves as a complete mitogen. Acute avian retroviruses containing oncogenes of the v-*src* family can render v-*myb*-transformed myeloblasts and v-*myc*-transformed macrophages independent of the exogenous supply of growth factors for their proliferation.[151] In rodent myeloid cells transformed by the v-*myb* oncogene, independency of the supply of growth factors is achieved via the secretion of a growth factor activity. Introduction of a retroviral vector containing the v-*myc* and v-*raf* oncogenes can abrogate the requirement of cell lines for IL-3 or IL-2 for growth, and this abrogation is associated with suppression of endogenous c-*myc* proto-oncogene expression.[152] Induction of estrogen-independent tumorigenicity in the human breast cancer cell line MCF-7 by introduction of a v-H-*ras* oncogene may occur by secretion of mitogen which would function in an autocrine manner.[153] These observations, and the fact that several oncoproteins are structurally and functionally similar to growth factors or their receptors, led to the suggestion that the relatively autonomous growth of oncogene-transformed cells, and perhaps also the growth of other types of neoplastic cells, may be due to the endogenous production of growth factors that would act in an autocrine fashion.[154] However, c-H-*ras* gene activation and growth factor independence may be associated with independent pathways to malignancy in rat mammary carcinogenesis.[155] Thus, although *ras* oncogenes are potent initiators of mammary carcinogenesis in rodents, other alterations are probably responsible for the appearance of the independence of growth factors that occurs during tumor progression.

Cooperation between two different oncoproteins is required in some cases for the production of autocrine growth factors in cells expressing viral oncogenes. The acute retrovirus MH2 carries two oncogenes, v-*myc* and v-*mil*, and is capable of inducing *in vitro* the transformation of hematopoietic cells with a macrophage-like phenotype.[156] Some mutants of the MH2 virus lack the v-*mil* oncogene.[157] Although the v-*mil* oncogene alone cannot directly transform macrophages, it induces v-*myc*-transformed macrophages to produce a specific type of growth factor that can be used by the cells to stimulate their own growth in an autocrine manner.[156] Furthermore, v-*mil* markedly enhances the capacity of v-*myc* to induce monocytic neoplasms *in vivo*. Reversion to a normal phenotype induced by particular agents may be accompanied by a restored ability to respond to specific growth factors.[44] Monocytic cells from mouse marrow are immortalized by a recombinant retrovirus carrying the combination of v-*raf* and v-*myc* oncogenes, and the proliferating cells do not require the addition of a specific growth factor supplement any more.[158] Both v-*myc* and v-*raf* oncogenes have to be expressed to induce proliferation of the bone marrow cells in standard medium, because viruses carrying either oncogene alone do not promote cell growth. These results suggest that the expression of both oncogenes may overrun the need for exogenous growth factors. These results may be interpreted on the basis of a model in which a competence signal is originated by v-*myc* expression and a progression signal by v-*raf* expression.[159]

Growth factors may have an important role in regulating the growth of oncogene-transformed cells. Factors contained in serum may contribute to regulation of the progression of K-MuSV-induced transformation in populations of freshly explanted rat adrenal cells, in the continued presence of the v-Ras oncoprotein.[160] Revertants of K-MuSV-transformed cells, although expressing elevated levels of v-K-Ras and producing TGFs, may express most of the properties associated with the nontransformed state.[161] In murine myeloid leukemic cell lines maintained in suspension cultures, no relationship was found between the capacity of a line to grow autonomously and its capacity to release autostimulating growth factors.[162] Thus, the expression of a viral oncoprotein is not necessarily associated with a transformed phenotype and, at least in some cases, oncogene-induced transformation may be independent of the endogenous production of growth factors by an autocrine type of growth stimulation.

The study of temperature sensitive (*ts*) mutants of acute transforming retroviruses may help to define the possible relations between the oncoproteins produced by these viruses and specific growth factors. The acute leukemia virus ALV E26 contains two oncogenes, v-*myb* and v-*ets*, and is capable of inducing malignant transformation of various cell types *in vitro* (myeloblasts, erythroblasts, fibroblasts) and causes a mixed erythroid/myeloid leukemia in chicks.[163] E26 virus mutants that are *ts* for myeloblast transformation have been isolated.[164] At the permissive temperature, *ts* E26-transformed myeloid cells resemble macrophage precursors and proliferate rapidly, provided the medium contains a specific growth factor, the chicken myelomonocytic growth factor. When shifted to the nonpermissive temperature, the cells stop growing, differentiate into macrophage-like cells, and lose the responsiveness to the particular growth factor, but still secrete the factor or a substance similar to the factor.[164]

Transformation of the same cells by different oncogenes may result in differences in their responses to specific growth factors. Chicken heart mesenchymal cells transformed with a c-*myc* gene and treated with heparin become quiescent and assume a normal morphology but are hypersensitive to stimulation with EGF, FGF, and PDGF.[165] These cells, however, do not respond to insulin. On the other hand, the same cells transformed with a v-*ras* oncogene and treated with heparin are refractory to mitogenic stimulation with EGF or FGF but are hypersensitive to insulin or IGFs. It is thus clear that transformation by different viral oncogenes may result in altered responses of the cells to specific growth factors and that in some cases the transformed cells become hypersensitive to the stimulation by particular growth factors.

Neoplastic transformation induced by viral oncogenes or activated proto-oncogenes is not necessarily dependent on the endogenous production of growth factors and operation of autocrine mechanisms. A replication-defective murine retrovirus expressing the human *trk* oncogene can abrogate the growth factor requirement of hematopoietic cells through a nonautocrine mechanism.[166] The Abelson murine leukemia virus (A-MuLV), which contains the v-*abl* oncogene, is capable of inducing transformation of growth factor-dependent myeloid cell lines into growth factor-independent, tumorigenic cell lines; however, v-*abl*-induced transformation is apparently not associated with endogenous synthesis of growth factors capable of determining autocrine stimulation.[167-169] A-MuLV-derived transformants of mast cells do not express or secrete detectable levels of the specific growth factor nor is their growth inhibited by an antiserum against the factor.[169] Neither the autonomous proliferation nor the spontaneous differentiation of A-MuLV-induced leukemic cells may involve autocrine secretion of growth factors.[170] Such results, as well as the results obtained in other systems,[171,172] argue strongly against the general validity of an autocrine mechanism of oncogenesis.

REFERENCES

1. **Pimentel, E.,** *Hormones, Growth Factors, and Oncogenes,* CRC Press, Boca Raton, FL, 1987.
2. **Westphal, M. and Herrmann, H.-D.,** Growth factor biology and oncogene activation in human gliomas and their implications for specific therapeutic concepts, *Neurosurgery,* 25, 681, 1989.
3. **Bonneterre, J., Peyrat, J. P., and Demaille, A.,** Growth factors and oncogenes in human solid tumors: clinical aspects, *Biomed. Pharmacother.,* 44, 25, 1990.
4. **Tahara, E.,** Growth factors and oncogenes in human gastrointestinal carcinomas, *J. Cancer Res. Clin. Oncol.,* 116, 121, 1990.

5. **Frauman, A. G. and Moses, A. C.,** Oncogenes and growth factors in thyroid carcinogenesis, *Endocrinol. Metab. Clin. N. Am.,* 29, 479, 1990.

6. **Thompson, T. C.,** Growth factors and oncogenes in prostate cancer, *Cancer Cells,* 2, 345, 1990.

7. **Powis, G. and Kozikowski, A.,** Growth factor and oncogene signalling pathways as targets for rational anticancer drug development, *Clin. Biochem.,* 24, 385, 1991.

8. **Laiho, M. and Keski-Oja, J.,** Growth factors in the regulation of pericellular proteolysis: a review, *Cancer Res.,* 49, 2533, 1989.

9. **Osborne, C. K. and Arteaga, C. L.,** Autocrine and paracrine growth regulation of breast cancer: clinical implications, *Breast Cancer Res. Treat.,* 15, 3, 1990.

10. **Manni, A.,** Endocrine therapy of metastatic breast cancer, *J. Endocrinol. Invest.,* 12, 357, 1989.

11. **Walker, M. J.,** Role of hormones and growth factors in melanomas, *Semin. Oncol.,* 15, 512, 1988.

12. **Pimentel, E.,** Hormones as tumor markers, *Cancer Detect. Prevent.,* 6, 87, 1983.

13. **Pimentel, E.,** Peptide hormone precursors, subunits and fragments as human tumor markers, *Ann. Clin. Lab. Sci.,* 15, 381, 1985.

14. **Pimentel, E.,** Growth factors as tumor markers (abstract), *J. Tumor Marker Oncol.,* 7, 28, 1992.

15. **Russell, P. J., O'Mara, S. M., and Raghavan, D.,** Ectopic hormone production by small cell undifferentiated carcinomas, *Mol. Cell. Endocrinol.,* 71, 1, 1990.

16. **Zirvi, K. A., Chee, D. O., and Hill, G. J.,** Continuous growth of human tumor cell lines in serum-free media, *In Vitro Cell. Dev. Biol.,* 22, 369, 1986.

17. **Brizzi, M. F., Avanzi, G. C., Veglia, F., Clark, S. C., and Perogaro, L.,** Expression and modulation of IL-3 and GM-CSF receptors in human growth factor dependent leukaemic cells, *Br. J. Haematol.,* 76, 203, 1990.

18. **Rizzino, A., Ruff, E., and Rizzino, H.,** Induction and modulation of anchorage-independent growth by platelet-derived growth factor, fibroblast growth factor, and transforming growth factor-β, *Cancer Res.,* 46, 2816, 1986.

19. **Lee, K., Tanaka, M., Hatanaka, M., and Kuze, F.,** Reciprocal effects of epidermal growth factor and transforming growth factor β on the anchorage-dependent and -independent growth of A431 epidermoid carcinoma cells, *Exp. Cell Res.,* 173, 156, 1987.

20. **Van Zoelen, E. J. J.,** Phenotypic transformation of normal rat kidney cells: a model for studying cellular alterations in oncogenesis, *Crit. Rev. Oncogenesis,* 2, 311, 1991.

21. **Mordan, L. J.,** Induction by growth factors from platelets of the focus-forming transformed phenotype in carcinogen-treated C3H/10T1/2 cells, *Carcinogenesis,* 9, 1129, 1988.

22. **Heideran, M. A., Fleming, T. P., Bottaro, D. P., Bell, G. I., Di Fiore, P. P., and Aaronson, S. A.,** Transformation of NIH 3T3 fibroblasts by an expression vector for the human epidermal growth factor precursor, *Oncogene,* 5, 1265, 1990.

23. **Jirik, F. R., Burstein, S. A., Treger, L., and Sorge, J. A.,** Transfection of a factor-dependent cell line with the murine interleukin-3 (IL-3) cDNA results in autonomous growth and tumorigenesis, *Leukemia Res.,* 11, 1127, 1987.

24. **Rogelj, S., Weinberg, R. A., Fanning, P., and Klagsbrun, M.,** Basic fibroblast growth factor fused to a signal peptide transforms cells, *Nature,* 331, 173, 1988.

25. **Takahashi, J. B., Hoshimaru, M., Jaye, M., Kikuchi, H., and Hatanaka, M.,** Possible activity of acidic fibroblast growth factor as a progression factor rather than a transforming factor, *Biochem. Biophys. Res. Commun.,* 189, 398, 1992.

26. **Kaleko, M., Rutter, W. J., and Miller, A. D.,** Overexpression of the human insulinlike growth factor I receptor promotes ligand-dependent neoplastic transformation, *Mol. Cell. Biol.,* 10, 464, 1990.

27. **van Zoelen, E. J. J., van Oostwaard, T. M. J., and de Laat, S. W.,** Transforming growth factor-β and retinoic acid modulate phenotypic transformation of normal rat kidney cells induced by epidermal growth factor and platelet-derived growth factor, *J. Biol. Chem.,* 261, 5003, 1986.

28. **van Zoelen, E. J. J., van Oostwaard, T. M. J., and de Laat, S. W.,** The role of polypeptide growth factors in phenotypic transformation of normal rat kidney cells, *J. Biol. Chem.,* 263, 64, 1988.

29. **McGeady, M. L., Kerby, S., Shankar, V., Ciardiello, F., Salomon, D., and Seidman, M.,** Infection with a TGF-α retroviral vector transforms normal mouse mammary epithelial cells but not normal rat fibroblasts, *Oncogene,* 4, 1375, 1989.

30. **Rhlm, J. S.,** Neoplastic transformation of human cells *in vitro, Crit. Rev. Oncogenesis,* 4, 313, 1993.

31. **Herlyn, M., Clark, W. H., Rodeck, U., Mancianti, M. L., Jambroslc, J., and Koprowski, H.,** Biology of tumor progression in human melanocytes, *Lab. Invest.,* 56, 461, 1987.

32. **Hamburger, A. W., White, C. P., Lurie, K., and Kaplan, R.,** Monocyte-derived growth factors for human tumor clonogenic cells, *J. Leukocyte Biol.,* 40, 381, 1986.
33. **Pierpaoli, W., Haran-Ghera, N., Bianchi, E., Mueller, J., Meshorer, A., and Bree, M.,** Endocrine disorders as contributory factor to neoplasia in SJL/J mice, *J. Natl. Cancer Inst.,* 53, 731, 1974.
34. **Pierpaoli, W., Haran-Ghera, N., and Kopp, H. G.,** Role of host endocrine status in murine leukaemogenesis, *Br. J. Cancer,* 35, 621, 1977.
35. **Pierpaoli, W. and Meshorer, A.,** Host endocrine status mediates onocogenesis: leukemia virus-induced carcinomas and reticulum cell sarcomas in acyclic or normal mice, *Eur. J. Cancer,* 18, 1181, 1982.
36. **Mishkin, S. Y., Farber, E., Ho, R. K., Mulay, S., and Mishkin, S.,** Evidence for the hormone dependency of hepatic hyperplastic nodules: inhibition of malignant transformation after exogenous 17 β-estradiol and tamoxifen, *Hepatology,* 3, 308, 1983.
37. **Lupulescu, A.,** Glucagon control of carcinogenesis, *Endocrinology,* 113, 527, 1983.
38. **Henderson, B. E., Ross, R. K., Pike, M. C., and Casagrande, J. T.,** Endogenous hormones as a major factor in human cancer, *Cancer Res.,* 42, 3232, 1982.
39. **de Waard, F.,** Hormonal factors in human carcinogenesis, *J. Cancer Res. Clin. Oncol.,* 108, 177, 1984.
40. **Moolgavkar, S. H.,** Hormones and multistage carcinogenesis, *Cancer Surv.,* 5, 635, 1986.
41. **Kanazawa, T. and Hosick, H. L.,** Transformed growth phenotype of mouse mammary epithelium in primary culture induced by specific fetal mesenchymes, *J. Cell. Physiol.,* 153, 381, 1992.
42. **Moses, H. L. and Robinson, R. A.,** Growth factors, growth factor receptors, and cell cycle control mechanisms in chemically transformed cells, *Fed. Proc.,* 41, 3008, 1982.
43. **Levine, A. E., Crandall, C. A., Brattain, D., Chakrabarty, S., and Brattain, M. G.,** Retinoic acid restores normal growth control to a transformed mouse embryo fibroblast cell line, *Cancer Lett.,* 33, 33, 1986.
44. **Powers, S., Fisher, P. B., and Pollack, R.,** Analysis of the reduced growth factor dependency of simian virus 40-transformed 3T3 cells, *Mol. Cell. Biol.,* 4, 1572, 1984.
45. **El-Enany, T. M. and Dubes, G. R.,** The set of growth factors stimulatory for a transformed rat cell of line NRK-49F depends on the identity of the transforming virus, *Tumour Biol.,* 7, 49, 1986.
46. **Gilks, C. B., Bear, S. E., Grimes, H. L., and Tsichlis, P. N.,** Progression of interleukin-2 (IL-2)-dependent rat T cell lymphoma lines to IL-2-independent growth following activation of a gene (*Gfi-1*) encoding a novel zinc finger protein, *Mol. Cell. Biol.,* 13, 1759, 1993.
47. **Auersperg, N. and Calderwood, G. A.,** Development of serum independence in Kirsten murine sarcoma virus-infected rat adrenal cells, *Carcinogenesis,* 5, 175, 1984.
48. **Durkin, J. P. and Whitfield, J. F.,** Partial characterization of the mitogenic action of pp60$^{v\text{-}src}$, the oncogenic protein product of the *src* gene of avian sarcoma virus, *J. Cell. Physiol.,* 120, 135, 1984.
49. **Durkin, J. P. and Whitfield, J. F.,** The mitogenic activity of pp60$^{v\text{-}src}$, the oncogenic protein product of the *src* gene of avian sarcoma virus, is independent of external serum growth factors, *Biochem. Biophys. Res. Commun.,* 123, 411, 1984.
50. **Zhan, X. and Goldfarb, M.,** Growth factor requirements of oncogene-transformed NIH 3T3 and BALB/c 3T3 cells cultured in defined media, *Mol. Cell. Biol.,* 6, 3541, 1986.
51. **Fusco, A., Berlingieri, M. T., Di Fiore, P. P., Portella, G., Grieco, M., and Vecchio, G.,** One- and two-step transformations of rat thyroid epithelial cells by retroviral oncogenes, *Mol. Cell. Biol.,* 7, 3365, 1987.
52. **Uemura, N., Ozawa, K., Tojo, A., Takahashi, K., Okano, A., Karasuyama, H., Tani, K., and Asano, S.,** Acquisition of interleukin-3 independence in FDC-P2 cells after transfection with the activated c-H-*ras* gene using a bovine papillomavirus-based plasmid vector, *Blood,* 80, 3198, 1992.
53. **Segawa, K.,** Cooperation of mitogenic growth factors with polyoma virus middle T antigen in transformation of secondary cultured rat cells, *Biochem. Biophys. Res. Commun.,* 136, 921, 1986.
54. **Mörike, M., Quaiser, A., Müller, D., and Montenarh, M.,** Early gene expression and cellular DNA synthesis after stimulation of quiescent NIH3T3 cells with serum or purified simian virus 40, *Oncogene,* 3, 151, 1988.
55. **Ford, R. J., Davis, F., and Ramirez, I.,** Growth factors for human lymphoid neoplasms, in *Mediators in Cell Growth and Differentiation,* Ford, R. J. and Maizel, A. L., Eds., Raven Press, New York, 1985, 233.
56. **Salle, V., Souttou, B., Magnien, V., Israël, L., and Crépin, M.,** Facteurs de croissance des cellules épithéliales de mastopathies et de carcinomes du sein en primoculture, *Bull. Cancer (Paris),* 79, 133, 1992.

57. **Kaplan, P. L. and Ozanne, B.,** Cellular responsiveness to growth factors correlates with a cell's ability to express the transformed phenotype, *Cell,* 33, 931, 1983.

58. **Stern, D. F., Roberts, A. B., Roche, N. S., Sporn, M. B., and Weinberg, R. A.,** Differential responsiveness of *myc-* and *ras*-transfected cells to growth factors: selective stimulation of *myc*-transfected cells by epidermal growth factor, *Mol. Cell. Biol.,* 6, 870, 1986.

59. **Holzer, C., Maier, P., and Zbindin, G.,** Comparison of exogenous growth stimuli for chemically transformed cells: growth factors, serum and cocultures, *Exp. Cell Biol.,* 54, 237, 1986.

60. **Ethier, S. P. and Moorthy, R.,** Multiple growth factor independence in rat mammary carcinoma cells, *Breast Cancer Res. Treat.,* 18, 73, 1991.

61. **Albino, A. P., Davis, B. M., and Nanus, D. M.,** Induction of growth factor RNA expression in human malignant melanoma: markers of transformation, *Cancer Res.,* 51, 4815, 1991.

62. **Siegfried, J. M. and Owens, S. E.,** Response of primary human lung carcinomas to autocrine growth factors produced by a lung carcinoma cell line, *Cancer Res.,* 48, 4976, 1988.

63. **Singletary, S. E., Tomasovic, B., Spitzer, G., Tucker, S. L., Hug, V., and Drewinko, B.,** Effects and interactions of epidermal growth factor, insulin, hydrocortisone, and estradiol on the cloning of human tumor cells, *Int. J. Cell Cloning,* 3, 407, 1985.

64. **Medrano, E. A., Cafferata, E. G. A., and Larcher, F.,** Role of thrombin in the proliferative response of T-47D mammary tumor cells, *Exp. Cell Res.,* 172, 354, 1987.

65. **Furcht, L. T.,** Critical factors controlling angiogenesis: cell products, cell matrix, and growth factors, *Lab. Invest.,* 55, 505, 1986.

66. **Folkman, J. and Klagsbrun, M.,** Angiogenic factors, *Science,* 235, 442, 1987.

67. **Bouck, N.,** Tumor angiogenesis: the role of oncogenes and tumor suppressor genes, *Cancer Cells,* 2, 179, 1990.

68. **Boswell, H. S., Srivastava, A., Burgess, J. S., Nahreini, P., Heerema, N., Inhorn, L., Padgett, F., Walker, E. B., and Geib, R. W.,** Cellular control of *in vitro* progression of murine myeloid leukemia: progression accompanies acquisition of independence from growth factor and stromal cells, *Leukemia,* 1, 765, 1987.

69. **Peres, R., Betsholtz, C., Westermark, B., and Heldin, C.-H.,** Frequent expression of growth factors for mesenchymal cells in human mammary carcinoma cell lines, *Cancer Res.,* 47, 3425, 1987.

70. **Sporn, M. B. and Todaro, G. J.,** Autocrine secretion and malignant transformation, *N. Engl. J. Med.,* 303, 878, 1980.

71. **Sporn, M. B. and Roberts, A. B.,** Autocrine, paracrine and endocrine mechanisms of growth control, *Cancer Surv.,* 4, 627, 1985.

72. **Heldin, C.-H. and Westermark, B.,** Growth factors as transforming proteins, *Eur. J. Biochem.,* 184, 487, 1989.

73. **Bajzer, Z. and Vuk-Pavlovic, S.,** Quantitative aspects of autocrine regulation in tumors, *Crit. Rev. Oncogenesis,* 2, 53, 1990.

74. **Sporn, M. B. and Roberts, A. B.,** Autocrine secretion — ten years later, *Ann. Int. Med.,* 117, 408, 1992.

75. **Kawano, M., Kuramoto, A., Hirano, T., and Kishimoto, T.,** Cytokines as autocrine growth factors in malignancies, *Cancer Surv.,* 8, 905, 1989.

76. **Yee, D., Favoni, R. E., Lebovic, G. S., Lombana, F., Powell, D. R., Reynolds, C. P., and Rosen, N.,** Insulin-like growth factor I expression by tumors of neuroectodermal origin with the t(11;22) chromosomal translocation. A potential autocrine growth factor, *J. Clin. Invest.,* 86, 1806, 1990.

77. **Greig, R., Dunnington, D., Murthy, U., and Anzano, M.,** Growth factors as novel therapeutic targets in neoplastic disease, *Cancer Surv.,* 7, 653, 1988.

78. **Leslie, K. B. and Schrader, J. W.,** Growth factor gene activation and clonal heterogeneity in an autostimulatory myeloid leukemia, *Mol. Cell. Biol.,* 9, 2414, 1989.

79. **Ohmura, E., Okada, M., Onoda, N., Kamiya, Y., Murakami, H., Tsushima, T., and Shizume, K.,** Insulin-like growth factor I and transforming growth factor α as autocrine growth factors in human pancreatic cancer cell growth, *Cancer Res.,* 50, 103, 1990.

80. **Perosio, P. M. and Brooks, J. J.,** Expression of growth factors and growth factor receptors in soft tissue tumors — implications for the autocrine hypothesis, *Lab. Invest.,* 60, 245, 1989.

81. **Mizukami, Y., Nonomura, A., Yamada, T., Kurumaya, H., Hayashi, M., Koyasaki, N., Taniya, T., Noguchi, M., Nakamura, S., and Matusbara, F.,** Immunohistochemical demonstration of growth factors, TGF-α, TGF-β and *neu* oncogene product in benign and malignant human breast tissues, *Anticancer Res.,* 10, 1115, 1990.

82. **Roholl, P. J. M., Weima, S. M., Prinsen, I., De Weger, R. A., Den Otter, W., and Van Unnik, J. A. M.,** Expression of growth factors and their receptors on human sarcomas. Immunohistochemical detection of platelet-derived growth factor, epidermal growth factor and their receptors, *Cancer J., 4,* 83, 1991.

83. **Goustin, A. S., Betsholtz, C., Pfeifer-Ohlsson, S., Persson, H., Rydnert, J., Bywater, M., Holmgren, G., Heldin, C.-H., Westermark, B., and Ohlsson, R.,** Coexpression of the *sis* and *myc* proto-oncogenes in developing human placenta suggests autocrine control of trophoblast growth, *Cell, 41,* 301, 1985.

84. **Nistér, M., Liberman, T. A., Betsholtz, C., Pettersson, M., Claesson-Welsh, L., Heldin, C.-H., Schlessinger, J., and Westermark, B.,** Expression of messenger RNAs for platelet-derived growth factor-α and their receptors in human malignant glioma cell lines, *Cancer Res., 48,* 3910, 1988.

85. **Browder, T. M., Dunbar, C. E., and Nienhuis, A. W.,** Private and public autocrine loops in neoplastic cells, *Cancer Cells,* 1, 9, 1989.

86. **Keating, M. T. and Williams, L. T.,** Autocrine stimulation of intracellular PDGF receptors in v-*sis*-transformed cells, *Science,* 239, 914, 1988.

87. **Fleming, T. P., Matsui, T., Molloy, C. J., Robbins, K. C., and Aaronson, S. A.,** Autocrine mechanism for v-*sis* transformation requires cell surface localization of internally activated growth factor receptors, *Proc. Natl. Acad. Sci. U.S.A.,* 86, 8063, 1989.

88. **Davis, J. A. and Linzer, D. I. H.,** Autocrine stimulation of Nb2 cell proliferation by secreted, but not intracellular, prolactin, *Mol. Endocrinol.,* 2, 740, 1988.

89. **Wellstein, A., Lupu, R., Zugmaier, G., Flamm, S. L., Cheville, A. L., Delli Bovi, P., Basilico, C., Lippman, M. E., and Kern, F. G.,** Autocrine stimulation by secreted Kaposi fibroblast growth factor but not by endogenous basic fibroblast growth factor, *Cell Growth Differ.,* 1, 63, 1990.

90. **Xin, L. W., Jullien, P., Lawrence, D. A., Pironin, M., and Vigier,** Chemically and virally transformed cells able to grow without anchorage in serum-free medium: evidence for an autocrine growth factor, *J. Cell. Physiol.,* 131, 175, 1987.

91. **Ethier, S. P. and Cundiff, K. C.,** Importance of extended growth potential and growth factor independence of *in vivo* neoplastic potential of primary rat mammary carcinoma cells, *Cancer Res., 47,* 5316, 1987.

92. **Ove, P., Coetzee, M. L., Scalamogna, P., Francavilla, A., and Starzl, T. E.,** Isolation of an autocrine growth factor from hepatoma HTC-SR cells, *J. Cell. Physiol.,* 131, 165, 1987.

93. **Liotta, L. A., Mandler, R., Murano, G., Katz, D. A., Gordon, R. K., Chiang, P. K., and Schiffmann, E.,** Tumor cell autocrine motility factor, *Proc. Natl. Acad. Sci. U.S.A.,* 83, 3302, 1986.

94. **Kohn, E. C., Liotta, L. A., and Schiffmann, E.,** Autocrine motility factor stimulates a three-fold increase in inositol trisphosphate in human melanoma cells, *Biochem. Biophys. Res. Commun.,* 166, 757, 1990.

95. **Stracke, M. L., Engel, J. D., Wilson, L. W., Rechler, M. M., Liotta, L. A., and Schiffmann, E.,** The type I insulin-like growth factor receptor is a motility receptor in human melanoma cells, *J. Biol. Chem.,* 264, 21544, 1989.

96. **Lang, R. A., Metcalf, D., Gough, N. M., Dunn, A. R., and Gonda, T. J.,** Expression of a hemopoietic growth factor cDNA in a factor-dependent cell line results in autonomous growth and tumorigenicity, *Cell,* 43, 531, 1985.

97. **Bajzer, Z. and Vuk-Pavlovic, S.,** Quantitation of autocrine regulation of tumor growth: a general phenomenological model, *Cancer Res.,* 47, 5330, 1987.

98. **Sasada, R., Kurokawa, T., Iwane, M., and Igarashi, K.,** Transformation of mouse BALB/c 3T3 cells with human basic fibroblast growth factor cDNA, *Mol. Cell. Biol.,* 8, 588, 1988.

99. **Yamada, Y. and Serrero, G.,** Characterization of an insulin-related factor secreted by a teratoma cell line, *Biochem. Biophys. Res. Commun.,* 135, 533, 1986.

100. **Hamburger, A. W. and White, C. P.,** Autocrine growth factors for human tumor clonogenic cells, *Int. J. Cell Cloning,* 3, 399, 1985.

101. **Richmond, A., Lawson, D. H., Nixon, D. W., and Chawla, R. K.,** Characterization of autostimulatory and transforming growth factors from human melanoma cells, *Cancer Res.,* 45, 6390, 1985.

102. **Klein, B., Jourdan, M., Vazquez, A., Dugas, B., and Bataille, R.,** Production of growth factors by human myeloma cells, *Cancer Res.,* 47, 4856, 1987.

103. **Tsai, T.-F., Yauk, Y.-K., Chou, C.-K., Ting, L.-P., Chang, C., Hu, C., Han, S.-H., and Su, T.-S.,** Evidence of autocrine regulation in human hepatoma cell lines, *Biochem. Biophys. Res. Commun.,* 153, 39, 1988.

104. **El-Badry, O. M., Minniti, C., Kohn, E. C., Houghton, P. J., Daughaday, W. H., and Helman, L. J.,** Insulin-like growth factor II acts as an autocrine growth and motility factor in human rhabdomyosarcoma tumors, *Cell Growth Differ.,* 1, 325, 1990.

105. **Bradbury, D., Bowen, G., Kozlowski, R., Reilly, I., and Russell, N.,** Endogenous interleukin-1 can regulate the autonomous growth of the blast cells of acute myeloblastic leukemia by inducing autocrine secretion of GM-CSF, *Leukemia,* 4, 44, 1990.

106. **Hoosein, N. M., Kiener, P. A., Curry, R. C., and Brattain, M. G.,** Evidence for autocrine growth stimulation of cultured colon tumor cells by a gastrin/cholecystokinin-like peptide, *Exp. Cell Res.,* 186, 15, 1990.

107. **Upp, J. R., Singh, P., Townsend, C. M., and Thompson, J. C.,** Clinical significance of gastrin receptors in human colon cancers, *Cancer Res.,* 49, 488, 1989.

108. **Sethi, T. and Rozengurt, E.,** Gastrin stimulates Ca^{2+} mobilization and clonal growth in small cell lung cancer cells, *Cancer Res.,* 52, 6031, 1992.

109. **Moody, T. W., Russell, E. K., O'Donohue, T. L., Linden, C. D., and Gazdar, A. F.,** Bombesin-like peptides in small cell lung cancer: biochemical characterization and secretion from a cell line, *Life Sci.,* 32, 487, 1983.

110. **Moody, T. W., Carney, D. N., Cuttitta, F., Quattrocchi, K., and Minna, J. D.,** High affinity receptors for bombesin/GRP-like peptides on human small cell lung cancer, *Life Sci.,* 37, 105, 1985.

111. **Miller, Y. E.,** Growth factors, oncogenes, and lung cancer, *Am. Rev. Respir. Dis.,* 132, 178, 1985.

112. **Zachary, I. and Rozengurt, E.,** High-affinity receptors for peptides of the bombesin family in Swiss 3T3 cells, *Proc. Natl. Acad. Sci. U.S.A.,* 82, 7616, 1985.

113. **Cuttitta, F., Carney, D. N., Mulshine, J., Moody, T. W., Fedorko, J., Fischler, A., and Minna, J. D.,** Autocrine growth factors in human small cell lung carcinoma, *Cancer Surv.,* 4, 707, 1985.

114. **Cuttitta, F., Carney, D. N., Mulshine, J., Moody, T. W., Fedorko, J., Fischler, A., and Minna, J. D.,** Bombesin-like peptides can function as autocrine growth factors in human small-cell lung cancer, *Nature,* 316, 823, 1985.

115. **Trepel, J. B., Moyer, J. D., Cuttita, F., Frucht, H., Coy, D. H., Natale, R. B., Mulshine, J. L., Jensen, R. T., and Sausville, E. A.,** A novel bombesin receptor antagonist inhibits autocrine signals in a small cell lung carcinoma cell line, *Biochem. Biophys. Res. Commun.,* 156, 1383, 1988.

116. **Gaudino, G., Cirillo, D., Naldini, L., Rossino, P., and Comoglio, P. M.,** Activation of the protein-tyrosine kinase associated with bombesin receptor complex in small cell lung carcinomas, *Proc. Natl. Acad. Sci. U.S.A.,* 85, 2166, 1988.

117. **Lemoine, F. M., Krystal, G., Humphries, R. K., and Eaves, C. J.,** Autocrine production of pre-B-cell stimulating activity by a variety of transformed murine pre-B-cell lines, *Cancer Res.,* 48, 6438, 1988.

118. **Nakamoto, T., Usui, A., Oshima, K., Ikemoto, H., Mitani, S., and Usui, T.,** Analysis of growth factors in renal cell carcinoma, *Biochem. Biophys. Res. Commun.,* 153, 818, 1988.

119. **Bajzer, Z., Pavelic, K., and Vuk-Pavlovic, S.,** Growth self-incitement in murine melanoma B16: a phenomenological model, *Science,* 225, 930, 1984.

120. **Stocking, C., Löliger, C., Kawai, M., Suciu, S., Gough, N., and Ostertag, W.,** Identification of genes involved in growth autonomy of hematopoietic cells by analysis of factor-independent mutants, *Cell,* 53, 869, 1988.

121. **Lange, B., Valtieri, M., Santoli, D., Caracciolo, D., Mavilio, F., Gemperlein, I., Griffin, C., Emanuel, B., Finan, J., Nowell, P., and Rovera, G.,** Growth factor requirements of childhood acute leukemia: establishment of GM-CSF-dependent cell lines, *Blood,* 70, 192, 1987.

122. **Vellenga, E., Delwel, H. R., Touw, I. P., and Löwenberg, B.,** Patterns of acute myeloid leukemia colony growth in response to recombinant granulocyte-macrophage colony-stimulating factor (rGM-CSF), *Exp. Hematol.,* 15, 652, 1987.

123. **Jasmin, C., Georgoulias, V., Smadja-Joffe, F., Boucheix, C., Le Bousse-Kerdiles, C., Allouche, M., Cibert, C., and Azzarone, B.,** Autocrine growth of leukemic cells, *Leukemia Res.,* 14, 689, 1990.

124. **Singletary, S. E., Baker, F. L., Spitzer, G., Tucker, S. L., Tomasovic, B., Brock, W. A., Ajani, J. A., and Kelly, A. M.,** Biological effect of epidermal growth factor on the *in vitro* growth of human tumors, *Cancer Res.,* 47, 403, 1987.

125. **van Zoelen, E. J. J., Koornneef, I., Holthuis, J. C. M., Ward-van Oostwaard, T. M. J., Feijen, A., de Poorter, T. L., Mummery, C. L., and van Buul-Offers, S. C.,** Production of insulin-like growth factors, platelet-derived growth factor, and transforming growth factors and their role in the density-dependent growth regulation of a differentiated embryonal carcinoma cell line, *Endocrinology,* 124, 2029, 1989.

126. **Anzano, M. A., Rieman, D., Prichett, W., Bowen-Pope, D. F., and Greig, R.,** Growth factor production by human colon cell lines, *Cancer Res.,* 49, 2898, 1989.

127. **Cantrell, D. A., Collins, M. K. L., and Crumpton, M. J.,** Autocrine regulation of T-lymphocyte proliferation: differential induction of IL-2 and IL-2 receptor, *Immunology,* 65, 343, 1988.

128. **Muraguchi, A., Nishimoto, H., Kawamura, N., Hori, A., and Kishimoto, T.,** B cell-derived BCGF functions as autocrine growth factor(s) in normal and transformed B lymphocytes, *J. Immunol.,* 137, 179, 1986.

129. **Fant, M., Munro, H., and Moses, A. C.,** An autocrine/paracrine role for insulin-like growth factors in the regulation of human placental growth, *J. Clin. Endocrinol. Metab.,* 63, 499, 1986.

130. **Cook, P. W., Pittelkow, M. R., and Shipley, G. D.,** Growth factor-independent proliferation of normal human neonatal keratinocytes: production of autocrine- and paracrine-acting mitogenic factors, *J. Cell. Physiol.,* 146, 277, 1991.

131. **Coffey, R. J., Jr., Derynck, R., Wilcox, J. N., Bringman, T. S.,** Production and auto-induction of transforming growth factor-α in human keratinocytes, *Nature,* 328, 817, 1987.

132. **Bascom, C. C., Sipes, N. J., Coffey, R. J., and Moses, H. L.,** Regulation of epithelial cell proliferation by transforming growth factors, *J. Cell. Biochem.,* 39, 25, 1989.

133. **Coffey, R. J., Jr., Sipes, N. J., Bascom, C. C., Graves-Deal, R., Pennington, C. Y., Weissman, B. E., and Moses, H. L.,** Growth modulation of mouse keratinocytes by transforming growth factors, *Cancer Res.,* 48, 1596, 1988.

134. **Soma, Y. and Grotendorst, G. R.,** TGF-beta stimulates primary human skin fibroblast DNA synthesis via an autocrine production of PDGF-related peptides, *J. Cell. Physiol.,* 140, 246, 1989.

135. **Sjölund, M., Hedin, U., Sejersen, T., Heldin, C.-H., Thyberg, J.,** Arterial smooth muscle cells express platelet-derived growth factor (PDGF) A chain mRNA, secrete a PDGF-like mitogen, and bind exogenous PDGF in a phenotype- and growth-state-dependent manner, *J. Cell. Biol.,* 106, 403, 1988.

136. **Miano, J. M., Vlasic, N., Tota, R. R., and Stemerman, M. B.,** Smooth muscle cell immediate-early gene and growth factor activation follows vascular injury. A putative *in vivo* mechanism for autocrine growth, *Arterioscler. Thromb.,* 23, 211, 1993.

137. **Porter, S., Glaser, L., and Bunge, R. P.,** Release of autocrine growth factor by primary and immortalized Schwann cells, *Proc. Natl. Acad. Sci. U.S.A.,* 84, 7768, 1987.

138. **De Larco, J. E. and Todaro, G. J.,** Growth factors from murine sarcoma virus-transformed cells, *Proc. Natl. Acad. Sci. U.S.A.,* 75, 4001, 1978.

139. **Kryceve-Martinerie, C., Lawrence, D. A., Crochet, J., Julien, P., and Vigier, P.,** Cells transformed by Rous sarcoma virus release transforming growth factors, *J. Cell. Physiol.,* 113, 365, 1982.

140. **Roberts, A. B., Anzano, M. A., Lamb, L. C., Smith, J. M., Frolik, C. A., Marquardt, M., Todaro, G. J., and Sporn, M. B.,** Isolation from the murine sarcoma cells of novel transforming growth factors potentiated by EGF, *Nature,* 295, 417, 1982.

141. **Marquardt, H., Hunkapiller, M. W., Hood, L. E., Twardzik, D. R., De Larco, J. E., Stephenson, J. R., and Todaro, G. J.,** Transforming growth factors produced by retrovirus-transformed rodent fibroblasts and human melanoma cells: amino acid sequence homology with epidermal growth factor, *Proc. Natl. Acad. Sci. U.S.A.,* 80, 4684, 1983.

142. **Anzano, M. A., Roberts, A. B., Smith, J. M., Sporn, M. B., and De Larco, J. E.,** Sarcoma growth factor from conditioned medium of virally transformed cells is composed of both type α and type β transforming growth factors, *Proc. Natl. Acad. Sci. U.S.A.,* 80, 6264, 1983.

143. **Hirai, R., Yamaoka, K., and Mitsui, H.,** Isolation and purification of a new class of transforming growth factors from an avian sarcoma virus-transformed rat cell line, *Cancer Res.,* 43, 5742, 1983.

144. **Yamaoka, K., Hirai, R., Tsugita, A., and Mitsui, H.,** The purification of an acid- and heat-labile transforming growth factor from an avian sarcoma virus-transformed rat cell line, *J. Cell. Physiol.,* 119, 307, 1984.

145. **Massagué, J.,** Type beta transforming growth factor from feline sarcoma virus-transformed rat cells: isolation and biological properties, *J. Biol. Chem.,* 259, 9756, 1984.

146. **Anzano, M. A., Roberts, A. B., De Larco, J. E., Wakefield, L. M., Assoian, R. K., Roche, N. S., Smith, J. M., Lazarus, J. E., and Sporn, M. B.,** Increased secretion of type beta transforming growth factor accompanies viral transformation of cells, *Mol. Cell. Biol.,* 5, 242, 1985.

147. **Haas, M., Altman, A., Rothenberg, E., Bogart, M. H., and Jones, O. W.,** Radiation leukemia virus and X-irradiation induce in C57BL/6 mice two distinct T-cell neoplasms: a growth factor-dependent lymphoma and a growth factor-independent lymphoma, *Leukemia Res.,* 11, 223, 1987.

148. **Fasciotto, B., Kanazir, D., Durkin, J. P., Whitfield, J. F., and Krsmanovic, V.,** AEV-transformed chicken erythroid cells secrete autocrine factors which promote soft agar and block erythroleukemia cell differentiation, *Biochem. Biophys. Res. Commun.,* 143, 775, 1987.

149. **Giancotti, V., Panbi, B., D'Andrea, P., Berlingieri, M. T., Di Fiore, P. P., Fusco, A., Vecchio, G., Philp, R., Crane-Robinson, C., Nicolas, R. H., Wright, C. A., and Goodwin, G. H.,** Elevated levels of a specific class of nuclear phosphoproteins in cells transformed with v-*ras* and v-*mos* oncogenes and by co-transfection with c-*myc* and polyoma middle T genes, *EMBO J.,* 6, 1981, 1987.

150. **Durkin, J. P. and Whitfield, J. F.,** The mitogenic activity of pp60$^{v\text{-}src}$, the oncogenic protein product of the *src* gene of avian sarcoma virus, is independent of external serum growth factors, *Biochem. Biophys. Res. Commun.,* 123, 411, 1984.

151. **Adkins, B., Leutz, A., and Graf, T.,** Autocrine growth induced by *src*-related oncogenes in transformed chicken myeloid cells, *Cell,* 39, 439, 1984.

152. **Cleveland, J. L., Jansen, H. W., Bister, K., Fredrickson, T. N., Morse, H. C., III, Ihle, J. N., and Rapp, U. R.,** Interaction between *raf* and *myc* oncogenes in transformation *in vivo* and *in vitro, J. Cell. Biochem.,* 30, 195, 1986.

153. **Kasid, A., Knabbe, C., and Lippman, M. E.,** Effect of v-*ras*^{H} oncogene transfection on estrogen-independent tumorigenicity of estrogen-dependent human breast cancer cells, *Cancer Res.,* 47, 5733, 1987.

154. **Sporn, M. B. and Roberts, A. B.,** Autocrine growth factors and cancer, *Nature,* 313, 745, 1985.

155. **Chiodino, C., Jones, R. F., and Ethier, S. P.,** The role of Ha-*ras* oncogenes in growth factor independence in rat mammary carcinoma cells, *Mol. Carcinogenesis,* 4, 286, 1991.

156. **Graf, T., von Weizsaecker, F., Grieser, S., Coll. J., Stehelin, D., Patschinsky, T., Bister, K., Bechade, C., Calothy, G., and Leutz, A.,** v-*mil* induces autocrine growth and enhanced tumorigenicity in v-*myc*-transformed avian macrophages, *Cell,* 45, 357, 1986.

157. **Martin, P., Henry, C., Ferre, F., Bechade, C., Begue, A., Calothy, C., Debuire, B., Stehelin, D., and Saule, S.,** Characterization of a *myc*-containing retrovirus generated by propagation of an MH2 viral subgenomic RNA, *J. Virol.,* 57, 1191, 1986.

158. **Blasi, E., Mathieson, B. J., Varesio, L., Cleveland, J. L., Borchert, P. A., and Rapp, U. R.,** Selective immortalization of murine macrophages from fresh bone marrow by a *raf/myc* recombinant murine retrovirus, *Nature,* 318, 667, 1985.

159. **Rapp, U. R., Cleveland, J. L., Fredrickson, T. N., Holmes, K. L., Morse, H. C., III, Jansen, H. W., Patschinsky, T., and Bister, K.,** Rapid induction of hemopoietic neoplasms in newborn mice by a *raf(mil)/myc* recombinant murine retrovirus, *J. Virol.,* 55, 23, 1985.

160. **Auersperg, N., Siemens, C. H., Krystal, G., and Myrdal, S. E.,** Modulation of normal serum factors of Kirsten murine sarcoma virus-induced transformation in adult rat cells infected in early passage, *Cancer Res.,* 46, 5715, 1986.

161. **Salomon, D. S., Zwiebel, J. A., Noda, M., and Bassin, R. H.,** Flat revertants derived from Kirsten murine sarcoma virus-transformed cells produce transforming growth factors, *J. Cell. Physiol.,* 121, 22, 1984.

162. **Fichelson, S., Heard, J.-M., and Levy, J.-P.,** The *in vitro* autocrine secretion of CSFs alone does not account for the longterm growth of murine myeloid leukemic cells in suspension cultures, *J. Cell. Physiol.,* 124, 487, 1985.

163. **Moscovici, M. G., Jurdic, P., Samarut, J., Gazzolo, L., Mura, C. V., and Moscovici, C.,** Characterization of the hemopoietic target cells for the avian leukemia virus E26, *Virology,* 129, 65, 1983.

164. **Beug, H., Leutz, A., Kahn, P., and Graf, T.,** *Ts* mutants of E26 leukemia virus allow transformed myeloblasts, but not erythroblasts or fibroblasts, to differentiate at the nonpermissive temperature, *Cell,* 39, 579, 1984.

165. **Balk, S. D., Riley, T. M., Gunther, H. S., and Morisi, A.,** Heparin-treated, v-*myc*-transformed chicken heart mesenchymal cells assume a normal morphology but are hypersensitive to epidermal growth factor (EGF) and brain fibroblast growth factor (bFGF); cells transformed by the v-Ha-*ras* oncogene are refractory to EGF and bFGF but are hypersensitive to insulin-like growth factors, *Proc. Natl. Acad. Sci. U.S.A.,* 82, 5781, 1985.

166. **Katzav, S., Martin-Zanca, D., Barbacid, M., Hedge, A.-M., Isfort, R., and Ihle, J. N.,** The *trk* oncogene abrogates growth factor requirements and transforms hematopoietic cells, *Oncogene,* 4, 1129, 1989.

167. **Oliff, A., Agranovsky, O., McKinney, M. D., Murty, V. V. V. S., and Bauchwitz, R.,** Friend murine leukemia virus-immortalized myeloid cells are converted into tumorigenic cell lines by Abelson leukemia virus, *Proc. Natl. Acad. Sci. U.S.A.,* 82, 3306, 1985.

168. **Cook, W. D., Metcalf, D., Nicola, N. A., Burgess, A. W., and Walker, F.,** Malignant transformation of a growth factor-dependent myeloid cell line by Abelson virus without evidence of an autocrine mechanism, *Cell,* 41, 677, 1985.

169. **Pierce, J. H., Di Fiore, P. P., Aaronson, S. A., Potter, M., Pumphrey, J., Scott, A., and Ihle, J. N.,** Neoplastic transformation of mast cells by Abelson-MuLV: abrogation of IL-3 dependence by a nonautocrine mechanism, *Cell,* 41, 685, 1985.

170. **Hines, D. L.,** Viral oncogene expression during differentiation of Abelson virus-infected murine promonocytic leukemia cells, *Cancer Res.,* 48, 1702, 1988.

171. **Rein, A., Keller, J., Schultz, A. M., Holmes, K. L., Medicus, R., and Ihle, J. N.,** Infection of immune mast cells by Harvey sarcoma virus: immortalization without loss of requirement for interleukin-3, *Mol. Cell. Biol.,* 5, 2257, 1985.

172. **Coffey, R. J., Jr., Goustin, A. S., Soderquist, A. M., Shipley, G. D., Wolfshohl, J., Carpenter, G., and Moses, H. L.,** Transforming growth factor α and β expression in human colon cancer lines: implications for an autocrine model, *Cancer Res.,* 47, 4590, 1987.

INDEX